Springer Series on Atomic, Optical, and Plasma Physics

Volume 113

Editor-in-Chief

Gordon W. F. Drake, Department of Physics, University of Windsor, Windsor, ON, Canada

Series Editors

James Babb, Harvard-Smithsonian Center for Astrophysics, Cambridge, MA, USA

Andre D. Bandrauk, Faculté des Sciences, Université de Sherbrooke, Sherbrooke, QC, Canada

Klaus Bartschat, Department of Physics and Astronomy, Drake University, Des Moines, IA, USA

Charles J. Joachain, Faculty of Science, Université Libre Bruxelles, Bruxelles, Belgium

Michael Keidar, School of Engineering and Applied Science, George Washington University, Washington, DC, USA

Peter Lambropoulos, FORTH, University of Crete, Iraklion, Crete, Greece

Gerd Leuchs, Institut für Theoretische Physik I, Universität Erlangen-Nürnberg, Erlangen, Germany

Alexander Velikovich, Plasma Physics Division, United States Naval Research Laboratory, Washington, DC, USA

The Springer Series on Atomic, Optical, and Plasma Physics covers in a comprehensive manner theory and experiment in the entire field of atoms and molecules and their interaction with electromagnetic radiation. Books in the series provide a rich source of new ideas and techniques with wide applications in fields such as chemistry, materials science, astrophysics, surface science, plasma technology, advanced optics, aeronomy, and engineering. Laser physics is a particular connecting theme that has provided much of the continuing impetus for new developments in the field, such as quantum computation and Bose-Einstein condensation. The purpose of the series is to cover the gap between standard undergraduate textbooks and the research literature with emphasis on the fundamental ideas, methods, techniques, and results in the field.

More information about this series at http://www.springer.com/series/411

Isak Beilis

Plasma and Spot Phenomena in Electrical Arcs

Volume 2

Isak Beilis
Department of Electrical Engineering
Tel Aviv University
Tel Aviv, Israel

ISSN 1615-5653 ISSN 2197-6791 (electronic)
Springer Series on Atomic, Optical, and Plasma Physics
ISBN 978-3-030-44746-5   ISBN 978-3-030-44747-2 (eBook)
https://doi.org/10.1007/978-3-030-44747-2

© Springer Nature Switzerland AG 2020
This work is subject to copyright. All rights are reserved by the Publisher, whether the whole or part of the material is concerned, specifically the rights of translation, reprinting, reuse of illustrations, recitation, broadcasting, reproduction on microfilms or in any other physical way, and transmission or information storage and retrieval, electronic adaptation, computer software, or by similar or dissimilar methodology now known or hereafter developed.
The use of general descriptive names, registered names, trademarks, service marks, etc. in this publication does not imply, even in the absence of a specific statement, that such names are exempt from the relevant protective laws and regulations and therefore free for general use.
The publisher, the authors and the editors are safe to assume that the advice and information in this book are believed to be true and accurate at the date of publication. Neither the publisher nor the authors or the editors give a warranty, expressed or implied, with respect to the material contained herein or for any errors or omissions that may have been made. The publisher remains neutral with regard to jurisdictional claims in published maps and institutional affiliations.

This Springer imprint is published by the registered company Springer Nature Switzerland AG
The registered company address is: Gewerbestrasse 11, 6330 Cham, Switzerland

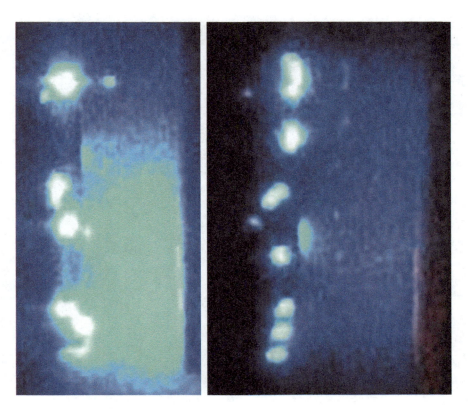

Photographs illustrating momentary distribution of the cathode spots and presence of an inter-electrode (gap 10 mm) plasma, which is detected after 45 s (left) and 70 s (right) of 200-A vacuum arc ignition with Zn cathode and Mo anode. In the left image, the glass of the observation window is weakly deposited and the plasma is observed better, whereas in the right image the window is strongly deposited and the plasma is more shielded

My special dedication of this book is to the blessed memory of my parents, Ioil and Sima, and my brother, Gregory

Preface

> ...*and you will hear it, and investigate thoroughly,*
> *and behold, the matter will be found true*
>
> (*Deuteronomy—Chap. 17*)

The electrical arc is commonly defined as an electrical discharge with relatively high current density and low gap voltage. Electrical arcs occur in vacuum, as well as in low- and high-pressure gases. An important issue is the origin of a conductive medium in the electrode gap. In vacuum, the conductive medium is generated by vaporization of electrode material during arcing. This arc may also be called an electrode vapor discharge. Even in most cases where a gas is present, electrode vaporization is the plasma source process in the near-electrode region. Many phenomena in these arcs are similar to those in vacuum arcs.

The electrical arc was first reported in the nineteenth century after the observation of a continuous arc between carbon electrodes while measuring carbon resistance by Petrov in 1802 in Petersburg and by Humphry Davy in 1808 in London, while experimenting with pulsed discharges. After some applications, preliminary investigations of the arc were conducted even in nineteenth century and systematic research was conducted from the beginning of twentieth century, firstly with graphite and then with metallic electrodes.

Different phenomena were observed in the arcs. Strongly constricted and highly mobile luminous forms appeared near the electrodes, later called cathode and anode "spots." In these regions, the electrical conductivity transits from very high in the metal to low in the plasma. Curious phenomena were observed, including fragmentation of the plasma in at the arc roots, plasma jet formation, and retrograde arc spot motion in a transverse magnetic field. Retrograde cathode spot motion in magnetic fields, supersonic plasma jet flow, extremely high electron emission density at the cathode, and the high current density in the arc root plasma could not be explained using earlier existing physical models. This was a consequence of the then limited knowledge of the plasma processes, leading to inadequate assumptions for the calculated models. Therefore, the behavior of the arc cathode region was considered mysterious for the long even though it was intensively studied. With time, experimental and theoretical investigations contributed knowledge that improved our understanding of some of these phenomena.

Improved diagnostic techniques were applied to the near-electrode region provided new information, but also required further explanations. Numerous investigators contributed experimental and theoretical knowledge. In the second half of nineteenth century, electrode repulsion, plasma fluxes, and the arc force on the electrodes were detected and studied by Dewar, Schuster, Hemsalech, Duffield Tyndall, and Sellerio. Langmuir, Tonks, and Compton developed the theory of the space charge sheath, and volt–current and plasma characteristics in the arcs in the first decades of twentieth century. Stark, and then Tanberg, Cobine, Gallagher, Slepian, Finkelnburg, Maecker, Gunterschulze, Holm, Smith, Froom, Reece, Ecker, Lee, Greenwood, Robson, Engel, Kesaev, Plyutto, Hermoch, Davies, Miller, Rakhovsky, Kimblin, Daalder, Juttner, Hantsche, Anders, Goldsmith, Boxman, the Mesyats group, the Shenli Jia group and many others significantly contributed to different aspects of the electrode and plasma characteristics of electrical arcs. The understanding of observed arc behavior, its characteristics in different regions of the electrode gap expanded with the progress of plasma science, especially after 1960 in low-temperature plasma, plasma fusion, and astrophysics.

This book consists of four main parts that describe progress in study of cathodic arc phenomena. It began with the early separate hypotheses of charged particle generation in the cathode region up to the present time including our own investigations extending over more than 50 years. The first introductory part contains five chapters and characterizes the basic plasma and electrode processes needed to study cathodic arc. Plasma concepts and particle interactions in the plasma volume and at the surface are presented in Chap. 1. Chapter 2 presents the kinetics of cathode vaporization and mechanisms of electron emissions. The mathematical formulation of heat conduction for different cathode geometries subjected to local moving heat sources are considered in Chap. 3. Chapter 4 presents the transport equations and mathematical formulation of particle diffusion, and mass and heat transfer phenomena in a multi-component plasma. The plasma–wall transition including cathode sheath formation is described in Chap. 5.

Part II presents a review of experimental studies of electrode phenomena in vacuum arcs. Chapter 6 describes the electrical breakdown of a vacuum gap and the methods of vacuum arc ignition. Chapter 7 reviews cathode spot dynamics, different spot types, and the problem of current density in the cathode spot. The spot types are described by their physical characteristics as determined experimentally in contrast to the previous confused numerical classification. The total eroded cathode mass loss is considered in Chap. 8, and cathode erosion in the form of macroparticles is considered in Chap. 9. Chapter 10 presents data concerning electrode energy losses and the effective voltage measured during the arc operation. The repulsive effect in the arc gap discovered in the second half of nineteenth century as a plasma flow reaction and the further measurements of the electrode force phenomena in vacuum arcs is reviewed in Chap. 11. Chapter 12 describes experimental investigation of the cathode jets, including jet velocity and ion current. Chapter 13 presents the state of the art of observations of spot motion in transverse and oblique magnetic fields. Current continuity in the anode region and the conditions for anode spot formation are described in Chap. 14.

Theory of the cathodic arc is presented in Part III. The evolution of cathode spot theory starting from the beginning of the twentieth century up to present time is described in Chap. 15. The progress and weakness of the developed theories are analyzed. The reason why the cathode spot was considered mysterious is discussed. The key issue is electrical current continuity from the metallic cathode to the plasma. In order to explain continuity, charge-exchange ion–atom interactions were examined, and it was shown that the ion current to the cathode is determined by the collisions in the dense near-cathode plasma. Consequently, a gasdynamic theory was advanced (Chap. 16), which consistently considers the cathode and multi-component plasma processes and determines the cathode spot parameters using a mathematically closed formulation. This approach describes the parameters and current continuity at cathode plasma region as functions of the cathode potential drop using measured cathode erosion rate and current per spot.

The cathode spot parameters for materials with a wide range of thermophysical properties were investigated. This study includes the cathode spot characteristics for three groups of materials. (1) Cathode materials with intermediate thermal properties (e.g., Cu, Al, Ni, and Ti). In cathodic arcs with these materials, the electron emission flux is comparable to the evaporated atom flux for a wide region of cathode temperatures. (2) Cathodes with refractory materials are characterized by larger flux of emitted electrons than the flux of evaporated atoms over a wide range of cathode temperatures. A modified model using a "virtual plasma cathode" allows obtaining an electron emission current density lower than the saturated density. (3) Volatile cathode material (e.g., Hg) forms an opposite case, which is characterized by an electron emission flux much lower than the vapor flux. The modified model considers an additional plasma layer adjacent to the cathode sheath, which serves as a "plasma cathode" supplying electrons for self-sustained arc operation.

A general physically closed approach is advanced considering the kinetics of cathode vaporization and gasdynamic of plasma expansion, which described as a kinetic theory in Chap. 17. The kinetic theory simultaneously considered atom and electron evaporation, taking into account conservation laws for particle density, flux, momentum, and energy in a non-equilibrium Knudsen layer. In the rarefied collisional Knudsen layer, the heavy particle return flux is formed at a length of few mean free paths from the cathode surface. A relatively high atom ionization rate is due to large energy dissipation which produced the high ion fraction in the back flux. The difference between direct evaporated atom flux and the returned flux of atom and ions determines the cathode mass loss, i.e., the cathode erosion rate G_k. It was obtained that the plasma velocity at external boundary of Knudsen layer is significantly lower than the sound speed. This very important result indicated that cathode spots operate self-sufficiently only when plasma velocity is lower than the sound speed, i.e., when the plasma flow near the cathode is impeded.

The cathode potential drop u_c was determined by considering the kinetics for emitted and plasma electrons. Electron emission was considered as electron evaporation. The electron beam relaxation zone (analogous to the Knudsen layer for atoms) depends on the electron beam energy, which is increased in the cathode sheath by eu_c. The returned electron flux depends on the plasma density and by the potential barrier in the sheath (u_c). Conservation of electron momentum and electron energy and the quasineutrality condition on the boundary of Knudsen layer allow determine u_c directly. *Two principle rules* based on necessity of impeded plasma flow near the cathode and satisfaction to a positive cathode energy balance at low spot current, which determine conditions of the cathode spot ignition and spot development. Chapter 18 describes formation of the cathode plasma jet. The jet begins as a relatively slow plasma flow, initially produced due to difference of plasma parameters (plasma velocity formation) in Knudsen layer. The plasma is further accelerated up to supersonic speed, determined by the cathode dense plasma, and the energy dissipated by electron beam relaxation due to gasdynamic mechanisms to the expansion region. The calculated plasma velocity agreed well with the measurements. The potential distribution is not monotonic, but rather has a hump with a height of the electron temperature.

Transient cathode spots were studied on the protrusions and bulk cathodes. The nanolife spot parameters, the mechanism of spot operation on film cathodes, specific spot parameters at large rate of current rise, and the nature of high voltage arc initiation were also studied.

Cathode spot motion in magnetic fields is studied in Chap. 19. Three important experimental phenomena are described: (1) cathode spot grouping and (2) cathode spot retrograde motion, i.e., in the anti-Amperian direction, when the arc runs in a transverse magnetic field, and (3) cathode spot motion in oblique magnetic fields. The retrograde spot motion model takes in account the kinetic principle rules of impeded plasma flow in the non-symmetric total pressure (gas−kinetic+magnetic) in the near-cathode plasma. The current per group spot was calculated taking into account that the plasma kinetic pressure is comparable to the self-magnetic pressure in the cathode plasma jet. The theory showed that the current per group spot and spot velocity increase linearly with the magnetic field and arc current, and these dependencies agree with experimental observations. Retrograde cathode spot motion in short electrode gaps and in atmospheric pressure as well as reversed motion in strong magnetic fields (>1 T) observed by Robson and Engel is described. The observed acute angle effect and spot splitting in oblique magnetic fields are explained. Robson's experimentally obtained dependencies of the drift angle on magnetic field and on the acute angle are described by this theory by using kinetic principle rules.

Chapter 20 presents a theoretical study of anode spots. An anode spot theory was developed using a kinetic treatment of anode evaporation and modeling the plasma flow in the plasma acceleration region. The plasma energy balance was described considering the Joule energy in the anode plasma, the energy dissipation caused by ionization of atoms, energy convection by the electric current, and the energy required for anode plasma acceleration. The anode surface temperature, current density, plasma density, plasma temperature, and plasma velocity were calculated

Preface

self-consistently. The main result is a significantly lower degree of anode plasma ionization (10^{-3} to 5×10^{-2}) than in the cathode plasma, due to mobility of electrons higher than the ions, which both supports current continuity in the anode and cathode regions, respectively.

The applications of cathode spot theory are discussed in Part IV. Chapter 21 describes unipolar arcs on the metal elements in fusion devices. The results of experimental and theoretical investigation (including details of the primary work of Robson and Thonenmann) are reviewed, beginning from the first reported data. The specifics of unipolar arcs on nanostructured surfaces in tokamaks are analyzed. Our own experimental results for a tokamak wall with a thin tungsten coating on carbon demonstrate the spot motion mechanism which is similar to spots on film cathodes.

Chapter 22 presents the application of vacuum arc plasma sources for thin film deposition. Novel arc plasma sources were developed in the last decades at Tel Aviv University. Low macroparticle density in the films was reached by converting cathode droplets to plasma through their re-evaporations on a hot refractory anode surface. Arc characteristics (plasma density, temperature) and deposition rate as functions of the distances to the substrate from the arc axis, gap distance, electrode configuration, and electrode materials are discussed. A theory was developed to describe self-consistently transient anode heating and anode plasma generation. The advantages of the method are compared with traditional deposition approaches.

Another vacuum arc application is microthrusters for spacecraft. Chapter 23 presents the general problems of vacuum arc plasma thrusters, their main characteristics, and their efficiency. Phenomena in arcs with small electrode gaps are considered, and processes in very short cathodic arcs relevant to thrusters are modeled. Simulation produced thrust efficiency results. Microthruster devices operating in space in the last decade are analyzed.

Chapter 24 applies cathode spot theory to laser–metal interaction and laser plasma generation. Experiments and previously developed theories of laser–target interaction are analyzed. It is shown that the plasma generated near the target by laser and the plasma generated in the cathode region of a vacuum arc have similar characteristics. Both plasmas are very dense and appear in a minute region, known as the laser spot and the cathode spot, respectively. The plasma interactions with matter in both cases are also similar. Taking into account a detailed analysis of the features of the laser spot, a physical model based on a previously presented kinetic approach of the cathode spot is described by modifying it for the particular features of laser-generated plasma. The model considers that the net electrical current is zero in the laser spot, in contrast to the current carried in a vacuum arc cathode spot. The specifics of target vaporization, breakdown of neutral vapor, near-target electrical sheath, electron emission from the hot area of the target, plasma heating, and plasma acceleration mechanism are detailed. The mathematical description and simulation for different target materials show an unusual new result. The laser power absorbed in the plasma is converted into kinetic and potential energy of the plasma particles and is returned to the target and the target surface temperature calculated, taking into account the plasma energy flux is significantly larger than that without this flux.

Finally, Chap. 25 reports about application of cathode spot theory to arcs formed in different technical devices: low-temperature plasma generators, plasma accelerators, devices using plasma of combustion products, and rail guns that accelerate the solid bodies. Cathode spots on the electrodes of these devices are explained using spot theory for film cathodes and considering a plasma flow consisting of lightly ionized dopants. Transition from a diffuse to an arc mode was modeled, considering electrode overheating instability with an exponential dependence of the electrical conductivity on temperature. Plasma accelerator experiments are discussed: A significantly larger electron emission current density than that described by the Richardson law was detected, and the models are analyzed. The contribution of ion current is demonstrated using the gasdynamic model of the cathode spot, taking into account the experimentally observed plasma expansion from the spot. A mathematical approach described the long-time problem of plasma column constriction in atmospheric pressure arcs, without recourse to Steinbeck's minimal principle. The arc mechanism and the observed arc parameters in rail gun are explained using cathode and anode spot models. Experimental reduction of the rate of solid body acceleration can be explained by the calculated arc velocity dependence on applied electrical power, which tends to saturate when the power increases, due to increasing arc plasma mass needed to support increasing arc current.

The text includes not only the cathode current continuity mechanism usually published in the spot theories, but also various ideas to understand many different spot aspects. The studies presented here cannot address all observed arc discharge phenomena—they cannot be completely explained by the developed theories. Nevertheless, I hope that the developed models systematically lead to further research, increase our understanding of different arc spot mechanisms, and reduce the mystery surrounding them. I hope this book will be useful for researchers in the physics of low temperature plasma and gas discharges, practicing engineers and students learning electrical engineering, plasma physics, and gas discharge physics.

I gratefully acknowledge colleagues for useful discussion, help, and support at various stages of this work. First, thanks to Vadim Rakhovsky who introduced me to this subject, worked with me during my postgraduate studies, and helped me to understand the early experiments on near-electrode phenomena conducted in widely known Veniamin Granovsky's laboratory. Thanks to laboratory colleagues N. Zykova and V. Kantsel for helpful direct discussion for the experiment at first stage of my spot study. I am grateful to Gregory Lyubimov for discussions and nice joint theoretical work on the electrode phenomena based on the hydrodynamics of plasma flow. Great thanks to Igor Kesaev, who explain the specifics of his world known experiments and the difficulties in my way by choosing this confused problem.

Many thanks to the supervisors of a number of All-Union Seminars: Academician Nikolay Rykalin (heat conduction in solids caused by a moving concentrated heat source), Academician Vitaly Ginsburg, Anri Rukhadze, Yuri Raizer (general physics and physics of gas discharges), and Sergey Anisimov (kinetics of solid vaporization). I appreciate Oleg Firsov, which noticed about the charge-exchange process during the discussions and to Boris Smirnov (elementary collisions of plasma particles), Alexey Morosov and Askold Zharinov (mechanisms of plasma acceleration).

I am grateful to Academician Vladimir Fortov, which initiated and supported the doctoral work in the department and Academician Gennady Mesyats for support the presentation of a thesis of Doctor of Science. As a result, I acquired much scientific experience and broad scientific knowledge, which played an essential role in my understanding of the many aspects of cathode spot science.

Since January 1992, my investigations were continued in the Electrical Discharges and Plasma laboratory at the Faculty of Engineering of Tel Aviv University. My great thanks to Samuel Goldsmith and Ray Boxman for very important support and efficient joint work at the first stage and for the excellent lab atmosphere as my work progressed. I gratefully acknowledge N. Parkansky, V. Zitomirsky, H. Rosenthal, V. Paperny A. Shashurin, D. Arbilly, A. Nemirovsky, A. Snaiderman, D. Grach, Y. Koulik, Ben Sagi, and Y. Yankelevich for their contributions at different stages of the investigation in the laboratory. A special thanks is addressed to Ray Boxman for helpful discussion of my cathode spot research and English help at the publication stage. I am grateful to Michael Keidar for our joint work that began at Tel Aviv University, continues presently and whose discussions give me great pleasure. Many thanks also for his help during processing of the manuscript. I appreciate Burhard Juttner, Erhard Hantzsche, Michael Laux, Boyan Djakov, Heinz Pursch, Peter Siemroth, Lorenzo Torrisi, Joachim Heberlein, and Emil Pfender for their help during my visits to their respective institutions.

Finally, I would like to thank my family, my loved children for being always patient with me. Greatly I appreciate my wife Galina, supporting my way of life in the science, for her understanding and taking care of all my needs.

Tel Aviv, Israel   Isak Beilis
2020

Contents

Part I Plasma, Particle Interactions, Mass and Heat Phenomena

1 Base Plasma Particle Phenomena at the Surface and in Plasma Volume 3
- 1.1 Plasma 3
  - 1.1.1 Quasineutrality 3
  - 1.1.2 Plasma Oscillations 4
  - 1.1.3 Interaction of Electron Beam with Plasma 5
  - 1.1.4 Plasma State 6
- 1.2 Particle Collisions with Surfaces 7
- 1.3 Plasma Particle Collisions 10
  - 1.3.1 Charged Particle Collisions 10
  - 1.3.2 Electron Scattering on Atoms 11
  - 1.3.3 Charge-Exchange Collisions 12
  - 1.3.4 Excitation and Ionization Collisions 13
  - 1.3.5 Electron–Ion Recombination 27
  - 1.3.6 Ionization–Recombination Equilibrium 30
- References 33

2 Atom Vaporization and Electron Emission from a Metal Surface 37
- 2.1 Kinetics of Material Vaporization 37
  - 2.1.1 Non-equilibrium (Kinetic) Region 37
  - 2.1.2 Kinetic Approaches. Atom Evaporations 39
  - 2.1.3 Kinetic Approaches. Evaporations Into Plasma 44
- 2.2 Electron Emission 46
  - 2.2.1 Work Function. Electron Function Distribution 46
  - 2.2.2 Thermionic or T-Emission 47
  - 2.2.3 Schottky Effect. Field or F-Emission 48
  - 2.2.4 Combination of Thermionic and Field Emission Named TF-Emission 50

		2.2.5	Threshold Approximation	53
		2.2.6	Individual Electron Emission	53
		2.2.7	Fowler–Northeim-Type Equations and Their Correction for Measured Plot Analysis	55
		2.2.8	Explosive Electron Emission	56
	References		....	62
3	**Heat Conduction in a Solid Body. Local Heat Sources**			69
	3.1	Brief State-of-the-Art Description		69
	3.2	Thermal Regime of a Semi-finite Body. Methods of Solution in Linear Approximation		70
		3.2.1	Point Source. Continuous Heating	70
		3.2.2	Normal Circular Heat Source on a Body Surface	71
		3.2.3	Instantaneous Normal Circular Heat Source on Semi-infinity Body	72
		3.2.4	Moving Normal Circular Heat Source on a Semi-infinity Body	73
	3.3	Heating of a Thin Plate		75
		3.3.1	Instantaneous Normal Circular Heat Source on a Plate	75
		3.3.2	Moving Normal Circular Heat Source on a Plate	76
	3.4	A Normal Distributed Heat Source Moving on Lateral Side of a Thin Semi-infinite Plate		77
		3.4.1	Instantaneous Normally Distributed Heat Source on Side of a Thin Semi-infinite Plate	77
		3.4.2	Moving Continuous Normally Distributed Heat Source on Thin Plate of Thickness δ	79
		3.4.3	Fixed Normal-Strip Heat Source with Thickness X_0 on Semi-infinite Body	80
		3.4.4	Fixed Normal-Strip Heat Source with Thickness X_0 on Semi-infinite Body Limited by Plane $X = -\delta/2$	80
		3.4.5	Fixed Normal-Strip Heat Source with Thickness X_0 on Lateral Side of Finite Plate ($X_0 < \delta$)	81
		3.4.6	Moving Normal-Strip Heat Source on a Later Plate Side of Limited Thickness ($X_0 < \delta$)	82
	3.5	Temperature Field Calculations. Normal Circular Heat Source on a Semi-infinite Body		83
		3.5.1	Temperature Field in a Tungsten Body	84
		3.5.2	Temperature Field in a Copper	86
		3.5.3	Temperature Field Calculations. Normal Heat Source on a Later Side of Thin Plate and Plate with Limited Thickness [25]	90
		3.5.4	Concluding Remarks to the Linear Heat Conduction Regime	93

3.6		Nonlinear Heat Conduction	94
	3.6.1	Heat Conduction Problems Related to the Cathode Thermal Regime in Vacuum Arcs	94
	3.6.2	Normal Circular Heat Source Action on a Semi-infinity Body with Nonlinear Boundary Condition	94
	3.6.3	Numerical Solution of 3D Heat Conduction Equation with Nonlinear Boundary Condition	97
	3.6.4	Concluding Remarks to the Nonlinear Heat Conduction Regime	98
References			99

4 The Transport Equations and Diffusion Phenomena in Multicomponent Plasma ... 101

4.1		The Problem	101
4.2		Transport Phenomena in a Plasma. General Equations	101
	4.2.1	Three-Component Cathode Plasma. General Equations	103
	4.2.2	Three-Component Cathode Plasma. Simplified Version of Transport Equations	105
	4.2.3	Transport Equations for Five-Component Cathode Plasma	109
References			112

5 Basics of Cathode-Plasma Transition. Application to the Vacuum Arc ... 113

5.1		Cathode Sheath	113
5.2		Space Charge Zone at the Sheath Boundary and the Sheath Stability	116
5.3		Two Regions. Boundary Conditions	120
5.4		Kinetic Approach	121
5.5		Electrical Field	126
	5.5.1	Collisionless Approach	127
	5.5.2	Electric Field. Plasma Electrons. Particle Temperatures	127
	5.5.3	Refractory Cathode. Virtual Cathode	128
5.6		Electrical Double Layer in Plasmas	132
References			137

Part II Vacuum Arc. Electrode Phenomena. Experiment

6 Vacuum Arc Ignition. Electrical Breakdown ... 143

6.1		Contact Triggering of the Arc	143
	6.1.1	Triggering of the Arc Using Additional Trigger Electrode	143

		6.1.2	Initiation of the Arc by Contact Breaking of the Main Electrodes	144
		6.1.3	Contact Phenomena	144
	6.2	Electrical Breakdown		148
		6.2.1	Electrical Breakdown Conditions	149
		6.2.2	General Mechanisms of Electrical Breakdown in a Vacuum	149
		6.2.3	Mechanisms of Breakdown Based on Explosive Cathode Protrusions	153
		6.2.4	Mechanism of Anode Thermal Instability	154
		6.2.5	Electrical Breakdown at an Insulator Surface	157
	6.3	Conclusions		160
	References			161
7	**Arc and Cathode Spot Dynamics and Current Density**			**165**
	7.1	Characteristics of the Electrical Arc		165
		7.1.1	Arc Definition	165
		7.1.2	Arc Instability	166
		7.1.3	Arc Voltage	168
		7.1.4	Cathode Potential Drop	170
		7.1.5	Threshold Arc Current	174
	7.2	Cathode Spots Dynamics. Spot Velocity		174
		7.2.1	Spot Definition	174
		7.2.2	Study of the Cathode Spots. General Experimental Methods	176
		7.2.3	Method of High-Speed Images	176
		7.2.4	Autograph Observation. Crater Sizes	188
		7.2.5	Summary of the Spot Types Studies	194
		7.2.6	Classification of the Spot Types by Their Characteristics	199
	7.3	Cathode Spot Current Density		201
		7.3.1	Spot Current Density Determination	201
		7.3.2	Image Sizes with Optical Observation	201
		7.3.3	Crater Sizes Observation	202
		7.3.4	Influence of the Conditions. Uncertainty	203
		7.3.5	Interpretation of Luminous Image Observations	204
		7.3.6	Effects of Small Cathode and Low-Current Density. Heating Estimations	206
		7.3.7	Concluding Remarks	207
	References			208

Contents

8 Electrode Erosion. Total Mass Losses 213
    8.1   Erosion Phenomena 213
        8.1.1   General Overview 213
        8.1.2   Electroerosion Phenomena in Air 216
        8.1.3   Electroerosion Phenomena in Liquid Dielectric Media 221
    8.2   Erosion Phenomena in Vacuum Arcs 226
        8.2.1   Moderate Current of the Vacuum Arcs 227
        8.2.2   Electrode Erosion in High Current Arcs 242
        8.2.3   Erosion Phenomena in Vacuum of Metallic Tip as High Field Emitter 245
    8.3   Summary and Discussion of the Erosion Measurements 249
    References 250

9 Electrode Erosion. Macroparticle Generation 255
    9.1   Macroparticle Generation. Conventional Arc 255
    9.2   Macroparticle Charging 264
    9.3   Macroparticle Interaction 269
        9.3.1   Interaction with Plasma 269
        9.3.2   Interaction with a Wall and Substrate 271
    9.4   Macroparticle Generation in an Arc with Hot Anodes 277
        9.4.1   Macroparticles in a Hot Refractory Anode Vacuum Arc (HRAVA) 277
        9.4.2   Macroparticles in a Vacuum Arc with Black Body Assembly (VABBA) 279
    9.5   Concluding Remarks 280
    References 280

10 Electrode Energy Losses. Effective Voltage 285
    10.1   Measurements of the Effective Voltage in a Vacuum Arc 285
    10.2   Effective Electrode Voltage in an Arc in the Presence of a Gas Pressure 292
    10.3   Effective Electrode Voltage in a Vacuum Arc with Hot Refractory Anode 295
        10.3.1   Effective Electrode Voltage in a Hot Refractory Anode Vacuum Arc (HRAVA) 296
        10.3.2   Energy Flux from the Plasma in a Hot Refractory Anode Vacuum Arc (HRAVA) 298
        10.3.3   Effective Electrode Voltage in a Vacuum Arc Black Body Assembly (VABBA) 300
    10.4   Summary 302
    References 304

11 Repulsive Effect in an Arc Gap and Force Phenomena as a Plasma Flow Reaction 307
- 11.1 General Overview 307
- 11.2 Early Measurements of Hydrostatic Pressure and Plasma Expansion. Repulsive Effect upon the Electrodes 308
- 11.3 Primary Measurements of the Force at Electrodes in Vacuum Arcs 320
- 11.4 Further Developed Measurements of the Force at Electrodes in Vacuum Arcs 331
- 11.5 Early Mechanisms of Forces Arising at Electrodes in Electrical Arcs 339
- 11.6 Summary 342
- References 343

12 Cathode Spot Jets. Velocity and Ion Current 347
- 12.1 Plasma Jet Velocity 347
- 12.2 Ion Energy 354
- 12.3 Ion Velocity and Energy in an Arc with Large Rate of Current Rise dI/dt 360
- 12.4 Ion Current Fraction 365
- 12.5 Ion Charge State 373
- 12.6 Influence of the Magnetic Field 393
- 12.7 Vacuum Arc with Refractory Anode. Ion Current 404
- 12.8 Summary 411
- References 415

13 Cathode Spot Motion in a Transverse and in an Oblique Magnetic Field 421
- 13.1 The General Problem 421
- 13.2 Effect of Spot Motion in a Magnetic Field 422
- 13.3 Investigations of the Retrograde Spot Motion 424
  - 13.3.1 Magnetic Field Parallel to the Cathode Surface. Direct Cathode Spot Motion 425
  - 13.3.2 Phenomena in an Oblique Magnetic Fields 475
- 13.4 Summary 483
- References 486

14 Anode Phenomena in Electrical Arcs 493
- 14.1 General Consideration of the Anode Phenomena 493
- 14.2 Anode Modes in the Presence of a Gas Pressure 494
  - 14.2.1 Atmospheric Gas Pressure 494
  - 14.2.2 Low-Pressure Gas 501
- 14.3 Anode Modes in High-Current Vacuum Arcs 503
  - 14.3.1 Anode Spotless Mode. Low-Current Arcs 505
  - 14.3.2 Anode Spotless Mode. High-Current Arcs 506

	14.3.3	Anode Spot Mode for Moderate Arc Current	506
	14.3.4	Anode Spot Mode for High-Current Vacuum Arc	509
14.4	Measurements of Anode and Plasma Parameters		523
	14.4.1	Anode Temperature Measurements	524
	14.4.2	Anode Plasma Density and Temperature	527
	14.4.3	Anode Erosion Rate	534
14.5	Summary		535
References			539

Part III Cathodic Arc. Theory

15 Cathode Spot Theories. History and Evolution of the Mechanisms 545

- 15.1 Primary Studies 545
  - 15.1.1 First Hypotheses 545
  - 15.1.2 Cathode Positive Space Charge and Electron Emission by Electric Field 546
  - 15.1.3 Mechanism of Thermal Plasma and Ion Current Formation 547
  - 15.1.4 Mechanism of "Hot Electrons" in a Metallic Cathode 548
  - 15.1.5 Mechanism of Local Explosion of a Site at the Cathode Surface 551
  - 15.1.6 Emission Mechanism Due to Cathode Surface Interaction with Excited Atoms 552
- 15.2 Mathematical Description Using Systematical Approaches 554
  - 15.2.1 The First Models Using System of Equations 555
  - 15.2.2 Models Based on Double-Valued Current Density in a Structured Spot 562
- 15.3 Further Spot Modeling Using Cathode Surface Morphology and Near-Cathode Region 566
  - 15.3.1 Modeling Using Cathode Traces and Structured Near-Cathode Plasma 567
  - 15.3.2 Cathode Phenomena in Presence of a Previously Arising Plasma 573
- 15.4 Explosive Electron Emission and Cathode Spot Model 579
  - 15.4.1 Primary Attempt to Use Explosive Electron Emission for Modeling the Cathode Spot 580
  - 15.4.2 The Possibility of Tip Explosion as a Mechanism of Cathode Spot in a Vacuum Arc 580
  - 15.4.3 Explosive Electron Emission and Cathode Vaporization Modeling the Cathode Spot 582

	15.5	Summary	588
	15.6	Main Concluding Remarks	591
	References		592
16	**Gasdynamic Theory of Cathode Spot. Mathematically Closed Formulation**		**599**
	16.1	Brief Overview of Cathodic Arc Specifics Embedded at the Theory	600
	16.2	Gasdynamic Approach Characterizing the Cathode Plasma. Resonance Charge-Exchange Collisions	600
	16.3	Diffusion Model. Weakly Ionized Plasma Approximation	601
	16.4	Diffusion Model. Highly Ionized Plasma Approximation	604
		16.4.1 Overall Estimation of Energetic and Momentum Length by Particle Collisions	604
		16.4.2 Modeling of the Characteristic Physical Zones in the Cathode Spot Plasma	605
		16.4.3 Mathematically Closed System of Equations	606
	16.5	Numerical Investigation of Cathode Spot Parameters	610
		16.5.1 Ion Density Distribution. Plasma Density Gradient	611
		16.5.2 Numerical Study of the Group Spot Parameters	612
		16.5.3 Low-Current Spot Parameters	615
		16.5.4 Nanosecond Cathode Spots with Large Rate of Current Rise	620
	16.6	Vacuum Arcs with Extremely Properties of Cathode Material. Modified Gasdynamic Models	624
	16.7	Refractory Material. Model of Virtual Cathode. Tungsten	625
		16.7.1 Model of Low Electron Current Fraction s for Tungsten	627
		16.7.2 Numerical Study of the Spot Parameters on Tungsten Cathode	630
	16.8	Vacuum Arcs with Low Melting Materials. Mercury Cathode	633
		16.8.1 Experimental Data of a Vacuum Arc with Mercury Cathode	633
		16.8.2 Overview of Early Debatable Hg Spot Models	634
		16.8.3 Double Sheath Model	638
		16.8.4 The System of Equations	639
		16.8.5 Numerical Study of the Mercury Cathode Spot	643
		16.8.6 Analysis of the Spot Simulation at Mercury Cathode	646
	16.9	Film Cathode	650
		16.9.1 Physical Model of Spot Motion on Film Cathode	650
		16.9.2 Calculation Results	652
		16.9.3 Analysis of the Calculation Results	654

	16.10	Phenomena of Cathode Melting and Macrodroplets Formation	656
	16.11	Summary	661
	References		663

17 Kinetic Theory. Mathematical Formulation of a Physically Closed Approach 669

	17.1	Cathode Spot and Plasma	670
	17.2	Plasma Flow in a Non-equilibrium Region Adjacent to the Cathode Surface	670
		17.2.1 Overview of the State of the Art	671
		17.2.2 Specifics of Kinetics of the Cathode Spot Plasma....	672
	17.3	Summarizing Conception of the Kinetic Cathode Regions. Kinetic Model	673
	17.4	Kinetics of Cathode Vaporization into the Plasma. Atom and Electron Knudsen Layers	674
		17.4.1 Function Distribution of the Vaporized and Plasma Particles	674
		17.4.2 Conservation Laws and the Equations of Conservation	676
		17.4.3 Integration. Total Kinetic Multi-component System of Equations	677
		17.4.4 Specifics at Calculation of Current per Spot	681
	17.5	Region of Cathode Potential Drop. State of Previous Studies	682
	17.6	Numerical Investigation of Cathode Spot Parameters by Physically Closed Approach	685
		17.6.1 Kinetic Model. Heavy Particle Approximation	685
		17.6.2 Spot Initiation During Triggering in a Vacuum Arc	689
		17.6.3 Kinetic Model. Study the Total System of Equations at Protrusion Cathode	690
		17.6.4 Bulk Cathode	696
	17.7	Rules Required for Plasma Flow in Knudsen Layer Ensured Cathode Spot Existence	711
	17.8	Cathode Spot Types, Motion, and Voltage Oscillations	714
		17.8.1 Mechanisms of Different Spot Types	714
		17.8.2 Mechanisms of Spot Motion and Voltage Oscillations	715
	17.9	Requirement of Initial High Voltage in Electrode Gap for Vacuum Arc Initiation	717
	17.10	Summary	718
	References		720

18 Spot Plasma and Plasma Jet 725
18.1 Plasma Jet Generation and Plasma Expansion. Early State of the Mechanisms 725
18.2 Ion Acceleration Phenomena in Cathode Plasma. Gradient of Electron Pressure 729
18.2.1 Plasma Polarization and an Electric Field Formation 729
18.2.2 Ambipolar Mechanism 730
18.2.3 Model of Hump Potential 731
18.3 Plasma Jet Formation by Model of Explosive Electron Emission 733
18.4 Ion Acceleration and Plasma Instabilities 734
18.5 Gasdynamic Approach of Cathode Jet Acceleration 735
18.5.1 Basic Equations of Plasma Acceleration 735
18.5.2 Gasdynamic Mechanism. Energy Dissipation in Expanding Plasma 736
18.5.3 Jet Expansion with Sound Speed as Boundary Condition at the Cathode Side 739
18.6 Self-consistent Study of Plasma Expansion in the Spot and Jet Regions 744
18.6.1 Spot-Jet Transition and Plasma Flow 744
18.6.2 System of Equations for Cathode Plasma Flow 745
18.6.3 Plasma Jet and Boundary Conditions 747
18.7 Self-consistent Spot-Jet Plasma Expansion. Numerical Simulation 748
18.8 Anomalous Plasma Jet Acceleration in High-Current Pulse Arcs 753
18.8.1 State of the General Problem of Arcs with High Rate of Current Rise 753
18.8.2 Physical Model and Mathematical Formulation 753
18.8.3 Calculating Results 756
18.8.4 Commenting of the Results with Large dI/dt 759
18.9 Summary 760
References 763

19 Cathode Spot Motion in Magnetic Fields 769
19.1 Cathode Spot Motion in a Transverse Magnetic Field 769
19.1.1 Retrograde Motion. Review of the Theoretical Works 770
19.2 Spot Behavior and Impeded Plasma Flow Under Magnetic Pressure 795
19.2.1 The Physical Basics of Impeded Spot Plasma Flow and Magnetic Pressure Action 796
19.2.2 Mathematics of Current–Magnetic Field Interaction in the Cathode Spot 796

	19.3	Cathode Spot Grouping in a Magnetic Field	798
		19.3.1 The Physical Model and Mathematical Description of Spot Grouping	798
		19.3.2 Calculating Results of Spot Grouping	801
	19.4	Model of Retrograde Spot Motion. Physics and Mathematical Description	801
		19.4.1 Retrograde Spot Motion of a Vacuum Arc	802
		19.4.2 Retrograde Spot Motion in the Presence of a Surrounding Gas Pressure	804
		19.4.3 Calculating Results of Spot Motion in an Applied Magnetic Field	805
	19.5	Cathode Spot Motion in Oblique Magnetic Field. Acute Angle Effect	807
		19.5.1 Previous Hypotheses	807
		19.5.2 Physical and Mathematical Model of Spot Drift Due to the Acute Angle Effect	809
		19.5.3 Calculating Results of Spot Motion in an Oblique Magnetic Field	813
	19.6	Spot Splitting in an Oblique Magnetic Field	815
		19.6.1 The Current Per Spot Arising Under Oblique Magnetic Field	815
		19.6.2 Model of Spot Splitting	816
		19.6.3 Calculating Results of Spot Splitting	818
	19.7	Summary	819
	References		823
20	**Theoretical Study of Anode Spot. Evolution of the Anode Region Theory**		**829**
	20.1	Review of the Anode Region Theory	829
		20.1.1 Modeling of the Anode Spot at Early Period for Arcs at Atmosphere Pressure	830
		20.1.2 Anode Spot Formation in Vacuum Arcs	842
		20.1.3 State of Developed Models of Anode Spot in Low Pressure and Vacuum Arcs	855
		20.1.4 Summary of the Previous Anode Spot Models	860
	20.2	Anode Region Modeling. Kinetics of Anode Vaporization and Plasma Flow	862
	20.3	Overall Characterization of the Anode Spot and Anode Plasma Region	862
		20.3.1 Kinetic Model of the Anode Region in a Vacuum Arc	865
		20.3.2 Equations of Conservation	866
		20.3.3 System of Kinetic Equations for Anode Plasma Flow	867

20.4	System of Gasdynamic Equation for an Anode Spot in a Vacuum Arc	871
20.5	Numerical Investigation of Anode Spot Parameters	874
	20.5.1 Preliminary Analysis Based on Simple Approach. Copper Anode	875
	20.5.2 Extended Approach. Copper Anode	875
20.6	Extended Approach. Graphite Anode	880
	20.6.1 Concluding Remarks	885
References		886

Part IV Applications

21 Unipolar Arcs. Experimental and Theoretical Study ... 895
- 21.1 Experimental Study ... 895
  - 21.1.1 Unipolar Arcs on the Metal Elements in Fusion Devices ... 896
  - 21.1.2 Unipolar Arcs on a Nanostructured Surfaces in Tokamaks ... 903
  - 21.1.3 Film Cathode Spot in a Fusion Devices ... 908
- 21.2 Theoretical Study of Arcing Phenomena in a Fusion Devices ... 911
  - 21.2.1 First Ideas of Unipolar Current Continuity in an Arc ... 912
  - 21.2.2 Developed Mechanisms of Unipolar Arc Initiation ... 914
  - 21.2.3 The Role of Adjacent Plasma and Surface Relief in Spot Development of a Unipolar Arc ... 920
  - 21.2.4 Explosive Models of Unipolar Arcs in Fusion Devices ... 924
  - 21.2.5 Briefly Description of the Mechanism for Arcing at Film Cathode ... 925
- 21.3 Summary ... 926
- References ... 928

22 Vacuum Arc Plasma Sources. Thin Film Deposition ... 933
- 22.1 Brief Overview of the Deposition Techniques ... 934
  - 22.1.1 Advanced Techniques Used Extensive for Thin Film Deposition ... 934
  - 22.1.2 Vacuum Arc Deposition (VAD) ... 935
- 22.2 Arc Mode with Refractory Anode. Physical Phenomena ... 938
  - 22.2.1 Hot Refractory Anode Vacuum Arc (HRAVA) ... 938
  - 22.2.2 Vacuum Arc with Black Body Assembly (VABBA) ... 940
- 22.3 Experimental Setup. Methodology ... 940
- 22.4 Theory and Mechanism of an Arc with Refractory Anode. Mathematical Description ... 945

		22.4.1	Mathematical Formulation of the Thermal Model	946
		22.4.2	Incoming Heat Flux and Plasma Parameters	947
		22.4.3	Method of Solution and Results of Calculation	949
	22.5	Application of Arcs with Refractory Anode for Thin Films Coatings	952
		22.5.1	Deposition of Volatile Materials	952
		22.5.2	Deposition of Intermediate Materials	967
		22.5.3	Advances Deposition of Refractory Materials with Vacuum Arcs Refractory Anode	983
	22.6	Comparison of Vacuum Arc Deposition System with Other Deposition Systems	988
		22.6.1	Vacuum Arc Deposition System Compared to Non-arc Deposition Systems	988
		22.6.2	Comparison with Other Vacuum Arc-Based Deposition Systems	989
	22.7	Summary	992
	References	994

23 Vacuum-Arc Modeling with Respect to a Space Microthruster Application 1003

	23.1	General Problem and Main Characteristics of the Thruster Efficiency	1003
	23.2	Vacuum-Arc Plasma Characteristics as a Thrust Source	1005
	23.3	Microplasma Generation in a Microscale Short Vacuum Arc	1007
		23.3.1	Phenomena in Arcs with Small Electrode Gaps	1007
		23.3.2	Short Vacuum Arc Model	1008
		23.3.3	Calculation Results. Dependence on Gap Distance . . .	1010
		23.3.4	Calculation Results. Dependence on Cathode Potential Drop	1012
	23.4	Cathodic Vacuum Arc Study with Respect to a Plasma Thruster Application	1015
		23.4.1	Model and Assumptions	1016
		23.4.2	Simulation and Results	1017
	23.5	Summary	1020
	References	1023

24 Application of Cathode Spot Theory to Laser Metal Interaction and Laser Plasma Generation 1027

	24.1	Physics of Laser Plasma Generation	1027
	24.2	Review of Experimental Results	1028
	24.3	Overview of Theoretical Approaches of Laser–Target Interaction	1034
	24.4	Near-Target Phenomena by Moderate Power Laser Irradiation	1037

		24.4.1	Target Vaporization 1037
		24.4.2	Breakdown of Neutral Vapor................... 1038
		24.4.3	Near-Target Electrical Sheath 1039
		24.4.4	Electron Emission from Hot Area of the Target 1039
		24.4.5	Plasma Heating. Electron Temperature............ 1039
		24.4.6	Plasma Acceleration Mechanism 1040

24.5 Self-consistent Model and System of Equations of Laser Irradiation... 1042

24.6 Calculations of Plasma and Target Parameters 1046
    24.6.1 Results of Calculations for Copper Target 1046
    24.6.2 Results of Calculations for Silver Target 1052
    24.6.3 Calculation for Al, Ni, and Ti Targets and Comparison with the Experiment................ 1054

24.7 Feature of Laser Irradiation Converting into the Plasma Energy and Target Shielding 1056

24.8 Feature of Expanding Laser Plasma Flow and Jet Acceleration 1058

24.9 Summary 1061

References 1062

25 Application of Cathode Spot Theory for Arcs Formed in Technical Devices 1067

25.1 Electrode Problem at a Wall Under Plasma Flow. Hot Boundary Layer 1067

25.2 Hot Ceramic Electrodes. Overheating Instability 1068
    25.2.1 Transient Process 1068
    25.2.2 Stability of Arcing and Constriction Conditions 1069
    25.2.3 Double Layer Approximation 1070
    25.2.4 Thermal and Volt–Current Characteristics for an Electrode-Plasma System 1077

25.3 Current Constriction Regime. Arcing at Spot Mode Under Plasma Flow with a Dopant..................... 1080
    25.3.1 The Subject and Specific Condition of a Discharge 1080
    25.3.2 Physical Model of Cathode Spot Arising Under Dopant Plasma Flow 1082
    25.3.3 Specifics of System of Equations. Calculations...... 1083

25.4 Arc Column at Atmospheric Gas Pressure 1087
    25.4.1 Analysis of the Existing Mathematical Approaches Based on "Channel Model".................... 1087
    25.4.2 Mathematical Formulation Using Temperature Dependent Electrical Conductivity in the All Discharge Tube 1089
    25.4.3 Results of Calculations 1092

25.5	Discharges with "Anomalous" Electron Emission Current		1095
25.6	High-Current Arc Moving Between Parallel Electrodes. Rail Gun		1098
	25.6.1	Physics of High-Current Vacuum Arc in MPA. Plasma Properties	1099
	25.6.2	Model and System of Equations	1100
	25.6.3	Numerical results	1103
	25.6.4	Magneto-Plasma Acceleration of a Body. Equations and Calculation	1104
25.7	Summary		1107
References			1109

Conclusion 1113

Appendix: Constants of Metals Related to Cathode Materials Used in Vacuum Arcs 1117

Index 1123

Part III
Cathodic Arc. Theory

Chapter 15
Cathode Spot Theories. History and Evolution of the Mechanisms

An electrical discharge in vacuum or in low pressure ambient gas can be supported if enough charge particles are generated in the electrode gap. The main question is continuity of the electrical current from the metallic cathode to an electro-conductive media in the gap. The problem is the current transition from the high conductive metallic cathode to relatively low conductive gap plasma. According to the experiment, this transition proceeds by the strong plasma contraction at the cathode surface in the form of so-called cathode spot. However, the phenomena at the cathode surface-plasma spot are very complicated due to transition from phenomena of heat-mass transfer in solid body to these phenomena in the plasma including dependence on elementary processes. In order to understand the nature of this subject, numerous theoretical studies were conducted in period from the past century up to present time. Below an analysis of the spot models, beginning from the separate hypotheses of charge particles emission by the cathode up to the present state of the art is provided.

15.1 Primary Studies

15.1.1 First Hypotheses

Some of primary works were reviewed in [1–5]. The details will be discussed in the works, which only referenced or even not considered in order to report the main published results. The first attempt to understand the electrical discharge phenomena are related to the beginning of the past century provided by Stark. The electro-conductivity of positive column produced in discharges in different gases he was studied in 1901 [6]. The presence of a minimal arc voltage was studied by Stark in 1902 [7] considering the atom ionization processes in different gases. The role of cathode potential drop and the influence of the current and gas pressure were discussed in 1903 [8]. Also in 1903, Stark [9] investigated the cathode electron emission. It was noted that since the cathode potential drop in the spot is low, the electrical force cannot influence the electron emission from the cathode (cite: "so kann es nicht die direkte wirkung der elektrischen kraft sein" [9]). According to Stark, the electron emission occurred due to Richardson law by thermionic mechanism, who studied it for platinum and graphite [10]. Indicating that the cathode temperature in the spot can be between 2500 and 3500 K the estimations electron emission current density was obtained at significant values 10^8 and 10^9 A/cm^2 for graphite and platinum cathodes, respectively. This result seems to be doubly considering modern knowledge. Also, the thermionic mechanism of electron emission was proposed by Mitkevich in 1905 [11] using experiments with graphite cathode.

15.1.2 Cathode Positive Space Charge and Electron Emission by Electric Field

However, the arc appearance at different cold cathodes finds effects that cannot be explained in frame thermionic electron emission. For example, it was not understandable the transient spots behavior at cold cathodes, the arcs at mercury cathode, and the transition from an arc with hot W to the arc with cold tungsten cathode, which was detailed subsequently by Finkelburg and Maecker [12] and by Tikhodeev [13]. Therefore, the further study of the cathode spot was continued with searches of the electron extraction mechanism from cold metals. Langmuir in 1913 [14], 1923 [15], 1929 [16] developed theory of a positive space charge, which produced by the ions concentrated at the cathode surface. This positive space charge causes the cathode potential drop in a sheath, which may exert an electric field at the cathode surface described by Poisson's equation. Langmuir [17] has suggested that the electric field can be strong enough to cause a large electron field-current from the cathode that later was named as field emission, F-emission.

Compton in 1927 [18] has developed this idea further showing that the "electrons from the cathode were supplemented by a 'pulling out' of electrons by the electric field which is concentrated at the cathode surface." His analysis based on an energy balance considerations assuming that in the mercury arc the electrons emanating from the cathode due to a high electric field and not to thermal emission. As Compton 1923 [19] noted, the field to electron extraction from the cathode should be sufficiently large and it is equivalent to a reduction in the ordinary value of the metal work function. The experiment showed that the electron extraction at fields lower than 10^6 V/cm was relatively small. However, as Compton was noted [19], taking into account the high temperature of the cathode, the "electron cloud should extend farther from the cathode surface" an effect of the sort suggested may be appreciable. This means the mechanism of thermal and field or T-F emission of the electrons (see Chap. 2).

Further, Compton [20] extended the study of heat balance at the mercury cathode by the introduction of the accommodation coefficient for neutralized ions and by evaluation of the energy gained by electrons in the cathode fall space and due to neutralization of the positive ions at the surface. It was indicated that existing experimental data shows that the Langmuir's theory of extraction of electrons by the field at a Hg arc cathode can consistent with the cathode heat balance, but this result cannot be uniquely proved by them due to of uncertainty in the order of magnitude of the energy fraction dissipated by different plasma particles. It was obtained that the temperature of mercury cathode spot does not exceed 200 °C using the relationship between the vapor pressure and temperature and the net rate of evaporation.

The electric field at the surface of the cathode for any given current density of electrons and positive ions, and for a given cathode drop was determined by Mackeown [21] using Poisson s equation (see Chap. 5). The calculated model developed by an assumption that the total cathode drop occurs in a distance less than one mean free path from the cathode. The solution was analyzed for mercury arc experiment with the current density of 4000 A/cm^2 and cathode drop of 10 V. The electric field at the surface of the cathode was determined as 5×10^5 V/cm if the current carried by positive ions was given as 5% of the total current at the cathode. There should be also mentioned in the study of the cathode space charge layer conducted by Ecker [22] (see Chap. 5).

The Langmuir theory of field emission also meets difficulties when the relatively low-current densities were measured in the cathode region. This theory periodically was considered again as the measurement techniques were improved indicating higher spot current density [2]. It should be noted that this is the case even up to date. Alternative mechanisms of electron emission were developed also to explain the new measurements of growth of current densities [23, 24]. As mentioned above, the difficulties appeared also at explanation the current continuity in an arc with mercury cathode. The last fact stimulated to develop some models of the cathode phenomena, which take in account separate processes. Let us consider these mechanisms consequently.

15.1.3 Mechanism of Thermal Plasma and Ion Current Formation

A thermal model of cathode spot was developed by Slepian [25], Slepian and Haverstick [26], Weizel et al. [27], and by Slepian et al. [28]. It was assumed that the current continuity is maintained due to just of ion flux and ion neutralization by the electrons at cathode surface. This model assumed that due to relatively low temperature of the cathode at copper and mercury so that the thermionic emission from the cathode is not essential. It was suggested to take in account thermal ionization of atoms in the gap obtained in accordance with Saha's equation to the formation of the ion current to the cathode. In this case, it is no longer necessary for the cathode to have a large electron emission or to be heated to a higher temperature than its boiling point. The respective current density as dependence on plasma temperature was presented in Fig. 15.1. The calculations were conducted for thermodynamic equilibrium plasma,

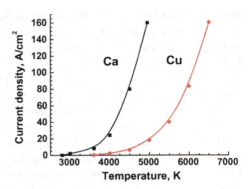

Fig. 15.1 Random ion current density to the cathode as function on temperature of the thermally ionized vapor [25]

and according to Langmuir the current density was determined by random ion flux to the cathode surface.

It can be seen that for equilibrium plasma the current density is small and not exceed 160 A/cm^2. The authors noted that in order to base their approach the ionized vapor was in an electric field and therefore can be deviated from thermodynamic equilibrium. Since the electrons take energy from the electric field and maintain themselves with a large energy. Thus, according to Langmuir in a mercury arc the electrons have a temperature of over 10^4 K. This result used to conclude that for a gas in an intense electric field degree of ionization will be expected much larger.

It can be noted that the Slepian's model yield not only limited values of current densities, but also arise the main question which was avoided in the work: it was not determined the mechanism of plasma heating in order to reach large temperatures and enough degree of ionization. In essence, it was arbitrary stated, that the vapor was heated by the arc current at low voltage in a strongly collisional quasineutral ionized gas at the cathode.

Later Bauer [29, 30] considered the thermal mechanism of vapor ionization by analysis of the cathode energy balance in a simple form using the experimental data of spot current density. It was assumed that the electron energy gained in the cathode space sheath is equal to the energy requested for atom ionization. As a result, the ratio electron to ion current densities was calculated of 1.5. Using this value, the sizes of space charge sheath, ionization regions, and vapor pressure were obtained. It was showed that the returned atom flux is 90% of the total flux vaporized from the cathode. It should be noted that the obtained results used primitive assumption using given arbitrary parameters, and therefore they can be considered only as an estimation of the spot parameters.

15.1.4 Mechanism of "Hot Electrons" in a Metallic Cathode

Effect of overheating of the electrons in a metal was considered previously by Ginsburg and Schabansky [31], and by Kaganov et al. [32]. Smith [33] developed a theory

15.1 Primary Studies

of a Hg arc cathode spot considering bombardment of the cathode by the energetic electrons from "tail" of electron Maxwellian distribution near the surface. It was intended to include an efficient transfer of energy to conduction electrons in the liquid mercury, raising the temperature of these electrons to relatively high values ("hot electrons"). This effect causes a thermionic electron emission great enough to yield the observed arc current density (of 4000 A/cm^2), while the atom remained at a very much lower temperature of the liquid of near 150 °C. It was claimed that non-equilibrium between electron and ion energy distributions inside of the metallic cathode takes place, and this effect was characterized by the hot electrons with temperature T_{ec}. To a quantitative analyze, Smith [33] considered a mathematical system including the equation of total current density j_0:

$$j_{em} = j_0 + j_{ret} \qquad (15.1)$$

where j_{em} is the electron emission current density determined by electron temperature in the metal T_{ec} and j_{ret} is the returned current density of the energetic electrons from "tail" of electron Maxwellian distribution with temperature T_{ep}. This current determined by a cathode charge space layer (sheath) with potential drop u_c.

Smith [33] used the cathode energy balance assuming the cathode heated by the energy flux from the plasma electrons and cooled by energy loss due to the electron emission from the cathode in the following form:

$$j_{ret} 2 k_B T_{ep} + j_{ret} \varphi = j_{em} \varphi + j_{em} 2 k_B T_{ec} \qquad (15.2)$$

where φ is the work function and k_B is the Boltzmann constant. Using (15.1) and (15.2) can be converted to:

$$j_{ret} 2 k_B T_{ep} = j_0 \varphi + j_0 2 k_B T_{ec} + j_{ret} 2 k_B T_{ec} \qquad (15.3)$$

Equation (15.3) was analyzed by arbitrary assumption that $T_{ec} = 4000$ K and for $\varphi = 4.26$ eV the relation was obtained in form:

$$\frac{j_0}{j_{ret}} = 3.495 \times 10^{-5} T_{ep} - 0.131 \qquad (15.4)$$

One (15.4) consists of 3 unknown. It was then concluded that in order to reach the condition of $j_{et} = j_0$, the plasma electrons should be about 4×10^4 K. Naturally, this conclusion is arbitrary and of course cannot characterize the real spot mechanism.

The above model of "hot electrons" was updated in the works of [34, 35]. It was proposed that the emitted from the cathode electrons accelerated in cathode sheath with potential drop of u_c were randomized in the near-cathode region with temperature of u_c and the discharge current density is

$$j_0 = en_e\sqrt{\frac{2eu_c}{m_e}} \tag{15.5}$$

The electron current density of the returned electrons to the cathode was depend on coefficient α and given as

$$j_{\text{ret}} = \alpha en_e\sqrt{\frac{2eu_c}{m_e}} \tag{15.6}$$

The cathode energy balance:

$$j_{\text{ret}}(2u_c + \varphi) = j_{\text{em}}(2k_B T_{\text{ec}} + \varphi) + \frac{\lambda}{l_e}(T_{\text{ec}} + T) \tag{15.7}$$

The last term is the energy loss due to cathode heat conduction with characteristic length l_e and coefficient of heat conductivity λ. The thermionic electron emission was calculated taking into account the effect of a temperature gradient [36]. It should be noted that this effect can be influenced only at heat flux density larger than 10^9 W/cm^2. This condition is far larger than that for mercury cathodic arc. As it can be seen, the number of unknown exceeds the number of equations. Therefore, two other parameters were introduced by some characterization of the ratios of j_{ret}/j_0 and $j_{\text{ret}}/j_{\text{em}}$. These ratios were given arbitrary for calculation j_0, n_e and different terms of the cathode energy balance as a function of the electron temperature T_{ec} in the metal, which varied again as given parameter for $u_c = 10$ V, $\varphi = 4.5$ eV. It is obvious that the calculated results are have not physical sense because they determined by the given arbitrary parameters. The other problem that cannot be understood is the assumption that no was taken energy loss during electron beam relaxation and that the plasma electron temperature equal to eu_c/k_B. Also, the model of "hot electrons" assumed that the electron current is significantly larger than that of the ion. This assumption meets a contradiction due to quasineutrality of the considered plasma. Application of this model to usual cathode materials is difficult due to their equilibrium state during the heating at measured current density, when the energy relaxation time in the metal will be lower than the arc pulse time. In this case, the mass loss and other processes depended on the metal temperature should be taken in account.

Thus, unrealistic assumptions, using arbitrary parameters, neglecting or not correct consideration of energy losses due to metal heat conduction, evaporation, and radiation as well as other problems indicate that the model of "hot electrons" cannot be accepted for description of the cathode spot phenomena. Finally, it should be noted that the relaxation time between electrons and lattice is very short (from ps to ns region) [37] and the mechanism of non-equilibrium metal heating with high power heat flux still not clear and the possibility of an experimental study of this effect was discussed in [38].

15.1.5 Mechanism of Local Explosion of a Site at the Cathode Surface

Another group of work based on an assumption that current density in the cathode spot can be so much large causing the explosive plasma generation (ions and electrons) due to Joule heating in a local cathode volume. One of the first Nekrashevich and Bakuto [39, 40] developed an approach, named a "migration theory." According to this model, the metal under the spot was overheated, and then explosive removal of the cathode material, current stopping and the spot operation fast shifted at a new area. According to this model, the spot is consequence of explosive migration, which can be after the arcing observed as number of erosion craters and traces.

Il'in and Lebedev [41] studied the Joule energy dissipation in the cathode showing that this heating can explain the cathode material removal at spot area. The model was based on works reported about spot current densities of 5×10^6–10^7 A/cm^2 for which the experimental data of spot lifetime was measured. The authors [41] assumed that at the end of spot life a vanishing of the heat conductivity for cathode material takes place and this effect associated with appearing of a "hot spot." As a result, a condition for metal explosion and formation of a new spot on the neighboring area was occurred. A dispersed layer at the place arises at the place of hot spot exploding, which forms a flare with relatively large velocity. As example, the removal mass calculated according to the developed thermal model was 0.55 μg, while the mass of 0.98 μg for iron cathode at time 450 μs and current 50 A indicating good agreement. There should be mentioned the work of Rich [42] where was shown that Joule heating in the arc cathode spot can be important at current density larger than 10^7 A/cm^2.

Rotshtein [43] suggested a model of "dense vapor cathode." He assumed that a region possibly 10 μm thick of very dense metallic vapor exists immediately adjacent the cathode surface in the spot. The high density perturbs the atomic fields so that the normally sharp energy levels were spread into the conduction zone. Metallic conduction is then possible from the cathode to this region, which is at a sufficiently high temperature to emit thermionically into the plasma. The ions bombarding the cathode serve to maintain the high local density. As evidence in favor of the conduction theory, Rotshtein indicated Smith's experiment [44] which observed a continuous spectrum originated within 10 μm from the cathode surface. Similarly to the above-described model, Rotshtein [45] considered an analogy between wire explosion and the spot as exploding conductor modeling the spot plasma generation as local metal explosion in the spot area. In essence, the problem reduced to the problem of metal state that discussed by Loeb in 1939 [46].

15.1.6 Emission Mechanism Due to Cathode Surface Interaction with Excited Atoms

Considering the above arc cathode theories an important conclusion was presented by Robson and Engel [47]. They noted, "no satisfactory theory exists to account for the current transfer at the cathode of the material having a low boiling point, such as mercury or copper". Whereas thermionic emission can account for the mechanism of arcs with cathodes of high boiling point, such as carbon and tungsten, this seems highly improbable for substances, which boil far below their emitting temperature. Field emission of electrons [17], current transfer by positive ions only [25], and other suggestions [33, 43] have been made; but they cannot account for both the observed high-current densities of order 10^6 A/cm^2 and the low cathode fall, which can be smaller than the ionization potential of the cathode material [48]."

Therefore, Engel and Robson [49] developed another theoretical model taking into account an excited atom flow to the cathode. According to this theory, the radiation from the excited atoms diffuses toward the cathode. The diffusion mechanism was determined by absorption and re-emission in the vapor and was ultimately absorbed by atoms, which finally strike the cathode. The electrons ejected from the cathode by the energy impact of excited atoms. The cathode potential drop in the space charge accelerated these electrons in space charge layer and new excited atoms in the dense vapor produced by their collisions with the neutrals. Positive ions are formed in the vapor by collisions between excited atoms, and by ionization due to stepwise processes of electron colliding with the atoms. The positive ions accelerated in space charge layer and supply energy for cathode heating and evaporation.

The model was developed for mercury arc. It was assumed that in the vicinity of the cathode surface the vapor density was many orders of magnitude higher than that far from the cathode because of the intense local evaporation. An electron ejected from the cathode and accelerated across the space charge layer with potential greater than 5 V will have sufficient energy to excite a Hg atom into the resonance state of 4.9 eV. Since the ionization energy of mercury is 10.4 eV, electrons, which have enough energy to excite atoms to the resonance state will also be capable to ionize the excited atoms according to the reaction:

$$Hg^*(4.9\,eV) + e(5.5\,eV) > Hg^+(10.4\,eV) + 2e. \tag{15.8}$$

At another hand, when relative high density of excited atoms is present, the collisions between them result in the formation of molecular ions according to the reaction:

$$Hg^*(4.9\,eV) + Hg^*(4.9\,eV) > Hg_2^+(9.65\,eV) + e(0.3\,eV) \tag{15.9}$$

To understand the energetic characteristic of the model, the energy balances for electrons and ions were studied. The electron energy balance:

15.1 Primary Studies

$$f(u_c + V^* - \varphi)j = F^*\text{eV}^* + j(1-\varphi)u_i + jR + j\varepsilon \quad (15.10)$$

where f is the electron fraction of the current j at the cathode F^* is the photon flux to the cathode and u_i is the ionization potential. R is the energy lost by radiation, and ε is the mean energy of the electrons, which carry by the electron.

The ion energy balance at the cathode surface by the de-excitation of excited atoms, which do not release electrons, by the neutralization and kinetic energy of positive ions:

$$(1-f)\text{eV}^*F^* + j(l-f)(u_c + u_i - \varphi) = j(C + P) \quad (15.11)$$

where C is the lost from the cathode heat conduction and P is the energy lost by evaporation. Taking in account definition of the yield in electrons/excited atom η it was assumed that

$$\eta e F^* = f_i. \quad (15.12)$$

Considering (15.10)–(15.12) a relation between η and f_i was obtained that was used to base the proposed model of excited atom flow to the cathode. It was noted that the mercury arc is the best illustration of the proposed mechanism, since all the excited states of the Hg atom have energies greater than the work function of this metal, and are therefore capable of electron emission. This model, however, consists also of arbitrary assumptions (regarding to determination of C, P, F^* and cathode temperature) that cannot base the above conclusion. Also according to Lee [50] the "difficulty with this theory is that one has to postulate a η of unity which seems to be too high."

The idea of Engel and Robson [49] taking into account the excited atoms in the cathode spot theory was further developed by Holmes [51]. He suggested mechanisms of electron emission by excited atom bombardment, as well by thermionic field emission. The relative role of these two processes was studied. It was noted that the proposed approach based on the research for transition to mercury arc spot from a stage of glow discharge published little later in [52]. According to the model, the atom flow with density N in the direction to the cathode was described by a diffusion equation in the region of cathode potential drop h. The atom flux to the cathode was determined by the momentum transfer from ions to neutral atoms by ion–atom collisions and the diffusion in every element of volume. The current density in region h was a sum of ion j_i and electron j_e with a fraction of $s = j_i/j_e$. Considering these definitions, the diffusion equation was solved and a dependence of atom density on distance x from plasma to the cathode was obtained. A Poisson equation using the relation between ion current density and the ion velocity determined by the expression of ion mobility depended on the atom density $N(x)$. Therefore, the cathode electric field obtained also as dependence on $N(x)$. Arbitrary was assumed that the atom density is determined by the saturated vapor pressure at the temperature T of the cathode. The electrons excite the atoms and ionize it by collisions with already excited atoms. This step ionization makes the major contribution to the total ionization, since normal ionization by direct electron impact is not effective due to the low voltage of the arc. The photons from the excited atoms diffuse to the cathode transferring the energy to it.

The energy balance was studied in two regions including the quasineutral plasma outside the spot itself, and the cathode fall region. There it was assumed equal flow of energy from one region to the other. The input flux is represented by the bombardment of the surface volume element by excited atoms, ions, and photons. The excited atoms use only part of their total energy in heating the cathode, the remainder being expended in extracting electrons from the cathode. The ions and photons expend almost all their energy in heating the cathode. Also, conduction will be neglected in this analysis. In addition to thermal dissipation in the cathode, there is an energy flux loss from the cathode caused by the cooling effect of the T-F emission current. The energy flux losses from the cathode caused by evaporation and by cathode heat conduction were neglected.

Two addition equations were derived from the following conditions. The first assumed that each electron emitted from the cathode must produce sufficient excited atoms and such electric field that one new electron should be ejected from the cathode. The second assumed that the flux of these atoms during the production of excited atoms could not exceed the total flux of neutral atoms, which impinge on the cathode. It was noted that as the motion of the arc was dependent on the speed of its component atoms, the spot velocity was assumed equal to the thermal velocity of the atoms within the cathode fall.

The model and the obtained results cannot be considered as realistic. The above-mentioned additional limiting conditions are arbitrary assumed. Firstly, no basic was that at the cathode that will produce a requested density of excited atoms and moreover the requested field. Secondly, it cannot be based on that every atom returned to the cathode was excited because of the very high-current density. No experiment confirmed that the spot velocity is equal to the thermal velocity of the atoms. The coefficient of electron emission by the excited atoms γ was given as arbitrarily large. Another problem is that the possibility of particle diffusion in very small thickness of space charge layer with cathode potential drop, specially, for current densities in the range of 10^6–10^7 A/cm^2. This problem was not only not justified, but also even not discussed in the frame of data from previously published works.

15.2 Mathematical Description Using Systematical Approaches

The absence of a possibility to describe the cathode mechanism by modeling separate processes stimulated an attempt to detailed analysis of a number of the cathode processes. The models developed taking into account a combination of different equations to determine main parameters of the spot are considered below.

15.2.1 The First Models Using System of Equations

Lee [53] presented a calculation using an equation of cathode T-F electron emission in general form (Chap. 2) for Cu and Ag taking into account the equation of electric field E_c [21] with cathode potential drop of 10 V. The results were presented as dependencies of electrical field on the varied electron current fraction s with current density j and the cathode temperature given as parameter. Other calculations were presented as dependencies of current density on varied electrical field with the cathode temperature given as parameter. The energy components were analyzed considering the input power per unit area carried to the cathode by the positive ions and energy losses by cathode vaporization and electron emission. It was concluded that positive ions should carry between 20 and 50% of the total current to the cathode in case when the cathode spot current density is between 10^5 and 10^6 A/cm^2. Lee concluded that for a copper arc and typical values of $j = 10^6$ A/cm^2, $u_c = 10$ V, $s = 0.5$, at a temperature of 3500°K, the field $E = 2 \times 10^7$ V/cm is sufficient to produce an electron current density (T-F electron emission) of 10^6 A/cm^2 without the use of a roughness factor. The above works were developed in the next more complicated model.

Lee and Greenwood in 1961 [54] published a comprehensive description of the cathode spot of the vacuum arc in a basic work, which is the first analysis of common cathode processes based on cathode evaporation model. Ecker in 1961 [1] also summarized the works used cathode evaporation models. Lee and Greenwood formulated a system of equations describing the cathode heating and the cathode electron emission. Let us consider their approach. The five dependent variables include (1) temperature T_s of the cathode spot, (2) electric field E at the cathode surface, (3) current density, (4) electrons current fraction s, and (5) radius of the circular spot r_s. These variables have been treated as constants over the spot area. Four equations were used:

(1) Mackeown's equation [21] determined a relation between E, j, and s at the cathode.

$$E = \left\{ \frac{4}{\varepsilon_0} j(u_c)^{0.5} \left[(1-s)\left(\frac{m_i}{2e}\right)^{0.5} - s\left(\frac{m_e}{2e}\right)^{0.5} \right] \right\}^{0.5} \quad (15.13)$$

(2) The current density of electrons emitted from the cathode is given (Sect. 2.2) by

$$sj = e \int_{-\infty}^{\infty} N_x(\varepsilon_e, T) D(\varepsilon_e, E) d\varepsilon_e \quad (15.14)$$

where $D(E,\varepsilon_e)$ is the probability that an electron with energy ε_e will emerge from the metal, $N(\varepsilon_e, T)$ is the supply function.

(3) The energy balance equation is written as

$$P_i = P_c + P_e + P_v + P_r \qquad (15.15)$$

$$P_i = (1-s)j(u_c + u_i - \varphi)$$

where P_i is the power input to the cathode spot by ion kinetic and potential energy, and the terms on the right are the various sinks for dissipating this energy. P_v is the power dissipated by vaporization determined Dushman equation [55], P_e is the power dissipated by the cooling effect of electron emission, is a complex function of E, φ, and T_s. To determine this power, the results of Lee [56] were used. P_r is the power dissipated by radiation. P_c is the power dissipated by thermal conduction (λ_T is thermal conductivity) into the cathode, which was determined by steady-state solution of heat conduction equation with temperature in the spot center as.

$$P_c = \lambda_T \sqrt{\pi} \frac{T_s - T_0}{r_s} \qquad (15.16)$$

(4) Assuming of uniform distribution of the current density in the cathode spot, the following equation of total spot current I was used:

$$I_i = j\pi r_s^2 \qquad (15.17)$$

Two limiting conditions were used to determine possible values for the fifth unknown. One f they is criterion for atom–ion balance, i.e., the ion flux cannot exceed the evaporated atom flux Γ_a and was written as

$$\frac{1-s}{e} j \leq \frac{\Gamma}{m_i} \quad \text{or}$$

$$s \geq 1 - \frac{e\Gamma}{jm_i} \qquad (15.18)$$

The second limiting condition was on the magneto-hydrodynamic (MHD) equation. Assuming that the vapor pressure is p and radial plasma expansion is absent this equation is

$$\frac{dp}{dr} = (j \times B)_r \qquad (15.19)$$

Taking into account zero ambient pressure an integration of (15.19) gives:

$$p_{\max} = 10^{-8} I j \qquad (15.20)$$

The dependence of electron current fraction s on varied cathode temperature is presented in Fig. 15.2. This figure demonstrates the common solution (pointed by

15.2 Mathematical Description Using Systematical Approaches

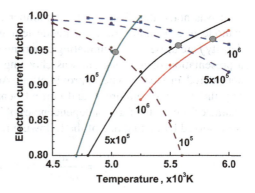

Fig. 15.2 Electron current fraction as dependence on cathode temperature: solid lines are the solution obtained from system (15.13)–(15.15), (15.17); dotted lines are the solution obtained from condition of (15.18) with current density given as parameter [54]

an arrow) obtained from system (15.13)–(15.15), (15.17), and condition of (15.18) with current density given as parameter.

A minimum current of the order of 2–10 A that can be supported by the spot processes was calculated. The current density obtained for Cu is 10^5–10^6 A/cm^2 with $s = 0.5$–0.7. The authors indicated that for the cathode material like copper, the atom–ion balance more limiting factor than the MHD equation. The physical reason for this is that in the case of copper, sufficient ion flux to the cathode is requested due to the T-F mechanism for which a strong electric field is needed. For materials like molybdenum and tungsten, the spot temperature may be so high that the electrons can be emitted thermally. In this case, a very high-current density that apparently necessary to satisfy energy balance at the cathode spot on a refractory material will increase the magnetic pinch force, increasing the importance of the MHD criterion. To this end, a question is appeared: since to reach the high-current densities the ion flux should be also important taking into account balance (15.15), then why the atom–ion balance is unimportant. The main limitations of this work consist of absent any mechanism of ion flux formation, the approach being steady state, and the system of equations being not closed and the solutions are multiple-valued.

Another mathematical model used the limiting conditions were published in a series of works by Ecker in 60–70 years of the past century (see references below). At this time, he noted that the exact information about physical processes was absent, and moreover, the parameters cannot be with high precision obtained for such mysterious subjects like cathode spot of an electrical arc. As one of the examples, Ecker indicated that the ion current from the plasma to the surface of the electrode is uncertain due to the lack of knowledge about the radial losses and the uncertainty of the electron temperature, which in turn is a consequence of the uncertainty of an energy balance in the plasma. Taking into account the mentioned difficulties and uncertainties, he suggested to determine the important spot parameters in some existing numerical ranges. To find the ranges, Ecker suggests using limiting relationships that limited a region of possible variation of the spot parameters, i.e., a region of limited existing spot characteristics. Such approach named "existence diagram" or "E-diagram" was in first used to study an effect of discharge contraction at the electrodes [1].

It was noted that from the large number of dependent variables describing the cathode in a vacuum arc, the cathode temperature in the spot T and the current density j are those of predominating experimental and theoretical interest. Therefore, a presentation of the diagram was choosing in a T-j plane. The spot model was considered in steady-state approximation. According to Ecker's study 1971 [57, 58], the spot temperature T and current density j were limited by three limiting characteristics in the form of dependencies of T on j. To this end, (15.13) and (15.14) were solved by adding one of the following relations:

$$\frac{j_i}{j} = \frac{u_c}{u_c + u_i} \quad or \quad j_i = (1-s)j \tag{15.21}$$

$$j_i = eW(T) \tag{15.22}$$

$$j\left\{(u_c + u_i)\frac{j_i(j,T)}{j} - \left[\varphi - 0.32\sqrt{\frac{Ij}{\pi}}\rho(T)\right]\right\} = \lambda_s W(T) + T\lambda_T(T)\frac{j\pi}{I} \tag{15.23}$$

where s is the electron current fractions, φ is the work fraction, and $W(T)$ is the rate of cathode evaporation in g/cm^2/s according to Langmuir–Dushman equation [55]. The first relation (15.21) indicates that the ion flux from the plasma is determined by the number of ions produced in the plasma in case, when all energy of the emitted electrons acquired in the sheath is spent on the vapor ionization. This means that calculated by (15.21) ion current density is maximal and this relation determined the low limit of the temperature (T_{cc}). The second relation (15.22) is, in essence, the atom–ion balance used already in [54] and it determines the upper limit of the ion current density (T_{ec}). According to Ecker [57, 58] the third relation (15.23) is the cathode energy balance that used in a simple form, and the calculated temperature (T_{nc}) is the upper limit of its values. It is due to the used power loss in form (15.16) was determined by steady-state solution of heat conduction equation with maximal temperature in the spot center and by the energy accommodation coefficient used equal to unit. The "E-diagram" calculated in the above-mentioned work [57], which turned out very small and that for Cu cathode can be seen in Fig. 15.3. Figure 15.4 shows that two E-areas were fund solutions denoted as 0-mode with low values of T-j and 1-mode with high values of T-j. The second mode was produced due to the presence of the term with resistive heating in cathode energy balance. The area of 0-mode significantly reduced when the spot current I decreased and approaches to a point at $I = 30$ A. Ecker 1971–1973 [59, 60] presented the mathematical formulation and calculations for cathode materials with a wide range of thermophysical properties (Cd, Zn, Bi, Sb, Ag, Sn, Cu, Cr, Ni, Fe, Ta, Mo, and W).

It should be noted that here the small area in "E-diagrams" was obtained erroneously and which was detected during analysis of spot phenomena in 1975 [61]. This error Ecker [62] then explained by "an error in the evolution of T that had occurred because in the relation $j_i u_i = j_e u_c$ the subscripts had been confused; thus, in previous

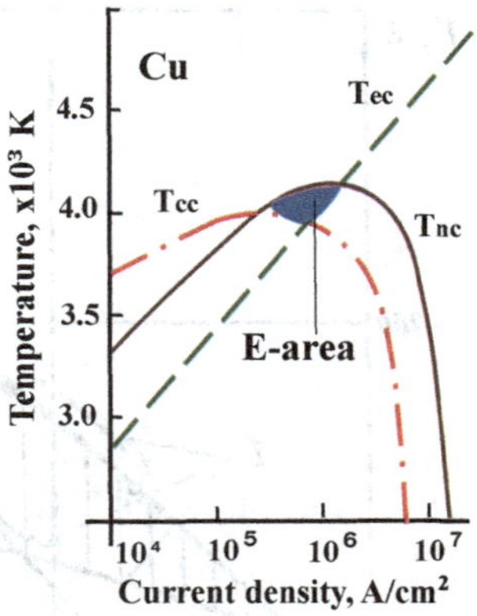

15.2 Mathematical Description Using Systematical Approaches

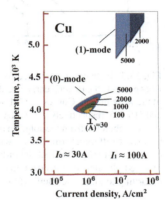

Fig. 15.3 Example of the existence diagram for the vacuum arc copper cathode. T temperature, j current density in the spot at the cathode surface [57]

Fig. 15.4 Existence areas of the 0-mode and the 1-mode shown as a function of the spot current. I_0 is the minimal current [58]

evolutions, we had used erroneously the relation $j_e u_i = j_i u_c$." The future Ecker's 1976–1978 [63, 64] similar investigations calculations were corrected, however, the principal point of view commented in [61] to the method of "E-diagram" remained. Therefore, let us discuss the main limitations and interpretation of the results of this method. Figure 15.5 shows the results of previous Ecker's calculation [57] denoted by dotted lines a small region (thickly hatched), and for comparison, the real solution with a wide region (bounding with solid lines) obtained using corrected equations but for a different limiting form of cathode balance:

$$j\left[(u_c + u_i)\frac{j_i(j,T)}{j} - \varphi_{ef}\right] = \lambda_s G \frac{j}{I} + 2\lambda_T(T-300)\left(\frac{j}{I}\right)^{1/2} \quad (15.24)$$

It can be seen that the curve I is close to the curve IA obtained using Lee and Greenwood data [54] while these both curves significantly different from the dotted (Ecker's) curve extending the area of "E-diagram." The dependences II are weakly

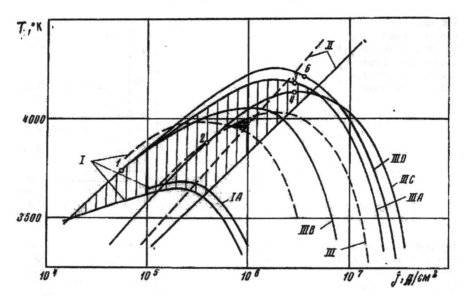

Fig. 15.5 E-diagram calculated for Cu cathode at $I = 200$ A. Thickly hatched region with dotted lines presents the Ecker's early calculations. Rare hatched region with solid lines presents calculation of the work in [61]. Curves I and II present the solutions of (15.13) and (15.14) together with (15.21) and (15.22), respectively. Curve IA was obtained using the corresponded result from [54]. Curves III were obtained from the solution of (15.13) and (15.14) together with relations for cathode energy balance in different forms: IIIA—with (15.24); IIIB—with (15.24), but $\varphi_{ef} = \varphi$; IIIC—with (15.24), but $\varphi_{ef} = \varphi - (e^3 E)^{1/2}$ (effect Schottky); IIID—with (15.24), but taking into account the energy by returned plasma electrons back to the cathode from [65]. The points (∘) indicate the exact solution according to the model Beilis [65] which will be described in a chapter below (see [61])

different. The cathode energy balance (15.23) is similar to that used in [54]. Therefore, now the minimal current should be this same as obtained from the solution [54] and it should be 2–10 A instead 30 A. However, even these values cannot be accepted because at such currents can be observed in transient spots. In additional, the used (15.23) is also not correct. Comparison of (15.23) and (15.24) indicate that in (15.23) was not taken in account effect Schottky and the rate of cathode mass evaporation was described by Langmuir–Dushman equation $W(T)$ that do not consider the returned mass flux in the kinetic layer near the cathode (see Chaps. 2 and 17). The calculations show that $W(T)$ significantly different from the measured cathode erosion rate G, which was taken in account in (15.24).

Comparison of the curves III in Fig. 15.5 calculated for different forms of cathode energy balance shows that these curves not only different, but it is also not possible to indicate how the corresponded dependence can be changed. This means that it is impossible to indicate how the E-diagram will be changed using an improved limiting relation (15.23). An example can be also illustrated by curve IIID that take in account the energy by returned plasma electrons back to the cathode calculated in [65]. It should be noted that that the last improving demonstration can stimulate

15.2 Mathematical Description Using Systematical Approaches

to find a more accurate form of the energy balance for E-diagram. But in this case, the solution will be presented at the curve III limited by curves I and II and therefore such approach contradicts to the idea of E-diagram.

In frame of mentioned above limitations of E-diagram (especially related to the curve III), it is doubtful to have positive opinion regarding seeking an achievement in the later Ecker's investigation [62–64, 66]. The developed method took in account the non-stationary of spot operation, the spot velocity and spot appearing at a roughness cathode surface. The non-stationary spot and the spot velocity were accounted by a local spot lifetime using a term in (15.23) describing the non-stationary heat conduction equation in 1973 [67]. It was obtained that the results of calculations for moving spot are quite close to the localized stationary spot and it was noted that is because the residence time is about the same as the thermalization time. The E-diagram for spot on roughness cathode surface was described using to calculate some arbitrary coefficient of local enhancement of the electric field. The calculations in all above-referenced works were conducted for constant $u_c = 15$ V.

An attempt to study the influence of the cathode potential drop Ecker [66] conducted in, so-called, unified analysis of the method, taking into account that the sheath voltage consists of the potential drop in Langmuir layer l_{sh} and in a quasineutral transition region (see Chap. 5). The potential drop in Langmuir layer was determined by law of "3/2", and in transition region u_1 from the following limiting condition (note, difficult to understand it):

$$j_e u_1(l_{sp}) \geq j_{eT} u_i$$
$$j = \sigma_{el} E_0 \qquad (15.25)$$

where j_{eT} is the returned current of plasma electrons to the cathode depended on electron temperature T_e, E_0 is the plasma electric field, σ_{el} is the plasma electrical conductivity, and length l_{sp} is the difference between sheath length and length of the plasma transition region. Using additional assumptions (including l_{sh} equal ion mean free path) in an ionization region, a resulting potential (depended on the potential drop) as a function of the parameter l_{sp} was calculated. This dependence passed through a minimum at $l_{sp} = 1$. This minimal value was used as the cathode potential drop u_c. Figure 15.6 presents u_c as dependence on varying T and j assuming $T_e = 10^4$ K and using stationary E-diagram. It was indicated that according to Fig. 15.6 the value of u_c changed weakly inside of the E-area and it is in range of values used in previous calculations [66]. Similarly, as of value u_c, it were calculated also dependencies on T_e and on the ion current fractions.

However, none of the obtained results can be interpreted as real characteristics because many assumptions are not justified and some not understandable assumptions were taken and the comparison of these dependences was provided with small areas from old not corrected E-diagram. Here, we should note that assumption of the ion mean free pass equal to length of Langmuir space charge layer was also not corrected. This assumption was used as a condition to calculate the cathode erosion rate instead uses its experimental value in 1973 [68]. However, our calculations using the model

Fig. 15.6 Cathode potential drop as function of the spot temperature and the current density for the example of the Cu vacuum arc. Shaded areas are existence region for 0- and 1-modes calculated in work [58, 59]

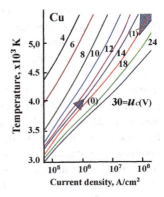

of [66], show an unrealistic result, indicating how critical is to choose the relations between different lengths in the cathode electrical sheath in order to obtain a solution of system of equation describing spot parameters.

An important question is validity of the Langmuir–Dushman equation [55] for $W(T)$ in order to calculate the cathode mass loss or erosion rate. Even not going into details, as a large particle flux is returned to the cathode, at least by the ions, this equation cannot be justified. In general, this question was discussed in [61] and previously [69] using balance of neutrals near the surface. Our detail investigations, considering kinetics of the evaporating mass flow, show that a significant returned flux was due to particle collisions in a non-equilibrium layer near the surface (see Chaps. 2 and 17). Figure 15.5 demonstrates the quantitative result comparing the solutions (1) using cathode energy balance in form close to (15.24) and solution (2) for which was used the balance of (15.23). The cathode temperature T according to the solution (1) is 3737 K, while from solution (2) this value is higher. The rate mass loss calculated according to equation [55] is obtained as $W(3737 \text{ K})I/j = 0.8$ g/s [61], but the experimental rate of erosion is about 100 μg/C, i.e., about 0.01 g/s that significantly different. For solution (2) it is just larger.

15.2.2 Models Based on Double-Valued Current Density in a Structured Spot

In the present section, we describe the spot models, in which it was assumed the presence of two values of current densities simultaneously during the spot lifetime to satisfy to different spot processes and observed parameters. So, in order to understand the electron emission mechanism and the cathode energy balance, Hull [70] was assumed presence of two values of current densities that were given arbitrary. On the one side, he assumed that to satisfy the cathode energy balance at the cathode a relatively large size cathode spot with average current density of 10^5 A/cm^2 should occur. On the other hand, this spot consists of a number of small spot randomly

15.2 Mathematical Description Using Systematical Approaches

moved with current density of 10^7 A/cm^2 that allowed to describe the cathode electron F-emission mechanism. The cathode energy balance was taken in form

$$f_i(V_c + V^* - \varphi)j = Tr_s/\lambda_T + W_s\lambda_s \tag{15.26}$$

where λ_T is the heat conductivity, r_s is the spot radius, W_s is the rate of evaporation of the cathode material, and λ_s is the latent heat of vaporization. To satisfy the Fowler-Nordheim equation a roughness factor of 2.5 was given that required electron emission of 10^7 A/cm^2. Furthermore, the ion current fraction was given by arbitrary value of $f_i = 0.05$. Thus, the number of given parameters leads to uncertainty of the results and of the model.

Fursey and Vorontsov-Vel'yaminov [71] studied the cathode processes after local explosion of a protrusion on the cathode surface. They developed a model of spot initiation assuming that an expanding plasma sphere, named "plasmoid," appeared after the explosion. During the expansion of the plasmoid in direction to the anode, an uncompensated positive charge appeared at the plasma front, setting up a strong electric field near the cathode, which stimulated an electron F-emission locally at the cathode. The process may be repeated several times extending at the surface and forming a plasma cloud near the cathode. It was noted that the important factor is the possibility creation of the localized field sufficient to electron F-emission. It can be realized assuming local enhancement of the average field at the protrusions by β-factor under extremely high plasma density. This assumption meets a problem because the space charge sheath thickness around is much lower than size of the protrusion and no enhancement can be.

Mitterauer 1972–1973 [72–74] developed a theory of dynamic field emission (DF-emission) to modeling an origin of the electrical current in the cathode spot of a vacuum arc. It was taken in account the results of breakdown phenomena studies of field emission from protrusions reported by Brodie [75] and from multi-protrusions by Tomaschke and Alpert [76] as well the effect of explosion electron emission studied by Mesyats [77]. Tomaschke and Alpert showed that field emission current originates from a number of separate localized emission sites on the cathode. These sites were shown to be as whisker projections on the cathode surface at which the electric field is greatly enhanced.

The DF-emission model suggested explosive evaporation and electron emission from different protrusions caused by a Joule heating. This process appeared during a spot life by a number of protrusions on a large cathode area. A plasma cloud was generated due to protrusion explosions inside of, so named "macrospot," in which the atom ionization, space charge sheath, and cathode bombardment by the ions support the spot current. The plasma regeneration was occurred during the macrospot lifetime by explosive of new tips produced after the tip explosion due to returned droplets, or plasma condensations. Mitterauer claimed that DF-emission model described explosive spot initiation and plasma regeneration by migration of the explosion from the tip to tip ("emission transfer") appeared at the cathode large area. It was noted that DF-emission model is only qualitative and it is very complicated to obtain a numerical solution.

Mitterauer [78] indicated the spot model evidence by experimental study of a cathode traces. Also, experimental results which relevant to the model of DF-emission of the cathode spot of cold cathode arc were reported in 1973 [79]. In particular, the observed micrograph is shown an average minimum current density derived for the macrospots of about 2.5×10^6 A/cm^{-2}. This result was in accordance with values obtained for macrospots on mercury and copper estimated previously.

The further development of the model related to a macrospot spot description and was presented by Mitterauer and Till in 1987 [80]. A system of equations including the Mackeown emission and transient heat conduction equations was solved for a given arbitrary the cathode electric field E and spot radius r_s. A hemispherical spot in the cathode body, and a spherical space charge sheath and ionization layer were assumed. The solution was presented for the condition of full atom ionization (degree, $\alpha = 1$). Also, it was assumed that the energy of electron emission accelerated in the space charge region was spent only on ionization. This assumption of course cannot be justified for the cathode spot plasma. It is difficult also to understand that the ion current density ratio to the ion current density can be equal to 2. The calculating results were obtained as dependencies of different given parameters. The cathode surface time-dependent temperature T was calculated as a function on E and r_s, which were given as parameters. It was obtained that T infinity increases for some certain time t (thermal runaway). Also, a critical spot radius was defined that determined at the thermal runaway condition.

In order to explain the measured low-current density spot ($\approx 10^4$ A/cm^2) that moved with a large velocity of about 10^3 cm/s, Beilis et al. [81] reported about an explosive model in 1973 and then the material was published in 1975 [82]. The model was based on a distribution of main spot parameters at some cathode area determined by non-uniformity of the cathode surface geometry. The initiation of the spot was due to Joule energy dissipation and Nottingham effect. The electron emission from the cathode entered into the cathode plasma and electrons from the plasma supported the electrical current to the anode. A new explosion under the expanding plasma supported the near-cathode plasma cloud. It was mentioned that Mitterauer [73] formulated an analogical point of view but for relatively large spot current density. The main goal that can be learned from this qualitative work is to illustrate that observed large luminous plasma size determined by high plasma jet velocity of 10^6 cm/s with the time scale of about 5–10 ns. With these parameters, the spot luminous characteristic size can be about of 100 μm, which exactly agrees with the experiment published in 1972 by Golub et al. [83].

Hantzsche [84] studied a thermal regime of the cathode with a heat conduction equation consisting of a volume energy source due to Joule energy dissipation. It was also considered a heat source at the surface by ion energy and the Nottingham effect. Different analytical solutions very obtained for specific cases such as for given surface temperature, just for volume heat source, etc. The maximal spot current was calculated for given spot current density of 10^7 A/cm^2 for different cathode materials.

A cathode spot calculation model considered rough surface conditions, i.e., the spot area composed of plane basis and superimposed protrusions was present by Hantzsche in 1974 [85]. The calculations provided for given area of the protrusions

15.2 Mathematical Description Using Systematical Approaches

top radius of about 0.1 μm as $0.2\pi a^2$ within the spot with radius $r_s = a$, and some other given parameters. The calculation shows that value of I/a increased by factor about 2 and the ratio of Joule energy to the energy by ion impact rise by order of magnitude from 10^{-2} to 10^{-3} with cathode temperature varied up to 4000 K. This result demonstrated the relatively role of the Joule cathode heating leading to possible of a thermal runway.

Hantzsche et al. in 1976 [86] developed a theory describing experimental data measured when the high-voltage pulses (up to 60 kV, pulse length of ns) was applied to an UHV diode with a needle-shaped cathode. The model considered the basic spot formation include processes of breakdown and arc spot initiation, heating, and melting of the cathode metal using a solution of the heat conduction equation determined the temperature $T(z, t)$ at a depth z below the center of the spot [87]. As the current I increases linearly with time t (according to used experiment), the surface heat flux per unit area and time in the spot surface caused mainly by ion impact, which assumed also linearly dependent on time like $Q = Q_o t$ (Q_o = constant). It was taken in account the droplet ejection and emission point reproduction, which was possible by the mechanism of the ejection of liquid droplets from the top of rising metal fountains. It is produced because of surface tension forces and by protrusion formation during solidification mainly at the crater rims.

The result shows three distinct stages of crater formation on a nanosecond time scale. Initially, micropoints explode without strong erosion, then melting occurs with pronounced smoothing of the surface and finally craters are formed by the action of ion pressure. The latter process is connected with the production of a few sharp micropoints at crater boundaries, and of numerous droplets. Micropoints and droplets maintain arcs and the following high-voltage breakdowns and during the discharge, they are displaced rapidly over the cathode surface. The characteristic values of time, current densities, and power input during heating, evaporation, and explosion of protrusions with radius 0.1 μm for a Cu cathode are presented in Table 15.1 [86]. The spot initiated at time of 0.01 ns (too small) with current density of 10^{10} A/cm^2 (too large) and develop to stationary cathode evaporation at time 10–100 μs with current density 3×10^7 A/cm^2.

The condition of thermal runaway was derived that showed unlimited increase of the temperature when a temperature dependent of resistive heat generation was accounted in the heat conduction equation 1979–1983 [88, 89]. This means that the heat conduction as linear process cannot remove the incoming resistive heat if a mean current density exceeds a critical value that depends only on the electrical and thermal conductivities. However, this thermal runway is not possible if electron emission cooling or in transient spot have time life is shorter than those necessity for thermal runaway.

In the next work, Hantzsche in 1981 [90] take in account that the observed arc spot is an exceptional non-stationary subject. Therefore, he indicated that a "dynamic" model (like that above developed by Mitterauer and other authors) must be developed taking into account the experience from investigation of vacuum breakdown including the surface effects. The inhomogeneity of the real surface

Table 15.1 Characteristic parameters of Cu spot development in time varied from protrusion (radius 0.1 μm) heating and explosion to a stationary state [86]

	Explosion or evaporation time constant, ns	Current density, A/cm^2			Power input, W
Joule heat generation dominant	0.01	5×10^{10}	Completely inertia-controlled explosion Fast explosion, heat losses inessential	Explosive (dynamic) evaporation Decreasing time constant	100
	0.1	7×10^9	Explosion time = heat conduction time		10
			'Slow' explosion, controlled by heat conduction	Melting	1
Ion impact and electron emission heating	1 100 1000 10,000	5×10^8	Limit of stationary, very rapid evaporation (>3000 K)		0.1
			Considerably stationary heating essential	Stationary heating	
	10^5 10^6	3×10^7	Slow evaporation Inessential heating (100 K)	Increasing temperature and evaporation rate	0.01

consists of its roughness, the existence of protrusions, and ridges or different inclusions. Hantzsche argued that appropriate spot description should use a combined model, which complete the conventional model in time dependent form by typical dynamic processes (resistive heating, new protrusion generation, metal melting, plasma expansion, and others) but not only with explosive points.

15.3 Further Spot Modeling Using Cathode Surface Morphology and Near-Cathode Region

Attempts to understand the cathode spot phenomena were conducted also using the data of cathode surface traces after the arc extinguish. Another way was associated with modeling of evolution of the surface craters under previously generated plasma near the cathode. Below the respective studies will be described.

15.3.1 Modeling Using Cathode Traces and Structured Near-Cathode Plasma

Osadin [91] studied a high-current vacuum arc (160 kA) between coaxial electrodes (Cu, Pb, Zn, Al and W) by calorimetric and optical measurements. He showed that the most of the arc energy was spend to production of the electrode erosion and ejection of the eroded material in form of plasma jet with high velocity up to $(1–3) \times 10^7$ cm/s. In the next work, Osadin [92] developed a cathode spot model using crater diameter d and depth h. Gurov et al. [93] tested the crater sizes previously. According to this experiment, the cathode spot consists of a number of craters with the size relation as $d \gg h$. Using this relation, it was assumed that the time of crater formation is much lower than the lifetime τ and therefore the process is quasistationary. The crater assumed in form close to hemisphere. If the charge passed though the crater is q_{ch}, the current density suggested determining as

$$j = \frac{4q_{ch}}{\pi d^2 \tau} \qquad (15.27)$$

The value of q_{ch} was given by the source capacity charge taking into account the effective area of total number of craters. The temperature T was suggested determine using Langmuir–Dushman formula and erosion rate Γ from the crater volume with mass density of cathode material ρ as

$$\Gamma = \frac{\rho h}{\tau}; \quad \Gamma = W(T) \qquad (15.28)$$

The effective voltage u_{ef} was calculated taking into account the energy with erosion mass and neglecting the heat conduction energy loss in form using (15.23):

$$j u_{ef} = \frac{\rho h \lambda_s}{\tau}; \quad \text{or} \quad u_{ef} = \frac{\pi d^2 \rho h \lambda_s}{4 q_{ch}} \qquad (15.29)$$

The next equation was atom–ion balance to calculate minimal value of electron current fraction s_{min}:

$$s_{min} \geq 1 - \frac{e \pi d^2 \rho h}{4 m q_{ch}} \qquad (15.30)$$

The system of (15.27)–(15.30) was solved varying τ in range from 0.1 to 10 μs. the results shows low values of u_{ef} (0.1 V for Cu) indicating low-energy loss with cathode material evaporation from the crater. The current density was changed from 1.6×10^8 to 1.6×10^6 A/cm^2 for Cu and from 5×10^7 to 5×10^5 A/cm^2 for Zn. The surface temperature was varied from 5300 to 3200 K for Cu and from 3300 to 1600K for Zn. Relatively high values of s_{min} were calculated: 0.97 for Cu, 0.95 for Mg, and 0.9 for Zn.

Additionally, in order to compare the measured spot current density, Mackeown equation [21] for electric field at the cathode surface and equations of electron emission in different forms were also calculated with given the experimental cathode potential drop. The results show that electron current density obtained by T-emission or by T-F electron emission does not satisfied the experimental value. It was concluded that an agreement with the measured current density can be obtained only in frame of emission mechanism taking into account a local fluctuation of electric filed, which was described by Ecker and Muller [94, 95] (see Chap. 2).

Also, the mathematical model was developed for high-current arc. Thereupon, a contradiction was obtained between calculated low values of u_{ef} (low-energy loss by the erosion rate) and result of the previous work [91] where the most of the arc energy was spent to production of the electrode erosion. Note, this theory, not reported of any value of arc current or some dependence on current. The results of the work [92] cannot be considered as useful due to many other contradictions including assumption that the process is stationary, disregarding the energy loss by cathode heat conduction, use Langmuir relation for calculating the mass evacuated from the crater and for determining the cathode surface temperature, atom–ion balance, etc.

Goloveiko [96, 97] further developed the spot model from Lee and Greenvood work [54] also using Mackeown equation for electric field, cathode energy balance, and equation to determine the electron emission current density j_e, in form presented by Murphy and Good [98]. Frenkel equation was used to determine the rate of mass flux G_n by cathode evaporation, that, in essence, this same as Langmuir–Dushman equation. The cathode heating was accounted by energy ion flux accelerated in the sheath, by resistive heating and cooling due to electron emission and cathode evaporation. The difference from the [54] consists in study the influence of ratios between atom and ion fluxes γ_{ni}, as well between atom and electron γ_{ne} fluxes, which were expressed as:

$$\gamma_{ni} = \frac{eG_n}{j_i}; \quad \gamma_{ni} = \frac{eG_n}{j_e} \quad s_i = \frac{j_i}{j_e} \qquad (15.31)$$

where j_i is the ion current density and s_i is the ion current fraction. The energy loss in the cathode body was determined solving one-dimension heat conduction equation taking into account the Joule energy dissipation and a moving the front surface of the cathode due to the cathode evaporation. The boundary conditions at the cathode surface were formulated considering the cathode energy balance. The cathode energy balance in steady-state approximation was obtained by integrating the mentioned heat conduction equation. The details of study of the cathode heat regime were described previously by Goloveiko [99]. The system consists of three equations for main four unknown (j_e, j_i, T, E), and therefore the solution was obtained by varying the electric field in range from 10^6 to 10^8 V/cm.

The results of calculations for Cu cathode ($u_c = 16$ V, $u_i = 7.7$ eV) are presented in Fig. 15.7. The cathode temperature decreased from about 8000 to 4000 K when E varied from 2×10^6 to about 4×10^7 V/cm, respectively. Using the obtained dependencies for γ_{ni} and γ_{ne}, Goloveiko [96] presented the following interpretation

Fig. 15.7 Ratios of atom to ion and atom to electron fluxes as dependence on given electric field at the cathode surface [96]

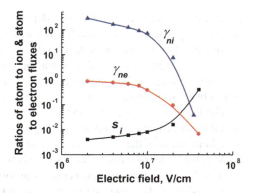

of the spot evolution. The spot was initiated at cold cathode at large electric field with F-emission with absent some cooling effects. At this stage, a large heat flux by the ions ($s_i \approx 1$) together with Joule source heated the cathode surface. Therefore, the spot operation was transited from cold to hot regime with decreasing of the electric field and increasing the heat source due to Joule energy dissipation in cathode volume. The volume heat source can stop the future development of the spot lifetime causing its migration.

The presented picture seems to be strange because the cathode spot evolution was analyzed using steady-state solutions. Other deficiencies of the above study include use equation of Murphy and Good, which is available in limiting cases, Langmuir equation that does not consider returned particles to the cathode, free parameters, like depth of liquid zone and heat conduction equation that was taken in one-dimension approximation. In the model, no any values of arc current or dependence on it was mentioned. Therefore, the model and calculated results cannot be considered as adequate description realistic spot behavior.

As in [96], Kulyapin [100] also used Mackeown equation for electric field, cathode energy balance and equation to determine the electron emission current density j_e, in form presented by Murphy and Good [98] and the rate of mass loss by cathode evaporation by Langmuir–Dushman equation. The solution was also found by varying the electric field E, as a free parameter, in a range from 5×10^6 to 3×10^7 V/cm. Also, the surface heat source and resistive heating with usual cooling terms were accounted. However, the cathode energy balance was presented in a different form using the heat conduction equation studying the liquid and evaporated phase boundaries of the metallic cathode. The heat conduction equation considering the moved liquid and evaporated phase boundaries were studied by Kulyapin [101] separately by given heat flux density in range of $q_0 = 10^3$ to 10^{10} W/cm². The expression for the surface temperature T and for moved boundaries as function on time were derived. These expressions were used solving the cathode energy balance together with the mentioned above system of equations.

The current densities j_i, j_e, T, n_i/n_e and $s_i = j_i/j_e$ in the spot were calculated (Ag, Cu, Au, Fe, and Ni) depending on electric field at the cathode surface varied as parameter. As an example the time dependent calculations were demonstrated for Ag showing that j_i, T are linearly increase, s_i was decreased and ratio n_i/n_e mainly remain constant (behavior of n_i/n_e and s_i are different from that result in Fig. 15.7) with E up to a critical value E_{cr}. It was reported that before E_{cr} the spot parameters were determined by the ion energy accelerated in the sheath. The calculated results were obtained ignoring any information about the spot current.

The following spot evolution was discussed. At the spot initiation, the incoming heat flux was equal to the possible energy losses. As the cathode surface was melted, a slight increase of the Joule energy dissipation was appeared and it is increased with E. At $E > E_{cr}$ the resistive heating exceeds the energy losses possible in the cathode body and at its surface, named as non-equilibrium heat regime. This situation causes an explosion of the metal under the spot and existence of the spot was stopped. A new spot was initiated. These phenomena explain the discrete spot motion. The spot lifetime was calculated as dependence on E and the value of E_{cr} was determined at point when this time approaches to zero. For the investigated cathode materials, E_{cr} was obtained in range $(1.2–2.3) \times 10^7$ V/cm and the spot life in range from 2.7 to 53.6 μs.

In the next work, Kulyapin in 1971 [102] it was determined the rate of cathode erosion for Sn, Ag, Al, Mo, Nb, Fe, W, and Ni using the results of heat flux, spot lifetime, and other parameters obtained in his previous work [100]. Note that in this work the influence of the erosion rate on the spot parameters was not considered. The expressions for material loss by melting and evaporation were derived separately before the moment when the spot should be dead. It was assumed that all melting material was evacuated in a volume determined by liquid boundary. The obtained spot lifetime was used to calculation of the spot velocity v_s by taken into account the spot diameter as a size at which the spot is fixed. This velocity was calculated for Cu, Au, Ag, Fe, and Ni in range of 117–850 cm/s. An agreement between the calculations and the measurement was indicated. Taking into account that this study is not self-consistent and other lacks that similar as was mentioned in above-referenced works the agreement can be considered as a criterion that necessary but not enough.

Kozlov and Khvesyuk [103] described a theoretical work with a specific spot model and an analysis of the cathode plasma structure. The near-cathode region was modeled consisting by different zones: (i) the first is the zone with large electric field E in which the emitted electrons acquired energy for atom excitation or ionization and the ions moved to the cathode with atom collisions; (ii) in the second zone, the atoms were ionized by impact collisions. No any collisions between generated low-energy electrons and accelerated beam electrons as well with fixed heavy particles was considered; (iii) zone of electron beam relaxation to the Maxwell electron distribution. In the first and second zones, the electric field was assumed significantly large. The Poisson equation was taken together with equations expressing momentum from modeled functions of velocity distribution of the particles. Although the Boltzmann kinetic equation was described, this equation was not solved and some numerical analysis of influence the particle collisions on sheath parameters was not conducted. The problem was considered in one-dimension and steady-state approximation. Also,

15.3 Further Spot Modeling Using Cathode Surface Morphology ...

the degree of atom ionization was assumed significantly small to neglecting the charge particle collisions, which are not real, case (see Chaps. 16 and 17).

The above model included the system of equations: (1) the heat conduction equations with volume heat source and heat flux at the cathode surface as a boundary condition to determine cathode temperature T; (2) equation of electron emission from the cathode depending on E and T; (3) equation of atom flux by the rate cathode evaporation; (4) Poisson equation. These equations were added with momentum equations described the beam loss of electrons due to direct impact or stepwise ionizations. The effective spot area was determined by normalization of the integral of electron current density assuming that a 90% of the current was occupied by this area. However, this approach seems to be adequate only in case of presence of two-dimension solution, which was not provided in this work.

The authors stated that both electron groups emitted and generated after ionization have mono-velocity distributions due to large electric field. While this assumption can be accepted for zones 1, the assumption of strong electric field in zone 2 indicates that its length should be about one mean free path of ions, otherwise, a large collisions leads to produce a state of quasineutrality with low electric field. Thus, a contradiction was appeared due to the assumptions about strong electric field in second zone, which is easy to show as collisional region with low field by an estimation using measured spot data.

The second part of the work of Kozlov and Khvesyuk [104] presented solutions according to the mathematical model discussed in their first paper [103]. Analytical solutions were provided for Poisson equation and the equations for electron, ion current, and particle densities in near-cathode regions. The cathode potential drop u_c calculated as dependence on given varied density of heavy particle n_g. The current density $j = 10^6, 10^7$, and 10^8 A/cm^2 and electron current fraction $s = 0.97, 0.98$, and 0.99 were used as parameters in the dependencies calculated for Hg and Cu cathodes. The results for Hg cathode were obtained by stepwise atom ionization and for Cu by direct electron impact mechanisms of atom ionization. The calculation shows that the u_c decreased with n_g at constant value of j and s. But u_c increased with rise of j and decreasing of s for constant n_g.

Note, these calculations provided using some arbitrary data and, therefore, indicate results, which can be not reflected to that obtained by complete analysis. Namely, it was reported that the total equations for cold cathodes were solved using data published for electron emission obtained by Lee [52], and data described the kinetic of vaporization and vapor expansion obtained by Afanas'ev and Krokhin [105], i.e., given data that not based their relation to the studied subject, especially for Hg. Also other arbitrary parameters as, n_g, etc. A number of spot parameters obtained by the authors [104] are presented in Table 15.2

No any data reported about cathode temperature, electric field, spot, or arc current. The results are presented in such confusing that impossible to trace the way of the solution conducted with free parameters and not by self-consistent method. The weakness of the electric field description and the contradictions of that model were detailed by Luybimov [68]. Another strange fact was related to the significant large $j = 7 \times 10^7$ A/cm^2. At electron current fraction $s = 0.96$ the ion current density is

Table 15.2 Spot parameters [104]

Material	n_g, cm^{-3}	j, A/cm^2	s	u_c, V
Hg	8×10^{20}	7×10^7	0.96	9
Cu	10^{20}	10^7	0.97	20

2.8×10^6 A/cm^2 and the heat flux 2.8×10^7 W/cm^2. At such large power density, the mercury will be at an unrealistic state and this critic state should be additionally explained. However, no any data regarding to the cathode temperature in the spot and for spot current was reported. It was informed that the current density agree by order of magnitude with the experiment, but it is a problem, because the measured by Kesaev [2, 3] j is about 5×10^4 A/cm^2, which is far from the data in Table 15.1. Additional weak point is related to an input parameter $\lambda = 4an_g \sigma_{ig}$ to obtain a series expansion of some expressions. However, estimation of $\lambda = 4 \times 10^{-6}$cm $\times 8 \times 10^{20}$ cm$^{-3} \times 10^{-14}$cm$^2 = 32$, i.e., this assumption was not fulfilled.

It should be noted that Beilis and Rakhovsky in 1969 [106] were first developed a spot model basing on *charge-exchange* of ion–atom collisions. A model considering characteristic plasma zones near the cathode taking electron impact mechanism of ionization was developed early in 1969, [107], and in 1970, [108, 109], from which the weaknesses of the above works were obviously followed (see Chap. 16). The same can be stated regarding the work of 1971, [110], were arbitrary estimated the probability of atom ionizations and an electric field (up to 5×10^7 V/cm) at the cathode using charge-exchange ion–atom collisions. A discussion of the published previously works reported in [68].

Harris and Lau [111] presented an advanced work. The plasma processes were considered taking into account different zones of the near-cathode region and Persson's theory of ambipolar particle diffusion in plasma for a plane-parallel case with a wall sheath [112]. The phenomena at the cathode surface including the atom evaporation and electron emission as well as the adjacent plasma were studied. The plasma region divided by two parts, the first at cathode side is acceleration zone (space charge) and then the next is atom ionization zone. The consideration includes the conservation laws expressed by equations of particle fluxes, momentum, and energy. The calculating (for Cu, Ca, and Mo) plasma parameters include the voltage–current characteristics, the electric field, current density, the voltage drop, and the potential distribution indicated a hump potential presence. However, the model used assumptions consisting of a number of principal lacks, which limited an adequate plasma description. A part of them is discussed below.

It was assumed that the ion and electron currents from the plasma to the cathode equal one to other and are determined by mechanism of ambipolar diffusion with zero current condition. The total spot current is determined by the current of electron emission from the cathode. Therefore, for current carrying plasma, the condition of zero current at the cathode plasma cannot be fulfilled. The energy of ion flux to the cathode was calculated without energy of the ion acceleration, which acquired in the sheath with cathode potential drop. It was also assumed that ion flux to the cathode and ion flux from anode side of the ionization zone are equal one to other and each of them equal to half of the atom flux evaporated from the cathode. These assumptions also could not be justified. The ion velocity at the anode and cathode side of the ionization zone is assumed as the sound velocity that also may be not fulfilled. The electric field at the cathode was determined by electron temperature ratio to the Debye length, but not by cathode potential drop, as usually.

The ion charge number and the effective potential of ionization were approximated by coefficients of proportionality to the plasma electron temperature, which was given arbitrary. No dependences on the atom density were presented. A potential hump of 20–30 V was obtained in the short ionization zone, while this hump should be arising in the plasma-expanding zone at distance of few spot radii from the cathode surface (see an analysis below). The current density of electron emission of about 10^8 A/cm^2 was obtained for Cu by electric field 4.6×10^7 V/cm (T_e= 5.7 eV). Estimation shows that in according to author's assumption for the cathode plasma it can be only at Debye length of 10^{-7}cm. It should be noted, the plasma processes were not analyzed in order to determine the ionization degree, ion density, ion flux, returned electrons to the cathode and the influence of the correctly determined space charge sheath.

Nemchinsky in 1979 [113] and 1983 [114] calculated the cathode spot parameters using, in essence, the gasdynamic model of the spot and the vaporization approach early modeled [65]. An additional modification was conducted considering a relation, that expressed the voltage in the quasineutral plasma, together with a "minimal principle" condition, which determined by dependence of the total arc voltage on the cathode temperature. The minimal voltage in this dependence was taken as a cathode potential drop. It should be noted that different minimal voltage dependences could be obtained by varying other parameters (current, current density, and other). So, an uncertainty of the studied parameters can be obtained by use such approach. The erosion rate was calculated using the saturation vapor pressure corrected by a coefficient of the returned particles, which was approximated from results studied in another work. Other models presented later [115, 116] also used input free parameters and arbitrary conditions, similar to work in [54], and the plasma processes were not always analyzed.

15.3.2 Cathode Phenomena in Presence of a Previously Arising Plasma

Prock [117] studied a dynamics of crater formation on the cathode of a metal vapor arc. The work consisted of number assumptions that could be not related to the real process. It was assumed that the initial plasma and initial crater parameters initiated by an electrical breakdown. Further crater grow was caused by the energy supply from the plasma. This process was stopped due to limitation of the energy for further melting because the leaved energy part not enough for electron emission or the evaporation, which necessary to supporting the discharge. The model takes into account phase changes, Joule and ionic heating, electron emission, and mass loss due to evaporation and ejection of molten metal. The equation of heat conduction was solved in conjunction with expansion of the crater (the Stefan problem) using a spherical one-dimensional model. Furthermore, Prock assumes that the configuration of the crater surface was determined by boiling phase if the vapor pressure exceeds the ion pressure. The limiting atom–ion balance was used as in work of [54]. Therefore, the calculation shows a lower bound for the spot lifetime and the final crater radius. The cathode material evaporation was calculated according to Dushman law [55]. As the initial crater radius was given then for used constant spot current I, in essence, the initial current density was given arbitrary. The current density was assumed decrease inversely of square of the radius in the cathode bulk. The cathode potential drop was expressed by the following arbitrary approximation $u_c = u_i + f(\varphi)$. It was indicated that the function $f(\varphi)$ was choose depending on mechanism of electron emission.

One of the main conclusions of Prock's work is the statement "because of energetically reasons, crater formation is only possible for current densities on the order of 10^8 A/cm^2, as shown for Cu." This conclusion is doubtful because it was based on the calculation for very small time of 0.4 ns and was obtained using the above-mentioned arbitrary assumptions. Prock compared the contribution of ionic and Joule heating at the end of crater formation for different metals. His conclusion that the contribution of Joule energy is large comparing to that by ionic energy source for different materials is also doubtful. Note, the ion current density j, is defined as difference between total current density (equal arbitrary to crater size) and current density of electron emission. So, this comparison was provided without consideration of plasma phenomena, origin of ion density and ion flux. It was estimated in large range j of $(2–11) \times 10^8$ A/cm^2 and also not in self-consistent manner.

Klein [118] extended Prock's model to investigate time development of the crater. He used an accumulation of the plasma in front of the spot to determine of the spot lifetime. In contrary to Prock's assumption of boiling temperature of the crater surface, Klein assumed that below given initial crater surface in the cathode an overheated phase is developed that sharply increase the crater radius by an explosion at the end of the overheating process. The temperature field is calculated via the transient heat conduction equation in radial symmetry with the Joule heating. The current density j at the cathode surface was defined as $I/(2\pi r_0)^2$ and $j(r)$ in the cathode body falls off proportional to $(r_0/r)^2$. The ion heating, evaporation, and radiation cooling were taken in account as surface boundary conditions. Also, the equations for electron emission and electric field at the cathode surface were used. The crater current, I and size r_0 were given and this means that the current density was also given. The calculation conducted for $r_0 = 0.4$ μm and $I = 50$ A for Cu showed that critical cathode temperature and explosion phase was reached at too small time 0.1 ns. It was then assumed that the whole plasma is returned to the cathode. And the further increase in heating rate leads to the development of a new temperature maximum below the surface after the last surface layer is removed. If the temperature maximum finally reaches the critical temperature, then a new surface layer is removed and the entire evaporated mass is increased. After a lifetime of about

15.3 Further Spot Modeling Using Cathode Surface Morphology ...

2 ns, the regime of thermal runaway changes to the regime of melt ejection, where the cooling effects compensate the Joule heating.

It should be noted that the calculated condition of the local cathode explosion phase is obvious and not surprise as a large current density of $j = 1.6 \times 10^{10}$ A/cm^2 was given corresponding to sub-micron initial size of current area. Also, it is difficult to understand how the crater plasma can stepwise to be returned after the explosions. It seems that such arbitrary assumption was provided to increase the crater size in order to satisfy to the measured data.

He and Haug [119] analyzed the thermal phenomena in a cathode as a possible mechanism for spot formation. The temperature field in the electrode is calculated with a 2D heat conduction equation in a cylindrical symmetry. The energy balance at the cathode surface according to Mitterauer et al. [80] determined the boundary conditions. The volume Joule source was defined by the electric field in the cathode body. This field was determined through the continuity equation, derived from Ohm's law (in a cylindrical coordinates) by neglecting the net charge source. It is due to the applied voltage on the electrode gap. However, in a real case, even with spot ignition in vacuum by a high voltage, there is often an ion flux to the cathode surface. It can be result from the previous electric discharge. The influence of the ion flux was accounted as an energy carrier for the cathode. When the surface field is not strong enough to generate heat flux to the electrode, an ion flux was considered as a supplementary heat source. Then, the electric field at the surface was used twofold. The electric field E and the ion current density j_i at the cathode surface are considered as two independent external parameters. At some significant value of j_i, the space charge was taken in account and additionally the electric field at the cathode surface was calculated according to Mackeown equation [21].

The calculations were conducted for copper cathode, cathode potential drop of u_c = 15 V and for planar or smooth surface, on which the electric field was applied at a circular zone with given radius of 10 μm. The results show that, without the ion flux, a high electric field is needed. In this case, at the surface and at the beginning, the only efficient source is the Joule heating due to electron emission. With a surface field equal to or stronger than 7.5×10^7 V/cm, the surface temperature increases from its initial value and then a cooling effect occurs with increasing emitted current density and of the metal vapor. After the surface temperature has reached a critical value, the sign of the thermal flux changes and the overheating phenomenon below the surface were calculated. The calculation showed that the surface temperature remains limited between 3400 and 3800 K, when a thermal runaway is occurred and the maximum temperature below the surface increases up to the critical value for Cu (8000 K). The time delay for the thermal runaway was about 10 μs for $E = 7:5 \times 10^9$ V/cm and 20 ns for $E = 9 \times 10^9$ V/cm.

According He and Haug [119] the time dependent evolution of the temperature at the surface center and of its maximum in the cathode bulk for different external electric fields is shown in form indicated in Fig. 15.8. To start spot formation on a cold cathode without ion impact, a strong surface electric field is necessary to about 10^8 V/cm. This required field was obtained due to the voltage applied across the electrodes and it enhanced by local surface rugosity. The influence of the ion

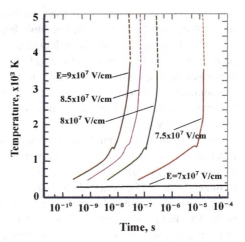

Fig. 15.8 Influence of the electric field on time dependent spot temperature evolution calculated without ion flux with value of electric field as parameter in range $(7-9) \times 10^7$ V/cm [119]

current was studied at minimal value $E = 7 \times 10^7$ V/cm. It was shown that $j_i = 5 \times 10^6$ A/cm^2, the spot ion current $I_i = 0.157$ A, and the temperature at the center is low, about 800 K at 100 μs. However, the temperature significantly increased at $j_i = 2.5 \times 10^7$ A/cm^2, $I_i = 0.8$ A that was considered as critical at which the spot ignition becomes possible. Thus, the intensity of the field required for spot initiation decreases with increasing ion flux. If the latter is sufficient, the electric field due to the space charge is sufficient to initiate a cathode spot.

It was indicated that the calculated overheating could be considered as the origin of the microexplosion, which can be a possible mechanism for the crater formation on the cathode. It should be noted that to reach the mentioned j_i the ion density should be 10^{21} cm^{-3} that is enough high. In additional, the used u_c is arbitrary value at stage before of spot ignition.

A 2D numerical model has been developed by Schmoll [120] to investigate the interaction of a cathodic spheroidal microtip apex with a low-temperature plasma. The voltage drop between the plasma and the cathode was used as a free parameter varied in the range 15–1000 V. Size of the tip base radius was 0.5 and 1 μm. Three time stages of heat conduction equation were discussed. The first is heating of the tip by energy of the ions accelerated in a sheath. The second is the tip melting with process of removing the liquid metal when (arbitrary assumed) the melting depth exceeds 0.1 μm also arbitrary changing the tip geometry. The third stage considers in addition to the solid and fluid phases at the tip, the metal vapor, which is generated by high temperatures at the protrusion apex. The end of this phase usually is the generation of dense metal vapor plasma in front of the microtip. The weakness of this study consists in using the following erroneous relations. (1) The force acting on the liquid surface due to the electric field strength, which is not based. (2) The flux of metal vapor particles from the tip and the evaporation tip cooling was calculated by means of the Langmuir formula at boiling temperature of the tip. (3) The ion pressure was determined by momentum of ion flux accelerated in the sheath at the

15.3 Further Spot Modeling Using Cathode Surface Morphology ...

protrusion surface without consideration of the force balance in the sheath. The plasma parameters were determined without consideration of atom ionization and plasma energy analysis. The electron emission was described in limiting approach of Murphy and Good formalism. No any data and discussion was presented related to the discharge current influence.

Niayesh [121] investigated an influence the electrode surface irregularities as a possible mechanisms of arc re-ignitions after interruption of high-frequency currents. The plasma state during the postarc period was investigated using the equations of conservation of mass and momentum in the space charge region in 2D approximation. The results (for 6 kV/μs and protrusion size 15 μm) indicate a considerable increase in ion density at the points with high electrical fields. No heat or other processes in the solid protrusion were analyzed and reported.

The instability phenomena in a copper vacuum arc were analyzed by Mungkung et al. [122] solving the traditional cathode spot system of equations including equations of (1) total spot current, (2) electron emission, (3) cathode energy balance, and (4) electric field at the cathode surface, i.e., in approach of [54]. This system was added by an arbitrary equation of heavy particle conservation and a relationship to determine the plasma electron temperature T_e determined by the Joule energy dissipation due to plasma jet electric field. To determine eight unknown, the ion current fraction f in the plasma jet and the effective cathode voltage were used as input parameters by independent varying of the spot current within 19–70 A. It was shown that a common solution of the equations for spot current below 20 A is absent. This fact was interpreted as instability of the vacuum arc. It was concluded that this instability is due to the electrons returning flux from the plasma region, which were taken in account only in the equation for electrical sheath neglecting of this electron flux in the other equations without any discussion. Therefore, it is understandable why the calculated values of electron current fraction of 0.1 and cathode potential drop of about 9 V were obtained so low. In contrary, the resulting value of $T_e \geq 2$ eV is large, while the estimated potential drop in the plasma is ≤ 1 V for relatively small the plasma electric field, about of 100 V/cm (using even a large $j = 10^5$ A/cm^2) and at a characteristic spot radius ~0.01 cm. The cathode jet mass was determined by *fjm/e* that intend fully atom ionization in the jet, while fluxes of the evaporated atoms and of the ions moving toward the cathode were assumed as independently different in the plasma near the sheath.

The cathode jet mass was defined by *fjm/e* that intend fully atom ionization in the jet, while this mass flux on the other hand determined by difference of the fluxes between evaporated atom Γ_{ev} and the ion (1-s)j moving toward the cathode and determined as independently from one to other near the cathode. As a result $\Gamma_{ev} = mj(1 + f\text{-}s)/e$, which indicated that for used $f = 0.1$ and at calculated s < 0.1 (see Fig. 3 in [122] for current > 40 A) the ion current exceeds the evaporated atom flux (s is the electron current fraction). Any analysis of ionization state of the vapor in mentioned region was absent.

The above considered thermal approaches and, in particular, the thermal model of [119], was taken in account by Uimanov [123] to investigate heat regime of a cathode surface of a microprotrusion. A numerical simulation was conducted using

2D non-stationary heat conduction equation with volume Joule heat source. The ion impact heating and electric field of the space charge zone near a protrusion at the cathode surface were taken in account. Joule heating of the microprotrusion at some certain values of the geometric parameters produced an electron emission current density followed by an explosion of the protrusion.

The shape of the microprotrusion surface was specified by the Gauss function with the height varied in range from 0.5 to 5 μm and the microprotrusion base radius of $r_m = 0.5$ μm. The explosion of the protrusions was investigated assuming an enhancement of the current density of the ions moving from the plasma to the cathode. The current enhancement parameter was determined by the ratio of the protrusion lateral surface area to its base area. The simulation was performed until the temperature in the protrusion reached a metal critical temperature given arbitrary as 8390 K for Cu. At this temperature, the calculations were stopped due to the assumption that the process went to the explosive phase and development of an electron emission center (see below).

Computations were performed for a copper cathode with cathode potential drop of 16 V and potential of atom ionization of 18 eV, which take in account ion charge of 2. The ion density current j_i varied in given large range from 10^7 to 10^8 A/cm^2 taking into account enhancement coefficient of the current determined according to the protrusion geometry. Considering such large values of j_i it is not surprise that the protrusion can reach the overheating temperature in range of ns even (at 10^8 A/cm^2) without Joule heat source by electron emission. Note, the result of calculation is very sensitive to the chosen input values, which early demonstrated in [119]. Namely, for $j_i = 10^6$ A/cm^2, the calculated overheating temperature is in range of μs. Additionally, the used initial ion density was taken as 10^{20} cm^{-3} at the sheath edge is strongly questionable because the previous plasma expands from the surface at distance of 10–100 μm in time of 1–10 ns. The arbitrary chosen small protrusion size limits the heat and current conduction. Also, the ion charge of 2 for Cu cathode was detected far from the cathode surface. The used potential of atom ionization of 18 eV is overstated.

Further development of Uimanov's 2D model to analyze the pre-explosion processes in a cathode protrusion has been described by Barengolts et al. [124]. The model includes a calculation of the cathode temperature in view of the surface heat fluxes carried by electrons and ions during the interaction of the cathode surface with the cathode plasma, and the Joule heating of the cathode. The calculations conducted with the following given plasma and protrusion parameters: ion density $n_{i0} \sim 10^{19}$–10^{20} cm^{-3}, electron temperature $T_e \sim$ 2–4 eV, cathode potential drop $U_c =$ 16–20 V. The used effective potential of ionization of $eV_i = 18$ eV is overstated as mentioned above. The shape of the protrusion surface was specified also by the Gauss with given characteristic sizes of height $h = 2$ μm, $d = 0.312$ μm, and $r_0 = 0.5$ μm [124]. The calculation domain of the plasma was 10 μm in size, the calculation domain near protrusion at the cathode was 3 μm. The enhancement parameter of ion current was taken as 4.

Calculations were carried out up to the point in time at which the maximal temperature in the cathode reached the critical temperature for copper also of 8390 K. As a difference from the approach of [123], in present model, the potential drop u_p across the cathode plasma was calculated. Using this result, the potential drop across the space charge sheath was determined as difference between the potential at the plasma boundary u_p, and the potential at the microprotrusion u_m. The calculated result showed a thermal instability (thermal runaway) in a cathode microprotrusion and that the critical temperature was reached within some tens of nanoseconds.

Of course, this result is a consequence given small geometry of the protrusion and of specific plasma parameters (at fixed cathode potential drop), indicated as parameters in the spot plasma, although that the model described the pre-explosion process. Also, it is not understandable how can be determined the sheath potential drop u_c as mentioned above and in the work (see [124]). For this a Laplace equation was used to calculation the potential drop u_p in the plasma adjacent to the protrusion, while the potential u_m was obtained from Laplace equation for body of the protrusion. These independent subjects not related to the value of u_c, which should be determined by analyze the space charge (Poisson equation) producing the sheath.

15.4 Explosive Electron Emission and Cathode Spot Model

The electron emission from a plasma plume generated by explosion of a cathode tip due to it resistive heating was defined as electron explosion emission (EEE) phenomenon [125–128] (see also Chap. 2). Using the fact that, the processes in a cathode spot are extremely complicated in many respects the authors of the EEE phenomenon indicated that explosive emission could have decisive significance in the functioning of the spot. The main motivation was based on transient character of EEE phenomenon. According their opinion (not correctly reflected the publications), the existed previous spot models were stationary and have not explained such experimental facts as the rapid displacement of the spots, fluctuations of the arc voltage, and the existence of multiply charged ions. Also, the roughness of the cathode surface and the inhomogeneity of its surface composition have not been taken into account.

So, these investigators asserted that *only the explosive process* could characterize the transient behavior of the cathode spot in a vacuum arc. This conclusion, however, ignore the real transient essence in the other models considered already above. In additional the transient spot description based on evaporation models (without explosions) was arbitrary referenced always as stationary approaches. Below this opinion and the details of the EEE approach, used to describe the spot mechanism, is analyzed and its relation to the other models is considered.

15.4.1 Primary Attempt to Use Explosive Electron Emission for Modeling the Cathode Spot

Mesyats [129], Bugaev et al. in 1975 [130], Litvinov et al. [131] first and then Mesyats and Proskurovsky [132] specified the phenomena of explosive electron emission (mainly as electrical breakdown mechanism) as a process that can be related to the cathode spots. In essence, the primary attempt was to use explosive electron emission to explain the cathode spot behaviour was qualitative. The authors' conclusion was based on comparing the observed EEE phenomenological results (in breakdown) with those experimental cathodic arc findings [130, 132]. It was indicated that such observations include the presence of a dense plasma near the cathode, high-current density of 10^7–10^8 A/cm^2, arc voltage fluctuation, craters, plasma expansion velocity, and the presence of multiply charged ions. Litvinov et al. in 1983 [133] discussed a crater formation and motion of the explosion location in framework of EEE. This result was obtained using thermal model with the non-stationary Joule cathode heating. Neglecting the energy loss by temperature gradient the heat conduction equation was solved for cathode of conical geometry. A coordinate r_{cd} or the cross section corresponding to the volume of cathode heated to the state of destruction was calculated. The time of destruction was determined when the input energy exceed the specific heat of sublimation. The mass of cathode material in this volume was assumed ejected away resulting in cathode erosion and crater formation. It was indicated that the initial explosion occurs as the result of field emission with a field $\geq 10^8$ V/cm, high-current density (~10^9 A/cm^2), and then at the surface form some conditions for appearance of a new explosion under the plasma. So, the erosion process consists of a series of microexplosions in the vicinity of the initial explosion center. The calculated results were compared with those results obtained for a high-voltage (30 kV) experiment. However, in case of vacuum arc, the cathode spot occur under only about 20–30 V. Therefore, the question is how the conditions for explosions can be occurred during vacuum arc operation with low voltage. Let us consider some calculations to understand the conditions.

15.4.2 The Possibility of Tip Explosion as a Mechanism of Cathode Spot in a Vacuum Arc

The model of the explosive emission and its application to explain operation of the relatively long-life spots (~1 μs) can be considered through the following scheme. Let us assume that the spot plasma with density n_i is generated due to a number of individual explosions of separate microscopic protrusions on the cathode surface. A space charge sheath is formed at the cathode surface with measured potential drop of about 10 V (Hg) and 15(Cu). It is taken in account that the sheath thickness d thus defined with reasonable accuracy by the 3/2-power law (Chap.2) and the ion density by the continuity equation, $j_i = e n_i v_{iT}/4$, where v_{iT} is ion thermal velocity.

15.4 Explosive Electron Emission and Cathode Spot Model

The value of the ion current density is given and the equation of electric field E can be used (see Chap.2). To support the further spot existence by explosion model, let us assume that after an explosion of a single protrusion of mass m_{ex} the plasma is uniformly distributed within a hemisphere with a radius equal to the spot radius r_s corresponding to given current density. As the explosion plasma expands with velocity v_j of about 2×10^6 cm/s [125, 130] one obtains the plasma density n_{ex} due to one explosion as

$$n_{ex} = \frac{m_{ex} - m n_{ex} v_j S_s t}{V_w m} \quad V_w = \frac{2}{3}\pi r_s^3; \quad S_s = 2\pi r_s^2; \quad r_s = \sqrt{\frac{I}{\pi j}} \quad (15.32)$$

or

$$n_{ex} = \frac{3 m_{ex}}{2\pi r_s^3 m} \frac{1}{1 + \frac{4}{3}\frac{v_j}{r_s}t} \quad (15.33)$$

where V_w and S_s are the volume and surface area of the hemisphere, respectively, m is the atom mass. A protrusion with radius r_b of the base and height h_h has mass $m_{ex} = \pi r_b^2 h_h \gamma$. Now, let us define the number of simultaneously explosions N needed to provide the ion density n_i corresponding to the current density j_i as the ratio of the density n_i to the density n_{ex} obtained from a single explosion. Then, the expression for N is:

$$\frac{n_i}{n_{ex}} = N = \left(\frac{I}{\pi j}\right)^{3/2} \frac{8 m_i j_i}{3 e v_{iT} r_b^2 h_h \gamma} \left(1 + \frac{4}{3}\frac{v_j}{r_s}\tau\right) \quad (15.34)$$

τ-is the explosion time of about (1–5) ns. Now to determine the protrusion sizes the critical explosive electric field is taken to be $E_{cr} = 10^8$ V/cm [125, 130]. The critical field is related to the average field by $E_{cr} = \beta E$ where β is the field enhancement factor. The field will be enhanced, if the sheath thickness h significantly exceeds the characteristic protrusion size. The magnitudes β and d will thus determine the maximum size of the protrusion, which under the conditions considered can be exploded. The calculations for Cu cathode show that $E = 10^7$ V/cm and $d \sim=(0.6–1) \times 10^{-6}$cm when u_c increases from 15 to 50 V, therefore, the explosion can be realized for $\beta = 10$. In this case, at least should be the $h_h = (0.6–1) \times 10^{-7}$cm. Let us consider that after explosion a spot with lifetime of $t = 10$ ns, total current density $j = 10^8$ A/cm^2, with ion current density $j_i = 10^6$ A/cm^2 will be appeared [134–136]. Using (15.34), $r_b = h_h = 10^{-7}$cm, the value of simultaneously explosions is $N = 3 \times 10^{14}$ with explosion time $\tau = 1$ ns. For spot time life of $t = 1$ μs, $j = 10^8$ A/cm^2, and $j_i = 10^6$ A/cm^2, the value of N increase to $\sim 10^{17}$. These values of N are fantastically large and difficult to justify. In additional, the characteristic sizes of the protrusions, on which, the electric field can be enhanced to the critical value estimated as a size of the order of the lattice constant. This fact should be taken into account by calculating the electron current by field emission, which defines the explosive emission conditions and its parameters. The theoretical analysis made by Ecker [62] leads to the same conclusion. Now let us study the explosion phenomena using the approach named as emission center.

15.4.3 Explosive Electron Emission and Cathode Vaporization Modeling the Cathode Spot

The further attempt to use explosive electron emission consists in definition and modeling a self-sustained emission center (EC) in 1983 [137], [138, 139] that later summarized by Mesyats [140]. Below, different approaches of cathodic arc study using the explosive phenomena are considered. It will be shown that further development of initially simple multi-explosions (only by Joule energy) the further explosion theory development was combined with the traditional emission vaporization mechanism of cathode processes.

15.4.3.1 First Approach to Use Cathode Vaporization in the Explosive Emission Center

Qualitative description of the emission center (EC). The scenario of the EC production was proposed as follows [140]. Initially, the current density reaches of about 10^9A/cm^2 (again *not clear how it can be in an arc with voltage of* 20 V) and then fast heating occurs of cathode material in a microvolume and its explosion resulting in efficient explosive emission. As the explosion develops, the emission zone increases in size and heating is dissipated by cathode heat conduction, evaporation, and ejection of heated cathode material. All these processes reduce the temperature in the EC operation zone and the thermionic electron emission current density.

The decrease in emission current density leads to reduction of the Joule heating. Therefore, the current ceases, ***but during this time a short-living portion of electrons*** is generated. Thus, the electron emission proceeds for short time, and then ceases because of the cooling due to energy loss. This portion ***ejection of electrons is defined as an EC***.

Quantitative description of the EC [137, 139, 140]. To determine the lifetime and other parameters of the EC, the cathode was modeled as a tip in form of conus with half angle θ at the top. To study of a crater at a plane surface was used $\theta = 90°$ (see Fig. 15.9). A system of equations was considered (in essence, in approach of model Ref [54] with different cathode geometry) including: (1) heat conduction equation in spherical coordinates with Joule heating as volume source, (2) Mackeown equation for electric field, (3) equation of thermionic electron emission with effect Schottky, and (4) total current $j \sim I/2\pi R_0^2$. The boundary condition for heat conduction equation described by the energy balance at the cathode surface that involves the energy fluxes transferred by evaporating atoms, the electron emission and the ion coming from the plasma to the surface. It was supposed that the evaporation zone coincides with the emission zone and both were represented by radius R_0 increased during the heating from an initial value. The time dependent Joule heating of the cathode was due to the time dependent current of thermionic electron emission enhanced by the electric field. The current extracted from the cathode was specified to be linearly rising with time ($dI/dt = $ const) or was given as constant value. When the electron emission current

15.4 Explosive Electron Emission and Cathode Spot Model

Fig. 15.9 Geometry of an emission center on the top of a point cathode and on the surface of a plane cathode

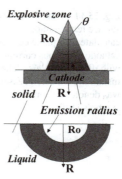

is high the current was limited by its given values. In this case, it was indicated that the work function should be increased due to virtual cathode appearance. However, no any theory of virtual cathode operation or its parameters (including an increase of the work function) was presented. The measured arc voltage was interpreted as the cathode potential drop. The rate of cathode evaporation was given by an exponential dependence on the surface temperature.

The solution showed that initial temperature reaches more than 30000 K and lifetime of the EC changes from few ns to 10 ns depending on cathode materials and given arc currents. The results for different cathode materials are presented in Fig. 15.10 and for Cu in Fig. 15.11. It was indicated that the EC at plane cathode can be developed at some critical rate of current rise and that cannot be lower than 10^{11} for Cu, 6×10^9 for Mo, and 4×10^9 A/s for W. For a conical tip, the critical dI/dt depends on angle θ and can lower up to $\sim 10^8$ A/s. For EC on a plane copper cathode at the moment of emission termination, the current density was decreased as 5.5×10^9, 3×10^9, 1.2×10^9, and 7×10^8 A/cm^2 when the current increased from as 10, 20, 50, and 100 A, respectively [140, 141]. This result indicates that the current density remains extremely high during the lifetime before the EC is dead.

Fig. 15.10 Time dependent surface temperature calculated for plane cathode at various metals and current of 20 A. The instant the emission ceases was denoted by circular symbols [138, 140]

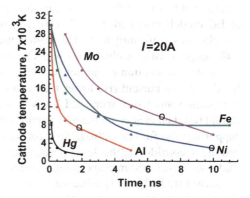

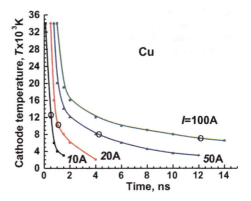

Fig. 15.11 Time dependent surface temperature calculated for plane cooper cathode at various current with dI/d$t = 10^{11}$ A/s. The instant the emission ceases was denoted by circular symbols [140]

The main debatable points of EC model. The initial radius was given $R_0 = 0.1$ μm for current carrying area (3×10^{-10} cm^2). The given values of R_0 and the currents, in essence, equal to given current density that for currents 10–100 A are 10^{10}–10^{11} A/cm^2. Therefore, it is not surprise that such fantastic cathode temperatures and, respectively, large electron emission appeared at times of about few ns. It was indicated that the ion current to the cathode was negligible low, but was not indicated its influence on the electric field and the electron emission enhancement. It is also not clearly described the procedure of accounting the cathode heating by the electron emission current with simultaneously given the values of rate of current rise and the current as a constant value. According to the explosive model, the EC was dead when the energy losses (due to increase of R_0 with time) by the material heat conduction, evaporation, and electron emission significantly cooled the cathode and cease of the Joule energy dissipation. This is critical point of the approach based on use of the EC. Let us explain this point.

In essence, this approach used system of four equations (for five unknown) as in [54] with main difference using transient heat conduction equation requesting initial conditions. The fifth unknown was given arbitrary as ion flux to the cathode determined by half of the evaporating atom flux from the cathode. Another problem of EC model presented in [137, 140] consists in the given initial current density that chosen significantly high. This high value consequently determines the substantially large resistive cathode heating. The result showed that for this condition the self-sustained solution could not be supported on time larger than few ns and also for the rate of current rise lower than some certain large values. However, as the evaporation atom flux increased his role was not only in cathode cooling but it should be taken in account the new phenomena producing additional plasma density supporting the further cathode spot formation. Therefore, the solution should be continued considering the dense plasma generated by the cathode evaporation. This study was developed in evaporating models. The mentioned system of equation was successful studied in [54] using some limiting conditions. For non-stationary cases, the solutions were demonstrated by gasdynamic mathematical formulation (see [65]) considering the plasma phenomena allowed to determine the fifth unknown, i.e., the

15.4 Explosive Electron Emission and Cathode Spot Model

ion current to the cathode and later by kinetic approach of the cathode vaporization (Chaps. 16 and 17).

15.4.3.2 Development of the Explosive Approach and EC Model

The further consideration of EC was concerned on its study using a different name of the explosive approach. To the name "EC" was added word "TON." Thus, the new name is "ECTON" assuming that the tip explosion is an elementary act. It can be assumed that the authors interpreted the "ECTON" act similarly to the acts from elementary particles "PROTON, ELECTRON" etc. Nevertheless, this approach could not be considered as elementary act and as independent subject similar to an elementary particle. First of all, this is a process depended on discharge conditions. The process duration and intensity determined on a given initial size of a protrusion or crater, which was measured in wide ranges. The size distribution depended on the cathode surface condition, cathode material (low- or high-melting temperatures), and arc parameters (current, gap length, gap plasma, voltage oscillations, and others). Thus, if a tip is explodes, the process is ambiguous in the arc conditions and determined by the mentioned parameters of the surface and of the arcs and cannot be interpreted as an elementary act. Moreover, the essence of the used and described above approach not depends on the name and therefore below it will be used the first definition of "EC."

To justify the given initial small sizes, the authors of [142] used, respectively, Daalder's [143] craters distribution on their sizes. According to these measurements Daalder calculated the (most probable) current density (in range of 2×10^7–1.9×10^8 A/cm^2) as a function of the current assuming that the crater having the most probable diameter is formed on the average by the total discharge current. Using Daalder's maximal current density ~10^8 A/cm^2 obtained at 50A, the authors of [142] provided some estimation in order to show the validity of the EC theory as a collective multi-EC process. Here, however, should be noted: (i) Daalder pointed that "to calculate the current density of a single discharge at the cathode one has to know if there are one or more emission sites active at the same time, a fact which is still unclear." This means that, in general, the current per crater (especially for small size) is an unknown parameter that requested further study.

According to the experiment [143] the total number of craters was in range of 170–440 depending on current; (ii) the collective multi EC process should be accompanying by, respectively, number of initial explosion of a protrusion that can be occurred at ~10^9 A/cm^2 and rate of current rise of 10^{11} A/s for Cu [140]. These parameters are far for those parameters in experiments for Cu in [143]. Using crater size, the relatively lower current density was measured at the moment of tungsten cathode spot death of 1.2×10^7 A/cm^2 for a cold cathode and 3×10^6 A/cm^2 for a hot one [144]. For closure gold electrical contacts, the current density of about $(2–6) \times 10^6$ A/cm^2 was reported for the crater sizes (5–10 μm) [145]. Also, the large range of current densities (of 10^6–10^8 A/cm^2) and spot lifetimes at the cathodes was obtained using the crater and trace sizes in vacuum arcs with presence of oxide and metallic films (see Chap. 7). Thus, even considering only the data of crater sizes measured in wide range of values, it is difficult accept a spot appearance as an elementary act satisfied to condition of initial explosion (crater 0.1µm) and of EC short lifetime with extremely large current density [137, 140].

Rossetti et al. [146] developed a numerical model of the predischarge heating process encountered by a microprotrusion on the cathode surface of a vacuum gap, due to the field emission producing high-current density flowing throughout the protrusion. The model is one-dimensional and non-stationary. The protrusion is sketched as a truncated cone and the material considered is tungsten, whose physical and thermal properties have been assumed temperature dependent. Microprotrusion height has been fixed to 10 µm; cone angles of 10° and 20° and tip radius of 0.1 and 0.05 µm have been investigated. Two different initial temperatures have been considered: 300 and 2000 K, as well as two different surface work functions: 4.5 and 2.7 eV.

The calculations showed that the melting temperature was reached inside of the tip for 4.5 eV, at $E = 10^8$ V/cm. For 2.7 eV, the melting temperature was reached at $E = 5 \times 10^7$ V/cm. For 2000 K, the critical electric fields was also $E = 10^8$ V/cm but yield the melting at the point immediately under the tip. The calculations confirmed the previous results that the explosion of protrusions occurred in a time of the order of 1–10 ns with current densities as 10^8 A/cm^2, while the corresponding electric field E at the tip can reach the value of 10^8 V/cm. Note, this conclusion was obtained again using very small protrusion size. The low value of work function can be obtained by the addition of thorium, barium, lanthanum, etc., in the base material. But the question what is the state of such doping material at temperatures significantly higher than the melting temperature.

The finite lifetime of an EC, obtained previously, was used to explain the cyclic character of a cathode spot appearing [147]. It has been indicated that protrusion explosions generate the arc plasma. The explosion occurs at the cathode surface by the Joule heating due to the high emission current density calculated by given small crater size. It was also indicated that "a cycle consists of two stages: the first stage lasts the time during which the EC is operative and the second stage of shorter duration during which a new EC is initiated by the ion current from the cathode plasma." However, no theoretical analysis was present how the first stage was initiated and how production of the new EC can be supported by ion current to the cathode and how the value of explosive current density of 10^9 A/cm^2 (to satisfy the specific action $h = 10^9$ As/cm^4) can be reached at the both stages.

The geometric shape and size of a solidified jet was considered by Mesyats et al. [148, 149] as a way for initiation a new explosive emission center (spot cell) during the interaction of the jet with the dense cathode plasma. An experiment was conducted with a piece of tungsten wire 150–200 µm in diameter used as a cathode and an arc current pulse of duration 700 ns, arc current ranged from 4 to 20 A. The arc triggered with a trigger electrode to which a pulsed positive voltage of 20 kV and duration 40 ns was applied. The experimental data showed that the track of a cathode spot contain a number of the solidified liquid metal protrusions of like jet-shaped. The authors represent separate drops of micrometer size jets with a neck of lower size

15.4 Explosive Electron Emission and Cathode Spot Model

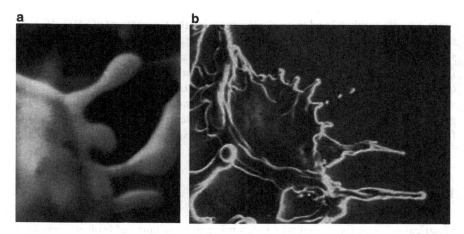

Fig. 15.12 Solidified liquid-metal in form of a protrusion: **a** protrusions formed at the edge of a crater observed in [150], **b** part of series of protrusions formed at the spot trace observed in [151]

(Figs.1–3 in [148]). It should be noted that such jet-shaped liquid metal protrusions was already detected previously after craters observations, for example, in Fig. 15.12 presented by Juttner [150, 151] (Fig. 15.12).

Recently [152] expanded the calculation of the explosive spot ignition at the neck configuration of a liquid metal at the periphery of a crater. Arbitrary initial geometry $R_0 = 0.1$ μm, plasma density, current of $I = 1$ mA and very large critical temperature varied from $(5-25) \times 10^3$ K were used. It should be noted, the real crater temperature after extinguish of the previous spot is significantly lower that used in the calculations. Estimations using energy balance for initially cold metallic neck (isolated cylinder with height L) taking into account the mentioned conditions, show that the neck temperature of 1000C can be reached only during 10–100 μs which is far from the obtained explosion period of ns.

The authors [148] described a scenario, mainly qualitative, of the initiation of a new explosive emission center (named also as a spot cell) occurred during the interaction of a liquid protrusion with dense cathode plasma proceeding in two stages. During the first stage, the protrusion is heated by the plasma ions. The second stage begins when the surface temperature reached the point of 3500–4000 K, and then the resistive heating becomes substantial as the emission current density increases with temperature. The current density at the protrusion base or current density through the neck was assumed geometrically enhanced. The degree of enhancement is assumed to be determined by a ratio between the total protrusion area to the base, or to it neck area.

However, as the ion current and the heat flux come to the total protrusion surface, the neck can be melted early in time and also distracted. Note, the total resistance of the protrusion with the neck (Fig. 15.12) can be considered as a number of series resistances, for which the electrical current is determined by the smaller resistance. As result, the electrical current for such configuration will be not enhanced but contra verse, become weaker. In additional, after reaching the temperature of 3500–4000 K, the material is melted (even for W), and of course, the geometry of the protrusion significantly changed. before any explosion emission can be occurred. Also, the fraction of the current to the protrusion is not known. This fraction can be very small to initiation of the heating in comparison with the total current at the local spot area where the protrusion arise. Besides, the mechanism of spot initiation on a melting protrusion occurred due to ion bombardment was reported previously [153, 154] in frame of vaporization model (without attract any explosion phenomena).

15.5 Summary

Electrical arc was discovered while measuring the resistance of broken graphite in early of nineteenth century, and then first its applications became important [134, 155]. In years after that, arc research and an attempt to understand the arc behavior was conducted mainly from the primary studies in second half of nineteenth century (see Chap. 11, [156]). In this chapter, the results of cathode spot study reviewed from the beginning of twentieth century up to the present time. The evolution of the main data and ideas are presented, starting from those, which developed for separate phenomena, up to systematic models inclusive and related to the cathode spot mechanism. For vacuum arc, different approaches were considered in order to describe the mechanisms of generating a plasma comprised of cathode material, based on cathode vaporization and local explosions. The description considers different stages of the studies showed evolution of the developed theories dependent on improvement of the diagnostic technique and cathode materials. The primary investigations were consisted in understanding the mechanism of electron emission to support the electrical current in the cathode region. The proposed mechanisms significantly varied depending on phenomena observed on such different cathodes as carbon or mercury, which involved thermionic or electron field emission, respectively.

Theory of field emission taking in account a high electric field at the cathode surface was developed due to necessity to explain how the current can be supported in a low-temperature spot of Hg arc. The further observation of arc on metallic cathodes and results of current density measurements stimulate new ideas of electron ejection from the cathode including de-excitation of excited atoms, presence of hot electrons in the cathode body, current supporting only by ion flux, and by primary models of local explosion at the cathode surface. To summarize of the early research works, at this stage, Robson and Engel noted [49]: "Many theories have been proposed in the last 15 years to explain the observed high-current densities associated with the cold cathode arcs. All of these theories are somewhat inadequate in themselves, but a combination of some of the more recent theories can give a satisfactory explanation of these high-current densities."

The problem of the cathode spot is not only the mechanism of electron emission but also the mechanisms of cathode plasma generation, especially when the arc arises in vacuum. An attempt to common (coupled) analysis of a number of

15.5 Summary

the cathode processes to determine of the main spot parameters was developed in the research works considered as second stage of the present review. The first such approach was given by Lee and Greenwood and by Ecker considering cathode vaporization, vapor atom ionization, and cathode space charge sheath. The mathematical formulation was presented by the main system of equations including equation of electron emission, electric field at the cathode surface, cathode energy balance, and the total current equations. Some limiting conditions used to close the mathematical formulation. In essence, this approach characterized as "vaporization model" used in the further models developed from about 1970th. To close the mentioned system of equations, except the limiting conditions, different modifications were used considering including sizes of the traces after the arcing, combination of plasma generation by explosions and by vaporization with given local and average values of current densities and other discharge parameters. The non-stationary models studied a time development of cathode crater considering a cathode thermal regime, but arbitrary giving an initial microsize of the crater. The heat flux to the cathode was determined in frame of the vaporization approach.

Beginning from about 1975 the approach of local cathode explosion became new interpretation using phenomena of explosion electron emission (EEE). This phenomenon produces both the cathode plasma and electron emission due to high-current density and overheating of a protrusion of the rough cathode surface at the breakdown stage at initiation of an electrical discharge. Firstly, it was assumed that after EEE the exploded plasma interact further with the cathode surface at sub-micron area that was a given parameter in the model explaining possibility of new explosions. Hereafter, to the cathode spot operation, the plasma generation was described in frame of vaporization model. A transient heat conduction equation was solved giving initial a high-current density and local Joule heating due to given microareas (crater or protrusion). As a result, the time of such interaction is very short (few nanoseconds) which was determined by time of breaking the heating process due to increase of the energy loss by evaporation, heat conduction, and others. Such localized plasma generation and ejected a package of electrons due to emission was named, first as emission center (EC).

Later this name was converted by formula "EC + TON = ECTON." The definition of EC was renamed as "ECTON" in order to indicate that a tip explosion can be considered as an elementary act. However, the *process* of transient vaporization using Joule heat source cannot be interpreted as an elementary act. Many factors influence on formation of the cathode surface morphology and on the microgeometry sizes (taken as input parameter in the EC model) during the arcing and new spot development. Moreover, the arbitrary assumptions limited the main conclusion about presence of a portion electrons. It is first of all the given very small initial characteristic size (10^9-10^{10} A/cm^2) leading to preferable role only the Joule heating. The model is not mathematically closed from which the size can be calculated and therefore cannot indicate presence another solution.

It should be noted that the used transient vaporization model can be solved for larger time when the ion flux to the cathode will be determined correctly and the EC size that is spot size will be not given, but will be determined by self-consistent study (such solution will be presented in the next chapters). One can meet other weak points of the explosion model considering the phenomena at film cathodes. The experiment showed a continued spot motion due to spot shifting after complete evaporation of the Cu film, which was observed as clear not discontinued tracks in each branches of spot moving. (see Chap. 7 and calculations below). Also, new EC can be produced by EEE when the condition of high electric field (≥ 100 MV/cm) will be formed at the cathode surface. This is problematic case in low-voltage arc discharge. The difficulties of produce such fields and possibility of current enhancements were discussed in [157]. Other problem of the EC model could by meet for Hg cathode. No usual mechanism of electron emission is applicable for Hg that used for the EC formation.

Taking into account that the EEE phenomenon requested a high electric field at the cathode surface during the arc with a multi-cathode cathode spot regime, the explosive model was modified at the last decade. An attempt to consider the local plasma interaction with small tips was considered without EEE. According to this approach [158, 159], the cathode protrusion was locally heated under previously generated plasma by energy of hot electrons (like in fusion devices), by the ion flux acquired in the cathode sheath and by Joule heating due to increase of the electron current density j_e due to T-F emission with increase of the protrusion temperature T. In essence, a vaporization model was used with the Mackeown, electron emission and balance equations, etc. The main difference from the EC model, here, the explosion occurred not as initial process but, vice versa, the process ended by the protrusion explosion. It was arbitrary assumed that the protrusion exploded when its temperature reached 20000 K and $j_e = 100$ MA/cm^2 (arbitrary points) for W and Cu cathodes.

To reach these critical values of j_e and T at time of about few ns, a specific range of high plasma densities and temperatures and of tip geometry were chosen. However, due to very small protrusion size, the products of explosion may not be reproduced the needed plasma parameters. Really, some above estimation showed that the mass of the exploding protrusion provided to be not enough, in order to support the input plasma density. Therefore, the calculating process can be not self-sustained. It is obvious that in all cases, the explosion can occur due to the high-current density. Although in recent work Mesyats [160] already indicate that "However, in an electric arc, the plasma-induced explosion of a liquid-metal jet occurs not at the jet tip but throughout its surface. In this case, Joule heating does not play a dominant part."

Thus, it should be noted some important points even not considering the detail of weakness of the explosion approach. The calculations in frame of vaporization model show that the electric field is reduced to support the electron emission at needed level due to surface heating. The characteristic times of both mechanisms are comparably short when chosen very dense of the initial plasma as input parameter. The protrusion geometry can be changed or completely destroyed due to melting and intense vaporization for a number of materials. The process of destroying can be occurred already in nanoscale time before reaching their overheating at mentioned above enormous temperature. The presence of intense cathode evaporation was shown even in the EC model (see also calculations in chapters below). So, it is difficult to understand the statement "according to the ecton model" with which often begin the developments by the authors of explosion model. Sometimes it seems as 'black box'?

15.6 Main Concluding Remarks

1. **The vaporization models** used a system of equations with arbitrary assumptions and arbitrary parameters including initial crater sizes, ion current density, electric field, electron current fraction, cathode surface temperature, and electron temperature. Therefore, the systems were not mathematically closed. Most of models calculate the rate of evaporation of cathode material according to the Langmuire–Dushman law that not takes in account the returned heavy particle flux. As result, the cathode mass loss was overestimated. A self-consistent study of the cathode plasma is needed.
2. **Explosion-EC models** used initial radius, angle, current or rate of current rise at certain large values. The lifetime of EC was determined using evaporating model with Joule heating. The **modified** explosion model used local plasma (arbitrary varied parameters) interaction with small microprotrusions.
3. **General point of view**. One can emphasis a specific point regarding to the above-reviewed papers. Most of published work, as usually, begins with an introduction reported like that "the cathode spot is complicated subject and up to present time the phenomena are not understand." At the same time, most of authors even not analyzed the positive results obtained in the previous works, further their description was presented as a new model using most of previous knowledges. In many cases, the previous work mentioned often only by citation. The calculated data are approved by comparison with experiment, from which difficult obtain exact numerical data. It is well known that Experiment is necessary but not enough by Cathode Spot study. As result, some new works describe models that very close to already published with repeating already well-known processes indicating as a new model with new arbitrary assumtions and weak points. Therefore, the cathode spot problem arises in a more confused form.

In some works, the cathode spot was studied numerically using commercial software which is "black box" with respect to understanding the adequacy of the results. Sometime, in such studies, multi-parametric initial characteristics were used. In this case, the resulting time dependent spot data can be varied with these input parameters and also not relate to a real spot behavior. Sometimes, the authors compare their results with that from other work with different mathematical formulation of the problems (for example, with different given an arbitrary initial parameters). All above-mentioned problems only confuse but not promote the understanding of the spot phenomena.

However, the cathode spot is really enough complicated subject at the point of view of its experimental study and observed unclear data. It is relevant to mention the spot characteristic provided by Hantsche in 2003 "*spots on the cathode surface are strange and fascinating phenomena. Spot is the purest form of electrical discharges are most surprising and very difficult to comprehend and to disclose at first glance. It remains an exciting problem and a challenge for future physical investigations, mainly due to the highly complex nature of arc spot operation, which calls for joint endeavors of several disciplines of physics, mathematics, chemistry, and material science."*

It should be noted that the role of the cathode plasma was not considered up to beginning of 70th years. In particular, the main issue—the ion flux formation, the mechanism of ions motion toward the cathode was not studied, and therefore the ion current fraction remained unclear. Also, it remains unclear what is the cathode plasma structure and what is the quantitative data of plasma density and electron temperature obtained in self-consistent approach. The approach that addresses aforementioned weaknesses will be described below in the next chapters. The advantages of the cathode spot theory described below consist in modeling of clear plasma structure, understanding the nature of particle and energy transfer as well of the ion flux generation, ion motion. It was shown that applicability of hydrodynamic and kinetic approaches to the investigation of the plasma flow beginning from the cathode surface up to plasma jet formation and jet expansion. The main difference between various models is that in some, the cathode potential drop u_c was assumed, whereas in later kinetic models u_c was calculated as part of a self-consistent set of equations [161, 162]. These last models represent a new glance regarding to role of the arc voltage at the moment of arc initiation and spot development. In this case, the spot temperature and the current density not rise significantly with time as in case of given constant u_c at which the calculated unlimited current rise (according EC model) was interpreted to show an explosive point.

References

1. Ecker, G. (1961). Electrode components of the arc discharge. *Ergebnisse der exakten Naturwissenschaften, 33*, 1–104.
2. Kesaev, I. G. (1964). *Cathode processes in the mercury arc*. NY: Consultants Bureau.
3. Kesaev, I. G. (1968). *Cathode processes in electric arcs*. Moscow: NAUKA Publishers. (in Russian).
4. Rakhovsky, V. I. (1973). *Physical bases of the commutation of electric current in a vacuum.* Translation NTIS AD-773 868 (1973) *of Physical fundamentals of switching electric current in vacuum.* Moscow: Nauka Pres.
5. Lafferty, J. M. (Ed). (1980). Vacuum *arcs*. In *Theory and applications*. New York: Wiley.
6. Stark, J. (1901). Berechnung der Leitfahigkeit durchstromenter Gase in der positiven Lichtsaule. *Annalen_der_Physik, 309*(1), 215–224.
7. Stark, J. (1902). Uber Ionisirung von Gasen durch Ionenstoss. *Annalen_der_Physik, 312*(2), 417–439.
8. Stark, J. (1903). Der Kathodenfall des Glimmstromes als Funktion von Temperatur, Stromstarke und Gasdruck. *Annalen_der_Physik, 317*(1), 1–30.
9. Stark, J. (1903). Zur Kenntnis des Lichtbogens. *Annalen der Physik, 317*(12), 673–713.
10. Richardson, Q. W., Coutt, B. A. (190.). On the negative radiation from hot Platinum. *Proceedings of the Cambridge Philosophical Society, 11*, 286. Richardson, Q. W. (1903). *Proceedings of the Cambridge Philosophical Society, 71*, 415.

References

11. Kapzov, N. A. (1947). *Electrical phenomena in gases and in vacuum*. In Gostekhisdat. M. (Ed.), (p. 500). (in Russian).
12. Finkelnburg, W., & Maecker, H. (1956). Elektrische bogen und thermisches plasma. *Handbuch der physic. Bd.*,22, 254–444(1956).
13. Tikhodeev, G. M. (1961). *Energetical properties of electrical welding arc*. Publisher Academy of Science of UdSSR. (In Russian).
14. Langmuir, I. (1913). Effect of space charge and residual gases on thermionic currents in high vacuum. *Physical Review*, 2(6), 450–486.
15. Langmuir, I. (1923). Positive ion currents from the positive column of mercury. *Science*, 58(1502), 290–291.
16. Langmuir, I. (1929). The interaction of electron and positive ion space charges in cathode sheaths. *Physical Review, 33*, 954–989.
17. Langmuir, I. (1923). Positive ion currents in the positive column of the mercury arc. *General Electric Review, 26*, 731–735.
18. Compton, K. T. (1927). The electric arc. *Journal of the American Institute of Electrical Engineers, 46*(11), 1192–1200.
19. Compton, K. T. (1923). Theory of the electric arc. *Physical Review, 21*(3), 266–291.
20. Compton, K. T. (1931). On the theory of the mercury arc. *Physical Review, 37*(9), 1077–1090.
21. Mackeown, S. S. (1929). The cathode drop in an electric arc. *Physical Review, 34*(4), 611–614.
22. Ecker, G. (1953). Die Raumladungszone an der Grenze des Bogenplasmas. *Zeitschrift für Physik, 135*, 105–118.
23. Froome, K. D. (1948). The rate of growth of current and the behaviour of the cathode spot in transient arc discharges. *Proceedings of the Physical Society, 60*(5), 424.
24. Froome, K. D. (1949). The behaviour of the cathode on an undisturbed mercury surface. *Proceedings of the Physical Society Section B, 62*(12), 805–812.
25. Slepian, J. (1926). Theory of current transference at the cathode of an arc. *Physical Review, 27*(4), 407–412.
26. Slepian, J., & Haverstick, E. J. (1929). Arcs with small cathode current density. *Physical Review, 33*(1), 52–54.
27. Weizel, W., Rompe, R., & Schon, M. (1940). Zur theorie der kathodischen eines entladungsteile lichtbogens. *Zeitschrift für Physik, 115*(3–4), 176–201.
28. Slepian, J., Berkey, W. E., & Kofoid, M. J. (1942). Arc cathodes of low current density at high amperage. *Journal of Applied Physics, 13*(2), 113–116.
29. Bauer, A. (1961). Zur Feldbogentheorie bei kalten verdampfenden Kathoden I. *Zeitschrift für Physik, 164*(5), 563–573.
30. Bauer, A. (1961). Zur Feldbogentheorie bei kalten verdampfenden Kathoden II. *Zeitschrift für Physik, 165*(1), 34–46.
31. Ginsburg, V. L., & Schabansky, V. P. (1955). Kinetic theory of the electrons and anomaly electron emission. *Dokl. AN SSSR, 100*(3), 445–448. (in Russian).
32. Kaganov, M. I., & Lifshiz,I. M., & Tanamarov, L. V. (1956). Relaxation between electrons and lattice. *Soviet Physics—JETP, 31, 2*(8), 232–237. (in Russian).
33. Smith, C. G. (1942). The mercury arc cathode. *Physical Review, 62*(1–2), 48–54.
34. Nevsky, A. P. (1970.) About electron temperature at a metal surface under action of a high power heat fluxes. *High Temperature, 8*(4), 898–899. (in Russian).
35. Nevsky, A. P., Scharakhovsky, L. I., & Yas'ko, O. I. (1982). Interaction of an arc with electrodes of a plasma torch. In Kisilevsky, L. I. (Ed.), Minsk. Nauka and Techinka.. (in Russian).
36. Bowers, H. C., & Wolga, G. J. (1966). Effect of a temperature gradient on thermionic emission. *Journal of Applied Physics, 7*(5), 2024–2027.
37. Morgan, W. L., Pitchford, L. C., & Boisseau, S. (1993). The physics of ion impact cathode heating. *Journal of Applied Physics, 74*(11), 6534–6537.
38. Kucherov, R. T., & Malhozov, M. F. (1981). About possibility of experimental determination a separation of electron temperature from the temperature of the crystal grid in a metal. *High Temperature, 19*(4), 878–881.

39. Nekrashevich, I. G., & Bakuto, I. A. (1955). On a question of the mechanism of electrical erosion of metals. Bulletin Byelorussian Academy of Sciences *Physical-Technical Institute, 2*, 167–177.
40. Nekrashevich, I. G., & Bakuto, I. A. (1959). On the mechanism of the emission of material from electrodes in a pulsed electrical discharge. *Inzh Fiz Zh, 11*(8), 59–65.
41. Il'in, V. E., & Lebedev, S. V. (1963). Destruction of electrodes by electric discharges of high current density. *Soviet Physics-Technical Physics, 7*(8), 717.
42. Rich, J. A. (1961). Resistance heating in the arc cathode spot zone. *Journal of Applied Physics, 32*(6), 1023–1031.
43. Rothstein, J. (1948). On the mechanism of electron emission at the cathode spot of an arc. *Physical Review, 73*, 1214.
44. Smith, C. G. (1946). Cathode dark space and negative glow of a mercury arc. *Physical Review, 69*(3–4), 96–100.
45. Rothstein, J. (1964). The arc spot as a steady state exploding wire phenomena. In Chace W. G., & Moore H. K. (Eds.), *Exploding wires* (Vol. **3**, pp. 115–224). New York: Plenum.
46. Loeb, L. B. (1939). *Fundamental processes of electrical discharges in gases.* New York: Jon Willey.
47. Robson, A. E., & von Engel, A. (1955). Excitation processes and the theory of the arc discharge. *Nature, 175*(4458), 646.
48. Lamar, E. S., & Compton, K. T. (1931). Potential drop and ionization at mercury arc cathode. *Physical Review, 37*(9), 1069–1076.
49. Engel, A., & Robson, A. E. (1957). The excitation theory of arcs with evaporating cathodes. *Proceedings of the Royal Society A. Mathematical, Physical and Engineering Science, A243*(1233), 217–236, (1957).
50. Lee, T. H. (1957). On the mechanism of electron emission in arcs with low boiling point cathodes. *Journal of Applied Physics, 28*(8), 920–921.
51. Holmes, A. J. T. (1974). A theoretical model of the mercury and copper vapor arcs. *Journal of Physics D. Applied Physics, 7*(10), 1412–1425.
52. Holmes, A. J. T., & Cozens, J. R. (1974). The abnormal glow discharge in mercury vapor and xenon. *Journal of Physics D. Applied Physics, 7*(12), 1723–1739.
53. Lee, T. H. (1959). T-F Theory of electron emission in high-current arcs. *Journal of Applied Physics, 30*(2), 166–171.
54. Lee, T. H., & Greenwood, A. (1961). Theory for the cathode mechanism in metal vapor arcs. *Journal of Applied Physics, 32*(5), 916–923.
55. Dushman, S. (1949). *Vacuum technique.* New York: Wiley.
56. Lee, T. H. (1960). Energy distribution and cooling effect of electrons emitted from an arc cathode. *Journal of Applied Physics, 31*, 924–927.
57. Ecker, G. (1971). The existence diagram a useful theoretical tool in applied physics. *Zeitschrift für Naturforschung, 26a*, 935–939.
58. Ecker, G. (1971). Zur theorie des vakuumbogens. *Beitrage aus der Plasmaphysik, 11*(5), 405–415.
59. Ecker, G. (1971). Theory of the vacuum arc II. The stationary cathode spot. *General Electric R-D. Report N71-C-195.*
60. Ecker, G. (1973). Modern development of the theory of electrode regions of an electric arc. *High Temperature, 11*(4), 773–779.
61. Beilis, I. I., & Lyubimov, G. A. (1975). Parameters of cathode region of vacuum arc. *High Temperature, 13*(6), 1057–1064.
62. Ecker, G. (1980). Theoretical aspects of the vacuum arc. In Lafferty J. M. (Ed.), *Vacuum arcs. Theory and application* (pp. 228–320). New York: Willey.
63. Ecker, G. (1976). The vacuum arc cathode,-a phenomenon of many aspects. *IEEE Transactions on Plasma Science, 4*(4), 218–227.
64. Ecker, G. (1978). Theoretical investigation of the cathode spot in a vacuum discharge. *High Temperature, 16*(6), 1111–1119.
65. Beilis, I. I. (1974). Analysis of the cathode in a vacuumarc, Sov. Phys. Tech. Phys *19*(2), 251–256.

66. Ecker, G. (1973). Unified analysis of the metal vapour arc. *Zeitschrift für Naturforschung A, 28*(3–4), 417–428.
67. Ecker, G. (1973). The non-stationary metal vapor arc. *Zeitschrift für Naturforschung, 28a*(3/4), 428–437.
68. Lyubimov, G. A. (1973). Cathode region of a high current arc. *Soviet Physics. Technical Physics, 18*(4), 565–569.
69. Lyubimov, G. A. (1970). Consumption of electrode material in cathode region of an arc. *Journal of Applied Mechanics and Technical Physics, N5,* 709–715.
70. Hull, A. W. (1962). Cathode spot. *Physical Review, 126*(5), 1603–1610.
71. Fursey, G. N., & Vorontsov-Vel'yaminov, P. N. (1968). Qualitative model of initiation of vacuum arc II. Field emission mechanism of vacuum arc onset. *Soviet Physics. Technical Physics, 12*(10), 1377–1382.
72. Mitterauer, J. (1972, September). Dynamic field emission (DF emission): A new model of cathode phenomena in cold cathode arcs. In *Proceedings of 2nd International Conference on Gas Discharges* (pp. 215-217).
73. Mitterauer, J. (1973). Dynamishe feldemission. Eine neue modellvorstellung des kathodenflecks an kalten kathoden. *Acta* Physica *Austriaca, 37,* 175–192.
74. Mitterauerer, J. (1976). Cathode processes in vacuum-breakdown and vacuum arcs. In *Proeedings of 4th International Conference on Gas Discharges* (p. 439).
75. Brodie, I. (1964). Studies of field emission and electrical breakdown between extended nickel surfaces in vacuum. *Journal of Applied Physics, 35*(8), 2324–2332.
76. Tomaschke, H., & Alpert, D. (1967). Field emission from a multiplicity of emitters on a broad-area cathode. *Journal of Applied Physics, 38,* 881–883.
77. Mesyats, G. A. (1971, September). The role of fast processes in vacuum breakdown. In Phenomena in Ionized Gases, Tenth International Conference, Invited Papers (p. 333).
78. Mitterauer, J. (1973). Experimental evidence of dynamic field emission (DF emission) within cathode spots of cold cathode arcs. *Nature Physics Science, 241*(113), 163–165.
79. Mitterauer, J. (1973). The surface dependence emission modes Hg arc film cathodes. *Journal of Physics. D. Applied Physics, 6*(8), L91–L93.
80. Mitterauer, J., & Till, P. (1987). Computer simulation of the dynamics of plasma-surface interactions in vacuum arc cathode spots. *IEEE Transactions on Plasma science, 15*(5), 488–501.
81. Beilis, I. I., Kantsel, V. V., & Rakhovsky, V. I. (1973). On explosive model of fast moving cathode spot. In *Proceedings All Union Conference on Electrical Contacts*. Moscow, U.S.S.R.: Nauka.
82. Beilis, I. I., Kantsel, V. V., & Rakhovsky, V. I. (1975). On explosive model of fast moving cathode spot. In *Chapter in book Electrical Contacts* (pp. 14–17). Moscow, U.S.S.R.: Nauka. (in Russian).
83. Golub, V. I., Kantsel, V. V., Rakhovsky, V. I. (1972). Current density and behaviour of cathode spots on copper during an arc discharge. In *Proceedings 2nd International Conference on Gas Discharges (London)* (pp. 224–226).
84. Hantzsche, E. (1972). Die maximale stromstarke eines bogen-brennflecks. *Beitr. Plasma Phys., 12*(5), 245–266.
85. Hantzsche, E. (1974). Modellrechnugen zum catodenfleck des vacuumbogens Kurzmitteilung. *Beiträge aus der Plasmaphysik, 14*(4), 135–138.
86. Hantzsche, E., Juttner, B., Puchkarov, V. F., Rohrbeck, W., & Wolff, H. (1976). Erosion of metal cathodes by arcs and breakdowns in vacuum. *Journal of Physics. D. Applied Physics, 9*(12), 1771–1781.
87. Carslaw, H. S., & Jaeger, J. C. (1959). *Conduction of heat in solids* (2nd ed.). Oxford: Clarendon Press.
88. Hantzsche, E. (1979). On the heat sources of the arc cathode spot. *Beiträge aus der Plasmaphysik, 19*(2), 59–79.
89. Hantzsche, E. (1983). Thermal runaway phenomena prevention in arc spots. *IEEE Transactions on Plasma Science, PS-11*(3), 115–122.
90. Hantzsche, E. (1981). Theory of cathode spot phenomena. *Physica, North-Holland Publishing Company, 104C,* 3–16.

91. Osadin, B. A. (1965). Energy release in high current vacuum discharge. *Soviet Physics-Technical Physics, 10*(7), 952–956.
92. Osadin, B. A. (1967). Contribution to the theory of the cathode spot in a heavy current vacuum arc. *Soviet Physics-Technical Physics, 12*(11), 1516–1521.
93. Gurov, S. V., Dzhafarov, T. A., Malinin, A. A., Osadin, B. A., & Tainov, Yu F. (1964). Electrode processes in a high current discharge in vacuum. *Soviet Physics–Technical Physics, 9*, 665–670.
94. Ecker, G., & Muller, K. G. (1959). Electron emission from the arc cathode under the influence of the individual field. *Journal of Applied Physics, 30*(9), 1466–1467.
95. Ecker, G., & Müller, K. G. (1959). Der Einfluß der individuellen Feldkomponente auf die Elektronenemission der Metalle. *Zeitschrift für Naturforschung A, 14*(5–6), 511–520.
96. Goloveiko, A. G. (1968). Elementary and thermophysical properties at the cathode in the case of a powerful pulse discharge. *Journal of Engineering Physics, 14*(3), 252–257.
97. Goloveiko, A. G. (1969). Limited processes at the cathode in the case of a powerful pulse discharge. *Journal of Engineering Physics, 16*(6), 1073–1081. (in Russian).
98. Murphy, E. L., & Good, R. H. (1956). Thermionic emission, field emission, and the transition region. *Physical Review, 102*(6), 1464–1473.
99. Goloveiko, A. G. (1967). Pulse action on metals of high power heat fluxes and volume het sources. *Journal of Engineering Physics, 13*(2), 215–224. (in Russian).
100. Kulyapin, V. M. (1971). Quantitative theory of cathode processes in an arc. *Soviet Physics-Technical Physics, 16*(2), 287–291.
101. Kulyapin, V. M. (1971). Some problems of heat conduction with phase transformations. *Journal of Engineering Physics, 20*(3), 497–504. (in Russian).
102. Kulyapin, V. M. (1972). Cathode erosion in an arc discharge. *Soviet Physics-Technical Physics, 17*(4), 622–625.
103. Kozlov, N. P., & Khvesyuk, V. (1971a). Cathode processes in electrical arcs I. *Soviet Physics. Technical Physics, 16*(10), 1691–1696.
104. Kozlov, N. P., & Khvesyuk, V. (1971b). Cathode processes in electrical arcsII. *Soviet Physics. Technical Physics, 16*(10), 1697–1703.
105. Afanas'ev, Yu V, & Krokhin, O. N. (1967). Vaporization of matter exposed to laser emission. *Soviet Physics JETP, 25*(4), 639–645.
106. Beilis, I. I., & Rakhovskii, V. I. (1969). Theory of the cathode mechanism of an arc discharge. *High Temperature, 7*(4), 568–573.
107. Beilis, I. I., Lubimov, G. A., & Rakhovskii, V. I. (1969). Electric field at the electrode surface at the cathode spot of an arc discharge. *Soviet Physics Doklady, 14*(3), 897–900.
108. Beilis, I. I., Lubimov, G. A., & Rakhovskii, V. I. (1970). Dynamics of the varation of the fraction of electron current in the region of an arc near the cathode. *Soviet Physics Doklady., 15*(2), 254–256.
109. Beilis, I. I., Lyubimov, G. A., & Rakhovskii, V. I. (1970, September). Dynamics of the variation of the fraction of electron current in the region of an arc discharge near the cathode. In *Proceedings of International Conference on Electrical Contact Phenomena* (pp. 31–33). Munich, Germany.
110. Nagaibekov, R. B. (1971). Ionization and ion charge exchange in the cathode spot of a vacuum arc. *Soviet Physics-Technical Physics, 16*(11), 1865–1867.
111. Harris, L. P., & Lau, Y. Y. (1974). Longitudinal flows near arc cathode Spots, New York: General electric Co, Schenectady. N.Y, Rep. **74**, CRD 154.
112. Persson, K.-B. (1962). Inertia-controlled ambipolar diffusion. *Physics of Fluids, 5*(12), 1625–1632.
113. Nemchinsky, V. A. (1979). Theory of the vacuum arc. *Soviet Physics-Technical Physics, 24*(7), 764–767.
114. Nemchinsky, V. A. (1983). Comparison of calculated and experimental results for stationary cathode spot in a vacuum arc. *Soviet Physics-Technical Physics, 28*(12), 1449–1451.
115. Lefort, A., Parizet, M. J., El-Fassi, S. E., & Abbaoui, M. (1993). Erosion of graphite cathodes. *Journal of Physics. D. Applied Physics, 26*(8), 1239–1243.

116. Coulombe, S., & Meunier, J.-L. (1997). Importance of high local cathode spot pressure on the attachment of thermal arcs on cold cathodes. *IEEE Transactions on Plasma Science, 25*(5), 913–918.
117. Prock, J. (1986). Time-dependent description of cathode crater formation in vacuum arcs. *IEEE Transactions on Plasma Science, 14*(4), 482-491.
118. Klein, T., Paulini, J., & Simon, G. (1994). Time-resolved description of cathode spot development in vacuum arcs. *Journal of Physics. D. Applied Physics, 27*(4), 1914–1921.
119. He, Z. J., & Haug, R. (1997). Cathode spot initiation in different external conditions. *Journal of Physics D: Applied Physics, 30*(4), 603.
120. Schmoll, R. (1998). Analysis of the interaction of cathode micro protrusions with low-temperature plasmas. *Journal of Physics. D. Applied Physics, 31,* 1841–1851.
121. Niayesh, K. (2000). Influence of electrode microstructues on the state of short vacuum gaps after interruption of high-frequency current. *Journal of Physics. D. Applied Physics, 33,* 2189–2191.
122. Mungkung, N., Morimiya, O., & Kamikawaji, T. (2003). An analysis of the instability phenomena of a low-current vacuum arc for copper cathode. *IEEE Transactions on Plasma Science, 31*(5), 963–967.
123. Uimanov, I. V. (2003). A two-dimensional nonstationary model of the initiation of an explosive center beneath the plasma of a vacuum arc cathode spot. *IEEE Transactions on Plasma Science, 31*(5), 822–826.
124. Barengolts, S. A., Shmelev, D. L., & Uimanov, I. V. (2015). Pre-explosion phenomena beneath the plasma of a vacuum arc cathode spot. *IEEE Transactions on Plasma Science, 43*(8), 2236–2240.
125. Mesyats, G. A., & Proskurovskii, D. I. (1971). Explosive electron emission from metallic needlesExplosive electron emission from metallic needles. *Soviet Physics JETP Letters, 13*(1), 7.
126. Mesyats, G. A. (1971, September). The role of fast processes in vacuum breakdown. In *Phenomena in Ionized Gases, Tenth International Conference,* Invited Papers (p. 333). Oxford.
127. Fursey, G. N., Antonov, A. A., & Zhukov, V. M. (1971). Explosive emission accompanying transition field emission into vacuum breakdown. *Vestnik LGY,10,* 75–78.
128. Fursey, G. N. (1985). Field emission and vacuum breakdown. *IEEE Transaction on Electrical Insulation, 20*(4), 659–670.
129. Mesyats, G. A. (1974). Electron explosive emission and electrical discharge in vacuum. *Proc. VI Int. Symp. Discharges and Electrical Insulation in Vacuum, 2,* 21. Swansea, UK.
130. Bugaev, S. P., Litvinov, E. A., Mesyats, G. A., & Proskurovsky, D. (1975). Explosive electron emission. *Soviet Physics Uspekhi, 18*(1), 51–61.
131. Litvinov, A., Mesyats, G. A., Proskourovsky, D. I., & Yankevich, E. B. (1976). Cathode processes at electron explosive emission. In *Proceedings of 7th International Symposium on Discharges and Electrical Insulation in Vacuum* (pp. 55–69). Novosibirsk.
132. Mesyats, G. A., & Proskurovskii, D. I. (1989). *Pulsed electrical discharge in vacuum.* Springer-Verlag.
133. Litvinov, E. A., Mesyats, G. A., & Proskurovsky, D. (1983). Field emission and explosive electron emission processes in vacuum discharges. *Soviet Physics Uspekhi, 26*(2), 138–159.
134. Anders, A. (2008). *Cathodic Arcs: From Fractal Spots to Energetic Condensation.* Springer.
135. Anders, A., Anders, S., & Juttner, B. (1992). Brightness distribution and current density of vacuum arc cathode spots. *Journal of Physics. D. Applied Physics, 25,* 1591–1599.
136. Jüttner, B. (2001). Cathode spots of electric arcs. *Journal of Physics. D. Applied Physics, 34,* R103–R123.
137. Litvinov, A., Mesyats, G. A., & Parfenov, A. G. (1983). The nature of explosive electron emission. *Soviet Physics—Doklady, 28,* 272–273.
138. Litvinov, E. A., Mesyats, G. A., Parfenov, A. G. (1984). The origin of the cyclical nature of explosive electron emission. *Soviet Physics—Doklady, 28,* 1019–1020.
139. Litvinov, E. A., Mesyats, G. A., Parfenov, A. G. (1984). Numerical modelling of the cathode processes at high current electron emission. Chapter in book. In *Emission high current electronics* (pp. 69–79). Nauka, Sibirian Brunch. Novosibirsk.

140. Mesyats, G. A. (2000). *Cathode phenomena in a vacuum discharge. The breakdown, the spark and the arc.* Moscow, Nauka.
141. Litvinov, E. A. (1985). Theory of the explosive electron emission. he *IEEE Transactions on Dielectrics and Electrical Insulation, 20*(4), 683–689.
142. Mesyats, G. A., & Barengol'ts, S. A. (2000, December). A high-current vacuum arc as a collective multiecton process. In *Doklady Physics* (Vol. 45, No. 12, pp. 640–642). Nauka/Interperiodica.
143. Daalder, J. E. (1974). Diameter and current density of single and multiple cathode discharges in vacuum. *IEEE Transactions on Power Apparatus and Systems,* (6), 1747–1757.
144. Puchkarev, V. F., & Murzakayev, A. M. (1990). Current density and the cathode spot lifetime in a vacuum arc at threshold currents. *Journal of Physics. D. Applied Physics, 23*(1), 26–35.
145. Boyle, W. S., & Germer, L. H. (1955). Arcing at electrical contacts on closure. Part VI. The anode mechanism of extremely short arcs. *Journal of Applied Physics, 26*(5), 571–574.
146. Rossetti, P., Paganucci, F., & Andrenucci, M. (2002). Numerical model of thermoelectric phenomena leading to cathode-spot ignition. *IEEE Transactions on Plasma Science, 30*(4), 1561–1567.
147. Barengolts, S. A., Mesyats, G. A., & Shmelev, D. (2003). Structure and time behavior of vacuum arc cathode spots. *IEEE Transactions on Plasma Science, 31*(5), 809–815.
148. Mesyats, G. A., Bochkarev, M. B., Petrov, A. A., & Barengolts, S. A. (2014). On the mechanism of operation of a cathode spot cell in a vacuum arc. *Applied Physics Letters, 104,* 184101.
149. Gashkov, M. A., Mesyats, G. A., Uimanov, I. V., & Zubarev, N. M. (2019). Molten metal jet formation in the cathode spot of vacuum arc. *IEEE Transactions on Plasma Science, 47*(8), 3456–3461.
150. Jüttner, B. (1979). Erosion craters and arc cathode spots in vacuum. *Beiträge aus der Plasmaphysik, 19*(1), 25–48.
151. Jüttner, B. (1979). Erosion craters and arc cathode spots in vacuum., Preprint 81-8, Akademie der Wissenschaften der DDR Zentralinstitut fur Electronenphyik, 19p. Berlin.
152. Tsventoukh, M. M. (2018). Plasma parameters of the cathode spot explosive electron emission cell obtained from the model of liquid-metal jet tearing and electrical explosion. *Physics of Plasmas, 25,* 053504.
153. Beilis, I. I. (1977). Cathode spots on metallic electrode of a vacuum arc discharge. *High Temperature, 15*(5), 818–824.
154. Beilis, I. I. (2001). State of the theory of vacuum Arcs. *IEEE Transactions on Plasma Science, 29*(5), 657–670.
155. Boxman, R. L. (2001). Early history of vacuum arc deposition. *IEEE Transactions on Plasma Science, 29*(5), 759–761.
156. Beilis, I. I. (2018). Vacuum arc cathode spot theory: history and evolution of the mechanisms. *IEEE Transactions on Plasma Science, 47*(8), 3412–3433.
157. Puchkarev, V. F., & Bochkarev, M. B. (1994). Cathode spot initiation under plasma. *Journal of Physics D: Applied Physics, 27,* 1214–1219.
158. Barengolts, S. A., Mesyats, G. A., & Tsventoukh, M. M. (2008). Initiation of ecton processes by interaction of a plasma with a microprotrusion on a metal surface. *Soviet Physics, JETP, 07*(6), 1039–1048.
159. Barengolts, S. A., Mesyats, G. A., & Tsventoukh, M. M. (2011). Explosive electron emission ignition at the "W-Fuzz" surface under plasma Power Load. *IEEE Transactions on Plasma Sciences, 39*(9), 1900–1904.
160. Mesyats, G., & Mesyats, V. (2018, November). The sequence of processes in the ecton cycle of a vacuum arc. In *Journal of Physics: Conference Series* (Vol. 1115, No. 2, p. 022020). IOP Publishing.
161. Beilis, I. I. (2011). Continuous transient cathode spot operation on a microprotrusion: Transient cathode potential Drop. *IEEE Transactions on Plasma Sciences, 39*(6), 1277–1283.
162. Beilis, I. I. (2013). Cathode spot development on a bulk cathode in a vacuum arc. *IEEE Transactions on Plasma Science, 41*(8), 1979-1986.

Chapter 16
Gasdynamic Theory of Cathode Spot. Mathematically Closed Formulation

According to the considered analysis in above chapters, the cathode spot is a complicated subject that is difficult to investigate not only experimentally, but also hardly studied theoretically. Therefore, the previously involved approaches used arbitrary parameters that are preferable to calculate. The main problem consists in perception for long time by the researches that the cathode spot is completely non-equilibrium formation and of extremely material state, which cannot be considered by traditional approaches [1]. The goal of any complete spot theory is to indicate mechanism that can determine the difference between non-equilibrium and equilibrium regions and to understand the charge particles' motion near the surface. As a result, in order to determine the current structure and especially the ion current, it can be large at the cathode surface. Below a system of equations is presented (based on author's contribution [2, 3]) modeling the ion motion to the cathode allowed to close the problem in comparison to previously published approaches. The mathematical formulation allowed using parameters relatively well measured to determine the spot parameters, which cannot be measured (without any data given previously arbitrary). Such model and respectively system of equations will be defined *as mathematically closed*.

Let us introduce the following definition. By the term of "cathode spot," the entire small region in both the plasma and the metal in which these two minute sub-regions support the current continuity between the highly conductive cathode body and low conductive adjacent dense plasma will be understood. The specifics of the *arc spot we now define as phenomenon, which arise as result of a local intense heating of the cathode*, in comparison to other discharges, for example glow discharge.

16.1 Brief Overview of Cathodic Arc Specifics Embedded at the Theory

In this chapter, the main experimental facts will be used as a base to modeling of the cathode processes in vacuum arc. The most significant and reliable among them is current contraction with relatively high density (>10^5 A/cm^2) in the separate luminous spots and thermal character of the metal evacuation, usually appearing due to influence of local high-intensive heat source. Thus, modeling of the self-supporting problem consists in searching for the sequence of the main common processes at the cathode and in the plasma, supplied local high power of the heat source and regeneration of the charges.

The results of previous studies of thermal regime show that heat inflow toward the cathode surface can be executed by the particles' bombardment, heat conductivity, radiation, and Joule dissipation of the electrical energy in the metal volume. In addition, these studies indicated that formation of the main energy sources is accompanied by current density produced by cathode electric field, a high cathode temperature (~4×10^3 K) and consequent high heavy particle density up to 10^{20} cm^{-3} with electron temperature ≥ 1 eV. The indicated parameters characterize the near-cathode region as a region with a plasma plume of enormous gaskinetic pressure (~10 atm). The question of any physical model is *how the ions in such plume moves to the cathode surface and what is the structure of electron and ion currents in the total spot current?*

16.2 Gasdynamic Approach Characterizing the Cathode Plasma. Resonance Charge-Exchange Collisions

Let us consider elementary collisions in the plasma, separated from the cathode surface by an electrical sheath, in order to understand the characteristic length of mean free path of the particles in the cathode vapor (Chap. 1). Atom ionization determines ion generation by electron impact. The ionization length depends on electron energy through a cross section σ_i, which is about 10^{-16} cm^2 for electrons, accelerated in a sheath with potential drop of 15 V. The plasma can consist of low or high ionized vapor. The model and calculations of the spot processes for metals of low ionized vapor are presented in the next section. Here, as example, let us consider the most typical case for Cu cathode [2, 3]. As the Cu plasma is dense and therefore not fully ionized, it can be assumed for estimation the neutral density n_a of 10^{20} cm^{-3}, and charge density n_i is lower about 10^{19} cm^{-3}. The Coulomb cross section of electron beam with charge particles is about 10^{-15} cm^2. Therefore, the mean free path $(n\sigma)^{-1}$ of beam electrons by atom ionization is about 10^{-4} cm at atom density of 10^{20} cm^{-3}, which determines the length of the ionization layer (ionization zone) in the cathode plasma. This same characteristic length for electron beam is determined the by interaction with the charge particle. The interaction between low

16.2 Gasdynamic Approach Characterizing the Cathode Plasma …

energy ions and other heavy particles is presented by ion–ion, electro–electron, and ion–atom collisions. The charge particle collisions are characterized by Coulomb cross section, which order of magnitude about 10^{-13} cm^2 for low-energy ions and electrons (~1 eV). For the ion–atom collisions, the charge-exchange cross section is relatively large ~10^{-14} cm^2 because in the cathode vapor for same type of particles, the interaction is characterized by resonance process. Therefore, the ion mean free pass is about 10^{-6} cm for n_i of 10^{19} cm^{-3} or for n_a of 10^{20} cm^{-3} in dense plasma near the surface.

Thus, the analysis of elementary particle collisions showed that the ion mean free path is significantly shorter than the mean free path of the energetic electron beam emitted from the cathode. This means that many collisions of low energy electron–electron, ion–ion, and ion–atom occurred in the relatively large ionization zone. As result, the particle flow in this region can be described in gasdynamic approximation. This enables to consider the mechanism of ion motion by the diffusion producing an ion flux toward the cathode surface. The diffusion model characterizes formation of the ion j_i and electron j_e current densities in space charge sheath and in ionization zone of the cathode plasma in vacuum arc; first was formulated by Beilis, Rakhovsky, and Lyubimov [2, 3]. The most important parameter is the electron current fraction s defined as:

$$s = \frac{j_e}{j_e + j_i} \qquad (16.1)$$

Let us consider different approximations of the diffusion model of the cathode dense plasma.

16.3 Diffusion Model. Weakly Ionized Plasma Approximation

The low ionized vapor near the cathode surface can take place in an initial stage of spot initiation and for arc initiated on volatile cathode materials with low melting temperatures. In this case, the ion generation can be accounted by direct impact of emitted electron beam with the atoms in the ionization layer in the cathode plasma. A model with characteristic plasma zones near the cathode, which considered electron impact ionization, was developed in 1970 [4, 5]. As the spot is a transient subject, the time-dependent electron current fraction $s(t)$ is an important issue.

This approach was mathematically formulated as follows. An ion flux Γ_i to the cathode surface can be formed due to the ion density gradient near the surface and is expressed by taking into account additionally the ion motion by plasma electric field E_{pl} in form:

$$\Gamma_i = \mu_i n_i E_{pl} - D_i \frac{dn_i}{dx}; \quad D_i \sim \frac{v_{iT}}{n_a \sigma_{ia}}; \qquad (16.2)$$

The ion diffusion equation with N_{e0} at $x = 0$ of the density of emitted electron flux (cm^{-2}s^{-1}):

$$\frac{dn_i}{dt} = \frac{d\Gamma_i}{dx} + F_i(x); \qquad (16.3)$$

The ion generation is described by ion source $F(x)$, determined by the atom ionization probability, which is dependent on distance from the cathode surface x as:

$$F_i(x) = n_a \sigma_i N_{e0} \text{Exp}(-n_a \sigma_i x); \qquad (16.4)$$

The boundary condition at plasma side before the electric sheath and faced to the cathode surface ($x = 0$):

$$\frac{dn_i}{dt} = \frac{n_i v_{iT}}{2D_i} = b n_i \qquad (16.5)$$

Here n_i and n_a are the ion and atom densities, respectively; D_i and μ_i are the diffusion coefficient and mobility of the ions, respectively; σ_{ia} and σ_i are the charge-exchange and atom ionization cross sections, respectively, v_{iT} is the ion thermal velocity. In order to determine the principle time-dependent behavior of the $s(t)$, let us consider the case of constant D_i, n_a, and N_{e0} and an initial certain density of neutral atoms and so:
Initial condition:

$$n_i(0, x) = 0 \qquad (16.6)$$

Considering that the plasma electric field is relatively small (satisfied to the condition $eE_{pl} \ll n_a \sigma_{ia} kT_i$), the solution of system (16.2)–(16.6) is obtained as following dependence of dimensionless plasma density on number of mean free path:

$$N_{\text{nor}} = \frac{n_i D_i a b}{N_{e0}(a+b)}$$

$$= \begin{bmatrix} \dfrac{\text{Exp}(\xi^2 t - ax)}{2(1+M)} \Phi^*\left(\xi\sqrt{t} - \dfrac{ax}{2\xi\sqrt{t}}\right) - \dfrac{\text{Exp}(\xi^2 t + ax)}{2(1-M)} \Phi^*\left(\xi\sqrt{t} + \dfrac{ax}{2\xi\sqrt{t}}\right) - \\ -\dfrac{\text{Exp}(-ax)}{1+M} + \dfrac{M^2 \text{Exp}\left(\dfrac{\xi^2 t}{M^2} - \dfrac{ax}{M}\right)}{1-M^2} \Phi^*\left(\dfrac{\xi\sqrt{t}}{M} - \dfrac{ax}{2\xi\sqrt{t}}\right) + \Phi^*\left(\dfrac{ax}{2\xi\sqrt{t}}\right) \end{bmatrix} \qquad (16.7)$$

where $M = \frac{2\sigma_i}{3\sigma_{ia}}$; $a = n_a \sigma_i$; $\Phi^* = 1 - \Phi(\varsigma)$; $\Phi(\varsigma) = \frac{2}{\pi} \int_0^{\varsigma_0} \text{Exp}(\varsigma^2) d\varsigma$; $\xi^2 = \frac{n_a v_{iT} \sigma_i^2}{3\sigma_{ia}}$.

As it is followed from the solution of (16.7) presented in Fig. 16.1, the distribution of normalized ion density N_{nor} is not monotonic and reached the maximum at distance

16.3 Diffusion Model. Weakly Ionized Plasma Approximation

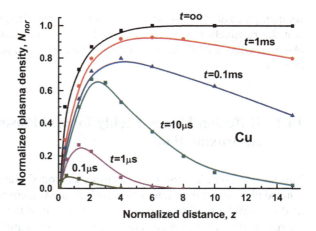

Fig. 16.1 Dimensionless electron current fraction as function on dimensionless distance from the cathode surface $z = ax$, expressed as number of electron beam mean free path for electron beam [4]

of about 1–3 mean free path depending on time. It can be seen that the plasma density reduced at cathode side due to ion motion toward the cathode surface.

At steady state, the normalized value of N_{nor} saturate to about unit and the distance of this region in general case can be determined by relation between cross sections of ionization and recombination, which will be considered in next section. Taking into account that the electron current fraction (16.1) can be obtained as $s = N_{e0}/[N_{e0} + \Gamma_i(0)]$ and using (16.7), the time-dependent relation of s at $x = 0$ can be expressed in the form:

$$s(\tau_d) = \left[2 - \frac{1 - \Phi(\sqrt{\tau_d})}{1 - M} \text{Exp}(\tau_d) + \frac{1 - \Phi(\sqrt{\tau_d/M^2})}{1 - M} M^2 \text{Exp}\left(\frac{\tau_d}{M^2}\right)\right]^{-1} \quad \tau_d = \frac{tnv\sigma_i^2}{3\sigma_{ia}}; \quad (16.8)$$

Figure 16.2 shows that s decreased with time and at steady state is equal 0.5 as minimal value. The characteristic time to reach the steady-state value is about 10 μs

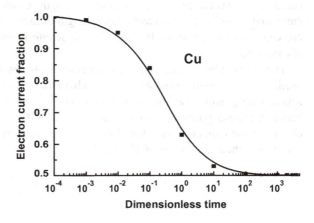

Fig. 16.2 Electron current fraction s as function on dimensionless parameter τ_d [4]

for mentioned above cross section of ionization and charge exchange. It can be noted that the relatively low value of s can be reached at transition to T-F emission of electron, when a large cathode temperature can be supported by ion component of current.

16.4 Diffusion Model. Highly Ionized Plasma Approximation

This section considers the slow electrons generated after atom ionization in highly ionized plasma and their interaction with emitted electron beam. The corresponding analysis will be used to modeling the formation of the fluxes of the particles and energy in the cathode multi-component and non-equilibrium cathode plasma in accordance with the model developed by Beilis, Lyubimov, and Rakhovsky [6].

16.4.1 Overall Estimation of Energetic and Momentum Length by Particle Collisions

Using analysis of Sect. 16.2, let us consider the characteristic parameters for partially ionized Cu cathode plasma. Besides the atom ionization, the accelerated electron beam also interacts with the generated slow plasma electrons and delivers them the energy and momentum. As the electron beam energy is 15 eV and respectively Coulomb cross section is $\sim 10^{-15}$ cm^{-2}, then the beam relaxation length is $l_r \approx 10^{-4}$ cm for plasma density of 10^{19} cm^{-3}. This same relaxation length of 10^{-4} cm is for energetic beam which arises with ionization collision at atom density of 10^{20} cm^{-3}. Since the interactions are electron–electron (equal masses), then the times (lengths) of the plasma electron relaxation for momentum and energy (Maxwellization) are equal. If the plasma electron temperature ~ 1 eV, the Coulomb cross section is $\sim 10^{-13}$ cm^{-2} and therefore the Maxwellization length of these electrons is about $l_{ee} = 10^{-6}$ cm with plasma density of 10^{19} cm^{-3}. This length is also for interactions between plasma ions.

The electron beam relaxation can be occurred also by the interaction with plasma oscillations. An estimation showed that characteristic time of electron beam interaction with plasma oscillations is about 10^{-12} s, while that time for interactions between plasma particles is significantly lower $\sim 10^{-14}$ s [6]. Therefore, this type of interaction cannot be developed due to high-intensive interaction of the plasma particles at their large level of 10^{19}–10^{20} cm^{-3}.

16.4.2 Modeling of the Characteristic Physical Zones in the Cathode Spot Plasma

The overall model can described accounting the typical plasma characteristics [6]. According to the above estimations, the cathode plasma may be divided into few main zones. The description can begin from the *cathode surface*, which is in a condensed state and assumed flat in first approximation. The cathode material far from the cathode spot is solid, while close to the cathode spot the material can be molten. Inside the cathode body, the Joule energy dissipation and the heat conduction away from the cathode spot take place. At the cathode surface, electron emission and atom evaporation are the processes occurred due to cathode heating and electric field at the cathode surface. The electrons emitted from the cathode surface transverse some short distance in the *ballistic zone* before suffering collisions, the length of which is defined by the mean free path (Fig. 16.3). The ballistic zone is a collisionless sheath in which the motion of the various particles can be described by collisionless equations (Chap. 5).

A strong electric field is present within the *ballistic zone* (sheath with cathode potential drop u_c), which accelerates the electrons away from the cathode. The electrons emitted from the cathode also suffer collisions, but because they are accelerated by the electric field in the *ballistic zone*, and the collision cross section decreases with the electron's kinetic energy, the mean free path for the beam electrons l_r is much larger than that of the plasma particles l_{ee}. Many collisions of low energy plasma electron occur in a region defined as the *electron beam relaxation zone l_r*. Thus, in the general case, the model of cathode region represents the adjacent to the surface ballistic zone. Also, the electron beam relaxation zone served as ionization region, in which the ions are generated due to direct electron impact and thermal ionization. There are two groups of electron inside the l_r region: monoenergetic accelerated

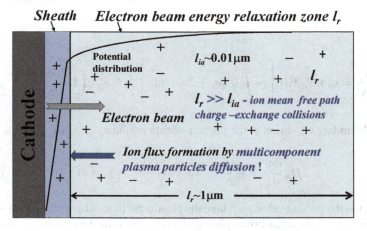

Fig. 16.3 Schematic presentation of the characteristic zones in the cathode spot plasma

electrons emitted from the cathode and plasma electrons, which velocity distribution may be considered as Maxwell's. This assumption follows from the fact that condition $l_r \gg l_{ee}$. The Debye length here is much smaller than the length (or width) of the plasma zone, and thus the plasma therein is quasineutral.

The dense plasma is optically thick, i.e., radiation emitted in the interior of the plasma is reabsorbed by the plasma, and is thus trapped within. The dense plasma is of a high thermal conductivity. The thermal diffusion time is much shorter than the spot lifetime, and hence, the temperature is relatively uniform throughout the zone. The energy, obtained by plasma electrons in the plasma electrical field E_p at length of l_{ee}, is small as compared with their thermal energy $eE_p l_{ee}/kT_e \ll 1$. As the spot radius r_s is considerably bigger than ionization region l_r, one-dimensional approximation can be used for its description. The flowing ions from the ionization region are accelerated by the electric field within the ballistic zone toward the cathode, and the energy carried by them is a significant source of cathode spot heating. Also the cathode heated by the energetic plasma electrons that returned to the cathode through the ballistic zone.

16.4.3 Mathematically Closed System of Equations

Assuming only singly ionized atoms for simplicity, the near-cathode dense plasma consists of three components that are ions, electrons, and atoms. Thus, the flowing ions toward the cathode can be obtained considering the analysis of multi-component and partially ionized plasma conducted in Chap. 5). The following diffusion equations were derived by studying (4.12)–(4.12) from Sect. 4.22.

1. $\Gamma_i = -(1+\theta)D_{ia}\dfrac{dn_i}{dx} + n_i v + \alpha D_{ia}\Psi; \quad \theta = \dfrac{T_e}{T}; \quad D_{ia} \sim \dfrac{0.4 v_{iT}}{(n_a + n_i)\sigma_{ia}}$ (16.9)

$$\frac{d\Gamma_i}{dx} = n_a \sigma_i N_{e0} \mathrm{Exp}\left[-\int_0^x (n_a\sigma_i + 2n_i\sigma_c)dx\right] + \beta_r n_0^2 n_i \left(1 - \frac{n_i^2}{n_0^2}\right) \quad (16.10)$$

The boundary condition at the plasma–sheath boundary ($x = 0$):

$$D_{ia}\frac{dn_i}{dt} = \frac{n_i v_{iT}}{2}; \quad v_{iT} = \sqrt{\frac{8kT}{\pi m}}; \quad j_i = e\Gamma_i \quad (16.11)$$

Here v is the plasma velocity, where the plasma particles originate from erosion of the cathode, n_0 is the equilibrium plasma density determined by Saha equation, σ_c is the Coulomb collision cross section, $\alpha = n_i/(n_i + n_a)$ is the degree of ionization, β_r is

16.4 Diffusion Model. Highly Ionized ...

the recombination coefficient, j_i is the ion current density, and Ψ is a term describing the particle flow due to the pressure and temperature gradients of electrons and ions. The details can be found in [6, 7].

2. Equation of continuity and momentum:

$$\rho v = G/m = \text{const}; \quad \rho v \frac{dv}{dx} = -\frac{dp}{dx} \quad n_a = \frac{p}{kT} - (1+\theta)n_i \quad (16.12)$$

3. Equation of state:

$$p = n_T T + n_e T_e; \quad n_T = n_a + n_i; \quad p_s(T_s) = n_s k T_s; \quad p_s(T_s) = A_s - B_s/T \quad (16.13)$$

where $p_s(T_s)$ and n_s are the equilibrium pressure determined by constants A_s, B_s from [8] and heavy particle density as function of cathode temperature T_s in the spot, respectively, p is the plasma pressure, n_T is the plasma heavy particle density, m is the heavy particle mass, and G- is the cathode erosion rate.

4. Equation of energy for the plasma electrons obtained from (4.18) of Sect. 4.22:

$$j_{em}\left(u_c + \frac{kT}{e_s}\right) - u_i(\Gamma_{iw} - n_i v) + \frac{j^2 l_r}{\sigma_{el}} - Q^*$$
$$= j_{et}\left(u_c + \frac{2T_e}{e}\right) + 3T_e\left(n_i v + \frac{j}{e}\right) - Q_r \quad (16.14)$$

$$Q^* = \int_0^{l_r} \frac{3m_e n_e}{m\tau}(T_e - kT)dx; \quad \frac{1}{\tau} = \frac{n_a}{n_i \tau_{ae}} + \frac{1}{\tau_{ie}}; j_{eT}$$
$$= \frac{1}{4}en_{ew}v_{eT}\text{Exp}\left(-\frac{eu_c}{T_e}\right); j = j_{em} - j_{eT} + j_i$$

Here k is Boltzmann constant, Q^* is the energy losses by elastic collisions, τ_{ei} is the effective time for elastic collisions between electrons and heavy particles, σ_e is the plasma electrical conductivity, u_c is the cathode potential drop, Q_r—plasma radiative energy flux, j_{em} is the emitted electron current density from the cathode, j is the spot current density, and j_{eT} is the returned electron current density toward the cathode. The electron temperature T_e in eV is determined by the balance between the energy flux carried by the electrons emitted from cathode, energy dissipation caused by ionization of atoms, and energy convection brought about by the electric current and electrode material losses in form of ions.

5. Equation of energy to determine the heavy particles' temperature $T = T_i = T_a$ (see (4.20) of Sect. 4.22):

$$\frac{5k}{2}\frac{dT}{dx}v(n_i + n_a) = -\frac{d}{dx}\left(-\lambda_a\frac{dT}{dx} + \mu_{ai}I_i\right) + Q^* \quad (16.15)$$

$$\mu_{ai} \approx 0.17\frac{kT_i}{m_e}; \quad \lambda_a = \frac{5}{2}\tau_{ai}\frac{k^2 n_a T}{m}$$

The estimations [7] provided by comparison of different terms in (16.15) showed that the temperature distribution in the relaxation zone depends on the plasma parameters and in most cases the condition of $T = T_s$ is fulfilled.

Let us consider the equations for phenomena at the cathode surface (electron emission, electric field) and in the cathode body (energy fluxes).

6. Equation of the electron current [7, 9] (see Sect. 2.2.4):

$$j_e = \frac{4em_e kT_s}{h^3} \int_{-\infty}^{\infty} \frac{\ln\left[1 + \text{Exp}(-\frac{\varepsilon}{kT_s})\right]d\varepsilon}{\text{Exp}\left[\frac{6.85 \times 10^7 (\varphi - \varepsilon)^{1.5}\theta(\tilde{y})}{E}\right]} - j_{eT} \quad (16.16)$$

$$\theta(\tilde{y}) = 1 - \tilde{y}\left[1 + 0.85\sin(\frac{1-\tilde{y}}{2})\right]; \tilde{y} = \frac{\sqrt{e^3 E}}{|\varepsilon|};$$

$$j_e = j_{em} - j_{eT}; s = j_e/j$$

In (16.16) the first term of the right part corresponds to emission current of electrons from the cathode, $\theta(\tilde{y})$ is the Nordheim function, φ is the work function, and ε—is the electron energy in metal.

7. Equation of cathode electrical field E is obtained by integration of Poisson's equation, in which the right part is the space charge resulting from the presence of positive charged ion current and negative charged electron emission and returned plasma electron currents derived in Sect. 5.52 (See also previous work [10]):

$$E^2 - E_{pl}^2 = 16\pi\sqrt{\frac{m_e}{2e}}\sqrt{u_c}\left\{j_i\left(\frac{m}{m_e}\right)^{0.5}\left[\left(1 + \frac{kT}{4\pi eu_c}\right)^{0.5} - \sqrt{\frac{kT}{4\pi eu_c}}\right.\right.$$
$$\left. - \frac{\sqrt{\pi}kT_e}{\sqrt{eu_c kT}}\left(1 - \exp\left(-\frac{eu_c}{kT_e}\right)\right)\right]$$
$$\left. - j_{em}\left[(1 + \frac{kT_s}{4\pi eu_c})^{0.5} - \sqrt{\frac{kT_s}{4\pi eu_c}}\right]\right\}$$

(16.17)

16.4 Diffusion Model. Highly Ionized ...

According to (16.17), the sheath will be stable when $T > \frac{\pi (kT_e)^2}{u_c}$ and with this condition, the monotonic potential distribution can be obtained near the sheath entrance. Taking the initial ion velocity according to the Bohm condition, i.e., $(kT_e/m)^{0.5}$, the cathode electric field can be obtained in form:

$$E^2 - E_{pl}^2 = 16\pi \sqrt{\frac{m_e}{2e}} \sqrt{u_c}$$

$$\left\{ \begin{array}{l} j_i(\frac{m}{m_e})^{0.5} \exp(0.5)\left[(1 + \frac{kT_e}{2eu_c})^{0.5} - \sqrt{\frac{kT_e}{2eu_c}} - \frac{kT_e}{\sqrt{eu_c}}(1 - \exp(-\frac{eu_c}{kT_e}))\right] \\ -j_{em}\left[(1 + \frac{kT_s}{4\pi eu_c})^{0.5} - \sqrt{\frac{kT_s}{4\pi eu_c}}\right] \end{array} \right\} \quad (16.18)$$

According to study of the cathode thermal regime (see Chap. 3) and for different conditions of Gauss distribution of a heat source and time of its action, there was produced a nearly flat portion of the temperature profile near the source center, which is close to the temperature T_s calculated from the time-dependent cathode energy balance written for the spot center (3.43). Let us take into account this result to develop a simple description of the thermal regime of the cathode spot.

8. Cathode energy balance:

$$j(1-s)u_{ef} + j_{et}(2T_{ec} + \varphi_{sh}) + q_j = q_T + \left(\lambda_s + \frac{2kT_s}{m}\right)G\frac{j}{I} - \sigma_{sB}T_s^4 \quad (16.19)$$

$$(1-s)u_{ef} = (1-s)(u_c + u_i) - \varphi_{sh};$$

$$q_T = 2\lambda_T \sqrt{\frac{j}{I}}(T_s - 300)\left(\frac{2}{\pi}\arctg\sqrt{\frac{4\pi ajt}{I_s}}\right)$$

$$\varphi_{sh} = \varphi - 3.79 \times 10^{-4}\sqrt{E}; \quad q_j = 0.32I^{3/2}\sqrt{\frac{j}{\pi}\rho_{el}}$$

where λ_s is the specific heat of evaporation of the cathode material, ρ_{el} is the cathode specific electrical resistance, q_T is the heat flux into the cathode due to heat conductivity, and I is the spot current.

It is assumed that the temperature is uniform within the spot and equal to T_s, the calculated temperature in the spot center (Chap. 3). It is further assumed that time dependence of spot parameters is determined only by the time dependence of the cathode heat conductivity as being the slowest process [7]. Cathode heating takes place by ion and electron bombardment of the cathode, and by Joule heating of the cathode body q_j according to Rich [11]. The terms of the left part of (16.19) correspond to these processes. Cathode cooling results from thermal conductivity, electron emission, evaporation of the electrode (erosion rate G in g/s), and radiation from the spot according to Stefan–Boltzmann law. The cathode thermal approach assumes that the heat flux to the cathode is constant during spot lifetime of the spot operation, and its value is determined by the self-consistent solution of the system equations. The cathode thermal regime taking into account the time-dependent heat flux will be considered below by calculation using the kinetic model of the spot development.

9. Equation of total spot current I_s for cathode spot assumed as circular form at the cathode surface:

$$I = jF_s; \quad F_s = \pi r_s^2 \quad (16.20)$$

The system of (16.9)–(16.20) represents nine main equations consisting of 12 unknowns and therefore requesting to define additional three unknowns. They can be chosen considering the spot parameters obtained relatively correctly from the experiment. First is the arc current, which should be given as a characteristic of the electrical circuit with the arc. The arc observation presents a certain number of spots, and therefore, the spot current I can be determined enough properly as one of the unknowns. The unknown characterized cathode mass loss is determined by the rate of cathode evaporation that depends on the cathode temperature and, in general, can be determined by Langmuir–Dushman formula [8]. However, a numerical analysis [12] shows the calculated result significantly exceeds the respectively measured data, due to presence large heavy particle flux returned toward the cathode and that not accounted by the mentioned formula. Therefore, parameter indicated the cathode mass loss could be characterized by the experimental value of erosion rate G. The next parameter well measured is the cathode potential drop u_c. It should be noted that this parameter was used in all published previously models reviewed above.

Thus, the above system allows to determine the following nine unknowns: j, E, T_s, r_s, s, n_T, n_i, T_e, and v_s. This system of equations describes in a self-consistent manner the mutual processes in the near-cathode plasma as well as in the cathode bulk and on the cathode surface. Therefore it is self-consistent closed system of equation, which determined the plasma parameters including the electron current fraction s which directly calculated, i.e., without arbitrary given or varied parameters as it was studied in previous publications.

16.5 Numerical Investigation of Cathode Spot Parameters

In this section, the application of the above described mathematically closed system of equation will be used to study the spot parameters for different cathode materials and arc conditions.

16.5.1 Ion Density Distribution. Plasma Density Gradient

One of the important issues of the cathode plasma study is to understand an influence of different plasma processes and plasma parameters on formation of the diffusion ion flux. To this end in first, a simulation of the (16.9)–(16.11) was conducted taking into account that the ion generation was provided by direct impact, stepwise, and thermal ionization [7]. The emitted electron beam intensity was reduced with distance from the surface by stepwise and direct impact ionization, as well due to scattering on the charge particles. In order to perform a general analysis, the mentioned equations were presented in following dimensionless form:

$$\frac{d^2 f}{dy^2} = A_1 \big[(1+\theta)(1-f) - A_2 \big]$$

$$\mathrm{Exp}\left\{ -\left[(1+\theta)(1-f) + \frac{f\sigma_c}{\sigma_i + \sigma^*} - A_2\left(1 - \frac{\sigma_i^*}{\sigma_i + \sigma^*}\right)\right] B_f y \right\}$$

$$- A_2 A_3 \mathrm{Exp}\left\{ -\left[(1+\theta)(1-f) + \frac{f\sigma_c}{\sigma_i + \sigma^*} - A_2\left(1 - \frac{\sigma_i^*}{\sigma_i + \sigma^*}\right)\right] B_f y \right\}$$

$$+ A_4 \frac{df}{dy} - f(1 - f^2) \tag{16.21}$$

Boundary conditions: $\frac{df}{dy}(y=0) = A_5 f_w$; $\quad \frac{df}{dy}(y=\infty) = 0$.
Here

$$f = \frac{n_i}{n_0}; \quad f = \frac{n_{iw}}{n_0}; \quad y = \left(\frac{\beta_r n_0^2}{D_{ia}}\right)^{0.5} x; \quad A_1 = \frac{\sigma_i j_{em}}{e\beta_r n_i^2}; \quad A_3 = \frac{\sigma_i^* j_{em}}{e\beta_r n_i^2}$$

$$A_4 = \frac{Gj}{Imn_0^2(\beta_r D_{ia})^{1/2}}; \quad A_5 = \frac{v_{iT}}{2n_0(\beta_r D_{ia})^{1/2}}; \quad B_f = (\sigma_i + \sigma^*)\left(\frac{D_{ia}}{\beta_r}\right)^{0.5};$$

The plasma parameters varied in wide ranges as following:
$j_{em} = 10^4$–10^6 A/cm^2, $n_0 = 10^{18}$–10^{20} cm^{-3}, $D_{ia} = 0.01$–10 cm^2/s; $A_2 = n^*/n_0 = 0.01$–1, n^* and n_0 are the density of excitation atoms and ion equilibrium density determined by Saha equation. According to the simulation, the ion density at the wall f_w is significantly different from the equilibrium ion density at large values of $\theta \approx 10$ and small D_{ia}. Figure 16.4 shows the dependence of normalized ion density distribution in plasma on distance from the cathode. When D_{ia} decreases from 0.1 to 0.01 cm^2/s, the value of f_w decreases from 0.55 to 0.25.

The simulation shows the results of ion density distribution, and the level weakly depends on processes of atom excitation, direct impact or stepwise atom ionizations, and ion outflow by the convection term due to cathode mass loss. This fact indicated on unessential role of the electron beam and on dominant influence of the thermal ionization in cathode plasma generation. It can be seen from (16.21) that if a condition

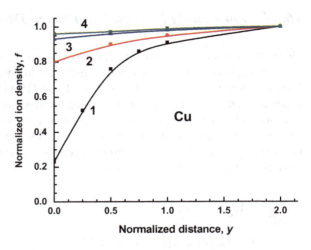

Fig. 16.4 Distribution of normalized ion density in the plasma as dependence on distance from the cathode surface. **4** $n_0 = 10^{19}$ cm^{-3}, $D_{ia} = 0.1$ cm^2/s, $j_{em} = 10^5$ A/cm^2, $\theta = 10$; **2** $n_0 = 10^{20}$ cm^{-3}, $D_{ia} = 1$ cm^2/s, $j_{em} = 10^5$ A/cm^2, $\theta = 10$; **3** $n_0 = 10^{19}$ cm^{-3}, $D_{ia} = 0.1$ cm^2/s, $j_{em} = 10^6$ A/cm^2, $\theta = 10$; **4** $n_0 = 10^{19}$ cm^{-3}, $D_{ia} = 10$ cm^2/s, $j_{em} = 10^4$ A/cm^2, $\theta = 3$ [7]

$$\frac{(1+\theta)}{\beta_r n_0^2 l_i} \ll 1 \qquad (16.22)$$

is fulfilled, then the ion density gradient is produced at distance from the cathode equal to the ion mean free path l_i. This means that the ion density is not changed and is equal to n_0 when the distance increased. In this case, the ion current toward the cathode surface can be determined by the random ion flux. The condition (16.22) was used to find a solution of the system of (16.9)–(16.11) taken in account the erosion rates measured by Kantsel et al. [13] as dependent on current. The saturation pressure at cathode temperature in the spot determines the heavy particle density, and it was assumed constant together with other plasma parameter in the relaxation zone. The developed approach here is very useful to determine the principle dependences of the above mentioned parameters on interrelated spot processes by mathematically closed approach. Note that a general approach taking into account the plasma parameters change near-cathode surface will be considered below using the kinetic model of the spot.

16.5.2 Numerical Study of the Group Spot Parameters

The system of (16.9)–(16.11) was first solved for copper cathode for slow group spot (SGS) by Beilis in [14] and then in [7] providing the current and time-dependent analysis. To understand the current influence, the study was conducted for currents in range of $I = 100$–500 A. For this range, the calculations showed that the current density decreased from 1.5×10^5 to 2×10^4 A/cm^2, degree of ionization from 0.2 to 0.07, electron temperature from 1.1 to 0.8 eV, and cathode temperature from 3820 to 3640 K with the current. The parametric simulation indicated that variation of the thermal conductivity, heat evaporation, cathode potential drop work function weakly

16.5 Numerical Investigation of Cathode Spot Parameters

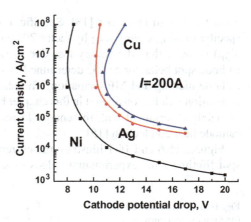

Fig. 16.5 Dependence of the spot current density on cathode potential drop at $I = 200$ A [15]

influences results, in particular, current density and electron current fraction. All spot parameters were weakly changed for spot lifetime larger than 1 ms.

The future detailed study of the group spot and spot fragments with lower currents was conducted for spots with lower currents and with cathode material from Cu, Ag, and Ni by Beilis [15]. The current density j as function on cathode potential drop u_c is presented in Fig. 16.5, for $I = 200$ A. The curves for Cu and Ag are close to one another, while some different dependences are for Ni. The dependences are characterized by a minimal values of u_c, which is ~10 V for Cu, Ag and 8 V for Ni, which are not equal to the experimental value, which are 15,13, and 18 V [16], respectively. In general, the calculated data are different from the potential of ionization although it is close for Ni.

The dependence of u_c on spot parameters can be obtained using the above energy balances to analyze the energy dissipation in the cathode of vacuum arc in form (see 1977 [15]):

$$eu_c = \varphi - 3.79 \times 10^{-4}\sqrt{E} + 3(1+\alpha\varsigma)T_e - 2sT_e + \left(\lambda_s + \frac{2kT_s}{m}\right)\frac{eG}{I}$$

$$+ Q_T + \alpha\varsigma u_i + 2T_e j(1-s)\left(\frac{mT_e}{m_e T}\right)^{3/2}\mathrm{Exp}\left(-\frac{eu_c}{T_e}\right); \quad \varsigma = \frac{eG}{mI} \quad (16.23)$$

Equation (16.23) shows that u_c depends on contribution of different terms. For example for Cu, the term $\alpha\varsigma u_i$ (energy to atom ionization) is about 0.1 eV, term due to the cathode erosion is about 0.4 eV and both terms weakly depend on spot current. Contribution of energy loss by electron emission is significant ~6 eV, and little larger is the energy loss by cathode heat conduction Q_T ~ 7–8 eV. This value of Q_T corresponds to Daalder's measurements of 6.2 V for 100 A [17]. The energy loss for cathode melting and MPs removing was neglected due to their low value in comparison to material loss by the erosion. It follows from condition $G_{MP}\lambda_{\mathrm{mel}}/G\lambda_s \ll 1$ which is fulfilled taking into account the removed melting material from the spot area by the macrodroplets G_{MP} ~ 60 μg/C and G ~ 30–40 μg/C, measured in accordance with Daalder's [18]. Specific atom evaporation energy λ_s exceeds the specific melting energy λ_{mel} by factor 20, and the average MP kinetic energy ($G_{MP} V_{MP}^2$) is lower than the energy loss due to vaporization. However, the dynamic of cathode spot behavior can be determined by cathode surface profile changed due to cathode melting and MP formation. The phenomena of cathode melting and MPs generation will be considered in this chapter below. Thus, the cathode potential drop is mainly determined with the energy losses by the atom evaporation energy and cathode heat conduction [15].

Figures 16.6 and 16.7 illustrate the current density dependences on current and spot lifetime t for experimental values of u_c (see above). It can be seen that the

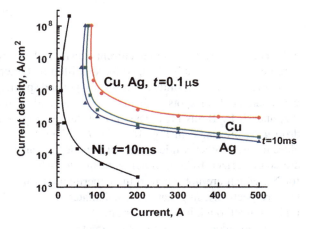

Fig. 16.6 Current density as function on spot current at t-10 ms and 100 µs (curve Cu, Ag) [15]

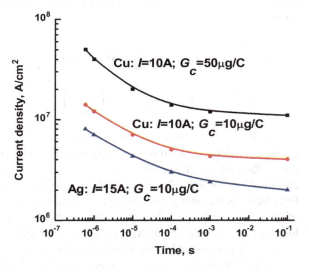

Fig. 16.7 Time-dependent result of spot current density at different spot current and cathode erosion rate for Cu and Ag cathodes [15]

dependences are not monotonic, and two-valued solutions with minimal value of I were obtained different for each metal. The current density increases when the spot lifetime decreased.

The curve for Ag is very close to that for Cu which sometimes not shown separately. The minimal current at which the system of equation can be solved is ~100 A for Cu and Ag and ~10 A for Ni at $G = 10^{-2}$–5×10^{-3} g/s. The calculations show that the dependence j on t for $I = 15$ A and dependences for Ag, Cu at $I = 10$ A presented in Fig. 16.7 can be obtained only for $G < 10^{-4}$ g/s and $j > 10^7$ A/cm^2. This result indicated that the spot with minimal current and small value of $t \sim 10^{-6}$ s which characterized the fast moving spot types can be described in frame of the present model. Nevertheless, the existence of such spots at $t \ll 10^{-6}$ s should be taken in account an increase of u_c at stage of their formation.

As illustration, the spot parameters calculated at 10 ms were presented in Table 16.1. The parameters corresponded to solutions "1" and "2" are related to the bottom and upper points on the calculated dependencies in the figures, respectively. It can be seen that the results for Ni at $I = 15$ A are similar to that obtained for Cu, Ag at large current of 200 A.

16.5.3 Low-Current Spot Parameters

A wide experimental study of group spots for Cu and CuCr for arc current range of 40–1500 A was provided in work of [19]. The spot number was found to increase linearly with arc current. The experimental results were discussed in frame of a theoretical study conducted for current of 40 A corresponding to the minimum current that sustained spots in the experiments after about 3 ms. The kinetic approach (see below) was used only for understanding the formation of the plasma flow at the cathode surface and to determine the velocity of the flow. For other parameters, the calculations were based mainly on above described model, i.e., using experimental u_c and G. The values u_c were varied in the range 14–17 V, being some volts smaller than the burning voltage measured at the electrodes. For convenience, we chose a constant erosion rate of 30 g/C [18]. The solutions were obtained as a function of the elapsed time from 10 ns to 3 ms. The upper boundary was given by the maximum delay of the optical measurement with respect to the arc ignition. This variation of the time was due to an uncertainty of the optically determined spot dynamics (displacement velocity and lifetime).

Let us consider the calculation results that shown in Figs. 16.8, 16.9, 16.10, 16.11, 16.12, 16.13 and 16.14. Generally, the curves for Cr essentially differ from those for Cu, as shown for example in Fig. 16.8 for the surface temperature T_s (t) and in Fig. 16.9 for the dependence of the density of heavy particles on the spot lifetime t. It can be seen that the cathode temperature is similar for the two materials in the time range 10–100 µs. Significant differences in the temperature are obtained for short lifetimes, such as 100 ns for which the difference may be as much as 500 K. In

Table 16.1 Spot parameter for two types of solutions named "1" and "2" for different cathode metals [15]

Metal	I (A)	Type solution	α	s	T_s (K)	T_e (eV)	E (MV/cm)	j (MA/cm^2)	n_i (10^{18} cm^{-3})	n_T (10^{18} cm^{-3})	v_s (10^3 cm/s)
Cu	200	1	0.1	0.450	3723	0.873	5.4	0.053	6.5	6.5	0.84
		2	0.86	0.977	5110	3.473	57.5	160	710	82	200
Ag	200	1	0.067	0.487	3530	0.880	6.97	0.078	11.8	17.6	0.38
		2	0.55	0.864	3921	1.873	30.7	585	220	40.2	12.4
Ni	200	1	0.017	0.356	3475	0.573	11	0.0018	0.25	1.46	–
		2	0.86	0.943	4877	4.507	51	45.50	480	56	49
Ni	15	1	0.083	0.401	3890	0.820	4.9	0.038	4.7	5.7	2.35
		2	0.72	0.746	4347	1.947	27	2.69	135	19	50
		$t = 1$ μs, $G = 10^{-4}$ g/s	0.787	0.51	4050	2.8	19	0.70	70	8.9	0.55

16.5 Numerical Investigation of Cathode Spot Parameters

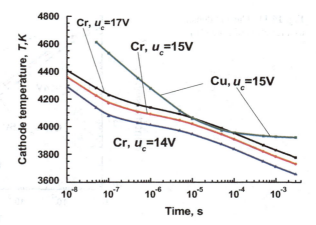

Fig. 16.8 Cathode temperature as a function of spot lifetime with cathode potential drop as parameter for Cr (u_c = 14, 15, 17) and for Cu (u_c = 15 V) [19]

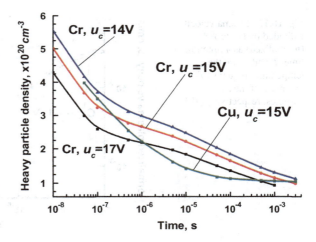

Fig. 16.9 Plasma density as a function of spot lifetime with cathode potential drop as parameter for Cr (u_c = 14, 15, 17) and for Cu (u_c = 15 V) [19]

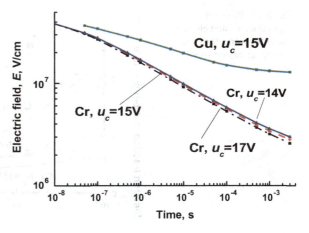

Fig. 16.10 Electric field strength as a function of time [19]

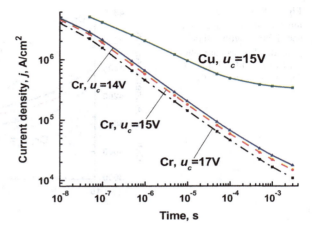

Fig. 16.11 Current density as a function of time [19]

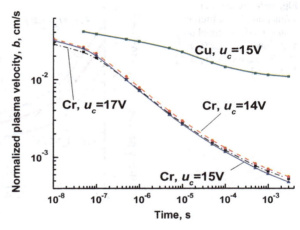

Fig. 16.12 Plasma velocity v divided by $(kT/m)^{0.5}$ (normalized) as a function of time. T and m are the temperature (equal to T_s) and mass of the heavy particles, respectively [19]

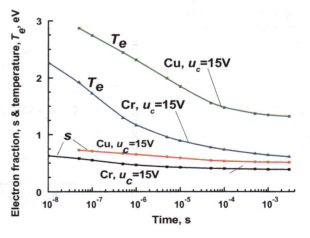

Fig. 16.13 Electron temperature T_e and electron current fraction s for Cr and Cu as functions on spot lifetime [19]

16.5 Numerical Investigation of Cathode Spot Parameters

Fig. 16.14 Degree of atom ionization α for Cr and Cu as functions on spot lifetime [19]

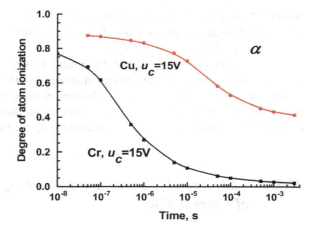

addition, it is interesting to note, that, for Cu, the temperature becomes practically constant for times >100 μs, whereas for Cr the temperature remains non-stationary throughout the time range.

Because the vapor pressure is somewhat higher for Cr than it is for Cu at the same temperature, a considerable difference in the particle densities is observed in the relatively broad time range of 2–100 μs (Fig. 16.9) for which the temperatures are similar for the two materials.

As it is shown in Fig. 16.10, the difference in the electric field strength increases monotonically with increasing time. The current density shows still more differences (Fig. 16.11). For Cu, the millisecond spot surpasses the Cr values of j by more than one order of magnitude. The plasma velocity near the cathode surface in the sub-sonic region of the dense plasma jet (Fig. 16.12), as discussed by Beilis [20], practically corresponds to the behavior of the current density, in agreement with the conservation of particle fluxes according to $G =$ constant.

The above calculated results show that the influence of the cathode falls on the spot parameters for both materials in the range 14–17 V appeared to be small. All parameters decreased monotonically with increasing lifetime and increased monotonically with decreasing cathode fall. Especially characteristic is the time dependence of all parameters for Cr, whence the non-stationary spot behavior. Thus, the most decisive parameter is the spot lifetime. Although for Cu and Cr the temperature and density of heavy particles are interrelated in a complex way, for both materials the current density varies monotonically with time. This might be connected with the experimentally observed behavior of the spot number versus current, this being similar for Cu and for the complex composite CuCr.

Figure 16.13 shows that the electron temperature T_e and electron current fraction s decrease with spot lifetime at $u_c =$ 15 V for Cr and Cu. It can be seen that in considered range of spot lifetime, T_e decreases by factor about 3 for Cr and by factor 2 for Cu, while s decreases from about 0.6 to 0.4 for Cr and from 0.7 to 0.5 for Cu. Another interesting result indicated that the degree of atom ionization significantly values at about 10 ns and α decreases with lifetime (Fig. 16.14). So the α decreases from 0.77 to 0.02 when t increased from 10 ns to 3 ms for Cr, and α decreases from 0.77 to 0.02 when t increased from 10 ns to 3 ms for Cr and from 88 to 0.4 when t increased from 50 ns to 3 ms for Cu.

Thus, the calculations show that the current density and the temperature do not decrease substantially when the current increases to 100–200 A, but increase considerably for currents <40 A. From the theoretical point of view, the basic physical parameter that influences the spot evolution is the heat flux density toward the cathode. This flux density is mainly determined by the ion current density.

16.5.4 Nanosecond Cathode Spots with Large Rate of Current Rise

Dynamics of nanosecond cathode spots in vacuum arc was investigated experimentally [21–24] (see details in Chap. 7). Shortly, the cathode surface luminosity for currents up to 150 A was investigated with high spatial (~μm) and temporal (~ns) resolution using laser absorption photography with a high-speed image converter and streak camera. The periodic fluctuations of the spot brightness in a nanosecond time scale were associated with the lifetime of the spot fragments. A fragment size of about 10 μm with a lifetime of about 10 ns at a given position and a current per fragment of about 10 A were found for a Cu cathode. The below model is considered in order to understand the nanosecond cathode processes in the spots calculating the cathode spot parameters (current density, plasma temperature, etc.). The main specifics of such spot is the current increase up to about 10 A in nanosecond duration, i.e., the model is considered for the arcs when spots are appeared at large rate of current rise $\geq 10^9$ A/s. This means that the model should be modified taking into account the heat flux rise to determine the cathode temperature from the cathode energy balance [25].

16.5.4.1 Model Modification for Nanosecond Spot Current Rise

As the spot current is $I = 10$ A during time $t = 10$ ns then the rate of spot current rise dI/dt is equal or higher than one GA/s. Thus in the nanosecond spot the value of dI/dt is relatively large. Therefore, the incoming heat flux to the cathode surface changes in time and depends on the current change during the spot life. In order to investigate the electron emission mechanism, the emitted electron beam current density from the cathode was calculated using Richardson's equation of thermionic electron emission with Schottky's correction factor (T-Sh-emission), and the equation for thermofield (T-F) emission; see above (16.16). We consider the case when the time of current increase is comparable with the spot lifetime. In this case, the non-stationary heat conduction equation may be studied with a time-dependent heat source in the following form:

16.5 Numerical Investigation of Cathode Spot Parameters

$$\frac{dT}{dt} = k_t \Delta T + \frac{j^2}{c\rho\sigma_{el}} \quad (16.24)$$

Boundary condition:

$$\lambda_T \frac{dT}{dx}(x=0) = q(y, z, t)$$

where $q(y, z, t) = F(y, z, dI/dt)$ is the time-dependent distribution function F of the heat flux at the cathode surface $x = 0$; y and z are Cartesian coordinates in the directions parallel to the cathode surface, T_s is the cathode body temperature, k_t λ_T, ρ, σ_{el} and c are the thermal diffusivity, thermal conductivity, density, electrical conductivity, and thermal capacity, respectively, of the cathode material.

In order to obtain a simple solution, we make a few assumptions. It is assumed that the current density is constant throughout the spot lifetime, i.e., spot area increases proportionally to the spot current. The Joule energy dissipation in the cathode bulk is described according to Rich's approximation [8] in the cathode energy balance. Assuming also a Gaussian distribution of the heat flux and that the spot current $I = t(dI/dt)$, the solution of the (16.24) can be obtained as [25]:

$$T_s(0, 0, 0, t) = \frac{u_{ef}}{2\pi\lambda_T}\sqrt{\frac{t}{\pi k_t}}\frac{dI}{dt}\left[\frac{dI}{dt}\frac{1}{4\pi k_t j} - 1\right]^{-1} \quad (16.25)$$

where $T_s(0, 0, 0, t)$ is the time-dependent cathode temperature in the spot center, u_{ef} is the effective cathode determined above, and j is the spot current density.

The cathode spot system of equations was solved together with (16.25), which determined the spot temperature in the cathode energy balance. An estimation of the non-stationary electron energy balance shows that, in the case of $j = $ const, the electron temperature T_e (eV) reaches it steady-state value during a characteristic time $t_R = 3.10^{-3} T_e \alpha / j u_c (1 - \alpha)$, where α is the degree of atom ionization and u_c is the cathode potential drop. When $u_c = $ 15–20 V, α is about 0.5, $T_e = $ 2–3 eV, j is in range 10^7–10^9 A/cm^2, and we obtain $t_R \sim 10^{-10}$–10^{-12} s. As the time t_R is small with respect to the spot lifetime (1–10 ns), then the steady-state electron temperature may use in the calculations below. Also the solution is obtained for simply case, taking in account the heavy particle flow in the kinetic layer. The cathode potential drop u_c and the erosion coefficient G(g/C) as well as the spot lifetime and the rate of spot current rise dI/dt are given.

16.5.4.2 Numerical Investigation of Nanosecond Spot Parameters

In order to investigate the electron emission mechanism and the role of tunneling effect, the emitted electron beam current density from the cathode was calculated using Richardson's equation of thermionic electron emission with Schottky's correction factor (T-Sh-emission), and the general equation for thermofield (T-F) emission

(16.16). In the calculations shown below, the measured erosion coefficient is given as 40 μg/C [18]. The spot current is 10 A. Taking in account that for small currents and large values of dI/dt, the cathode potential drop fluctuates [26], and thus the values of u_c are given in a wide range 15–50 V. In the proposed model, the spot radius r_s as well as T_s, j, T_e, α, plasma density n, electric field at the cathode surface E and electron current fraction s, are all calculated parameters.

Figure 16.15 presents the spot current density dependence on the cathode potential drop for dI/dt = 10 GA/s and t = 10 ns. It may be seen that the cathode potential drop has a minimum value of 15 V for T-Sh emission and about 18 V for T-F emission. A T-F current density was higher than that with T-Sh emission, but the difference was small for u_c> 25 V (see Fig. 16.1). However, the T-F current density was higher than that with T-Sh emission by factor 2 when u_c< 20 V. The numerical analysis shows that the contribution of field electron emission due to tunneling mechanism increases by relative high cathode electric field when the spot current density exceeds 3×10^7 A/cm^2.

Figure 16.16 demonstrates the current density dependence on the rate of current rise. The current density, in general, increases linearly with dI/dt, and for characteristic cathode potential drop 16–20 V, the T-F current density is about twice that of the T-Sh emission.

Figure 16.17 shows that the electron temperature increases linearly with the cathode potential drop. The value of T_e reaches 8–10 eV when the cathode potential drop reaches 50 V. The electron temperature was about 1 eV higher when the T-F electron emission was used in the calculations instead of T-Sh emission. It should be noted that according to the calculations, the electron temperature changes in range of 3.5–5 eV when the rate of spot current rise is in range 10^9–10^{11} A/s and u_c= 20 V.

The calculation of the various terms in cathode energy balance shows that the contribution of Joule heating is not large, about 1–5% of the total energy input. The main energy flux is due to returned electron (50–75%) and ion (20–45%) fluxes.

For a spot current of 10 A, the spot current density reaches a relatively high level (>10^8 A/cm^2). The calculation for case of T-F electron emission and dI/dt = 10 GA/s shows that in the above given range of cathode potential drop the electron current

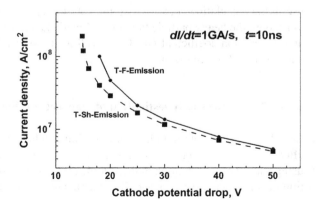

Fig. 16.15 Spot current density dependence on the cathode potential drop for Richardson–Schottky (T-Sh) electron emission and for thermofield (T-F) electron emission mechanisms. Spot current is 10 A [25]

16.5 Numerical Investigation of Cathode Spot Parameters

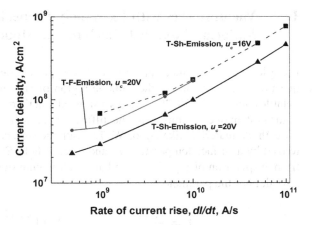

Fig. 16.16 Spot current density versus the rate of spot current rise (dI/dt) in cases of T-Sh-and T-F-Emission. Spot current is 10 A [25]

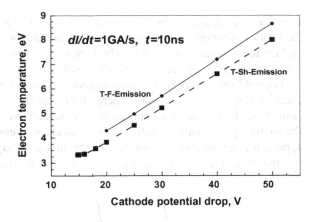

Fig. 16.17 Plasma electron temperature as a function on the cathode potential drop calculated for T-Sh-and T-F-Emission [25]

fraction changes from 0.8 to 0.97, current density increase from $5 \cdot 10^6$ to 2×10^8 A/cm^2 and spot radius decrease $r_s = 7.9$–1.3 μm. In this range of current density, the spot temperature is about 4300–6000 K, the plasma density is 2×10^{20}–2×10^{21} cm^{-3}, the electric field at the cathode surface is $(3$–$4) \times 10^7$ V/cm and the ionization fraction is 0.93–0.99.

The calculation indicates also that the current density can reach about 10^9 A/cm^2 when dI/d$t = 10^{11}$ A/s (see Fig. 16.16). The comparison with the experimental data shows that the measured spot radius $r_s = 1$–10 μm [21–24], measured plasma density $(3$–$6) \times 10^{20}$ cm^{-3} [21], and the electron temperature about 4–5 eV [27] are in the range of above calculated values.

16.6 Vacuum Arcs with Extremely Properties of Cathode Material. Modified Gasdynamic Models

For a long time, the mechanism of cathode spot operation has been an important question in the physics of the cathode region of vacuum arcs with refractory and volatile cathodes [7, 28, 29]. Analysis showed that there is no mathematical solution of the system of equations using the diffusion approach (gasdynamic model (GDM)) for cathode materials like Hg or W. The cathode surface conditions can be characterized by a surface temperature T_s and electric field E, resulting in an atomic flux from evaporation of $\Gamma_{ns}(T_s, E)$ and an electron emission flux of $\Gamma_{es}(T_s, E)$. Their ratio defines the parameter

$$Y = \frac{\Gamma_{es}(T_s, E)}{\Gamma_{ns}(T_s)} = AT_s^{5/2} \mathrm{Exp}\left[\frac{\varphi_{sh}(1-\chi)}{kT_s}\right] \quad (16.26)$$

where $\chi = \varphi_{sh}/\varphi_{ev}$, A is the cathode material-dependent constant, $\varphi_{ev} = \lambda_s$ (eV) is the heat of evaporation, and φ_{sh} is the work function with the Schottky factor. The surface temperature for several cathode materials was calculated as a function of current density by two approaches. The results are presented in Fig. 16.18.

The bold set of curves is a simultaneous solution of the cathode energy balance equation (16.19), conservation of energy for the electrons in the electron beam relaxation zone (16.14), electric field equation in the ballistic zone (16.17), together with the ion transport equation (16.9), i.e., without electron emission (16.16). The second approach is a simultaneous solution of the first three equations mentioned above, together with the electron emission (16.16), i.e., without ion transport equation (16.9).

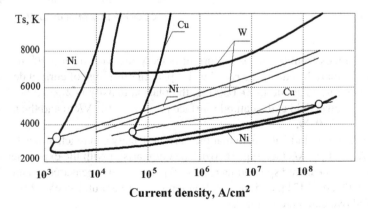

Fig. 16.18 Dependences of cathode surface temperature on spot current density for $I = 200$ A. The dependences marked by thin lines illustrate the solution of equations for electric field, balances for cathode and plasma, solved together with the equation of electron emission, while the solid lines obtained with the same equations but together with equation for ion current (i.e., without equation for electron emission). Circular points that can be also two-valued illustrate the solution presence

16.6 Vacuum Arcs with Extremely Properties of Cathode Material … 625

The diagram in Fig. 16.18 shows the presence or absence of a solution by GDM system of equations. It may be seen that the two curves have an intersection, and hence a solution, in the case of Cu and Ni, but do not intersect in the case of W. The curves of Ag have a solution and practically overlap the Cu curves, and therefore, they are not shown for clarity. The curves of Hg that likewise do not have an intersection will be presented below. Circular points mark the solution presence. The GDM calculations show that the materials having intermediate thermophysical properties (e.g., Cu, Ag, and Ni) have values of χ of around unity, and solutions are readily found. However, when $\chi < 1$, i.e., for refractory metals (e.g., W, see Tables in Appendix of the book), increasing the surface temperature results in $Y \to \infty$. Under these circumstances, there is high electron emission, causing plentiful electron emission cooling, but there is no ion current and hence insufficient ion bombardment heating, and thus a solution cannot be found for the cathode energy balances (16.14) and (16.19). Likewise, the GDM has no solution if $\chi \gg 1$, e.g., for Hg. There is insufficient electron emission, while the vaporized atom flux is too much to reach a concordance between plasma particle and energy balances. Thus, also for the volatile materials, the plasma energy balance cannot be established with the present model, due to insufficient plasma heating by electron beam energy. In both cases there, the particle and energy balances had some contradictions.

Thus, Fig. 16.18 indicates that solutions can be obtained for materials having a mid-range of thermophysical parameters, while modified models are necessary for extremely refractory or extremely volatile materials. To overcome these contradictions, new ideas were developed for the near-cathode sheath that determines the spot mechanism. An adjacent layer to the cathode was hypothesized, which served as a "virtual cathode" for W material and a "plasma cathode" supplying the required electron flux for Hg.

16.7 Refractory Material. Model of Virtual Cathode. Tungsten

For the refractory materials, the electron space charge near the cathode produced an electric field in the direction that reduced the large thermionic electron emission, thus acting as a virtual cathode [30, 31]. The potential distribution is schematically illustrated in Fig. 16.19. Two characteristic regions are formed in the sheath. In first region at some distance from the cathode surface, a minimal potential u_m is produced (zero electric field) due to negative charge of large electron emission, and the emitted low energy electrons are returned toward the cathode. After the minimum, the energetic electrons are accelerated. In the second region, the electron acceleration continues due to the potential increase by positive charge of the ions flux to the cathode from the plasma.

Integrating Poisson's equation from $u = u_m$ to $u = 0$ taking into account the simple case of single charged ion current extracted from the plasma boundary toward the

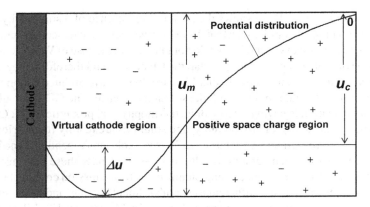

Fig. 16.19 Schematic diagram of the electric sheath with a virtual cathode

cathode the following expression for electric field E_{pl} at the plasma side is obtained (see Chap. 5, 5.33):

$$\varepsilon_0 E_{pl}^2 = 4en_{i0}\sqrt{u_m kT}\left[\left(\sqrt{1+\frac{kT}{u_m}}-\sqrt{\frac{kT}{u_m}}\right)-\sqrt{\pi}s_r\sqrt{\frac{kT_e}{u_m}}\left(1-e^{-\frac{u_m}{kT_e}}\right)- \right.$$
$$\left. -\sqrt{\pi}s_e\left(1+\frac{kT_s}{2u_m}\exp\left(\frac{u_m}{kT_s}\right)\Phi^*\left(\sqrt{\frac{u_m}{kT_s}}\right)-\sqrt{\frac{\pi u_m}{4kT_s}}\right)\right]$$

(16.27)

$$s_e = \frac{j_e}{j_i}\varepsilon; \quad s_r = \varepsilon j_{eT0}\exp(-u_c/kT_e); \quad \Phi^*(\xi)$$
$$= (1-\Phi(\xi)); \quad j_{eT0} = n_{er0}\sqrt{\frac{T_e}{2\pi m_e}}; \quad \varepsilon = \sqrt{\frac{m_e}{m}}; \quad (16.28)$$

where j_{eT0} is the random current density of the returned electrons from the plasma. The hydrodynamic model was solved with the above modifications and using experimental values for u_c, I, and G in the cases of W, Mo, and Gd cathodes. In the case of W, using ($u_c = 22$ V, $I = 20$ A, and $G = 10$ μg/C), a solution was found with the following spot parameters: $j = 2 \times 10^6$ A/cm², $T_s = 7400$ K, $T_e = 6$ eV, $s = 0.99$, $\alpha = 1$ (with 50% of the ions doubly charged), $n_T = 10^{19}$ cm⁻³, and $\Delta u < 1$ V.

An examination of the heat balance at the cathode surface reveals that the largest heating factor is from bombardment by the hot plasma electrons, while Joule heating accounts for about 30% of the equivalent heat flux, and ion bombardment an even smaller fraction. This should be contrasted to the case of intermediate metals, where ion bombardment was the dominant heating mechanism. It should be noted that the electron temperature calculated here is significantly higher than that calculated for intermediate metals (e.g., for Cu, $T_e \leq 2$ eV). This is a consequence of the high electron current fraction s that leads to a high heat flux into the electron beam relaxation zone, and due to low energy loss resulting from a low backflow of ions to the cathode. The predicted value of the current density j is close to values determined experimentally: $j = 8 \times 10^5$ A/cm^2 (from an experimentally measured cathode spot diameter of 40 μm) [32], and about 10^7 A/cm^2 [33]. At elevated temperatures, the calculated current density decreases by a factor of approximately 2 for initial temperatures of up to 1500 K, while the experimental results also show a decrease, by a factor of 5 for initial temperatures of 1800 K [33]. The model results are similar for Mo, but with a smaller value of Δu.

The hydrodynamic model was also applied for Gd, using the following input parameters: $u_c = 13$ V, $I = 20$ A, and $G = 100$ μg/C. It was found that: $j = 3 \times 10^3$ A/cm^2, $T_s = 2700$ K, $T_e = 3$ eV, $s = 0.97$, and $\Delta u = 0$. Thus, Gd, with $\chi = 0.75$, represents the marginal case with $E = 0$ and divides the refractory metals from the intermediate metals. The electron temperature and electron current fraction resemble the typical values for the refractory metals, while the current density and surface temperature are more typical of the intermediate metals.

The calculations in framework of above approach show that the electron current fraction $s = j_e/j$ is very large, approaching 0.99, and consequently, a relatively large total spot current density $j > 10^6$ A/cm^2 is obtained. However, in vacuum arc that slowly move and has steady-state spots with $j < 10^6$ A/cm^2 could be also observed on tungsten [34–36] and molybdenum [37] cathodes. At the same time, the above calculations show that lower j is consistent with lower $s < 0.8$. Consequently, two questions arise: *(i) what is the mechanism controlling the relatively small electron to ion current ratio in cathode spots on refractory cathodes? (ii) How does the spot on refractory cathodes operate in a self-sustained manner with a low s?* Thus, the nature of the spot operation with relatively low electron current fraction ($s < 0.9$) and the relation between the spot parameters and s are important to be clarified for refractory cathodes. Let us consider the model of the spot on refractory vacuum arc cathodes in order to study the low s and the corresponding spot plasma parameters.

16.7.1 Model of Low Electron Current Fraction s for Tungsten

According to GDM cathode spot theory, the electron and ion currents are controlled in a positive space sheath formed in a transition layer between the cathode and a quasineutral plasma. For refractory cathodes, the virtual cathode approach used which is described in (16.27) for E_{pl} (at right side in Fig. 16.19). The above solution for the virtual cathode was obtained using zero electric field at which the plasma side of the sheath E_{pl} is also assumed. According to Langmuir [38] theory in this case, the ratio of emitted electron current density j_e to the incident ion current density j_i is proportional to the square root of the ion mass m to the electron mass m_e ratio, i.e.,

$j_e/j_i \sim (m/m_e)^{1/2}$, [31]. Thus, j_e is much larger than j_i. Mathematical analysis of the sheath structure will show that the ratio j_e/j_i can be lower if $E_{pl} \neq 0$.

Recently, the two-scale problem (Bohm condition) in the plasma–wall transition layer was investigated [39]. It was shown that the plasma conditions could be matched to the sheath conditions when a certain dependence is maintained between the ion velocity v_i and electric field E_{pl} at the presheath–sheath boundary. The calculation shows that E_{pl} reaches a finite value lower than $E_d = T_e/r_d$ and v_i is about 0.9 of Bohm velocity v_B for a small electron to ion currents ratio (~1), where r_d is the Debye radius.

Taking into account this result, a nonzero electric field at the plasma sheath interface is assumed in the developed below model and a near-cathode quasineutral plasma consisting of ions with different charges will be considered [40]. The mathematical formulation is based on the gasdynamic cathode spot theory. The following system of equations is considered.

(1) Equation of quasineutrality: $n_e = \sum Z n_{iz}$, where n_{iz} is the ion density with ion charges $Z = 1, 2, 3, 4$, n_e is the electron density.
(2) The ion current density j_{iz} is calculated using n_{iz} and assuming the Bohm ion velocity for simplicity.
(3) Plasma energy balance.
(4) Cathode energy balance.

The expressions of (3) and (4) were accounted for the multi-charged ions. In the following calculations, these equations are modified taking in account the ionization energy loss for ions with different ionicity as

$$\Psi_e = n_e v \left(2T_e + \sum_1^z \left(\frac{f_z}{Z} \sum_1^z u_{iz} \right) \right) \quad (16.28)$$

$$\Psi_s = j_i \left(u_c + \sum_1^z \left(f_z \sum_1^z u_{iz} - f_z Z \varphi \right) \right) \quad (16.29)$$

where Ψ_e is the term related to the *plasma energy balance* and in the energy flux toward the cathode by ions with different ion charge z and Ψ_s is the term related to the *cathode energy balance*, where u_{iz} is the energy of ionization of ions with charge Z, $f_z = j_{iz}/j$ is the current density fraction for ions with charge Z and φ is the cathode work function, v is the plasma mass velocity determined by the cathode erosion rate G. u_c is the cathode potential drop and $u_c = u_m - \Delta u$.

(5) System of four Saha equations used for calculation of the density of four types of ions n_{iz} with $Z = 1$–4.

$$\frac{n_z n_e}{n_{z-1}} = \frac{2g_z}{g_{z-1}} \left(\frac{2\pi m_e T_e}{h^2} \right)^{3/2} \text{Exp}\left(-\frac{u_{iz}}{T_e} \right) \quad (16.30)$$

16.7 Refractory Material. Model of Virtual Cathode. Tungsten

where g_z is the statistical weight, n_0 is the neutral atom density, and h is the Planck constant. The Saha equations for the multi-ionization case were described previously [41].

(6) The plasma pressure is $p = nkT_e + n_T kT$, k is the Boltzmann constant, and here temperatures are in °K. The heavy particle density n_T is calculated from the saturation cathode vapor pressure determined by the cathode temperature T_s, assuming equality with the heavy particle temperature T. Here, the pressure correction from non-ideality of the plasma gas is neglected due to small non-ideality parameter $(z^2 e^2 n_e^{1/3}/T_e < 1$ [42], e is the electron charge, see calculation below).

(7) The electric field E_{pl} is obtained from Poisson's equation, taking into account the space charge from ions with different charges, plasma electrons back flowing toward the cathode, electrons emitted by the virtual cathode and the quasineutrality condition (1) at the sheath-plasma interface. In general, quasineutrality is violated at the sheath–plasma boundary due to $E_{pl} \neq 0$, yielding a net charge density of Δn. However, an estimation shows that Δn is much smaller than n_e, and therefore, condition (1) can be used in order to determine the ion to electron density ratio at the sheath edge. Setting the potential at the plasma–sheath boundary to zero and the electric field at the potential minimum plane (i.e., the virtual cathode) to zero the solution of Poisson's equation is obtained in the following form (Chap. 5, 5.39):

$$E_{pl}^2 = \frac{4j_i}{\varepsilon_0}\sqrt{\frac{eu_m m}{2}} \left\{ \begin{array}{l} e_N^{0.5} \sum \left[\frac{f_z}{\sqrt{z}} \left(\sqrt{1 + \frac{T_e}{2Zeu_m}} - \sqrt{\frac{T_e}{2Zeu_m}} \right) \right] \\ - \beta_{re0}\sqrt{\frac{\pi T_e}{eu_m}} \left(1 - e^{-\frac{eu_m}{T_e}} \right) \\ - \beta_{eb} \left[1 + \frac{1}{2}\sqrt{\frac{\pi kT_s}{eu_m}} e^{\frac{eu_m}{kT_s}} \Phi^* \left(\sqrt{\frac{eu_m}{kT_s}} \right) - \frac{1}{2}\sqrt{\frac{\pi kT_s}{\varphi_m}} \right] \end{array} \right\}$$
(16.31)

$$f_{iz} = \frac{Zn_{iZ0}}{\sum_Z Zn_{iZ0}}; \quad \beta_{re0} = \frac{\varepsilon j_{eT0}}{j_i}; \quad \beta_{eb} = \frac{\varepsilon j_{eb}}{j_i}; \quad j_{eb}$$

$$= j_{em} \exp\left(-\frac{\Delta u}{kT_s}\right); \quad j = j_{eb} + j_i \quad s = \frac{j_{eb}}{j}$$

ε_0 is the permittivity of vacuum, e_N is the root of the natural logarithm, β_{eb} is the fraction of electrons current density emitted by the virtual cathode j_{eb}, β_{re0} is the fraction of returned electrons in the current density, K_d is the ratio of E_{pl} to the presheath electric field E_d determined as T_e/r_d.

(8) The potential drop Δu can be calculated using the thermionic emission electron current density j_{em} and the electron current density j_{eb} emitted by the virtual cathode as $\Delta u = T_e Ln(j_{em}/j_{eb})$.

(9) The total spot current is given by $I = \pi r_s^2 j$ assuming a circular spot, where r is the spot radius.

For a given I and u_c (or minimal potential), the unknowns n_T, n_e, j, j_{eb}, n_{iz}, r_s, Δu, E_p, T_s, and T_e can be calculated from (1) to (9) when the parameter K_d is determined by relation of $E_d = T_e/r_d$. To understand the influence of this parameter, the above system of equations was studied as dependence on s as a varied parameter.

16.7.2 Numerical Study of the Spot Parameters on Tungsten Cathode

Below the results of calculations are described according to the above mathematical formulation and to the work [40] performed for a tungsten cathode with $I = 20$ A and a cathode erosion coefficient in the range of $G \sim 10$–100 μg/C. The cathode potential drop has not been measured for a tungsten vacuum arc. Reece [43], however, found a correlation between the arc voltage and the cathode boiling temperature, showing that for refractory materials the voltage is a few volts higher than that for intermediate materials (Cu, Ag). Therefore, in the following calculation $u_m = 20$ V is used as an example. The general calculation shows that the cathode vapor is highly ionized, T_s is in region of 6300–7500 K, the density n_T is 10^{19}–10^{18} cm^{-3}, and T_e is 2–4 eV when the s is in region of 0.5–0.9.

Figure 16.20 shows the calculated parameter K_d as dependence on s. It can be seen that K_d decreases with s from 0.3 to 0.15 when s varied from 0.5 to 0.98. In this case, the spot current density j is enough low and is about 10^5 A/cm^2 for $s < 0.9$.

Figure 16.21 shows the dependence of the electric field E_{pl}, on s, with the cathode erosion rate a parameter. It can be seen that the value of E_{pl} is about 1 MV/cm, only varies slightly with s when $s < 0.9$ and after a maximal value at about $s = 0.99$ sharply decreases to zero. E_{pl} increases with G. The calculation also shows that the ratio K_d is in region 0.28-0.13 when $s < 0.9$ and then for $s > 0.9$ sharply decreases to zero.

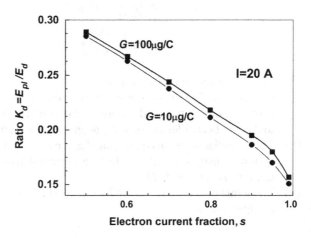

Fig. 16.20 Parameter K_d indicates the ratio of the plasma electric field to the field E_d in the Debye layer at sheath–plasma boundary as dependence on s

16.7 Refractory Material. Model of Virtual Cathode. Tungsten

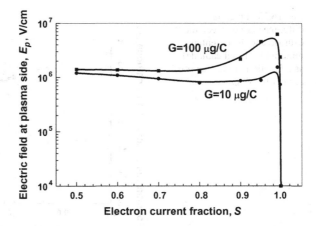

Fig. 16.21 Dependence of the electric field at the sheath–plasma interface, $E_{pl}=E_p$, on electron current fraction

The current density j is about $(0.1–1)$ MA/cm^2 for $s < 0.85$ and strongly increases with s when $s > 0.85$ (Fig. 16.22). The value j increases with G.

The potential difference between potentials at the cathode surface and the virtual point (minimum) is presented in Fig. 16.23. According to the these calculations, the Δu decreases from ~2.5 to zero when s increases up to 0.95.

For higher s, the potential Δu becomes negative indicating that the virtual cathode disappears. The cathode erosion rate has negligible effect on the positive Δu region of the Δu-s curve and only slightly affects the negative Δu region.

The dependence of the ion state distribution, as expressed by current fraction for ions with different charge f_z, on the electron current fraction is shown in Fig. 16.24 (for $G = 10$ μg/C). When s increases, the first ion current f_1 decreases while f_4 increases by several orders of magnitude. The ion current fractions f_2 and f_3 are maximal at $s = 0.7$ and $s = 0.85$, respectively.

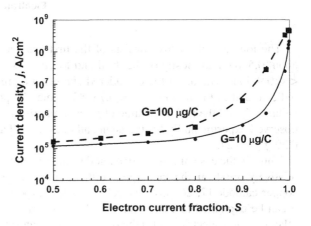

Fig. 16.22 Spot current density j as a function of s

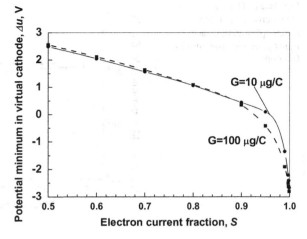

Fig. 16.23 Potential difference Δu between the cathode and the potential minimum (virtual cathode) as a function of the electron current fraction

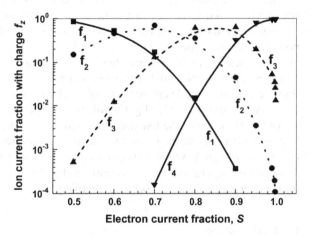

Fig. 16.24 Influence of the electron current fraction on the ion current fraction f_z with ionicity Z ($G = 10\ \mu g/C$)

The ion current mostly consists of the first and second charged ions when s is about 0.5–0.6 and mostly of the third and fourth ions for $s > 0.9$. In the region $0.6 < s < 0.9$, ions with all of the considered charges are present. The calculated results indicate that in the region of $s = 0.75$–85, it may be possible to explain the small value of f_1 and the maximal value of f_3 in comparison to other fractions, which was observed experimentally for tungsten cathodes in [44] as: $f_1 = 0.03, f_2 = 0.25, f_3 = 0.39, f_4 = 0.27$.

Thus, in the case of refractory cathodes, highly charged ions can be generated for relatively small s in the spot region, rather than in the plasma jet as occurred for the copper cathode [7, 26, 45]. The relatively low experimental value of j shows that it can be explained in case when the electron current fraction is controlled by the relatively large electric field at the sheath–plasma interface E_{pl}. Although this electric field is large (~1 MV/cm), it is smaller than the field determined as $E_d = T_e/L_d$ by factors between 3 and 8 when $s < 0.9$. The plasma region with E_d, in essence formed to accelerate the ions to a velocity, needed to stable sheath operation. This important point indicated that the theory of virtual cathode for refractory materials cannot be applied to the quasineutral (not disturbed) plasma region with the low electric field and use zero field as a boundary condition at the right side of the sheath.

16.8 Vacuum Arcs with Low Melting Materials. Mercury Cathode

In this section, a double sheath model developed by Beilis [46] is presented. The early studies devoted to discharges on mercury cathodes which are (in contrast to other materials) always unstable [16, 47–50] were motivated by the importance of this discharge in switching, rectification, and lighting [16, 51, 52] technology. Due to the instability of mercury spots, refractory rods are usually used to anchor them in practice [53–56]. An analysis of published theoretical attempts to explain the behavior of the mercury cathode spot was presented in Chap. 15 and in [46]. Let us first consider the experimental data.

16.8.1 Experimental Data of a Vacuum Arc with Mercury Cathode

Kesaev [16, 49] presented most detailed characteristics of the mercury cathode spot. In contrast to other materials, the near-cathode plasma in a Hg vacuum arc consists of two regions [16, 48, 49]. There is a dark space directly adjacent to the cathode surface and then a brightly luminous region. The size of the dark space is nearly 10^{-3} cm, which is much larger than the calculated thickness of a cathode drop sheath in the Langmuir approximation.

The cathode temperature Ts as determined by continuum spectrum measurements is in the range of about 1000–2000 K. The current density j in the cathode spot is in the range of 10^4–10^5 A/cm^2, [49, 54]. Kesaev's measurements [49] for liquid mercury gave the spot size as about 10 μm, which corresponds to $j \sim 5 \times 10^4$ A/cm^2. Kesaev [16] noted that the mercury discharge had two forms: the *fundamental form* and the *transitional form*. He conducted some interesting experiments on the arc discharge stability and found (1) a threshold current of $I_d \sim 0.1$ A below which the spot disappears and (2) the discharge voltage oscillates with periods of 0.1 ms and 0.1–3 μs, respectively, for the fundamental and transitional arc forms. The cathode drops for these arcs are ~9–10 V and ~18–21 V, respectively. When the current is close to the threshold value, both forms can be observed. However, at larger currents, only the fundamental form is observed. High-frequency oscillations are connected with the extinction of a spot and the creation of a new spot. Observation of the transient high-frequency bursts provided a convenient method of measuring the spot lifetime t.

Kesaev described the mercury arc as a dynamic process, during which spots spontaneously would divide into two spots, and some spots would spontaneously extinguish. For currents above a value of 0.1 A, usually more than one spot coexisted on the cathode surface. Kesaev explained spontaneous spot extinction with an *internal* spot instability, the mechanism of which he did not detail. If a spot extinguished, the total arc current would be redistributed among the remaining spots. According to Kesaev, the redistribution is caused by a *reformation* mechanism, which is also unknown. The discharge ceased to burn only when all spots disappeared at the same time.

The threshold current has been reported by Khromoy [53, 54] and Eckhardt [55] to be in the range of 0.6–2.5 A and according to their results depends on the electrode temperature, preparation, and surface cleanliness. According to the Kesaev's experimental measurements (Tables XIX and XX from [49]), the erosion coefficient is equal to ~4 × 10^{-3} g/C in the case of free spot and ~6 × 10^{-4} g/C in the case of an anchored spot. Generally, in vacuum arc this erosion is in the form of droplets and vapor. In the mercury case, much of the erosion occurs by droplets [16, 49, 57] and determination of the vapor fraction is found to be difficult due to the considerable contribution of evaporation from the mercury surface that is outside of the spots. The difference of the erosion rate between free moved and fixed spots Kesaev [49] is explained by smaller droplets in case fixed spot. However, the arc current and arc time in arcing with fixed spot (1–2 A, as example t ~1800 s) exceed those values for moved spot (10–18 A, t ~180 s) about order of magnitude. Eckhardt [58] has tried to take into account this effect. She developed a novel measuring technique. The method applied permits a separation of these two contributions, thus yielding results that are more accurate. The measured vapor erosion rate obtained as ~10^{-5} g/C for the anchored spot in experiments for Hg arc of 15, 30 and 90 A.

16.8.2 Overview of Early Debatable Hg Spot Models

A review of the existing spot theories was presented in Chap. 15. Mainly was discussed the mercury cathode spot mechanisms, and some of their controversies are summarized. Let us consider their main approaches in order to understand the new idea developed by the author to describe one of most mysterious case of current continue of Hg spot.

Two general cases can be emphasized. In the first, the temperature T_s ~(3–4) × 10^3 K necessary for sufficient electron emission considerably exceeds the mercury critical point T_{cr}. With this temperature, a highly volatile material such as Hg remains in the gas state, however, with atom density approximately that of a solid. It is obvious that in this case processes of classical emission cannot occur. In the second, when the temperatures are low $T_s < T_{cr}$, deviation from equilibrium depends on the conditions

16.8 Vacuum Arcs with Low Melting Materials. Mercury Cathode

of the vapor flow in the Knudsen layer. The minimum flow of returning particles is 18% of the flow of atoms evaporated into vacuum [59] (see Chap. 2). Under these conditions, in spite of the accordingly large deviation from equilibrium, the level of atom density with regard to the electrons remains very large according to GDM calculations.

Another approach [16, 60] assumes the presence of a large electrical field (~10^8 V/cm) on the cathode surface. A field of this magnitude can only be realized by an ion space charge, which by Mackeown's equation [61] is associated with an ion current flux of $j_i \geq 10^7$ A/cm^2. Under this value of j_i an analysis of the cathode surface energy balance, which only takes into consideration the heat flux associated with energy of ion neutralization at the cathode surface (potential of ionization, u_i), i.e., $u_i j_i$, results in an unrealistically high cathode surface temperature of (3–4) × 10^4 K for Hg. Thus, it is clear that the use of the field emission theory has certain difficulties.

The contradiction in the field emission theory for the Hg cathode spot can be seen more clearly if the equations for the cathode spot are written and solved in two alternative dependencies, for the current density j versus electron current fraction s (Fig. 16.25): (1) a simultaneous solution of the cathode energy balance, conservation of energy of the plasma electrons, equation for the electric field, and the ion transport equation, but without taking into account the equation for electron emission; (2) these same equations, together with the Fowler–Nordheim equation for electron emission,

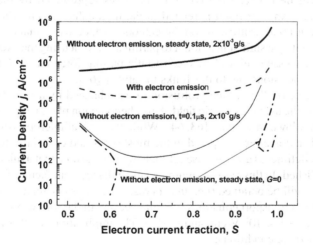

Fig. 16.25 Dependencies of the current density on the electron current fraction ($I = 1$ A) obtained from the solution of the traditional system of equations, considering the equation of electron emission, but not considering the equation of ion flow to cathode from the plasma (bold curve), and also, in reverse, considering equation of the ion flow, but not considering the equation of electron emission (other curves). The dashed curve corresponds to a time from the spot ignition of $t = 0.1$ μs and an erosion rate $G = 2 \times 10^{-3}$ g/s. The thin solid curve (with erosion rate $G = 2 \times 10^{-3}$ g/s) and (stroke-dotted) curve ($G = 0$) correspond to steady-state conditions

but without taking into account the ion transport equation. A true solution of the entry problem should occur at the intersection of these curves, but it may be seen that the two types of curves do not intersect.

While it might seem that an intersection might occur of the scale of Fig. 16.25, detailed analysis indicates that unrealistic values for various plasma parameters occur for large and small values of s. For example, for small values of s, when the electron current fraction decreases to 0.55, the cathode temperature will increase, and hence the vapor concentration adjacent to the Hg cathode will increase until the density becomes unreasonably high. For example, assuming an erosion rate of $G = 2 \times 10^{-3}$ g/s, for $s = 0.53$ (thin solid curve in Fig. 16.25), we obtain T_s ~2500 K and, consequently, an unrealistic neutral density of n_0 ~10^{31} cm^{-3} is produced. This phenomenon precludes the application of case (2) curves for small values of s. Similarly, at large values of s, the energy of the plasma ion flux becomes value smaller than the energy required for cathode evaporation. For $0.88 < s < 0.96$, the predominant positive term in the cathode energy balance is Joule heating in the cathode, and a solution is still possible. However, for $s > 0.96$ there is insufficient total energy input to maintain the evaporation, and no solution was found.

When we considered the transient cases, the current density is constant over a wide range of s (see dashed curve). When $G = 0$, the character of the curve changes so that the function $j(s)$ has a break and the solution exists only when $s < 0.617$ and $s > 0.95$ (stroke-dotted curves, Fig. 16.25). When $0 < G < 2 \times 10^{-3}$ g/s, all the curves will lie between the two limiting cases represented by the stroke-dot and thin curves. Another point is related to the field enhancement. Electrical field enhancement at microprotrusions on the cathode surface will enhance the electron emission, but not by a sufficient amount to permit a solution using the existing model, as it will be discussed below. The electrostatic pulling force on the liquid surface causes microprotrusions due to the Tonks instability [62].

The local electric field at the protrusion tip is enhanced by a factor of β: $E_{loc} = \beta E$, where E is the average electric field. This phenomenon plays a significant role in electrical breakdown in vacuum [63, 64]. While the typical microprotrusion growth time [62] (~1 ms) is large compared to the most types and especially for super-fast cathode spot lifetime, as a worse case estimate we will assume that a microprotrusion is fully established. In this case, assuming a field enhancement factor of order β: ~10 a full solution will be obtained (i.e., the curves, which was calculated from the field emission and ion transport equations will intersect) with the following parameters: j ~10^6 A/cm^2, E ~5×10^6 V/cm, and s ~0.9. The implications of β ~10, however, lead to unrealizable conditions:

(a) The requirements for the microprotrusion necessary to realize β ~10 are that [61] h/r ~10, where h is the height of the protrusion and r is the radius of curvature at the tip.
(b) The tip must be within the space charge sheath where a high average field is present.

16.8 Vacuum Arcs with Low Melting Materials. Mercury Cathode

(c) Given that the sheath is only $\sim 10^{-6}$ cm thick, we must require that the microprotrusion height is say an order of magnitude less, i.e., $h \sim 10^{-7}$ cm, and hence $r \sim 10^{-8}$ cm.

(d) Thus, we see that the radius tip must be about one atom thick, and thus even if this model were correct on this size scale, the total current would be miniscale ($\sim j_{rh} = 10^{-9}$ A) and not consistent with experimental observations.

(e) Furthermore, the size scale would be about the surface potential barrier, which prevents the use of the relation $E_{\text{loc}} = \beta E$ and expressions for electron field emission. Rodnevich [65] showed that the influence of field enhancement is not dominated when $\varphi/Eh \gg 1$ as in the present case.

Thus, the present model for electron emission does not lead to a consistent solution in the case of the Hg cathode spot. The further considered mechanisms of mercury cathode spot were discussed in Chap. 15, but below some details can be discussed.

1. The electron pulling from the cathode may be occurred by electron explosive emission [61], when the cathode surface is irregular (caused by mentioned above liquid surface instability). However, this phenomenon also initiates with electron field emission from microprotrusion and realizes with critical electric field $E_{cr} \sim 10^8$ AV/cm for which the thermal instability of microprotrusion is taken place. Such large value of E_{cr} requires more larger β and smaller microprotrusion size, as well necessity of large j_i as it is above discussed in case of field emission theory. The other weakness of mechanism of explosion electron emission was discussed in Chap. 15.

2. Electron emission may be the result of various secondary electron emission processes. The most effective of these is the bombardment by excited Hg atoms, which have a metastable level with an excitation energy of $u^* = 4.9$ eV widely discussed in [66–69]. The theory of this process is characterized by a coefficient for secondary electron emission [70] of $\gamma_m = 10^{-2}$. This process will be important if the electron current produced is comparable to the ion current. An expression for the ratio of electron to ion currents can be formulated by using the Boltzmann relationship [71] to calculate the excited atom density, and Saha's equation to calculate the electron density:

$$\frac{j_e}{j_i} = \gamma_m \frac{j^*}{j_i} \approx 3 \times 10^{-14} \gamma_m \text{Exp}\left(\frac{u_i - 2u^*}{2T_e}\right) \sqrt{n_a} T^{-3/4} \qquad (16.32)$$

where $j^* = e\Gamma^*$ is the equivalent excited atom current, while Γ^* is the excited atom flux and e is the electron charge. It may be calculated that $j_e/j_i \approx 1$ when $n_a \approx 10^{24}$ cm^{-3} even if T_e is on the order of 1 eV, or in another words, the atomic density in the plasma must be unrealistically larger than solid density for secondary emission to yield the requisite electron emission.

3. According to the thermal theory [16, 72], current continuity at the cathode surface is maintained by a flow of ions from the near-cathode plasma, i.e., $j = j_i$. The Hg atoms are ionized by the plasma electrons, which are heated to a temperature of T_e by Joule dissipation in a hemispherical region of the plasma adjacent to the cathode surface having a radius r approximately equal to the radius of the cathode spot. On

the other hand, j cannot be greater than that value which will lead to a heat flux that can be conducted into the interior of the cathode. If we have a cathode surface temperature of T_e ~1000 C, we have that $j \approx (2\lambda_T T_s/I^{0.5})^2$ ~10^4 A/cm^2, where λ_T is the thermal conductivity of the cathode, and u_{ef} is the effective heating voltage. If all of the Joule heating goes only into ionization and heating of the electrons, a simple relationship for the electron temperature may be obtained: $T_e \approx (u_{pl} - u_i)/3$ for the condition when $u_{pl} > u_i$, where u_i is the ionization potential of the Hg atoms, and the potential drop in the plasma is given by $u_{pl} \sim jr_s/\sigma_{el} = (jI/\pi)^{0.5}/\sigma_{el}$.

However, using the previous relationship for j, it can be seen that $u_{pl} = 2\lambda_T T_s/(\pi u_{ef}) \leq 1$ V, where this estimation relies on the Spitzer [73] calculation of the plasma conductivity σ_{el}. Thus, it may be seen that $u_{pl} < u_i$ and there is insufficient Joule heating to heat the plasma up to even the cathode surface temperature. Under these low plasma temperatures, the fractional ionization is very low, and an absurdly large ion density (>10^{22} cm^{-3}) is required to maintain the ion current due to the low ion mobility (diffusion constant D_{ia} ~10^4 cm^2/s). Neither taking account of transient thermal processes nor the non-ideal nature of the cathode spot plasma [74] alters the above conclusion.

Thus, the above overview shows that the published various conventional approaches lack of information or contradicted assumptions to explaining the cathode spot in the case of the Hg cathode. In the following paragraphs, a modified model in which an additional "plasma cathode" supplying electrons is hypothesized. This plasma cathode is adjacent to the cathode surface and separated from the bulk plasma by a double layer [75].

16.8.3 Double Sheath Model

As can be seen in the above analysis, existing models cannot consistently explain cathode spot operation on mercury. A new model is proposed [46], in which the near-cathode region is modeled with two distinct regions, divided by a double layer. A double layer in a high-voltage discharge with a low concentration of charged particles was observed experimentally [76]. Moreover, double layer also appears over a considerable range of the particle densities even when sharp changes of the discharge channel cross section occur [77]. Generally, the thickness of the plasma double layer is of the order of the mean free path.

Referring to Fig. 16.26, the plasma cathode (region I) is in direct contact with the cathode surface (boundary 1) and shares boundary 2 with the double sheath layer (region II). Electrons are created in region I by electron impact, and boundary 2 thus serves as a virtual cathode for the rest of the discharge, while current continuity is maintained at boundary 1 mostly by the ion current. Thus, there is no requirement for the cathode surface to emit an electron current equal to the circuit current—the electrons are generated in the near-cathode plasma, similar to what occurs in the cathode region of a glow discharge.

16.8 Vacuum Arcs with Low Melting Materials. Mercury Cathode

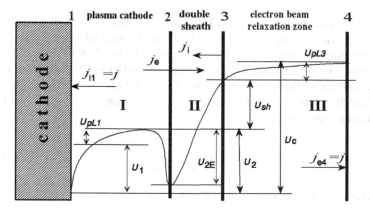

Fig. 16.26 Model of the near-electrode plasma with electrical double sheath and the schema of the potential distribution in the near-cathode region

Region II is a collisionless space charge double layer having a relatively large potential drop (see Fig. 16.26), and bounded by boundary 2, which it shares with region I, and boundary 3, which it shares with the external plasma (region III). The region III extends to boundary 4 for a length in the anode direction of the relaxation length of the electron beam emitted from the region I. Both electrons emitted from region I in the direction of the anode and ions extracted from region III move through the double layer (region II). They are accelerated by the potential drop in the double layer at the way in their respective destinations. In particular, the additional energy imparted to the ions on their way to region I is critical in providing the power that ultimately is necessary for ionizing atoms, and hence producing the necessary electron flux in region I.

While in reality there will be a lateral, as well as axial, variation of the parameters (indeed the radius of the cathode spot is microscopic while the typical cathode is macroscopic), the treatment in this model will be one dimensional, which considerably simplifies the analysis.

16.8.4 The System of Equations

The basic assumption of the model is that the potential distribution will be such that the energy balance and current continuity in all regions is maintained. Region I is heated by the ion flux from the external plasma (region III) which is accelerated as it passes through the double layer (region II). It is assumed that the density and thickness of region I are such that these ions undergo many collisions, and thus the energy of this flux is deposited in region I. A further consequence of the collision dominated nature of region I is that it may be described to a reasonable approximation as being uniform and isothermal with a temperature $T_1 = T_{el} = T_{il}$.

The bulk of the plasma in region I will have a positive potential with respect to the cathode surface, as illustrated in Fig. 16.26, with a sheath having a potential drop of u_1 separating the two. This sheath will both brake the flow of thermal electrons from region I to the cathode surface, as well as accelerate the ions from region I onto the cathode surface. Labeling the ion flux at the cathode surface as j_{il} and the thermal electron flux there as j_{etl}, we may define their ratio as $s = j_{etl}/j_{il}$ and obtain the following expression for u_1:

$$u_1 \approx T_1 Ln\left(s_T^{-1}\sqrt{\frac{m}{2\pi m_e}}\right) \tag{16.33}$$

According to definition, the total current density on cathode depends on the value of s_T and given by:

$$j_1 = j_{il}(1 - s_T) \tag{16.34}$$

In region I, the electrical current transport mechanism gradually changes from ion motion to the cathode surface that collects the impinging ions, to electron motion away from the cathode, as we progress from the cathode surface into the volume of region I. The plasma density in region I is quite high so that the diffusion layer will be smaller than the mean free path of the charged particles, and thus the ion current at the cathode surface will be given by:

$$j_{i1} = en_{i1}v_{iT}/4 \tag{16.35}$$

where v_{iT} is the thermal velocity of the heavy particles. At boundary 2 of region I, the random electron current j_{et2} considerably exceeds the discharge current. A potential barrier appears in region I adjacent to boundary 2 with a height u_2 such that the electron flow is retarded to the extent that it is equal to the discharge current. Thus

$$j_{e2} = j_{et2}\text{Exp}\left(-\frac{u_{2E}}{kT_{e1}}\right); \quad u_{2E} = u_2 - r_D E \tag{16.36}$$

where r_D is the Debye radius. The value of the electric field in the double layer (region II) can be determined from Mackeown's equation [61]. The ion current entering the double layer from region III is given by:

$$j_{i3} = en_{i3}v_{iT3}/4 \tag{16.37}$$

where n_{i3} is the ion density in region III near boundary 3, and $v_{iT3} = v_T$.

The electrons' current density fraction s in the double sheath can be expressed as:

$$s = j_{e2}/(j_{e2} + j_{i3}) \tag{16.38}$$

16.8 Vacuum Arcs with Low Melting Materials. Mercury Cathode

The conservation of energy equation for regions I and III, taking into account the energy flux by convection into and out of each region by electrons, ions, and excited mercury atoms, as well as energy gained by the charged particles within each region from the electric field, are given respectively by:

$$(u_i^1 + 2T_1)\frac{j_{i1}}{e} + \left(u_1 + \frac{2T_1}{e}\right)j_{et1} + q_p$$

$$= j_{i2}\left(u_{sh} + \frac{(u_i + 2T_3)}{e}\right) - j_{i2}\left(u_{2E} + \frac{2T_1}{e}\right) + \Gamma_{a2}^* u^* + j u_{pl1} \quad (16.39)$$

$$\left(u_{sh} + u_{2E} + \frac{2T_1}{e}\right)j_{e2} + j u_{pl3} = j_{i3}\frac{u_i}{e} + j_{e4}\frac{2T_{e2}}{e} + \Gamma_{a3}^* u^* + \varsigma j\alpha\left(\frac{u_i + 2T_{e2}}{e}\right) \quad (16.40)$$

where q_p is the energy flux, which is transferred by the neutral atoms on the boundaries 1 and 2:

$$q_p = 2T_1 G_{a1} - 2T_s G_{as} - 2T_1 G_{a2} + 2T_3 G_{a3}$$

ς is a dimensionless parameter characterized the cathode erosion rate by $\varsigma = eG/(m_i I)$, where G (g/s) is the cathode mass evaporation rate, u_{pl1} and u_{pl3} are the potential drops in the bulk of the plasmas in regions I and III, respectively determined by the plasma electrical conductivity, T_{e3} is the electron temperature in region III, and T_3 is the temperature of the heavy particles in regions I and III and is assumed to be equal, Γ_{a1} is the atom flux at boundary 1 from the plasma region I to cathode. Γ_{as} is the atom flux from the cathode due to evaporation when the surface temperature at the cathode spot is Γ_s, Γ_{a2}, Γ_{a3}, Γ_{a2}^*, and Γ_{a3}^* are the random fluxes of atoms and excited atoms through boundaries 2 and 3, respectively, α_1, α_2 are the degrees of ionization of the mercury atoms in regions I and III, respectively, u_1 is an effective ionization potential in region I which takes into account that a significant fraction of the atoms from region III is excited to the metastable level.

Because of continuity, the electron and ion current densities j_e and j_i at the boundaries are related by $j_i = j_{e4} = j$, $j_{e2} = j_{eT3}$ and $j_{i2} = j_{i3}$ and $T_1 = T_3$. The heavy particle concentrations are determined from the equations of particle conservation on boundaries 1 and 2:

$$\frac{n_1}{n_s} = \frac{T_s}{T_1} - \frac{4\varsigma j}{e n_s v_{iT}}; \quad (16.41)$$

$$\frac{n_3}{n_1} = (1 - \alpha) - \frac{4\varsigma j}{e n_1 v_{iT}}; \quad (16.42)$$

where n_s is the equilibrium Hg vapor density corresponding to a cathode surface temperature of T_s [78], n_1 and n_3 are the density of mercury heavy particles in regions I and III. Equations (16.39) and (16.40) take into account that electrons accelerated in region 2 produce step ionization of the mercury atoms in region III, because of the long-lived Hg metastable state at $u^* = 4.9$ eV. The concentration of the excited atoms in the metastable level, n_a^*, depends on the energy of the electron beam and the concentration of the neutral atoms n_a. As the excited atoms flow from region III to region I, the effective ionization potential u_{a1} in region I will be decreased and can be expressed in following form:

$$u_1^i = n_i - T_1 \ln\left[1 + \frac{n_a^*}{n_{a1}} \text{Exp}\left(\frac{u_a^*}{T_1}\right)\right]; \qquad (16.43)$$

Assuming that the cathode spot is circular, then the total current will be:

$$I = p r_s^2 j \qquad (16.44)$$

$$j = j_{e2} + j_{i3} \qquad (16.45)$$

The time-dependent equation of heat conduction in the cathode may be expressed as:

$$\frac{dT_s}{dt} = a\Delta T_s; \qquad (16.46)$$

Boundary condition:

$$\lambda_T \frac{dT_s}{dt}(z=0) = -q_T \qquad r < r_s$$

where a and λ_T are the cathode thermal diffusivity and thermal conductivity, respectively, t is time, and q_T is the net heat flux delivered to the cathode surface, and is determined by:

$$q_T = q_i + q_a + q_j + q^* - q_{ev} - q_j \qquad (16.47)$$

where q_i is the heat flux delivered by the impinging electrons and ions from the plasma and is given by:

$$q_i = j_{i1}(u_{i1} + u_1 + 2T_1) + 2T_1 j_{et1}$$

q_a is the net heat flux of convection by the impinging unexcited atomic flux (after correcting for the evaporated atomic flux) and is given by:

$$q_a = 2T_1 G_{a1} - 2kT_s G_s,$$

q^* is the heat flux of convection by impinging excited Hg atoms

16.8 Vacuum Arcs with Low Melting Materials. Mercury Cathode

$$q^* = (U^* + 2T_1)G_1^*$$

and q_j is the equivalent heat flux generated by ohmic heating within the cathode [11, 79]. The heat losses from the cathode surface include the energy of the electrons emitted to neutralize the impinging ion flux, $q_\varphi = j_i \varphi$, and the energy of the evaporated atoms, $q_{ev} = j_i \varphi_s Gj/I$.

The system of equations is completed with the addition of Saha's equation in regions I and III. The solution of the equations depends on the cathode spot condition. If the spot is burning on a film of Hg anchored to a refractory post, then the thermal model is similar to the film cathode model (see below). If the spot moves freely on the surface of a thick liquid Hg volume, then the thermal model of the massive cathode is appropriate.

16.8.5 Numerical Study of the Mercury Cathode Spot

In the present analysis, the mobile cathode spot will be considered without explicitly taking into account excited atoms (i.e., without calculating its flux and concentration). This considerably simplifies the problem and concentrates our attention on the essence of the main processes. The presence of excited atoms is taken into account implicitly only through the decreasing of the ionization potential u_1^i in accordance with (16.43). Besides, the decrease in u_1^i caused by the presence of a dense plasma [80], which may be reach ~1 eV, is also taken into account. Thus, we use for $u_1^i = 8$ eV. In the region III, where there is an influence of the dense plasma on decrease of the potential of ionization ui, we use ui = 9.4 eV.

In addition, the case where the ionization potential $u_1^i = 4.9$ eV was investigated. Due to presence of metastable Hg atoms, various processes leading to ionization may occur [81–83] including: (1) formation of atomic and molecular ions due to collisions of two excited atoms, (2) electron collision step ionization, and (3) molecular ion formation by collision of an atomic ion with a neutral atom (conversion process [79]). Besides, it should be noted that secondary emission of excited atoms from the cathode surface following neutralization of impinging ions might occur [84]. The significance of this process is that they lead to a reduction of the effective potential for mercury ionization, u_1^i The results obtained here, namely that it is necessary to take into account energy transport by the metastable Hg atoms, coincide with Kesaev's [5, 16] assumption that excitation of the mercury atoms plays a significant role in this kind of discharge.

The system of (16.32)–(16.47) and two Saha's equations contain 20 unknown variables: $j, j_{i1}, j_{e2}, j_{i3}, I, u_1, u_2, s_T, u_{sh}, s, T_s, T_1, T_{e2}, n_1, n_2, q_T, G, r_s, \alpha_1, \alpha_2$, while we have 16 equations. The problem is not closed because we need four additional equations. The value of s_T ~0.1 can be obviously chosen to satisfy the conditions $j_{et1} \ll j_{i1}$, i.e., ion current at the cathode is equated to the circuit current. In order to

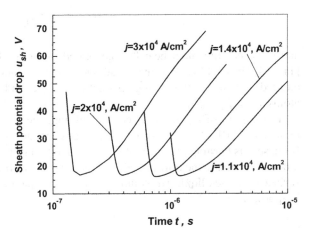

Fig. 16.27 Dependence of the sheath potential drop on the time from the spot ignition ($I = 0.1$ A; $G = 0$)

close the problem, the erosion rate G and spot current I will be set by experimental values [16, 53–55]. Because the cathode erosion is both in droplet and vapor form the parameter, G will be varied over a wide range.

The solution was obtained using a two-stage iteration method. In the first stage, the current density (fourth unknown) was the varied parameter. In the second stage, the current density was determined from additional conditions (see below). The solution was analyzed in the time interval of 0.1–1 μs from cathode spot initiation, which corresponds with the high-voltage oscillation transitional mode of the arc, and at 10^{-4} s, which corresponds to the fundamental mercury arc mode. The numerical study is presented in accordance with results of work in [46].

Figure 16.27 shows the dependence of the sheath potential drop u_{sh} in the double sheath on the time from spot initiation t, with different current densities j for $I \sim 0.1$ A, $G = 0$. It may be seen that in the beginning, u_{sh} decreases with t, as occurs for the metals such as Cu, Ag, and Ni (with $j < 10^6$ A/cm^2 see Sect. 16.4.4 [15]). However, with further increases of t the value of u_{sh} passes through a minimum and then grows without limit. Moreover, the time of the u_{sh} minimum becomes smaller with increasing current density.

An analysis showed that the existence of the unstable branch in the dependence of u_{sh} on t is caused by the relatively large neutral atom flow, which transports a lot of energy from the plasma (q_p). In the calculation, whose results are shown in Fig. 16.27, the role of the atom flow was considered only in the plasma energy balance (and not in the cathode energy balance). In other calculations, the atom energy was not accounted at all. The results are shown in Fig. 16.28. In this case, an unstable branch of u_{sh} was not observed, but rather u_{sh} falls monotonically to a minimum.

The results of the calculations at $t = 0.3$ μs in the case where energy transported by the atomic flow is accounted everywhere are shown in Fig. 16.29. It is shown that with increasing spot current, the sheath potential drop decreases. Increasing of the spot current density also leads to a decrease of u_{sh}. However, as shown in Fig. 16.27, the influence of j depends on t and at longer times this dependence may become reversed, i.e., u_{sh} will increase with j.

16.8 Vacuum Arcs with Low Melting Materials. Mercury Cathode

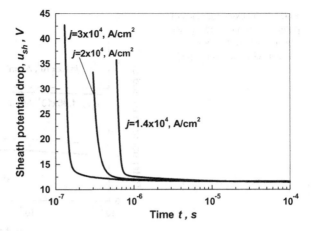

Fig. 16.28 Dependence of the sheath potential drop on the spot lifetime without considering atom energy flux from the plasma region I, ($I = 0.1$ A; $G = 0$)

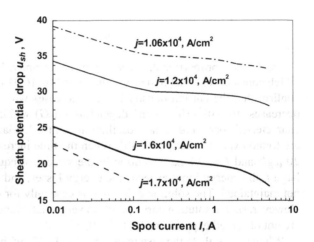

Fig. 16.29 Sheath potential drop as function on spot current ($G = 0; t = 0.3$ μs)

It is important to note that in all above calculations, the potential drop in the reflecting sheath near cathode u_1 and potential drop u_2 on the outside boundary of the region I are very close to each other (Fig. 16.30). The spot parameters obtained from the system (16.33)–(16.47) solution for the time t_m for which u_{sh} is minimized for $I = 0.1$ A, $j = 1.4 \times 10^4$ A/cm², $G = 0$ (Fig. 16.27) are shown in Table 16.2.

Using the fact that $u_{pl1} = u_2 - u_1$ we can estimate the plasma region I width as Ohm's law $L_1 = \sigma u_{pl1}/j$. Taking in account that $u_{pl1} = 0.5$ V (see Table 16.2), the calculation gives $L_1 = 1.4 \times 10^{-3}$ cm for $T_1 = 1$ eV.

To confirm the previous conclusion about the decreasing of the current density with time, it is necessary to calculate the current density, and not set it as a parameter. To make such a calculation we need another equation or condition. Therefore, we will take into account the above-mentioned fact that u_1 approximately equal u_2 and assume that:

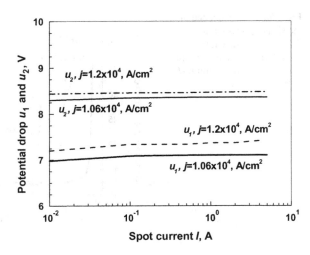

Fig. 16.30 Dependencies of potential drop u_1 and u_2 (see Fig. 16.26) on the spot current ($G = 0$; $t = 0.3$ μs)

$$u_1 = K_u u_2, \qquad (16.48)$$

where K_u is a proportionality coefficient about unity. Results of the calculations, which consider relation (16.48), are shown in Fig. 16.31 for $K_u = 1$ and $K_u = 1.1$ and confirm that the current density indeed decreases with time. Moreover, j strongly decreases (to ~10^2–10^3 A/cm² depending on G) within t ~10^{-4} s, which is the characteristic period of voltage oscillations for the fundamental form of the mercury arc found experimentally [16, 49]. When the time increases ($t > 10^{-4}$ s), with $G = 20$ μg/s and $I = 0.1$ A, a solution of the system of equations was not found: For large t, the energy input exceeds the energy losses, and thus, the energy balance is not maintained. The solution exists in this case only for $G < 20$ μg/s and $I < 0.1$ A. However, such a solution corresponds to very small values of $j < 10^2$ A/cm²), which are not observed experimentally [16, 49, 53–55].

When $u_i^1 = 4.9$ eV, the system of (16.33)–(16.47) with various spot currents I and erosion rates G ($K_u = 1$) was solved for u_{sh} and the results are presented in Fig. 16.32. It may be seen that u_{sh} decreases with time, and that for various combinations of (I, G) reaches 10 V at $t = 0.1$ ms. Moreover, u_{sh} slightly depends on I, if the erosion coefficient G/I is constant. Here, the characteristic spot parameters are varied $T_{e2} = 1.3$–0.9 eV, $T_1 = 0.77$–0.56 eV, $s = 0.45$–0.46, $u_1 = 6.5$–5 V, j ~2×10^4–5×10^3 A/cm², respectively if the spot lifetime varies from 10^{-6}–10^{-4} s ($I = 0.1$ A, $G = 20$ μg/s). An estimation of the value u_{pl3} gives <0.1 V.

16.8.6 Analysis of the Spot Simulation at Mercury Cathode

The mechanism of the u_{sh} instability on the mercury cathode may be understood as follows. Initially, the surface temperature is low, and the Hg vapor density is small.

16.8 Vacuum Arcs with Low Melting Materials. Mercury Cathode

Table 16.2 Parameters of the Hg cathode spot with $I = 0.1$ A, $j = 1.4 \times 10^4$ A/cm^2, $G = 0$

T (μs)	u_{sh} (V)	u_1 (V)	u_2 (V)	u_{2E} (V)	T_s (K)	T_1 (eV)	T_{e2} (eV)	n_s (10^{21} cm^{-3})	n_1 (10^{21} cm^{-3})	n_2 (10^{21} cm^{-3})	s	α_1	α_2
0.8	16.3	9.3	9.8	8.0	675	1.0	1.2	2.94	0.69	0.35	0.33	0.5	0.6

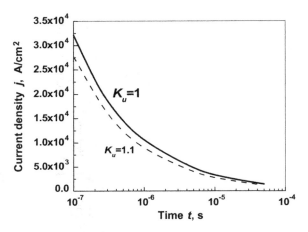

Fig. 16.31 Spot current density as dependence on the spot lifetime ($I = 0.1$ A; $G = 200$ μg/s), considering relation (16.48) from the text

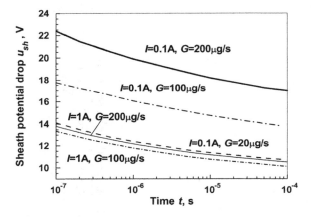

Fig. 16.32 Dependence of the sheath potential drop on the spot time considering relation (16.48) from the text ($K_u = 1$)

The degree of ionization must be relatively high to supply the necessary ion flow to the cathode. Consequently, the plasma energy losses in region I caused by the neutral atom outflow are small, because the atom density is relatively small. At the same time, relatively large power is needed to support the high degree of ionization in region I (a relatively high plasma temperature is required) under these conditions. In this case, the spot is sustained with large values of u_{sh}.

As the time increases, however, the cathode temperature and the heavy particle density increase while the degree of ionization falls, and consequently the power losses fall as well. As a result, u_{sh} decreases. This process continues until the energy flux associated with the neutral atom outflow will be the main power outflux, due to the decreasing of the degree of plasma ionization. In this case, the plasma energy losses continue to grow leading to an increasing of the sheath potential drop. Moreover, the fact that growth of the heavy particle concentration is a strong function of the temperature explains the unstable behavior of the right branch in the variation of u_{sh} with t, as seen in Fig. 16.27. If we take into account the atom energy contribution

16.8 Vacuum Arcs with Low Melting Materials. Mercury Cathode

to the cathode energy balance, the results will look more complicated. Due to the more intensive cathode heating, the rate of growth of the right section of the curves in Fig. 16.27 will be even stronger.

Comparing the theoretical results in Fig. 16.27 with Kesaev's experiments [16, 49], we conclude that a cathode spot with a given current density reaches a critical point in its temporal development at that time when it reaches the minimum value of $u_{sh}(t)$. Either the spot extinguishes, or the spot must increase in area so that j is decreased. The latter possibility is equivalent to moving from one curve in Fig. 16.27 to another curve further to the right. This process can occur continuously, such that either the spot extinguishes, or j continues to decrease. The transition from one curve to another may be accompanied by oscillations in u_{sh} with frequency of 10^6–10^7 Hz, which is characteristic of transitional mode of the Hg arc [16, 49]. The process continues until j reaches some value, after which spot extinguishes. Duration of the process may be compared with the time of fundamental spot mode.

The sheath potential drop is assumed in the above calculation to be close to the minimal value seen in the $u_{sh}(t)$ curves in Fig. 16.27. Generally, u_{sh} depends on a complex of processes, which take place in the near-cathode plasma of the mercury arc. For an estimate, it is possible to obtain an expression for $u_{sh} = u_1^i + 2(T_e + sT_i) + u_n$, by adding (16.39) and (16.40) and neglecting the terms connected with small parameter s_T, where $u_n = q_p/j$ is the effective potential for the power transported by the neutral atoms from the plasma. Considering that the plasma temperature is approximately 1 eV and taking into account that the discharge voltage is $u_c = u_{sh} + u_1 + u_{pl1} + u_{pl2}$ (Fig. 16.26), we can see from above-mentioned relation that the low value of u_c for the mercury discharge can be obtained when the values of u_n and u_1^i are low as well.

It should be noted that the temperature in region I could be lower than the plasma temperature in region III (Fig. 16.26). This result makes it possible to understand the experimentally observed "dark space," which appear immediately adjacent to the cathode surface, and signifies a lesser intensity of visible radiation originating there as compared to the intensity of the visible radiation in the region further from the cathode surface. The width of plasma region I calculated here agrees with the observed "dark space" width. An additional factor, which would enhance the excitation, and hence the radiation, in region III relative to region I, is the acceleration of the electrons in region II prior to their entry into region III, thus imparting a directed energy to the electrons in addition to their thermal energy. The fact that there are electrons of relatively high energy in the near-cathode plasma has been established experimentally [16], by studying the excitation of neon atoms that were especially introduced into the mercury arc gap. The lowest neon excitation level is 16 eV. From the Fig. 16.26, it can be seen that the potential drop in region II is $u_{sh} + u_{2E}$ (~24 V, see Table 16.2) and, consequently, accelerates the electrons enough to reach the neon excitation energy what explains experiment [16, 49].

The numerical analysis here shows that the spot on clean mercury can burn with threshold currents of about ~0.1 A, which agrees with experiments [16, 49]. It also should be noted that the sheath potential drop depends considerably on the process of ionization in the plasma and on thermal processes at the electrode. The difference in these processes, depending on arc burning conditions such as arc current, cathode cooling and geometry, as well as surface cleanness, may cause a wide interval of values (up to ~2.5 A) of the threshold currents, as observed experimentally [16, 49, 54, 55]. It is also important to measure arc voltage, to which threshold current corresponds in the given conditions.

Thus, the model predicts a temporal evolution of the discharge voltage which at constant current (until $t = 0.1$–1 μs) and current density would initially decrease, and then increase without limit. The latter behavior is avoided in practice by either spot extinction, or by a decrease in the current density. Repetitive extinction and re-ignition at constant current density corresponds to the transitional Hg arc mode ($t = 0.1$–1 μs), while longer lifetime spots (until $t = 0.1$ ms) in which the current density decrease with time corresponding to the fundamental Hg arc mode. The arc voltage unlimited increasing with time for transitional arc form connects with maintaining of plasma energy balance, while the temporal unlimited current density decreasing for fundamental arc form connects with disturbing of cathode energy balance. Physical explanations for these modes were previously unavailable.

16.9 Film Cathode

The detailed experimental study of the spots (cells) behavior in an arc burning on film cathodes was conducted by Kesaev [16] (see Sect. 7.2.4). The thin Cu films deposited on a glass and Bi films deposited on a glass and Mo substrates were studied. Kesaev shows that for different film thickness df and spot current If, the spot moved with a velocity vs and left a clear erosion track of width δ in which all the metal thin film was removed from the substrate. The arc duration t was up to 100 μs, and the erosion rate can be calculated as $G = \gamma df \delta vs$, where γ is the specific metal density.

The thickness of copper thin films was varied in the range of df ~0.01–0.1 μm and arc duration up to $t_f = 100$ μs. For the observation of the Cu tracks, the following parameters were determined: spot velocity vs ~10^3–10^4 cm/s, erosion rate Gf ~46–64 mg/C, spot current If ~0.1–0.9 A, and track width δ ~2–15 μm for $df < 0.14$ μm.

16.9.1 Physical Model of Spot Motion on Film Cathode

In order to understand the spot mechanism on thin film cathodes, the GDM was modified [85] taking into account the equation of heat conduction for thin films in the cathode energy balance and by using the experimental dependence between the film erosion rate and track size in the form $G = v\delta \rho d$ (Fig. 16.33).

The model was used experimental fact that the spot moves when the underlying film is evaporated [16]. It can be shown by an estimation of the distance over which

16.9 Film Cathode

Fig. 16.33 Schematic presentation of metallic film deposited on glass substrate and the track with width δ produced due spot motion

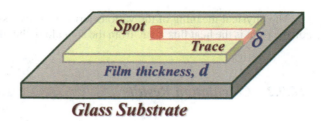

the metal can be heated to a temperature comparable to the spot temperature T_s over the lifetime t_f yields $l \sim (Q_T t_f/T_s c\gamma\pi d)^{0.5}$, where Q_T is the heat which flows into the film from the plasma, and c is the specific heat.

As for typical case $Q_T \approx 0.3$ W for $d_f \approx 0.02$ μm, $t_f \approx 0.1$ μs, $T_s \approx 3500\text{--}4000$ K, then $l \approx 10$ μm. On the other hand, the width of the signature δ is a function of the film thickness, lying in the range 1–10 μm. Therefore, while the spot was at a given position it heats the metal over a distance comparable or more to the spot diameter. In the evolution of the temperature rise of the film due to Joule energy dissipation, it is assumed that the current flows away radially from the spot and that the profile of the current over the film thickness is uniform. According to [85], the temperature increases at a point $R \gg r_s$ at which the spot arrives at this point was obtained (v_s = constant) as:

$$\Delta T = \frac{I_f^2}{\sigma_{el} c \gamma v_s r_s (2\pi d_f)^2}; \quad (16.49)$$

As it was shown in [85], the value of ΔT at points on the spot trajectory is slight regardless of the film thickness. Since the estimation is upper limit the Joule heating, in thermal regime of the metal can be neglected. The spot was assumed a circular over which the profiles of the spot parameters are uniform except the film thermal regime for which a Gauss heat flux to the film was taken in account (see Chap. 3). As it is modeled above, the temperature of the spot surface is taken to be the temperature at the center of the spot. Assuming that the erosion occurs in the vapor phase (experiments [16] showed that the erosion in droplets was small) and neglecting the heat loss to the glass substrate, the energy balance equation for the film cathode was obtained in following form [85]:

$$(1-s)u_{ef}I_f + I_{et}(2T_e + \varphi_{sh}) - (\lambda_s + \frac{kT}{m})G = \frac{4\pi\lambda_T d_f (T_s - T_0)\text{Exp}\left(-\frac{v_s t_0}{2a}\right)}{\int_{t_0}^{t_0+t} \frac{dt'}{t'} \text{Exp}\left(-\frac{v_s^2 t'}{4a} - \frac{v_s^2 t_0^2}{4at'}\right)}$$
(16.50)

$$(1-s)u_{ef} = (1-s)(u_c + u_i) - \varphi_{sh};$$

$$\varphi_{sh} = \varphi - 3.8 \times 10^{-4}\sqrt{E(\text{V/cm})}; \quad t_0 = \frac{4a}{r_s^2} \quad (16.51)$$

The physical meaning of the various terms was discussed above, and expression on the right is the heat flux away from the spot along the film due to heat conduction.

16.9.2 Calculation Results

The calculations were conducted using Kesaev's data for I_f and $G(g/s)$ [16]. Kesaev does not reported data for u_c and for its dependence on the film thickness.

Therefore, the value of u_c was varied over a broad range. Note there was presented a first approach to understand the prnciple spot parameters on the film cathode. At next approach (Chap. 17) the study will be conducted calculating u_c. The results calculated for Cu films of various d_f are shown in Fig. 16.34 as a plot of the current density j as a function on u_c [85]. The calculations were carried out for spot lifetime and for films $d_f < 0.5$ μm, since the experimental determination of the erosion rate becomes unreliable for thicker films. Figure 16.35 presents also the calculations for bismuth deposited on the glass substrate. As in case of bulk cathode, the solutions for film cathodes are also two-valued. However, the spot diameter obviously cannot be larger than the transverse dimension of the signature and this condition, i.e., $2r_s = \delta$, correspond to higher second solution (small circles in figure) and to minimal current density j_{min}.

On the other hand, it can calculate the possible maximal current density j_{max} (minimal r_{smin}) at the spot. The transverse dimension of the signature obviously cannot exceed the transverse dimension of the melting isotherm δ_m so that condition $\delta_m = \delta$ sets an upper limit on the range of possible current densities. The value of r_{smin} (j_{max}) is determined from the system of equation with heat conduction equation using melting temperature T_m:

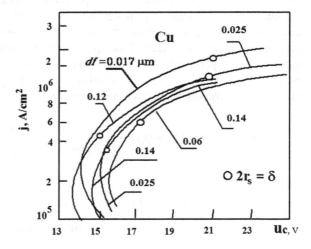

Fig. 16.34 Current density as function on cathode potential drop for different Cu film thicknesses

16.9 Film Cathode

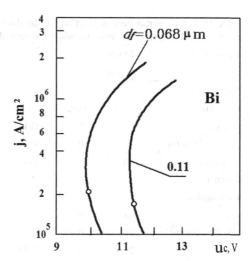

Fig. 16.35 Current density as function on cathode potential drop for different Bi film thicknesses

$$T_m\left(\frac{\delta_m}{2}\right) = \frac{Q_{TL}}{4\pi \lambda_T d_f} \text{Exp}\left(\frac{v_s^2 t_0}{2a}\right) \int_{t_0}^{t_0+t} \frac{dt'}{t'} \text{Exp}\left(-\frac{v_s^2 t'}{4a} - \frac{v_s^2 t_0^2 + \left(\frac{\delta_m}{2}\right)^2}{4at'}\right) \quad (16.52)$$

To evaluate the properties of the cathode region the data of heavy particle $n_h = n_a + n_i$, degree of ionization, the electron current fraction s, electric field E electron temperature T_e, velocity of the plasma expansion v cathode temperature, the energy required for evaporation Q_s, for heat conduction Q_T for case $2r_s = \delta$ are calculated and presented in Table 16.3. The calculated results show that the vapor near the cathode is highly ionized and that the fraction of the electron current is relatively not large ($s = 0.75$). The energy components Q_s, Q_T and total Q are comparable in order of magnitude. Comparison of Q_s with total energy loss Q estimated by Kesaev [16] showed that $Q_T > Q$.

On this basis, Kesaev concluded that THE energy for evaporation could not be important because it was not energetically favorable and that most of the energy supplied was expended in melting the metal in the volume of the signature by heat conduction. This estimation was provided by using $s = 0.95$ obtained calculating the metal thermal regime considering heat point source approach (assuming $r_s \ll \delta$, i.e., significantly large j) and condition $\delta = \delta_m$. The above calculations using general thermal approach and two limited conditions ($r_s = \delta$ and $\delta = \delta_m$) allow obtained $j_{\max}$ and $j_{\min}$ or $r_{s\min}$ and $r_{s\max}$.

Table 16.3 Spot parameters as function on film thickness corresponded to cathode potential drop $u_{c\delta}$, at which r_s equal to the experimental transverse dimension of the signature

Parameters	Material						
	Copper					Bismuth	
Experiment							
d_f (μm)	0.17	0.025	0.06	0.12	0.14	0.068	0.11
I_f (A)	0.1	0.15	0.4	0.9	1.2	0.06	0.08
$v_s \times 10^3$ (cm/s)	13	10.1	0.57	0.34	0.34	0.32	27
G (μg/s)	46	86	260	560	820	130	230
δ (μm)	2.5	4	9	16	20	6	8
Calculations							
$u_{c\delta}$	21.3	19	17	15	15.5	10	11.25
$j \times 10^5$ (A/cm^2)	20	10.7	5.8	4.1	3.3	2.0	1.33
T_e	2.13	1.6	1.33	1.2	1.16	0.86	0.83
T_s	4000	3980	3955	3977	3945	3780	3690
$E \times 10^7$ (V/cm)	2.4	1.9	1.5	1.25	1.16	1.1	1
$v\ 10^2$	720	490	310	200	200	10	10
$n_T \times 10^{20}$ (cm^{-3})	1.2	1.2	1.16	1.2	1.1	11.1	10.3
α	0.83	0.58	0.38	0.28	0.256	0.029	0.025
s	0.761	0.696	0.641	0.616	0.6	0.568	0.52
Q_s (W)	0.268	0.492	1.513	3.258	4.763	0.149	0.262
Q_T (W)	0.156	0.279	0.8	1.692	2.425	0.103	0.192
Q_T/I_f (V)	1.56	1.86	2.0	1.88	2.0	1.03	1.92

16.9.3 Analysis of the Calculation Results

The calculations indicate that the obtained values for spot radii reveal a difference by factor about two, i.e., r_s can be comparable with δ. It can be seen from results of Chap. 3 that also for bulk cathode, the temperature decreases to about the melting temperature for Cu at distance about $r = r_s$. In addition, certain experimental data directly show that the difference between spot radii, and the transverse dimension of the signature is clearly smaller than the difference between r_{smin} and r_{smax}. Indeed, Kesaev [16] reported that the edge of the signature is melted and he assumed that the metal could contract to the edge of the signature due to surface tension force. Thus, on the base of the calculated and experimental data, it can be concluded that the transverse dimension of the signature can be used as a scale dimension of the spot. Also the energy loss by film heat conduction (with an effective voltage for heating

16.9 Film Cathode

of 1–2 V) is smaller than in bulk cathodes (6–7 V [17]). This result, and the small spot size for thin film cathodes (<0.1 μm), indicates that the spot operation on the thin film cathode is different than on the bulk cathode, due to different heat loss and spot motion mechanisms.

The energy loss in the films is relatively small due to very thin thickness and as result, the low values of spot current. This result corresponds to the measured relatively small cathode potential drops in range of (8–12) V for film cathodes with metal substrates [86] which turn out to be lower than the cathode potential drops for large number of materials of bulk cathodes [16]. An exception to this rule is small value of u_c bismuth due to specific low its heat conductivity and heat of atom evaporation. Therefore, it is not surprise that the measured bulk cathode (8.55 V) and film cathode (8.3 V) are equal one to other for Bi. The small energy loss also justly for Cu film while the calculated large value of $u_{c\delta}$, lies above the measured value for Cu film (11.2 V) deposited on Cu bulk substrate (15 V). The calculated value of $u_{c\delta}$ for Cu was obtained here because the condition $r_s = \delta$ was fulfilled at the second solution (high current density) of the system of equations. It will be shown below that the agreement of calculation with the measured cathode potential drop for Cu film cathode will be reached in frame of kinetic model in which the two-value solution is excluded because the cathode potential drop will be determined.

Let us discuss the possibility of studying the physical properties operating in a spot on a bulk cathode through the use the film cathodes. One of the way indicated such possibility is that as the film thickness was increased, the spot on the film cathode converted into the corresponding spot on a bulk cathode. This assumption was used in the formulation of the Kesaev's experiment [16]. If the spots on film correspond to the above model, i.e., if they are spots, in which thermal and electron emission processes are governing, and the erosion results from evaporation, then it can be confidently asserted that the nature and properties of the spots should be similar by the transition to a bulk cathode. Really such mechanism can explain the low-current, low lifetime, and fast-moving spots on bulk cathodes but taking into account the peculiarity of the cathode surface such as cleaning, presence of an impurities, roughness with protrusions.

The surface properties allow understand the spot motion in frame of material expense like expense of the film, protrusion evaporation, and others. Low cathode potential drop occur when the cathode heat loss can be balanced by a current-dependent heat source. Therefore, both the film cathode and rough bulk cathodes (with protrusions, or preliminary heating) have spot currents, which are much smaller than the spot current for the flat bulk cathode. The details of spot behavior of the different spot types will be analyzed considering the kinetic model below. However, accordingly, an arc spot on film cathode is apparently an "independent" physical subject clearly deserving study. This study contributes much to the understanding of the physical processes operating in cathode spots of different types.

16.10 Phenomena of Cathode Melting and Macrodroplets Formation

During cathode spot operation, the metal is melted and the forces due to electrostatic and hydrodynamic plasma pressure, convection of the liquid as well surface tension determined the fractions of remained the liquid metal and removal of droplets. Different mechanisms were developed to explain the behavior of the melted metal under the cathode spot action in vacuum arcs. One of first McClure [87] considered the mechanism of droplets ejection is due to pressure on the molten area of the cathode surface produced by cathode plasma jets using measured the current density j and the cathode force F per unit current (see Chap. 9).

Hantzche [88] obtained a simple solution of the hydrodynamic equations describing the cathode molten area taking in account the forces acting on the cathode arc spot surface which composed of the ion pressure, the neutral gas pressure, and the evaporation recoil while electrostatic forces diminish the effective pressure. The melting front velocity was determined by a solution of the heat conduction equation, assuming a constant heat flux density per second the temperature increase It was stated that an acceleration of the liquid layer toward the borders of the spot, mainly due to pressure forces, leads to an excavation of the melt from the spot and to crater formation. The overflow of the metal at the rims of the crater, to the ejection of splashes, causes the rise of emission tips.

The calculations for Mo cathode shown the ejection velocities at the spot rims of $(2–3) \times 10^4$ cm/s and the layer depth of about 0.4 μm. An expression was derived describing the spot motion indicating that the corresponding velocity is about equal to the melting front velocity, but because of the random character of this motion, the measured values were significantly smaller. The obtained results did not take in account the effect of tension force. In addition, the used momentum flux of accelerated ions in the sheath can prove to be insignificant because of its compensation by the momentum flux from the electric field near the cathode surface.

The metal convection and tension force of thermocapillary effect in frame hydrodynamic mechanism were studied by Zharinov and Sanochkin [89, 90]. It was considered a single spot behavior on a metal liquid layer with size l_L that much larger than the spot size r_s and the liquid thickness layer h_L. As between hot spot and the peripheries of the liquids a temperature gradient arise, a force is produced due to gradient of tension coefficient α'_t. As a result in the liquid with a free convection regime, two symmetric cells form on two sides of heating (by spot) line. These cells characterized by a certain length and maximal velocity depended on α'_t. This situation is unstable because some small fluctuations of spot parameters cause an asymmetry of the cells and therefore motion of the liquid together with the current carried spot.

The calculations showed relatively large velocity that is used to explain the measured data of spot motion [89] as well as to claim that significant value of plasma pressure can be compensated by the dynamic flow strength [90]. As the cells approach each other and characteristic distance between cells becomes smaller than its length,

16.10 Phenomena of Cathode Melting and Macrodroplets Formation

the thermocapillary vortices begin interact. This is the regime of hindered convection and allows to discuss the mechanism of spots repulsion as cathode spots draw together and divide. Sanochkin and Filippov [91, 92] provided a detailed study of the thermocapillary mechanism of cathode spot convergence and phenomenon of mutual repulsion.

The phenomenon of fixation of the cathode spot based on some self-regulating hydrodynamic mechanism of liquid motion was also considered using formation the thermocapillary vortex cells produced in liquid layer during point heating near a vertical wall [93, 94]. If the spot is far from the wall, the convection pattern will be symmetric on both sides of the heating point. However, near the wall, the asymmetry of the convection will be directed and a state of equilibrium is violated. When the spot reached the vicinity of the wetted wall, it will move toward this wall until that it becomes fixed.

Finally, Sanochkin [95] summarized his work and the works with his co-authors establishing some connections between different observed cathode spot phenomena and the results of a set of Navier–Stokes and energy equations solved numerically in case of liquid layer local heating by an energy from the cathode spot with size much lower than the length of the layer. It was concluded that the thermocapillary convection approach makes possible to study the dynamics of the cathode spot including the equilibrium between plasma pressure and pressure due to velocity head, mechanism of the spot motion, the spot fixation, etc.

It should be noted that cathode melting around the spot and droplets ejection are experimental fact as well as spot fixation in mercury arc with Mo rod. However, the assumption related to the cathode state and use developed theory of thermocapillary convection are far from that observed during spot operation. First of all the spot as a heat source develop in time and therefore no symmetric pattern of the thermocapillary vortex can be expected. According to the observation and to the calculations, the liquid area arising around the spot is of size of the spot (see Chap. 3, [96]). The size of isotherm of melting temperature exceeds the spot diameter by factor lower than 2, and therefore, the requested for the model condition $l_L \gg r_s \gg h_L$ is not fulfilled. The liquid zones from single fragments can be overlapped in a group spots, but this spot type is practically rest spots and the small moving fragments can be explained by non-uniformity of the surface profile during their function. The character and velocity of super fast spot SFS motion were also determined by the cathode surface profile consisted of a number protrusions or contaminations, which supplied appropriate heat conduction condition for relatively small energy requested for spot development at small current per spot (See Chap. 17).

Let us consider the cathode spot moving on mercury cathode that modeled as heat source acted on a liquid layer. The cathode temperature inside of the spot is relatively small. About few 100 °C and therefore the assumption regarding the presence of temperature difference of 1000 °C [88] necessary to generation convection flow in viscose Hg is very doubtful. The mechanism of spot operation for mercury arc in presence of Mo rod can be explained by spot burning in vapor of the Hg film wetting around the Mo rod. Therefore, it is clear appearance of spot fixation at the Mo rod and its linear form. However, the main problem of Hg arc is the mechanism of electron emission from the cathode at very low temperature, which the method of thermocapillary convections was not solved.

Of course, in general case the thermal model of spot on metallic cathode should consider the cathode melting and the possibility of convective heat transfer. The question is how it influences all spot parameters considering the self-consistent solution of the total system of equation describing the complex of spot phenomena simultaneously. Such solution was obtained by Buchin and Zektser [97] considering the melting pool around the spot and effect of liquid convection taking into account the GDM spot model and system of equations [7, 98]. They concluded that if one takes into account the liquid metal pool of melt in the region of the cathode spot, one obtains only small quantitative change in the parameters of the spot, in comparison with calculation without the liquid metal, within the framework of the used model [7, 96].

Mesyats and Uimanov [99] described the microcrater formation during the postexplosion hydrodynamic stage of the operation of an emission center named cathode spot. The plasma experimental data for pressure and effective cathode voltage with given spot current were used to determine heat generation in the cathode due to electric current by Joule effect. Two-dimensional axisymmetric equations for the incompressible fluid flow and for the charge and heat transfer in a cathode including the melt pool and the solid cathode were considered. The surface heat flux and current density spatial distribution were assumed dependent of a spot radius that, in essence, also as a given parameter. It has been shown that for the spot current ranging between 1.6 and 7 A and the time of current flow through the cell ranging between 15 and 60 ns, the crater diameter is 3–7 μm and the maximum current density in the spot center equal to $(1–3) \times 10^8$ A/cm^2.

Nachshon, Beilis, and Ashkenazy [100] studied the cathode melting and crater formation using system of equations including heat and electrical conduction equations and hydrodynamic of melting material flow taking into account the plasma pressure and the pressure due to surface tension. A two-phase heat transfer for the solid and liquid phase was considered. The solid is given as a liquid of very high viscosity, and only the liquid phase is allowed to move. The flow is considered compressible, non-isothermal, non-turbulent, and sub-sonic described by the following equations:

1. Heat equation (conduction and convection):

$$\rho c \frac{\partial T}{\partial t} + \rho c \vec{u} \cdot \vec{\nabla} T = \vec{\nabla}(\lambda_T \vec{\nabla} T) + q_J \qquad (16.53)$$

Here ρ is the density, c is the heat capacity, λ_T is the heat conductivity

2. Momentum equation for material flow with it viscosity:

$$\rho \frac{\partial \vec{u}}{\partial t} + \rho(\vec{u} \cdot \vec{\nabla})\vec{u} = \vec{\nabla} \cdot \left(-p\hat{u} + \mu\left(\vec{\nabla}\vec{u} + (\vec{\nabla}\vec{u})^T\right) - \frac{2}{3}\mu(\vec{\nabla} \cdot \vec{u})\hat{u}\right) + \vec{F}$$

16.10 Phenomena of Cathode Melting and Macrodroplets Formation

$$\vec{F} = \frac{(1-\theta)^2}{\theta^3 + \varepsilon} F_0 \vec{u} \tag{16.54}$$

where F is the volume force due to liquid flow, μ is the viscosity.

3. Continuity equation

$$\frac{\partial \rho}{\partial t} + \vec{\nabla} \cdot (\rho \vec{u}) \tag{16.55}$$

4. Quasistatic electric conduction

$$\vec{\nabla} \cdot \vec{j} = 0 \tag{16.56}$$

$$\vec{j} = \left(\sigma_{el} + \varepsilon \frac{\partial \rho}{\partial t} \right) \vec{E}$$

$$\vec{E} = -\vec{\nabla} V \tag{16.57}$$

We define a phase field θ for each phases where $\theta = 1$ is solid and $\theta = 0$ is liquid. The simulation evaluates θ at each step based on the temperature:

$$\theta(T) = \begin{cases} 0 & T < T_L \\ \frac{T-T_L}{T_H - T_L} & T_L < T < T_H \\ 1 & T < T_H \end{cases} \tag{16.58}$$

where $T_H = T_m + \Delta T$, $T_L = T_m - \Delta T$ are temperature range to allow for a continuous calculation of the phase transition and the melting temperature T_L and $T_m - 1358$ K.

The thermophysical parameters are:

$$\rho = \theta \rho_{sol} + (1-\theta) \rho_L$$
$$\lambda_T = \theta \lambda_{Ts} + (1-\theta) \lambda_{TsL}$$
$$\mu = \theta \mu_{sol} + (1-\theta) \mu_L \tag{16.59}$$

$$\alpha_m = \frac{1}{2} \frac{(1-\theta)\rho_L - \theta \rho_{sol}}{(1-\theta)\rho_L + \theta \rho_{sol}} \tag{16.60}$$

where α_m is the fraction of liquid and solid for each calculated element.

The boundary conditions.

For the heat conduction equation taken in account that the heat flux in the spot is Gauss distributed and energy loss due to cathode evaporation

For pressure:

$$P = P_v \left(1 + \alpha \frac{T_e}{T} \right) + \sigma_{ten} \vec{\nabla} \cdot \hat{n} \tag{16.61}$$

The second term of (16.61) is the electron pressure and is determined by ionization fraction, α and electrons temperature T_e. The last term is the contribution of the surface tension. The surface tension gives a correction to the pressure inside, which is the surface tension divided by the curvature with radius R of the surface $\Delta P = \sigma_{ten}/R$. As the curvature is $1/R = \vec{\nabla} \cdot \hat{n}$ then $\Delta P = \sigma_{ten} \vec{\nabla} \cdot \hat{n}$.

The initial conditions in the bulk and on the boundary, $t = 0$: $V = 0$, $T = 300$ K, $u = 0$, $\partial z_{boundary}/\partial t = 0$.

The problem was simulated a 2D axial symmetric approximation. The system of equations were solved using a finite element method in COMSOL. A moving mesh boundary condition setting the normal velocity was l considered.

As example, the results (reported in [100]) of simulation of the liquid phase and crater formations at current of 10 A, spot radius of 10 μm, and for mesh size 0.25 μm are presented in Figs. 16.36 and 16.37. The results are given for Cu by an axial representation. It was shown that the crater rim edge might form subsequent droplets. Liquid layer was formed during arcing, and the liquid was expelled outward. The calculations indicate a resolidified region below the crater as well as material at the edge of the crater, which could form detached droplets.

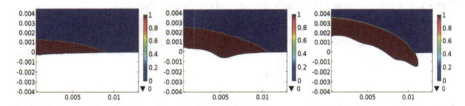

Fig. 16.36 Evolution of liquid phase fraction over time for 75, 150, and 250 ns shown from left to right, respectively. Left side is $r = 0$, bottom is plasma surface boundary. Dimensions are in mm. Brown color is liquid, blue is solid, and white is the plasma [100]

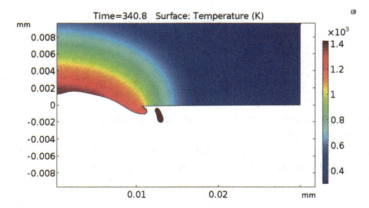

Fig. 16.37 Temperate map of the simulation after time 340.8 ns and small droplet formation. Left side is $r = 0$, bottom is plasma surface boundary. The color at right show the temperature. White color is the plasma [100]

16.11 Summary

As it was shown in Chap. 15, the early models considered various processes and structures in the cathode spot separately. However, new experimental spot data and developed methods of thermal solids and plasma studies stimulated further theoretical thinking. Although the developed models somewhat advanced, the models were not mathematically closed due limitation in available investigations of essential features of the cathode plasma. The present chapter reports further author's ideas using the modern developments in the plasma science, e.g., approaches of multi-component particle and energy flow in partially ionized gas, plasma–wall transition studies and plasma particle collision cross-sectional data. These ideas allowed to explain not only the current continuity (spot plasma and cathode parameters), but the different measured arc phenomena detected experimentally.

The developed gasdynamic model (GDM) first used the resonance charge-exchange collisions to base the diffusion mechanism of ion flux formation toward the cathode. By taking into account processes occurring in the near electrode plasma, a closed system of equations was formulated. The equations which may be solved simultaneously and yield the values of various cathode and plasma parameters, some of which can be measured experimentally (cathode potential drop u_c, Erosion rate G), and some of which cannot. The mathematically closed system of equations for the cathode plasma was solved self-consistently with the equations for cathode heat conduction. This enabled understanding and determining directly the early unknown electron current fraction s and analyzing the main cathode plasma parameters.

An interesting result of calculation is that the model predicted considerable differences between a relatively small thermal conductivity of materials such as Ni, and relatively high thermal conductivity materials such as Cu and Ag. The calculations indicate a minimal values of spot current and cathode potential drop at which the solution of system of equations exist that for Ni is low ~10 A, while for Cu, Ag it is ~100 A. Similarly, u_c for Ni = 8 V, and 10 V for Cu and Ag. Because of the intrinsically low thermal conductivity in the Ni arc, there will be less of a tendency to form group spots than with Cu and Ag. Experimentally, a group spot sub-structure is observed in relatively high-current (~600–1000 A) arcs on Cu, Ag, and W cathodes, while group spots (with relatively large current~100–200A) are not observed on Ni [101]. It is due to lower input heat flux requested to reach sufficiently temperature for self-consistent processes supported the Ni arc existing with separately moving spots. These spot can be supported by thermal mechanism and by evaporation as independent spot at the bulk cathode. Therefore, caution should be exercised in inferring details about the mechanism of the cathode spot based on macroscopic phenomenological classification—spots which have similar macroscopic characteristics (fast-moving spot for Ni and for Cu, W with low spot current ~10 A) may differ in their microscopic structure, and in the physical mechanisms of their operation.

The principal mechanism of a self-sustained nanosecond cathode spot operation take in account the specifics of the arcs when spots are appeared at large rate of current rise $\geq 10^9$ A/s. In this case, the cathode temperature depends not only on spot current, but also on current rise during the non-stationary spot operation. According to the calculation, the large current density (10^7–10^8 A/cm^2) can be supported by very high electron current fraction (~0.9). Therefore, the incoming heat flux in the plasma energy balance is large and, as results, the electron temperature reaches a high value (a few eV). In this case, the plasma generates a large flux of hot electrons, which reach the cathode surface. Their energy flux together with that of the ion energy flux, heats the cathode, and thus supports cathode evaporation. When the plasma electron energy is not taken in account, the Joule and ion bombardment heating are comparable for j ~10^8 A/cm^2.

Existence of the cathode spot demands material flux continuity at the cathode plasma interface, and current of electron emission, which is sufficient at produced cathode surface temperature. It was established that for cathodes constructed from materials having intermediate thermophysical properties, the electrons are emitted from the cathode surface by Schottky-enhanced thermionic emission—at the high temperatures necessary for sufficient evaporation of these materials. This emission mechanism is sufficient, and the effect of tunneling is negligible, except of nanosecond spot appearing with rate of current rise larger 10^9 A/s.

Thus, the above developed model shows that, while a high electric field directed into the cathode surface requires Schottky enhancement of the thermionic emission on intermediate materials, the field is still weak in the case of volatile cathodes for emission of the electrons by known mechanisms. In the case of refractory materials, the electric field is actually directed outward from the cathode surface, i.e., in the direction that retards electron emission.

For refractory materials, which require an even higher temperature to maintain material flux continuity, thermionic emission is over sufficient. An new idea was developed for which was taking in account the effect of the electron space charge near the cathode produces an electric field in the direction that reduces the large thermionic electron emission, thus acting as a virtual cathode. It was shown that range of electron current fractions can be increased using condition $E_p \neq 0$ at the boundary of sheath–presheath interface and the solution showed E_p/E_d, where E_d calculated as ~T_e/r_d, r_d is the Debye length. The calculation shows that for refractory cathode, a nonzero electric field at the plasma-sheath interface and the quasineutral plasma can be consisted of ions with different charge state in the cathode region.

In the case of the volatile materials, an additional plasma double layer adjacent to the cathode surface is hypothesized, which serves as a "plasma cathode," supplying electrons to the remainder of the discharge. The electrons are emitted from a *plasma cathode* produced by thermal ionization of the evaporated material. So, the developed model of the near-cathode region of the arc indicate three characteristic plasma regions, which in order of their distance from the cathode surface are called the first plasma region, the double sheath, and the second plasma region. The arc current at the cathode surface is closed by ion flux from the adjacent cathode plasma. Electrons emitted from the first plasma region are accelerated through the double sheath region into the second plasma region. Likewise, ions flowing from the second ion region are accelerated through the double sheath into the first plasma region where they serve as a significant heat source. Two types of time-dependent solutions were found,

16.11 Summary

with characteristics times of 0.1–1 μs and 100 μs corresponded to the experimentally observed spot lifetimes for the transitional and fundamental discharge forms, respectively.

An important feature of the model is the key role played by the flux of excited Hg atoms and the outflow of neutral atom in energy balance of the plasma layer adjacent to the cathode. The model also predicts that the atoms further from the cathode would be preferentially excited in comparison to the plasma layer adjacent to cathode, due to both a higher plasma temperature and high directed energy of the free electrons in that region. This preferential excitation may explain the experimentally observed dark space in the Hg arc discharge and the relatively small cathode potential drop in a basic form having a cathode drop of 8.5–10 V, which usually appears when there is sufficient arc current for several cells to exist simultaneously. The observed transient form, having a cathode drop of 18 V, appears at low current, when the spot is on the verge of extinction. A possible interpretation of the model results is that the step ionization describes the basic form, while the model with direct ionization describes the transient form. Acceleration of electrons from the first plasma region through u_{sh} explains the observed fast electrons by Ne atom excitation in Hg arc.

It should be noted that after publication GDM (~1970), a number of researchers studied the cathode spot, in essence on base of the evaporation model, with some modifications using different some arbitrary given parameters (except the u_c from experiment [16]) in order to formulation a closed the mathematical approach.

The above results were obtained by closed mathematical approach, which traditionally used experimental value of u_c and in addition the erosion rate G. An extended information will be obtained from physically closed mathematical approach based on kinetics of the cathode evaporation and plasma generation. This approach (next chapter) allows calculate the time-dependent values of u_c and G and develop mechanisms explained different cathode spot phenomena such as motion, spot types, splitting, and nature of magnetic field influence.

References

1. Loeb, L. B. Fundamental processes of electrical discharge in gases, New York (1939) and also (1947) (Russian translation, Gostekhisdat, M-L, 1950).
2. Beilis, I. I., & Rakhovskii, V. I. (1969). Theory of the cathode mechanism of an arc discharge. *High Temperature, 7*(4), 568–573.
3. Beilis, I. I., Lyubimov, G. A., & Rakhovskii, V. I. (1969). Electric field at the electrode surface at the cathode spot of an arc discharge. *Soviet Physics Doklady, 14*(3), 897–900.
4. Beilis, I. I., Lyubimov, G. A., & Rakhovskii, V. I. (1970). Dynamics of the variation of the fraction of electron current in the region of an arc discharge near the cathode. *Soviet Physics Doklady, 15*(2), 254–256.
5. Beilis, I. I., Lyubimov, G. A., Rakhovskii, V. I. (1970). Dynamics of the variation of the fraction of electron current in the region of an arc discharge near the cathode. In Proceedings of V International Conference on Continnum Phenomena (pp. 31–33). Munich, Germany.
6. Beilis, I. I., Lyubimov, G. A., & Rakhovskii, V. I. (1972). Diffusion model of the near cathode region of a high current arc discharge. *Soviet Physics—Doklady, 17*(1), 225–228.

7. Beilis, I. I. (1974). Analysis of the cathode in a vacuum arc. *Soviet Physics—Technical Physics, 19*(2), 251–256.
8. Dushman, S. Scientific foundation of vacuum technique, J. M. Lafferty (Ed.) (p. 691). NY: Wiley.
9. Beilis, I. I. (1974). Emission process at the cathode of an electrical arc. *Soviet Physics—Technical Physics, 19*(2), 257–260.
10. Beilis, I. I. (1977). Theoretical analysis of cathode phenomena in vacuum arc discharge. *Chapter in book: Part I., 3rd International Symposium, Switching Arc Phenomena 1977*, Lodz, Poland: Theh. Univ. (pp. 194–200).
11. Rich, J. A. (1961). Resistance heating in the arc cathode spot zone. *Journal of Applied Physics, 32*(6), 1023–1031.
12. Beilis, I. I., & Lyubimov, G. A. (1975). Parameters of cathode region of vacuum arc. *High Temperature, 13*(6), 1057–1064.
13. Kantsel', V. V., Kurakina, T. S., Potokin, V. S., Rakhovskii, V. I., & Tkachev, L. G. (1968). Thermophysical parameters of a material and electrode erosion in a high current vacuum discharge. *Soviet Physics—Technical Physics, 13*, 814–817.
14. Beilis, I. I. (1972). Determination of the parameters of the near-cathode region of arc discharges. *Chapter in book: Electrical contacts* (M. Nauka, pp. 104–106) (in Russian).
15. Beilis, I. I. (1977). Cathode spots on metallic electrode of a vacuum arc discharge. *High Temperature, 15*(5), 818–824.
16. Kesaev, I. G. (1968). *Cathode processes in electric arcs*. Moscow: NAUKA Publishers. (in Russian).
17. Daalder, J. E. (1977). Energy dissipation in the cathode of a vacuum arc. *Journal of Physics. D. Applied Physics, 10*, 2225–2234.
18. Daalder, J. E. (1976). Components of cathode erosion in vacuum arcs. *Journal of Physics. D. Applied Physics, 9*, 2379–2395.
19. Beilis, I. I., Djakov, B. E., Juttner, B., & Pursch, H. (1997). Structure and dynamics of high-current arc cathode spots in vacuum. *Journal of Physics. D. Applied Physics, 30*, 119–130.
20. Beilis, I. I. (1986). Cathode arc plasma flow in a Knudsen layer. *High Temperature, 24*(3), 319–325.
21. Anders, A., Anders, S., Juttner, B., Botticher, W., Luck, H., & Schroder, G. (1992). Pulsed dye laser diagnostics of vacuum arc cathode spots. *IEEE Transactions on Plasma Science, 20*(4), 466–472.
22. Juttner, B. (1995). The dynamics of arc cathode spots in vacuum. *Journal of Physics. D. Applied Physics, 28*(4), 516–522.
23. Juttner, B. (1997). The dynamics of arc cathode spots in vacuum: new measurements. *Journal of Physics. D. Applied Physics, 30*(2), 221–229.
24. Juttner, B. (1999). Nanosecond displacement times of arc cathode spots in vacuum. *IEEE Transactions on Plasma Science, 27*(4), 836–844.
25. Beilis, I. I. (2001). A mechanism for nanosecond cathode spot operation in vacuum arcs. *IEEE Transactions on Plasma Science, 29*(N5), Part 2, 844–847.
26. *Handbook of vacuum arc science and technology*, R. L. Boxman, P. J. Martin, & D. M. Sanders (Eds.). (1995). N.J.: Noyes Publ. Park Ridge.
27. Puchkarev, V. (1991). Estimating the electron temperature from fluctuations in vacuum arc plasma: cathode spot operation on contaminated surface. *Journal of Physics. D. Applied Physics, 24*(5), 685–692.
28. Bek-Bulatov, I Kh, Borukhov, M Yu., & Nagaibekov, R. B. (1973). Cathode spot in a vacuum arc with refractory metals. *Soviet Physics—Technical Physics, 18*, 1401–1402.
29. Hantzsche, E. (1974). Model calculations for the cathode spot of the vacuum arc. *Beitr. Plasmaphysics, B14*(4), 135–138.
30. Beilis, I. I. (1988). Model of steady-state arc spot on a refractory cathode in vacuum. *Soviet Physics—Technical Physics, 14*, 494–495.
31. Beilis, I. I. (1988). On a mechanism of vacuum arc spot function on refractory cathodes. *High Temperature, 14*, 1224–1226.

References

32. Zykova, N. M., Kantsel, V. V., Rakhovsky, V. I., Seliverstova, I. F., & Ustimets, A. P. (1971). The dynamics of the development of cathode and anode regions of electric arcs I. *Soviet Physics—Technical Physics, 15*(11), 1844–1849.
33. Puchkarev, V. F., & Murzakayev, A. M. (1990). Current density and the cathode spot lifetime in a vacuum arc at threshold currents. *Journal of Physics. D. Applied Physics, 23,* 26–35.
34. Finkelnburg, W., & Maecker, H. (1956). Electrishe bogen und thermishes plasma. *Handbuch der Physik, Bd. XXII*, 254–444.
35. Rakhovsky, V. I. (1976). Experimental study of the dynamics of cathode spots development. *IEEE Transactions on Plasma Science, 4,* 87–102.
36. Selicatova, S. M., & Lukatskaya, I. A. (1972). Initial stage of a disconnection vacuum arc. *Soviet Physics—Technical Physics, 17,* 1202–1208.
37. Persky, N. E., Sysun, V. I., & Kromoy, Y. D. (1989). The dynamics of cathode spots in a vacuum discharge. *High Temperature, 27*(6), 832–836.
38. Langmuir, I. (1929). The interaction of electron and positive ion space charges in cathode sheaths. *Physical Review, 33*(N6), 954.
39. Beilis, I. I., & Keidar, M. (1998). Sheath and presheath structure in the plasma wall-transition layer in an oblique magnetic field. *Physics of Plasmas, 5*(5), 1545–1553.
40. Beilis, I. I. (2004). A mechanism for small electron current fraction in a vacuum arc cathode spot on a refractory cathode. *Applied Physics Letters, 84*(8), 1269–1271.
41. Allen, C. W. (1973). *Astrophysical quantities.* London: Athlone Press.
42. Fortov, V. E., & Yakubov, I. T. (1990). *Physics of nonideal Plasma.* N.Y.: Hemisphere Publ.
43. Reece, I. P. (1963). The vacuum switch. *Proceedings IEE, 110*(4), 793–811.
44. Brown, I. G., Feinberg, B., & Galvin, J. E. (1988). Multiply stripped ion generation in the metal vapor vacuum arc. *Physics of Plasmas, 63*(10), 4889–4898.
45. Beilis, I. I. (2001). State of the theory of vacuum arcs. *IEEE Transactions on Plasma Science, 29*(5), 657–670.
46. Beilis, I. I. (1996). Current continuity and instability of the mercury vacuum arc cathode spot. *IEEE Transactions on Plasma Science, 15,* 488–501.
47. Engel, A., & Steenbeck, M. M. (1934). *Elecktrishe Gasentladungen Ihre Physic Und Technik.* Berlin: Springer.
48. Smith, G. (1942). The mercury arc cathode. *Physical Review, 62,* 48–54.
49. Kesaev, I. G. (1964). *Cathode processes in the mercury arc.* N.Y.: Consultants Bureau.
50. Ecker, G. (1961). Electrode components of the arc discharge. *Ergeb. Exakten Naturwiss, 33,* 1–104.
51. Ecker, G. (1980). Theoretical aspects of the vacuum arc. In J. M. Lafferty (Ed.) *Vacuum arcs: Theory and application* (pp. 228–320). N.Y.: Wiley.
52. Waymouth, I. J. (1971). *Electrical discharge lamps.* London: MIT Press.
53. Khromoi, Yu. D., Zemskova, L. K., & Korchagina, Yu. I. (1978). Pinning of a cathode spot with pulsed discharge current. *Soviet Physics—Technical Physics, 23*(N8), 920–922.
54. Poroshin, S. N., & Khromoi, Yu. D. (1988). The current threshold of anchored cathode spot. *Teplofiz. Vys. Temp., 26*(N6), 1226–1228 (in Russian).
55. Eckhardt, G. (1980). Properties of anchored cathode spot of a DC mercury vacuum arc. *IEEE Transactions on Plasma Science, PS-4*(N4), 295–301.
56. Sena, L. A. (1959). The appearance of a cathode spot at the boundary of a dielectric and mercury in an ionized gas. *Soviet Physics—Technical Physics, 4*(1), 3–9.
57. Udris, Ya. Ya. (1958). *Trudy VEI, Investigation in the field of electrical discharges in gases. N63* (pp. 107–128). Moscow: Gosenergoizdat (in Russian).
58. Eckhardt, G. (1971). Efflux of atoms from cathode spots of low-pressure mercury arc. *Journal of Applied Physics, 42*(13), 5757–5760.
59. Anisimov, S. I., Imas, Yu A, Romanov, G. S., & Khodyko, Yu V. (1971). *Action of high-power radiation on metals.* Springfield, Virginia: Nat. Tech. Infor. Serv.
60. Hull, A. W. (1962). Cathode spot. *Physical Review, 12b*(N5), 1603–1610.
61. Mackeown, S. S. (1929). The cathode drop in an electrical arc. *Physical Review, 34,* 611–614.

62. Tonks, L. (1935). A theory of liquid surface rupture by a uniform electric field. *Physical Review, 48,* 562–568.
63. Mesyats, G. A., & Proskurovsky, D. I. (1989). *Pulsed electrical discharges in vacuum.* Heidelberg: Springer.
64. Batrakov, A. V., Popov, C. A., & Proskurovsky, D. I. (1993). Observing of the dynamics of the electrohydrodynamic instability on the fluids cathode surface. *Letters Journal of Technical Physics, 19*(19), 66–70.
65. Rodnevich, B. B. (1978). The structure of the potential barrier on boundary of metal. *Journal of Applied Mechanics and Technical Physics, N5,* 79–86. (In Russian).
66. Robson, A. E., & Engel, A. (1955). Excitation processes and the theory of the arc discharge. *Nature, 175*(N4458), N9, 646.
67. Robson, A. E. (1959). The new theory of cathode spot. *Radiotechnika and Elektronika, 4*(N8), 1295–1298 (In Russian).
68. Engel, A., & Robson, A. E. (1958). Mechanism of electron emission from the arc cathode. *Journal of Applied Physics, 19*(N4), 734; Lee, T. H. On the discussion of the T-F theory by Robson and Von Engel, ibid. (pp. 734–735).
69. Engel, A., & Robson, A. E. (1957). The excitation theory of arcs with evaporating cathodes. In *Proceedings of the royal society A. Mathematical, physical and engineering science*, A243 N1233 (pp. 217–236).
70. Engel, A. (1955). *Ionized gases.* Oxford: Clarendon Press.
71. Biberman, L. M., Vorobyev, V. S., & Yakubov, I. T. (1987). *Kinetics of non-equilibrium, low temperature plasmas.* N. Y: Consultant Bureau.
72. Slepian, J. (1926). Theory of current transference at the cathode of an arc. *Physical Review, 27*(4), 407–412.
73. Spitzer, L. (1956). *Physics of fully ionized gases.* N.Y.: Interscience.
74. Alekseev, A., Fortov, V. E., & Yakubov, I. T. (1983). Physical properties of high-pressure plasmas. *Soviet Physics Uspekhi, 26*(2), 99–115.
75. Beilis, I. I. (1990). The nature of an arc discharge with a mercury cathode in vacuum. *Soviet Technical Physics Letter, 16*(5), 390–391.
76. Lutsenko, E. I., Sereda, N. D., & Kontsevoi, L. M. (1976). Investigation of the charge volume sheath creation in plasmas. *Soviet Journal of Plasma Physics, 2*(N1), 39.
77. Andrews, J. G., & Allen, J. E. (1971). Theory of a double sheath between two plasmas. *Proceedings of Royal Society, London, A320,* 459–472.
78. Tablitzy Physicheskih Velychin, I. Kikoyn (Ed.) (1976). Moscow: Atomizdat (in Russian)
79. Ecker, G. (1980). Theoretical aspects of the vacuum arc. In J. M. Lafferty (Ed.), *Vacuum arcs: Theory and application* (pp. 228–320). New York: Wiley.
80. Sutton, G. W., & Sherman, A. (1965). *Engineering magnetohydrodynamics,* N.Y.: McGrawn-Hill Book Company.
81. Raizer, Yu. P. (1991). *Gas discharge physics,* Berlin: Springer.
82. Babucke, G., Hartel, G., & Kloss, H. G. (1991). On the energy balance in the core electrode-stabilized high-pressure mercury discharges. *Journal of Physics. D. Applied Physics, 24,* 1316–1321.
83. Bashlov, N., Zissis, G., Charrada, K., Stambouli, M., Milenin, V., & Timofeev, N. (1994). Simulation of the high-pressure mercury discharge lamp during the middle phase of start-up (medium mercury pressure). *Journal of Physics. D. Applied Physics, 27,* 494–503.
84. Massy, H., & Burhop, E. (1952). *Electronic and ionic impact phenomena.* Oxford: Clarendon Press.
85. Beilis, I. I., & Lyubimov, G. A. (1976). Theory of the arc spot on a film cathode. *Soviet Physics—Technical Physics, 21*(6), 698–703.
86. Grakov, V. (1967). Cathode fall in vacuum arcs with deposited cathodes. *Soviet Physics—Technical Physics, 12,* 1248–1250.
87. McClure, G. W. (1974). Plasma expansion as a cause of metal displacement in vacuum-arc cathode spots. *Journal of Applied Physics, 45*(5), 2078–2084.

88. Hantzsche, E. (1977). A new model of crater formation by arc spots. *Beitr. aus der Plasmaphysik, 17,* 65–74.
89. Zharinov, A. V., & Sanochkin, Yu V. (1982). Possible explanation of the mechanism of cathode spot motion. *JETP Letters, 36*(5), 182–185.
90. Zharinov, A. V., & Sanochkin, Yu V. (1983). Meniskus shape and equilibrium condition of the liquid metal surface in a cathode spot in vacuum arc. *Soviet Physics—Technical Physics, 9*(23), 629–630.
91. Sanochkin, Yu V, & Filippov, S. S. (1985). Hydrodynamic mechanism for the repulsion of current cells as cathode spots draw together and divide. *Soviet Technical Physics Letter, 11,* 305–306.
92. Sanochkin, Yu V, & Filippov, S. S. (1986). Interaction of thermocapillary cells during convergence and phenomen of mutual repulsion of cathode spots. *High Temperature, 24,* 706–711.
93. Sanochkin, Yu V. (1986). Hydrodynamic nature of the fixation of a cathode spot. *Soviet Technical Physics Letter, 12,* 180–181.
94. Sanochkin, Yu V. (1988). Cathode spot location and thermocapillary convection in horizontal liquid layer with point free-surface heating near a vertical wall. *High Temperature, 26,* 384–391.
95. Sanochkin, Yu V. (1987). thermocapillary convection and dynamics of cathode spot at the liquid cathode. *Nuvo Cimento, 9D,* 941–958.
96. Beilis, I. I., & Lyubimov, G. A. (1976). "Signature" determination of current density at the cathode spot in an arc. *Soviet Physics—Technical Physics, 21,* 1280–1282.
97. Buchin, V. A., & Zektser, M. P. (1990). The effect of cathode melting in the spot region of a vacuum arc. *Soviet Physics—Technical Physics, 35*(4), 456–459.
98. Beilis, I. I., Lyubimov, G. A., & Zektser, M. P. (1988). Analysis of the formulation and solution of the problem of the cathode jets of a vacuum arc. *Soviet Physics—Technical Physics, 33*(10), 1132–1137.
99. Mesyats, G. A., & Uimanov, I. V. (2017). Semiempirical model of the microcrater formation in the cathode spot of a vacuum arc. *IEEE Transactions on Plasma Science, 45*(8), 2087–2092.
100. Nachshon, I., Beilis, I. I., & Ashkenazy, Y. (2017 March). Crater evolution and droplet generation in vacuum arcs during cathode spot evolution. In *6th international workshop mechanisms of vacuum arcs, MeVArc 2017,* Jerusalem, Israel.
101. Zykova, N. M. (1968). *Investigation of the dynamics of the development of cathode and anode spots of electric arcs.* Academy of Science Siberian Brunch, Institute of Physics, Krasnoyarsk: Thesis.

Chapter 17
Kinetic Theory. Mathematical Formulation of a Physically Closed Approach

Most of published works (Chap. 15) used a system of equation, in which the number of unknown not equal to the number of equations. In this case, a number of arbitrary parameters, characterizing the cathode plasma, the cathode surface, and the spot current structure, added to the mathematical description. The cathode potential drop u_c was used as input parameter traditionally referred to its measurement with high accuracy. Another parameter is the cathode erosion rate G was calculated by evaporation rate using Langmuir–Dushman (or Frenkel) formula that is incorrect not only by describing the mass loss in vacuum (with rarefied collisions), but also, it does not account the returned mass flux to the surface in the cathode dense plasma. Nevertheless, the study of transient spot parameters by gas dynamic model (GDM, Chap. 16, [1]) shows that the values u_c and G can influence the spot parameters during its transient development. The presence of two values in solution when these parameters are fixed requested a physical reason to choose the correct data. In addition, the observed voltage oscillations indicate that there is more complicated behavior of cathode spot at low arc current. Thus, an investigation of the cathode spot behavior requires use of mathematical formulation in a physical closed approach.

Let us define the understanding of term "physical closed approach". The physically closed self-consistent approach we define as an assembly of equations described the physical processes by a mathematical formulation, which does not require any arbitrary and experimentally given parameters, except the material constant. To reach such approach formulation, two main subjects should be investigated. In first, the returned mass flux to surface during cathode evaporation should be determined. In second, the energy and momentum of the plasma particles should be analyzed in the charge space sheath at the cathode surface in order to determine the mutual dependence with the cathode potential drop formation. Both phenomena occur near the cathode surface at length of mean free pass of the heavy particles, i.e., essentially non-equilibrium, and therefore should be considered in a kinetic approximation. The formulation of the cathode spot phenomena in a physically closed approach was developed and investigated by Beilis [2]. This extended description of the approach named as "kinetic theory" presented below.

17.1 Cathode Spot and Plasma

The specifics of a vacuum arc is the presence of large current density in the cathode spot. The cathode spot description connected with a mechanism forming a conductive media in order to support the observed current density at low value of cathode potential drop. As the cathode spot is a self-consistent subject of physical processes including the cathode heating by plasma interaction, the cathode plasma parameters and its flow are an important issue. The first accounting of mass and energy transfer in multi-component plasma and its mutual interaction with the cathode in the spot theory was developed by GDM, (Chap. 16, [1]) by the assumption of constant plasma parameters in the region immediately near the cathode surface. This means that the distribution of plasma density is not changed in the relaxation zone beginning from the cathode surface, and the density was determined by equilibrium vapor condition at the cathode temperature. This assumption was undertaken due to a low mass velocity in the relaxation zone collaborated by the GDM calculations indicating that the heavy particle density is close to that determined by the equilibrium conditions. This is the result of large part of particles (as ions and neutral atoms) are returned back to the cathode.

Thus, the Gasdynamic approach allowed understanding the mechanism and origin of ion current, electron emission fraction, and characteristics of the plasma at given experimental cathode potential drop. However, this approach cannot be accepted when it is necessary to determine the mass velocity of the plasma flow (important parameter), which produced by a jump changes of the heavy particle density near the surface that different from the equilibrium condition. The net of cathode mass loss is a difference between evaporated and returned heavy particles. This difference of heavy particles density could be determined by studying a non-equilibrium (kinetic) layer produced immediately near the cathode surface. The calculations using the kinetic approach show that the heavy particle density remains at level similar to that calculated by the GDM, but an additional very important information (except the mass velocity) can be obtained, regarding to the amount of cathode erosion rate and to the cathode potential drop. This chapter presents details of the kinetic model, system of equations, and the results of calculation, developed by Beilis [2, 3] considering together evaporation of the heavy particles and electrons as a common process.

17.2 Plasma Flow in a Non-equilibrium Region Adjacent to the Cathode Surface

Here, the previously published works are reviewed indicating the state of the problem. The specifics of the kinetic cathode evaporation into the dense plasma are analyzed.

17.2.1 Overview of the State of the Art

The kinetics of matter vaporization in general terms was reviewed in Chap. 2. The main problem consists in study of non-equilibrium (Knudsen) layer generated at the wall surface during its vaporization due to rarefied collisions at a length of mean free path of the evaporated particles. The laser–target interaction is one of phenomena, which is similar to plasma–cathode interaction in a cathode spot because both act with highly power density. When the laser is of moderate intensity (the energy not absorbed in the vapor), the evaporated material expands free in the form of neutral atoms, which characterized by sound speed. Such approach was studied by Anisimov [4] considering the Knudsen layer as a discontinuity surface in the hydrodynamic treatment of the vapor expanding from laser irradiation of a solid, similar to earlier treatments of the shock wave problem [5]. Assuming that the velocity at the external boundary of the Knudsen layer is the sonic velocity $v = v_{sn}$ (free expansion of the vapor, i.e., Mach number $M = 1$), Anisimov [4] obtained that 18% of the evaporated atom flux are returned to the target even in vacuum.

In the literature, a few works studied the kinetics of cathode vaporization in a vacuum arc in which the analysis was limited to consideration of a separate plasma processes and by assumptions limiting the study. Kozlov and Khvesyuk [6] presented some kinetics of atom ionization by emitted electron beam. However, they did not consider the kinetics of heavy particles flow and a change of gasodynamic parameters at the cathode surface.

Nemchinsky [7] conducted such analysis considering the conservation law in simple form. He assumed that the cathode vaporization should be minimal, and therefore, the velocity of the heavy particles at the external boundary of the non-equilibrium layer should be equal to the sound velocity like to such assumption in case of laser–target interaction. A relation for cathode erosion rate was obtained, which arises as a dependence on the electron pressure. This dependence was obtained studying of plasma flow only in the quasineutral cathode plasma without consideration of the influence of the ion flow, returned plasma electrons in expression of an electric field in the sheath. However, it is an error approach because the momentum of electrical field in the space charge sheath compensates the action of the mentioned plasma pressure at the sheath-plasma boundary (see [2, 3] and Eq.(17.33) below). Physically its mean, that the force due to electron pressure at the sheath-plasma boundary acts the cathode through the electrical field.

The future analysis [8] studied the parameters of the metal kinetic vaporization into vacuum using Monte Carlo numerical approach accounting the atom ionization, particle interactions, and possible energy dissipation. This work approved again a transition of the plasma flow velocity to the sound speed occurred at the external boundary of the Knudsen layer (or at an exit from zone of energy dissipation) and as in the previous work limited by the electron pressure gradient. Note that the mentioned analysis was conducted parametrically for the quasineutral plasma region, i.e., without self-consistent consideration. In addition, taking into account the fact of ignoring momentum from the electric field, the used approach, in essence, described free vapor flow that is close to that formulated by laser study.

In contrast, Djakov [9] took in account the momentum from electric field but only partially in the presheath layer with parameters of this layer *given arbitrary*. In essence, the cathode spot was studied by system of equation like that formulated in GDM [10] (Chap. 16) but only using an additional equation of momentum conservation of the particle flow. However, this equation again determines the spot parameters, by the influence of the electron pressure gradient as in work [7]. The obtained both results cannot be accepted because this influence is compensated by the momentum from electrical field that was ignored by consideration of the conservation law in the quasineutral plasma region near the cathode surface. In addition, the system of equation was solved using the current density as free input parameter, i.e., by not closed formulation. The change of the gas dynamic parameters in the Knudsen layer was not considered although a ratio between values of the equilibrium density and density in plasma near the cathode was calculated using mentioned incorrect momentum equation.

17.2.2 Specifics of Kinetics of the Cathode Spot Plasma

In the cathode spot, the ionized vapor flows not free because its velocity is significantly lower than the sound speed. This result obtained previously from the GDM (see also below). This fact indicates that the coupled cathode processes in the spot could be realized at certain regime of the plasma, in which parameters are determined by solution of a self-consistent problem. Intense energy dissipation in the cathode body produces extremely dense particle fluxes, in which flow corresponds to flow regime of continue media. However, before the flow reaches this regime at a certain length, the particles transfer their momentum and energy from each to other, and as a consequence, the particles significantly change their trajectory. As a result, the near-cathode plasma consists of two characteristic regions. In the first region adjacent to the surface (Knudsen layer), the velocity function distribution is changed with distance and acquires some equilibrium state at the external boundary of Knudsen layer. The parameters at this boundary can be used as boundary condition to the second, gasdynamical relaxation region as well for the further plasma expanding region.

To solve the problem of establishment of the ionized vapor flow in the Knudsen layer in immediate vicinity of the cathode, one must use the system of Boltzmann kinetic equations for particles of different types. The asymptotic solution of this problem gives the parameters on the boundary of Knudsen layer. The physical processes at the surface and in the near-cathode region determine the boundary conditions for such system of equations. In the arc spot, the cathode heats up so strongly that electron emission from the surface is important along with intense vaporization of its atoms. As in the GDM, it can be taken into account that the electron beam acquiring energy in the region of cathode potential drop relaxes over a distance of about 1 μm [1], where the beam energy becomes equal to the energy of plasma electrons. Therefore, two groups of electrons take place. The plasma is so dense that the characteristic mean free path proves to be of the order of or somewhat greater than the region of the potential drop (Chap. 16).

17.3 Summarizing Conception of the Kinetic Cathode Regions. Kinetic Model

Thus, several zones divide the cathode spot vicinity, where the material flow in each zone can be mathematically described. Let us characterize them qualitative. The first zone is the *cathode* itself, which close to the cathode plasma. Within the cathode, the heat generation by Joule heating as well the heat conduction away from the cathode spot in the body is taken into account. At the cathode surface, electron emission and evaporation are the key processes.

The particles emitted from the cathode surface are accelerated at some short distance in the *ballistic zone* before suffering collisions, the length of which is defined by the mean free path. Poisson equation describes the motion of the various particles in the ballistic zone that is a collision less sheath. A strong electric field is present within the ballistic zone. A backflow of electrons and ions from the plasma transverse the ballistic zone in addition to the particles emitted from the cathode. The back flowing ions are accelerated by the electric field within the ballistic zone toward the cathode, and the energy carried by them is a significant source of cathode spot heating.

Atoms emitted from the cathode surface will collide with each other and with the backflow of heavy particles, in the *heavy particle relaxation zone*, which begins at the cathode, overlaps the *ballistic zone,* and extends a distance of a few times the heavy particle mean free path, whereas the heavy particles evaporated from the cathode surface have a half Maxwell velocity distribution with cathode temperature, at the external boundary of the *heavy particle relaxation zone (Knudsen layer)*, sufficient collisions have occurred such that a shifted Maxwell distribution with new particle temperature may be used to describe the emitted heavy particles.

The electrons emitted from the cathode also suffer collisions, but because they are accelerated by the electric field in the *ballistic zone*, and the mean free path for the electrons is longer than that of the heavy particles. Electron collisions occur in a region defined as the *electron beam relaxation zone*. Here, the electrons undergo numerous collisions, including ionizing collisions with the emitted atoms of sufficient frequency such that a highly ionized plasma is generated. The *electron beam relaxation zone* begins at the cathode surface, overlaps the *ballistic zone* and the *heavy particle relaxation zone,* and extends for a distance defined by that required for the energy of the fast entering electrons to equilibrate with the electrons created by electron impact ionization within this zone. This equilibration is by means of numerous electron-electron collisions. This distance is typically a hundred times larger than the mean free path for ionization.

At the beginning of the *electron beam relaxation zone,* the electron distribution function has two distinct components, one describing the entering beam electrons emitted from the cathode surface, and another component, which is isotropic and describes the plasma electrons. Both types of electrons may collide with and ionize the neutral atoms in this region. In principle, multiple ionizations are possible. At the end of the *electron beam relaxation zone,* a single, shifted Maxwellian distribution may be used to describe all of the electrons. The Debye length here is much smaller than the length (or width) of the plasma zone, and thus, the plasma therein is quasineutral. The zone has a high conductivity, and thus, both the electric field strength and the potential drop are so small that the electron energy gain from the field is much smaller than their thermal energy. Because of the high plasma density, this region is optically thick, i.e., radiation emitted in the interior of the plasma is reabsorbed by the plasma and is thus trapped within. Another consequence of the high electron density and temperature is a high thermal conductivity. The thermal diffusion time is much shorter than the spot lifetime, and hence, the temperature is relatively uniform throughout the zone. The plasma exits the far side of the *electron beam relaxation zone* in the form of an expanding plasma jet into what we will call an *expansion zone.*

17.4 Kinetics of Cathode Vaporization into the Plasma. Atom and Electron Knudsen Layers

The neutral atoms and electron emission from the cathode were considered as common process of vaporization of the atoms and electrons taking into account that the electron flux is enhanced by the electric field in the ballistic zone. Therefore, two relaxation (Knudsen) lengths appear near cathode.

17.4.1 Function Distribution of the Vaporized and Plasma Particles

Let us detail the above description, according to what four characteristic boundaries can be distinguished in the near-cathode region, Fig. 17.1. The origin of the coordinate system is defined at the cathode surface, and the x-direction coincides with the direction of the vapor flow. Boundary 1 is at the cathode surface at $x = 0$. Boundary 2 is at the external boundary of space charge (ballistic) zone. Boundary 3 is at a distance of the Knudsen layer length from the cathode surface for the heavy particles and plasma electrons. Boundary 4 is located at a distance equal to the electron beam relaxation length (electron beam Knudsen region) from the cathode surface.

17.4 Kinetics of Cathode Vaporization into the Plasma. Atom and Electron ...

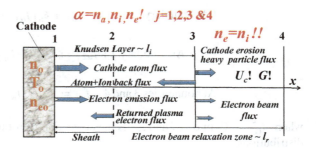

Fig. 17.1 Model of cathode evaporation in a vacuum arc. Diagram of kinetic layers in the cathode current carrying plasma

The gasdynamic parameters (density, velocity, and temperature) at these boundaries are denoted as $n_{\alpha j}$, $v_{\alpha j}$, $T_{\alpha j}$, where indices α- e, i, a are the electrons, ions, atoms, respectively, and $j = 1, 2, 3, 4$ indicates the boundary number, and n_{e0} and n_0 are the equilibrium electron and heavy particle densities determined by the cathode surface temperature T_0.

The velocity function distribution (VDF) is half-Maxwellian (positive velocity $v_{\alpha x}$) with a temperature at the first boundary in accordance with particle emission from the cathode determined by evaporation for atoms [11] and for electrons [12]. The evaporated atoms are significantly ionized within one mean free path [10], and therefore, the reverse flux consists of atom and ions, whose velocity distribution (negative velocity v_x) at boundary 1 can be approximated as shifted half-Maxwellian ones by the plasma parameters determined at this boundary from the problem solution [13].

At the external boundary 3 of the Knudsen layer, the heavy particles are in equilibrium. Therefore, the heavy particle distribution function is a shifted full Maxwellian with parameters $n_{\alpha 3}$, $u_{\alpha 3}$, and $T_{\alpha 3}$. For electrons, the VDF consists of a shifted full Maxwellian for the plasma electrons and a shifted half-Maxwellian for the electron emission beam. At boundary 4, the parameters of all particles are in equilibrium, and therefore, there the distribution functions are shifted full Maxwellians with parameters $n_{\alpha 4}$, $u_{\alpha 4}$, and $T_{\alpha 4}$. Between the boundaries, 1 and 2 is the space charge sheath. At boundary 2, the plasma electrons are returned, while the ions and emitted electrons are accelerated with energy eu_c, where u_c is the cathode potential drop. These effects determine the charged particle shift velocities and change the parameters of the VDF at boundary 2.

The velocity distribution function of the particles for $v_{\alpha x} < 0$ is denoted by $f_{\alpha j}^-$ and for $v_{\alpha x} > 0$ is by $f_{\alpha j}^+$, where

$$f_{\alpha j}^{\pm} = n_{\alpha j} \left(\frac{m_\alpha h_{\alpha j}}{\pi}\right)^{3/2} \exp\{-m_\alpha h_{\alpha j}[(v_{\alpha x} - v_{\alpha j})^2 + v_{\alpha y}^2 + v_{\alpha z}^2]\} \quad (17.1)$$

and where $h_{\alpha j} = (2kT_{\alpha j})^{-1}$, $v_{\alpha y}$, $v_{\alpha z}$ are the random particle velocity components in the y- and z-directions, m_α is the particle mass and k the Boltzmann constant.

Next, the conditions at the boundaries are written as

$$\text{At boundary 1}: f_{\alpha 1} = f_{0\alpha 1}^+ + \xi_{\alpha 1}^- f_{\alpha 1}^-, \quad f_{\text{oil}}^+ = 0 \tag{17.2}$$

$$\text{At boundary 2}: f_{\alpha 2} = \xi^+ f_{\alpha 2}^+ + \xi_{\alpha 2}^- f_{\alpha 2}^- \tag{17.3}$$

$$\text{At boundaries 3 and 4}: f_{\alpha j} = f_{\alpha j}^+ + f_{\alpha j}^- \quad j = 3, 4 \tag{17.4}$$

where $f_{0\alpha 1}^+ = 0$ for $\alpha = i$ and $\xi_{\alpha j}$ is the correction factor to the Maxwellian function distributions.

The condition at boundary 2 is the condition for joining the parameters at the external boundary of the electrical sheath with the region where the particle motion is practically determined by collisions. The conditions at boundaries 3 and 4 reflect the fact of establishment of equilibrium parameters of the plasma particles. In contrast to boundary 3, at boundary 4, the electron beam disappears, and then, the electron current is defined only by the electrical conductivity of the plasma (in essence, boundary 4 serves as the external boundary for the electron Knudsen layer). If the dependence of the particle distribution function on the coordinates is known, the plasma parameters are determined through relations expressing the laws of conservation of particle number, momentum, and energy. In writing the respective conservation laws, one must allow for the input of momentum and energy to the plasma from the electromagnetic field. Note, for the problem of plasma flow in cathode spot, it is only important to know the parameters in the established regions and not their dependence on the coordinates so that the problem comes to the integration of equations expressing the laws conservation between the control boundaries.

Let us define the coefficients $\xi_{\alpha j}$. Henceforth, we assume that the corresponding distribution functions are close in form to a Maxwellian function and it can neglect the influence of their departure on the parameters of the established region. The basis for such an assumption is the large reverse flux of the particles under the condition of the arc spot. In addition, the use of a Maxwellian distribution function for f^- at boundary 1 under the condition of free vaporization during laser heating [13] also gives good agreement with experiment. With allowance for this assumption below, it will be taken $\xi_{\alpha j} = 1$.

17.4.2 Conservation Laws and the Equations of Conservation

Thus, the problem is reduced to integrating the conservation equations at the above boundaries, as moments of known velocity distribution functions in the following form:

$$\int v_x f(\mathbf{v}) d\mathbf{v} = C_1 \quad \int v_x^2 f(\mathbf{v}) d\mathbf{v} = C_2 \quad \int v_x v^2 f(\mathbf{v}) d\mathbf{v} = C_3 \tag{17.5}$$

17.4 Kinetics of Cathode Vaporization into the Plasma. Atom and Electron ...

where $f(\mathbf{v})$ is the velocity distribution function at the boundaries, $\mathbf{v}$ is the velocity vector, vx is the velocity component in the x-direction, and $C1$, $C2$, and $C3$ are the results of the integration.

The (17.5) at the above boundaries are expressed in following form [4, 13]:

$$\int f(x, v_\alpha) dv = n_\alpha(x); \quad \int v_{\alpha x} f(x, \mathbf{v}_\alpha) d\mathbf{v}_\alpha = n_\alpha(x) u_\alpha(x);$$

$$\frac{m}{2} \int [(v_{\alpha x} - u_\alpha)^2 + \mathbf{v}_y^2 + \mathbf{v}_z^2] f(x, \mathbf{v}_\alpha) d\mathbf{v}_\alpha = \frac{3 n_\alpha(x) k T_\alpha(x)}{2} \qquad (17.6)$$

where $f(x, \mathbf{v}_\alpha)$ is the velocity distribution function at the control boundaries, $\mathbf{v}_{\alpha x}$ is the velocity vector, $\mathbf{v}_{\alpha x}$ is the particle velocity component in the x-direction, and $\mathbf{v}_\alpha$ is the velocity of the ionized vapor.

17.4.3 Integration. Total Kinetic Multi-component System of Equations

Using the definitions of VDF at mentioned boundaries and (17.1)–(17.6), the procedure of integrating brings to following system of equations describe the heavy particles and electrons flow after their vaporization from the cathode and taking into account specifics of atom ionization and plasma generation during the flow of the particles [2].

Equation of heavy particles conservation:

$$n_1 = \frac{n_0}{2} - \frac{n_{a1}}{2}[1 - \Phi(b_{a1})] - \frac{n_{i1}}{2}[1 + \Phi(b_{i1c})] \qquad (17.7)$$

Equations of heavy particles continuity:

$$n_1 v_1 = n_j v_j \quad j = 2, 3, 4 \qquad (17.8\text{--}17.10)$$

$$n_2 v_2 = \frac{n_0}{2\sqrt{\pi d_0}} - \frac{n_{a1}}{2\sqrt{\pi d_{a1}}} [\exp(-b_{a1}^2) - b_{a1}\sqrt{\pi}[1 - \Phi(b_{a1})]] - A_{i1} \qquad (17.11)$$

$$A_{i1} = \frac{n_{i1}}{2\sqrt{\pi d_{i1}}} [\exp(-b_{i1c}^2) + b_{i1c}\sqrt{\pi}[1 - \Phi(b_{i1c})]]$$

$$= \frac{n_{i2}}{2\sqrt{\pi d_{i2}}} [\exp(-b_{i2}^2) - b_{i2}\sqrt{\pi}[1 - \Phi(b_{i2})]] \qquad (17.12)$$

Equations of heavy particles momentum flux:

$$\frac{n_0}{4 d_0} - \frac{n_{a1}}{2 d_{a1}} \left[(0.5 + b_{a1}^2)[1 - \Phi(b_{a1})] - \frac{b_{a1}}{\sqrt{\pi}} \exp(-b_{a1}^2) \right]$$

$$+ \frac{n_{i1}}{2d_{i1}}\left[(0.5 + b_{i1}^2)[1 - \Phi(b_{a1})] - \frac{b_{i1}}{\sqrt{\pi}}\exp(-b_{i1}^2)\right]$$
$$= B_2 + B_{e2} + B_{ec} + F(E_{12}) \tag{17.13}$$

$$B_j + B_{ej} = B_{j+1} + B_{ej+1}; \quad j = 2, 3; \qquad B_{ej} = \frac{n_{ej}}{2d_{ej}}(1 + 2b_{ej}^2); \qquad j = 3, 4$$
$$\tag{17.14–17.15}$$

Equations of heavy particles energy flux

$$\frac{n_0}{\sqrt{\pi}d_0^{3/2}} + \frac{n_{a1}}{\sqrt{\pi}d_{a1}^{3/2}}[(2.5 + b_{a1}^2)[1 - \Phi(b_{a1})]b_{a1}\sqrt{\pi} - (2 + b_{a1}^2)\exp(-b_{a1}^2)]+$$
$$+ \frac{n_{i1}}{\sqrt{\pi}d_{i1}^{3/2}}[(2.5 + b_{i1}^2)[1 - \Phi(b_{i1})]b_{i1}\sqrt{\pi} - (2 + b_{i1}^2)\exp(-b_{i1}^2)] + A_{i1}u_c = W_2$$
$$\tag{17.16}$$

$$W_j = W_{j+1}; \quad j = 2, 3 \tag{17.17–17.18}$$

Equations of electron flux continuity:

$$\frac{n_{e1}}{2\sqrt{\pi d_{e1}}}[\exp(-b_{e1}^2) + b_{e1}\sqrt{\pi}[1 - \Phi(b_{e1})]]$$
$$= \frac{n_{e2}}{2\sqrt{\pi d_{e2}}}[\exp(-^-b_{e2c}^2) - b_{e2}\sqrt{\pi}[1 - \Phi(^+b_{e2c})]] = A_{e1} \tag{17.19}$$

$$n_{e1} = \frac{n_{e0}}{2} + \frac{n_{e1}}{2}[1 - \Phi(b_{e1})]; \quad n_{ej} = n_{ij} \quad j = 3, 4 \tag{17.20–17.22}$$

$$e_0 n_{e1} v_{e1} = e_0 n_{e2} v_{e2} + j_{e0} - j_i = j_e \tag{17.23}$$

$$j_{e0} + e_0 n_{ej} v_{ej} = e_0 n_{e4} v_{e4}; \quad j = 2, 3 \tag{17.24–17.25}$$

$$n_{e1} v_{e1} = \frac{n_0}{2\sqrt{\pi d_{e0}}} - A_{e1} \tag{17.26}$$

Equation of electron momentum flux:

$$B_{e1} = B_{e2} + B_{ec} + F(E_{12}) - j_i(2mu_c/e_0);$$
$$B_{e2} = B_{e3} - F(E_{12});$$
$$B_{e3} + B_{ec} = B_{e4} - F(E_{12}) \tag{17.27–17.29}$$

Equations of electron energy flux:

17.4 Kinetics of Cathode Vaporization into the Plasma. Atom and Electron ...

$$\frac{n_{e0}}{\sqrt{\pi}d_{e0}^{3/2}} + \frac{n_{e1}}{\sqrt{\pi}d_{e1}^{3/2}}[(2.5+b_{e1}^2)[1-\Phi(b_{e1})]b_{e1}\sqrt{\pi} - (2+b_{e1}^2)\exp(-b_{e1}^2)]$$

$$= \frac{n_{e2}}{\sqrt{\pi}d_{e2}^{3/2}}\{[(2.5+b_{e2}^2)[2-2\Phi(b_{e2})+\Phi(^-b_{e2c})-\Phi(^+b_{e2c})]b_{e2}\sqrt{\pi} -$$

$$-3\sqrt{\frac{d_{e2}}{d_{ep}}}\exp(-^-b_{e2c}^2)b_{e2}b_{ec} + 2(2+b_{e2}^2)\exp(-b_{e2}^2)$$

$$-(2+^-b_{e2}^2)\exp(-^-b_{e1}^2) - (2+^+b_{e2}^2)\exp(-^+b_{e1}^2)\}$$

$$+ q_{ec} - j_e u_c = W_{e2} \tag{17.30}$$

$$W_{e2} - q_{ec} + j_e u_c = W_{e3} - q_{23}^g + q_{23}^e + q_{23}^n \tag{17.31}$$

$$W_{e3} + q_{ec} = W_{e4} - q_{34}^g + q_{34}^e + q_{34}^n \tag{17.32}$$

Definition of the following expressions:

$$B_{e1} = \frac{n_{e0}}{4d_{e0}} - \frac{n_{e1}}{2d_{e1}}\left[(0.5+b_{e1}^2)[1-\Phi(b_{e1})] - \frac{b_{e1}}{\sqrt{\pi}}\exp(-b_{e1}^2)\right];$$

$$B_{ej} = \frac{n_{ej}}{2d_{ej}}(1+2b_{ej}^2); \quad j = 3, 4$$

$$B_{e2} = \frac{n_{e2}}{2d_{e2}}\left\{\begin{array}{l}(0.5+b_{e2}^2)[2+2\Phi(b_{e2})+\Phi(^-b_{e2c})-\Phi(^+b_{e2c})]+\\ +\frac{b_{e2}}{\sqrt{\pi}}[2\exp(-b_{e2}^2)-\exp(-^+b_{e2}^2)-\exp(-^-b_{e2}^2)]\end{array}\right\}$$

$$B_{ec} = \frac{n_{eb}}{2d_{eb}}\left\{\frac{b_{ec}}{\sqrt{\pi}}\exp(-b_{ec}^2) + (0.5+b_{ec}^2)[1+\Phi(b_{ec})]\right\}$$

$$q_{ec} = \frac{n_{eb}}{\sqrt{\pi}d_{eb}^{3/2}}[(2.5+b_{ec}^2)[1+\Phi(b_{ec})]b_{ec}\sqrt{\pi} + (2+b_{ec}^2)\exp(-b_{ec}^2)]$$

$$W_j = \frac{n_j}{\sqrt{b_j^2/d_j^3}}(2.5+b_j^2) \quad j = 2, 3, 4$$

$$h_{\alpha j} = (2kT_{\alpha j})^{-1}; \quad m_\alpha h_{\alpha j} = d_{\alpha j}; \quad m_\alpha h_{\alpha 0} = d_{\alpha 0}; \quad m_e h_{eb} = d_{eb};$$
$$d_{ij+1}^n = (n_{ij+}v_{j+1} - n_{ij}v_j)u_i; \quad j = 2, 3$$

$\alpha - e, i, \alpha$ are the electron, ion, and neutral atom, m_α—mass of the particles

$$^-b_{ejc} = (v_{\alpha c} - v_{\alpha j})^2 d_{\alpha j}; \quad ^+b_{ejc} = (v_{\alpha c} + v_{\alpha j})^2 d_{\alpha j}; \quad b_{\alpha j}^2 = v_{\alpha j}^2 d_{\alpha j};$$
$$b_{\alpha c}^2 = v_{\alpha c}^2 d_{\alpha b}; \quad v_{\alpha j}^2 = \frac{2eu_c}{m_\alpha}$$

q^g_{jj+1}, $F(E_{jj+1})$ are the inputs of energy and momentum from the electric field in the corresponding regions, q^{el}_{jj+1} and q^{in}_{jj+1} are the energy losses by plasma electrons in elastic and inelastic collisions with heavy particles. Estimations using calculated data from GDM show that these terms (except for inelastic) are important in regions 1–2 where they are determined by u_c. j_{e0} is the emission current density, $\Phi(b_{\alpha j})$ is the probability integral function, and u_i is the ionization potential. In obtained these equations, it was assumed that the ion j_i and electron j_e current densities are constant in the space charge ballistic region, while the total current density is constant everywhere. In describing the charged particles in the space charge region, we allowed for corresponding shift of the distribution function and the partial reflection of plasma electrons at boundary 2 connected with the presence of a potential barrier (u_c). To determine the input momentum from the electric field E_{12}, one must solve the Poisson equation in the layer 1–2 with the given density distribution of the charged particles. The obtained pf such a distribution is connected with an analysis of a complicated kinetic equation. Since the particle velocity in region of the cathode potential drop significantly exceeds the thermal velocity during expansion is insignificant, the space charge due to fluxes of ions and emitted electrons can be obtained from the condition of constancy of the corresponding current densities. If the distribution of the plasma electrons can be taken as Boltzmann distribution, then the momentum will have

$$F(E_{12}) = j_i\sqrt{\frac{2mu_c}{e}} - j_e\sqrt{\frac{2m_e u_c}{e}} - kn_{e2}T_{e2}\left[1 - \exp\left(-\frac{eu_c}{kT_{e2}}\right)\right] \quad (17.33)$$

In general, three main gasdynamic parameters at each four boundary and for each four boundaries and for each of three particles there are 37, unknowns (including the cathode potential drop u_c) that follow: $n_{\alpha j}$, $T_{\alpha j}$, $v_{\alpha j}$, u_c. While the number of equations is only 26, this number of unknowns can be reduced to the number of equations taking into account the following conditions. It can be taken in account that the mean free paths for ion-ion and ion–atom collisions are equal and the effective momentum and energy relaxation lengths for heavy particles are determined by ion-ion collisions and are about one mean free path. Therefore, it can be assumed that $T_{ij} = T_{\alpha j}$ ($j = 1, 2, 3$) and $v_{ij} = v_{\alpha j}$ ($j = 3$) and the number of unknown is reduced by 6. It should be noted that at boundary 2, the ion velocity is determined by the condition at the sheath–plasma interface, and at boundary 1, the ions are accelerated toward the cathode in the space charge layer, reaching a velocity of $v_c = (eu_c/m)^{1/2}$. This velocity is used as the shifted velocity for the ion distribution function at boundary 1. In frame of the assumptions, the number of unknowns is reduced to 29, and it is n_{ej}, n_{ij}, n_j, v_j, v_{ij}, T_{ej}, T_j, u_c, where $n_j = n_{ij} + n_{aj}$. The above system consists of 26 equations! To determine the next three unknowns, it can be used the specifics of plasma flow in contrast to only neutral atom flow by laser–target vaporization.

Since the space charge layer thickness is less than the Knudsen layer thickness, we have two quasineutrality conditions, which in case of single ionized ions, for example, for Cu [9] is $n_{ej} = n_{ij}$ ($j = 2, 3$). The quasineutrality condition for another case will take a special consideration. Finally, if the ionization–recombination length and plasma velocity are small in comparison with the length of the relaxation zone and to the thermal velocity, respectively, then at boundary 4, Saha's equation is fulfilled [9], i.e., $n_{e4} = n_{es}$, where n_{es} is the equilibrium electron density. In this case, an additional diffusion equation (Chap. 16) determines the ion density at boundary 3, closing the above system of equations. In order to determine the equilibrium parameters (n_{e0}, n_0) at the cathode surface, it is necessary to supplement the kinetic equations with equations for emission of electrons and heavy particles from the cathode as well as with equation for the electric field (E) at the surface and the cathode energy balance (T_0) [3, 10]. The electron energy balance is determined by energy influx from the emitted electrons and energy dissipated by atom ionization, convective transport by the electric current, and ion outflow in the relaxation zone. The cathode and plasma parameters at the mentioned boundaries including the cathode potential drop u_c and rate of cathode mass loss may be determined as a part of the self-consistent solution.

17.4.4 Specifics at Calculation of Current per Spot

In this formulation, the above system of equations is physically closed in framework of description of the cathode and plasma parameters. One problematic parameter is the current per spot. Sometimes, this parameter was determined by given arc voltage as input parameter, which, in essence, should be also calculated. In general, the current per spot can be obtained by two ways. One way is integrating the ion and electron distribution using appropriate study of the cathode thermal regime determining the 2D temperature distribution at the cathode surface. In this case, there a problem arises related to gradients of plasma parameters and plasma fluxes in radial direction at 2D approximation of cathode evaporation, electron emission, and gasdynamic flow of current carrying cathode plasma discussed previously [14]. So, it is very complicate approach, which sometimes solved using different arbitrary conditions but ignoring the problem with plasma radial gradients. In this case, the obtained result cannot be correct although the authors compared it with the experiment. The second way determines the current per spot is by taking into account the near-cathode plasma electrical conductivity and the spot area. However, the experiments show that the current per spot critically depends on the cathode geometry and cathode surface state. The last factor, as a rule, is impossible to take in account as a geometrical condition because the surface change occurs by material melting and force variation resulting in plasma–surface interaction during the spot operation. Certain difficulties also arise by solving the problem of cathode melting and liquids motion in frame of self-sustained transient kinetic formulation of the cathode spot development. So, this way meets also other difficulties. Nevertheless, the spot current will be calculated below for some typical cases in order to understand the overall state of this parameter.

Also, the plasma flow consists of two basic regions, kinetic before boundary 3 and hydrodynamic after boundary 3. As results of the kinetic solution, the relations between the equilibrium parameters of n_{e0}, n_0, T_0, and the non-equilibrium plasma parameters at boundary 3 serve as boundary conditions for hydrodynamic equations of mass, momentum, and energy in the expanded plasma jet (Chap. 18). Examples of calculations of partial equations as well as total calculating results will be presented below showing the role of the kinetic approach to understand the physics of cathode current continuity, different spot types, appearing of which depends on cathode geometry, surface state, cathode materials, and other conditions. There an important issue is the possibility of calculation of the cathode potential drop u_c. Therefore, let us consider the state of previous approaches used for calculation of this parameter.

17.5 Region of Cathode Potential Drop. State of Previous Studies

The preliminary estimations show that although of relatively weak change of the pressure and density in the non-equilibrium layer, the plasma velocity and, consequently, the erosion rate can reach significant values [4]. As these parameters are determined on the dissipated energy, they also significantly dependent on the cathode potential drop. The relation between u_c and G determined the other spot parameters at different arc conditions. Therefore, the spot calculation using the given experimental minimal value of u_c can limit an interpretation of the results. This limitation also caused by uncertain structure of the measured u_c, which included of two regions. One of these regions is the drop immediate in the adjacent sheath (really determined the thermal and ionization processes), while the second region reflects the fall in the cathode plasma. Taking into account the transient character of vacuum arcs, observations of the different spot types, and voltage oscillations, it is obvious the necessity to obtain u_c by a direct method. Let us consider the previous different attempts and approaches to determine the important spot parameter, cathode potential drop.

An approach [6] was developed in which u_c searched as an additive to the potential of atom ionization, by assumption of linear distribution of the electrical field in the ionization zone. The procedure of solution the Poisson equation assumed presence a small parameter. However, this assumption cannot be approved using the results of calculations [15]. In essence, it was considered approach of weakly ionization and relatively large electric field in region of ion generation that not corresponded to the boundary condition. The formulation of the mathematical description is not closed. Similar approach was conducted [16] by calculating the space charge region assuming atom ionization by electron beam but in the absence of interactions between the charge particles of dense cathode plasma.

A number of research [17–20] used, so-called, minimal principle to determine value of u_c. The essence of such method consists in calculation of a minimal value of u_{cmin} in a dependence of u_c on one of spot characteristics (cathode temperature, spot radius, current density, spot current, etc.) given parametrically. Such dependence taking into account also the potential drop in a quasineutral region was obtained at variation of cathode temperature T_s [19, 20]. As the dependence T_s on spot radius

17.5 Region of Cathode Potential Drop. State of Previous Studies

r_s is identical to a dependence of T_s on j, then the dependence u_c on T_s is analogy to dependence u_c on j (see Chap. 16 and [3]). Adding the potential drop in plasma region does not change the result of minimal value of u_c.

Ecker [18] obtained the minimum of u_c by variation of a thickness of the near-cathode layer including Langmuir sheath and the plasma presheath (L_{sh}), in which mostly occurred breaking of returned plasma electrons. The condition of $L_{sh} \ll l_{ia}$ was assumed as "commonly used", where l_{ia} is the ion mean free path. The minimum of cathode potential drop was found as minimum of the energy spend to the returned electrons breaking for given values of the total current density $j = 10^6$ A/cm^2 and the cathode spot surface temperature T. However, the calculation of u_{cmin} [18] was obtained exactly at condition of $L_{sh} = l_{ia}$ and was demonstrated that the potential u_c rises very rapidly (to about 60 V) as soon as the parameter L_{sh} little deviated from the minimum value $(L_{sh})_{min}$. The additional weakness of Ecker's [18] approach is also due to arbitrary input parameters of j and T_s as well as by arbitrary assumptions used by derivation of the final expression for the potential u_{cmin} as a function of the parameter L_{sh}. The condition of $L_{sh} = l_{ia}$ used as an expression to calculate the cathode erosion rate G [15]. However, the calculations for Cu in frame of GDM [10] show that the obtained values of G significantly exceed that measured for spot currents I varying up to 500 A and become negative values at $I < 70$ A. This result was obtained by the use of only one input parameter $u_c = 15$ V and shows critic sensitivity to weakly deviation from condition $L_{sh} = l_{ia}$.

In general, the weakness and lacks of approaches based on "minimal principle" widely discussed [21–23]. For cathode spot, the mentioned dependences usually obtained at keeping other as constant (current, erosion rate, cathode or electron temperatures, etc.). However, the value of u_c is not only sensitive to the given parameters but also depends on the model selected. As example, in the cited above works, the u_c was obtained for different minimal values in range from 10 to 16 V. In addition, the calculations indicate the presence of minimal values for other parameters, I_{min}, t_{min}, j_{min} (Chap. 16). So, what minimum should be chosen? Therefore, the use of "minimal principle" at keeping constant other parameters cannot be accepted. In some cases, the minimal parameter can be absent [19]. In case of accounting the total arc voltage variation, arise a question related to uncertainty of the plasma expansion geometry.

In the kinetic approach, a direct calculations u_c and G can be conducted using the kinetics of the evaporated cathode product flow. The physics of this approach consists in consideration of the electron emission as a process evaporation together with the atom vaporization with the following atom ionization. According to the laser interaction, the theory could be closed by given sound speed of the atom flow at the Knudsen layer. In case of cathode spot it is not flow of neutrals but rather it is plasma flow. Therefore considering the flow of electrons and ions an additional condition of quasineutrality can be used to study the plasma velocity at the external boundary of Knudsen layer. In contrast to the metal evaporation by laser–target interaction, in the arc spot, the returned flux of the charge particles depends on value of potential barrier in the space charge sheath, which in turn produced by the charge particles generated due to high power dissipation in the current carrying cathode region. The sheath potential drop determined by charge particle motion which in turn determined the potential barrier for the returned plasma electrons. As the charge particles motion is determined by the relations of energy and momentum of charge particles flow the analysis show that these relations allow express and calculating the corresponding potential height of the barrier, i.e., u_c.

The rate of mass flow G of the ionized vapor at the external boundary of heavy particle relaxation zone (the Knudsen layer) expressed as

$$G = mn_3v_3F_s, \quad F = \pi r_s^2 j \quad m = m_i = m_a, \quad n_3 = n_{a3} + n_{i3} \qquad (17.34)$$

The above model considered the cathode mass loss by the evaporation ignoring the MPs generation. The energy loss for cathode melting and MPs generation can be neglected due to their energy loss is low in comparison with energy loss by the vaporization (see Chap. 16).

The second stage of calculations the total system of (16.7–16.32) includes description of the kinetics and gasdynamics flow in quasineutral ionization zone taking into account the influence of the inflow of the momentum and energy of the electrons also in the non-equilibrium layers. This system was added by the equation for the cathode body and for electrical field at the cathode surface. After relaxation zone, the hot plasma expands away in the form of the cathode plasma jet. The physics of plasma jet will be considered in the next Ch.18. Here, we present the integral equations of particle momentum and energy to understand the character of the jet velocity change calculated below together with the parameters in the kinetic and ionization regions. Integral equations for the plasma jet represented in the following form:

Momentum Conservation:

$$p_3 F_3 + G v_3 = p(x) F(x) + G v(x) \qquad (17.35)$$

Energy Conservation:

$$G v_3^2 + 3I \frac{T_{e3}}{e} = G v^2(x) + 2I \frac{T_e(x)}{e} + I u_{pl} \qquad (17.36)$$

Ohm's law:

$$u_{pl} = \frac{I}{2\pi \sigma_{el} r_s} \qquad (17.37)$$

where the subscript 3 refers to parameters at boundary 3. The values $p(x)$, $v(x)$, and Te(x) are the pressure, velocity, and electron temperature at the external jet boundary at x distance. Equation (17.37) obtained by the solution of Laplace's equation for the potential in the quasineutral region of the near-cathode plasma under the assumption that its electrical conductivity σ_{el} is a weak function of x, and the current, at boundary of the electrode, is uniformly distributed in the circular spot area. In order to obtain the

17.5 Region of Cathode Potential Drop. State of Previous Studies

jet velocity far from the cathode, the conservation of momentum and the assumption of steady state require that the net force on this expansion zone is zero.

17.6 Numerical Investigation of Cathode Spot Parameters by Physically Closed Approach

The system of (16.7)–(16.32) is sufficiently complicated. This system can be significantly simplified by taking into account a weakly difference in the parameters between boundaries 2 and 3 because the cathode potential drop mainly occurs in zone 1–2. This fact allows not determining the vapor parameters directly at boundary 2 but only taking into account the effect of ballistic zone on motion of the charge particles. Therefore, it seems appropriate to study kinetics of cathode vaporization in two stages. As a first stage, it can obtain relations between parameters of the plasma flow (erosion rate, jump of gasdynamic values) in the non-equilibrium layer and spot parameters (current density, temperature) at given value of u_c. These relations can be obtain using the kinetic system of equation only for heavy particle and the result at this stage represents deviation from the equilibrium. The result at this stage represents deviation from the equilibrium. The second stage is study of the total kinetic system of equation (including electron flow) which allow to determine the mentioned above plasma flow parameters in couple with the cathode potential drop and cathode erosion rate.

17.6.1 Kinetic Model. Heavy Particle Approximation

The first stage of the problem can be considered by investigation of the heavy particle flow. The corresponding simplistic system of equation can be obtained ignoring terms related to the vaporization of the electrons in following form [24, 25].

Equation of heavy particles conservation:

$$n_1 = \frac{n_0}{2} - \frac{n_{a1}}{2}[1 - \Phi(b_{a1})] - \frac{n_{i1}}{2}[1 + \Phi(b_{i1c})] \qquad (17.38)$$

Equations of continuity:

$$n_1 v_1 = n_3 v_3 \qquad (17.39)$$

$$n_3 v_3 = \frac{n_0}{2\sqrt{\pi d_0}} - \frac{n_{a1}}{2\sqrt{\pi d_{a1}}} \left[\exp(-b_{a1}^2) - b_{a1}\sqrt{\pi}[1 - \Phi(b_{a1})] \right] - A_{i1} \qquad (17.40)$$

$$A_{i1} = \frac{n_{i1}}{2\sqrt{\pi d_{i1}}} \left[\exp(-b_{i1c}^2) + b_{i1c}\sqrt{\pi}[1 - \Phi(b_{t1c})] \right]$$

$$= \frac{n_{i3}}{2\sqrt{\pi}d_{i3}}[\exp(-b_{i3}^2) - b_{i3}\sqrt{\pi}[1 - \Phi(b_{i3})]] \quad (17.41)$$

Equations of momentum flux:

$$\frac{n_0}{4d_0} - \frac{n_{a1}}{2d_{a1}}\left[(0.5 + b_{a1}^2)[1 - \Phi(b_{a1})] - \frac{b_{a1}}{\sqrt{\pi}}\exp(-b_{a1}^2)\right]$$
$$+ \frac{n_{i1}}{2d_{i1}}\left[(0.5 + b_{i1}^2)[1 - \Phi(b_{i1})] - \frac{b_{i1}}{\sqrt{\pi}}\exp(-b_{i1}^2)\right] = B_3 \quad (17.42)$$

Equations of energy flux

$$\frac{n_0}{\sqrt{\pi}d_0^{3/2}} + \frac{n_{a1}}{\sqrt{\pi}d_{a1}^{3/2}}[(2.5 + b_{a1}^2)[1 - \Phi(b_{a1})]b_{a1}\sqrt{\pi} - (2 + b_{a1}^2)\exp(-b_{a1}^2)]$$
$$+ \frac{n_{i1}}{\sqrt{\pi}d_{i1}^{3/2}}[(2.5 + b_{i1}^2)[1 - \Phi(b_{i1})]b_{i1}\sqrt{\pi} - (2 + b_{i1}^2)\exp(-b_{i1}^2)] + A_{i1}u_c = W_3$$
$$\quad (17.43)$$

$$W_3 = \frac{n_3}{\sqrt{b_3^2/d_3^3}}(2.5 + b_3^2); \quad B_3 = \frac{n_3}{2d_3}(1 + 2b_3^2) + F(E_{12});$$

$$b_3 = v_3\sqrt{\frac{m}{2kT_3}} \quad \Phi(b_3) = \frac{2}{\sqrt{\pi}}\int_0^{b_3} e^{-\xi^2}d\xi$$

The system of (17.38)–(17.43) was solved together with equations of electron emission, electric field, cathode energy balance in quasistationary formulation for long lifetime Cu group spot with current of 200 A (see Chap. 16). The cathode erosion rate G(g/s) was calculated as dependencies on the current density j (Fig. 17.2) or as dependencies on normalized plasma velocity b_3 (Fig. 17.3) at the external boundary of the heavy particle Knudsen layer. u_c is taken as a parameter to obtain a family of curves. The effect of the ion motion in the ballistic zone is presented in Fig. 17.4. The

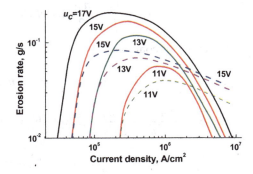

Fig. 17.2 Evaporation rate versus current density in the *heavy particle relaxation zone* with $I = 200$ A and Cu cathode and cathode potential drop u_c varied as parameter. *Solid lines* are for calculation that not considered the returned electron flux to the cathode surface. Calculation dashed lines included these fluxes

17.6 Numerical Investigation of Cathode Spot Parameters by Physically ...

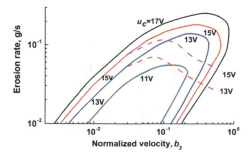

Fig. 17.3 Evaporation rate versus normalized plasma velocity on the external boundary of the *heavy particle relaxation zone* with $I = 200$ A and cathode potential drop u_c varied as parameter. Solid lines are for calculation that not considered the returned electron flux to the cathode surface. Calculation dashed lines included these fluxes

results show that the normalized flow velocity b_3 and current density j are double-valued functions of the net erosion rate G. The lower values of b_3 and j correspond to mode 1, while the higher values correspond to mode 2 (Chap. 16). The calculations show that mode 1 heavy particle flow at the end of the heavy particle relaxation zone is always subsonic. The dependences of j and b_3 indicate a maximal value of G_{max}, and for $G > G_{max}$, no solution can be obtained. Influence of different factors on the solution was studied by their excluding or including during procedure of the solution. In general, j and b_3 significantly increased for moderate change of G. Increase of u_c leads to decrease of b_3 at fixed G. So, the solid lines were calculated assuming that all heavy particles flow as neutrals atoms, i.e., in the equations an influence of space charge sheath on the ion motion and an influence of the returned plasma electrons were ignored. Nevertheless, the heat flux to the cathode from accelerated in the sheath ions were taken into account. The dashed lines indicate the dependences calculated accounting the energy flux due to returned plasma electron flux (marked by symbol "*" in Table 17.1). The results, indicated by the dashed lines, show that the mode 2 solution is modified, leading to lower net erosion rates and higher flow velocities even approached to sonic velocity.

For $I = 200$ A case presented, the calculation indicates a mode 1 with current density of 5×10^4 A/cm^2. The ratio of the heavy particle temperature adjacent to the surface temperature, T_1/T_s, is 0.989 (Table 17.1), i.e., the heavy particles are slightly "cooler" than the surface, while the ratio of the heavy particle density at that location to the Langmuir density, n_1/n_0, is 0.991.

The normalized flow velocity for the mode 1 solution varies from 0.008 at the beginning of the heavy particle relaxation zone, to 0.009 at its exit, i.e., very little change occurs in this subsonic region. The total heavy particle density declines by a factor of 0.838 from the beginning to the end of the zone, while the heavy particle temperature increases by approximately the same factor (Table 17.1). So, the flow is subsonic, and the returned heavy particle flux to cathode is comparable with the direct atom flux from the cathode.

Table 17.1 Plasma and jump parameters at the heavy particle relaxation zone at $u_c = 15$ V $I = 200$ A, Cu [25]

G g/s	$j10^5$ A/cm^2	s	T_0, K	$E10^6$ V/cm	α	$n_0 10^{19}$ cm^{-3}	T_e, eV	b_1	b_3	n_3/n_1	n_1/n_0	T_3/T_1	T_1/T_0	p_1/p_0	Type
0.21	0.533	0.456	3770	5.4	0.11	7.3	0.88	0.008	0.009	0.838	0.991	0.837	0.989	0.981	1
0.3	180.0	0.926	5232	35.8	0.90	110	2.90	0.474	0.965	0.441	0.668	0.809	0.530	0.193	2*
0.5	0.763	0.473	3820	6.30	0.13	8.28	0.94	0.0225	0.0226	0.833	0.975	0.832	0.971	0.948	1*
0.5	23.0	0.766	4344	23.0	0.97	25.0	3.23	0.377	0.486	0.654	0.711	0.708	0.605	0.397	2
0.5	25.3	0.777	4432	23.7	0.76	29.0	2.0	0.314	0.380	0.704	0.745	0.725	0.696	0.503	2*

17.6 Numerical Investigation of Cathode Spot Parameters by Physically ...

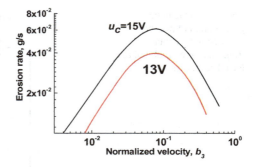

Fig. 17.4 Evaporation rate versus normalized plasma velocity on the external boundary of the *heavy particle relaxation zone* with $I = 200$ A and cathode potential drop u_c varied as parameter. In additional, the effect of the ion motion was taken into account in the ballistic zone

Concluding Remarks Thus, the obtained results indicate an important fact consist in significantly lower plasma velocity in the Knudsen layer in comparison with the sound velocity. For the measured erosion rate 0.02 g/c [26] of the group spot, the normalized velocity b_3 is about 0.01. With increase of G from 0.01 to 0.1 g/s, the solution of mode 1 corresponds to b_3 in range 0.04–0.1. As an exception, the solution indicates mode 2, at which higher flow velocity is calculated even approached to sonic velocity. However, this flow arises for mode 2 when the cathode heating mainly supported by the energy of the returned plasma electrons to the cathode. This solution is accompanied with an extremely parameters which is possible at very high current density $j > 10^7$ A/cm^2, electron temperature ~4 eV, and very high cathode temperature that can exceed the critical temperature of copper cathode and which not realized in a group spot of 200 A [10, 27].

The further spot study by using total system of equation will take into account the transient character of spot development. The important point of spot development is the initial plasma that produced by triggering the discharge gap. It should be taken into account that the cathode surface state as the spot can be developed on a protrusion or located on a smooth area of a bulk cathode depending on the current. Before these phenomena are studied, let us consider previously the different types of the arc triggering. This knowledge is necessary to formulate the initial conditions of the spot initiation.

17.6.2 Spot Initiation During Triggering in a Vacuum Arc

The initial near-cathode plasma supports the spot initiating. The spot development depends on the kind of arc triggering or on the plasma of a previously extinguished spot. So, the initial conditions were determined by the parameters of the primary plasma. In the experiments, this plasma was initiated by a triggering mechanism [28]. Figure 17.5 illustrates schemes of plasma production, contact triggering, and by high voltage local breakdown. Obviously, for some certain initiating plasma parameters, the plasma density and heat flux to the cathode during a lifetime τ could support the future spot development. The primary plasma cloud supports initial cathode

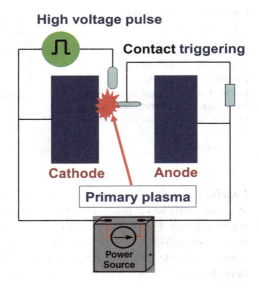

Fig. 17.5 Schematic presentation of plasma production by contact triggering or by high voltage local breakdown

spot formation in order to develop an arc. The primary plasma exists for a small characteristic lifetime τ. During this time, the applied voltage is mostly distributed in a sheath near the cathode with a small part in the high conductivity plasma that fills the gap. When a spot is formed during this time, a self-sustained process between cathode heating, electron emission, and plasma regeneration in the cathode vapor determines the self-sustained primary plasma parameters.

As in this case, the initiated plasma should be self-sustained, and the parameters of this plasma can be determined by calculating self-consistently from the above model using the corresponding cathode system of equations (Chap. 16). Assuming that during this very small time, the plasma parameters (density and heat flux q to the cathode) do not vary significantly as the spot is formed, and the self-sustained primary plasma parameters (density, temperature, ionization) and cathode temperature are obtained by solving the system of equations for a time duration τ. *The question is how the behavior of spot parameters during the spot development time depends on τ?*

17.6.3 Kinetic Model. Study the Total System of Equations at Protrusion Cathode

In this section, the spot parameters are studied for non-uniform surface geometry consisted of a number of protrusion placed on the surface as well as during the spot appeared at the flat surface of a bulk cathode.

17.6.3.1 Heating Model

Let us consider a cylindrical protrusion with radius R and height h placed on planar Cu cathode surface (Fig. 17.6.) The spot parameters self-consistently calculated from primary plasma conditions at $t = 0$ until the parameters reach a state and when the protrusion vanishes [29, 30].

In general, the cathode heat regime should take into account heat conduction both in the protrusion and the cathode body [31]. However, the transient problem for such cathode configuration is greatly complicated numerically by nonlinear and multi-parametric dependencies for the case of a self-consistent solution. For simplicity, in order to obtain the principal character of the transient plasma and parameters as the spot develops a uniform temperature distribution assumed in the protrusion surface. The energy flux q_T is considered as energy loss be heat conduction in the protrusion volume. Indeed, the heat flows a length L of about the protrusion size when the characteristic time is less than $t_p = L^2/a$, where a is the thermal diffusivity. For $L = 2R_0$ and $R_0 = 5$ μm with a Cu cathode $t_p = 1$ μs [30]. Thus, the cathode energy balance can be used in form as previously defined ([3, 10] and Chap. 16) and is:

$$q_i(t) + q_J(t) + q_{eth}(t) = q_T(t) + q_r(t) + q_{er}(t) + q_{em}(t) \qquad (17.44)$$

The following transient heat conduction equation was used to express the cathode energy balance:

$$T(t) = \int_0^t \frac{q_T(t')(1 + 2\frac{h(t')}{R(t')})}{c\rho h(t')} dt' + T_{in} \qquad (17.45)$$

where c is the heat capacity, ρ the cathode material density, T_{in} the initial temperature of the protrusion (room temperature), and $R(t')$ the time-dependent protrusion radius. The energy flux $q_T(t')$ (W/cm^2), which is nonzero, is the time-dependent energy dissipation by heat conduction into the body of the cathode determined by

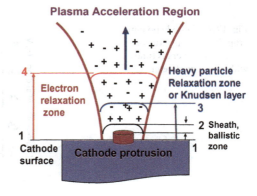

Fig. 17.6 Schematic presentation of the cathode protrusion and the plasma surrounding it

$$q_T(t') = q_i(t') + q_J(t') + q_{eth}(t') - q_r(t') - q_{er}(t') - q_{em}(t') \qquad (17.46)$$

In order to simplify the solution, it is also assumed that the heat flux is from the plasma that is uniformly distributed around the protrusion and that $R(t) = h(t)$. The Ohmic energy flux q_J is $j^2(t)R(t)/\sigma_{el}$, where σ_{el} is the highly ionized plasma electrical conductivity which depends on electron temperature, and j is the current density.

17.6.3.2 Methodology

The solution begins with an initial protrusion radius R_0. Plasma and cathode parameters are calculated, and the cathode evaporation rate $G(t)$ (g/s) is obtained by solving the equations with a small time step Δt. Then, a new protrusion radius R_1 is calculated taking into account the protrusion mass decrease of rate $G(t)$ during Δt and assuming that the erosion is uniform over the whole protrusion surface. In the next time step, the next (and smaller) radius R_2 is calculated using the protrusion radius R_1. This procedure is repeated n time steps, each time calculating q_n, R_n, and the other variables. The run is completed after n steps when the protrusion radius approaches zero, i.e., when $R_n \ll R_0$ or at $t = n\Delta t = 1$ μs. The time step Δt is chosen sufficiently small so that all cathode spot parameters are independent of Δt during t. Also, the time step is small enough that $q_n(t)$ is assumed to be constant during each Δt.

17.6.3.3 Calculation Results for Protrusion Cathode

The calculation using total system of equations (see Sect. 17.4.3) was conducted for a Cu cathode protrusion in a cylinder form with radius (and also height) $R_0 = 5$ μm, as example, and for primary characteristic plasma ignition times of $\tau = 35, 50, 100,$ and 200 ns [28, 30]. Figure 17.7 shows the dependence of u_c and the size as a function of time. It can be seen that initial large value of u_c decreases significantly with time in a time interval of about 2–3 times τ and then quickly approaches an asymptotic value that is independent of τ.

Fig. 17.7 Cathode potential drop (left) and protrusion size R (right) versus time with τ as a parameter

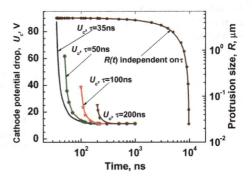

17.6 Numerical Investigation of Cathode Spot Parameters by Physically ...

Fig. 17.8 Spot current and cathode temperature versus time with τ as a parameter

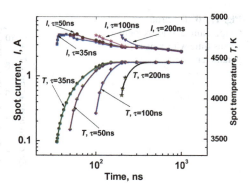

It can be seen that u_c is significantly larger when τ is 35 ns ($u_c = 90$ V) than that for 200 ns ($u_c = 25$ V). The protrusion size slightly decreases up to 1–2 μs and then decreases significantly until the protrusion disappears. This dependence is only shown for times >1 μs in order to demonstrate the tendency of the protrusion size to decrease with time. The additional energy loss in the cathode (in a general problem formulation) can change the rate of protrusion size decrease for times >1 μs (Fig. 17.8).

Initially, N_0 and G_k depend on τ and sharply increase with time, reaching a steady state at 200 ns that was independent of τ. The large values of G(g/C) in comparison with the experiment are due to large size of protrusion and indicate that the steady-state values cannot be reached (Fig. 17.9).

Initially, the electron temperature T_e and jet velocity depend on τ and sharply decrease with time, reaching at 200 ns steady state, at which time they are independent of τ (Fig. 17.10). T_e and V agree with the measured values [3, 32]. The plasma jet velocity decreases from about $(2–3) \times 10^6$ to 7×10^5 cm/s with spot time with $T_e \sim 1$ eV. The total degree of ionization α_{itotal} approaches 1 (Fig. 17.11 and 17.12) indicating that the cathode spot plasma is fully ionized at all times. The fraction of atoms in the 1st through 5th ionization states is presented in Fig. 17.11 with $\tau = 35$ ns.

Fig. 17.9 Heavy particle density N_0 and erosion rate G versus time with τ as a parameter

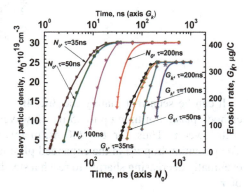

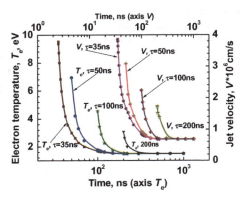

Fig. 17.10 Electron temperature and jet velocity versus time with τ as a parameter

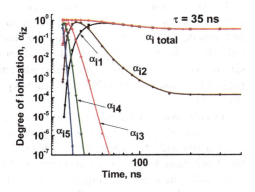

Fig. 17.11 Degree of ionization α_{itotal} of the cathode plasma as well as ion state fractions α_{iz} for the ions with a Z ionization level (Z = 1, 2, 3, 4 and 5) versus time ($\tau = 35$ ns)

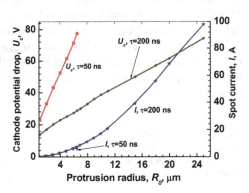

Fig. 17.12 Cathode potential drop and spot current versus protrusion radius with τ as a parameter

It can be seen that mainly the 3rd, 4th, and 5th charged ions contribute to the cathode plasma in time up to 45 ns. However, the ion charge is significantly lower with lower τ. Thus, for $\tau = 200$ ns, mainly the 1st through 3rd ions are present. The potential drop u_{pl} in the expanded plasma varies non-monotonically with time, eventually decreasing sharply to reach a steady-state value of $u_{pl} \sim 5$ V, independent of τ.

Thus, the sum of u_p with the u_c in steady state agrees with the measured Cu arc voltage (~20 V) [3, 32]. The influence of the initial protrusion radius R_0 on u_c and I calculated for $\tau = 50$ and 200 ns is presented in Fig. 17.12, showing that both substantially increase with R_0. Note, the dependences of I on R_0 for $\tau = 50$ and 200 ns coincide one to other. These calculations show that for small protrusion sizes (~1 μm), the value u_c remains about 20 V, while the protrusion current can be significantly small about 0.1 A. The calculated cathode spot current density is $j \sim$ 1–2 MA/cm², and the electron current fraction is ~0.75. The main contribution to the cathode electron current is from thermionic electron emission enhanced by the Schottky effect. The cathode electric field E is $\sim 10^7$ V/cm.

The calculated results show that the steady-state spot parameters are reached below t_p when the energy balance of the protrusion determines the cathode temperature. When the spot life exceeds t_p, the energy loss in the cathode body should be also taken into account in the energy balance [31]. The cathode evaporation fraction K_{er}, defined as the ratio of the cathode evaporation flux to the Langmuir evaporation flux in vacuum, was calculated to be 0.7–0.4, and decreases with time. This result indicates that near the cathode strong non-equilibrium heavy particle flow takes place, and therefore, the plasma velocity is lower than the sound speed (Langmuir flux) in the Knudsen layer.

17.6.3.4 Cyclic and Continuous Models of Spot Operations

Let us compare the character of the time-dependent cathode and plasma parameters on a protrusion obtained by the two models, the continuously developing spot, and the cyclic spot operation models. The previously developed model [29, 30] was used to calculate the parameters of a spot that cyclically appears on a cold protrusion. This model was developed due to observations of cyclic spot operation [27 (chap. 7)]. According to the model [29, 30], a value of Δt was chosen as a cyclic time of spot operation at a protrusion, and at which, the plasma parameters were assumed constant. However, after time Δt, the protrusion size reduced by its evaporation with rate of the calculated erosion. Thus, transient character of spot operation was determined by the time dependence of protrusion size that cyclically decreases due to its erosion.

In continuous transient model, the plasma and cathode parameters are determined self-consistently as a continuous function of time t with given initial conditions at $t = 0$. The transient character of spot development is determined by all plasma and cathode parameters including continuous decrease of the protrusion size. Therefore, all spot parameters were obtained as continuous dependences on time from the moment of spot initiation.

While the spot parameters u_c, u_p, and T_e decrease and T, n_0, and G_k increase with continuous model from some initial value to their steady-state level with spot time, these dependencies were reversed for the cyclic model after a number of spot cycles with significant decrease of the protrusion size. Indeed, the initial T, n_0 and G are approximately constant with time up to about 1 μs for cyclic spot operation [30]. This difference is because the plasma parameters are determined by the cathode temperature development for a continuously operating spot, while for cyclic operation, these parameters are determined by the protrusion size decrease with each cycle when the protrusion is initially cold or at an uncertain cathode temperature that not self-consistent with the plasma. The above results indicate that the parameters of the spot are continuously dependent on time during spot development in each cycle even when observed cyclic spot operation. Thus, spot initiation and development can be described consistently by a vaporization mechanism with the kinetic and gasdynamic models without involving any phenomena of explosion of irregularities at the cathode surface in spite of the cathode spot was observed as a cyclic subject.

17.6.3.5 The Main Result of Solution of the Kinetic Approach

The spot (and the arc) continues burning when a spot can be formed at the first stage of ignition, which is supported by relatively large initial value of u_c. The large u_c occurred due to small time τ of triggered plasma, when a *large heat flux* is requested to sufficiently heat the initially cold cathode in order to obtain a *temperature* needed for spot development. At each next moment, the cathode is heated, and its temperature increases. As a result, the requested self-sustained heat flux from the plasma to the cathode decreases and consequently decreases the requested cathode potential drop which thereby determines this flux. Thus, the calculation predicts plasma parameter variation and the decreasing of u_c due to continued cathode heating with time. This means that in order to form a spot, the u_c initially must be significantly larger than that value during the development stage with subsonic plasma flow in the relaxation zone.

17.6.4 Bulk Cathode

Let us study a vacuum arc with solid bulk cathode and plain surface at which a spot arises [33, 34]. Figure 17.13 shows the dense cathode plasma adjacent to the surface with characteristic kinetic layers and the further region of expanding accelerated plasma.

17.6.4.1 Transient Model of Cathode Heating

The transient thermal model takes into account that the plasma plume and consequently the heating are concentrated on the cathode surface. The solution of the time-dependent heat conduction equation for a solid body heated by a concentrated heat source has the following general form:

17.6 Numerical Investigation of Cathode Spot Parameters by Physically ...

Fig. 17.13 Schematic diagram of main plasma regions presented in accordance with the kinetic model of adjacent plasma to cathode and a region with the gasdynamic plasma expansion

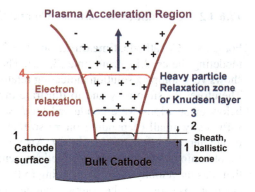

$$T(t) = \int_0^t q(T, t') f(t' t_0) dt' + T_{00} \qquad (17.47)$$

where $T(t)$ is the time dependent cathode temperature in the spot (see Chap. 3), $q(T,t')$ is the time-dependent heat flux density to the cathode surface, T_{00} is the surface temperature before spot initiating, $f(t', t_0)$ is a time-dependent function resulting from integrating the heat conduction equation in differential form, and $t_0(a, r_s)$ is a parameter indicated the heat flux concentration, determined by thermal diffusivity coefficient a and by spot radius $r_s = I/\pi j$. Equation (17.47) is the relationship between the transient temperature and different terms of the cathode energy balance (Chap. 16). Using this relationship and the above kinetic system of equations that described the cathode plasma phenomena, the time-dependent heat flux density $q(T,t')$ and spot temperature $T(t')$ were determined. It was assumed that the dynamics of plasma parameters and $q(t)$ determined by the transient character of $T(t')$ (see below).

The temperatures $T(t')$ and $q(T,t')$ are appeared as implicit parameters due to nonlinear effects of the cathode spot phenomena. Therefore, the cathode plasma system of equations was solved numerically using an iteration method at each time step of integration of (17.47). Thus, the solution of the cathode system of equations will determine the dependences $q_n(T,t), f_n(t)$ at each time step Δt and consequently the temperature $T(t)$. For the first time step, the initial triggered plasma lifetime τ was used to determine the initial temperature T_{in} that supported reproduce a secondary plasma density self-consistently. These parameters were used for future integration of (17.47). A relatively small time step Δt was chosen for integration so that calculated $q_n(\Delta t)$ can be considered as mostly constant during Δt. The solution continues up to time $t = n\Delta t$ for which the $T(t)$ as well as the cathode plasma parameters reached steady state. The calculations were conducted for different cathode materials—Cu and Cr, as well partially for Ag and Al which have intermediate thermophysical properties, and also for W, a refractory material.

17.6.4.2 Calculation Results for Cu and Cr Cathode

For Cu and Cr cathode, **time durations** $\tau = 2, 7.5, 10, 50$, and 100 ns are chosen for modeling the arc triggering time [3, 35]. The preliminary study of the total model shows very complex solution indicating relative small current per spot, around 10 A. Therefore, for simplicity and to illustrate the main characteristics of the spot, the below calculations were conducted for $I = 10$ A. The time step Δt was chosen sufficiently small (depending on τ) so that the results of calculation for all spot parameters were independent of Δt. Also, in order to overcome the complications for obtaining the solution of the kinetic system of equation, it was taken into account that plasma velocity is low comparing to the thermal velocity, i.e., b_3 is small. The calculations show that the solution can be significantly simplified if the ion density obtained without convective term due to small velocity b_3 determines the quasineutrality condition. In some cases, however, this quasineutrality condition was considered with convective term. It is sensitive by calculating the difference between direct and returned heavy particles fluxes in the Knudsen layer that determined the cathode erosion rate and parameter K_{er}. This case of the solution will be specially indicated as "with b_3".

The calculated time dependence of u_c is presented in Fig. 17.14 (without b_3) in Fig. 17.15 (with b_3), which show the small influence of the quasineutrality condition with convective term. This result shows that the time dependence of u_c and the

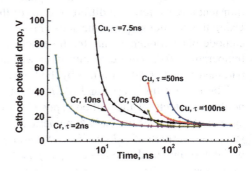

Fig. 17.14 Dependence of u_c on time t with τ as parameter for Cu and Cr (without b_3)

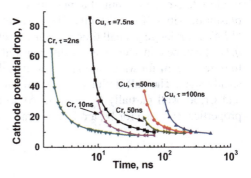

Fig. 17.15 Dependence of u_c on time t with τ as parameter for Cu and Cr (with b_3)

17.6 Numerical Investigation of Cathode Spot Parameters by Physically ...

maximal u_c are determined by the values of τ, i.e., on the type of spot triggering in the experiment. The large $u_c \sim 90$–100 V (Cu) and 65–70 V (Cr) were calculated with the small τ for which solutions of the total system of equations were obtained for bulk cathodes at 7.5–8 ns (Cu) and ~2 ns (Cr), respectively.

It can be seen that u_c significantly decreased with τ and with time t and converged to a steady-state value ~13–15 V, while the smallest τ determined by the metal properties. At each step, the heat flux from the plasma to the cathode was obtained which in turn determined the calculated dependences of the u_c and other plasma parameters on time during the spot development. $T(t)$ time dependences are shown in Fig. 17.16.

The cathode temperature increased from some initial value T_{in} (obtained at time τ) depending on τ to a steady-state temperature that did not depend on the initial conditions. T_{in} increased with τ for both cathode materials. T_{in} was lower for Cu (about 3500 K) for $\tau = 7.5$ ns than for Cr (3700 K) even with lower $\tau = 2$ ns.

The calculations showed that the equilibrium neutral atom density (Fig. 17.17) produced by cathode evaporation increased with time from an initial low values depending on τ and reached the steady-state level of 2×10^{20} cm^{-3} (Cu) and 8×10^{20} cm^{-3} (Cr) independent of τ. Figure 17.18 shows the time-dependent electron temperature for Cu and Cr with τ as parameter. T_e decreased with time from initial

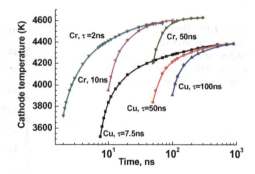

Fig. 17.16 Cu and Cr cathode spot temperature as function of time with τ as a parameter

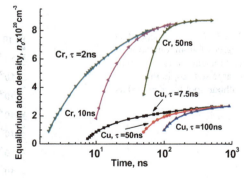

Fig. 17.17 Equilibrium neutral atom density at Cu and Cr cathodes as a function on time with τ as the parameter

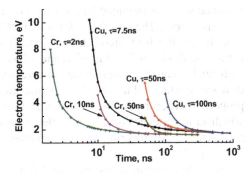

Fig. 17.18 Electron temperature for Cu and Cr as a function on time with τ as the parameter

values depending on τ to a steady state of about 1.5 eV, independent of the initial condition.

The initial T_e was larger for Cu than for Cr cathodes. Figure 17.19 for Cu and $\tau = 7.5$ ns and Fig. 17.20 for Cr and $\tau = 2$ ns illustrate the time dependence of the atom ionization fractions with charge state α_{ij} in the cathode spot where $j = 0$, 1, 2, 3, 4, 5 and α_{i0} is the total degree of ionization. The cathode spot plasma was fully ionized for all initiating plasma parameters considered (<30–40 ns), i.e., $\alpha_{i0} = 1$ and then decreased to about 0.5–0.6 with time. The ion charge state with the 1st

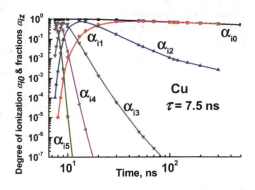

Fig. 17.19 Time-dependent degree of atom ionization α_{i0} and ionization fractions with charge state α_{ij} in the Cu cathode spot for $\tau = 7.5$ ns

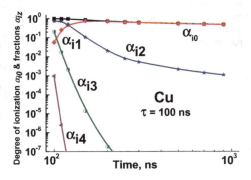

Fig. 17.20 Time-dependent degree of atom ionization α_{i0} and ionization fractions with charge state α_{ij} in the Cu cathode spot for $\tau = 100$ ns

17.6 Numerical Investigation of Cathode Spot Parameters by Physically ...

Fig. 17.21 Time-dependent degree of atom ionization α_{i0} and ionization fractions with charge state α_{ij} in the Cr cathode spot for $\tau = 2$ ns

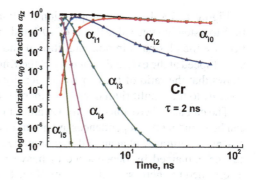

through 5th ions was present, though mainly the 3rd and 4th, for $\tau = 7.5$ ns (Cu) and $\tau = 2$ ns (Cr). For $\tau = 100$ ns, the plasma mostly consisted of the 1st–3rd ions (see Fig. 17.21).

An important parameter is the normalized plasma velocity $b_3 = \sqrt{\frac{v_3^2 m}{2kT_3}}$ indicated a ratio of the plasma velocity at the external boundary 3 of Knudsen layer to the thermal velocity, sound speed (see 17.4.3). The calculation showed that b_3 is very small (see Fig. 17.22). Another important calculated parameter is the fraction of the evaporated cathode material K_{er} is the ratio of the net evaporation rate into the adjacent spot dense plasma, to the Langmuir evaporation rate into vacuum.

This parameter is important because in many published previously works, the erosion rate was erroneously studied by cathode evaporation according to the Langmuir formula. The kinetic system of equation is very complicated in order to obtain its solution. As b_3 is small, for simplicity, in the above calculations, the quasineutrality conditions were considered using the ion density determined from the ion flux without convective term due to velocity b_3. However, the convective term influences by calculating the difference between direct and returned heavy particles fluxes in the Knudsen layer that determined the cathode erosion rate and also on parameter K_{er}.

Fig. 17.22 Normalized plasma velocity at the external boundary 3 of Knudsen layer as function on time (with b_3)

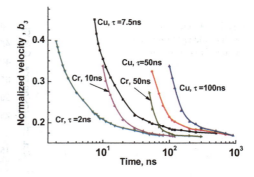

Figure 17.23 shows the evaporation fraction K_{er} decreased with time from about 0.25 to a steady-state value that is of ~0.1 and lower. This parameter is lower than unit because the plasma is in non-equilibrium and the plasma velocity is lower than the sound speed at the external boundary of the Knudsen layer. Therefore, the calculation shows that the ratio of heavy particle density at the external boundary of Knudsen layer n_3 to the equilibrium density n_0 increased during the spot development.

The calculated erosion rates for Cu and Cr cathodes are presented in Fig. 17.24. It can be seen that the dependences are not monotonic initially increase with time then passing a saturated plateau and then continue to increase with time. This behavior can be explained by dependence of heavy particle flux from the cathode, which determined by increase of the density (Fig. 17.17) but with simultaneously decrease of the convective velocity b_3 (Fig. 17.22). The obtained values of G are in range from about 20 to 50 μg/C and are in accordance with Daalder's data [36].

The calculation shows that for Cu, the ratio n_{03} of heavy particle density at the external boundary of Knudsen layer n_3 to the equilibrium density n_0 increased during the spot development. The ratio n_{03} for Cu increased from initial values of 0.49–0.55 dependent on τ, to a steady-state value of 0.66, while this ratio was 0.31 when the cathode evaporated in vacuum (with the sound speed at boundary 3) [4]. For a Cr cathode, this ratio varied in the range of 0.52–0.61 at initial stage, dependent on τ, to

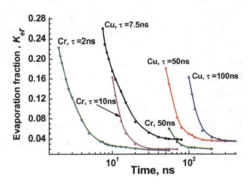

Fig. 17.23 Cu and Cr cathode evaporation fractions as a function of time with τ as the parameter (with b_3)

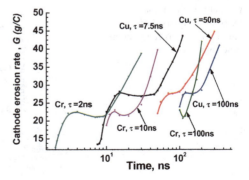

Fig. 17.24 Dependence of the cathode erosion rate for Cu and Cr on time (with b_3)

17.6 Numerical Investigation of Cathode Spot Parameters by Physically ...

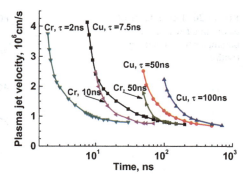

Fig. 17.25 Cu and Cr cathode plasma jet velocity as a function of time with τ as the parameter

a steady-state value of 0.67. The cathode spot current density was $j \sim 1$–3 MA/cm^2, and electron current fraction s varied between 0.7–0.8.

The cathode plasma jet velocity (see Fig. 17.25) for Cu and Cr initially was relatively large (4–2) × 10^6 cm/s, and it decreases with time to the measured values $\sim 1 \times 10^6$ cm/s, independent on τ, i.e., on the conditions of spot initiation [32, 37]. The calculations show that the potential drop in the expanding jet, u_{pl}, decreased with time from about 13 V at the initially to 5–6 V at steady state for both cathode materials. The main contribution to the cathode electron current was from thermionic electron emission (due to the large cathode temperature), enhanced by the Schottky effect. The tunneling effect also weakly contributed. The contribution of this effect is determined by the electric field at the cathode surface E that not exceed $\sim$(2–3) × 10^7 V/cm. The Ohmic heating of the cathode was low in comparison with the cathode heating due to the ion and returned electron heat fluxes.

17.6.4.3 Calculation Results for Ag and Al Cathodes

For Al and Ag cathodes, the time of spot initiation was chosen in range $\tau = 6$–100 ns, and the spot current was 10A. Δt was chosen sufficiently small so that the results of calculation for all cathode spot parameters were independent on Δt. The calculated time-dependent u_c for Al and Ag is presented in Fig. 17.26. The calculation shows that u_c dependence on time was very sensitive near the value of τ. The larger $u_c \sim 70$ V was at first step with $\tau = 10$ ns. For Al cathode, the u_c was relatively lower and increases from 20 to 50 V when τ decreased from 100 to 6 ns.

Figure 17.27 shows the cathode evaporation fraction that is comparatively low with respect to Langmuir evaporation flux into a vacuum. K_{er} decreases from about 0.25 to about 0.04 with the spot time in range shown in Fig. 17.27. For Al: $K_{er} = 0.2$–0.05. Figure 17.28 shows the heavy particle increase with time for different initial values of τ. It can be seen that this density increases from 6×10^{19} to about 1.5×10^{19} cm^{-3} when increased from 6 to 50 ns.

When the spot time increases from a few ns to about 100 ns, the spot parameters are changed for Ag cathode as: the heavy particle density $n_0 = (0.4$–$2) \times 10^{20}$ cm^{-3},

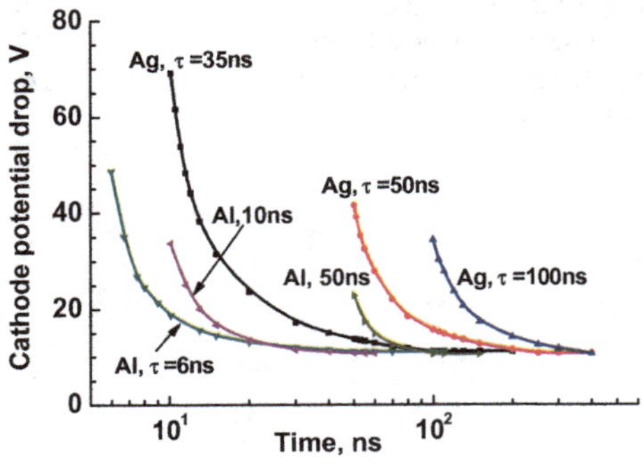

Fig. 17.26 Dependence of u_c on time t with τ as parameter for Al and Ag (with b_3)

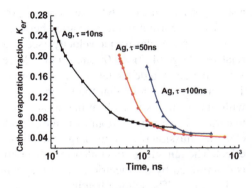

Fig. 17.27 Ag cathode evaporation fractions as a function of time with τ as the parameter (with b_3)

Fig. 17.28 Equilibrium neutral atom density at Al cathode as a function on time with τ as the parameter (with b_3)

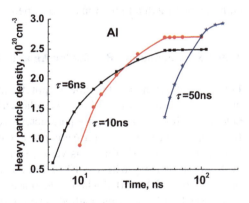

electron current fraction $s = 0.75 - 0.63$, current density $j = 2.5 - 1.3$ MA/cm^2, electron temperature $T_e = 8 - 2$ eV, total degree of ionization $\alpha_{10} = 1 - 0.7$. Cathode temperature $T = 3040-3800$ K, the erosion rate $G = 30-60$ μg/C. The characteristic parameters for Al were $G = 7.4-12$ μg/C, while the measured value is 28 μg/C [38]); $V \sim 2 \times 10^6$ cm/s (1.5 $\times 10^6$ cm/s—experiment [38]), $s = 0.75 - 0.65$; $j \sim (4 - 2) \times 10^6$ A/cm^2, $T_e = 6 - 1.5$ eV.

17.6.4.4 Calculation Results for W Cathodes

This section applies the proposed model of the virtual cathode concept [39, 40] to describe the spot on refractory cathodes. The calculations were conducted with a W cathode to demonstrate a presence of a self-consistent solution. Therefore, the spot parameters were determined only at the initial times τ, showing the possible solution that then allows the spot development in time. The calculations were conducted for a wide range of triggered initial plasma lifetimes τ, from 2 ns to 100 μs.

The spot current was 10 A. As it was mentioned above, the virtual cathode concept requires the given electrical field at the sheath–plasma boundary which was characterized by coefficient k_d. Previous calculations showed that for W cathode, k_d was mostly constant for widely varying spot plasma parameters determined as dependence on electron current fraction and the coefficient $k_d = 0.1$ [39]. Below this value was taken for calculation in the expected range of plasma parameters.

The calculations show that when τ increased from 2 ns to 100 μs, then the CPD decreased from 45 to 25 V, the current density from 2×10^7 to 5×10^5 A/cm^2, and cathode temperature in the spot from about 10,000 to 7600 K (Fig. 17.29). The electron temperature decreased from 5.7 to 3 eV, the plasma jet velocity from 1.5×10^6 to 1.1×10^6 cm/s (Fig. 17.30), and the atom density n_0 decreased from 8×10^{20} to 2×10^{19} A/cm^2, and K_{er} weakly increased from about 0.49 to 0.52 (Fig. 17.31) with τ in mentioned range.

The cathode plasma was mostly fully ionized and consisted mainly of ions with charge 2 and 3 (Fig. 17.32). Also, the potential drop in the expanding plasma jet U_p decreased from 8 to about 1 V, k_m varied from 2.5 to 2.9 and U_m from 2.6 to 1.9 with τ. The erosion rate weakly varied in range 170–130 μg/C, the ratio n_{30} as 0.63–62, and electron current fraction varied between 0.74 and 0.80.

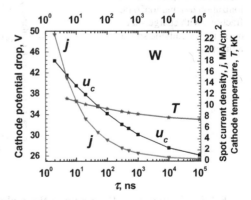

Fig. 17.29 u_c, T, and j for W cathode as a function of τ

Fig. 17.30 T_e, and V for **W** cathode as a function of τ

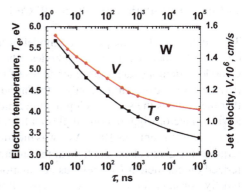

Fig. 17.31 K_{er} and n_0 for **W** cathode, as a function of τ

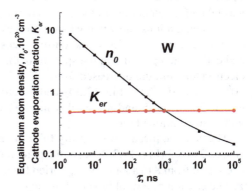

Fig. 17.32 Degree of atom ionization α_{i0} and ionization fractions with charge state α_{ij} in the **W** cathode spot as a function of τ

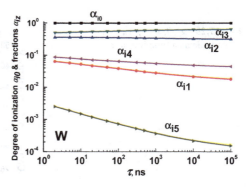

17.6.4.5 Concluding Remarks to the Calculated Results for the Bulk Cathode

First of all, it should be noted in this Chapter it was considered the case when the characteristic particle collision times (electron-electron collisions—t_{ee}, ionization—t_i, resonance charge exchange—t_{ia}) are shorter than that the primary plasma lifetime

17.6 Numerical Investigation of Cathode Spot Parameters by Physically …

at τ (even in few ns range) in order to generate ***the secondary near-cathode plasma which should support the spot development***. Also, the case when the characteristic times for dense plasma in the spot are shorter than that for solid thermal time was considered, and therefore, these plasma parameters in the transition period were determined by the target heating time [13]. The calculations show that the cathode spot plasma parameters strongly vary in the first 10 ns for Cr ($\tau = 2$ ns) where the density is 10^{20}–10^{21} cm^{-3}. For Cu, the density is 4×10^{19}–10^{20} cm^{-3} in range of 10–20 ns. T_e changed from its maximal value to the level of 2 eV and lower very rapidly (in range from few ns to 10 ns after ignition in the nanoscale of τ). The estimations show that $t_{ee} \sim 10^{-13} - 10^{-14}$ s, $t_i \sim 10^{-10} - 10^{-11}$ s, and $t_{ia} \sim 10^{-10} - 10^{-11}$ s [41–43] for $T_e \sim 2$–10 eV and plasma density of 10^{19} cm^{-3} (and lower for larger density), i.e., relatively low indicating that above-mentioned condition is fulfilled for spot development even in first 10–20 ns. Note that the observed fast dynamic cathode processes in range 20 ns–100 μs were detected only due to average periods of brightness fluctuations [35, 44].

Let us discuss the obtained dependencies of the plasma and cathode parameters. Firstly, the minimal temperature for each considered τ was determined by a minimal vapor density, which was fully ionized and needed to support the particle and energy fluxes for further spot development. Secondly, for smaller τ, larger energy flux needed to sufficient heat the initially cold cathode. Thus, when the initial plasma lifetime τ was significantly small, a relatively large energy flux to the cathode was obtained from the solution to reach the required cathode temperature for self-consistent producing a new portion of secondary cathode plasma. The required energy flux can be supported through a corresponding high value of u_c. With time, the cathode temperature increased as well the intensity of cathode evaporation increased, and consequently, the plasma density grows. In this case, the self-sustained heat flux from the plasma to the cathode can be supported by lower value of u_c. As a result, the plasma parameter variation and also the decrease of u_c with time depend on the cathode heat regime, which demonstrated by the dependences on the cathode material properties (see results for Cu and Cr).

The calculation showed that at relatively large spot duration and at steady state, the sum of u_c and u_{pl} agrees well with the measured arc voltage of 20 V for Cu and Cr [3, 32]. But at arc initiation, the calculation showed the necessity of a relatively large initial power supply voltage (50–100 V), depending on initial plasma parameters, to insure arc ignition. The voltage u_p can be estimate from relation like is $Iu_p \sim GV^2$. In essence, Ohmic and electron pressure effects determine u_p. The, respectively, contribution of these effects was discussed in [45] and can be obtained calculating the potential distribution with distance from the cathode in the jet. Such calculation will be presented in Chap. 18. According to the calculation, the ion charge was larger at the beginning of spot formation due to larger energy dissipation with larger u_c and smaller for the developed spot. This result agrees well with experimental measurements of ion charge state for Cu and Cr [46] as well as for other cathode materials—Mo [47], Pt [48], and Au [49].

The rate of cathode mass loss (cathode erosion) was determined by the difference between the evaporated and returned heavy particle fluxes. Here, the other momentums including the momentum flux from the near-cathode electric field (firstly shown in [2]) compensated the returned relatively large momentum of electron plasma pressure acted at the external sheath boundary. Therefore, it was not influenced on the heavy particles flow directly in the Knudsen layer (except the electron momentum by factor $\exp(-u_c/T_e)$. However, the plasma electron energy and momentum (with small mass comparing to heavy particle mass) were transferred to heavy particles at some distance from the cathode influencing on the cathode evaporation rate due to plasma heating as well as producing a non-free (impeded) plasma flow. The subsonic flow in the Knudsen and electron beam relaxation regions accelerates to supersonic plasma jet due to energy dissipation and electron pressure gradient discussed in [50]. The external pressure can influence on the location point of sonic velocity. The jet velocity and jet potential drop were calculated by integral equations using the results of plasma parameters (density and temperature) in the spot to obtain the estimation of the mentioned jet values. The phenomena related to the cathode melting and MPs generation are discussed in previous Chap. 16.

The most important result for the tungsten cathode is the obtained solution using the vaporization model with a virtual cathode for electron emission. The calculated sum of u_c and u_p for a W cathode for large $\tau \geq 1$ μs reaches arc voltage of 25–26 V. As this voltage in development stage is weakly different from the initial stage (for large τ), the calculated value agrees well with the measured arc voltage [1, 32]. Taking into account that the calculated effective cathode voltage is one third of the total arc voltage, the resulting energy dissipated in the cathode body is enough to reach an initial temperature, which promote the future spot development with $j \sim 10^7$ A/cm^2 (low Joule heating). When the triggered plasma appears for a longer time ($\tau > 100$ ns), the calculated cathode temperature can be relatively low (7500–8000 K), and the low density cathode vapor is fully ionized with maximal ion flux fraction. This range of temperatures was determined by the minimum fully ionized plasma density, when the spot can develop self-consistently, supporting the plasma with necessity parameters. There the tungsten vapor density reached $\sim 2 \times 10^{19}$ cm^{-3} at relaxation length ~ 1 μm even at time of 2 ns, while thermionic electron emission current density was $\sim 6 \times 10^6$ A/cm^2.

The potential minimum u_m in the virtual cathode model is also proportional to the cathode temperature by factor k_m which was calculated relatively small (2.5–2.9) and therefore u_m is low (2.6–1.9 V). The evaporation fraction K_{er} (the ratio of the calculated net evaporation rate to the evaporation rate into vacuum) is important. Note, large of the heavy particle flux is returned to the cathode due to rarefied collisions in the Knudsen layer and impeded plasma flow and thus $K_{er} < 1$.

17.6.4.6 Kinetic Model. Study of the Total System of Equations for Film Cathode

Kesaev experimentally studied spot and cell behavior of arcs on Cu thin film cathodes deposited on glass substrates [51]. He showed that the spot moved with velocity v_s and left a clear erosion track of width δ, in which all the film was removed. For arc durations was up to 100 µs, the erosion rate was calculated using measured δ and v_s as

$$G(\text{g/s}) = \gamma d_f \delta v_s, \qquad (17.48)$$

where γ is the specific metal density and d_f is the film thickness. The spot parameters and spot motion were studied using gasdynamic model (Chap. 16) [52] for different d_f by taking into account the energy loss in thin film by its heat conduction and erosion $G(\text{g/s})$, varying u_c. The calculations show agreement between the modeling mechanism of spot motion due to film evaporation under the spot and that spot behavior observed by its motion in film deposited on a glass substrate. Also, another important result indicates that the energy loss due to film heat conduction is comparable with that loss by film evaporation, which cannot be neglected as it was done previously by the estimation of electron current fraction [51]. The calculation also shows that the track width observed by spot motion or width of melting isotherm does not exceed the spot diameter more than by factor 2.

This section will further develop the film spot study using the kinetic approach calculating not only the parameters G and u_c but also the spot velocity from (17.48) for different d_f using the calculated G and δ. The kinetic model also uses the experimental finding that the spot moves when the underlying film is evaporated and took into account the current and d_f dependence of the spot velocity [51]. The calculations take into account the film energy balance considering the specifics of heat conduction process for the metallic film deposited on a glass substrate described in Chap. 16. Figure 17.33 shows the calculated cathode potential drop u_c as a function of d_f. This result is comparable with that measured by Grakov for Cu films deposited on a Cu substrate (~11 V) [53].

This low value of u_c for Grakov's measurements was explained by low energy losses in the deposited film determined the mechanism of spot motion due to film evaporation [52]. The normalized plasma velocity at the first boundary $b_1 = \sqrt{\frac{v_1^2 m}{2kT_1}}$;

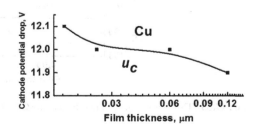

Fig. 17.33 Cathode potential drop as function on film thickness for Cu film cathode

and at external boundary $b_3 = \sqrt{\frac{v_3^2 m}{2kT_3}}$ of the Knudsen layer is shown in Fig. 17.34, which demonstrated a subsonic plasma flow.

The measured and calculated dependences of tracks width on film tickness (Fig. 17.37) demonstrate their equality for $d_f = 0.017$ μm, while small difference between them can be seen with increase of d_f.

The dependence of film cathode erosion rate G in g/C on film tickness is presented in Fig. 17.35. The calculated result indicates mostly constant value of G, while weakly dependence on d_f is obtained in the experiment.

The obtained value of G is used to calculate the spot velocity dependence on d_f, which presented in Fig. 17.36. It can be seen that while some difference between calculated and measured value takes place for very small thickness (0.017 μm), these values approached on to other with d_f up to the be equal value at $d_f = 1.4$ μm. Figure 17.38 shows the comparison of energy losses due to film heat conduction Q_T and vaporization Q_s. It can be seen that $Q_s > Q_T$ for all range of considered film thicknesses. Table 17.2 presents the range of characteristic parameters of the cathode spot for the calculated range of d_f (Fig. 17.37).

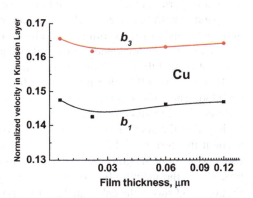

Fig. 17.34 Normalized plasma velocity at the internal b_1 and external b_3 boundaries of Knudsen layer as function of film thickness

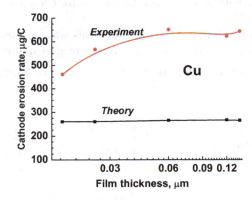

Fig. 17.35 The erosion rate of the film cathode G(μg/C) as function of film thickness

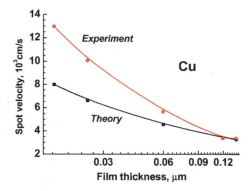

Fig. 17.36 Dependences of calculated and measured spot velocities on film thickness

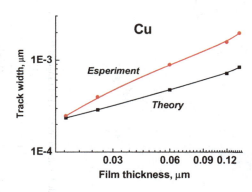

Fig. 17.37 Track width as dependence on film thickness for calculated and measured data

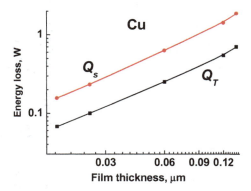

Fig. 17.38 Energy losses due to film heat conduction Q_T and vaporization Q_s as dependences on film thickness

17.7 Rules Required for Plasma Flow in Knudsen Layer Ensured Cathode Spot Existence

The above study of the partial kinetic mode using equations describing only heavy particles flow and the calculations of the plasma flow according to the total system

Table 17.2 Characteristic parameters of the appeared on film Cu cathode for above-mentioned range of film thicknesses

T_0, K	s	j, MA/cm^2	T_e, eV	E, MV/cm	α	n_0, 10^{20} cm^{-3}	V, 10^6 cm/s	K_{er}	n_2/n_0	T_2/T_0
4700	0.76	1–2	1.6–1.7	20	0.4–0.5	4–5	0.9–1	0.38–0.4	0.68–0.69	~1

of equations for microprotrusion as well for bulk cathode show that in the Knudsen layer the evaporated atom flux is comparable with the flux of the returned atoms and ions. The evaporation fraction K_{er} decreased with time, reaching 0.1 in some cases. The calculated velocity of the dense plasma b_3 is significantly smaller than the "sound velocity" ($b_3 \ll 1$). Thus, the ion current to the cathode formed at the boundary 3 is characterized by a convective term, in essence, like as $j_{i3} \sim j_{i0}(1 - b_3)$, where j_{i0} is the ion current density determined by the thermal velocity at boundaries 3 and at outer boundary 2 of the ballistic zone. It is clear that near the sound speed when $b_3 \to 1$, the ion current $j_{i3} \to 0$, and as result, no solution of the closed system of equation could be found. As the sound velocity characterizes free gasdynamic gas flow, the cathode plasma flow in the Knudsen layer is not free but rather impeded due to high plasma density and intense energy dissipation by the electron beam in the relaxation zone. The impeded character of near-cathode plasma in a vacuum arc is followed from the presence of number of common phenomena, which provided the energy flux to the cathode, heating of the cathode and plasma as well energy dissipation between the cathode body and plasma in coupling with plasma motion and expansion.

On the other hand, the calculations show that the spot cannot be developed due to its short lifetime in the case of small spot current. In this situation, power is not sufficient to support the requested cathode surface temperature due to relatively large heat conduction loss. Such case is occurred in attempt to describe the nano or microsecond high-speed spots at moving on flat surface of a bulk cathode [10, 54]. At the same time, spot with similar characteristic can be studied in frame of described here model, as example on film cathodes (*high velocity, small lifetime*), or on protrusion cathode (*small lifetime*) with current per spot up to ~0.1 A due to very low heat conduction loss. Understanding of these factors was obtained by numerous calculations of self-consistent mathematical study, the common phenomena for different observed spot types.

Accordingly, **two main Principle Rules of the Kinetic Model** for the cathode spot in vacuum arcs were derived [25] and discussed [3, 31]:

I. **Plasma flow in the region adjacent to the cathode surface should be impeded in order to ensure the cathode spot ignition and further self-consistent spot development;**

II. **Cathode heat loss must be smaller than the inflow power to a surface of the cathode with a given geometry** [25]. This is equivalent to a presence of the effective cathode voltage (defined by a ratio of the cathode heat conduction loss to the current) at least being smaller than the cathode potential drop in the ballistic zone.

The relatively small subsonic plasma flux in Knudsen layer is the result of a difference between large direct evaporated atom flux and returned atom and ion fluxes. Therefore, the formation of this small flux can explain why the observed cathode spot behavior (with high-pressure plasma near the surface) is sensitive to the variation of a lower surrounding pressure. The derived rules represent two principles, which can be used to explain the multiformity of cathode spot types depending on arc conditions.

17.8 Cathode Spot Types, Motion, and Voltage Oscillations

Let us consider the spot behavior and their experimental characteristics presented in Table 7.10 in Chap. 7. Based on the above Principle Rules, mechanisms of transient spot dynamics, spot motion, and arc voltage oscillation are discussed below. These non-stationary spot characteristics are explained in frame of thermal cathode vaporization without any explosions. The main accounted point that the cathode spot located on the cathode surface with some irregularities, film, and protrusions requires low input energy to reach sufficiently high temperature for intense evaporation.

17.8.1 Mechanisms of Different Spot Types

Experiments indicate that the spot type and their characteristics are determined by the arc current, heat loss in the cathode, and ambient vapor or gas. Furthermore, different type spots occur on bulk and film cathodes (Chap. 7, Table 7.10).

When a bulk cathode surface is contaminated by an oxide film and other impurities, super fast spots (SFS) with velocities up to 10^4 cm/s were observed at relatively low arc current (~100 A) [55, 56]. It can be understood taking into account that SFS spots appear due to a rapid rise of the temperature and intense vaporization of the relatively thin oxides or contamination layer in times from a few tens to hundreds of nanosecond. This time determined the local spot lifetime and the high spot velocity (Fig. 17.39, picture 1). An estimation show that heat penetrates into a 0.1 μm thickness film in ~1 ns. The low spot current is due to low heat loss in the thin surface layer [52] that in accordance with Principle Rule II is fulfilled. Usually, SFS spots were observed until that the cathode cleaned by the arc [55]. The non-stationary heat conduction of the cathode body determines the transient cathode spot operation and its cyclic extinguishing and reigniting. The spot lifetime is associated with a characteristic thermal time. In the case when the spot current is relatively small (~10 A), the spot can operate on the cathode micro-area with low cathode heat losses balanced by the low heat flux from the spot. The thermal time and subsequently the velocity of the spot on clean surfaces depend by the cathode roughness (Fig. 17.39, picture 2). The spot lifetime is determined by the thermal time constant and exhaustion of relatively larger metal irregularities in range from 0.1 to ~1 μm in comparison with the thin contaminated layers <0.1 μm). Due to this difference in the conditions of the cyclic spot operation, transient moderately fast spots (MFS) were observed in experiments on cleaned cathodes [55, 57]. Sometimes a new spot can be observed at large distance from the previously dead spot. This

17.8 Cathode Spot Types, Motion, and Voltage Oscillations

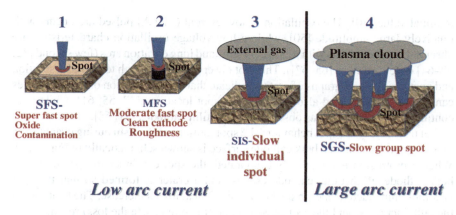

Fig. 17.39 CS types. Schematic drawing of spot types: 1. Super fast spot, 2. Moderate fast spot, 3. Individual spots, 4. Slow grouping spot

new spot can be appeared due to interaction of a protrusion with declined plasma jet before the old spot was dead.

According to the experiment [2, 27, 58], the spot types influenced by the presence of ambient gas of pressure significantly lower than the cathode plasma pressure. This effect can be understood taking into account change of the flow regime formed in Knudsen layer due to the accumulation of density along the expanding cathode plasma after the interaction of its rarefied part with the ambient gas [25]. This is similar to the passing reflected wave from a target, and as result, the plasma velocity, super sound transition point, and the plasma density adjacent to the cathode can change.

In low-pressure ambient gas, slow individual spots (SIS) can exist separately [27, 58] because the Principle Rule I is fulfilled due to the gas presence (Fig. 17.39 picture 3). At higher spot currents (<1 kA) on bulk cathodes in vacuum, the heat loss from the individual spots is compensated by the heat supplied by the large current and other nearby spots (Principle Rule II). Furthermore, these individual spots tend to group, thus creating a common vapor cloud that satisfies Principle Rule I (Fig. 17.39, picture 4). This group spot is favorable, and new spots can be ignited under the common plasma cloud under impeded plasma flow condition. Slow group spots (SGS) with a relatively long lifetime (≤ 1 ms) were observed [27, 59–61]. In very high current arcs ($\gg 1$ kA) with intense vaporization, the SIS or SGS types occur because Principle Rule I is automatically fulfilled [10].

17.8.2 Mechanisms of Spot Motion and Voltage Oscillations

Spot operation and motion are characterized by voltage fluctuations of arcs occurred with different amplitude and the rate of cathode voltage growth relative to some minimal value [50]. The oscillation in low-current (~10 A) pulsed arc occurs with relatively large amplitude [50], while a low voltage oscillation characteristics are observed in experiments with relatively stable and long duration arcs (few second [32, 62–64] or few minutes [65–67]). The spot observation with high temporal resolution and post arcing photographs of craters indicate that the spot motion on bulk cathodes can be continuously and also with discrete spot locations [27, 55, 61], while only continuous spot motion was observed on thin film cathodes [51, 68].

Let us consider the spot behavior. The spot motion mechanism during evaporation of a microprotrusion on a bulk cathode surface is shown schematically in Fig. 17.40. When a protrusion is completely evaporated, the spot continues to operate on the bulk cathode. With time, cathode erodes, and a crater is formed so that the spot becomes embedded deeper in the body and 3D heat losses increase. The heat loss in the bulk increases, and the requested voltage (to compensate the loss) reaches such high values that the old spot dies because Principle Rule II is no longer satisfied, and a new spot appears at the nearest neighboring protrusion, where the voltage and heat conduction energy can be smaller. This causes apparent spot motion: i.e., one spot dies, and a new spot is ignited at a nearby location. The characteristic time for protrusion evaporation can be relatively small, $\leq 1\mu s$, and it determines the local spot lifetime and the consequential high spot velocity (~10^3 cm/s). As the ignition direction has equivalent probability, the spot motion resembles a random walk [35, 55].

When a spot extinguishes and reignites at a new location, u_c fluctuates (Fig. 17.40). In essence, voltage oscillation in low current pulsed arcs [51] is associated with time-dependent u_c. When the spot extinguishes (usually before reaching steady state) and then the next spot appears at a neighboring location, the voltages u_c increases

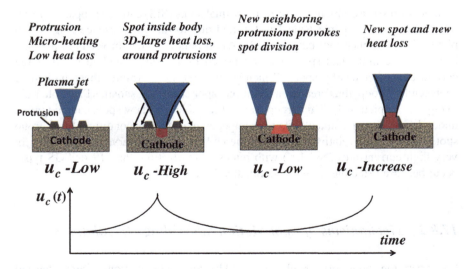

Fig. 17.40 Clean cathode. Moderate fast spot motion. Voltage oscillation

depending on temperature at the cathode place which is lower than that in the developed spot. The power supply must be able to support this voltage fluctuation. In experiments with relatively long duration and higher current arcs [32, 69], cathode energy loss can be compensated by dissipation even in bulk cathodes [10].

17.9 Requirement of Initial High Voltage in Electrode Gap for Vacuum Arc Initiation

Let us consider a requirement of initial high voltage in the interelectrode gap for arc development [70]. According to the experiment, while the developed electrical arc is a relatively high current discharge which operates at low voltage, arc initiation requires higher initial voltage, and an appropriate power supply will be required [28, 67, 68, 71, 72]. The arc voltage in the developed stage is low and depends on the cathode materials, ranging from 10 V (Hg) [51] to 26 V (Mo) [32] and for Cu cathodes is about 20 V. However, after arc triggering or in the presence of an external initial plasma, the voltage of the power supply should be also much larger (~100 V) than the burning arc voltage in order to prevent the arc extinction when the voltage oscillates [73]. This experimental result is in accordance with above calculations of $u_c(t)$ [70]. Really, as the lifetime of triggered initial low dense plasma is short, the above calculations show the necessity of a relatively high u_c in order to continue the spot development.

After triggering, secondary plasma is generated at the cathode surface during time τ. So, the secondary cathode plasma parameters (density, ionization fraction, temperature, etc.) should reach parameters that are suitable to reproduce the electroconduction medium in order to support the current continuity in the gap. The total arc voltage u_g for moderate arc currents (~10^2A) consists of two main parts: (i) the cathode potential drop which is u_c in the electrical sheath, whose thickness is on the order of the plasma Debye length and (ii) the potential drop in the cathode plasma jet, u_{pl}. Thus, the main goal of the problem is to calculate u_c and u_{pl} as functions of time.

During a primary plasma plume appearing, the cathode temperature must increase from the room value to a value allowing the rate of vaporization needed to sustain the plasma for cathode spot initiation. As the lifetime of τ is small, the required initial heat flux to the cathode is relatively large [69] to reach appropriate temperature. This heat flux can be supported by relatively high initial voltage of the power supply, which falls in the sheath near the cathode surface. With time, the cathode temperature increases to a steady-state level, and the energy losses due to thermal conduction into the bulk of the cathode, electron emission, and cathode evaporation are balanced by the incoming energy flux from the plasma to the cathode. In steady state, the main energy losses are minimal—this reduces the required energy flux from the plasma and therefore the cathode voltage. Thus, dependence of the arc voltage time is a result from the dynamics of the spot parameters, which evolve along with the cathode temperature.

Also, when the spot extinguishes (before reaching steady state) and then the next triggering spot appears at a neighboring location, the voltages u_c and u_{pl} can be increased depending on temperature at the cathode place, which is lower in comparison with that in the developed spot. At this moment, the power supply should be able to support this voltage fluctuation. This effect explains the use at least the welding power supplies whose open circuit voltage is typically about 80–100 V. Thus, the relatively large experimental voltage requested for arc ignition was caused by the mechanism of the initial spot appearing which is initiated by the interaction of the triggered low life primary plasma with the cold cathode bulk. The nature of furthermore voltage decreasing caused by the heat flux decrease with cathode temperature during the time of spot development.

17.10 Summary

The above solution of a system of equations represents a self-consistent investigation describing the evolution of different parameter characterized spot phenomena. In contrary, at last decade in the published works, an attempt to study an influence of coupling arc conditions, as a rule, connected with accounting one effects, but with some not adequate assumptions regarding to others, although that the authors indicate their studies as self-consistent. Sometimes, the authors of the published works presented models as self-consistent while using arbitrary input parameters or due to use a certain physical effect, which can be important for certain arc conditions. On the other hand, some authors take into account specific effects (or condition) and claim on completeness of the used model. The examples of different such works were reviewed in Chap. 15. Thus, it should be noted that the definition of the calculating approach as a self-consistent meets some uncertainty. In the present chapter a self-consistent physically closed model is developed which allows obtain the solution without parameters given arbitrary or given from an experiment except the thermophysical material parameters. The calculations for given spot current (observed average value) for demonstration the spot mechanism was conducted due to uncertainty of this value even during the arc operation discussed above.

The model is applied to different conditions indicating the initial plasma parameters that also calculated self-consistently taking into account the cathode thermal regime for times characterizing experimental methods of triggering arc. It is well known that the cathode spots appear in various forms depending on unpredictable conditions, also produced by the arc burning. Therefore, an attempt to completely spot description can meet unpredictable difficulties. Nevertheless, consequently, the developing spot theory should be consisted of principle ideas, which is possible to explain the mechanisms of typical spot phenomena, its characteristics, and behavior.

An advanced physically closed approach is the present kinetic model developed by the author. The kinetics of cathode evaporation adjacent to the Knudsen layer together with gasdynamic plasma flow in the dense plasma region and in the expanding plasma jet was investigated. The model describes the returned fluxes of heavy particles and determines directly the net erosion of cathode material.

17.10 Summary

The main difference between the various published models is that the cathode potential drop u_c was a given parameter, whereas in the kinetic model u_c, was calculated as part of a self-consistent set of equations. The present kinetic model newly examines the role of the arc voltage at the moment of arc initiation and spot development. With increase in self-consistently calculated u_c, the spot temperature and the current density do not increase with time unlimited. This, however, occurs for planar cathode in case of given plasma spatial (crater) and temporal distribution, when a constant u_c or the constant arc voltage was assumed, for example, in work [74]. The important issue is the understanding that u_c is not constant value in a transient arc when the first spots were initiated. In the developed arc characterized by multi-spot regime, the u_c decreases to the steady-state value.

For stationary arcs (with hot electrodes or high current arcs), the spot can operate relatively long time, for example, up to milliseconds. In this case, the cathode temperature increased with time reaching a state when the input energy is compensated by the energy losses. Then, the temperature and the self-consistent spot parameters weakly changed approaching to mostly steady state. The above model allows to obtain such solutions indicating that the time to reach this state is relatively small (at order of initial plasma triggered time) of around 10–100 ns depending on cathode metal and spot parameters.

Another advantage of the present physically closed approach is its ability to derive two fundamental principle rules. According to the rules, an understanding of the mechanisms appearing of different spot types observed over long period of time was developed. Considering the above mechanisms of spot types, it is obvious that the cathode surface irregularities, the metal melting, and self-organized energy due to the plasma cathode interaction limit the spot lifetime and its parameters, which of course can be shorter than the steady-state values. The observed arc voltage oscillations, which are significantly larger at low (near a threshold) current, become an explanation in frame of the two principle rules derived by the kinetic model of cathode phenomena of a vacuum arc. The parameters including the low value cathode potential drop for film cathodes (special type of spots) were calculated, and an understanding a mechanism of operation of spots at such cathodes is presented. The kinetic model demonstrates a presence of an adequate theoretical description of cathode spot on tungsten using virtual cathode eliminating the contradiction between the large electron emissions while low rate of material evaporation.

The above model is formulated assuming atom ionization equilibrium at the jet origin. This assumption can be estimated by equation:

$$\frac{d(nv)}{dx} = \beta_r n_i n_0^2 \left(1 - \frac{n^2}{n_0^2}\right) \qquad (*)$$

From (*), the following condition should be fulfilled $\frac{v}{\beta_r n_0^2 l_r} \ll 1$, n_0 is the equilibrium charge particle density. This condition for Cu is 10^{-3} and therefore is fulfilled due to the large plasma density $n_0 > 10^{19}$ cm^{-3}, low velocity near the cathode $v < 10^4$ cm/s, electron beam relaxation length $l_r \sim 10^{-4}$ cm, and $\beta_r = 10^{-27}$ cm^6/s.

However, in the general case, when the plasma density deviates from the equilibrium value (e.g., for metals when there is a large convective flux and low degree of ionization), the ion current to the cathode should be determined by the ion density near the cathode surface in region l_r in accordance with the diffusion model (Chap. 16).

Some remarks related to the Kesaev's experimental data [51]. The cell structure of cathode spots was observed using simple photographing the image of spots on moving film only for low melting metals as Hg, In, and Pb with melting point of −38.83 °C, 157 °C, and 356.7 °C, respectively, and therefore, his result cannot be applied to wide range of high melting materials. For these materials, other researchers studied the phenomena of spot splitting which obtained different result dependent on thermophysical properties [75]. Another point related to the current per spot, for which one used Kesaev's measurements threshold current I_d or $2I_d$. For Cu was used for study spot in vacuum the threshold current [51] of 1.6 and 3.2 A, respectively, [76–79]. It should be noted that Kesaev's measurements of the threshold current were conducted for arcs in air atmosphere arcs in order to reach stability data, but not in vacuum [80]. In addition, the measured spot current significantly depends on cathode material properties and surface state and can be much different from the threshold current [35, 75].

References

1. Beilis, I. I., Lubimov, G. A., & Rakhovskii, V. I. (1972). Diffusion model of the near cathode region of a high current arc discharge. *Soviet Physics-Doklady, 17*(1), 225–228.
2. Beilis, I. I. (1982). On the theory of erosion processes in the cathode region of an arc discharge. *Soviet Physics-Doklady, 27,* 150–152.
3. Beilis, I. I. (1995). Theoretical modeling of cathode spot phenomena. In R. L. Boxman, P. Martin, & D. Sanders (Eds.), *Vacuum arc science and technology*. Park Ridge, NJ: Noyes Publications.
4. Anisimov, S. I. (1968). Vaporization of metal absorbing laser radiation. *JETP, 37*(1), 182–183.
5. Mott-Smith, H. M. (1951). The solution of the Boltzmann equation for a shock wave. *Physical Review, 2*(6), 885–892.
6. Kozlov, N. P. & Khvesyuk, V. (1971). Cathode processes in electrical arcs. I, *Soviet Physics—Technical Physics, 16* (N10), 1691–1696.
7. Moizhes, B. Ya., & Nemchinsky, V. A. (1980). Erosion and cathode jets in a vacuum arc. *Soviet Physics—Technical Physics, 25,* 43–48.
8. Moizhes, B. Ya., & Nemchinsky, V. A. (1982). Formation of a jet during vaporization in vacuum. *Soviet Physics—Technical Physics, 25,* 438–441.
9. Djakov, B. E. (1983). A model for the cathode mechanism in low-current metal vapor arc. *Journal of Physics. D. Applied Physics, 16,* 343–355.
10. Beilis, I. I. (1974). Analysis of the cathode spots in a vacuum arc. *Soviet Physics—Technical Physics, 19*(2), 251–260.
11. Knake, O., & Stranski, I. N. (1956). Mechanism of vaporization. *Progress in Metal Physics, 6,* 181–235.
12. Dobretzov, L. N., & Homoyunova, M. V. (1966). *Emission electronics*. Moscow: Nauka.
13. Anisimov, S. I., Imas, Yu A, Romanov, G. S., & Khodyko, Yu V. (1971). *Action of high-power radiation on metals*. Springfield, Virginia: National Technical Information Service.
14. Beilis, I. I. (2003). The vacuum arc cathode spot and plasma jet: Physical model and mathematical description. *Contributions to Plasma Physics, 43*(3–4), 224–236.

References

15. Lyubimov, G. A. (1973). Cathode region of a high current arc. *Soviet Physics—Technical Physics, 18*(4), 565–569.
16. Belkin, G. S., & Kiselev, V. Ya. (1973). About limit regime of cathode spot in a vacuum arc. *Soviet Physics—Technical Physics, 18*(7), 764–767.
17. Ecker, G. (1961). Electrode components of the arc discharge. *Ergebnisse der exakten Naturwissenschaften, 33*, 1–104.
18. Ecker, G. (1973). Unified analysis of the metal vapor arc. *Zeitschrift für Naturforschung, 28a*, 417–428.
19. Nemchinsky, V. A. (1979). Theory of the vacuum arc. *Soviet Physics—Technical Physics, 24*(7), 764–767.
20. Nemchinsky, V. A. (1983). Comparison of calculated and experimental results for stationary cathode spot in a vacuum arc. *Soviet Physics—Technical Physics, 28*(12), 1449–1451.
21. Raizer, Y. P. (1997). *Gas discharge physics*. Berlin, New York: Springer © 1991.
22. Nedospasov, A. V., & Khait, V. D. (1991). *Base of physical processes in a devices with low temperature plasma*. Moscow: Energoatomisdat (in Russian).
23. Beilis, I. I., Sevalnicov, A. Y. (1991). The column of electrical arc at atmospheric pressure. *High Temperature, 29* (5), 856–863 (1991). (in Russian).
24. Beilis, I. I. (1985). Parameters of kinetic layer of arc discharge cathode region. *IEEE Transactions on Plasma Science, PS-13* (N5), 288–290.
25. Beilis, I. I. (1986). Cathode arc plasma flow in a Knudsen layer. *High Temperature, 24*(3), 319–325.
26. Kantsel', V. V., Kurakina, T. S., Potokin, V. S., Rakhovskii, V. I., & Tkachev, L. G. (1968). Thermophysical parameters of a material and electrode erosion in a high current vacuum discharge. *Soviet Physics—Technical Physics, 13*, 814–817.
27. Zykova, N. M., Kantsel, V. V., Rakhovsky, V. I., Seliverstova, I. F., & Ustimets, A. P. (1971). The dynamics of the development of cathode and anode regions of electric arcs, I. *Soviet Physics—Technical Physics, 15*(11), 1844–1849.
28. Boxman, R. L. (1977). Triggering mechanisms in triggered vacuum gaps. *IEEE Transactions on Electron Devices, ED-24*, 122–128.
29. Beilis, I. I. (2007). Transient cathode spot operation at a microprotrusion in a vacuum arc. *IEEE Transactions on Plasma Science, 35*, N4, Part 2, 966–972.
30. Beilis, I. I. (2011). Continuous transient cathode spot operation on a microprotrusion: Transient cathode potential drop. *IEEE Transactions on Plasma Science, 39*, N6, Part 1, 1277–1283.
31. Beilis, I. I. (2001). State of the theory of vacuum arcs. *IEEE Transactions on Plasma Science, 29*, N5, Part 1, 657–670.
32. Davis, W. D., & Miller, H. C. (1969). Analysis of the electrode products emitted by dc arcs in a vacuum ambient. *Journal of Applied Physics, 40*, 2212–2221.
33. Beilis, I. I. (1988). The cathode potential drop of an arc in the vapor of the electrode discharge. *Soviet Physics-Doklady, 33*, 125–127.
34. Beilis, I. I. (2013). Cathode spot development on a bulk cathode in a vacuum arc. *IEEE Transactions on Plasma Science, 41*, N8, Part II, 1979–1986.
35. Jüttner, B. (2001). Cathode spots of electric arcs. *Journal of Physics. D. Applied Physics, 34*, R103–R123.
36. Daalder, J. E. (1975). Erosion and the origin of charged and neutral species in vacuum arcs. *Journal of Physics. D. Applied Physics, 8*(14), 1647–1659.
37. Yushkov, G. Y., Anders, A., Oks, E., & Brown, I. G. (2000). Ion velocities in vacuum arc plasmas. *Journal of Applied Physics, 88*(10), 5618–5622.
38. Polk, J. E., Sekerak, M. J., Ziemer, J. K., Schein, J., Qi, N., & Anders, A. (2008). Theoretical analysis of vacuum arc thruster and vacuum arc ion thruster performance. *IEEE Transactions on Plasma Science, 36*, N5, 2167.
39. Beilis, I. I. (1988). On the mechanism of the vacuum arc spot operation on refractory cathodes. *High Temperature, 26*(6), 1224–1226.
40. Beilis, I. I. (2004). Mechanism for small electron current fraction in a vacuum arc cathode spot on a refractory cathode. *Applied Physics Letters, 84*(8), 1269–1271.

41. Braginskii, S. I. (1965). Transport properties in a plasma. In M. A. Leontovich (Ed.), *Review of plasma physics* (p. 1). Consultants Bureau, 205–311.
42. Zeldovich, Y. B., & Raizer, Yu P. (1967). *Physics of shock waves and high temperature hydrodynamic phenomena.* N.Y.: Academic Press.
43. Biberman, L. M., Vorob'ev, V. S., & Yakubov, I. T. (1987). *Kinetics of nonequilibrium low-temperature plasmas.* New York: Consultants Bureau.
44. Juttner, B. (1997). The dynamics of arc cathode spots in vacuum: New measurements. *Journal of Physics. D. Applied Physics, 30*(2), 221–229.
45. Beilis, I. I., Rakhovsky, V. I., & Zektser, M. P. (1985). Current density in the cathode spot of a vacuum arc. Gradients of electron temperature and density. *Soviet Physics-Doklady, 30*, 476–478.
46. Oks, E., Yushkov, G. Y., & Anders, A. (2011). Ion charge state distribution in high current vacuum arc plasmas in a magnetic field. *IEEE Transactions on Plasma Science, 24*, N3, Part 1, 1174–1183.
47. Oks, E., Yushkov, G. Y., & Anders, A. (2008). Temporal development of ion beam mean charge state in pulsed vacuum arc ion sources. *Review of Scientific Instruments, 79*, 02B301.
48. Yushkov, G. Y., & Anders, A. (2008). High charge state ions extracted from metal plasma in the transition regime from vacuum spark to high current vacuum arc. In *Proceedings of 23rd International Symposium on Discharges and Electrical Insulation in Vacuum*, Romania, Bucharest, September, pp. 341–344.
49. Adonin, A., & Hollinger, R. (2012). Development of high current Bi and Au beams for the synchrotron operation at the GSI accelerator facility. *Review of Scientific Instruments, 83*, 02A505.
50. Beilis, I. I., & Zektser, M. P. (1991). Calculation of the parameters of the cathode stream an arc discharge. *High Temperature, 29*(4), 501–504.
51. Kesaev, I. G. (1968). *Cathode processes in electric arcs.* Moscow: Nauka Publishers. (in Russian).
52. Beilis, I. I., & Lyubimov, G. A. (1976). Theory of the arc spot on a film cathode. *Soviet Physics—Technical Physics, 21*(6), 698–703.
53. Grakov, V. (1967). Cathode fall in vacuum arcs with deposited cathodes. *Soviet physics—Technical physics, 12*, 1248–1250.
54. Beilis, I. I. (1977). Cathode spots on metallic electrode of a vacuum arc discharge. *High Temperature, 15*(5), 818–824.
55. Achtert, J., Altrichter, B., Juttner, B., Peach, P., Pusch, H., Reiner, H. D., et al. (1977). Influence of surface contaminations on cathode processes of vacuum discharges. *Beitrage Plasma Phys, 17*(6), 419–431.
56. Anders, S., & Anders, A. (1991). Emission spectroscopy of low-current vacuum arcs. *Journal of Physics. D. Applied Physics, 24*(11), 1986–1992.
57. Anders, A., Anders, S., Juttner, B., Botticher, W., Luck, H., & Schroder, G. (1992). Pulsed dye laser diagnostics of vacuum arc cathode spots. *IEEE Transactions on Plasma Science, 20*(4), 466–472.
58. Rakhovsky, V. I. (1976). Experimental study of the dynamics of cathode spots development. *IEEE Transactions on Plasma Science, 4*, 87–102.
59. Djakov, B. E., & Holmes, R. (1974). Cathode spot structure and dynamics in low current vacuum arcs. *Journal of Physics. D. Applied Physics, 7*, 569–580.
60. Djakov, B. E., & Holmes, R. (1971). Cathode spot division in vacuum arcs with solid metal cathodes. *Journal of Physics. D. Applied Physics, 4*, 504–509.
61. Beilis, I. I., Djakov, B. E., Juttner, B., & Pursch, H. (1997). Structure and dynamics of high-current arc cathode spots in vacuum. *Journal of Physics. D. Applied Physics, 30*, 119–130.
62. Plyutto, A. A., Ryzhkov, V. N., & Kapin, A. T. (1965). High speed plasma streams in vacuum arcs. *Soviet Physics JETP, 20*, 328–337.
63. Kimblin, C. W. (1973). Erosion and ionization in the cathode spot regions of vacuum arcs. *Journal of Applied Physics, 44*(7), 3074–3081.

64. Schulman, M. B., Slade, P. G., & Bindas, J. A. (1995). *IEEE transactions on components, packaging and manufacturing technology*, P.A, 18, N2.
65. Merinov, N. S., Ostretsov, I. N., & Petrosov, V. A. (1976). Anode processes with negative potential drop at the anode. *Soviet Physics—Technical Physics, 21,* 467–472.
66. Beilis, I. I., Shnaiderman, A., & Boxman, R. L. (2008). Chromium and titanium film deposition using a hot refractory anode vacuum arc plasma source. *Surface and Coatings Technology, 203*(5–7), 501–504.
67. Messik, S., Kampf, J., Allen, R., Chan, C., & Sroda, T. (1992). Investigation of the properties of type 303 stainless steel thin fils deposited with the anodic arc. *Material Letters, 14,* 63–66.
68. Guile, A. E. & Jüttner, B.(1980). Basic erosion processes of oxidized and clean metal cathodes by electric arcs. *IEEE Transactions on Plasma Science,* PS-8, 259–269.
69. Plyutto, A. A., Ryzhkov, V. N., & Kapin, A. T. (1965). High speed plasma streams in vacuum arcs. *Soviet Physics—JETP, 20,* 328–337.
70. Beilis, I. I. (2010). The nature of high voltage initiation of an electrical arc in a vacuum. *Applied Physics Letters, 97,* 121501.
71. Lafferty, J. M. (1966). Triggered vacuum gaps. *Proceedings of the IEEE, 54,* 23–32.
72. Anders, A., Brown, I. G., MacGill, R. A., & Dickinson, M. R. (1998). 'Triggerless' triggering of vacuum arcs. *Journal of Physics. D. Applied Physics, 31,* 584–587.
73. Ehrich, H., Karlau, J., & Muller, K. G. (1986). Initiation of arcing at a plasma-wall contact. *IEEE Transactions on Plasma Science,* PS-14, N5, 603–608.
74. Barentgolts, S. A., Shmelev, D. L., & Uimanov, I. V. (2015). Pre-explosion phenomena beneath the plasma of a vacuum arc cathode spot. *IEEE Transactions on Plasma Science, 43*(8), 2236–2240.
75. Boxman, R. L., Martin, P. J., & Sanders D. M. (Eds.). (1995). *Handbook of vacuum arc science and technology.* Park Ridge, N.J.: Noyes Publ.
76. Mesyats, G. A. (2000). *Cathode phenomena in a vacuum discharge. The breakdown, the spark and the arc.* Nauka: Moscow.
77. Mesyats, G. A., & Mesyats, V. (2018). Structure of the ecton cycle of a vacuum arc. In *Proceeding of 28th International Symposium Discharges and Electrical Insulation in Vacuum,* Germany, Greifswald, September, pp. 333–336.
78. Gashkov, M. A., Mesyats, G. A., Uimanov, I. V., & Zubarev, N. M. (2019). Molten metal jet formation in the cathode spot of vacuum arc. *IEEE Transactions on Plasma Science, 47*(8), 3456–3461.
79. Gashkov, M. A., Mesyats, G. A., Uimanov, I. V., & Zubarev, N. M. (2018). Model of the formation of liquid-metal jet in the cathode spot of vacuum arc discharge. In *Proceeding of 28th International Symposium Discharges and Electrical Insulation in Vacuum,* Germany, Greifswald, September, pp. 337–340.
80. Kesaev, I. G. (1976). Laws governing the cathode drop and the threshold current in an arc discharge on pure metals. *Soviet Physics—Technical Physics, 9*(8), 1146–1154.

Chapter 18
Spot Plasma and Plasma Jet

Metallic plasma formed in the cathode spots of a vacuum arc expands into vacuum as highly ionized and considerably accelerated plasma jets. The jet properties determine its wide use for different applications [1–4]. Therefore, knowledge of the mechanism of jet origin is an issue, which is important not only as scientific subject, but also has great practical interest. In this chapter, the progress of the models reported in the literature and their analysis are presented. Also our own developed ideas allowed to understanding the nature of the cathode jet formation and its expansion are considered.

18.1 Plasma Jet Generation and Plasma Expansion. Early State of the Mechanisms

The presence of vapor or plasma flow was discovered in nineteenth century appeared as result of repulsion effects produced in electrical arcs. The early experimental studies (including Tanberg's [5] measurements of super sound plasma velocity) as well as the first discussion of the causes of the force between the electrodes and of the produced plasma jets in the arcs were provided in Chap. 11. The later experimental investigation of the vacuum-arc plasma jet and jet velocities was detailed in Chap. 12. To understand the jet origin, let us summarize the basic early published hypothesizes in first.

The early attempt to understand of the effects of the supersonic flow was consisted from separate propositions. The first is the assumption of unrealistic extremely high cathode temperature up to 5×10^4 K with following intense cathode vaporization [5]. Compton 1931 [6, 7] suggested for Hg cathode that an interpretation of the high pressure is to be found in the existence of an "accommodation coefficient" for ions which accelerated in the cathode potential drop layer, strike, and were neutralized at the surface. Risch and Ludi [8] assumed that high-energy plasma flow can be realized due to potential energy of multiply ions by their neutralization at the cathode surface causing the observed force.

Tonks [9] developed the most realistic model. He considered the electron pressure $p = nkT_e$ in a cathode spot plasma taken in account a momentum change due to the rebounding of plasma electrons from the positive ion sheath covering a negatively charged electrode. In essence, Tonks used the electron pressure to explain Kobel's measurements [10] of the force at the cathode of the Hg arc. However, the plasma density, electron temperature, and ion current fraction were determined arbitrary, and he did not discussed the measured supersonic jet velocity as result of the explained cathode force.

Finkelnburg [11] considered a thermal model of the electrode vapor jets by arcs. He assumed that the energy W_j (W/cm^2) for vapor production was transferred to the surface of either electrode by the incident electrons or positive ions accelerated by the anode or cathode fall, respectively. The velocity of the electrode vapor jet ejected perpendicularly from the electrode surface was determined by following expression

$$V = \frac{W_j}{\lambda_s \gamma} \qquad (18.1)$$

where λ_s (Ws/g) and γ (g/cm^3) are the specifics heat evaporation and density of the electrode material. It was indicated that the calculated vapor jet velocities are in agreement with those measured for mercury sparks by Haynes [12] and for the high-current carbon arc for which the measurements were conducted by the author. It should be noted that the energy W_j was calculated using certain values of the anode and cathode falls and with unrealistic assumption that the total incident energy was spent to the cathode evaporation neglecting the other losses, for example, the cathode heat conduction loss.

Robson and von Engel [13] considered the ions which moved through dense vapor taking into account that a fraction of their momentum should be transferred by the returned particles to the cathode. Using equation of particle conservation with returned particle flux and measured cathode force, Robson and von Engel estimated the particle velocity as 10^4 cm/s that shows rather lower than measured by Tanberg's 10^6 cm/s. Also, Robson and Engel [14] indicated that electrons from the cathode gain energy in the cathode fall, and then their random velocity becomes significant value due to frequent scattering. As a result, an electron pressure was produced in the plasma, and this pressure acts the cathode with force calculated as $F = 50$ dyn/A at $j = 10^5$ A/cm^2. The electrons interact with atoms by elastic collisions, with the result that the evaporating atoms are accelerated away from the cathode by the gradient of the electron pressure. The atoms move into regions of lower density, and in this way, a beam of fast atoms was formed, i.e., the jet. Thus, in essence, the authors obtained also that the plasma jet was produced due to gradient of the electron pressure, close to that assumed by Tonks [9].

Maecker [15] considered a contraction zone due to self-magnetic axial pressure p, which produces an axial transport of plasma forming a plasma jet. The mathematical description included Maxwell equations for magnetic field H and equation of Lorentz

18.1 Plasma Jet Generation and Plasma Expansion. Early ...

force F_L in form:

$$\text{rot} H = -\frac{4\pi j}{c}; \quad F_L = \frac{1}{c} j \times H \tag{18.2}$$

The pressure is obtained assuming

$$\frac{\partial p}{\partial x} = F_L \tag{18.3}$$

For a uniform channel of radius r_0 with current I, and current density $j = I/\pi r_0^2$ as radial dependence

$$p(r) = \frac{Ij}{c^2}\left(1 - \frac{r^2}{r_0^2}\right) \tag{18.4}$$

To calculate the plasma velocity v, the equation of hydrodynamics plasma flow was used in form:

$$\frac{4\pi j}{c} = \text{grad}\, p + \rho v \, \text{grad}\, v \quad \text{and}$$

$$(v\, \text{grad})v = \text{grad}\, \frac{v^2}{2} - v \times \text{rot}\, v \tag{18.5}$$

Taking into account that electromagnetic force for motion along the axis, i.e., rot v = 0, the following expression is obtained:

$$\rho \text{grad}\, \frac{v^2}{2} = \text{grad}\, p \tag{18.6}$$

And for maximal magnetic pressure [from (18.4) at $r = r_0$], the equation for plasma velocity is:

$$\frac{Ij}{c} = p_{\max} = \rho_j \frac{v^2}{2}; \quad v_{\max} = \sqrt{\frac{2Ij}{c\rho_j}} \tag{18.7}$$

where ρ_j is the plasma mass density in the expanding jet. The estimation of $v_{\max}$ was conducted for carbon arc assuming arbitrary $\rho_j = 10^{-5} \text{g/cm}^{-3}$. For $I = 200$ A, $2r = 2.4 \times 10^{-5}$cm, the velocity $v_{\max}$ was obtained as 3.5×10^4 cm/s. This value is about the thermal particle velocity determined taking into account the relatively large carbon cathode temperature, and therefore, it does not explain the plasma expansion in a vacuum.

Ecker [16] also developed model of cathode jet formation by plasma contraction due to self-magnetic field. It was indicated that the effect appeared by small cylindrical symmetry caused by axial pressure gradient. Ecker extended Maecker's model taking into account also the sputtering, neutralization, and reflection of the accelerated ions, as well that the emitted electrons accelerated in the space charge and deliver their momentum to the contraction zone. These processes were denoted by last term in the following equation:

$$\frac{\bar{\rho} v_{max}^2}{2} = -\frac{Ij}{c^2} + \bar{\rho} \int \left(\frac{dv}{dt}\right)_0 ds \qquad (18.8)$$

where $\bar{\rho}$ is the constant average value of density ρ. The solution of (18.8) was obtained as function of current I, electron current fraction f, ionization potential u_i, cathode potential drop u_c, current density j in form

$$v_{max}^2 = \frac{2j}{\bar{\rho}} f(I, j, f, u_i, \varphi, u_c, \bar{\rho}) \qquad (18.9)$$

Equation (18.9) was obtained for reflected ions with following two cases of ion coefficient neutralization from the surface a_+, and atom accommodation coefficient a_n:

(a) $a_+ = a_n = 0$;
(b) $a_+ = a_n = 1$;

The maximal velocity was calculated as function of ratio $j/\bar{\rho}$, and the results are presented in Fig. 18.1. According to Ecker,s description, the thick lines refer to the C cathode. For this case at $j/\bar{\rho} < 10^9$ A cm/g the thermal vaporization of cathode material has lower effect on the plasma jet in comparison with other effects. For lower range of $j/\bar{\rho}$, the determining influence comes from the collisions of the electrons emitted from the cathode and the ions reflected at the cathode. The effect of the self-magnetic field is subordinate and becomes noticeable only with currents >100 A. With increasing $j/\bar{\rho}$, the vaporization eventually predominates completely in the range >10^9 A cm/g. The lines for $I = 1$ and 500 A are separated from one to other at small values of $j/\bar{\rho}$, while join together at large values of $j/\bar{\rho}$.

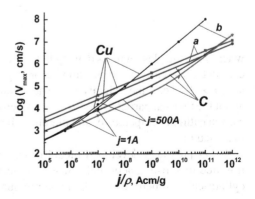

Fig. 18.1 Velocity v_{max} of the plasma jet of Cu and C cathodes for two limiting cases **a** and **b** plotted versus the ratio of current density j and average mass density $\bar{\rho}$ in the jet [16]

For Cu cathode, the dependencies are quite similar except that overall picture of the phenomena which is shifted to smaller values of $j/\bar{\rho}$. The vaporization already comes into effect when $j/\bar{\rho}$ is about 10^5 A cm/g. Below this range, the main influence is to be found in the collisions of electrons emitted and ions reflected at the cathode. The self-magnetic field produces deviations noticeable with high current >100 A. In case, for 1 and 500 A, the curves join in range >10^7 A cm/g.

Summarizing Ecker's work, it can be emphasized a fact of his positive attempt to study most of published ideas together in order to compare different effects. However, a detailed analysis of this model indicates that the calculated results can be considered as only qualitative dependencies. The model consists of a number of arbitrary assumptions. It is followed from derivation of some expressions by using the second term in right side of. (18.8) and by using arbitrary input parameters, such as $f = 0.9$. It was not discussed what is the range of values of j and others. The influence of the emitted electrons and space charge layer was provided without details of plasma parameters. The argument $j/\bar{\rho}$ of calculated dependences is completely not understandable due to uncertainty of the value $\bar{\rho}$ in the jet through its expansion.

18.2 Ion Acceleration Phenomena in Cathode Plasma. Gradient of Electron Pressure

This section considers further modeling of the plasma acceleration by a force due to *gradient of electron pressure* produced in the cathode spot plasma. Based on this force, the models are distinguished by their interpretation.

18.2.1 Plasma Polarization and an Electric Field Formation

Zeldovich and Raizer [17] have studied the effect of plasma polarization and an electric field production during a shock-wave generation. In the expanding binary plasma, small change of ion and electron densities due to their diffusion causes generation of a high volume charges and high plasma polarization, which prevent the further separation of the charge particles. The characteristic length of quasineutrality violation is Debye radius r_d. The charge particle separation has supported thermal motion, and therefore, the potential drop is about of electron temperature. The electron gas is strongly compressed in strong jump at a length of mean free pass, which is occurred due to strong compression of the ion gas. In this case, the potential drop is determined by the compression work of the electron gas in form [17].

$$e(\varphi_2 - \varphi_1) = kT_e Ln \frac{\rho_2}{\rho_1} \qquad (18.10)$$

According to (18.10) potential difference is about electron energy.

18.2.2 Ambipolar Mechanism

A model of ambipolar acceleration of the ions was presented by Plyutto [18] for a spherically expanding plasma of the vacuum sparks and arcs. The mode considers a high rapid motion of electrons, which is bounded with the heavy ions by the Coulomb force, produced by charges separations. The plasma cloud expands due to electron gas, and the ions are radially accelerated. This collective process transfers the energy from electrons to the ions similar to plasma polarization [17] and to the ambipolar diffusion produced in a sheath at contact of plasma isolated wall [19].

Equations continuity and momentum with electron pressure p_e:

$$v\frac{\partial \rho}{\partial r} + \rho\frac{\partial v}{\partial r} + \frac{2v\rho}{r} = 0$$

$$\rho v \frac{\partial v}{\partial r} + \frac{\partial p_e}{\partial r} = 0 \qquad (18.11)$$

After conversation:

$$\frac{\partial(\rho v)}{\partial r} = -\frac{2v\rho}{r}$$

$$\frac{\partial(mv^2)}{2\partial r} = -\frac{\partial(nkT_e)}{n\partial r} \qquad (18.12)$$

After integration from r_0 to r and $n_0 v_0$ to nv assuming isothermal expansion:

$$\frac{\partial(\rho v)}{\partial r} = -\frac{2v\rho}{r}$$

$$\frac{\partial(mv^2)}{2\partial r} = -\frac{\partial(nkT_e)}{n\partial r} \qquad (18.13)$$

$$Ln\frac{\rho v}{\rho_0 v_0} = -2Ln\frac{r}{r_0} \text{ or } Ln\frac{n}{n_0} = -2Ln\frac{r}{r_0} - Ln\frac{v}{v_0}$$

$$\frac{mv^2}{2} - \frac{mv_0^2}{2} = -kT_e Ln\frac{n}{n_0} \qquad (18.14)$$

$$\frac{mv^2}{2} - \frac{mv_0^2}{2} = kT_e(2Ln\frac{r}{r_0} + Ln\frac{v}{v_0}) \qquad (18.15)$$

Taking in account $v \gg v_0$, the ion energy W_i is:

$$\frac{mv^2}{2} = W_i = 4.6\,kT_e \, Log\frac{r}{r_0} \qquad (18.16)$$

Plyutto (see formula (1) in work [18]) obtained formula (18.16). Using electron energy $kT_e = 25$ eV, $Log(r/r_0) = 2$, he explain the experimental values of W_i as

230–460 eV. It was indicated that the isothermal expansion was supported by Joule energy dissipation and due to high heat conduction of the small dense plasma cloud. The region in which the heavy ions effectively accelerated is close to the cathode [20].

Hendel and Reboul [21] has developed similar mechanism. They indicated that during the short period (~1 μs) of ion acceleration by space charge, only a negligible amount of heat could be delivered to the ions by collisions with electrons due to limited transfer of momentum. Therefore, the energy transfer was assumed to be adiabatic, and thermodynamic equation was used $p_e V_v = nkT_e$ and $p_e V^{5/3} = c$. As a result, the ion energy is:

$$W_i = \int_{V_{v1}}^{V_{v2}} p_e dV_v \tag{18.17}$$

And the ion drift velocity was deduced as:

$$v_i = \left(\frac{3kT_e}{m}\right)^{1/2} \tag{18.18}$$

Thus, this approach shows that the final ion energy being approximately equal to the electron energy. Equation (18.18) allows to explain the parabolic decrease of the ion energy from about 2×10^6 to 10^5 cm/s with the ion mass for materials Al, Sn, Zn, Sb, and Pb, respectively, from a pulsed arc with $T_e = 25$ eV. Tyulina [22] also explained the observed plasma velocity by acceleration of the ions at the front of a plasma cloud by space charge field created by the fast electrons.

18.2.3 Model of Hump Potential

Plyutto et al. [23] developed another interpretation of the mechanism of ion acceleration in vacuum arc due to gradient of electron pressure. They assume the presence of a large electron density in the cathode spot region, in which the vapor of the cathode metal is ionized and multiply charged ions are generated. The electron density n_e decreases in all directions away from the peak density, causing a potential peak, or hump, of significant value to occur in the cathode spot plasma region. The ions are accelerated then by ambipolar mechanism with hump potential that is mathematically formulated by following equations.

The equation of motion of the electrons is:

$$m_e n_e v_e \frac{dv_e}{dx} = -\frac{d(n_e k T_e)}{dx} + en_e \frac{d\varphi_p}{dx} - R_{\text{sor}} \tag{18.19}$$

Using isothermal condition, neglecting the convective terms and the ion–electron friction R_{sor}, it can be obtained

$$\frac{kT_e}{e}\frac{d(n_e)}{n_e dx} = \frac{d\varphi_p}{dx} \qquad (18.20)$$

Since the plasma density decreased in both directions to the cathode and from the cathode as plasma jet, maximal electron density n_{e0} is produced near the surface. After integration in the jet direction, the formula presented in work [23] was derived as:

$$\varphi_{p0} - \varphi_p = \frac{kT_e}{e} Ln \frac{n_{e0}}{n_e} \qquad (18.21)$$

Formula (18.21) indicated the maximal plasma potential φ_{p0} that can exceed significantly the cathode potential drop. According to the work [23] for $kT_e \sim 5\text{--}8$ eV and $Ln(n_{e0}/n) = 3$, $\varphi_{p0} - \varphi_p = 20\text{--}40$ V.

Davis and Miller [24], also postulate a mechanism for the acceleration of ions, which move under potentials greater than the arc voltage. They presented a model of hump potential with following model. The electrons gain energy from the electric field and ionize the emitted neutrals from the cathode at an average electron energy which reached a certain value. The onset of ionization produces a sharp rise in the local positive ion density. The rate of production of ions reaches a peak at some location and then falls with distance from the cathode. The decrease of ionization rate occurs because the electrons have yielded up much of their energy and also due to decreasing of the density of neutrals with distance from the cathode spot.

As the electrons have a much higher average velocity than the ions, they tend to flow out of the region of production much faster than the ions. This results in a net local surplus of positive ions, which creates a potential maximum (hump) in this region. As this potential hump increases in magnitude, the electrons are slowed and the positive ions are accelerated until the net charge loss rates of ions and electrons are equal (i.e., an ambipolar diffusion-controlled process). Thus, in the steady state, new ions are produced in location of the potential hump, and therefore, these ions must possess energies at least as great as the potential at which they were created. The authors indicated that the model of a potential hump explains their observation that the ion energy distributions of the various ions are quite similar if the energies are given in electron volts/unit charge.

18.3 Plasma Jet Formation by Model of Explosive Electron Emission

According to study of the phenomena of explosive electron emission, the speed of glow boundary motion appeared to be independent on voltage pulse duration and remained approximately constant during plasma motion within range of (2–3) × 10^6 cm/s. To describe this plasma motion, Mesyats [25] used the model of adiabatic expansion in vacuum [17]) for 1 ns after explosion. In this case, the velocity determined by the energy accumulated in the cathode protrusion derived in form:

$$v = \sqrt{\frac{4\gamma}{\gamma - 1}\eta\varepsilon_0} \qquad (18.22)$$

where γ is the adiabatic exponent, η is the metal superheat factor, and ε is the specific heat of metal sublimation. Using $\eta = 2$–5, $\gamma = 5/3$, the calculation show $v = (1$–$2) \times 10^6$ cm/s, which is in good agreement with the measured value. Litvinov [26] used Zeldovich model and (18.22) to describe the flare explosion. Also, Litvinov et al. [27] calculated this same value of $v \sim 2 \times 10^6$ cm/s for Al, Cu, Mo, and W and $v = (0.5$–$1) \times 10^6$ cm/s for Pb studying (18.21) and (18.22).

The further study [28, 29] of the explosive plasma expansion into a vacuum takes in account the influence of the electron pressure gradient, which is in accordance with the Plyutto model [18]. In this case, instead of the energy accumulated in the condensed phase, the Joule energy dissipation in the plasma plume was taken into account. As a result, an increase of the electron plasma pressure is occurred. It was assumed that the velocity of expanding plasma is due to the thermal explosion of the emission center (EC), and it is equal to the velocity of the vacuum-arc cathode jet. The final expression for the jet velocity was obtained in form [28, 30]:

$$v = \frac{2}{\gamma - 1}\sqrt{\gamma\frac{kT_i + ZkT_e}{m}} \qquad (18.23)$$

It can be seen that (18.23) describes the plasma sound speed at electron temperature. Using results of magneto-hydrodynamic calculation for exploding wires [31]. Loskutov et al. [32] studied the expanding plasma produced by explosion of an emission center. The initial stage of the explosion of a field electron emitter has been studied considering transition of the state of the material through several phase stages. 2D plasma expansion was described using combination of 1D equations of MHD flow for a cylindrical symmetry with r-axis, and 1D hydrodynamic equation for planar case in direction of z-axis coincides with the emitter direction, divided into layers along the z-axis, in which the current flows. The calculations for Cu include a number of input parameters such as tip radius, taper angles, rate of current rise, and others. According to the calculations, the velocity of plasma expansion was obtained in range of (2–3) × 10^6 cm/s, and with rate of current rise of 10^9 A/s, the explosion of the tip occurs in 1.5 ns.

18.4 Ion Acceleration and Plasma Instabilities

Some hypothesizes were published, in according to which the cathode jet formation in an arc occurs due to collective plasma processes. Aref'ev et al. [33] considered an anomaly scattering of emission of the electron beam accelerated in the space charge sheath by its interaction with the plasma oscillations. Two kinds of instabilities develop by such interaction: (i) ion-sound caused by beam motion relatively the plasma ions and (ii) beam instability caused by beam motion relatively the slow plasma electrons. The study of the instabilities conducted using gasdynamic system of equation taking into account the turbulent friction force in the momentum equation and Poisson equation for self-consistent electric field. However, the analysis conducted without accounting collisions between the charged particles, which significantly change the condition of electron beam relaxation, and which are important in the dense plasma of the cathode spot.

Aksenov et al. [34, 35] studied another mechanism of cathode plasma acceleration caused by Buneman instability. Buneman [36] initially showed that directed electron energy is dissipated in a plasma into random energy by "collective interactions" with the ions. A mechanism for the buildup of bunches from small fluctuations was described briefly by estimating that the initial electron drift can be destroyed within some tens of plasma periods. In the further detailed study, Buneman [37] showed that the growth of a local disturbance in the mentioned phenomena takes place without effective propagation. The turbulent flow pattern created under nonlinear conditions was calculated numerically, demonstrating the tendency toward randomization of the initial drift energy for wide plasma conditions. The effect stops process of "runaway" in about 100 plasma periods, after which there is "plasma heating" by "collective interaction." This instability [38] occurs when the current exceed a critical value, and then, the electrostatic field perturbations increase exponentially. The authors [34, 35] try to explain the large kinetic energy and wide energy spectrum of the ions, which were measured in the discharge with a long interelectrode gap.

Alterkop et al. [39] conducted a consistent theoretical analysis of the conditions for excitation of the Buneman instability and its influence on parameters of the cathode plasma jet. The analysis shows that the perturbation leads to the relative change of drift velocity which is 5×10^{-2} of electron temperature is 0.2 which can be understood due to Buneman instability. However, as the condition of exciting this kind of instability occurs at relatively low plasma density, this mechanism can be considered only at distance far from the cathode at which the plasma jet already was accelerated.

18.5 Gasdynamic Approach of Cathode Jet Acceleration

As the plasma pressure in the spot and in the initial part of expanding cathodic jet is significantly large (according to the experiment and calculations), the plasma flow is allowed to be considered in a gasdynamic approximation. In first, the model considered in simple form of integral equations, and then, the study of plasma flow described the distribution of plasma parameters using a system of differential equations.

18.5.1 Basic Equations of Plasma Acceleration

Let us consider the gasdynamic system of equation neglecting the influence of thermal and ionization terms for simplicity and principle analyzing of the acceleration mechanism in following form:
Equation of mass conservation:

$$\frac{d}{dx}[S_a(m_i n_i v_i + m_e n_e v_e)] = 0 \tag{18.24}$$

Momentum equations for ions and electrons

$$m_e n_e v_e S_a \frac{dv_e}{dx} = -\frac{d(S_a n_e k T_e)}{dx} + e n_e S_a \frac{d\varphi_p}{dx} - R_{sor} \tag{18.25}$$

$$m_i n_i v_i S_a \frac{dv_i}{dx} = -\frac{d(S_a n_i k T)}{dx} - Z_i e n_i S_a \frac{d\varphi_p}{dx} + R_{sor} \tag{18.26}$$

where is the ion charge state, R_{sor} is the ion-electron friction, T_e, T are the temperatures of electrons and ions, respectively. Equations (18.24)–(18.26) should be closed by a quasineutrality condition. When adding (18.24)–(18.26) and also neglecting the small terms due to $m_e \ll m$ and $T_e \gg T$, we obtain:

$$m n_i v_i S \frac{dv_i}{dx} = -\frac{d(S n_e k T_e)}{dx} + e(n_e - Z n_i) S \frac{d\varphi_p}{dx} \tag{18.27}$$

Taking into account the quasineutrality $n_e = Z n_i = n$, $v_i = v$ and $G = mnvS$, where S is the cross-sectional area of the expanding jet, after integration from the spot (index "0") up to distance x, the equation is

$$G(v_x - v_0) = -(S_x p_x - S_0 p_0) \tag{18.28}$$

As $v_x \gg v_0$, due to small velocity v_0 in the dense spot plasma with pressure p_0 and $p_0 \gg p_x$, the expansion in vacuum (18.28) is:

$$G v_x = S_0 p_0 \tag{18.29}$$

Equation of energy conservation in integral form:

$$G\frac{v_x^2}{2} = Iu_p \qquad (18.30)$$

Taking into account the parameter $\varsigma = \frac{eG}{mI}$ (see (16.23) from Chap. 16), (18.30) can be presented as:

$$\frac{mv_x^2}{2} = W_x = \frac{eu_p}{\varsigma} \qquad (18.31)$$

Equations (18.30) and (18.31) allow to calculate the velocity of accelerated cathode material eroded in plasma form. Thus, the cathode spot can be considered as a microplasma accelerator, and the value of ς is the exchange parameter which indicated the cathode erosion rate ratio to the electron current, i.e., the degree of electron energy transfer to the ions [40].

18.5.2 Gasdynamic Mechanism. Energy Dissipation in Expanding Plasma

Equation (18.31) was used by Lyubimov [41, 42] to explain the mechanism of plasma acceleration in cathode jet of a vacuum arc considering the plasma region between external boundary of space charge sheath and a boundary at which the plasma particle collisions are disappeared. The analysis shows that the jet can be accelerated due to energy dissipation current I in the considered region with plasma voltage of u_p, and with relatively small u_p ~1 V. To understand this point, let us calculate the plasma voltage u_p for cathode materials analyzed in [41] using (18.31). The results are presented in Table 18.1 mainly for the experimental data of Plyutto [23] as well as for data obtained by Tamberg [5] and Eckhardt [43, 44]. The calculation shows that the request values of u_p for jet acceleration to the measured energy are very low and are in range of 0.3–2.6 V for different cathode materials. These values of u_p significantly lower even than that measured and which can be obtained by difference between arc voltage u_{arc} and the cathode potential drop as $u_p = u_{\mathrm{arc}} - u_c$. The estimation, for example, for Ag, Cu, Al, Zn, and Mg is 4, 5, 5, 2.5, and 3 V, respectively [45].

The significant difference of calculated u_p from the measured values is not understandable and indicates a necessity of a further studies. Moreover, some estimations conducted by Lyubimov [42] indicate that the momentum equation like (18.29) cannot be satisfied using the parameters calculated for dense plasma in the cathode spot. It can be assumed that the calculated values u_p can be different from real values due to some additional energy losses appeared through the arcing. All this should be studied considering the cathode phenomena by self-consistent approach.

18.5 Gasdynamic Approach of Cathode Jet Acceleration

Table 18.1 Results of calculation of parameter ζ and plasma voltage u_p using Plyutto et al. data [23] for $I = 300$ A

Material	Arc voltage, u	Erosion plasma G × 10^{-5} μg/C	Velocity × 10^5 cm/s	Energy W, eV	A_i	Parameter $\varsigma = \dfrac{10^5 G(g/C)}{A_i}$	$u_p = \zeta W$, V
Mg, 170 A	15.0	4.2	8.8	9.5	24	0.175	1.66
Al	20.8	6	6.5	5.8	27	0.222	1.3
Ni	19.6	5	7	15	59	0.085	1.27
Cu	20.0	6.5	7.8	20	64	0.102	2.0
Ag	17.8	7.2	8.4	39	108	0.0667	2.6
Zn	12.5	16	2.3	2.2	65	0.244	0.54
Cd, 170 A	11.1	31	1.8	1.9	112	0.277	0.53
Hg	10	1	7.5	56.25	200	0.005	0.28
Cu*, I = 32A		1.53	13.1	53.6	64	0.024	1.28

The data for Hg are measurement observed by Eckhardt [43, 44]. Data for Cu* obtained by Tanberg [5]

The contradiction followed using momentum equation, stimulated Zektser and Lyubimov [46] to consider the gasdynamic system of equations to describe the distribution parameters of the plasma jet expanding from the region of cathode spot. The system equations include the particle flux, momentum, and energy fluxes conservations presented in differential form. The plasma parameters were calculated as dependencies on distance x from the cathode surface. The spot assumed as circular form, and a truncated cone approximated the jet geometry. The boundary conditions were pressure $p = 0$, density $n = n_a + n_i + n_e = 0$, electric field $E = 0$ at distance far from the cathode $x = \infty$.

The initial cross section of the jet ($x = 0$) was located at outer boundary of the space charge sheath of the cathode. The system of equations characterized the spot parameters that were taken according to Beilis model [47]. The solution for $I = 300$ A, $u_c = 15$ V, and $G = 1.6 \times 10^{-2}$ g/c is presented in Table 18.2. The author concluded about agreement between calculated and the experimental jet data. However, detailed analysis of the model and results of calculation presented by Zektser and Lyubimov [46] for Cu brings us to the conclusion that this work consists of a number of contradictions, which is important to discuss:

Table 18.2 Results of calculations obtained by the model of work [46]

Material	j, A/cm²	T_s K	n_T, cm^{-3}	α	T_e, eV	v_0, cm/s	v_∞, cm/s	u, V
Cu	3×10^4	3573	2.4×10^{17}	0.95	3.84	6.2×10^4	0.97×10^6	19

(1) First of all, estimation of ion current fraction s_i from used formula $s_i = en_T \alpha v_{iT}/4j = 0.03$, i.e., significantly low and not consist with that calculated from the spot theory.

(2) At this low s_i, the cathode temperature calculated from cathode balance (18.16 [46]) does not exceed 260 °C which $\ll T_s$ in Table 18.2. Also, the electric field at the cathode surface is also very low at this s_i and cannot influence in the cathode sheath.

(3) The plasma velocity in the spot v_0 was calculated to be significantly lower than thermal velocity, and therefore, the near-cathode pressure should correspond to the saturated pressure at the cathode temperature. However, the calculated pressure of saturated vapor at $T_s = 3573$ K [46] is about 2×10^7 bar which exceeds the pressure obtained from the calculated plasma data, $n_T k T_s = 2 \times 10^5$ bar, i.e., by two orders of magnitude.

(4) At the calculated plasma density of about 10^{17} cm^{-3}, the electron beam relaxation length is ~10^{-2} cm, while at this length the jet structure is highly non-uniform according to Fig. 3 from work [46]. It was indicated that the jet is accelerated to the sonic velocity in distance ~5×10^{-4} cm, and over the same distance, there are sharp changes in the electron temperature and the particle density. This means that the significant plasma acceleration and parameters change occur in region significantly lower than particle collisions of charge particles (~10^{-3} cm for $T_e = 3$–4 eV) or charge exchange (~5×10^{-4} cm), i.e., in almost collisionless plasma. The sound velocity was reached at distance ~5×10^{-4} cm and the measured velocity of 10^6 cm/s at distance ~10^{-2} cm from the cathode surface, i.e., inside of the cathode spot size (spot radius $r_s = 6 \times 10^{-2}$ cm).

Thus, the problem was not solved in work [46]. One of reasons of weakness in the mathematical formulation is related to the assumption of zero plasma parameters of the jet far from the cathode surface. At first, such conditions can be satisfied by different parameters occurred in the dense spot region depending on the character of plasma expanding in vacuum and location of the critical cross section. Secondly, the plasma parameters can be different from zero at significant distance from the cathode. Additional analysis of the work [46] brings to conclusion that model formulation was erroneous, and the solution consists of calculating errors.

In general case, the solutions found for the different regions such as spot and expanding plasma jet must be matched at some boundary between the regions. An attempt to study this matching was provided by Beilis [48] considering a consistent solution for system of equations which described the plasma in the cathode spot [47] together with (18.29)–(18.30). The possibility of such matching was demonstrated by an analysis of the space charge sheath, electric field at the cathode surface, and the thermophysical properties of the cathode material including the pressure of saturated vapor at the cathode temperature [48]. As a result, this analysis allows understanding the formulation of a correct boundary conditions at the cathode side as it is the main requirement defining the expanding plasma flow in the jet. Before formulation of these conditions, let us consider the published models in which a condition of sound speed of the plasma was taken at the cathode side to study the flow of the cathode jet.

18.5.3 Jet Expansion with Sound Speed as Boundary Condition at the Cathode Side

Harris [49] was first to use the sound speed as boundary conditions. He assumed sound velocity of the ions expanding from the ionization zone in direction to the anode (see Chap. 15). In contrast to the above conical case [46], Moizhes and Nemchinsky [50, 51] developed gasdynamic model using spherical approximation of the plasma jet expansion. The boundary conditions were given at hemisphere with a radius equal to the spot radius using experimental data or some arbitrary estimation. However, these conditions do not correspond to the cathode surface parameters solved with the equations for cathode body. In this case, a problem appeared, due to uncertainty between calculated and parameters given from experiment. It should be also noted that the measured parameters are determined by the experimental conditions, which cause low accuracy due to spread of the measurements. The sound speed was taken as condition in the hemisphere (spot), while it is known the plasma flow is subsonic in the spot near the cathode surface (Chaps. 16 and 17).

Wieckert [52] studied expansion of the cathode spot plasma in vacuum arc considering stationary, quasione-dimensional two-fluids model with isothermal or cold ion plasmas. The description starts outside the ionization region at constant area A_s of a cross section determined by spot radius r_s and by given function $A(x)$ in expansion direction x as

$$A(x) = \pi r_s^2 + \Omega x_s^2 \qquad (18.32)$$

The form (18.32) indicates that for large values of x, the expansion becomes asymptotically spherically symmetric in solid angle Ω. Equations (18.25) and (18.26) combined together with energy equation for electrons and ions were evaluated in order to determine the ion acceleration by the pressure gradient force. This force acts in two ways: (i) in form of electric field responding to the potential hump and (ii) via the electron–ion friction due to collisions of the charged particles. The electron pressure depends on electron temperature that is determined by Joule energy dissipation, i.e., Iu_p (similar to work [41]). The above equations were reduced to a form dependent on Mach number M describing the flow with a critical point given at the ion sound speed. The system of equations integrated taking into account that a critical point is existed.

The solution was studied for *given values* of I, j, j_i, Z and for a certain value of ratio of electron-to-ion temperature. As an example, the problem was solved for Cu cathode at $I = 20$ A, current density 5×10^5 A/cm^2, ion current of 1 A, and electron temperature $T_e = 1.5$ eV. The dependence of Mach number on distance x

was obtained with a given boundary value of M_0. It was shown that the transition of the plasma flow to supersonic regime can be realized for $M_0 > 0.5$. The calculations also show that M_0 decreased with T_e, which is used at the critical point as parameter. However, in all cases of the experimental value of super speed (10^6 cm/s), the plasma flow was reached at distance of about 1 cm where the collisions cease. This means that the Joule energy is relatively small in the region near the cathode. It reached the value requested for acceleration only at distance of 1 cm.

A more detailed analysis of above-formulated model was conducted in the next work by Wieckert [53] taking into account the ions of different charges and wider range of input parameter. The asymptotic mean ion charge state Z, ion energy $E_{i\infty}$ at $x = 1$ cm, hump potential location, and its height were studied for Cu, Al, and Ti in the region of $j = 5 \times 10^5$–10^7 A/cm^2 and current $I = 20$ and 200 A. It was shown that the Z and $E_{i\infty}$ are higher for Ti and lowest for Al and for larger I and j. The hump potential strongly increased with j and weakly depended on cathode material. The present research is useful not just for principle view of the basic dependences, but also for achieving the complete understanding of the self-consistent investigations of the spot parameters and cathode plasma expansion.

Krinberg et al. [54] considered the spherical expansion of fully ionized plasma in the supersonic region of the cathode plasma by the given critical conditions at a sphere of certain radius r_*. The system of equations like (18.24)–(18.26) and plasma energy balance were used with $T_i = 0$, taking into account the heat flux from the plasma. The solution was analyzed from the critical sphere chosen at r_* to a distance $x = \infty$ far from the r_*. A preliminary study was conducted for boundary condition at $x = \infty$ as $T_e \to 0$ and $n \to 0$ except the sound speed at r_*. The solution was provided at small region ($0.8 r_*$) in direction to the cathode and in direction of the expanding jet up to $10^5 r_*$. It was shown that some influence of the heat flux from the plasma on the electron temperature and weakly on the expanding ion energy. The plasma jet of a vacuum arc was described using additional boundary condition including the value of sphere radius r_0, (assumed equal to spot radius) solid angle Ω in which the plasma expanded, current I, the ratio of electron to the ion current γ_r, average ion charge Z, and also ratio $I/2r_0 = 10^4$ A/cm.

Using the above parameters and conditions, the calculations were conducted and a discussion with comparison to the experimental data was provided. The research can be useful regarding to the analysis of different near-cathode processes, estimation of their importance and to generally considering of the problem. However, the main deficiency is the uncertainty of the boundary conditions which most of them were given from measurements far from the cathode (γ_r, Z), and also wide scattering of data related to the spot parameters, such as spot radius, plasma density, and temperature. The assumption of zero values of the plasma jet parameters at distance far from the critical section can be not satisfied even at expanding into vacuum (see above). In addition, such condition may be not correct as they correspond to multiple and not a unique solution at the point, in which the jet is generated. Therefore, the calculations can be considered as qualitative results that sometime cannot be adequate to reflect the problem.

18.5 Gasdynamic Approach of Cathode Jet Acceleration

The above model was continued to be developed by Krinberg [55] in order to describe acceleration of different ion charges in cathodic plasma expansion. The mathematical formulation took in account the force due to gradient of electron pressure in the cathode plasma, which is heated by Joule energy dissipation. The ion heating is due to interaction with the hot electrons, which is proportional to the difference temperatures between electrons T_e and ions T and depended on ion charge state Z by standard following expression:

$$Q_T = \frac{3\langle Z^2 \rangle m_e n_e}{m \tau_e}(T_e - T) \tag{18.33}$$

The equations described the flow transition through the sound speed v_{sn} were derived taking the expression $v_{sn} = (5\langle Z\rangle Te/3\ m)^{0.5}$, i.e., also depend on Z. The resulting equations were obtained in form of dependences of Mach number M and T_e on distance (i.e., on ratio r/r_s) from the critical section at the spot radius r_s. The calculations show that M increases and are saturated from 3.42 to 3.57 when the ratio r/r_s increases from 10^3 to 10^5. The value of r_s was taken arbitrary as about 1–3 μm. Thus, even for this small value of r_s, the jet velocity increases to the measured value of 10^6 cm at distance of about 1 cm. Again, this result is consequent of assumption of electron heating by Joule energy dissipation.

According to the model [55], the main reason of deviation of each ion charge velocity from the plasma velocity is: (i) the effect of density gradient, (ii) the dependence of the accelerated rate in an electric field on difference ion charge Z_k, and (iii) the dependence of the ion–electron frictional force on Z_k^2. The calculated results show that the ion velocities of different species in low-current vacuum arcs were different by about 1%. This conclusion agreed with the measured data indicated weak dependence of the ion velocity on the ion charge [56, 57].

A similar spherical formulation of the problem was used to study the expanding cathode jet at the transition zone to supersonic flow [58]. It was described a region in the initial part of plasma flow where the charge particle density has a maximum and the ions move in the direction opposite to the plasma flow. In general, the ion flux from the plasma supports the spot existence, but the location where this flux is formed should be determined by immediately near the surface and not in the transition region, where the plasma velocity is equal or larger than thermal velocity of the ions. Also, using data (γ_r, Z) that measured far from the investigated point can bring to contradicted result.

In another work [59], the authors developed model, in which the cathode plasma parameters were estimated using data of multi-charged ion state. The cathode spot parameters were calculated using the measurements of ion charge state [60], Saha equations, and results of jet expansion indicating a relation between parameters in critical section and at external part of the jet. The procedure consists in following. First, the relation between the electron temperature T_e and ratio of I/d (d is the spot diameter), as well between plasma density n and ratio of I/d, was obtained in the critical section of expanding cathode plasma. Then T_e and n were calculated from Saha equation by substituting the measured charge-state distribution. Using these data and their above relation with I/d, the spot parameters including also the current density, spot sizes, and spot current were obtained for wide range of cathode materials in range of $T_e = 4$–10 eV, $n = 10^{20}$–10^{22} cm^2, $I = 0.5$–1 A, and $j = (1$–$30) \times 10^{12}$ A/cm^2. It should be noted, that correctness of the procedure of this research is not obvious due to assumption of freezing the charge state and T_e, as well due to number of assumptions by modeling the spherical cathode plasma expansion. Therefore, no wonder that the calculated values such as spot currents (1 A) or current density (10^9 A/cm^2) look doubtful for cathodes including Cu, Fe and for refractory materials as Ta, Nb, Mo, W (plasma density 10^{22} cm^{-3}) considering the steady-state plasma jet expansion.

Considering the published models of the cathode plasma jet, Hantzsche [61–64] summarized the remained open questions and emphasizes what is the strength of the different forces and what are their quantitative contribution to the final ion energy. In order to understand these questions, a mathematical model was developed assuming stationary spherical fully ionized plasma expansion characterized by an averaged charge state. The spot was modeled as a point source of plasma, which radially expand with radius $r \gg r_s$. The used equations are like the system (18.23)–(18.25) with dependence on r, to which added by the energy flux equations for electrons and ions. The system of equations was presented in form, which allows to estimate the contribution by electric field, by frictions due to different velocities of the electrons and ions, as well by energy transfer due to elastic electron–ion interactions, like (18.33). The system of equations was solved analytically for the jet region much larger than the critical section using an asymptotic power series. The distribution of plasma parameters and ion energy distribution were obtained from this solution using experimental data for Z, I_i/I and the asymptotic ion energy E_∞ for given value of current I. The electron and ion temperatures and ion energy distribution were calculated as dependence of parameter r/I (in range 0–10^4 μm/A). The results of these calculations were used to study the contribution of different forces in the ion acceleration considering the momentum equation of the ions in form:

$$\frac{dE_i}{dr} = ZeE - \frac{dT}{dr} - \frac{T}{n_e}\frac{dn_e}{dr} + e(v_e - v_i)Z\frac{e^2 n_e}{\sigma} + c_0 Z\frac{d(T_e)}{dr} \quad (18.34)$$

The analysis of (18.34) for copper ($E_\infty = 100$ V, $Z = 1.8$, $I_i/I = 0.08$, $c_0 = 0.71$) shows that the ion kinetic energy (left term) is increased by electric field (first term, ~30%), by the ion pressure gradient (second term ~15%, T is the ion temperature), and by the friction between electrons and ions (fourth and fifth terms ~55%). Note that, the electric field, which has the opposite direction in the plasma expansion zone, forming a potential hump (possible up to 10–15 V above the asymptotic potential in direction to the anode [64]) near the cathode spot. The percentages are indicated for the case without effect of plasma heat conduction and for $T_e \neq T$ [63]. Some redistribution of different forces was shown when taking into account the heat conduction, convection terms, and for isothermal condition. In order to investigate more carefully the dependence of the plasma parameters on arc current, the revised model [65] took in account the variability of the Coulomb logarithm and an influence of the boundary

18.5 Gasdynamic Approach of Cathode Jet Acceleration

conditions depended on spot current. The author indicated that although some influence of the new input parameter, the formerly derived results and conclusions relating to the expanding plasma remain valid with only minor numerical changes. It should be noted that the mechanism of ion acceleration due to gradient of particle pressures and friction is obviously understandable. However, the electric field mechanism of quasineutral plasma is requested to be explained.

Afanas'ev et al. [66–68] developed an analytical model of the vacuum-arc cathode plasma jet. A paraxial approximation was used to obtain the solution of hydrodynamic equation. This approach was developed for quasistationary axially symmetrical plasma jet and one-fluid isothermal hydrodynamic flow without taking in account the viscosity, heat conductivity, and radiation. The plasma jet has to pass through a critical cross section (with sound speed) using a certain relationships between the plasma parameters. The paraxial approximation means description of a narrow plasma jet. Formation of the jet boundary was assumed by the self-magnetic field. The boundary between the plasma flow and the vacuum is considered as a free one with condition: zero normal components of velocity, current density, and pressure at some boundary r. The condition of zero pressure assumed absence of a drop of magnetic field passing the boundary, i.e., absence of surface currents. Three independent parameters were given in critical cross section, which are the temperature, the average ion charge, and the potential drop between the cathode and the critical point in the jet.

A set of calculations from the critical point ($z = 0$) were present in direction to the anode, i.e., in direction of plasma acceleration, as well as in direction ($z < 0$) to the cathode. The integration of derived system of equations was conducted for given $I = 20$, 50, and 100 A and for series given with the mean ion charges and temperatures. The results of the axial values of density, temperature, and potential drop in direction to the cathode were obtained for $I = 50$ A, $T = 5.3$ V, and $Z = 3.5$. Some difficulties of comparison of the calculated distribution of the plasma parameters with the published measurements were discussed. The authors [66, 67] indicated that the model allows calculating distribution of the jet plasma parameters only at distance of several millimeters from the cathode surface, while the measurements were conducted at great distance, more than 10 cm. This same is related to the integral characteristics of the arc, in particular, using the voltage measured at low current ~30 A in a short arc. At $z < 0$, the calculations can be provided correctly only up to distance of half spot radius which is sufficiently far from the cathode surface. Although the mentioned lacks use of a number of free parameters, the model allowed to understand a general tendency of the plasma parameters dependencies, the critical cross section, the presence of a hump potential of several values of the temperature, and others.

A series of mathematical approaches to understand the arc plasma expansion was developed by Keidar et al. [69, 70] and by Beilis et al. [71]. 2D jet expansion of supersonic plasma was considered as modeling of the free boundary formed in radial direction of the expanding jet. It was studied that the arc plasma for disk electrodes and plasma flow through ring anode. It was shown that a conical shape of the jet is produced with parabolic radial profile, which depends on strength of

18.6 Self-consistent Study of Plasma Expansion in the Spot and Jet Regions

Let us consider the system of equations and the boundary conditions for self-consistent processes by modeling of the boundary spot-jet plasma and plasma flow in the accelerated region.

18.6.1 Spot-Jet Transition and Plasma Flow

Let us consider the problem taking into account the described spot models (Chaps. 16 and 17). It is known that the cathode spot is a source of a high-velocity plasma jet. Within the vaporization model for flat and smooth surface, high-intensity energy dissipation takes place in the plasma and at the surface. The near-cathode plasma flow is strongly collision dominated, and the ions in the expanding plasma capture a considerable fraction of the electron momentum and energy. The high-plasma flux is a result of the high pressure at the surface. As the energy dissipation region (i.e., zone of the electron beam relaxation) is much smaller than the spot radius, a reaction force is generated perpendicular to the cathode surface. Therefore, within the circular spot, the radial plasma gradient is smaller than the gradient in perpendicular direction. As a result, a one-dimensional flow of eroded cathode material is formed at a distance of order of spot radius from the cathode, and a jet is thus created. At increasing distance from the cathode, the plasma expands, and the velocity of the plasma increases. As justification of jet shape of the expanding cathode plasma, it can adduce the experimental fact that jet has an elongated shape [72].

The plasma is accelerated by several mechanisms [40]. Number of papers described the arc jet expansion [23, 47, 53, 55, 63]. Firstly, this is the pressure of the heavy particles near the cathode surface. Secondly, the electron pressure acts because of the difference between the electron and ion temperature. The first and the second terms are components of the thermal mechanism, in which the enthalpy of the particles is transformed into kinetic energy. The energy is acquired by the electrons in the electric field at the cathode surface and then transferred to the heavy particles by elastic collisions. Here, we should emphasize that, in the light of the above model, the boundary conditions defined by the presence of a high-intensity energy source at the cathode are of great importance in order to generate a jet, contrary to "classical" jet flow, which is determined by nozzle geometry [73].

18.6.2 System of Equations for Cathode Plasma Flow

Let us consider the problem of cathode jet issuing into vacuum in the gasdynamic approximation [74, 75]. The system of equations corresponds to an averaging of the parameters in the plane of a transverse section of the jet. Here, the distribution of the parameters assumed to be nearly uniform in the core of the section. The transition zone at the periphery of the jet is assumed narrow and is replaced by a discontinuity of the parameters, with the assumption that the pressure and the normal component of the electric field are zero on the outside of this discontinuity. For this formulation of the problem, the system of equations reduces to following differential equations [75]:

1. *Equations of continuity, momentum, and energy of heavy particle flow:*

$$\frac{d}{dx}(mn_T v F) = 0; \quad n_T = n_a + n_i; \quad T_a = T_i = T \quad (18.35)$$

$$\frac{d}{dx}(pF + Gv) = eE_{pl}F(n_i - n_e); \quad p = n_T T + n_e T_e; \quad (18.36)$$

$$2n_T v k \frac{dT}{dx} = vkT \frac{dn_T}{dx} + evn_i E_{pl} + \frac{3\varepsilon n_e}{\tau_e} k(T_e - T);$$

$$\varepsilon = \frac{m_e}{m}; \quad \tau_e^{-1} = \left[n_i v_{eT} \left(1 + \frac{n_a \sigma_{ea}}{n_i \sigma_{ei}}\right) \right] \quad (18.37)$$

2. *Electron energy:*

$$\left[\frac{3}{2}n_e v + \frac{3.2 j_e}{e}\right] k \frac{dT_e}{dx} = \frac{u_c}{F} \frac{d(j_b F)}{dx} + jE_{pl} - \frac{u_i}{F} \frac{d(n_i v F)}{dx}$$

$$- \frac{n_e k T_e}{F} \frac{d(vF)}{dx} - \frac{3\varepsilon n_e}{\tau_e} k(T_e - T); \quad (18.38)$$

3. *Ion continuity:*

$$\frac{1}{F} \frac{d_i}{dx}(n_i v F) = S_e; \quad S_e = \beta_i n_i n_a - \beta_r n_i n_e^2 \quad (18.39)$$

4. *Electron continuity:*

$$\frac{1}{F} \frac{d}{dx}(j_e F) = -eS_e + j_b; \quad j_b = j_{b0} \text{Exp}\left[-\int_0^x (n_a \sigma_i + (n_e + n_i)\sigma_c) dx\right]$$

$$(18.40)$$

5. *Equation of electric field*:

$$\frac{\varepsilon_0}{F}\frac{d}{dx}(E_{pl}F) = e(n_i - n_e) \qquad (18.41)$$

6. *Ohm's law in the plasma jet*:

$$j_e = \sigma_{el}\left[E_{pl} - \frac{1}{en_e}\frac{d}{dx}(n_e k T_e)\right] + ev n_e; \quad j = ev_e n_e - ev n_i + j_b \qquad (18.42)$$

7. *Equation of total current*:

$$I = j(x) \cdot F(x) \qquad (18.43)$$

here F is the cross section of the jet, which is always a function of the distance x from the cathode surface, j_{b0} is the beam of emission electron current density at $x = 0$, n_T is the heavy particle density in plasma jet, τ_e is the effective time collision between electron–ion and electron–atom, β_i and β_r are the coefficients of ionization and recombination, respectively, σ_i and σ_c–are the cross sections of atom ionization and Coulomb collisions, respectively, ε_0 is the vacuum permittivity, and u_i is the atomic ionization energy.

In writing the gasdynamic (18.38) and (18.40) for the electrons, it is accounted that there are two kinds of electrons, namely low-energy plasma electrons and electrons emitted from the cathode and accelerated in the space charge sheath. The beam electrons (j_b) are assumed to be monoenergetic with energy equal to the cathode potential drop. The scattered beam electrons are assumed to relax instantaneously into plasma electrons. This permits the use of the gasdynamic equations for the electrons [with allowance for the corresponding influx of energy from the accelerated beam electrons in (18.38)] everywhere outside the Knudsen layer for the heavy particles, even in the beam relaxation region, which is non-equilibrium with respect to the electron beam.

In general, the plasma near the electrodes is quasineutral, and only a small deviation from quasineutrality is possible. However, in the equations considered, the terms proportional to the space charge density were taken in account in order to have the possibility of matching the solution for the collisional region near the cathode with the solution in the region far from the cathode, where the particles move mostly without collisions. In the quasineutral ($n_e = n_i$) region, the E_{pl} can be determined from (18.42). For simplicity, only single ionization was assumed. In order to describe multi-charged ions, the system of equations should be extended by adding mass continuity equations for each higher ionized state. Estimation shows that the vapor is single ionized immediately near a Cu cathode surface [47], and the multi-charged ions appear far from the electron beam relaxation zone at a distance where the plasma density is sufficiently reduced.

18.6.3 Plasma Jet and Boundary Conditions

In formulating the boundary conditions for system (18.35)–(18.41), it is necessary to take into account the change in the parameters in Knudsen layer for the heavy particles. Under the investigated conditions, the Knudsen layer extends a distance of the order of the mean free path of an ion. Structurally, it consists of space charge sheath, smaller in extent than the mean free path of an ion, and a quasineutral outer part. Analysis of the distribution of the parameters in such layer shows that ultimately there exist a number of relations among the gasdynamic parameters of the plasma at the outer boundary of the layer, the temperature, and physical constants of the cathode, and the cathode's potential drop [76].

According to above kinetic model (see Chap. 17), the boundary conditions follow from analysis of the Knudsen layer for the near-cathode plasma. The parameters at outer boundary of this layer were obtained assuming that the degree of atom ionization at this boundary corresponds to an equilibrium at the electron temperature that satisfied the integrated energy balance in the electron beam relaxation zone (in essence, the Knudsen layer for emitted electron beam). Thus, the boundary conditions for the system of differential equations for the jet are formed at boundary 3, where the velocity distribution functions for the plasma electrons and heavy particles are in equilibrium. Six boundary conditions are needed for the six differential (35)–(40). As the equations are of the first order, values for the parameters sought should be prescribed at the left edge of the jet placed at the external side of the kinetic layer (boundary 3). There, the parameters (index "w") suffer a discontinuity, so that their relation to the equilibrium parameters (index "0") or to the other parameters at boundary 3 was derived as boundary conditions from Knudsen layer for heavy particles [75, 76]:

$$v_w = v_3 \tag{18.44}$$

$$T_w \equiv T_3 = k_T T_0 \tag{18.45}$$

$$n_{iw} \equiv n_{i3} = k_i(T_e, n_{a3}) \tag{18.46}$$

$$G_w \equiv G_3(g/s) = m(n_{i3} + n_{a3})v_3 F_3 \tag{18.47}$$

From Knudsen layer for electron beam (boundary 4, Chap. 17):

$$T_{ew} \equiv T_{e3-4} = k_e j_{e0} \tag{18.48}$$

$$j_{ew} \equiv j_{e4} = j_{e0} + j_i - j_{eT} \tag{18.49}$$

The ion current density j_i and the returned electron flux j_{eT} to the cathode are determined by the plasma density and quasineutrality condition. The kinetic system of equations is very complicated (Chap. 17), therefore the expressions for boundary conditions here expressed in simplicity in a concise form, with corresponding proportionality factors k_α. These factors, however, represent complicate functions of number of the plasma parameters (T, v, n). Each of six equations in the system (18.35)–(18.40) corresponds each of six conditions (18.44)–(18.49), respectively. The boundary condition for (18.41) is determined from this equation together with (18.42) through the jet expansion up to a point, at which the solution for the collisional region in the jet should be matched with the region, where the particles move without collisions.

The system of (18.35)–(18.43) with conditions (18.44)–(18.49) describes the x-dependence of the unknowns v, n_T, n_i, T, T_e, E_{pl}, j, j_e, and F in the jet for a given arc current I. One of simple approximation of the jet geometry in the initial weakly expanding region used specified rectilinear boundaries in form

$$F(x) = F_0 (1 + \frac{x}{r_s} tg\phi)^2; \quad F_0 = \pi r_s^2 \tag{18.50}$$

where ϕ is the divergence angle of the jet. When the plasma density reduces in a way that the radial plasma distribution in the jet volume became important, the 2D model [71] should be used in order to further consider the plasma expansion.

18.7 Self-consistent Spot-Jet Plasma Expansion. Numerical Simulation

Let us describe the results of coupled study that relates the parameters of the cathode spot, whose study is extremely complicated, to the parameters of the jet far from the cathode which have been determined experimentally with high accuracy. The method for calculating of the jet based on using boundary conditions that take in account the processes and parameters at the cathode and in the spot plasma [77]. This approach makes it possible to study the phenomena in self-consistent formulation that allows us to match the parameters at left side of the jet with the parameters of certain type of cathode spot. The jet parameters obtained in the calculations are determined, through the boundary conditions at the cathode, by processes in the cathode spot, which evidently corresponds to the physics of the phenomenon. To this end, in general, it should use the above-mentioned boundary condition that is consequent of kinetic approach. However, the common study of such conditions together with jet parameters distribution is dramatically complicated. In order to demonstrate the spot-jet self-consistent description, this study is conducted using, for simplicity, the gasdynamic model (Chap. 16) for calculating of the cathode spot parameters.

The above system of (18.35)–(18.43) can be reduced to main five ordinary differential equations for the electron T_e and heavy particle temperatures, the space charge, the flow velocity v, and degree of ionization (Chap. 16). This system should be added

18.7 Self-consistent Spot-Jet Plasma Expansion. Numerical Simulation

by equations for electron emission, electric field, metal saturated pressure, cathode energy, and electron energy balances. Also, the cathode potential drop u_c, the spot current I, and the cathode erosion rate G are given parameters in accordance with GDM. The geometry of the stream is the truncated cone in accordance with (18.50).

In addition, a condition was used which determined the transition through a sonic point ($v = v_{sn}$) at some point in the stream far from the cathode. The five equations can be transformed, so that three of them (T_e, T and v) will have a structure similar to that of the equation for reversal of response in ordinary gas dynamics

$$\frac{1}{v}\frac{dv}{dx} = \frac{C_1 - C_2}{1 - M^2}; \quad \frac{1}{T_e}\frac{dT_e}{dx} = \frac{B_2 a_2 - B_1 b_2}{1 - M^2}; \quad \frac{1}{T}\frac{dT}{dx} = \frac{T_e}{T}\frac{B_1 b_1 - B_2 a_1}{a_1 b_2 - b_1 a_2}$$
(18.51)

where:

$$C_1 = \frac{\frac{dT}{dx} + \alpha \frac{dT_e}{dx} + T_e \frac{d\alpha}{dx}}{1 + \alpha T_e}; \quad C_2 = \frac{\rho_e E_{pl} F M^2}{v G}:$$

$$a_1 = -\alpha\left(1 - \frac{1}{1 - M^2}\right): \quad a_2 = 2 + \frac{1}{1 - M^2}$$

$$b_1 = \frac{3}{2} + \frac{\alpha T_e}{\alpha T_e (1 - M^2)} - j_e \frac{mF}{e\alpha G}\left(4.3 - \frac{\alpha T_e}{(\alpha T_e + T)(1 + M^2)}\right);$$

$$b_2 = \frac{T}{(\alpha T_e + T)(1 + M^2)}\left(1 + j_e \frac{mF}{e\alpha G}\right)$$

$$B_1 = -\frac{d\alpha}{dx}\left(1 - \frac{1}{(1 - M^2)}\right) + \frac{1}{F}\frac{dF}{dx}\left(\frac{T}{T_e} + \alpha\right) - j_e \frac{A_{el} m F}{T_e G} - \frac{e\alpha}{T_e}\left(\frac{I}{F} - j_e\right)$$

$$B_2 = -\frac{d\alpha}{dx}\left[-\frac{T_e}{(\alpha T_e + T)(1 - M^2)} - \frac{u_i}{\alpha T_e} + \frac{mF}{e\alpha G}\left(\frac{1}{\alpha} - \frac{u_i}{\alpha T_e}\right)\right]$$

$$+ \frac{1}{F}\frac{dF}{dx}\frac{mF}{e\alpha G}\left(\frac{I}{F} - j_b\right) - \frac{j_e m F}{\alpha G \sigma_{el} T_e}\left(\frac{I}{F} - j_b\right)$$

$$- 1.8 \frac{m}{e\alpha}\frac{d(j_e F)}{dx_e} + \frac{A_{el} m F}{\alpha G T_e} - \frac{u_c m}{e\alpha G T_e}\frac{d(j_b F)}{dx_e}$$

There A_{el} is the term characterizing the exchange of energy between electrons, α is degree of atom ionization and heavy particles in elastic collisions [see (18.33) and in (18.37)]. The fourth and fifth equations, i.e., those for the ion component and the space charge density, do not have any special behavior at the sonic point. It is known that the sonic point is located at some distance from the cathode, and then, the numerators and denominators of (18.51) must go to zero simultaneously. An analysis shows that under these conditions, this requirement leads to two conditions on the parameters at sonic point:

$$M = 1; \quad C_1 - C_2 = 0 \tag{18.52}$$

The equation of motion (18.51) is reduced to a relation for velocity v that has two branches corresponding to subsonic $v = v_1(x)$ and supersonic $v = v_2(x)$ flows. We have solved the Cauchy problem in the numerical calculations with a subsonic branch $v = v_1(x)$ at the initial cross section and an electron temperature T_{e0} chosen such that at some point of the stream, with increasing distance x from the cathode the sonic transition condition (18.52) is satisfied with a given degree of accuracy. The location of this point is initially unknown and is determined in the process of solving the problem. The subsequent calculations were done along the supersonic branch $v = v_2(x)$ in accordance with measured value of the cathode erosion rate. At sonic point, it is $v_1 = v_2 = v_{sn}$ in accordance with condition (18.52).

Calculations have been done for a copper cathode with $u_c = 15$ V, an aperture angle of the stream of $\phi = 45°$, and $G = 100$ μg/C. The present calculations provided for developed spot when the cathode potential drop decreased to the low experimental value (see Chap. 17). The group spot is considered, and therefore, the spot current is $I = 100$ and 300 A. Although such spot consists of fragments, we assume that the average parameters of mixed jet not significantly differ from that for the single jets due to close distribution of the fragments (Chap. 7). Some difference can be expected due to smaller fragment current and nevertheless use the considered model here to substantially simplify study of the group spot including mixing process of the single plasma jets.

The dependences of the jet parameters were plotted along the jet on dimensionless parameter x/r_s (r_s is the spot radius) with the origin at the left hand side of the quasineutral region [77]. The jet velocity increases (passing the sound speed) within a region of size comparable to r_s and asymptotically reaches the experimental value (10^6 cm/s) over distances on the order of the 2–3 of spot diameters (Fig. 18.2).

The heavy particles density n_T varies mainly over distances greater than the spot size (Fig. 18.3a). Near the spot ($x < r_s$), n_T varies by slightly more than an order of magnitude (Fig. 18.3b). The temperature distribution (Fig. 18.4) has a maximum. T_e is relatively large already at the spot region, and therefore, this value of T_e and large

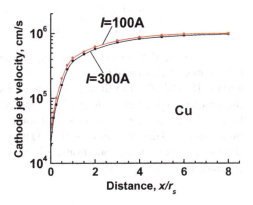

Fig. 18.2 Plasma velocity distribution along the expanding cathode plasma jet

18.7 Self-consistent Spot-Jet Plasma Expansion. Numerical Simulation

Fig. 18.3 **a** Distribution of the heavy particle density along the jet corresponding to the logarithmic scale of the length with the origin at the point $x/r_s = 1$. **b** Distribution of the heavy particle density along the jet corresponding to the standard scale of the length with the origin at the point $x/r_s = 0$

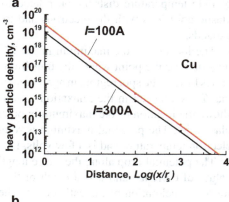

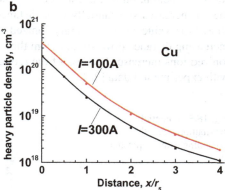

Fig. 18.4 Electron T_e and heavy particles T temperature distributions along the expanding plasma jet

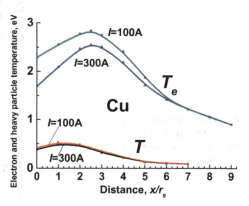

spot plasma density mainly (large spot plasma pressure) contribute in the formation of the accelerated jet at relatively low distance from the cathode surface. For the electrons, the peak is evidently by the increased contribution of gradient and convective terms to the corresponding energy equation. The peculiarities of the heavy particle temperature distribution related to the transfer of kinetic energy through elastic collisions with the electrons and to the conditions under which the stream expands.

The location of the maximum in the T_e distribution depends on the current and may lie near the point where the transition through the sonic speed takes place. T_e varies little in the spot region, in agreement with the earlier [78, Chap. 16] assumption that T_e is constant in the relaxation zone for electron emission beam. Figure 18.5 shows the location of the maximum in potential distribution along the length of the plasma jet. The potential maximum is roughly the same as that of the peak in the electron temperature and is a few volts, i.e., of the same order as T_e.

The potential drop along the entire length of the jet is 8–10 V for the considered values of currents, and it is a result of the plasma expansion, but it is not a cause for any acceleration of the cathode jet. The degree of ionization α of the vapor in the jet increases substantially over distances about the spot diameter. Figure 18.6 shows the value of α for singly ionized atoms. The calculations show that with increasing distance downstream from the cathode, the fraction of doubly and triply ionized ions increases because of the reduced heavy particle density in accordance with experimental data [23].

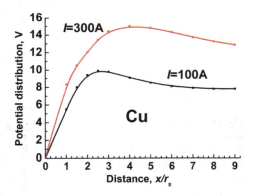

Fig. 18.5 Potential distribution along the expanding cathode plasma jet

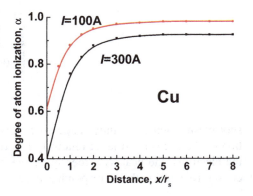

Fig. 18.6 Variation in the degree ionization of cathode vapor along the expanding plasma jet

18.8 Anomalous Plasma Jet Acceleration in High-Current Pulse Arcs

In this section, we will analyze the phenomena of large jet velocity in experiments and modeling the mechanism explained the measurements accounting the high rate of current rise

18.8.1 State of the General Problem of Arcs with High Rate of Current Rise

Recently in short pulsed high-current vacuum arcs [79, 80] and earlier for sparks [18, 20], a considerable increase of the ion energy in comparison with longer duration low-current arcs has been observed. The measured ion energy range in microsecond arcs with current $I > 1$ kA and rate of current rise $>10^8$ A/s is 1–10 keV (*high energy*), while for $I \sim 100$ A millisecond arcs, it is only about 100 eV (*low energy*) [20, 23, 24, 81]. Let us consider the acceleration mechanism of high-energy cathode plasma.

A mechanism of collective acceleration of *high energy* ions by electrons in the expanding plasma was proposed in [18]. Krinberg and Paperny [80] explain the *high energy* ions by an additional acceleration of the expanded plasma jet from the spot at some distance from the cathode due to magnetic plasma pinching. These authors [80] indicated that the supersonic flow of the cathode plasma is formed at distance about 2–3 mm and more. An influence the self-magnetic field was studied in theirs related works [82, 83]. While the self-magnetic field affects the jet geometry, Alferov et al. [84] concluded that self-magnetic pinching decelerates the plasma flow in the cathode jets.

Therefore, it is important to understand a possibility of the origin of high-energy ions. Let us consider the problem using gasdynamic mechanism of plasma acceleration, taking into account experimental specifics according to which the spot phenomena at high rate of current rise. Below a respective physical model and its mathematical formulation can explain the origin of the high-energy ions observed in vacuum arc with a high rate of current rise. The model will take into account the multi-charged ions generated in the spots. The influence of the spot parameters (cathode erosion rate, current, and cathode potential drop) on the ion energy and their dependence on the rate of current rise will be calculated and discussed.

18.8.2 Physical Model and Mathematical Formulation

Beilis [85, 86] developed the physical and mathematical approach. Let us consider this approach and take in account that the plasma accelerates in the cathode jets is due to plasma pressure gradient and by energy transfer from the electrons to the heavy plasma particles by collisions. The plasma heating is determined by the electron current fraction in the spot and Joule energy dissipation in the jet. Both processes depend on the spot parameters (e.g., current, cathode potential drop, plasma density, and temperature) and obviously on the rate of current rise dI/dt. The arc voltage can sufficiently increase with dI/dt [87]. To demonstrate the principle mechanism of unusual acceleration in the present analysis, we will use the kinetic (for heavy particles [88]) model of the cathode spot, modified for the particular features of a spot with high dI/dt (> 100 MA/s). The following assumptions: a circular spot was located on a smooth surface. Near the hot surface of the spot, highly ionized electrode vapor plasma exists, whose structure consists of the following partially overlapping regions: (1) a ballistic zone comprising a space charge sheath region, (2) a non-equilibrium kinetic plasma region, (3) an electron relaxation region, and (4) a plasma acceleration region (see Chap. 17). According to this model, the difference of heavy particle fluxes formed near the surface can be obtained that determines the velocity of the mass flow and the net cathode evaporation rate Γ. The ratio of this net cathode evaporation rate to the total evaporated mass flux from the cathode determines the cathode mass loss fraction $K_{er} = \Gamma/W(T_s)$.

In the relaxation region, the atoms are ionized by electrons emitted from the cathode as well as by plasma electrons with temperature T_e. The plasma could be sufficiently heated producing the ions with multi-charged state. The plasma is collision dominated, and its behavior can be treated by gasdynamic approach. Plasma quasineutrality requires that $n_e = \sum Z n_{iz}$, where n_{iz} is the density of ions with charge Z = 1, 2, 3, 4 and n_e is the electron density. The plasma energy flux, taking into account the ionization energy loss for ions, with different ion charge number is expressed as

$$n_e v \left[2T_e + \sum_1^Z \left(\frac{f_z}{Z} \sum_1^Z u_{iz} \right) \right]$$

where u_{iz} is the energy of ionization of ions with charge Z, $f_z = j_{iz}/j$ is the current density fraction for ions with charge Z, v is the plasma mass velocity in the spot region, and j is the spot current density.

The cathode body is heated ohmically and by the incident ion and electron fluxes from the near-electrode plasma. In order to understand the cathode energy balance for large dI/dt, let us consider the spot current change during the spot life. The incoming heat flux to the cathode surface changes in time and depends on dI/dt. We consider the case when the current rise time is comparable with the spot lifetime. The non-stationary heat conduction equation with a time-dependent heat source $q(x,y,t)$ = $f(x,y,dI/dt)$ should be solved in order to obtain the cathode body temperature. The term in the right side of this expression is the time-dependent distribution function of the heat flux onto the cathode surface $z = 0$, and x and y are Cartesian coordinates in the directions parallel to the cathode surface.

We assume that the spot current density j is constant throughout the spot lifetime, i.e., the current increases by the spot area increasing proportionally to the spot current. Assuming also, a Gaussian spatial distribution of the heat flux, and that $f(x,y,dI/dt)$

18.8 Anomalous Plasma Jet Acceleration in High-Current Pulse Arcs

is a linear function of time t, i.e., spot current $I = t(dI/dt)$ where dI/dt is constant, the solution of the heat conduction equations can be presented:

$$T_s(0,0,0,t) = \frac{u_{ef}}{2\pi \lambda_T} \sqrt{\frac{t}{\pi k_t} \frac{dI}{dt}} \left[\frac{dI}{dt} \frac{1}{4\pi k_t j} - 1 \right]^{-1} \qquad (18.53)$$

where $T_s(t)$ is the time-dependent cathode temperature in the spot center, k_t is the thermal diffusivity, λ_T is the thermal conductivity, and u_{ef} is the effective cathode voltage [89] and is dependent on the cathode potential drop u_c, ionization energy, work function φ, and cathode erosion rate $G(g/C)$. The ion energy flux toward the cathode is taken as

$$j_i \left[u_c + \sum_1^Z \left(f_z \sum_1^Z u_{iz} - f_z Z \varphi \right) \right] \qquad (18.54)$$

A system of four Saha equations was used to calculate the density of four types of ions n_{iz} with $z = 1$–4 according to [90]. The ion current density j_{iz} is calculated assuming that all of the ion velocities toward the cathode are equal. The plasma pressure in the spot p_s reaches a large value, 10–100 atm [90]. At increasing distance from the cathode, the plasma expands due to plasma pressure, and the velocity of the plasma increases. The plasma is accelerated by the mechanisms described and discussed above. These are components of the thermal mechanism, in which the enthalpy of the particles transformed into kinetic energy. As the goal of the model is to show the possibility of plasma accelerating in arcs with large dI/dt, instead of taking the complicated equations in differential form, the plasma acceleration was calculated using these equations in their integral form. The jet momentum equation and equation of energy conservation determine the ion velocity and its energy W_i. The simple expression in an integral form was used as [75]:

Momentum conservation:

$$p_{x=0} F_{x=0} + Gv = p_j F_j + G V_j \quad F_{x=0} = \pi r_s^2 \qquad (17.55)$$

Energy conservation:

$$Gv_{x=0}^2 + 3I \frac{T_{e,x=0}}{e} = Gv_{x=0}^2 + 2I \frac{T_{ej}}{e} + I u_{pl} \qquad (17.56)$$

where m is the ion mass, j is the spot current density, V_j is the plasma jet velocity far from the cathode, u_p is the plasma jet voltage, and $p_{x=0}$ is the plasma pressure calculated from the spot theory.

The system of equations for the Cu cathode spot (Chap. 16) and using kinetic approach for heavy particles flow in the Knudsen layer [Chap. 17, 88, 91] were solved together with (18.53)–(17.56), to calculate T_e, plasma density n, T_s, j, and other spot parameters. For this formulation, the input parameters are the cathode potential drop u_c and spot current I, or rate of spot current rise. Also, the jet velocity and the ion energy can be calculated when the erosion rate is given, and vice versa, the erosion rate can be calculated when the velocity or the jet energy is a given parameter. Below, we consider the results associated with the influence of the spot parameters on plasma acceleration as a function of dI/dt.

18.8.3 Calculating Results

It was experimentally observed [79, 80] that arc voltage oscillation increased from 30–50 V up to few hundred volts and the spot current can be also changed with dI/dt. However, mostly the high voltage was dropped not in the cathode spot plasma but rather across the cathode–anode gap outside the spot, and the voltage fluctuations generate new spots, which support the arc current rise with time.

Therefore, the calculations were provided for wide regions of I and u_c in order to understand the character of ion energy dependence on spot parameters, and here, the possible plasma acceleration mechanism in the spot region is considered with relatively low $u_c < 50$ V, which is assumed to increase with dI/dt. The dependence of the jet velocity on I for $dI/dt = 1$ GA/s and $u_c = 20$ V is presented in Fig. 18.7. It can be seen that the jet velocity increases linearly with spot current up to $I = 1000$ A, and then, this dependence is weaker. For $I \approx 100$ A and $G = 100$ μg/C, the velocity approaches the low-current experimental data obtained by Davis and Miller [24]. The velocity reaches 6×10^6 cm/s with $I = 1000$ A and $G = 70$ μg/C.

Increasing cathode potential drop also increases the plasma jet acceleration (Fig. 18.8). With $G = 70$ μg/C, the jet velocity increases by a factor of ~1.5 when the cathode potential drop increases from 15 to 30 V and then tends towards saturation with $u_c > 30$ V.

The calculation shows (Fig. 18.9) that plasma velocity increases at a distance from cathode of about few spot radii from 10^6 to 10^8 cm/s when G decreases from 100 to 1 μg/C for $I = 20$ A and $t = 10$ ns, i.e., the plasma can be significantly accelerated and V_j is sensitive to variation of G.

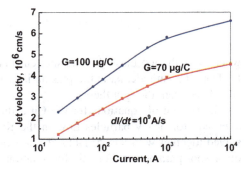

Fig. 18.7 Jet velocity V_j as a function of spot current I, with the cathode erosion rate as a parameter

18.8 Anomalous Plasma Jet Acceleration in High-Current Pulse Arcs

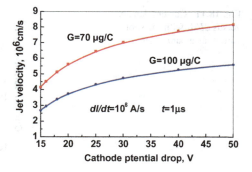

Fig. 18.8 Cathode jet velocity as a function of the cathode potential drop, with the cathode erosion rate as a parameter

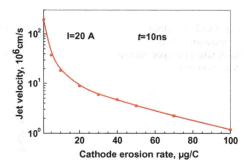

Fig. 18.9 Cathode jet velocity as a function of the cathode erosion rate

Thus, the erosion rate substantially affects the plasma jet acceleration and the ion energy. It is important to determine the region of G that corresponds to the measured ion energy values for different spot currents. To accomplish this, G was calculated using measured ion energy dependence on dI/dt data [79, 80]. The results of this calculation for $I = 0.02$–1 kA and $u_c = 20$ V are presented in Fig. 18.10. It can be seen that the cathode erosion rate decreases from about 300 to 10 μg/C when ion energy increases from 0.1 to 20 keV. Note that for this region of ion energy, the value of dI/dt increased from 0.01 to 10 GA/s, respectively [79, 80].

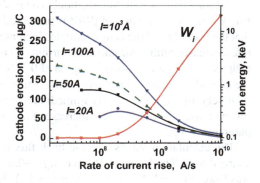

Fig. 18.10 Calculated cathode erosion rate (left axis) and measured ion energy [79, 80] (right axis) as functions of dI/dt for spot current 0.02–1 kA

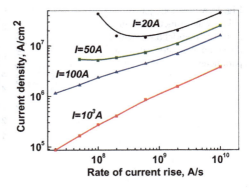

Fig. 18.11 Current density versus dI/dt with the spot current as a parameter

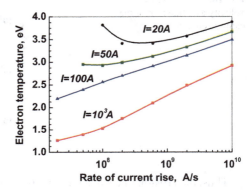

Fig. 18.12 Electron temperature versus dI/dt when the spot current is a parameter

Figure 18.11 shows j as function of dI/dt. The current density j increases from 10^5 to 5.10^7 A/cm^2 with dI/dt in the considered region of spot currents. The increase of j rises the electron temperature T_e. The dependence of T_e on dI/dt is presented in Fig. 18.12. It can be seen that the electron temperature dependence is similar to the current density dependence. T_e does not exceed 4 eV for $u_c = 20$ V in the considered spot current range.

However, the calculation shows that T_e reaches about 8 eV for $dI/dt = 1$ GA/s and $t = 10$ μs when the cathode potential drop is about 40–50 V. The cathode mass loss fraction K_{er} as a function of dI/dt is presented in Fig. 18.13. These results indicate that the back flux of the heavy particles increases, and therefore, the cathode erosion rate decreases with dI/dt and spot current. It can be seen (18.13) that the calculated cathode erosion mass flux is in region of 0.5–0.05 of total mass flux evaporated from the cathode. The heavy particle returned flux is large ($K_{er}<1$) which is high due to relatively high plasma density immediately near the cathode surface. This returned flux increases with dI/dt due to increase of the plasma density in the above-mentioned parameter range.

The calculation shows that the back flux consists mainly of singly and doubly charged ions. The ion current fraction f_1 ~0.6-0.7, f_2 ~0.4–0.3 and maximum of f_3 was lower than 10^{-3} from ions with $z = 1, 2, 3$, respectively ($u_c = 20$ V).

Fig. 18.13 Cathode mass loss fraction K_{er} as a function of dI/dt, with the spot current as a parameter

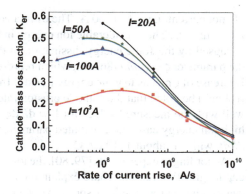

Fig. 18.14 Singly charged ion current fraction as a function of dI/dt with the spot current as a parameter

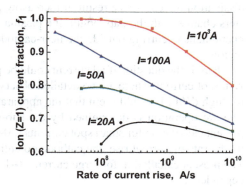

Figure 18.14 shows f_1 as a function of dI/dt. The ion flux towards the cathode mostly consists of singly charged ions for relatively small dI/dt ($< 10^8$ A/s) and large spot current, and this flux substantially decreases with increasing dI/dt and decreasing spot current. The degree of ionization α in the near-cathode plasma is in range of 0.8–0.85, and the electron current fraction s is 08–0.9 for $I \leq 100$ A. These values decrease to α ~0.7–0.8 and s ~0.6–0.7 for $I = 1$ kA. The plasma jet voltage u_p (for spot current >100 A) increases from about 10 to ~100 V when dI/dt increases from 10^8 to 10^{10} A/s in accordance with experiment. The heavy particle density is in region 10^{20}–10^{21} cm^{-3}.

18.8.4 Commenting of the Results with Large dI/dt

The model demonstrates that the observed high-energy ions can be obtained by calculating the pressure in the near-cathode region and the plasma energy balance in the case of a high level of dI/dt considering the integral conservation equations. The present calculations also demonstrate a possible mechanism for the observed high-energy plasma acceleration, and therefore, they were performed for a wide region of spot current, i.e., 20–1000 A. The estimated spot sizes are about 0.2–0.01 cm for large currents. The considerable ion energy in the cathode plasma jet (at large dI/dt) is caused by the Joule energy dissipation in the jet volume and by the decrease of the parameter $\zeta = eG/m$ [86, 88], which represents the ion flux-to-electron flux ratio. Decrease of ζ means that the energy transfer from the electrons to the ions increases. It should be noted that the cathode erosion rate calculated for low ion energy agrees well with the measured values for low dI/dt, which is ≤ 100 μg/C [23], and in the high ion energy case, the calculated parameter G is in the region measured for high dI/dt, which is about 1–10 μg/C.

According to experiments [79, 80], the ion energy increases with the interelectrode gap length. Two reasons could explain the measured results: (i) the cathode erosion can be changed as result of spot type changing with gap length; (ii) Joule energy dissipation increases as a result of increasing gap voltage. Thus, as the spot parameters change with the gap length, the plasma acceleration depends not only on the phenomena occurring outside the near-cathode region, but also in the near-cathode region.

There is the question if all current could be passing through one spot at used large values of current. There are no detailed observations of the cathode region structure for high dI/dt. It would seem that the appearance of a group spot is most probable. The arc operation is determined by the sub-spots in the group. As shown in the present calculation for small spot currents (20–100 A), the jet velocity is determined by the rate of current rise, the cathode potential drop, and the cathode erosion rate. The present calculation for large currents (~1 kA) just indicates the tendency of the dependence.

Another question appears when the large value of spot current used as a parameter of the model. The pressure due to the self-magnetic field P_m increases with I. Nevertheless, an estimation shows that for large currents, the ratio $P_m/P_k \sim 10^{-2}$, i.e., remain small, where P_k is the kinetic pressure in dense plasma of the spot.

18.9 Summary

An unusual electrode-vapor high-velocity flow is one the first phenomenon discovered while studying the electrical arcs. The cause of such flow was for first explained by the effect of force observed by repulsion of the electrodes. The consistent experimental studies (including Tanberg's 1930 measurements) showed production of the super sound plasma velocity in form of jets. The first hypothesizes subsequently presented to explain super sound plasma velocity were based on ion accelerating in region of the cathode potential fall region or accounting of the potential energy of multiply charged ions producing high ion pressure. An important assumption of large electron pressure $p = nkTe$ in a cathode spot plasma was provided for first, by Tonks to explain a nature of the cathode force. There a mechanism should be mentioned, according to which a contraction zone was produced due to self-magnetic axial pressure, and which increase an axial transport of plasma forming a plasma jet.

18.9 Summary

However, the preliminary models consisted of number of shortcomings, and the phenomenon remained not understandable. The further modeling of the plasma acceleration was developed using different mechanisms of the cathode forces. In essence, the nature of the forces comes to action of a force due to *gradient of electron pressure* produced in the cathode spot plasma. Based on this force, the models are distinguished by their interpretation. There are ambipolar mechanism due to high-speed motion of electrons, which bounded with the heavy ions by the Coulomb interaction, plasma polarization, and an electric field producing, presence of a the peak plasma density, causing a potential peak, or hump potential, of significant value which decreased in the anode direction accelerating the ions in the cathode plasma region.

As the plasma pressure in the cathode region is significantly large, the further development of plasma accelerating was based on a gasdynamic approach. In essence, this approach is based on the above-mentioned mechanisms (plasma pressure and charge particle friction), but the mathematical description become exact formulation using respective system of equations which expressed the mass, momentum, and energy conservation law. The preliminary estimations using the equations in an integral form showed that gasdynamic mechanism of plasma acceleration could explain the measured cathode jet velocity. The question is what is the value of jet voltage u_p and the length of the expanding plasma is requested to that jet to be accelerated. Another question is related to a possibility to describe the jet expansion coupled with cathode spot processes taking into account the plasma velocity transition to the sound speed through a critical cross section of the flow.

As a solution of this problem is very complicated, a number of authors considered the problem of solving the plasma flow beginning from the critical cross section using the ion sound speed as boundary condition at the left side of the jet. The numerical investigations (Wieckert) show that the typical length is relatively large, about 1 cm, at which the jet velocity become value comprised with that, obtained experimentally. A fundamental point related to calculating the parameters of the jet from the critical cross section rather than from the cathode spot should now be noted. The transition condition together with the existing relations for the current density, electric field, pressure, and heavy particle density form a set of equations at sonic point, which are not enough for determining all requested variables. Three parameters (arbitrary) requested are to be specified at the sonic cross section. In general, this way to choose and specify these parameters is incorrect. It is since they could not be independent, and moreover, they are determined by the conditions in the cathode spot. Even if we simplify the problem and consider a fully ionized isothermal plasma, the number of unknowns differs from the number of available formulas. The situation is actually more complicated, since the conditions not only contains the parameters of the problem but also their derivatives, which must also be specified at sonic point. This circumstance makes it impossible to use these conditions for a unique determination of the parameters at the sonic cross section separately from an analysis of the conditions on the cathode surface. Therefore, in general, it is impossible to determine the parameters of the cathode spot using calculations of the jet downstream from the sonic cross section without arbitrarily specifying a number of parameters at this cross section. Evidently, this applies equally to calculations of the supersonic region of the jet. In order to fully describe the problem, a self-consistent approach for plasma expansion in the spot and jet regions was formulated [85].

The peculiarity of this formulation consists in description of the boundary conditions determined by jump of the gasdynamic parameters at the cathode plasma transition in frame of kinetic model of the spot (see Chap. 17). The processes occurring in the Knudsen layer serving as the boundary conditions studied a substantial effect on gasdynamic flow of the cathode-ionized vapor. As the full formulation is very complex, for the first step, the investigations were provided using the cathode spot parameters as boundary condition calculated for simplicity in frame of GDM. A numerical simulation for long life group spot allow obtain the distribution of the heavy particle density, plasma velocity, electron temperature, electrical potential and degree of vapor ionization along of the expanding jet.

The analysis shows that the jet velocity reaches its measured value at a distance of about 1–2 spot radius, and the heavy particle density is about 10^{20} cm^{-3} and decreases by two orders of magnitude at distance of about three spot radii from the cathode surface. The electron temperature and electrical potential both have maximums. The location of the potential maximum is roughly the same as that of the peak in the electron temperature (both at about three radii of spot from the cathode surface), and it is only a few volts, i.e., of the same order as T_e.

In the considered model, a circular spot with a Gauss heat flux distribution is used. This approximation is easy to use in order to obtain an analytic solution for the cathode temperature distribution. It should be noted, however, that the calculated cathode temperature in the spot center differs weakly from the temperature calculated assuming uniform heat flux in the spot when the self-consistent cathode spot model (Chap. 3) is used. As found previously [47], the spot plasma consists mainly of singly and doubly charged ions, because of the large plasma density (10^{20}–10^{21} cm^{-3}) in the spot. The experimentally observed multi-charged ions are due to thermal ionization occurring during the plasma expansion. With increasing distance from the cathode, the plasma density significantly decreases, while T_e slightly changes causing multiple ionizations (Fig. 18.4). The typical distance in which the multi-charged ion fraction becomes significant is about 3–5 of spot radii, i.e., in the plasma acceleration region. The plasma jet expansion was described by gasdynamic equations in integral form, together with calculated model for plasma in the cathode spot region used to solve.

It was applied for understanding of the anomalies plasma expansion in arcs with high rate of current raise dI/dt taking into account generation of the multi-charged ions. The measured high ion energy of 10 keV in arcs with high dI/dt of 0.1–10 GA/s could be calculated using gasdynamic approach of plasma acceleration. The plasma jet acceleration is due to large plasma pressure produced in the cathode spot region. The mechanism related to measured parameters could be explained as a result of a decrease of the ion to the electron flux ratio (parameter ζ) produced in the spot with increasing dI/dt. The heavy particle back flux to the cathode surface consists mainly of singly and doubly charged ions.

The above formulation is relevant for plasma expansion into vacuum ($p = 0$). In the case of expansion in an interelectrode plasma or ambient gas of nonzero but small pressure, the flow can remain supersonic, but the ambient pressure causes a change in the plasma parameter distributions along the jet and in the energy production. This leads to a change in the electric current and voltage of the jet. At higher ambient pressures, however, the strong disturbance leads to submerged flow regime. Two-dimensional and effusion effects are now important. At a certain pressure, the flow will be subsonic, characteristic of the high pressure arc.

Conclusion. As the density and electron temperature in the spot plasma is very large, the main mechanism of the jet formation and acceleration is by the plasma pressure gradient of the spot produced due to the intense energy dissipation in the electron beam relaxation zone l_r (Chaps. 16 and 17) by emitted energetic electron beam indicated by Beilis [48, 92] studied this mechanism. The calculated *"potential hump," cannot be a cause of an electrostatic mechanism of the plasma acceleration, but rather is only the result of a plasma expansion* in which the electron pressure is larger than the heavy particle pressure [89, 93]. However, at distance $r > l_r$ and $r \geq r_s$ (T_e and u_p pass the peak), the jet continues to be accelerated by the plasma pressure gradient by energy Iu_p [41, 53, 63, 64] produced a plasma electric field, a transfer energy and friction by $T_e > T$, which that role was estimated by Hantzche. As the energy Iu_p can be relatively not large in vicinity of the cathode surface, the plasma jet acceleration by this mechanism to the velocity reported in the experiment occurred at significantly larger distance from the cathode surface (~1 cm) showed by Wieckert.

References

1. Lafferty, J. M. (Ed.). (1980). *Vacuum arcs. Theory and applications.* New York: Wiley.
2. Boxman, R. L., Martin, P. J., &. Sanders, D. M. (Eds.). (1995). *Handbook of vacuum arc science and technology.* Park Ridge, New Jersey: Noyes Publications,
3. Anders, A. (2008). *Cathodic arcs: From fractal spots to energetic condensation.* Springer.
4. Keidar, M., & Beilis, I. I. (2018). *Plasma engineering* (2nd ed.). London, New York: Academic Press, Elsevier.
5. Tanberg, R. (1930). On the cathode of an arc drawn in a vacuum. *Physical Review, 35*(9), 1080–1090.
6. Compton, K. T. (1930). An interpretation of pressure and high velocity vapor jets at cathodes of vacuum arcs. *Physical Review, 36*(4), 706–708.
7. Compton, K. T. (1931). On the Theory of the Mercury Arc. *Physical Review, 37*(9), 1077–1090.
8. Risch, R., & Ludi's, F. (1932). Die Entstehung des Strahles schneller Molekiile an der Kathode eines Lichtbogens. *Zeitschrift für Physik, 75*(11/12), 812–822.
9. Tonks, L. (1934). The pressure of plasma electrons and the force on the cathode of an arc. *Physical Review, 46*(9), 278–279.
10. Kobel, E. (1930). Pressure and high velocity vapor jets at cathodes of a mercury vacuum arc. *Physical Review, 36*(11), 1636–1638.
11. Finkelnburg, W. (1948). A theory of the production of electrode vapor jets by sparks and arcs. *Physical Review, 74*(10), 1475–1477.
12. Haynes, J. R. (1948). The Production of High Velocity Mercury Vapor Jets by Spark Discharge. *Physical Review, 73*(8), 891–903.
13. Robson, A. E., & von Engel, A. (1957). An explanation of the Tanberg effect. *Nature, 179*(4560), 625–625.

14. Von Engel, A., & Robson, A. E. (1957). The excitation theory of arcs with evaporating cathodes. *Proceedings of the Royal Society of London. Series A. Mathematical and Physical Sciences*, *243*(1233), 217–236.
15. Maecker, H. (1955). Plasmastromungen in lichtbogen infolge eigenmagnetischer kompression. *Zeitschrift für Physik, 141,* S198–S216.
16. Ecker, G. (1961). Electrode components of the arc discharge. *Ergebnisse der exakten Naturwissenschaften, 33,* 1–104.
17. Zeldovich, Ya. B., & Raizer, Yu P. (1966). *Physics of shock waves and high temperature hydrodynamic phenomena.* New York: Academic Press.
18. Plyutto, A. A. (1961). Acceleration of positive ions in expansion of the plasma in vacuum spark. *Soviet Physics JETP-USSR, 12,* 1106–1108.
19. Granovsky, V. L. (1952). *Electrical current in a gas.* Moscow: Gostechisdat. (In Russian).
20. Korop, E. D., & Plyutto, A. A. (1971). Acceleration of ions of cathode material in vacuum breakdown. *Soviet physics - Technical physics(Engl. Trans.), 15: No. 12, 1986–9 (Jun 1971).*
21. Hendel, H. W., & Reboul, T. T. (1962). Adiabatic acceleration of ions by electrons. *Physics of Fluids, 5*(3), 360–362.
22. Tyulina, M. A. (1965). Acceleration of ions in a plasma formed by breaking a current in a vacuum. *Soviet Physics—Technical Physics 10,* 396–399.
23. Plyutto, A. A., Ryzhkov, V. N., & Kapin, A. T. (1965). High speed plasma streams in vacuum arcs. *Soviet Physics JETP, 20*(2), 328–337.
24. Davis, W. D., & Miller, H. C. (1969). Analysis of the electrode products emitted by dc arcs in a vacuum ambient. *Journal of Applied Physics, 40*(5), 2212–2221.
25. Mesyats, G. A. (1971, September). The role of fast processes in vacuum breakdown. In *Phenomena in Ionized Gases, Tenth International Conference, Invited Papers* (pp. 333–357).
26. Litvinov, E. A. (1974). Kinetics of cathode flare at explosive electron emission. In *High-power pulsed sources of accelerated electrons* [in Russian] (pp. 23–34). Nauka, Novosibirsk.
27. Litvinov, E. A., Mesyats, G. A., & Proskurovskii, D. I. (1983). Field emission and explosive electron emission processes in vacuum discharges. *Soviet Physics Uspekhi, 26*(2), 138–159.
28. Mesyats, G. A., & Proskurovskii, D. I. (1989). *Pulsed electrical discharge in vacuum.* Springer-Verlag.
29. Mesyats, G. A. (2000). *Cathode phenomena in a vacuum discharge.* Moscow: Nauka.
30. Litvinov, E. A. (1985). Theory of explosive electron emission. *The IEEE Transactions on Dielectrics and Electrical Insulation, 20*(4), 683–689.
31. Bakulin, Y. D., Kuropatenko, V. F., & Luchinskii, A. V. (1976). MHD calculation for exploding wires. *Soviet Physics—Technical Physics, 21*(9), 1144–1147.
32. Loskutov, V. V., Luchinskii, A. V., & Mesyats, G. A. (1983). Magnetohydrodynamic processes in the initial stage of explosive emission. *Soviet Physics—Doklady, 28*(8), 654–656.
33. Aref'ev, V. I., Leskov, L. V., & Nevsky, A. P. (1973.) On anomaly scattering of electrons in cathode layers of gas discharges. *Soviet Physics—Technical Physics, 43*(8), 1660–1666. in Russian.
34. Aksenov, I. I., Konovalov, I. I., Padalka, V. G., Sezonenko, V. I., & Khoroshikh, V. M. (1985). Instabilities in a plasma of a vacuum arc with gas discharge gap: I. *Soviet Journal of Plasma Physics, 11,* 787–791.
35. Aksenov, I. I., Konovalov, I. I., Padalka, V. G., Sezonenko, V. I., & Khoroshikh, V. M. (1985). Instabilities in a plasma of a vacuum arc with gas discharge gap: II. *Soviet Journal of Plasma Physics, 11,* 791–794.
36. Buneman, O. (1958). Instability, turbulence, and Conductivity in current carrying plasma. *Physical Review Letters, 1,* 8–9.
37. Buneman, O. (1959). Dissipation of currents in ionized media. *Physical Review, 115*(3), 503–517.
38. Kadomtzev, V. V. (1970). *Collective phenomena in plasma.* Moscow: Nauka.
39. Alterkop, B., Beilis, I., Boxman, R., & Goldsmith, S. (1994, May). Influence of current instabilities on the parameters of the vacuum arc plasma jet. In *XVI international symposium on discharges and electrical insulation in vacuum* (Vol. 2259, pp. 76–81). International Society for Optics and Photonics.

40. Morosov, A. I. (1978). *Physical basis of space electro-reaction-propulsion units*. Moscow: Atomisdat. in Russian.
41. Lyubimov, G. A. (1975). Mechanism of acceleration of cathodic vapor jets. *Soviet Physics-Doklady, 20*(5), 830–832.
42. Lyubimov, G. A. (1977). Dynamics of cathode vapor jets. *Soviet physics—Technical physics, 22*(2), 173–177.
43. Eckhardt, G. (1971). Efflux of atoms from cathode spots of low-pressure mercury arc. *Journal of Applied Physics, 42*(13), 5757–5760.
44. Eckhardt, G. (1973). Velocity of neutral atoms emanating from the cathode of a steady-state low-pressure mercury arc. *JJournal of Applied Physics, 44*(3), 1146–1155.
45. Kesaev, I. G. (1968). *Cathode processes in electric arcs*. Moscow: NAUKA Publishers. (in Russian).
46. Zektser, M. P., & Liubimov, G. A. (1979). Fast plasma jets from the cathode spot in a vacuum arc. *Soviet Physics Technical Physics, 24*, 3–11.
47. Beilis, I. I. (1974). Analysis of the cathode spots in a vacuum arc. *Soviet Physics Technical Physics, 19*(2), 251–256.
48. Beilis, I. I. (1982). On vapor flow from cathode region of a vacuum arc. *Izvestiya Siberian branch of Academy Nauk SSSR, Seriya Thekhicheskih Nauk, N1,* 69–77. In Russian.
49. Harris, L. P., & Lau, Y. Y. (1974). *Longitudinal flows near arc cathode spots*. Rep. 74, CRD 154. Schenectady, New York: General electric Co.
50. Moizhes, B. Y., & Nemchinsky, V. (1980). Erosion and cathode jets in a vacuum arc. *Soviet Physics—Technical Physics, 30,* 34–37.
51. Nemchinsky, V. (1985). Gasdynamic acceleration of a cathode jets in a vacuum arc. *Soviet Physics—Technical Physics, 25,* 43–48.
52. Wieckert, C. (1987). Expansion of the cathode spot plasma in vacuum arc discharges. *Physics of Fluids, 30*(6), 1810–1813.
53. Wieckert, C. (1987). A multicomponent theory of cathodic plasma jet in a vacuum arcs. *Contributions to Plasma Physics, 27*(5), 309–330.
54. Krinberg, I. A., Lukovnikova, M. P., & Paperny, V. L. (1990). Steady-state expansion of current-carrying plasma into vacuum. *Soviet Physics—JETP, 70*(3), 451–459.
55. Krinberg, I. A. (2001). Acceleration of multicomponent plasma in the cathode region of a vacuum arc. *Soviet Physics—Technical Physics, 46*(11), 1371–1378.
56. Bugaev, A. S., Gushenets, V. I., Nikolaev, A. G., Oks, E. M., & Stachowiak, G. Y. (1999). Influence of a current jump on vacuum arc parameters. *IEEE Trans on Plasma Sci., 27*(4), 882–887.
57. Yushkov, G. Y., Anders, A., Oks, E. M., & Brown, I. G. (2000). Ion velocities in vacuum arc plasmas. *Journal of Applied Physics, 88*(10), 5618–5622.
58. Krinberg, I. A. (1994). Ionization and particle transfer in an expanding current carrying plasma. *Physics of Plasmas, 1*(9), 2822–2826.
59. Krinberg, I. A., & Lukovnikova, M. P. (1995). Estimating cathodic plasma jet parameters from the vacuum arc charge state distribution. *Journal of Physics. D. Applied Physics, 28,* 711–715.
60. Brown, I. G., & Godechot, X. (1991). Vacuum arc ion charge-state distributions. *IEEE Transactions on Plasma Sciences, 19*(5), 713–717.
61. Hantzsche, E. (1990). A hydrodynamic model of vacuum arc plasmas. In *Poceedings of XIVth Internernational Symposium on Dielectrics and Electrical Insulation in Vacuum*. Santa-Fe, USA.
62. Hantzsche, E. (1990). A simple model of diffuse vacuum arc plasmas. *Contributions to Plasma Physics, 30*(5), 575–585.
63. Hantzsche, E. (1991). Theory of the expanding plasma of vacuum arcs. *Journal of Physics. D. Applied Physics, 24,* 1339–1353.
64. Hantzsche, E. (1992). A hydrodynamic model of vacuum arc plasmas. *IEEE Transactions on Plasma Science, 20*(1), 34–41.
65. Hantzsche, E. (1993). A revised theoretical model of vacuum arc spot plasmas. *IEEE Transactions on Plasma Science, 21*(5), 419–425.

66. Afanas'ev, V., Djuzhev, G., & Shkolnik, S. (1992a). Hydrodynamic model of the plasma jet of the cathode spot in vacuum arc. Report of Sankt-Petersburg, A.F. Ioffe Phys.Tech. Inst. RAN, Russia, 44p..
67. Afanas'ev, V., Djuzhev, G., & Shkolnik, S. (1992b). Hydrodynamic model of the plasma jet of a vacuum arc cathode spot. I. Calculation of the jet in the critical cross section. *Soviet Physics—Technical Physics, 37*(11), 1085–1088.
68. Afanas'ev, V., Djuzhev, G., & Shkolnik, S. (1993). Hydrodynamic model of the plasma jet of a vacuum arc cathode spot. II. Calculation of the jet.. *Soviet Physics—Technical Physics, 38*(3), 176–183.
69. Keidar, M., Beilis, I., Boxman, R. L., & Goldsmith, S. (1996). 2D expansion of the low-density interelectrode vacuum arc plasma jet in an axial magnetic field. *Journal of Physics D: Applied Physics, 29*(7), 1973–1983.
70. Keidar, M., Beilis, I. I., Boxman, R. L., & Goldsmith, S. (1997). Voltage of the vacuum arc with a ring anode in an axial magnetic field. *IEEE Transactions on Plasma Science, 25*(4), 580–585.
71. Beilis, I. I., Keidar, M., Boxman, R. L., & Goldsmith, S. (1998). Theoretical study of plasma expansion in a magnetic field in a disk anode vacuum arc. *Journal of Applied Physics, 83*(2), 709–717.
72. Reece, M. P. (1957). The Tanberg Effect. *Nature, 180*(4598), 1347–1347.
73. Emmons, H. W. (Ed.). (1958). *Fundamentals of gas dynamics*. New Jersey: Prinston University Press.
74. Beilis, I. I., Lyubimov, G. A., & Zektser, M. P. (1988). Analysis of the formulation and solution of the problem of the cathode jets of a vacuum arc. *Soviet Physics—Technical Physics, 33*(10), 1132–1137.
75. Beilis, I. I. (2003). The vacuum arc cathode spot and plasma jet: Physical model and mathematical description. *Contributions to Plasma Physics, 43*(3-4), 224–236.
76. Beilis, I. I. (1982). On the theory of erosion processes in the cathode region of an arc discharge. *Soviet Physics—Doklady, 27,* 150–152.
77. Beilis, I. I., & Zektser, M. P. (1991). Calculation of the parameters of the cathode stream an arc discharge. *High Temparture, 29*(4), 501–504.
78. Beilis, I. I., Lubimov, G. A., & Rakhovskii, V. I. (1972). Diffusion model of the near cathode region of a high current arc discharge. *Soviet Physics—Doklady, 17*(1), 225–228.
79. Astrakhantsev, N. V., Krasov, V. I., & Paperny, V. (1995). Ion acceleration in a pulse vacuum discharge. *Journal of Physics. D. Applied Physics, 28*(12), 2514–2518.
80. Krinberg, I. A., & Paperny, V. (2002). Pinch effect in vacuum arc plasma sources under moderate discharge currents. *Journal of Physics. D. Applied Physics, 35*(6), 549–561.
81. J. Kutzner and C.H. Miller, Integrating ion flux emitted from the cathode spot region of diffuse vacuum arc *Journal of Physics D: Applied Physics, 25*(3), 686.
82. Krinberg, I. A., & Zverev, E. A. (2003). Additional ionization of ions in the interelectrode gap of a vacuum arc. *Plasma Sources Science and Technology, 12,* 372–379.
83. Krinberg, I. A. (2005). Three modes of vacuum arc plasma expansion in absence and presence of a magnetic field. *IEEE Transactions on Plasma Sciences, 33*(5), 1548–1552.
84. Alferov, D. F., Korobova, N. I., Novikova, K. P., & Sibiriak, I. O. (1990, September). Cathode spots dynamics in vacuum discharge. In *Proceedings of XIVth ISDEIV* (pp. 542–545).
85. Beilis, I. I. (2004). Nature of high-energy ions in the cathode plasma jet of a vacuum arc with high rate of current rise. *Applied Physics Letters, 85*(14), 2739–2740.
86. Beilis, I. I. (2005). Ion acceleration in vacuum arc cathode plasma jets with large rates of current rise. *IEEE Transactions on Plasma Science, 33*(5), 1537–1541.
87. Paulus, I., Holmes, R., & Edels, H. (1972). Vacuum arc response to current transients. *Journal of Physics. D. Applied Physics, 5*(1), 119–132.
88. Beilis, I. I. (1985). Parameters of the kinetic layer of arc-discharge cathode region. *IEEE Transactions on Plasma Science, 13*(5), 288–290.
89. Beilis, I. (1996). Theoretical modeling of cathode spot phenomena. In *Handbook of vacuum arc science and technology* (pp. 208–256). William Andrew Publishing.

90. Allen, C. W. (1973). *Astrophysical quantities*. London: Athlone Press.
91. Beilis, I. I. (1986). Cathode arc plasma flow in a Knudsen layer. *High Temperature, 24*(3), 319–325.
92. Beilis, I. I. (2013). Cathode spot development on a bulk cathode in a vacuum arc. *IEEE Transactions on Plasma Science, 41*(8), 1979–1986.
93. Beilis, I. I. (2018). Vacuum arc cathode spot theory: history and evolution of the mechanisms. *IEEE Transactions on Plasma Science, 47*(8), 3412–3433.

Chapter 19
Cathode Spot Motion in Magnetic Fields

The near-cathode phenomena in electrical arcs such as spot motion and spot dynamics, especially in a magnetic field, determine the arc performance and playing an important role in different applications. The current per spot and spot motion depend on the arc current, gap length, interelectrode pressure [1–3] and determine the stability of the vacuum arc. Several of the important vacuum-arc phenomena observed when the arc runs in a magnetic field. While cathode spot motion is normally random, in a magnetic field the motion is directed in the retrograde $-\mathbf{j} \times \mathbf{B}$ direction, i.e., in the anti-Amperian direction [4–6]. Cathode spot grouping, spot splitting, cathode spot motion in oblique magnetic field, and acute angle effect are the phenomena that indicate the complexity of the arc operation as a subject which up to present time is unclear and needed consecutive study. This chapter summarizes the main physical hypotheses and models published in the literature as well presented the author's own attempts to describe the spot behavior and spot motion in the presence of a magnetic field.

19.1 Cathode Spot Motion in a Transverse Magnetic Field

Proposed mechanisms for the "retrograde" spot motion were widely reviewed [1–9]. Numerous hypotheses were developed and published in the literature. Although each new mechanism put in a claim for more clear explanation, nevertheless the overall state of the art indicates that the problem is still open. Below, the published works are presented close to their original description with some critical analysis showing, as it was possible, the progress of the study and why the problem is still open. Also, description of the models is systematized, some addition point of view or analysis of works are provided that not considered previously.

19.1.1 Retrograde Motion. Review of the Theoretical Works

Analysis of the theoretical works below we begin from primary hypotheses and then the further approaches will be considered.

19.1.1.1 Primary and Developed Hypothesis

Some points related to understand the experimental works were already discussed in Chap. 13 and part of them will be just briefly mentioned here. One of first mechanism was indicated by Stark in [10], who also first observed this effect. He suggested, that the positive charged ions are shifted in the observed opposite direction under action of the voltage induced in parallel to the cathode surface due to plasma motion in a transverse magnetic field and due to Hall effect. This effect was used in further studies but using different interpretations which will be analyzed in this chapter.

Minorsky [11] explained the observed spot rotation assuming a circular electron currents produced by radially magnetic field. This mechanism was criticized by Tanberg's [12]. He expressed a doubt regarding the mentioned Minorsky's theory because that effect of retarding motion does not request a closed trajectory of the spot motion. In turn, Tanberg proposed model assuming a stream of positive ions moving away from the cathode that can be deflected the spot in reverse direction. However, this idea cannot explain the observations because the positive ions moved not separately but raiser in the form of quasineutral plasma and with relatively low velocity in the region adjacent to the cathode. Moreover, the supersonic flow in the cathode jet occurs at distance of about 1–2 spot radii (Chap. 18).

After more than ten years, Smith [13, 14] suggested to explain the retrograde spot motion in a mercury arc considering the thermomotive force produced due to negative Righi–Leduc effect. The conduction electrons along a vertical line in the liquid can appear a thermal gradient. In the presence of a transverse magnetic field to the direction of a temperature gradient in a conductor, a new temperature gradient is produced perpendicular to both the direction of the original temperature gradient and to the magnetic field. According to Righi–Leduc phenomenon, the magnetic field transverse to the thermal gradient gives a rise to a thermomotive force in the liquid that transfers heat and moves the spot in the observed direction of motion. To realize this mechanism, Smith estimated a large temperature gradient of 5×10^8 degrees/cm. This gradient was supported by taking unrealistic electron temperature of 4000 K near the emitting top Hg surface of the spot while the atomic constituent of the excited region is at approximately 150–200 °C.

Such an enormous thermal gradient in the liquid could cool the spot and extinguish the arc due to the thermal conduction. Therefore, Smith assumed that the Thomson heat counteracts the heat loss by thermal conduction. As the thermomotive force requests relatively large temperatures in the solids, it is small and also cannot explain the anomalous spot driving for arcs with different cathode materials. Yamamura [15] commented that the Smith explanation of spot motion using the negative Righi–Leduc effect does not seem to be plausible. He indicated that not only the metals

19.1 Cathode Spot Motion in a Transverse Magnetic Field

with a negative coefficient of the Righi–Leduc effect (for instance, Mg, Cu, and Ag), but also the metals with a positive coefficient (for instance, Fe) show reverse driving. It is impossible that the coefficient of the Righi–Leduc effect changes its sign from positive to negative with any change of air pressure. Later Yamamura [16] indicate that the Righi–Leduc effect is not the primary phenomenon for describing the magnetically induced arc motion on lead and zinc electrodes.

Significantly, later Kigdon [17] was also used the model of Smith [13]. He indicated that the Hall voltage seems to be too small to be significant, but estimates of the Righi–Leduc thermomagnetic effect are more promising. Kigdon developed a semi-quantitatively model of the retrograde motion of the spot in a magnetic field, which is due to the temperature gradient setup across the spot by the Righi–Leduc thermomagnetic effect. This model is based on assumption that the random motion of the cathode spot on mercury is caused by radial temperature gradients at the edge of the spot and due to fluctuations. It was also assumed that the ion diffusion time into the cathode material is limited to the surface layers and therefore affects the possible secondary electron emission from the cathode associated with the neutralization of the ions. Secondary emission of electrons caused by positive ions is known as an Auger effect [18–20]. This effect has been studied usually by the interaction ions of noble gas with metals and has been observed as an external Auger effect when the ion is immediately outside the metal surface. The last assumption using Auger effect is questionable in case of Hg arc because some difficulties were appeared with the energy balance between losses due to Auger electron emission and new ion generation at cathode potential drop that comparable with the potential ionization of Hg atom.

The conditions for the presence of positive or negative Righi–Leduc effect were discussed [17]. A positive Righi–Leduc coefficient will make the retrograde edge of the spot hot and will decrease the secondary emission there, so that the spot will move away from that side in the ponderomotive direction. Conversely, a negative Righi–Leduc coefficient will make the spot move in the retrograde direction. Estimations of the temperature gradient were conducted for solid Zn, Pb, Cu, Ag, Sn, and for liquid mercury. It was shown that some metals which do not show retrograde motion have positive Righi–Leduc coefficients, whereas all the metals which do show retrograde motion, except iron, have negative Righi–Leduc coefficients. It was concluded that the retrograde velocity on bulk liquid mercury is subject to the same velocity limit of about 10^4 cm/s as is the random velocity in zero field. The main problem of the above approach that the Righi–Leduc effect is small to shift the temperature maximum along some direction at the cathode surface [4, 5].

At the other side, Himler and Cohn [21] assumed that the electrons emitted from the cathode exert a mutual repulsion on each other, thereby causing some electrons to move transverse to the arc. This statement is supplemented also by the fact that electrons are emitted from the cathode in random direction. These authors indicated that at normal temperature and gas pressure, the mean free path of an electron is very low, and therefore, the electron diffusion from the region of the cathode spot is small also.

Under the mutual action of the electric and magnetic field, the electrons will tend to travel in a cycloidal path. With the very short mean free path at atmospheric pressure, the electrons can be considered to travel in short straight line segments, undergoing collisions with gas molecules. As the pressure is reduced, the electrons start to travel over larger portions of the cycloid before collision. At longer the mean free path, the greater will be the deviation of the electrons from straight line motion and the deviation will be affecting the behavior of the arc by following model.

It was assumed that one group of electrons can travel to the right, another one to travel to the left (note that is completely arbitrary statement). The force $F = q(vB)$ will tend to curve the motion of the electrons in different direction, as shown in Fig. 19.1. Those electrons traveling to the right will be forced back into the cathode to the right of where they were emitted, while those traveling to the left will be curved toward the anode. The electrons which are forced back toward the cathode form an electron cloud over part of the cathode spot, reducing further emission from that side of the cathode spot. Electrons traveling to the left, which are the electrons that finally reach the anode and contribute to the actual current flow in the arc, will collide with gas molecules to the left of the cathode spot. Thus, positive ions from the plasma will be formed to the left of the cathode spot and will bombard the cathode on that side. This will heat up a new region on the cathode to the left of the cathode spot so that emission can take place from the area just to the left of the initial cathode spot. This process causes the cathode spot to migrate to the retrograde direction.

The electron diffusion model of Himler and Cohn [21] was used by Yamamura [15]. He also assumed that electrons with energy corresponding to the cathode fall diffuse to the left side and are accelerated by the electric field, and thus these electrons efficiently ionize the gas molecules by collision. The electrons diffusing to the right side are decelerated by the electric field because these electrons are moving in an opposite direction to the electric field. Thus, the electrons moving to the right side cannot efficiently ionize the gas molecules by collision. In this way, the positive space charge cloud spreads to the left side producing the retrograde spot motion. The above hypotheses cannot be accepted as an explanation of a real process, because the electron beam emitted from the cathode is disappearing on the short length of the electron beam relaxation zone and then the quasineutral plasma flow is occurred (Chap. 16).

Longini [22, 23] assumes the effect of arc motion in the transverse magnetic field is due to electron field emission that depends on of space charge and therefore electric field distribution at the cathode surface. A qualitative model was developed showing that the space charge distribution can be characterized by a peak. Longini assumed

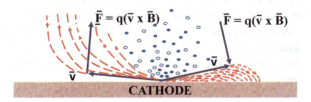

Fig. 19.1 Diagram of the cathode spot indicating the direction of electron motion at low pressure. Figure taken from [21]

19.1 Cathode Spot Motion in a Transverse Magnetic Field

that the total electric field distribution at the cathode is independently determined by ion space charge (ion field) and by electron space charge (electron field) ignoring interaction between these.

Both charge particles will be deflected in the same direction by the magnetic field with the electrons being deflected hundreds of times farther than the ions because of their small mass. When a tangential field is applied, the average instantaneous position of the ions and electrons will be displaced in the direction of their deflection. The electron field at the cathode will no longer partially neutralize the ion field symmetrically as it can see in Fig. 19.2, but will have a greater neutralizing effect in the direction of their displacement. Thus, while each of the component field is shifted to the right, the peak of the net field is shifted to left, accounting for the observed motion of the spot. According to Longini [22], this model demands that spot motion should reverse its direction as the magnetic field is greatly increased which has been observed. The absurdity of this hypothesis is obvious.

There should be mentioned work of Jerome Rothstein [24], who reported in 1950 a very doubt idea at the American Physical Society meeting. After an analysis of spot parameter, Rothstein proposed that as in the anomalous Hall effect the holes are directed in the retrograde direction, and a number of electron emission are following the effectively positive holes. Later the ions formed by collision outside the spot return bringing holes to a region shifted in the retrograde direction. Absurdity of assumption of hole conductivity in the dense plasma in a cathode spot also is obvious.

Ware [25] consider the Ettingshausen effect to explain the discharge behavior in a transverse magnetic field. He indicated that in the spot a direct transfer of energy from electrons in the gas to electrons in the metal lead to thermionic emission. Since the Ettingshausen effect causes one side of the discharge to become hotter than the other, the position of the arc spot will move in this direction. This is the direction of retrograde motion. The Ettingshausen effect is dependent on the gas pressure.

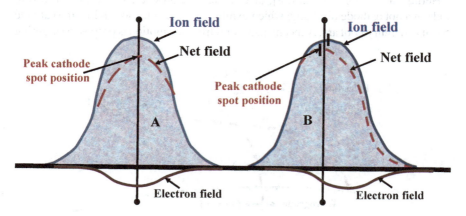

Fig. 19.2 Qualitative electric fields at the cathode surface with magnetic field into the paper. A-zero magnetic field; B- Under magnetic field the peak field is left shifted at the cathode surface producing left (reverse) spot motion. Figure taken from [22]

Eidinger and Rieder [7] criticize the work [25] and indicate two problems of this hypothesis such as necessity of only thermionic mechanism for electron emission and that the temperature maximum should be actually move by the Ettingshausen effect and very quickly. They also expressed their doubt asking questions: Does the effect not occur with sufficient heating of the whole cathode? Is the thermionic emission was observed at high spot velocities as well as for low boiling point cathode materials?

St. John and Winans model developed in [26] taken in account that the cathode spot consisted of a positive ion sheath separated by a dark space from the cathode and an electron cloud between the positive ion sheath and anode. The emitted electrons accelerated through the cathode dark space continuing his travel *by a curved path* due to the magnetic field and produced new positive ions on the forward side of the cathode spot.

These new positive ions would be attracted by the positive ion sheath. The new positive ions are drawn toward the cathode spot region by the forces exerted on them by the electrons in the cathode (f1), the positive ion sheath (f2), the electron cloud above the sheath (f3), and the electrons streaming toward the new positive ions (f3) (see Fig. 19.3). A resultant force would cause them to overshoot the positive ion sheath in the tangential to the cathode surface direction. When the traveling ions being pulled into the cathode, a new spot was started in the retrograde direction.

St. John and Winans in [27] extended the above mechanism of arc motion in a magnetic field to explain the rapid velocity rises with increasing magnetic field strength by associating them with the effect of Hg^{2+} and Hg^{3+} ions. The proposed model of retrograde spot motion is very qualitative because no any calculations of the forces f1, f2, f3, and f4 are presented. Therefore, a resultant force, which moves the generated ions in tangential direction and their overshoot the sheath region, seems to be stated arbitrary.

Hernqvist and Johnson, in [28], investigated the glow region of the ball-of-fire mode in a hot cathode discharge which exhibited retrograde motion like the cathode spot of a mercury pool arc. A mechanism for retrograde motion is proposed assuming

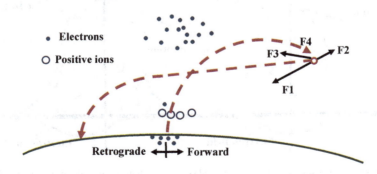

Fig. 19.3 Schematic presentation of a qualitative description of the cathode spot motion. Figure taken from St. John and Winans [26]

19.1 Cathode Spot Motion in a Transverse Magnetic Field

presence of a dark space between the cathode and the negative glow of the discharge. The author indicates that the sheath stability requires that the ratio of ion j_i to electron j_e current densities must everywhere along the boundary between the dark plasma and the glow region be equal $j_i/j_e = (m_e/m)^{1/2}$. It was assumed that an applied magnetic field bends the electron flow in the dark plasma. Thus, the density of the dark plasma is increased on one side of the glow region and decreased on the other side and the above condition for current ratio violated. As result, an electric field appeared, which will cause a decay of plasma density at the boundary, and the space charge sheath will seek to travel into the plasma deficient in density. Since the bending of the electron flow in the dark plasma is in the direction of electromagnetic force, the negative glow region will travel in the retrograde direction.

Hernqvist and Johnson indicated that the mechanism for retrograde motion is valid only for low values of the magnetic field strength because perturbation of the discharge was induced with increasing the magnetic field. An expression for the velocity of retrograde motion is derived based on the stability criterion for a space charge sheath connecting two plasmas. Conclusions from a possible analogy with the mercury pool arc were made. However, an analysis of cathode plasma taking into account the plasma particle collisions and their relaxation length was not provided. Therefore, no bending of the electron trajectory can be in the region significantly lower than Larmor radius. It is obvious that the above-mentioned ratio of ion j_i to electron j_e currents cannot be fulfilled in the space charge sheath. According to Eidinger and Rieder [7], this theory cannot explain the retrograde velocity with magnetic field as well the observed sudden increase of the arc velocity in the retrograde direction when a strong magnetic field was applied.

Robson and Engel proposed in [29] and then extended in 30] an idea based on some "resultant magnetic field" acting on the cathode region. In essence, to this end they were one of first to consider an influence of self-magnetic field. The authors suggested that the applied magnetic field deflects the positive column in the Ampere direction with respect to the cathode spot and the current path in the arc is curved in the vicinity of the spot. This deformation of the arc sets up a local self-magnetic field of the order i/R, where i is the arc current and R an effective radius of curvature of the arc which depends on the actual field and other discharge parameters. This field will act so as to oppose the applied field H_0 and so the resultant field H in the cathode spot region will be given by $H = H_0 - i/R$. Thus, the authors assumed that an arc bent over in the Amperian direction, forms a loop in which the magnetic field of the arc opposes the applied field to yield the retrograde motion. The resulting field is composed of the applied field and the "loop" field of the curved current path above the cathode spot. At low pressure, an increase of the former is accompanied by a rapid increase of the latter and thus the resultant field changes its sign. In general, the idea of take in account of the self-magnetic field is positive, but use of the curvature of the arc and calculation of the self-magnetic field with arbitrary value of $R = 10 \, \mu m$ to approve the model cause significant doubt.

Smith [31] provided the analysis of the force theory advanced by Robson and Engel [29] to explain retrograde motion. He detailed the problem by modeling the arc geometry as it is shown in Fig. 19.4. A magnetic field H0 is perpendicular to the paper.

Fig. 19.4 Schematic presentation of the arc geometry in applied transverse magnetic field. Copper arc deflected to right. Figure is similar to that from Smith [31]

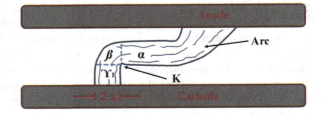

The arc path is bent over to the right. The arc was divided into three parts. Region α is a long, horizontal, round section, β is the elbow, γ is the remainder, including the current in the massive cathode. The self-magnetic field Hs was calculated for each arc region. The resultant force per unit area in the retrograde Fr and in Amperian Fa directions, respectively, were calculated in the point K (Fig. 19.4) as

$$F_r = \frac{(H_s - H_0)^2}{8\pi} \qquad F_a = \frac{(H_a - H_0)^2}{8\pi} \qquad (19.1)$$

For current of 3 A, and $a = 10\,\mu$m, then the value Hs is 1000 oersteds. In arriving at this value, the current density assumed of 10^6 A/cm^2. In the experiments, H_0 ranged upward to values near 10^4 oersteds. Smith commented that calculated H_s appears to be greater than was really existed, since to get it, a relatively high-current density was chosen. Smith also noted, whereas expansion of the arc path as it leaves the cathode is expected to make the radius larger than 10 μm. It seems that the resultant force is preponderantly in the Amperian direction, and the Robson and Engel theory is not sustained.

Thus, Smith concluded that the calculations indicate the magnetic field opposing the applied field is too weak to cause retrograde motion. Arcs apparently not having a bend in the sense nevertheless show the retrograde motion. Also according to St. John and Winans [26], this mechanism [30] requires that the retrograde velocity be nearly proportional to the square of the arc current, while measurements show it to be less than proportional to the first power of arc current.

Kesaev [4, 5, 32] was also taking in account the self-magnetic field, and for first time, he indicated a presence a maximum value of resultant magnetic field located in the opposite side of the spot with respect the Amperian direction. He used the superposition of self-magnetic field and the external magnetic field and as results detected the magnetic field asymmetric distribution around the spot (with maximum value). In order to explain the retrograde motion, Kesaev used the previous experimental fact that the cathode spot stability increased in the external magnetic field. These experiments Kesaev explained by the increase of the plasma density due to electron-optic an action of the magnetic field. Therefore, it was assumed that the spot moves toward the maximal magnetic field, located on the retrograde side. However, no similar action in the dense plasma can be occurred due to relation $\omega\tau \ll 1$. Therefore, although that experiment shows the spot motion in direction of

19.1 Cathode Spot Motion in a Transverse Magnetic Field

larger magnetic field [4, 5, 33], the physics of spot stability and the mechanism of the spot motion in the presence of a transverse magnetic field remained unclear and not specified.

Ecker and Muller [34, 35] developed theoretical model of "ion potential tube" in order to explain the retrograde rotation. They assumed that most of the field lines do not go directly to the cathode, but leave the space charge in radial directions and end at electrons in the neighborhood. From this, the authors also assumed that ions and electrons in the fall region form a kind of potential tube. The curvature of the tube axis is defined by the path of the heavy ions. The electrons emitted at the cathode end of the tube are forced to follow the tube axis until their energy has grown large enough to overcome the radial potential wall. The electrons coming from the cathode and following the ion path therefore must develop a velocity component in the retrograde direction. The electrons produce new ions on the retrograde side of the cathode spot and thus cause retrograde motion.

The radial velocity depends on the deflection of the potential tube. The curvature of the tube axis was calculated using the equations of continuity and momentum for ion flow together with the Poisson equation [35]. It was taken into account the voltage induced by $v \times B$ where v is the ion or electron velocities in magnetic field B. The results indicate that this curvature depends on the extension of the fall region, the mean free path of electron ionization, and the time to re-establish the new cathode region. As the theory cannot taken in account, many uncertainties of the experiment, only, one normalization experimental point was used to study. The calculated results are shown satisfactory agreement with the experimental dependences. According to Kesaev [5], the model consists of a number of absurd assumptions although the obtained results are in agreement with the experiment. Among those assumptions are the absurdity of inertial and direct motion of the electrons in the space charge region, the weaker curvature of the electron trajectory in comparison with that trajectory for ions in the presence of magnetic field, the model of cathode space charge sheath in the form of an "potential tube" contradict to very small sheath thickness in comparison with the spot radius and others.

A series of works studied the mechanisms of the spot motion in a transverse magnetic field were published by Guile, Secker, and co-authors. In one of the early work, Guile et al. [36] suggested that transverse galvanomagnetic and thermomagnetic forces modify the electron emission and are responsible for the cathodic arc motion. It appears that these forces also influence the motion of the anode spot. The idea appeared due to observed melting of the electrode surface occurred as the spot moved. So, it was assumed that electron emission has been proceeded by thermionic mechanism at the higher temperatures. It was assumed that within the cathode area there is a large electron current density, normal to the electrode surface and also a large temperature gradient in the normal direction of the surface. In all the experiments, the magnetic flux was perpendicular to this normal direction, so that transverse galvanomagnetic and thermomagnetic forces can arise. It was indicated that these forces (see above) have been mentioned by Smith and by Ware in connection with the retrograde motion of the mercury arc.

Later Guile and Secker [37] have demonstrated new experimental results (Chap. 13) for magnetically induced arc motion on lead and zinc electrodes indicate that the Righi–Leduc effect is not the primary phenomenon, as has been previously suggested. In this connection, the results of Yamamura [16] and Kingdon [17] were analyzed. According to authors' opinion [37], the retrograde motion occurred at reduced pressure is the result of enhanced emission on backside of the spot, either directly or as a result of the charging of surface oxide layers.

In another work, Guile and Secker [38] studied the published theories (including the cited above) in order to describe their own experiments (Chap. 13). The authors reported that since both continuous forward and retrograde movements of the cathode spot arise from conditions appeared at the cathode surface, then any theory proposed for retrograde motion must take this fact into account. Therefore, they reach the same conclusions as Smith [31] regarding the theory suggested by Robson and von Engel [29, 30]. The models do not seem to be adequate since surface conditions were not into account. Some corrections were conducted by Smith who calculated resultant magnetic field by providing more complicated analysis of the self-magnetic field for arc configuration in Fig. 19.4. A new interpretation of the forces F_r and F_a in (19.1) to, respectively, spot motion was given.

According to the review, Guile and Secker [38] concluded, that theories suggested for retrograde motion do not appear to be able to account for the dependence of retrograde cathode movement on the nature of the cathode surface. Similar conclusion was provided by comments in work [39]. In particular, it was noted that electrode influence can be significant (as one of examples, interaction of electrode and column) and should be accounted by the analysis of experimental results. Also noted that the spot movement has been found and was not necessarily caused only by electron emission or arc column friction effects, and that electrode vapor jets are probably involved [40].

Secker [41] considered a model to understand the influence of the transverse magnetic field set up by current flow in the electrodes. He showed that the greater velocity on the magnetic steel electrode caused by the increased transverse magnetic field set up due to a skin effect at the rapidly moving cathode spot. It was assumed that a skin effect should be present near the cathode spot, since the current paths to consecutive emitting sites are changed very rapidly within the electrodes. An expression was derived for determine the "skin depth," which accounted the properties of the conducting medium and the rate of change of current at the cathode spot.

It was obtained that if a certain critical velocity is attained, the skin depth is so great that there is no appreciable distortion of the current flow pattern in the cathode electrode. At the critical velocity, however, the reduced skin depth perturbs the current flow pattern and thus causes an increase in the transverse magnetic field just outside the cathode. This in turn increases the arc velocity so further enhancing the skin effect.

Lewis and Secker [42] described experiments which demonstrate clearly the effect of cathode oxide layers on arc velocity. A model of the cathode root region of the arc which is consistent with the experimental results is proposed. The model taken into account assumption that the electrons emitted from the cathode will be accelerated in the cathode fall region by the positive ion space charge field. As suggested by Ecker and Muller [34, 35], the positive ions were considered as restricted to a clearly defined "tube" which is bent by the action of the applied transverse magnetic field. The authors [42] stated that the predictions of the model, which is applicable to both Amperian and retrograde motion, are also in satisfactory agreement with results obtained by the previous investigators. It can be noted that this model has the same lacks as the model [34, 35], according to which the mechanism of retrograde spot motion remains not understandable.

The theory of Ecker and Muller [34, 35] was further extended by Guile [43] to the higher pressures where forward spot motion occurs. The ion potential tube model taking into account the electron–molecule and ion–molecule collisions in the presence of a gas were considered. In this case, at the shift of ion and electron tubes there will also be a shift of the region in which excited molecules also reach the cathode. The original site, receiving less positive ions and excited molecules, is likely to decline in emission efficiency. On the other hand, sites in the adjacent region to which the ions have been directed and will increase in emission efficiency as a result of the ion layer built up and the heat energy transferred to the surface.

The arc velocity (in the continuous mode) will depend, therefore, on the rates of decline of emission from the old sites and growth on the new. These were determined by surface conditions, arc current, gas pressure, as well as by the magnetic field in the cathode fall space. The author [43] widely analyzed the observed spot motion under various conditions in light of the potential tube mechanism. According to this mechanism, relationships between magnetic field and velocity were obtained for both Amperian and retrograde motion. Sometimes, as the cathode spot moves in the retrograde direction while the column and anode spot tend to continue in the Amperian direction. This leads to discontinuous jumping of the cathode spot in the forward direction.

19.1.1.2 Further Developed Ideas and Models

Bauer in [44, 45] provided an explanation for the irregular movement of the cathodic arc, by its motion in the magnetic field. Bauer analyzes the Hall effect and assumed that the direction of the electron drift deviates in the magnetic field from the direction of the electric field. The field emission current is deflected noticeably to the side by the Lorentz force. The temperature maximum is pushed aside by the cathode trap energy of this deflected electron current. The pressure $p(r)$ and $T(r)$ distributions indicated in Fig. 19.5 results drawn asymmetrical course of the space charge density. The limit temperature on the side, which corresponds to the Amperes rule, is reached at a lower pressure p_2 compared to p_1. Therefore, there is the higher ion density, i.e., a maximum on the opposite side. Since the spot zone is favored with the higher ion density, the spot is deflected noticeably sideways in the direction opposite to the Amperian force.

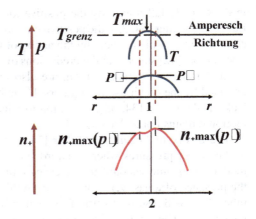

Fig. 19.5 Schematic profile of the effective temperature, the pressure, and ion density in the ionization region as a function from the center distance under the action of a transverse magnetic field. Figure is similar to that from [44]

When the current intensity increased, a splitting of the spot in single spots was occurred due to the self-magnetic field of the arc current. The average spot current was of 5 A. The above explanation notice that the temperature grows with reduced pressure and therefore asymmetrical localization of ionization rate occurred. This result is not understandable taking into account time dynamic of plasma parameters in the spot. Besides the conducted statements presented in qualitative manner and not validated even by some estimations of the physical processes (Chap. 16).

Hull [46] suggested that retrograde motion is caused by the drift of slow electrons produced after atom ionization by electron beam emitted from the cathode and accelerated in the cathode space charge sheath. The slow electrons move in the "correct" direction, parallel to the mercury surface, thus weakening the electric field drawing the positive ions to the surface, so that the ion current on the "retrograde"-side predominates. But further Hull used that the Tonks result, according to which the slow electrons are constrained by a magnetic field and they are drift at an angle to the direction which they would form an existing potential and concentration gradient. The drift at an angle depends on temperature of the slow electrons.

The spot motion under transverse magnetic field Hull was interpreted for parameters of Gallagher experiment [47] for Hg cathode. Hull suggested that at low ambient pressure, the drift of slow electrons tends to the right, nearly parallel to the mercury surface, and creating a negative space charge above the surface. This negative space charge reverses the electric field above the surface, so that the flow of ions to the surface on the "retrograde"-side predominates, causing the spot to move in the "retrograde" direction. On the other hand, for spot parameters at high pressure the drift of the slow electrons is nearly normal to the surface, and their space charge no longer produces a reverse field near the cathode. The ionization produced farther from the cathode, by the curved paths of the electron beam, predominates, and the spot moves in the "correct" direction.

The problem of Hall's theory is the absence of any analysis of the particle collisions in the plasma above the cathode surface and in the plasma quasineutrality. Another problem consists in the estimation of energy for slow electrons. The energy of these

19.1 Cathode Spot Motion in a Transverse Magnetic Field

slow electrons in case of mercury cathode is estimated taking into account that the electron beam first excites the metastable level at 5.44 V, then additionally the beam ionizes these excite atoms. An averaging of the resulting electron beam energy brings to the energy of slow electrons of about 0.3 eV. This very low value was derived due to erroneous assumption that the electron beam energy is a difference between cathode potential drop (10 V) and cathode work function (4.5 eV).

Hermoch and Teichmann [48] used magneto-hydrodynamic approach to explain the retrograde spot motion. They indicated that in a plasma moving through a transverse magnetic field and there induced an electric field determined as $E = v_0 \times B$. This field produces a transverse (Lorentz) and a Hall currents. This field E will further influence in such a way that either there is produced closed current in the direction of the field or the field induces the space charges, which can be a positive one and amplifies the electric field intensity in the direction of the retrograde motion, or a negative one—at the site of a spot corresponding to a normal motion. The most important influence is a displacement of the ion channel in a retrograde direction. The magnitude of such a displacement (provided a closing of the current) corresponds to a drift velocity of the ions in a given electric field.

Thus, the explanation consists of a plasma drift in retrograde direction. Some validation of the authors' hypothesis was conducted by qualitative discussion of the experimental data in light of above MHD description and a correlation between velocity of the retrograde motion and velocity of the cathode plasma jet [49]. After that it was arbitrary concluded that the plasma jets in a cathode region established a situation favorable for the origin of a retrograde motion of a cathode spot in a transverse magnetic field. However, it is not clear how the low MHD fields or ion drift induced in the quasineutral plasma of the jet can influence on the phenomena occur in the cathode space charge sheath with significantly strong electric field. The other problem arises by explanation that a significant displacement of the ions (by the field) under intensive convective plasma expansion with velocity measured in the highly ionized jet. The paradox consists in generation an electric field that can shift the source that produced it.

Murphree and Carter [50] explained the retrograde arc rotation between concentric cylindrical electrodes, produced due to the presence a transverse magnetic field, by the generation of electrical current in cathodic arc region caused due to temperature non-equilibrium. The analysis was based on generalized Ohm's law taking into account the gasdynamic forces (pressure gradients) and temperature non-equilibrium. The plasma was assumed as partially ionized consisting of three components including electrons, single-ionized ions, and neutrals. The viscous effects and spacial changes in the degree of ionization were neglected in the expressions for the electrical conduction and current density. It was shown that a reversal of radial current was induced on retrograde side of the cathodic arc region.

The current density produced by the non-equilibrium effect increased with reduction in pressure since the degree of non-equilibrium increased. The electrical current produced by non-equilibrium effect was on retrograde side of the arc, whereas the current produced by the electromagnetic field is in the positive direction. With reduction in pressure, a high degree of non-equilibrium is obtained the net current on retrograde side of arc region will increase, while the current in positive direction was reduced. The model describes such arc characteristics as the rotational speed, arc shape, and arc motion with reduction of the chamber pressure. It seems that the model used the Ohm's law was studied for a rest plasma and do not taken in account the influence of the high-velocity cathode plasma jet.

Seidel and Stefanik [51] indicate that the retrograde spot motions originates due to asymmetry of high-current density in the spot current of the vacuum arc. They assume a possibility of retrograde dissipations of the ions trending toward the cathode by means of a stream of neutral atoms. These ions produce something like an umbrella around the vapor stream and reach the surface of the cathode at the border of this of the cathode spot. Therefore, maximum current density should be obtained on the border of the spot. It was stated, when a transverse magnetic field is superposed, the ions will be reached the cathode on the retrograde side of the arc, i.e., a new spot will occur just here. This authors' opinion was based on dependences of spot current density and vapor pressure in the spot on current, but the calculations were presented without explanation of the method and provided independently on the process of retrograde motion.

Sherman and Webster [52] and later Agrawal and Holmes [53] estimated a saturation velocity of the spot using data of current density j and temperature T from Ecker's theory of "existing diagram." The saturation spot velocity v_{sm} was obtained using the cathode thermal regime according to Carslaw and Jaeger [54] by modeling the moving spot as a heat source at the cathode surface and taken the value of ju_{ef} as the power density input to the surface for circular spot $(I/\pi j)^{0.5}$:

$$v_{sm} = 1.44 a_t j^{1.5} I^{0.5} \left(\frac{u_{ef}}{kT}\right)^2 \tag{19.2}$$

The velocity calculated using effective voltage due to loss by cathode heat conduction u_{ef} taken according to Daalder's data. The authors noted that in spite of widely differing physical properties, the investigated non-refractory metals and the maximum retrograde spot velocities were obtained between 10 and 100 m/s, while the measured maximum spot velocities are within the same order of magnitude (10–65 m/s). The weakness of such approach considering thermal process is obvious because the used input values can be varied during the spot motion.

Djakov and Holmes in [55] explained the retrograde spot motion considering spot movement in a magnetic field formed by electrical current of other spots arise in radially expanding ring. The self-magnetic field was calculated as an inversely proportional dependence on radius R of the ring-shaped spots formation. The assumptions are based on arc stability. It is assumed that (i) the lifetime of elementary spot for any arc current depends on magnetic field in the same way as for a one spot arc; (ii) the displacement of the spots during their lifetime is neglected so that the spot displacement is considered to be the result only of spot rebuilding. As a result, a relation of ring displacement R on average current per spot I and lifetime t was derived. It was indicated that the model based on the experimental data for Hg arc, and therefore, it is first approach as a simple explanation, which should be improved.

The further Djakov and Holmes consideration [56] of the retrograde spot motion based on assumption, in which the heat conduction within the cathode plays an important role. The model takes into account that in transverse magnetic field the ion distribution is undisturbed, but an asymmetry is provoked in the electron distribution. A continuous emitting line was considered that produced by a number of cathode spots. In this case, a variable potential drop u_h (Hall potential) in direction perpendicular to the emitting line was generated in transverse magnetic field. As a result, the ionization and excitation of the cathode vapor are enhanced by the increased potential drop on the retrograde side of the spot. The power q delivered by the ions to the cathode surface per unit area is $q = j_i(u_0 + u_h)$, where j_i is the ion current density and u_0 includes u_c and other constant components in the power input per ampere. Due to the asymmetrical heating the temperature maximum at the cathode surface is displaced in the retrograde direction. The increase of Hall electric field enhances electron emission on the retrograde side of the spot, and the vapor jet follows the spot displacement. The calculated results indicated that at relatively low magnetic field, the spot velocity increased linearly, but at strong magnetic field, the dependence was declined from linear, which is in agreement with the observations.

Djakov [57] continue the consideration of asymmetry in the cathode heating due to Hall effect in quasineutral cathode plasma. The high-current arc of 3–7 kA and spatial distribution of the current I for Cu cathode as well as the Bi, Zn, and Pb materials for $I < 100$ A were analyzed. The Hall voltage expressed as $u_h = r_s jB/en_e$ was produced parallel to the cathode surface in plasma column, where the density reduced to $n_e = 10^{14}$ cm^{-3}. The Hall voltage estimated as several volts sufficient to modify the asymmetry of the heat flux to the cathode. It was assumed that the Hall voltage influences the heat flux distribution because the equipotential lines become deformed near the metal surface. There we should note, that this assumption is very doubt due to small Hall voltage arise in the quasinetral plasma and not influence on the heat flux to the cathode determined by the adjacent to the cathode surface a large cathode potential drop u_c in the sheath. Moreover, even the weak asymmetry in the heat source cannot determine some asymmetry in the temperature distribution in metallic cathode with relatively large thermal diffusivity (Chap. 3). This note related al to other works used the Hall voltage influence.

A similar explanation of the retrograde effect was considered by Moizhes and Nemchisky in [58] taking into account the increasing of the value of u_c by the generation of an additional voltage drop u_h in the near-surface quasineutral plasma due to Hall effect. It is also assumed that the heat flux to the cathode was determined by the ion acceleration to the value of $u_c + u_h$ at the retrograde site of the spot. However, in the dense plasma at the cathode surface the plasma density is so large that condition of $\omega\tau \ll 1$ is fulfilled, and, therefore, the effect is negligible small comparing to the large electric field in the cathode sheath.

Another explanation Nemchisky provided in [59] considering the cause of the spot motion by its jump from one to other location on the cathode surface. He assumed that such spot jump was stimulated by large energy loss on ohmic resistance of a plasma of long length produced due to depth crater formation in the cathode body during the spot operation. The presence of Hall effect determined the asymmetry of the resultant voltage. Also it was assumed that, when an increase of the voltage in the plasma reaches a critical value with increasing its length, the spot changes the location. A statistical treatment was used to obtain the condition of spot location, and therefore, the spot velocity as dependence on magnetic field and current. However, in addition to the above comments, the experiment [60] not always shows a depth crater filled with a plasma, the crater always is filled with the liquid metal or the crater with evacuated the liquid metal due to very large plasma pressure acted on the liquids in the spot.

Hantzsche [61] studied the forces acting on the cathode arc spot surface and removing the molten layer from the crater which composed mainly of the ion pressure, the neutral gas pressure, and the evaporation recoil. The liquid layer of thickness of 0.1 μm at the solid cathode (significantly smaller than the spot radius) was determined by treatment using the hydrodynamic approach. It should be noted that sometimes the craters were produced in form like to that presented in Fig. 19.6, i.e., mostly planar in which the diameter is much larger than the depth. In additional, the calculations [59] used some assumptions and experimental data that not understandable and should be discussed separately.

The above hypothesis relating to the increase of plasma voltage produced due to plasma electric resistance as well as the Hall effect was used to explain the retrograde spot motion on mercury cathode [62]. For mercury cathode, a spot deepening was assumed due to sunken of the spot into the liquid by the large pressure. In the later work of Moizhes and Nemchisky in [63], this same model was used and considering the dependence on electrode gap as well as an influence of the axial magnetic field. The Hall effect was also used by Emtage et al. [64] in order to explain the motion of a high-current arc in forward direction.

In contrary, Daybelge [65] indicated, that the Hall effect is very small, because the mean free path of electrons is shorter than their gyroradius. He studied the expansion of metal vapor from an emission center (EC) without consideration of the heating

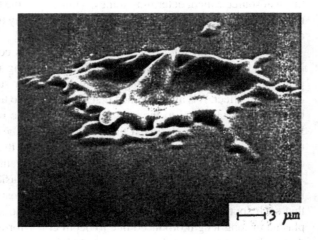

Fig. 19.6 Crater produced at Mo cathode according to Juttner [60]. Used with permission

19.1 Cathode Spot Motion in a Transverse Magnetic Field

problem itself. To analyze the metal plasma, the general magneto-hydrodynamic system was used including equations of continuity, momentum, and entropy in 2D approximation. The strong variation of electrical conductivity was taken into account as well the induced electric field caused by plasma motion across magnetic field. As boundary conditions the plasma was assumed to outflow from a cylinder of radius r_0 (equal to a crater size) with given initial Mach number M_0. All processes inside the crater region were ignored.

The calculations conducted for distance of $20r_0$ and for constant self-magnetic field B_s along the plasma boundary assuming constant arc current in the explosive electron emission. The influence of the expanding metal plasma on the electric field distribution around the emission center was studied. It was shown that expanding metal plasma strongly enhances the electric field in vicinity of the EC, which was symmetric in the absence of the external magnetic field. In this case, two maxima on both sides of at crater can activate of new EC on both sides with equal probability, which is the reason to random walk of the cathode spot. In the presence of the transverse magnetic field, even neglecting of the B_s, an anti-symmetric electric field distribution arises around the EC, so that the field was reversed on the Amperian side of the crater making any electron emission on this side unlikely, whereas strongly field enhancement on the retrograde side of the crater favoring the rise of new EC on this side.

Some not understandable questions related to the model [65] should be noted. The non-stationary processes of electron explosive emission and plasma expansion from a transient EC cannot be studied by assuming constant electrical current or current density. The asymmetry electric field distribution was originated in the quasineutral expanding plasma. Therefore, it is not clear how this field in such plasma can influence the electric field at a protrusion to support a new EC displacement. Also, it is not clear how the input parameters, given arbitrary in the model, are related to the real cathode spot phenomena.

Drouet in [66] and similarly in [67] also was used the asymmetric magnetic field distribution which results from the asymmetrical combination of the self and applied magnetic fields. He assumed that the plasma is *confined* at the retrograde side, and therefore, the *plasma density increase* influencing the *work function* of the cathode material in way that it is reduced enhancing the electron emission and thus favors the retrograde motion by the new spot ignition. Let us consider the details.

1. The fact of discharge confinement and asymmetric action in the presence of a transverse magnetic field was previously reported by Kesaev [4, 5, 31, 32] and by Emtage et al. and reference therein [68]. Also, Meunier and Drouet [69, 70] experimentally studied the arc behavior with the application of the 850 G magnetic field. The confirmed influence of the transverse magnetic field showing that the plasma light was confined along the field lines close to the cathodic plane. The detected amplitude of the ion flux was more than three times stronger along the field lines than perpendicular to the field, whereas without field the same flux intensity was measured in both directions. In the direction parallel to the field lines the plasma cloud is asymmetric, the luminous region going further away from the electrodes in the forward or Amperian direction while the arc

spot propagates in the retrograde direction. The appearance of ejected metallic droplets was largely reduced with the magnetic field.
2. The work function reduction was based by reference to the Leycuras work [71]. In this work, the lowering of the work function from a metal was explained by the polarization energy of the excess electrons in a dielectric. It was analyzed experimental results obtained by widely studied of the work function φ for a number metals immersed in a few dielectric liquids and in argon. The results showed that reduction of φ was determined by the presence of a dielectric. In case of cathode spot, Leycuras assumed that the role of a dielectric plays the presence of high dense vapor. This assumption contradicts to well establish opinion about the near-cathode high dense and highly electro-conductive plasma. In addition, the self-consistent solution of the cathode system of equation indicated that the plasma parameters and current density weakly depend on the variation of the work function [72]. Only the calculated cathode surface temperature weakly varied with φ in order to reach requested electron emission intensity to satisfy the cathode processes self-consistently.
3. The work of Fabre [73] was cited in order to base the plasma density increases with the magnetic field. He studied experimentally the capture and containment of a laser produced plasma with a magnetic field. Experimental results show the deceleration of the laser product expansion and a dependence of high energy particle losses due to the large target size. There should be noted that the plasma production in case of laser–target interaction is a result of action an independent heat source. Therefore, the near surface plasma generation is completely different from the plasma generation in case of self-consistent cathode processes, which are different in different spot types. The difference of both the phenomena can be considered from the works [74, 75].

Thus, although that Drouet's assumption about asymmetry of the resultant magnetic field and its influence on the arc plasma is not new in the literature, there are unanswered questions about the role of plasma density increasing (which also proposed by Kesaev [4, 5]) and the interpretation of published experiment about to reduction of the work function.

Fang in [76] developed a model retrograde motion of vacuum arcs in transverse magnetic fields considering the outflow of positive ions from the cathode spot with velocity v_z in z-direction perpendicular to the cathode surface. In transverse magnetic fields, the high-speed ion jets from the cathode to the anode will be deflected in the retrograde direction and thereby the positive space charge shifts to the same direction. It was reported about not understandable assumption according to which the separation of ion charge in plasma jet can occur at the moment of explosion. While the electrons will be bent in the forward direction, the high-speed ion jets will be deflected in the retrograde direction by the transverse magnetic fields. The retrograde deflection of the positive ions will cause a retrograde shift in the location of the maximum electric field at the cathode surface and thus a retrograde shift in the location of the high electron emission from the cathode and finally in the retrograde motion of vacuum arcs.

19.1 Cathode Spot Motion in a Transverse Magnetic Field

The ion deflection was assumed due to force like to expression $e(-u_a/d + vB)$, where u_a is the arc voltage, d is the gap distance, v is the ion velocity, and B is the magnetic field induction. It was obtained the expression for velocity v_y of ion deflection in y-axis indicated direction parallel to the cathode surface. This expression is depended on number of parameter such as cathode potential drop, ion mean free path, current density, electron current fraction, cathode surface temperature, and initial ion velocity in the spot region. The further correction of this expression was conducted in order to calculate the spot velocity as dependence on the cathode temperature variation due to the external cathode heating [77]. At numerical study part of the mentioned, parameters were given arbitrary or calculated using arbitrary assumptions. Also, doubtful using a force by the average electric field $e(-u_a/d + vB)$ generated in the quasineutral plasma (i.e., not related to the space charge sheath), in which vB is, as rule, small and was already commented at the analysis of the listed above similar works. Besides, not understandable problem of above-described model is the assumption of separate ion flow in the cathode quasineutral plasma jet caused due to an explosion process.

Harris [78] theorized the different phenomena occurred in group spot of a vacuum arc including splitting, grouping, and the retrograde motion of the fragment or a single cathode spot. He assumed that the retrograde effect is caused by the motion of the electrons to the edge of the spot. The electrons and heavy ions within the plasma react differently to the Lorentz force that is applied. In vacuum, the heavy ions could be considered stationary when compared to the low mass electrons. The velocity of the electrons is determined by their reaction with the magnetic field (Lorentz force) and the "drag" that is generated due to electron–ion interaction. The electrons that are ejected from the cell are displaced in the "forward" direction. Since the electron space charge is now located at the "forward" side of the cathode spot, the electron emission on that side is reduce and is increased on the retrograde side of the cell. As result, the electron emission center will move toward the retrograde side of the cell to the same degree that the negative space charge will move toward the "forward" side. And the electron cloud will return to its original position. Although this mechanism appears to be stable when considering the electron behavior, it is not the case for the ions. The location change of the electron emission center is accompanied by a change in the location of the atom evaporation center (after some thermal delay—which is required to heat the surface). Since the "forward" side of the cell now cools off, while the "retrograde" side heats up—the plasma is emitted from the retrograde side and the plasma is shifted backwards. Figure 19.7 shows the time dynamics of electron beam forward and retrograde new generated plasma altered the space charge fields displacement explained by the cycle of steps.

This cycle can be described in way reported by Harris. Figure 19.7a who indicated the situation in the absence of Lorentz forces parallel to the cathode surface. Here the electron and ion space charge clouds with positive charge approximately of a mean free path for atom ionization off the cathode surface. The arrows indicate that the electrons tend to flow away from the cathode perpendicular to the surface. Figure 19.7b illustrates the change in the electron trajectories in the presence of transverse Lorentz force that tends to push electrons forward toward the left. Further

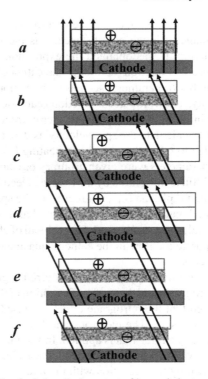

Fig. 19.7 Dynamical cycle of relative displacement of ion and electron clouds caused by Lorentz forces in a cathode plasma for **a** in the absence of the magnetic field. The presence of the magnetic field; **b** change of the electron trajectories; **c** electron space charge is shifted slightly to the left of its original position; **d** displacement of the mean locus of electron emission, the emitted electrons cover the exposed remained the positive space charge; **e** the electron deficiency on the backward edge is filled in, and the space charge cloud of the fragment is properly compensated; **f** the picture is identical to that of (**b**) except for the retrograde, shift of the fragment plasma and of the emission site at end of the cycle

after a few electron transits times, the electron space charge in the plasma region is located slightly to the left as indicated in Fig. 19.7c. Because the transit time of an emitted electron passing through the plasma region is very low, the estimated maximum transverse electron displacements were less than a Debye length, which is of the order of 10^{-6} cm. Therefore, it is not given rise to significant transverse electric fields opposing the electron drift. However, the sign of the longitudinal electric field is changed and that suppresses the electron emission at the edge of the fragment, while at the same time strongly enhancing electron emission at the opposite side of the fragment where the electron cloud tends to recede. So, the displacement of the mean locus of electron emission and the emitted electrons covers the exposed positive space charge, which is taken to be stationary during these changes.

Figure 19.7d shows the backward shift of the locus of electron emission which must be equal in magnitude to the forward displacement of the electron cloud. Thus,

19.1 Cathode Spot Motion in a Transverse Magnetic Field

within another few electron transit times, the electron excess on the forward side of the fragment flows away, the electron deficiency on the backward edge is filled in, and the space charge cloud of the fragment is properly compensated, as indicated in Fig. 19.7e. Then ions are created when atoms evaporated from the emitting surface undergo ionizing collisions with plasma electrons, as shown in Fig. 19.7e. After some thermal delay, by a corresponding shift in the locus of atom evaporation as the surface under the forward edge of the fragment cools and that under the backward edge heats up. There results a backward shift in the position of the entire plasma as old plasma at the forward edge of the fragment flows away and is not replaced, while new plasma is formed opposite the newly emitting area at the backward edge. This process results in the condition given in Fig. 19.7f, which is identical to that of Fig. 19.7b, except for the backward, or retrograde, shift of the fragment plasma and emission site. The cycle then runs again. Finally, a mathematical treatment was presented that allows calculate the parameters of retrograde spot motion as dependence on the magnetic field. The calculation results were in agreement with that obtained from the experiment.

Summarizing, it can be stated that the idea of emitted electrons displacement while the remained ions produced new positive space charge sheath at the retrograde site of the spot is used in a number above cited works. Harris referred to the similar, but very qualitative, work of Longini [22]. Again, the main problem that the effect of electron magnetization is produced at length significantly larger (>10 μm) than the length of the space charge sheath, or the mean free path (0.01 μm) or even the ionization zone (1 μm). In other word, all above processes of charge shifting cannot be occurred in the dense plasma of the spot (fragment). The mentioned charge separation at the edge of the spot cannot be accepted even if the plasma there is enough rarefied. This is due to, firstly, not clear position of this edge, and secondly, the intensity of the thermal and emission processes cannot dominate in the rarefied plasma in comparison with that in the nucleus and therefore determines the spot shifting. Finally, the quasineutrality of the plasma remains even in rarefied edge of the spot.

Sanochkin [79, 80] reported an attempt to explain the retrograde effect by asymmetric curvature of a molten metallic layer generated due to current carrying microspot surrounding in a transverse magnetic field. The used model considered a point heat source acted on a liquid metal layer which length is much larger than its thickness. A thermocapillary force caused a convective motion of the liquid, which is in general symmetric around the heat source in the absence of the field. In the presence of a magnetic field, the ponderomotive force is added to the gravitation in the Amperian direction and as result the level of the molten layer decreased at this side producing an asymmetric level of the liquid surface around the heat source. Asymmetry of the molten surface, according to the author's opinion, leads to asymmetry of the electrical field, which decreased on side of the pit and it is concentrated around the higher edge of the liquid meniscus. Therefore, the surface heating due to ion bombardment is also asymmetric and the heat flux is shifted in direction opposite to the Amperian force causing respectively shift of the spot in this direction. However, the observed cathode spot phenomena occur mainly at the solid cathodes and the liquid area around the spot is about of the spot radius and is not associated with the assumptions used in the theory of thermocapillary convection. Besides, the surface

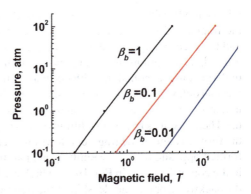

Fig. 19.8 Static pressure dependence on applied magnetic field with constant parameter β_b as a parameter. Using Schrade data [82]

heating due to ion bombardment cannot be asymmetric due to thickness of the space charge sheath is much lower than the characteristic scale of the melting region at the surface. At liquid cathode like Hg, the level of possible cathode temperatures in the spot is relatively low and not associated with the requested level of temperature difference corresponding to the modeled phenomena of metallic convection (see also Chap. 16).

Schrade et al. in a series of publication [81–86] developed an idea of magnetohydrodynamic instability, which provokes the ignition of new cathode spots in the retrograde direction. It was determined, that an arc is stable with respect to a certain disturbance or displacement if the arc can be restored to its original balanced configuration. In 1973, Schrade [82] indicated a parameter $\beta_b = P_{st} 8\pi / B^2$, which is the ratio of average static pressure P_{st} to magnetic pressure. The magnetic field 0.5 T corresponds to a pressure of 1 atm. If the applied magnetic field is much smaller than 0.5 T, the value of β_b becomes much larger than unit for an arc pressure more than 1 atm. In Fig. 19.8, curves of constant β_b are plotted in log-log scale as function of the gas pressure P_{st}, and the magnetic field B. The arc column can be stable when $\beta_b > 1$. According to Fig. 19.8, this condition can be fulfilled for cathode spot dense plasma. Thus, the value of β_b turns out to be a crucial parameter in the stability criteria of an arc in a magnetic field.

Tseskis work [87] should be mentioned as well. This work pointed that magnetohydrodynamic instabilities of the arc column lead to the ignition of new cathode spots in the retrograde direction. A kink instability as a result of interaction of the arc current with the magnetic field was studied. Ion current fraction in the arc as a constant parameter obtained experimentally has been utilized. The conducted study leads to some approximate result regarding to the dependence for the velocity of retrograde motion on the magnetic field and which include three constant parameters given arbitrary. The arc instability was studied to find a condition when the discharge axis will be kink, and this instability will be developed.

A general stability criterion for a discharge channel which emanates (arise) from a crater of a low-pressure arc cathode attachment was derived later Schrade et al. [84] in order to explain arc spot motion by a kink instability of the discharge channel

19.1 Cathode Spot Motion in a Transverse Magnetic Field

causing the channel to bend more and more. The model considers dissipation mechanisms of a non-stationary single spot by taking experimental data of the crater size on pure copper and molybdenum cathodes. It was assumed that the current density distribution over any cross-sectional area of such a discharge channel can be approximated by a so-called eccentric parabola of grade n. A finite channel segment of certain length and volume of any arbitrarily shaped current carrying plasma channel was considered. The forces which acted on the particles within this segment were: the Lorentz force, the centrifugal force, the Coriolis force, and the resultant gasdynamic force due to convection, pressure, and friction effects. The derived stability criterion was determined by the ratio of self-magnetic field to the external magnetic field. The criterion indicates conditions when the channel becomes less stable for small disturbances in the retrograde direction. The consequence of this fact is a kinking of the channel in the retrograde direction and an eventual contact with the cathode surface on the retrograde side of the original spot or crater.

Similarly to above model, Auweter-Kurtz and Schrade [83] studied the stability of a current carrying plasma channel. The model takes in account that one end is fixed to the metal surface. It was shown that transverse convective forces induced by magneto-gasdynamic effects play an important role in the dynamics of a plasma channel and that the vapor pressure of the evaporated electrode material must be larger than about twice of the average magnetic pressure across the spot area in order to obtain a stable arc channel.

Qualitative model of arc instability was described by the authors [88]. Let us discuss the main aspects of their mechanism. Firstly, the arc consists of an axially symmetrical current carrying channel having the origin in the cathode spot. The self-magnetic field of a force profile produces a pressure increase in the plasma channel. When a fluctuation curved the channel, then the magnetic lines on the concave side of the channel are condensed and on the convex side were spread out. Therefore, the radial forces on the concave side become larger than on the convex side and the force profile, distributed over the cross-sectional area of the channel, becomes asymmetrical. This asymmetry affects the cathode vapor plasma, which under high pressure and ohmically heated can now no longer be held together by magnetic force because it expands or stream out on the weak force side of the channel and was thereby diverted off the axial direction. So, the axial vapor and plasma jet flowing out from the cathode spot were deflected in the direction of smaller force fields, while the intrinsic current carrying channel was deflected in the opposite direction. This deflection of the current carrying channel can become so strong that a contact with a neighboring surface point vaporized the cathode and form a new cathode spot (Fig. 19.9).

The other direction of instability is related to the case when a transverse magnetic field is applied to the arc channel. With this instability development, the radial force profile can become asymmetrical without additional curving of the discharge channel. This asymmetry causes a deflection of the vapor and plasma jet in the Amperian direction, while a resulting gasdynamic surface force acts on the current carrying channel in the opposite retrograde direction. It can be shown that under certain condition this gasdynamic force in the retrograde direction predominates over Lorentz force in the

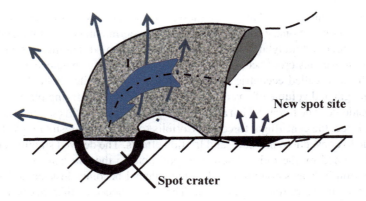

Fig. 19.9 Curvature of the arc plasma channel and new spot ignition due to the bending of the current channel. Figure is similar to that from Auweter-Kurtz and Schrade [83]

Amperian direction. The channel axis is thereby curved in the retrograde direction until the discharge channel contacts and heats up the cathode surface. As a result, a new cathode attachment arises on the retrograde side of the original spot.

Thus, the idea and mathematical treatment widely presented in the literature by Schrade and co-authors is based on study a discharge instability caused the discharge channel curving. A high degree of discharge channel curvature leads to new spot ignition due to plasma contact with the cathode. But there is a problem with this mechanism because any analysis provided by the authors of probability of such ignition is due to contact of rarefied expanding plasma with new cold point at the cathode surface was absent. Moreover, it is difficult to represent the plasma jet acceleration up to supersonic velocity at distance of about one spot radius (Chap. 18) that can be curved up to like to half circular configuration with a contact of dense plasma at a some new cathode surface. Also, a number of assumptions related to choose a certain segment, its length, parabolic density profile, and others sound arbitrary and not related to the real cathode plasma expanding in forward and radial directions at some distance from the surface.

An ionization–energy model was developed by Bobrov et al. [89] to explain the spot motion without assumption of new spot ignition at retrograde side of the old spot. The model was considered as a basic concept approving Kesaev's principle of "maximum magnetic field" and his respectively assumption of increase here the plasma density. This increase can be provided taking into account the neutral gas ionization in the plasma region by the absorption of the electromagnetic energy, which is maximal at the side of maximum magnetic field. So, the model describes the retrograde motion of a cathode spot as a process of plasma propagation to a direction of maximum dissipation of Poynting energy, which is in the direction opposite the Lorenz force. Although that the analysis of plasma heating and atom ionization was conducted, the model is qualitative. This model did not present any calculation regarding to the level of the plasma heating as dependence on distance from the cathode surface to understand where this approach is effective. However, it is known

19.1 Cathode Spot Motion in a Transverse Magnetic Field

(see below) that at near-cathode surface at which the plasma density is very high, the influence of the magnetic field is negligible low due to low value of $\omega\tau \sim 10^{-5}$. Also, the plasma properties are quietly equalized in the dense highly collisional plasma (with characteristic time of 10^{-13} to 10^{-14} s at electron–electron collisions) even of some local its parameter fluctuation in the nearest to surface plasma region. Thus, the proposed approach cannot be accepted, because the released energy can be far from the cathode surface, where it is actually important.

Jüttner and Kleberg [90–92] presented an excellent experiment in order to study the spot motion in a transverse magnetic field using an improved technique having high time and space resolution (see details also in Chap. 13). It has been found that a plasma cloud expanded in the form of plasma jets in the retrograde direction and the new spots are ignited in the direction of this plasma expansion. The jet ejection is assumed as a product of instability of spot plasma confinement in the magnetic field. The plasma expansion reach velocities of $v_{jet} = 5$ km/s on average. An attempt has been made to modeling the spot motion.

The model based on expanding jet motion crossing the magnetic field, producing an electric fields $E_{jet} = v_{jet} \times B$, which assumed is promote the ignition of a new spots in the retrograde direction. It should be reminded that the field E_{jet} and the plasma drift in retrograde direction using hydrodynamic approach was already studied by Hermoh and Teichmann [48, 49] and in other above discussed publications (see above comments). But here is a number of weak points in this approach as well. The electric field E_{jet}, in essence, is the field produced inside of the quasineutral plasma. An estimation shows that it is very low field, in range of 5–50 V/cm for B in range of 0.1–1 T, respectively. Corresponding potential drops even at size of spot ($r_s = 10\,\mu$m) consists of 5–50 mV. Therefore, it is not clear how such weak plasma field in the quasineutral plasma influences on any phenomena of spot ignition. In the above-cited works, such field was used to calculate a voltage which was considered as additional value to the cathode potential drop in the space charge sheath. Such adding is not relevant and not related to a process of spot ignition (see above). It should be noted regarding to the instability assumption. The assumption of plasma spot instability is not agreed with Kesaev [4, 5] observations, which indicated a vice versa fact, i.e., an increasing of spot stability observed in the presence of an applied magnetic field. Usually, the magnetic field influences the instability of intrinsic plasma parameters. So, the direction of plasma ejection can be formed randomly as result of energy dissipation due to the plasma instability (ion sound, acoustic, Buneman, and others) [93]. However, specific configurations of the magnetic field were used to prevent the random high-pressure plasma motion due to instabilities [94].

Lee [95] reported on a model of radial electric field formation by the gyrocenter shift due to the neutral density gradient caused by charge-exchange collisions of the ions and their elastic scatterings. This model, developed early [96] in the field of fusion plasma physics, was proposed to describe the arc discharges by the generalization of the theory resulting from a short mean free path λ_{ia} compared with the gyroradius r_L. When there is a magnetic field perpendicular to the gradient direction of ions and neutrals, the ion drifts perpendicular to the magnetic field such as $E \times B$ drift and diamagnetic drift contribute since ions lose their momentum at the charge-exchange collisions. When there is ion pressure gradient ∇p oriented in the same direction of electric field, the force acting on the ion is $eE - ev_D B - \nabla p/n_i$, where v_D is drift velocity. After the charge-exchange collisions of an ion with a neutral with a mean free path λ_{ia}, the ion starts a new acceleration since the motion of a neutral is independent from the electric and magnetic fields. The expression for the drift velocity was derived as [95]:

$$v_D = \frac{1}{1 + \frac{r_L^2}{\lambda_{ia}^2}} \left(\frac{E}{B} - \frac{\nabla p}{eBn_i} \right) \quad (19.3)$$

The expression (19.3) shows that when $r_L \gg \lambda_{ia}$, the velocity v_D tends to zero. This is the case for dense plasma in cathode spot. When the mean free path is comparable with gyroradius, the term with neutral density gradient in expression for radial current density becomes weak. It is because the gyrocenter shift was taken by the average over a circle of gyromotion, and many ions do not make a whole gyromotion in this case [95]. Therefore, only a limited number of ions fly longer than gyroradius. This point was accounted by the ratio of this number of ions to the whole ion number which is $\text{Exp}(-r_L/\lambda_{ia})$.

While the above model is not applicable to the cathode spot motion in a vacuum arc ($\lambda_{ia}/r_L \to 0$), the presented calculation for the case of experimental study using experiment [50] for discharge in an Argon indicate useful result. The neutral density of an Argon in front of the cathode sheath is assumed of 10^{15} cm^{-3}. It is also assumed that the gas pressure is proportional to the neutral density. The arc voltage dependence on the gas pressure follows the experimental measurement in which the arc current is constant, 120 A. The drift velocity was calculated at distance corresponding to 1/20 of the neutral decay length is plotted in Fig. 19.10 for the different gas pressures and magnetic fields. The results indicate how the ratio of the gyroradius to the mean free path of charge exchange varies the drift velocity since the gyroradius is a function of magnetic flux density and the mean free path is a function of gas pressure. It can be see that retrograde spot motion is appeared at low pressure of an Argon.

Pang et al. [97] discussed a hypothesis to explain their experiment for vacuum arc with van-type and spiral-type contacts in an applied transverse magnetic field. The hypothesis is based on Beilis theory [98, 99] according to which a spot or group spot is formed preferable under already existed plasma cloud impeded the expanding plasma plume and save the necessary plasma density for spot development. This model was confirmed by Pang and co-authors observations indicating that when the cathode spots are close, an overlapping region of plasma plumes is formed. So, according to the theory [98, 99] the overlap region impedes the plasma flow, increasing the near-cathode plasma density. New spots can be ignited more easily near these concentrated cathode spots in the overlap region, which is favorable for the formation of cathode spot groups.

They also indicate that for the contacts considered in the experiment both a transverse magnetic field and an axial magnetic field exist between the contacts. The collimating effect of axial magnetic field reduces the plasma density. Meanwhile,

19.1 Cathode Spot Motion in a Transverse Magnetic Field

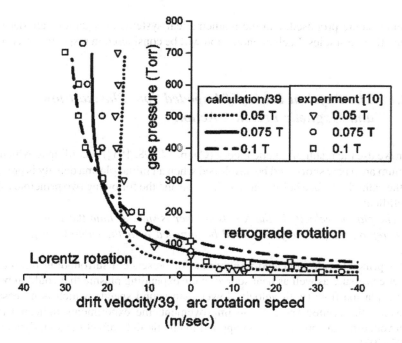

Fig. 19.10 Profiles of drift velocity calculated by gyrocenter shift and measured arc spot velocity in [50]. The calculated values in this figure were taken at the distance of 0.15 mm from cathode and divided by 39. Figure taken from [95]. Used with permission

the transverse magnetic field component drives the plasmas to move in a direction perpendicular to the magnetic field. The plasma density in this direction is increased and so is the possibility of the new cathode spots igniting, leading to a phenomenon in which the cathode spot group moves under the action of the transverse magnetic field. According to the model [99], a group spot and the retrograde spot motion were described for an arc in vacuum, while with increasing the gap pressure the spot tend to be separate individual passing to slow individual spot (SIS) type (see Chap. 7), which also was observed by Djakov and Holmes [100]. Some mixing situations between conditions in a vacuum and gap filled pressure were discussed in [97].

19.2 Spot Behavior and Impeded Plasma Flow Under Magnetic Pressure

The observed spot behavior includes the group spot formation, cathode spot motion, spot splitting, and effect of acute angle in the presence of a magnetic field. The nature of these phenomena is explained in the frame of the kinetic theory. The physical mechanisms are proposed, and the mathematical systems of equations are derived for quantitative studies. Each phenomenon will be considered in respective sections.

19.2.1 The Physical Basics of Impeded Spot Plasma Flow and Magnetic Pressure Action

Again we use the result of kinetic theory (Chap. 17, [98, 101]) according to which in a vacuum arc a new spot could be developed when an initial plasma density is present near the cathode surface before the spot ignition and the following two principle rules are fulfilled:

1. *The plasma velocity in the Knudsen layer is smaller than the sound velocity, and therefore, the plasma flow should be impeded near the cathode surface (first rule).*

This principle rule indicated requirement of presence a surrounding plasma or magnetic pressure as well as heat source in the expanding plasma that could impede the free plasma flow (with velocity lower than sound speed), which is necessary to develop the ignited spot. As an improvement, the experiments indicated that the surrounding gas pressure or an applied magnetic field affect the spot dynamics [5, 102–105].

2. *The heat losses due to heat conduction in cathode bulk should be smaller than the incoming energy to the cathode from the plasma for spot current and cathode temperature, which is necessary during the spot operation (second rule).*

Based on these two principle rules, the mechanisms of cathode spot motion, its splitting, and group spot formation explained taking into account the magnetic pressure action. A mechanism for spot motion consists in a new spot ignition under a plasma cloud or magnetic pressures in the direction where the pressure is sufficient to impede the plasma flow. But first, we describe the mathematical formulation of the forces and pressures due to current–magnetic field interaction.

19.2.2 Mathematics of Current–Magnetic Field Interaction in the Cathode Spot

Let us consider these phenomena when a magnetic field is present, taking into account the mentioned above two rules. An influence of the magnetic field on the plasma processes in the spot (electron diffusion, electron drift, etc.) depends on the Hall parameter $\beta = \omega\tau$, where ω is the electron cyclotron frequency and τ is the electron–ion collision frequency. The value of ω is $10^{10}-10^{11}$ s^{-1} for magnetic field 0.1–1 T. At the surface of a cathode spot, the plasma density is very high (about $10^{19}-10^{20}$ cm^{-3}); therefore, τ is about 10^{-15} s and β is very small ~10^{-5} to 10^{-4}. Thus, the influence of the external magnetic field on the spot plasma parameters (e.g., potential drop, space

19.2 Spot Behavior and Impeded Plasma Flow Under Magnetic Pressure

charge, heat and particle fluxes, etc.) in the dense spot plasma is negligible. However, the electrical current in the spot and in the cathode plasma jet in vacuum induces a self-magnetic field, which is superimposed with any applied external magnetic field. This interaction can influence the spot current and spot velocity. As the cathode plasma is highly ionized and highly electrically conductive, the magnetic diffusion time is estimated to be much shorter than the dwell time of the plasma in the vicinity of the spot. Thus, the magnetic field penetrates into the cathode plasma. In general, the plasma jet flow in a magnetic field is described by the momentum equation

$$\rho \frac{d\mathbf{v}}{dt} = -\nabla P + \mathbf{j} \times \mathbf{B} \tag{19.4}$$

and by Maxwell's equation

$$\mu \mathbf{j} = \nabla \times \mathbf{B} \tag{19.5}$$

where $\mathbf{B}$ is the magnetic field, $\mathbf{j}$ is the current density, $\mathbf{v}$ is the plasma flow velocity, p is the plasma kinetic pressure, ρ is the plasma mass density, and μ is the magnetic permeability.

Without external magnetic field in steady state in the case of small radial plasma expansion assuming radial plasma velocity $v_r = 0$, the force balance per unit volume in the radial direction r can be obtained from (19.4) to (19.5) in the following form:

$$\frac{dp}{dr} = -\frac{d(B_{sm}r)}{\mu r dr} B_{sm} \tag{19.6}$$

where B_{sm} is the self-magnetic induction and r is the radial coordinate. After integration of (19.6), the radial pressure balance is

$$p(r_0) - p(r) = -\frac{\mu I_{s0} j}{4\pi} \left(1 - \left(\frac{r}{r_0}\right)^2\right) \tag{19.7}$$

where r_0 is the radius of discharge cross section, I_{s0} is the spot current in the single spot. The difference between the value of the kinetic pressure on the discharge axis ($r = 0$) and its value at the edge of the discharge channel is:

$$p(r_0) - p(0) = -\frac{\mu I_{s0} j}{4\pi} \tag{19.8}$$

Equation (19.8) describes the pressure in the current channel when the magnetic pressure is also included. In the cathode spot, the kinetic pressure generally differs from the self-magnetic pressure p_{sm} ($p \gg p_{sm}$) and the difference between these terms depends on distance from the cathode in the expanding plasma. Taking into account that the plasma density decreases due to the expansion of the accelerated axial plasma, the self-magnetic pressure can become comparable to the kinetic pressure at a certain distance from the cathode.

As the pressure ratio is proportional to the current and the spot current grows, the plasma flow becomes unstable when the magnetic pressure exceeds the kinetic pressure [106]. In this case, the spot can extinguish, limiting the spot current rise. In addition, when an external magnetic field is applied, or produced by an adjacent spot, the non-uniform pressure difference in the radial direction determines the spot motion. Let us consider the effects of magnetic pressure on the cathode spots behavior.

19.3 Cathode Spot Grouping in a Magnetic Field

There the physical model [99, 107] and the results of calculations are presented demonstrating the spot current dependence on magnetic field which was observed in a vacuum arc.

19.3.1 *The Physical Model and Mathematical Description of Spot Grouping*

The experimental study [92, 100, 102, 103, 108, 109] shows that the spot current, and therefore, the current per single jet is limited to a material-dependent value. The cathode spot expansion on the cathode surface in a ring away from their origin as well as tendency of spot grouping with increase in magnetic field were also observed [51, 110, 111]. In order to understand the observations, it will be taken into account that the cathode plasma expands and accelerates to velocity V_s and that the plasma density, n, and the current density, j, subsequently decrease with distance along the *z-axis*-oriented perpendicular to the cathode surface. Also, the self-magnetic field increase with spot current for a single spot. Therefore, the relation between the pressure (p_{sm}) by self-magnetic field and kinetic pressure along the jet will depend on the plasma parameter distribution in the axially expanding cathode plasma.

Different models were used in order to calculate the dependence of the plasma distribution on the distance from the cathode. According to the calculation of the plasma jet expansion [101, 112, Chap. 18] for a given spot current in the range of 100–300 A, the plasma velocity passes the sound speed at a distance of about 2–3 spot radii r_s and then saturates at a distance of about 10 r_s. Also, this calculation shows that the plasma density decreases similarly to the current density after the distance of velocity saturation and both depends on the cross section of the expanding jet. A two-dimensional calculation of supersonic vacuum arc plasma jet expansion with a free boundary, i.e., without any assumption about the jet geometry [113], showed that an initially the jet remains conical near the origin and the expanded shape depends on the initial jet angle and spot current. According to the two-dimensional calculation [114]

19.3 Cathode Spot Grouping in a Magnetic Field

in high-current arcs the Joule energy dissipation in the jet is so small that the plasma temperature in the jet increases only weakly. Thus, the kinetic pressure p and pressure (p_{sm}) generated by the self-magnetic field can be comparable, for example, at a distance of plasma acceleration region, where the plasma density strongly decreases. In this case, the near-cathode plasma parameters will be disturbed, and as result, the spot can be converted by forming an additional new spot, which consistent with the changed self-magnetic field. Therefore, the self-consistent processes in the converted spots and in the plasma expansion region should be described for some individual spot or group spot current. Thus, the spot splitting and the increase in spots number will be sustained by an increase in arc current. When an external magnetic field is applied (induced a pressure p_{em}), the total magnetic pressure changes; hence, the spot current must also change to a larger spot current. In order to determine the spot current, two assumptions will be used in further quantitative modeling.

(1) The group spot splits if the self-magnetic pressure p_{sm} is comparable (for simplicity is equal) to the plasma pressure p anywhere in the accelerated cathode plasma, i.e., when the following condition is fulfilled:

$$\frac{p}{p_{sm}} = 1 \qquad (19.9)$$

(2) The external magnetic field affects group spot splitting when the self-magnetic pressure is equal to the pressure p_{em} generated by the external magnetic field, i.e., when:

$$\frac{p_{em}}{p_{sm}} = 1 \qquad (19.10)$$

Let us consider a single-group spot jet without an external magnetic field. The pressures decrease with distance from the cathode as the current density and plasma density both decrease and therefore relation (19.9) along the z-axis will be as:

$$p(z) = p_{sm}(z) \qquad (19.11)$$

where the axial dependence of the pressure (19.8) on the self-magnetic field $P_{sm}(z)$ is:

$$p_{sm}(z) = \frac{\mu I_{s0} j(z)}{4\pi}, \quad j(z) = \frac{I_{s0}}{A(z)} \qquad (19.12)$$

and where $A(z)$ is the cross-sectional area of plasma jet as a function of the distance z and $j(z)$ is the current density distribution along the z-axis, assuming that the expanding jet is uniformly distributed in the cross-sectional area $A(z)$. In the cathode plasma jet, the plasma kinetic pressure p is not an independent variable, but rather is determined by

$$p(z) = n_h(z)kT(z) + n_e(z)kT_e(z) \tag{19.13}$$

where n_e and n_h are the electron and heavy particle densities, respectively, and are functions of the distance z, k is the Boltzmann constant, $T_e(z)$ and $T(z)$ are the electron and heavy particle temperatures, respectively, and are likewise functions of the distance. The equation of mass conservation is

$$mV(z)n_h(z)A(z) = GI_{s0} \tag{19.14}$$

where m is the heavy particle mass, $V(z)$ is the plasma jet velocity, and $G(g/C)$ is the cathode erosion, i.e., the mass eroded per unit charge transfer. Using (19.12)–(19.14) and the fact that the plasma jet is highly ionized [101] (i.e., $n_h \sim n_e$), the relationship between the self-magnetic and kinetic pressures in the plasma jet can be obtained as:

$$\frac{p_{sm}}{p} = \frac{\mu I_{s0} m V(z)}{4\pi G k(T_e(z) + T(z))} \tag{19.15}$$

Taking into account the condition (19.11), and the asymptotic values of jet velocity V_j, and electron T_{ej}, and heavy particle T_j temperatures (see Chap. 18), the expression for the current per single spot can be obtained from (19.15) as

$$I_{s0} = \frac{4\pi G k(T_{ej} + T_j)}{\mu m V_j} \tag{19.16}$$

Let us consider large arc currents, i.e., $I \gg I_{s0}$, when numerous spots appear on the cathode surface. In this case, the spot current I_s depends on the relation between the self-magnetic and kinetic pressures in the plasma jet and the pressure from an external magnetic field P_{em} caused by superposition of the self-magnetic field of others spots [110] or by an applied magnetic field. The magnetic pressure can be expressed as:

$$p_{em}(z) = B_{em} j(z) r(z) \tag{19.17}$$

Taking into account (19.12), the ratio between the pressure caused by the external and the self-magnetic fields can be obtained as:

$$\frac{p_{em}}{p_{sm}} = \frac{4\pi B_{em} r(z)}{\mu I_s} \tag{19.18}$$

Using condition (19.10), the dependence of the spot current I_s on the external magnetic field is:

$$I_s = \frac{4\pi B_{em} r_{ef}}{\mu} \tag{19.19}$$

19.3 Cathode Spot Grouping in a Magnetic Field

where r_{ef} is the effective radius of the jet for which condition (19.10) is fulfilled, i.e., where the external magnetic field is comparable with spot's self-magnetic field.

19.3.2 Calculating Results of Spot Grouping

Without a magnetic field. The calculations are performed for a copper cathode and for plasma jet velocity $V_j \sim (1{-}2) \times 10^4$ m/s. The electron temperature is $T_{ej} \sim$ 1–2 eV, and the cathode erosion coefficient is $G \sim 30{-}100$ μg/C [101]. For these arc parameters, the calculation according to (19.16) gives the group spot current I_{so} of 50–200 A, in agreement with measured values [100, 102, 103, 108, 110]. The lower value of I_{so} is obtained for the low values of T_{ej} and G.

With a magnetic field. The current per group spot is calculated for copper cathode by (19.19). The calculated results are compared with the measurements of Chromoy et al. [110] where the linear dependence of the spot current on the magnetic field is clearly expressed. This dependence occurs, according to (19.19) some effective radius r_{ef} of the plasma jet for which condition (19.10) is fulfilled. Both measured and calculated spot current dependencies on the applied magnetic field are shown in Fig. 19.11, where it can be seen that good agreement is obtained when $r_{ef} = 0.01$ cm.

19.4 Model of Retrograde Spot Motion. Physics and Mathematical Description

In this section, the mechanism of retrograde spot motion developed by Beilis [99] for arcs in vacuum and in the presence of a surrounding gas pressure is described.

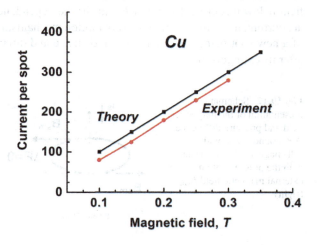

Fig. 19.11 Dependence of the group spot current on the magnetic field. Experiment [110], and the theory calculation according to (19.19) [99]

19.4.1 Retrograde Spot Motion of a Vacuum Arc

Let us consider a single spot on the cathode surface in a vacuum arc. In general, both the pressure by self-magnetic field and the plasma kinetic pressure are distributed axially symmetric (Fig. 19.12 left). In this case, according to the spot motion mechanism (Chap. 17), the spot moves randomly to a suitable nearby protrusion on the cathode surface in order to satisfy the second principle rule (low cathode heat loss). However, with an external magnetic field, the magnetic pressure resulting of the sum of external magnetic field B_{em} and the self-magnetic field B_{sm} is distributed asymmetrically and the larger pressure is on the side opposite to the Ampere direction (Fig. 19.12 right). According to the first rule mentioned above, ignition of a new spot is more probable under the larger pressure, and therefore, the spot appears to move in the retrograde direction.

As the magnetic pressure is $B^2/2\mu$, then the pressure difference between the two sides is $\Delta P = 2B_{em}B_{sm}/\mu$. Measurements [110] indicate that the spot current increases with the external magnetic field; therefore, the pressure difference ΔP and the pressure gradient will also increase with the external field in multi-cathode spot regime. Thus, the factors, which encourage selective ignition on the retrograde side, intensify with increasing external magnetic field strength, and hence the probability for new spot ignition on the retrograde side likewise increases with the external field, so that the directed retrograde spot velocity increases with the external field. However, this dependence saturates with the field because the maximal velocity is limited by the cathode surface state that determines the random spot velocity as well as by increased gap pressure. Also, the saturated velocity depends on the arc current. This is due to spot current dependence on arc current for certain external magnetic field [110]. According to observations [110], the arc current affects the group spot velocity in the dependence on B only near the saturated velocity. For magnetic field below a critical value of B, the velocity is independent of arc current but depends on the magnetic field. We consider two limited cases: (i) the dependence on magnetic field below this critical value of B and (ii) the dependence of saturated velocity on arc current, using the fact that the spot motion mechanism is based on the probability of a new spot being ignited at a given distance and direction from an existing spot after its extinguishing.

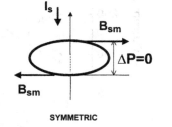 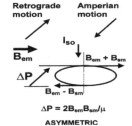

Fig. 19.12 Schematic presentation of the magnetic field and pressure difference ΔP around a spot with self-magnetic B_{sm} (left) and with the presence of an external magnetic field B_{em} (right)

19.4 Model of Retrograde Spot Motion. Physics and Mathematical ...

In order to calculate these dependencies, we assume that the new spot ignition probability is proportional to the magnetic pressure gradient across the spot, i.e., $\Delta P/2r$. Also, we consider a case when the new spot is ignited on the distance about radius of old spot r during the ignition time Δt and characterized by some plasma density (or mass density ρ_s). As the ratio $r/\Delta t$ is the spot velocity v_s, then according to the spot ignition probability the value of v_s^2 should be proportional to ΔP. Therefore, taking in account momentum equation and the above expression for ΔP, the spot velocity increase can be calculated formally using the following equation:

$$\rho_s v_s \frac{dv_s}{dr'} = \frac{2B_{em}B_{sm}}{\mu 2r_{ef}} = \frac{I_s}{2\pi r_{ef}^2} B_{em} \qquad (19.20)$$

where r' is the radial coordinate. Taking in account that the group spot current increases with magnetic field, as expressed by (19.19), (19.20) can be written as:

$$\rho_s v_s \frac{dv_s}{dr'} = \frac{2B_{em}^2}{\mu r_{ef}} \qquad (19.21)$$

The integration (19.21) taking in account that the new spot ignition is around r_{ef} where the velocity increases to v_s give:

$$\rho_s \frac{v_s^2}{2} = \frac{2B_{em}^2}{\mu} \qquad (19.22)$$

$$v_s = k_B B, \quad k_B = \left(\frac{4}{\mu \rho_s}\right)^{0.5} \qquad (19.23)$$

Equation (19.23) indicates that the spot velocity increases linearly with the magnetic field, in accordance with previous measurements [5, 52, 110].

According to the measurements, the dependence of the retrograde spot velocity on the magnetic field is saturated when the arc current I is lower or equal to the spot current, and therefore, the velocity depends only on the current and not on the magnetic field [110]. As the condition (19.10), i.e., $B_{em} = B_{sm}$ is fulfilled in case of $I \sim I_{so}$, then the new spot ignition probability is proportional to the self-magnetic pressure gradient around the spot, i.e., $2B_{sm}^2/\mu 2r$, and the saturated velocity dependence on the current is:

$$\rho_s v_s \frac{dv_s}{dr'} = \frac{1}{\mu r_{ef}} \left(\frac{\mu I}{2\pi r_{ef}^2}\right)^2 \qquad (19.24)$$

Taking in account that the new spot ignition is around r_{ef} where the velocity increases to v_s after integration (19.24):

$$\rho_s \frac{v_s^2}{2} = \frac{\mu I^2}{4\pi^2 r_{ef}^2} \qquad (19.25)$$

$$v_s = k_I I, \quad k_I = \frac{1}{\pi r_{\text{ef}}} \left(\frac{\mu}{2\rho_s}\right)^{0.5} \tag{19.26}$$

Equation (19.26) indicates that the spot velocity increases linearly with the arc current, in accordance with the measurements [110]. It should be noted that for relatively small values of arc current (<60 A) and magnetic field (<0.05 T) [5, 52], the velocity depends simultaneously on B and I (large transition range to saturated value) the case, which is not considered here.

19.4.2 Retrograde Spot Motion in the Presence of a Surrounding Gas Pressure

Let us consider surrounding gas pressure p_{sur} and the effect of a strong magnetic field. In the case of surrounding gas, there is a channel in the neutral gas that conducts the arc current between cathode and anode. This conduction channel is constricted due to thermal conduction into the surrounding gas, as only a constricted current flow can produce sufficient ionization and electrical conductivity. The ponderomotive force applied by the external magnetic field on the current I in the channel is proportional to the channel length L_{ch} and is given by the expression:

$$F = I B_{\text{em}} L_{\text{ch}} \tag{19.27}$$

According to the model proposed here, reversal spot motion occurs when the surrounding gas pressure p_{sur} in the discharge gap is larger than the pressure difference between the two spot sides generated by the self-magnetic field, i.e., when $\Delta p \ll p_{\text{sur}}$. Retrograde motion is not observed in pressures greater than 1 Torr for relatively large gap ~1 cm [78]. The estimation using (19.8) shows that self-magnetic field pressure p_{sm} will equal $p_{\text{sur}} = 1$ Torr when the current density is about 10^3–10^4 A/cm^2 for arc currents of about 100–10 A. According to the *first principle rule*, the new spot ignition without external magnetic field and large p_{sur} in different direction is equiprobable. However, when the magnetic field is present, the preferable motion is in the direction of the ponderomotive force acting on the conduction channel in the neutral gas. The velocity can be obtained from the equation of motion

$$m_{\text{ch}} \frac{dv_s}{dt} = F \tag{19.28}$$

where m_{ch} is the channel mass. Taking into account (19.27) and that $m_{\text{ch}} = \pi r_{\text{ch}}^2 L_{\text{ch}} m n_{\text{ch}}$, after integration of (19.28) the velocity is

$$v_s = \frac{I B_{\text{em}} k T_g t}{\pi r_{\text{ch}}^2 m p_{\text{sur}}} + C \tag{19.29}$$

where T_g is the surrounding gas temperature, m is the atom mass, r_{ch} is the channel radius, t is the time, C is the constant of integration. Equation (19.29) depends on relation r_{ch}^2/t, which is equivalent to the spot diffusion parameter [3].

Another circumstance occurred in Robson and Engel's experiment [30] when an arc burned in a gas-filled small electrode gap (e.g., 0.5 mm) and the conduction channel length, L_{ch}, was also small and the spot moved in the retrograde direction. This spot motion is due to asymmetric magnetic pressure distribution around the spot ($\Delta p/r \sim j B_{em}$). As the gap was small, the magnetic pressure was determined by larger current density near the cathode surface, than in the extended cathode jet. Therefore, the pressure of the surrounding gas can be lower than the magnetic pressure, which will impede the plasma flow during the spot ignition. In this case, the mentioned above *first principle rule* will be fulfilled because the asymmetric magnetic pressure although the surrounding gas is presented.

When the spot current I increase, the spot current density decreases [101] and then the magnetic pressure decreases as well as the magnetic pressure will be decreased when the external magnetic field decreases. This is the case in the experiment [30] where the reversal spot motion in short (0.5 mm) gas filled gap is observed when the current increases and the external magnetic field decreases. In a very strong external magnetic field (a few T), the ponderomotive force acting on the conduction channel grow and the spot moves according to (19.26) in the Amperian direction even in the arcs with gaps of about 2 mm [115].

Thus, according to the present model, the– transition of spot motion from one mechanism (ignition of a new spot under the asymmetric pressureretrograde motion) to the second mechanism (spot ignition due to the conduction channel displacement—Amperian motion) depends on the gas pressure, gap length, arc current, and magnetic field and agrees with the qualitative model in [67]. An excellent illustration of this transition is shown by Robson's experiment [115], where the arc moves in the retrograde direction when the external magnetic field B_{em} is lower than 1 T, while the motion changes to the Amperian direction when B_{em} increases in the range of 1.2–5 T.

19.4.3 Calculating Results of Spot Motion in an Applied Magnetic Field

19.4.3.1 Results of Retrograde Motion

In order to calculate the retrograde spot velocity, the proportional coefficients in (19.23) and (19.26) are required. Both these coefficients depend on the plasma density ρ_s. This parameter is calculated previously from spot theory [101, Chap. 17] and can be compared with the values obtained from (19.23) and (19.26). The calculation shows that using (19.23) the experimental value of the proportional coefficient $K_B =$

(2–10) × 10^4 cm/(sT) could be obtained when the plasma density ρ_s is about 0.01–0.1 g/cm^3. Using (19.26), the experimental value of $K_I = (0.2$–$0.25) \times 10^2$ cm/(sA) [110] can be obtained when the plasma density ρ_s is about 0.2 g/cm^3. As follows from these results, the density ρ_s in both cases can occur in the plasma of spot region [101, Chap. 17]. In order to obtain more exact results, the self-consistent solution of the spot theory adding (19.23) and (19.26) is necessary. Here the calculations are performed in order to demonstrate the measured linear dependencies of the spot velocities on B and I.

19.4.3.2 Strong Magnetic Field. Amperian Spot Motion

The calculation of the reversal spot motion in strong magnetic field is compared with Robson measurements (air + argon, $p_{sur} = 30$ Torr) [115] using (19.29). The constant C is obtained using the measured initial values of magnetic field and velocity when the reversal motion is appearing, i.e., C is equal 509 and 173 for two currents $I = 5$ and 10 A, respectively, and initial $B = 1.5$ T [115]. The parameter r_{ch}^2/t of spot motion in (19.29) depends on the conditions (cathode surface state, gas state, and others) in the experiment [115]. As the condition in the experiment [115] is unknown, this parameter cannot be determined by independent method.

Assuming that the relation r_{ch}^2/t weakly depends on B the calculated (solid lines, Tg ~ 1000 K) according (19.29) and the measured (points) results for arc currents 5 and 10 A show a linear dependence of v_s on B (Fig. 19.13). The good quantitative agreement between the experimental and calculated results (Fig. 19.13) was obtained for $r_{ch}^2/t = 3.1 \times 10^3$ cm^2/s when $I = 10$ A and for $r_{ch}^2/t = 2.7 \times 10^3$ cm^2/s when $I = 5$ A. These values are much higher than that measured without magnetic field for 100 A vacuum arc (20–30 cm^2/s) in [3, 92]. It should be noted that the parameter

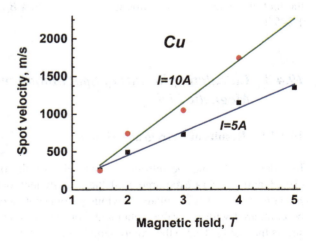

Fig. 19.13 Amperian spot velocity dependence on strong magnetic fields. The points are from Robson's experiment [115]. The lines are the calculated dependencies for 5 and 10 A

r_{ch}^2/t could be reduced taking into account the drag force. The channel radius r_{ch} of arc column in cold gas and strong magnetic field should be calculated from additional specific model.

19.5 Cathode Spot Motion in Oblique Magnetic Field. Acute Angle Effect

When a magnetic field obliquely intersected the cathode, a specific effect was observed for long time ago (1948). The effect consisted in a drift motion of cathode spot under some angle θ to the retrograde direction, known as effect of "acute angle". A unique experiment later was conducted by Robson [116], which demonstrated different drift angle dependencies. He measured the drift angle θ between the retrograde direction and the cathode spot motion direction (see Fig. 19.14) as a function of the acute angle of the magnetic field with respect to the cathode surface φ, the current I, and the magnetic field strength B. Below the drift mechanism and the mentioned dependencies were calculated using a model of retrograde spot motion crossed the normal component of magnetic field developed by Beilis [117].

19.5.1 Previous Hypotheses

There have been limited attempts to theoretically study the drift angle. An attempt to explain the mechanism of acute angle was conducted by Litvinov et al. [118]. The model of emission center (EC) (Chap. 15) was used, and the current of returned electrons from the plasma to the cathode surface at the periphery of the spot was taken into account. An interaction of assumed ring-shaped currents of the returned electrons with the magnetic field was considered. The author' noted that magnetic field tends turn the ring current in the anti-Amperian direction, while symmetry of the magnetic field of the spot current is not violated. A calculating formula was presented as a final result without details of it derivation. This relation limits the calculation only by a dependence between the drift angle and the angle inclination of the field. Even this final relation consists of arbitrary input parameters and assumptions (including condition $R \gg r_s$, R is the radius of an electron ring and r_s is the spot radius).

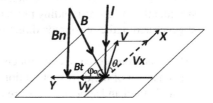

Fig. 19.14 Components of the oblique magnetic field and of the cathode spot velocity with on the cathode plane. Retrograde motion is in the x- direction

A similar model using the ring-shaped currents of the returned electrons is also considered by Barengolts et al. [119]. It was claimed that the occurrence of new EC is connected with the returned electrons from the plasma. The new EC is appeared at the location, at which the reverse electron current is larger. In addition, a more detail consideration of the above model and relation [118] was presented by the authors [119]. They take in account the asymmetry magnetic field distribution by the superposition with the self-magnetic fields due to electron ring current and EC current. The authors indicated the already known opinion that a higher magnetic field was found on the retrograde side. According to this fact was assumed that a higher magnetic field leads to a higher kinetic energy of the particles due to the electromagnetic energy [**E** × **H**] dissipation, which reproduce on impact with the surface of the cathode (see above [89]). Thus, the conditions for vaporization a tip on the retrograde side are more favorable than on the Lorentz side.

However, number of questions arise indicating strong weakness of both approaches [118, 119]. According to self-consistent solution of the cathode spot parameters [101, (Chaps. 16 and 17)], the current of returned electrons is significantly lower than the current of electron emission from the cathode, ion current, and the total current of the cathode spot, and therefore, this reverse current is insufficient for contribution any cathode heating, besides of some contribution for specific cases. Furthermore, the reverse electrons are formed in the thin space charge sheath (~0.01 μm) that is significantly lower than the Larmor radius (~10 to 100 μm). Also, it is not understandable how a large density reverse electrons can be occurred far from the spot location considered for case of large distance between two separately spots.

The drift cathode spot motion was considered by Garner [120]. He used the retrograde model of the work in [99] and described above. To reach the goal the author assumed that density of the ion space charge n_{sp} in the cathode sheath move with the spot velocity v_s and produce a current density j_{sp} in the retrograde direction (Fig. 19.12). It was also assumed that this current generate an azimuthal magnetic field B_{az} producing additional asymmetry of magnetic pressure in azimuthal direction by interaction with the self-magnetic field of spot current directed to the cathode surface. Just a relation between drift angle θ and the acute angle φ was derived.

However, the value of magnetic field B_{az} in Garner's model should be determined by a thickness of the space charge sheath and not by radius r_{ef} of the cross section of the expanding quasineutral plasma, as it was done by Garner (in the final published version [120]). The main lack of this point is that the sheath thickness is a size of a Debye radius r_D (~10^{-6} cm), i.e., very small. In this case, the calculation shows the value of $B_{az} = j_{sp}\mu r_D/2 \approx 10^{-5}$ T, i.e., is very small (by used j_{sp} 1.4 × 10^9 A/m^2), and consequently, the corresponding pressure is significantly smaller in comparison with that generated in the retrograde direction. As result, the drift angle is also negligible small.

Thus, the considered above models not presented an adequate understanding of the mechanism of the drift spot motion. Also, the dependencies of the drift angle on a magnetic field and spot current were never considered. In particular, the drift angle dependencies on magnetic field and acute angle measured by Robson [116] weren't

19.5 Cathode Spot Motion in Oblique Magnetic Field. Acute Angle Effect

analyzed previously and have not been understood. There are several remaining questions "What is the mechanism of spot drift in an oblique magnetic field?" and "How do the magnetic field components influence the drift spot motion?" Below an idea will be proposed which allow to study the mechanism for cathode spot drift in an oblique magnetic field and calculate the drift angle dependencies on the current, the magnetic field strength, and on the acute angle of the magnetic field with respect to the cathode surface.

19.5.2 Physical and Mathematical Model of Spot Drift Due to the Acute Angle Effect

The current carrying plasma in a magnetic field is subject to the momentum equation

$$\rho d\mathbf{v}/dt = -\nabla p + \mathbf{j} \times \mathbf{B} + n_{sp} e \mathbf{E} \tag{19.30}$$

and by Maxwell's equation

$$\mu \mathbf{j} = \nabla \times \mathbf{B} \tag{19.31}$$

where e is the electron charge, $n_{sp} = n_e - Zn_i$, n_e is the electron density, n_i is the ion density, and Z is the average charge of the ions, $\mathbf{B}$ is the magnetic field, $\mathbf{j}$ is the current density, $\mathbf{v}$ is the plasma velocity, p is the plasma pressure, ρ is the plasma mass density, and μ is the magnetic permeability.

The magnetic field vector $\mathbf{B}$, which obliquely intersects the cathode surface with an acute angle φ, may be decomposed into two components, B_n normal to the cathode surface (directed along the $-z$-axis) and B_t tangential (directed along the y-axis) to the cathode surface (Fig. 19.14). The cathode spot motion, as evidenced by the track left on the cathode surface [116, 121], may be described by a velocity vector $\mathbf{v}$ which may be decomposed into two components: v_x in the retrograde (+x) direction, and v_y in ($-y$) perpendicular to the retrograde direction. The angle θ between $\mathbf{v}$ and the retrograde direction (Fig. 19.14) can be defined as the drift angle and may be described by:

$$tg\theta = \frac{v_y}{v_x} \quad \text{and} \quad \theta = \text{arctg}\left(\frac{v_y}{v_x}\right) \tag{19.32}$$

In the following derivation, a model will be formulated for determining the velocity components v_x and v_y that can be used to predict the drift angle θ. In general, the mathematical problem formulated by (19.30)–(19.32) is complex and its analysis is complicated. Therefore, below a simple approach was developed in order to understand the physics of spot drift.

Let us consider the mechanism of the apparent cathode spot motion (see above two principle rules, Chap. 17). In first, to understand the idea let us again detail the mechanism of spot motion without and with magnetic field (retrograde case). Without a magnetic field, the cathode spot emits a plasma jet which flows significantly larger in the axial direction than in radial flow [122]. The axial flow produces the visible plasma in the interelectrode region at some distance from the cathode, but a portion of the plasma flows parallel and close to the cathode surface. The cathode spot is in a fixed location during its lifetime, and it evacuates the melting metal producing a crater at the cathode surface. As the crater deepens, the heat losses in the bulk solid, and therefore, the voltage necessary to sustain spot operation, increase.

At some time, operation of a new spot becomes favorable. The plasma close to the cathode surface concentrates the interelectrode voltage across a thin sheath and produces a sufficiently strong field, which form a heat flux, supporting the development of a new neighbor spot whose operation is more favorable at a tip than that of the old spot at the bulk. Thus, a new spot might ignite on some protrusion on the cathode surface, such as debris ejected earlier, or the rim of the crater evacuated by the old cathode.

Without any transverse magnetic field, the self-magnetic pressure is circularly symmetric around the cathode spot, and there is no preferred direction for ignition of a new cathode spot. The probability of dead the old spot and new ignition besides the above-described mechanism can be also determined by relation between self-magnetic and own plasma kinetic pressures [99]. Thus, the location of the new ignition is random. The sequence of new cathode spots igniting and old cathode spots dying produces random walk motion of the cathode spot location.

For a new cathode spot to ignite, the following processes must occur: (1) The radial (or lateral [91, 92]) plasma expansion should be such that there is sufficient plasma to ignite the new cathode spot location; and (2) the new spot must develop, and in particular the local cathode surface must be sufficiently heated for enough evaporation of cathode material and sufficient electron emission, so that the new spot can be sustained. Calculations [123] showed that the heating time is in range of 10–100 ns depending on the initial plasma density, current, and current density. The velocity of spot is in range of 10^2–10^4 cm/s (Chap. 7). The time required for a 10 μm spot to move 10 μm will be 0.1–10 μs, i.e., longer than cathode heating time. Thus, ignition of a new spot in a preferable direction is limited by condition (1), i.e., the time required for the direct plasma to expand.

Consider first the case where the magnetic field has only a tangential component, aligned in the ($-y$) direction and the model of the retrograde spot motion developed for an arc subjected to a tangential magnetic field. Superposition of this field with the azimuthal self-field generated by arc current flow from the plasma into the cathode produces a stronger magnetic field, and hence a larger magnetic pressure, on the ($+x$) side of the cathode spot. This larger magnetic field impedes the plasma flow at the ($+x$) side, producing a higher plasma density (non-free expansion) in this direction. In contrast, on the ($-x$) side, imposition of the external magnetic field weakens the total magnetic pressure so that the plasma flows freer in ($-x$) direction, and hence,

19.5 Cathode Spot Motion in Oblique Magnetic Field. Acute Angle Effect

the plasma density on the ($-x$) side is lower than on the ($+x$) side. New cathode spots are more likely to be ignited where the plasma is most impeded, and the plasma density is highest, namely in the ($+x$) direction, which is the $-\mathbf{j} \times \mathbf{B}$, or retrograde direction [99].

Now let us note the character of spot motion. As mentioned above, under the large magnetic pressure the plasma density distribution becomes asymmetric, with a larger density on the $+x$ side of the cathode spot. Thus, the cathode is heated more on this side, leading to more evaporation and eventually more new plasma production on this side, and finally igniting a new spot on this side. With this new spot, the process of producing another new spot in ($+x$) direction repeats. It should be noted that continuous or stepwise character of this motion depends on the magnetic field strength and surface relief [38]. We will consider this cycle of igniting new spots on the $+x$ side of previous spots assuming a continuous radial expansion of an accompanying plasma that also observed optically [32, 124] and by tracks [116].

This approach provides the basis for retrograde motion of the cathode spot, and it was found that the velocity of the cathode spot can be expressed by using (19.20) as

$$v_x = \left(\frac{I B_t}{\pi m n_h r_{ef}} \right)^{0.5} \quad B_t = B \cos \varphi \tag{19.33}$$

where r_{ef} is the distance in the x-direction from the center of the "old" cathode spot to the location, m is the atom mass, n_h is the heavy particle density, and I is the spot current.

The ***imposition of an oblique magnetic field*** modifies the above-described mechanism. Let us consider the case where the imposed magnetic field also has a normal component B_n, i.e., the field is oblique. It is well known that an electric field is induced when a conducting plasma moves across a magnetic field. In general, if we consider a fixed laboratory frame and designate the electric and magnetic fields therein as $\mathbf{E}$ and $\mathbf{B}$, and another frame, attached to the plasma, which has a velocity $\mathbf{v}$, that is much less than the speed of light, then the electric field in the plasma frame, $\mathbf{E'}$, is given by [106, 125]:

$$\mathbf{E'} = \mathbf{E} + \mathbf{v} \times \mathbf{B} \tag{19.34}$$

We apply (19.33) to the plasma moving in the x-direction (Fig. 19.14) from the "old" cathode spot to the site of the "new" spot, across the normal component B_n of the magnetic field. The electrical current in the cathode spot plasma is mostly normal to the cathode surface, i.e., in the $-z$-direction (Fig. 19.14). Accordingly, we neglect the current flow in the other directions, i.e., we take $j_x = 0$ and $j_y = 0$, and hence $E_x = 0$ and $E_y = 0$. Then the electric field has a y-component $E'_y = v_x B_n$ in the system coordinate moving with the plasma [125]. This electric field will exert a force on the positively charged particles whose force density in (19.30) is

$$F_y = n_{sp} e E'_y \tag{19.35}$$

A positive space charge n_{sp} can be formed not only in the sheath, but also in a transition region between quasineutral plasma and the sheath adjacent to the cathode and also named presheath [122] (Fig. 19.15). Its value depends on the degree of plasma ionization. The condition that $Zn_i > n_e$, ($\rho_{sp} > 0$ and $F_y > 0$) is fulfilled in the presheath region, where the ions begin to be separated from the electrons and where the returned electrons are formed. The ion velocity distribution function will be shifted in the y-direction by a velocity v_{oy} due to the F_y force (Fig. 19.15). Therefore, during the spot lifetime, an ion flux Γ_y in y-direction is formed due to asymmetric function distribution in according to

$$\Gamma_y = n_{sp}\left(\frac{m}{2\pi kT}\right)^{1.5} \int_{-\infty}^{\infty} v'_y e^{-\frac{m(v'_y-v_{oy})^2}{kT}} dv'_y \int_{-\infty}^{\infty} e^{-\frac{mv_x^2}{kT}} dv_x \int_{0}^{\infty} e^{-\frac{mv_z^2}{kT}} dv_z \simeq n_{sp}v_{oy}$$

(19.36)

This ion flux is important for new spot ignition near the old spot in the y-direction. Let us consider the probability of igniting the spot at a location which is shifted in the y-direction from the retrograde direction. Again, let us consider that the plasma expands from the old spot with smaller radial flow close to the cathode surface [122] than in the axial direction. Under the force F_y, (19.35), the peak of the plasma density distribution is shifted in the y-direction (19.36). Thus, the new cathode spot will be ignited at a location displaced in the y-direction from where it would have been with a purely tangential external magnetic field. This process of new spot ignition repeats and produces an apparent motion of the cathode spot which is displaced in the -y-direction from the +x retrograde direction.

We also assume that: (i) only the tangential component B_t determines the spot velocity in retrograde direction, v_x, which in turn determined the field E_y and (ii) the force produced by electric field component E_y' determines the component of the spot velocity perpendicular to the retrograde direction, v_y. The velocity v_y achieved between ignitions of cathode spots may be calculated by integrating the equation of motion using (19.35):

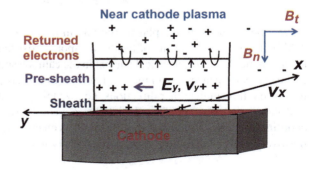

Fig. 19.15 Schematic presentation the cathode plasma in the spot including cathode surface, space charge sheath and presheath, returned plasma electrons and the near-cathode plasma

19.5 Cathode Spot Motion in Oblique Magnetic Field. Acute Angle Effect

$$mn_h v_y \frac{dv_y}{dy} = F_y = en_{sp} v_x B_n \quad B_n = B \sin \varphi \tag{19.37}$$

where m is the ion mass, n_h is the heavy particle density. Integrating (19.37) in the range of velocity from 0 to v_y and from $y = 0$ to $y = r_{ef}$, we obtain:

$$v_y = \sqrt{\frac{2ev_x B_n r_{ef} n_{sp}}{mn_h}} \tag{19.38}$$

This is the velocity at the end of the period of new spot appearing, i.e., when the space charge plasma arrives at the site of the new spot. Further, the velocity component of spot shifted in y-direction will be determined by (19.38). Using (19.32) and (19.38), the drift angle θ can be obtained as:

$$\theta = \text{arctg}\left(\frac{v_y}{v_x}\right) = \text{arctg}\left(\sqrt{\frac{2e B_n r_{ef} n_{sp}}{m v_x n_h}}\right) \tag{19.39}$$

19.5.3 Calculating Results of Spot Motion in an Oblique Magnetic Field

Equations (19.33) and (19.39) were calculated for an Al cathode to obtain the drift angle θ as functions of magnetic field inclination angle φ, magnetic field B, and the arc current I. The dependencies thus obtained were then compared to those measured by Robson [116]. Theoretical study [126] of the spot parameters on Cu and Al cathodes showed that the heavy particle densities n_h were around 5×10^{19}–5×10^{20} cm^{-3} with a high degree of ionization. Considering these data, the drift angle θ was calculated for $n_h = 10^{20}$ cm^{-3} and for highly ionized plasma with $n_{sp}/n_h = 0.5$.

It was assumed that arc currents used in Robson's experiments are equal to the spot current. The calculations of (19.39) in order to describe Robson's experimental dependencies should be conducted by choosing parameter r_{ef}. In general, this parameter can be obtained self-consistently using the closed theory of spot development and plasma expansion taking into account the cathode surface, kinetic, and hydrodynamic plasma phenomena [123, Chap. 17]. Such investigation is very complicated and is beyond the scope of demonstration the model. Therefore, r_{ef} was chosen to match Robson's experimental dependencies [116] with maximal accuracy and then obtained r_{ef} will be compared with spot radii previously calculated from spot theory. The calculated data are presented by dotted lines in the below figures while the solid lines present experimental data.

Figure 19.16 shows the drift angle θ as function of magnetic field inclination φ for magnetic field strengths of $B = 0.05$, 1.0 and 1.5 T and $I = 2.8$ A using $r_{ef} = 40$ μm. Both the calculated and measured drift angles increased with φ, and their values were close for $\varphi > 30°$. Figure 19.17 shows that with $I = 2.8$ A, θ increases with B and that the calculated and experimental results agree when $r_{ef} = 25$ μm was used at $\varphi = 15°$ and when $r_{ef} = 35$ μm was used at $\varphi = 45°$.

The drift angle θ as a function of the arc current is presented in Fig. 19.18 with φ as a parameter and $B = 0.1$ T. In this calculation, it was assumed that spot current density remains constant when the current increases, and therefore, the spot area increased proportionally with arc current. As a result, $r_{ef}(I)$ was determined by relation $r_{ef}(I) = r_{ef}(I_0)(I/I_0)^{0.5}$, where $r_{ef}(I_0)$ is the effective radius at $I_0 = 2.8$ A.

In general, Fig. 19.18 shows that the calculated and measured $\theta(I)$ agree relatively well when $r_{ef} = 20$ μm at $\varphi = 15°$ and $r_{ef} = 50$ μm at $\varphi = 45°$. However, the measured rate of rise of θ with I was larger than that calculated for $\varphi = 45°$ while these rates are in good agreement for $\varphi = 15°$. The matched r_{ef} varied with φ because the relation between axial and transverse components of magnetic field can influence spot characteristic such as spot type, spot stability, and the plasma distribution in the expanding region.

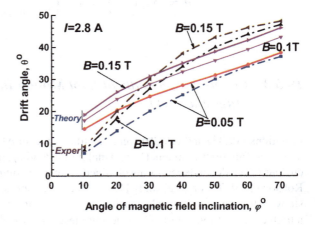

Fig. 19.16 Drift angle θ as function of the acute angle φ of the oblique magnetic field. Experiments [116] —solid lines, calculations—dotted lines

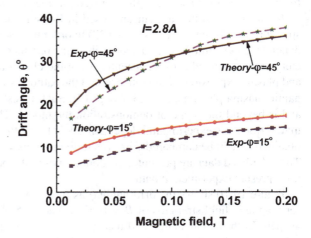

Fig. 19.17 Drift angle θ as function of the strength of oblique magnetic field with the acute angle φ of the oblique magnetic field as a parameter. Experiments [116]—solid lines, calculations—dotted lines

19.5 Cathode Spot Motion in Oblique Magnetic Field. Acute Angle Effect

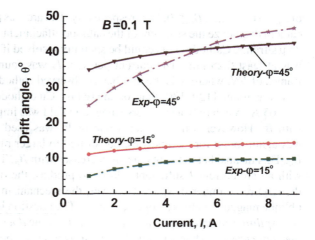

Fig. 19.18 Drift angle θ as function on the current I with the acute angle φ of the oblique magnetic field as a parameter. Experiments [116]—solid lines, calculations—dotted lines

Thus, the theory developed here agrees with Robson's results for r_{ef} in range of 20–50 μm for different dependencies, and over the relatively wide range of B and φ tested by Robson. Taking into account that the calculated [123] spot current density for the Al cathode is $(1–2) \times 10^6$ A/cm² ($I = 10$ A), and the spot radius r_s is in range of 13–20 μm and therefore r_s can be comparable with r_{ef}.

19.6 Spot Splitting in an Oblique Magnetic Field

Let us consider the spot behavior and its characteristics when a magnetic field obliquely intersected the cathode. The mechanism and calculated approach developed by Beilis [127] are described.

19.6.1 The Current Per Spot Arising Under Oblique Magnetic Field

As it was mentioned above, the experiments [92, 100, 102, 103, 108] indicated that the current per spot depends on the arc conditions, electrode material, arc time, magnetic field, etc. In a magnetic field parallel to the cathode surface (named as transverse or tangential), the current per spot increased with the magnetic field strength [52, 110]. In the last decade, the current per spot was investigated also in fields which obliquely intercept the cathode surface [128–130]. The arc was operated with constant current rectangular pulses with amplitude up to $I = 300$ A and durations up to 4 μs. Two pairs of coils produced a uniform magnetic field with components normal and tangential to the cathode surface, whose strengths could be varied independently in the ranges of $B_n \leq 0.35$ T and $B_t \leq 0.25$ T, respectively. The arc was photographed by a framing camera to analyze the structure of the cathode attachment of the arc with an exposure 25 μs/frame. Cathode spots could be spatially resolved if the distance between them was at least 0.25 mm. The current per spot I_s was found as the mean over many frames of I/N, where N is the number of observed cathode spots.

It was found [129, 130] that for an arc on a Cu cathode without a magnetic field, $I_s \approx 65$ A. When only a transverse magnetic field was imposed, I_s increased linearly with B_t. However, when a fixed value of B_n was used imposed, a linear relation was again found for I_s versus B_t, but shifted to larger ranges of B_t with increasing B_n. This means that the spot current decreases with B_n. These results were obtained with an arc current I sufficiently large to produce the observed dependences, i.e., the required number of spots. However, the mechanism of spot splitting under an oblique magnetic field is an open question. The question is: *what is the mechanism of spot splitting in an oblique magnetic field and how does their component influence the spot current?* Thus, an actual problem is how can be determine the mechanism for cathode spot splitting in an oblique magnetic field and to formulate a model which predicts the current per spot, I_s, and explains the influence of the tangential and normal field components, B_t and B_n.

19.6.2 Model of Spot Splitting

The model is based on the relation between the magnetic pressures produced by the self-magnetic field generated by current flow in the cathode spot and the plasma jet, and the applied external magnetic field [99]. The plasma jet generated in the cathode spot expands away from the cathode (in the z-direction). Calculations [101, 122] showed that the plasma density decreases similarly to the current density and both decrease with increasing cross section $A(z)$ of the expanding jet. The kinetic pressure P_k and pressure generated by the self-magnetic field p_{sm} can be comparable, for example, in the expanding plasma acceleration region, where the plasma density strongly decreases and reaches a relatively low value [99]. If p_{sm} is comparable to p, the inward magnetic pressure contracts the flow to some degree and disturbs the near-cathode plasma parameters. In this case, self-consistent processes of the spot (including cathode heating, emission, evaporation, and atom ionization) and in the plasma expansion region can be sustained at a particular spot current, which depends on the magnetic pressure. When the arc current increases, spots split and the number of spots increases. When an external magnetic field is applied, the magnetic pressure changes; hence, the spot current will be also changed.

According to the above description, the spot splits if anywhere in plasma jet $p_{sm} \geq p$, or if $p_{em} \geq p_{sm}$ where p_{em} is the external transverse magnetic field pressure. These criteria explain the observed spot current increase with external transverse magnetic field (see above). Following the idea of self- and external magnetic field interaction during the cathode plasma expansion, let us take in account the influence of magnetic

19.6 Spot Splitting in an Oblique Magnetic Field

pressures produced by the normal and tangential components of an oblique magnetic field.

In order to understand the observation [129], it should also be taken into account that the self-magnetic field in a single spot increases with spot current and the axial dependence of the pressure by the self-magnetic field is:

$$p_{sm}(z) = \frac{\mu I_s j(z)}{4\pi}, \quad j(z) = \frac{I_s}{A(z)} \tag{19.40}$$

where $j(z)$ is the current density along the z-axis in the expanded jet and is assumed uniform in the radial direction within the cross-sectional area $A(z)$. In the cathode plasma jet, the plasma kinetic pressure is determined by

$$P_k(z) = n_h(z)kT(z) + n_e(z)kT_e(z) \tag{19.41}$$

where n_h and n_e are the electron and heavy particle densities, respectively, and are functions of the distance. The heavy particle density was determined by the mass conservation equation:

$$mV(z)n_h(z)A(z) = GI_s \tag{19.42}$$

where m is the heavy particle mass, $V(z)$ is the plasma velocity, and G is the cathode erosion coefficient in g/C. The magnetic pressure produced by the tangential magnetic field, and the current density can be written as:

$$P_{tm}(z) = B_t j(z) r(z) \tag{19.43}$$

The magnetic pressure P_{nm} produced by normal magnetic field can be used in the following form

$$P_{nm}(z) = \frac{B_n^2}{2\mu} \tag{19.44}$$

Taking into account that the magnetic pressures due to self-magnetic and normal magnetic fields can be balanced by the kinetic pressure and pressure due to tangential magnetic field, the pressure balance can be written using (19.40)–(19.44) as:

$$\frac{\mu I_s^2}{4\pi^2 r_{ef}^2} - \left(M + \frac{B_t}{\pi r_{ef}}\right) I_s + \frac{B_n^2}{2\mu} = 0, \quad M = \frac{GDkT}{mV\pi r_{ef}^2} \quad D = (1 + \alpha\theta), \tag{19.45}$$

where α is the degree of ionization, $\theta = T_e/T$, M is a parameter characterizing the plasma kinetic pressure, and r_{ef} is the effective radius of the jet for which condition (19.45) is fulfilled. The solution of (19.45) is:

$$I_s = \frac{N + (N^2 - R^2)^{0.5}}{Q}$$

$$N = M + \frac{B_t}{\pi r_{ef}} \quad R = \frac{B_n}{\sqrt{2\pi} r_{ef}} \quad Q = \frac{2\mu}{4\pi^2 r_{ef}^2} \quad (19.46)$$

Thus, the different components of the external magnetic field and the magnetic pressure which they produce influence the spot current. Equation (19.46) determines the influence of the applied magnetic field on the spot current.

19.6.3 Calculating Results of Spot Splitting

In general, the complete cathode spot model (Chaps. 17 and 18) together with (19.45) can be used to determine self-consistently the parameters including V, G, plasma density and temperatures taking into account the applied oblique magnetic field. As this approach is very complicated, a simple approach is considered below, which uses measured values of these parameters to demonstrate the effect of the normal component of the oblique magnetic field [129]. The calculations were conducted for a copper cathode, and for plasma jet velocity $V \sim 1.5 \cdot 10^4$ m/s, $D = 4$, $G = 30$ μg/C [101]. The resulting dependences of I_s on B_t with B_n as parameter (dotted lines) and the measured results [129] (solid lines) are presented in Fig. 19.19, and it is shown that I_s decreases with B_n.

It can be seen that the spot current linearly increased with B_t for constant B_n which is in accordance also with early published experimental results [52, 110]. The calculated shifts with B_n and the slope of the I_s-B_t curves agree well with measured dependencies [128, 129] at some effective radius r_{ef} (shown in Fig. 19.19) of the cathode plasma for which condition (19.45) fulfilled. The values of r_{ef} decreased with B_n. This tendency agrees with the theoretical result [122] and with experimental

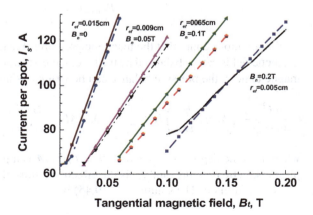

Fig. 19.19 Calculated (dotted lines) and measured in [129] (solid lines) dependencies of current passed by a cathode spot on the tangential magnetic field B_t with normal magnetic field B_n as a parameter for values indicated in the figure (from [129]). The respective effective radius r_{ef} for different B_n is indicated also

observation that the radial plasma jet cone angle decreased with the axial magnetic field [131].

19.7 Summary

In general, the arc behavior in the presence of a magnetic field undergo different changes including cathode spot retrograde motion, drift motion, spot splitting, and spot grouping. The phenomena of retrograde cathode spot motion in the presence of a transverse magnetic field were detected experimentally in 1903 and have been studied over the century. The numerous hypotheses were published, and the main attempts to explain the phenomena were reviewed and can be divided into few following groups.

(1) The first hypotheses were associated with following assumptions: (i) presence of a ring-shaped electrical current near the cathode surface; (ii) use a high-speed cathode jet consisted only of positive ions which can be deflected by Lorentz force in the Amperian direction while causing the spot motion in reverse direction; (iii) the thermomagnetic and galvanomagnetic forces produced due to negative Righi–Leduc effect and by the Ettingshausen effect according to which the heat flux and electrical current in a transverse magnetic field generated asymmetric temperature distribution at the cathode surface causing an asymmetric electron emission; (iv) curved line of the current density arising in the arc in the vicinity of the spot.
(2) The numerous works considered a shifting of positive space charge relatively to the negative charges at the cathode surface because (i) the different influences of the magnetic field on electrons and ions; (ii) pulled in retrograde direction of ions generated from atom ionization by the emitted electrons declined in the Lorentz direction; (iii) declined ion motion in a magnetic field resulting in produce a curvature potential tube, through which the emitted electron flows producing new ions due to the ionization in the retrograde side of the old spot.
(3) A number of model associated with effects of asymmetry (i) of resultant magnetic field taking into account superposition with self-magnetic field and assuming an enhancement of spot stability or an enhancement of electron field emission under maximal field at the retrograde side of the spot; (ii) of the neutral gas ionization in the plasma region by the absorption of the electromagnetic energy, which is maximal at the side of maximum magnetic field; (iii) of electric field and voltage generated in the quasineutral plasma above the cathode surface and caused by the Hall effect, by high velocity plasma motion ($\mathbf{V} \times \mathbf{B}$); (iv) bend of the plasma to the retrograde side due to magneto-hydrodynamic instability of the plasma column at some distance from the cathode surface; (v) of electron emission, which reduce in "forward" direction and was increased on the retrograde direction due to negative charge cloud produced by the electrons that are ejected from the spot in the "forward" direction; (vi) a displacement of hot area in a thin molten layer at the cathode due to thermocapillary effect.

The deficiencies of the published previously hypotheses were commented and also considered in above. However, there are principle common weaknesses.

(i) Mostly all listed mechanisms describe the behavior and processes of the cathode quasineutral plasma and not related to the plasma adjacent to the cathode surface in the spot, which is highly dense (10^{20} cm^{-3}) and strongly collisional. In this electron beam relaxation zone (Chap. 16) the electron Larmor radius $r_L = 6 \times 10^{-8} v_{eT}/B = 6 \times 10^{-3}$ cm ($B = 10^3$G, $v_{eT} = 10^8$ cm/s), i.e., significantly larger than the spot radius $r_s < 10^{-3}$ cm. So, the interaction between the dense plasma in the spot and magnetic field is negligible low due to low value of $\omega\tau \sim 10^{-5}$.

(ii) The arbitrary parameters were used to calculate the additional voltage due to Hall effect or considering mechanisms of appearance of asymmetric parameters in the plasma. Besides, the additional voltage is induced in the near-cathode electrically conducting plasma (not in the cathode sheath), and therefore, the breakdown mechanism is unclear. Also, it should be taken into account that the plasma properties are rapidly equalized in the dense highly collisional plasma (with characteristic time of 10^{-13} to 10^{-14}s at electron–electron collisions) even by arise of some local fluctuation of plasma parameters in the nearest to surface region. Therefore, an approach is used asymmetric energy dissipation (as example, due to asymmetric voltage and others) in the dense near-cathode plasma cannot be accepted.

Thus, the analysis of previous publications shows that existing explanations of the retrograde effect are confused, incomprehensible and the phenomena are remained unclear. The explanations of the spot behavior in an oblique magnetic field and spot splitting were considered by limited number of works and also were not consisted of unanimously accepted understanding. It should be noted that most of developed models indicated agreement with the experiment, while consisting contradicted assumptions. This fact stimulates the continued study of the problem.

Our new approach is presented in this chapter that we hope allowed more comprehension of the spot phenomena in a magnetic field. In this approach, a physical modeling of the phenomena and mathematical method was described, which developed using new concept based on kinetic model of the spot initiation and development (Chap. 17). The main aspect of the approach is analysis of the pressure gradient produced by the magnetic field as a force determined the spot motion, instead of consideration the influence on the plasma parameters by the magnetic field strength only. The approach also takes in account the self-magnetic field produced by the arc current and which first proposed by Robson and Engel 1956, Kesaev 1957, Smith 1957, Guile Secker 1958. Also some details were commented to "maximum principle" formulated by Kesaev 1957, who first indicated the presence of a maximum in the magnetic field distribution around the spot. Let us summarize our results.

Transverse magnetic field. Two main cathode spot phenomena in transverse magnetic field are considered: (1) the increase of spot current with magnetic field and (2) retrograde spot motion.

The spot current model was based on the comparison between the self-magnetic pressure and pressure generated by external magnetic field. It was shown that the

19.7 Summary

plasma jet flow, and hence, the spot operation will be stable when the self-magnetic and kinetic pressures are in equilibrium. The stable plasma flow in the plasma acceleration region and the self-consistent processes of cathode heating, evaporation, atom ionization, and plasma expansion determine the certain relation between the spot current and cathode erosion rate and plasma temperature for given cathode material. In case of multi-spot arc, the spot current linearly increases with the magnetic field and this dependence agree well with that observed. In high-current arcs, the different observed values of group spot currents explained by the dependence of the spot current on the cathode and electron temperatures and cathode erosion rate (see 19.16), which vary due to different experimental conditions. In general, the electron temperature in the jet can be increased with external magnetic field strength because of increase the Joule heating.

The retrograde spot motion model is based on the assumption that an initial pressure is required for new spot ignition, i.e., on the first principle rule (Chap. 17) that the plasma flow in the cathode region must be impeded. The superimpose of the self- and external magnetic fields produced a large magnetic pressure on the side of the retrograde motion, and hence, there is an increase in the probability of new spot ignition and a retrograde motion of the spot. Although the assumption in [4] (the electron diffusion coefficient is decreases by factor $1 + \beta^2$) cannot be fulfilled due to small the parameter β, the present model coincides with Kesaev's principle of "maximum of magnetic field" because it is in accordance with used here the maximum of magnetic pressure.

Drouet [67] also proposed an increase in plasma density (and doubtfully enhancement of electron emission) at the retrograde side of spot, assuming anisotropy in the confinement of the plasma expanding due to asymmetric magnetic field distribution. It should be noted that the plasma confinement (mentioned in [67]) could be realized far from the cathode surface, when the plasma density is sufficiently reduced. In this case, not clear whether a mechanism that postulates plasma density increases in the spot region is feasible. According to present model, the plasma flow should be not only confined but also rather impeded in the Knudsen layer at the cathode surface. This condition promotes non-free flow (sub-sonic) plasma flow, which can provide the ion flux to the cathode surface and support the necessity energy flux.

Our numerical analysis using the developed mathematical approach allowed describe two experimental observations: (i) the linear dependence of the spot velocity on the external magnetic field and (ii) the linear dependence of spot velocity on arc current. The developed model also allows understanding of some observation of the spots behavior using two rules from kinetic theory of the spot. According to the experiments, the retrograde spot velocity depends on the electrode gap, electrode state (roughness, temperature) and electrode material. The different velocities that were observed for different cathode materials [76, 110] are due to different spot currents and spot types. The retrograde spot velocity decreased with the number of arcing operations, i.e., after cathode cleaning by the arc, the roughness was reduced and the surface became smoother [76]. The spot velocity is lower on the clean cathodes because the spot mobility at the rough surfaces is determined due to rapid removing protrusion or film by evaporation. The difference in heat fluxes between bulk and rough or film cathodes is result of different cathode heating regime needed to the enough level for spot initiation and development (second principle rule Chap. 17).

Furthermore, the spot retrograde velocity is lower on hot cathodes [77], which also due to lower heat losses in bulk and film cathodes when the cathode is initially hot (second rule). Smith [132] observed the reversal of spot motion on a heated tantalum cathode, while Fang observed retrograde motion of spots on copper and aluminum hot cathodes (~900 K) [77], and Juttner and Kleberg on hot molybdenum (~2000 K) [91]. Juttner and Kleberg explained the different results by additional gap pressure caused by outgazing from the electrodes or as a transition to the spotless arc mode. The last conclusion (a thermionic arc mode) is also proposed by Smith [132]. Thus, the observed dependence of the retrograde spot velocity on the surface state can be understood on base of the second principle rule of kinetic model (see Chap. 17).

When the external gas is present, the gas discharge channel, created between the electrodes, moves in the Amperian direction due to the ponderomotive force. The discharge channel affects when the channel length (gap distance) and the magnetic field are relative large and the ponderomotive force (19.27) and (19.28) determines the direction of the spot motion. This was observed experimentally by Robson [30] in very strong magnetic fields (up to 5 T). When the gap distance is small the ponderomotive force can be negligible, then the new spot ignition and therefore the motion was determined by the asymmetric magnetic pressure to be in the retrograde direction. Likewise, Robson and Engel [30] observed retrograde spot motion in short (0.5 mm) gap atmospheric arcs.

Another circumstance is in vacuum when the gap distance increases. The spot moves in the retrograde direction due to the asymmetric magnetic pressure. When the gap distance is increased, the plasma losses also increase, and therefore, the plasma pressure in the gap decreases. Without a magnetic field and in the low-pressure gap, the spot mobility is larger (the plasma flow is impeded weakly) and, as a result, the random spot velocity is higher. With a magnetic field, the retrograde directed spot velocity also increases [77] because the magnetic field's role is to "straighten" the trajectory of random spot motion. The velocity saturates at some gap distance because the plasma pressure sufficiently low and its effect becomes negligible. It should be noted that in the small gap the anode secondary processes (atom desorption, sputtering, evaporation, etc.) also affect the interelectrode plasma expansion.

Juttner and Kleberg show [91] that in transverse magnetic fields the plasma propagates along the cathode surface in the retrograde direction before that the spot changes its location. As indicated by the authors, the plasma emanating from a spot and expand along the surface at some distance from the spot and, naturally, above the spot. This experimental result could be understood taking into account that at retrograde side of spot the plasma flow was impeded in the normal direction to the cathode surface due to the larger magnetic pressure. This fact stimulates the radial expansion, i.e., along the surface in the retrograde direction, while at direct side of the spot, the flow was not impeded. Therefore, at direct side, the plasma is accelerated and thus the energetic plasma (due to large plasma pressure) expands in direction normal to the cathode surface (second principle rule Chap. 17).

The further plasma propagation in this direction takes place in the absence of the pressure asymmetry (because it is far from the spot current). The mentioned plasma propagation can stimulate a new spot ignition far from the old spot, as it is observed [91], in the presence of favorable surface condition (spike, protrusion, film). Therefore, the observed spot displacement can be not straight and has a zigzag character. However, the new spot can arise when the plasma flow during the spot development will be impeded in the Knudsen layer at the cathode surface. The mechanism of spot ignition and dynamics of spot types is considered in Chap. 17.

The maximal direct velocity at a moderate magnetic field is determined by a random step velocity (in the absence of magnetic field) averaged for a number of steps. For large magnetic field, the velocity can be exceed the random velocity due to larger probability to new cathode spot ignition than that probability in the absence of magnetic field in usual random spot motion. Therefore, the straightened spot velocity can be larger than chaotically moving spot velocity at significantly high external magnetic field.

Finally, the model of spot motion in an oblique magnetic field allow to explain the acute angle effect, in particularly the Robson's measurements and also the mechanism of spot current splitting in under tangential and normal oriented magnetic fields.

References

1. Boxman, R. L., Martin, P. J., & Sanders, D. M. (eds.) (1995). *Handbook of vacuum arc science and technology*. Park Ridge, NJ: Noyes Publications.
2. Rakhovsky, V. I. (1976). Experimental study of the dynamics of cathode spots development. *IEEE Transactions on Plasma Science, PS-4*(N2), 81–102.
3. Juttner, B. (1997). The dynamics of arc cathode spots in vacuum: New measurements. *Journal of Physics. D. Applied Physics, 30,* 221–229.
4. Kesaev, I. G. (1964). *Cathode processes in the mercury arc*. New York, NY: Consultants Bureau.
5. Kesaev, I. G. (1968). *Cathode processes in electric arcs*. Moscow: Nauka Publishers. (in Russian).
6. Lafferty, J. M. (eds.) (1980). *Vacuum arcs. Theory and applications*. New York, NY: Wiley.
7. Eidinger, A., & Rieder, W. (1957). Das Verhalten des Lichtbogens im transversalen Magnetfeld. *Arch Elektrotech, 43*(2), 94–114.
8. Ecker, G. (1961). Electrode components of the arc discharge, Ergeb. *Exakten Naturwiss, 33,* 1–104.
9. Anders, A. (2008). *Cathodic arcs: From fractal spots to energetic condensation*. Berlin: Springer.
10. Stark, J. (1903). Induktionsercheinungen am Quecksilber-lichtungen im Magnetfield. *Physikalische Zeitschrift, 4,* 440–443.
11. Minorsky, N. (1928). Rotation of the electric arc in a radial magnetic field. *Journal of Physics Radium, 9*(4), 127–136.

12. Tanberg, R. (1929, September). Motion of an electric arc in a magnetic field under low gas pressure. *Nature, 7* (124 new volume), *7*, 371–372.
13. Smith, C. G. (1942, July 1). The mercury arc cathode. *Physical Review, 62*, 48–54.
14. Smith, C. G. (1943). Motion of the copper arc in transverse magnetic field. *Physical Review, 63*(N5-6), 217.
15. Yamamura, S. (1950). Immobility phenomena and reverse driving phenomena of the electric arc. *Journal of Applied Physics, 21*(3), 193–196.
16. Yamamura, S (1957). Immobility phenomena and reverse driving phenomena of the electric arc, driven by magnetic field. *Journal of the Faculty of Engineering (University of Tokyo), 25*, 59.
17. Kigdon, K. H. (1965). The arc cathode spot and its relation to the diffusion of ions within the cathode metal. *Journal of Applied Physics, 36*(4), 1351–1360.
18. Oliphant, M. L. (1930). Liberation of electrons from metal surfaces by positive ions. *Proceedings of the Royal Society (London), A127*, 373–387.
19. Hagstrum, H. D. (1954). Theory of auger ejection of electrons from metals by ions. *Physical Review, 96*(2), 336–365.
20. Kingdon, K. H., & Langmuir, I. (1923, August). *The removal of thorium from the surface of a thoriated tungsten filament by positive ion bombardment. Physical Review 22*, 148–160.
21. Himler, G. J., & Cohn, G. I. (1948). The reverse blowout effect. *Electrical Engineering, 67*, 1148–1152.
22. Longini, R. L. (1947). Motion of low-pressure arc spots in magnetic fields. *Physical Review, 72*(2), 184–185.
23. Longini, R. L. (1948). A note concerning the motion of arc cathode spot in magnetic fields. *Physical Review Letters Editor, 71*(9), 642–643.
24. Rothstein, J. (1950). Holes and retrograde arc spot motion in a magnetic field. *Physical Review, 78*(N3), 331.
25. Ware, A. A. (1954). Galvomagnetic and thermomagnetic effects in a plasma. *Proceedings of the Physical Society. Section A, 67*(10), 869–880.
26. St. John, R. M., & Winans, J. G. (1954). Motion of arc cathode spot in a magnetic field. *Physical Review, 94*(N5), 1097–1102.
27. St. John, R. M., & Winans, J. G. (1955). Motion and spectrum of arc cathode spot in a magnetic field. *Physical Review, 98*(N6), 1664–1671.
28. Hernqvist, K. G., & Johnson, E. O. (1955). Retrograde motion in gas discharge plasmas. *Physical Review, 98*(5), 1576–1583.
29. Robson, A. E., & Engel, A. (1954). Origin of retrograde motion of arc cathode spots. *Physical Review, 93*(6), 1121–1122.
30. Robson, A. E., & Engel, A. (1956). Motion of a short arc in a magnetic field. *Physical Review, 104*(1), 15–16.
31. Smith, C. G. (1957). Motion of an arc in a magnetic field. *Journal of Applied Physics, 28*(11), 1328–1331.
32. Kesaev, I. G. (1957). On the causes of retrograde arc cathode spot motion in a magnetic field. *Soviet Physics-Doklady, 2*(1), 60–63.
33. Kesaev, I. G. (1972). Direct motion of the cathode spot due to asymmetry conditions at its boundary. *Soviet Physics-Doklady, 203*(5), 1037–1040. in Russian.
34. Ecker, G., & Muller, K. G. (1958). Theory of the retrograde motion. *Journal of Applied Physics, 29*, 606–608.
35. Ecker, G., & Muller, K. G. (1958). Theorie der, Retrograde Motion. *Zeitschrift fur Physik, 151*, 577–594.
36. Guile, A. E., Lewis, T. J., & Menta, S. F. (1957). Arc motion with magnetized electrodes. *British Journal of Applied Physics, 8*(11), 444–448.
37. Guile, A. E., & Secker, P. E. (1965). Retrograde running of the arc cathode spot. *British Journal of Applied Physics, 16*(130), 1595–1597.
38. Guile, A. E., & Secker, P. E. (1958). Arc cathode movement in a magnetic field. *Journal of Applied Physics, 29*(12), 1662–1667.

References

39. Guile, A. E., Adams, V. W., Lord, W. T., & Naylor, K. A. (1969). High-current arcs in transverse magnetic fields in air at atmospheric pressure. *Proceedings of the Institution of Electrical Engineers, 116*(4), 645–652.
40. Guile, A. E., & Sloot, J. G. J. (1975). Magnetically driven arcs with combined column and electrode interactions electrical engineers. *Proceedings of IEE, 122*(6), 669–671.
41. Secker, P. E. (1960). Explanation of the enhanced arc velocity on magnetic electrodes. *British Journal of Applied Physics, 11*(8), 385–388.
42. Lewis, T. J., & Secker, P. E. (1961). Influence of the cathode surface on arc velocity. *Journal of Applied Physics, 32*(1), 54–63.
43. Guile, A. E., Lewis, T. J., & Secker, P. E. (1961). The motion of cold-cathode arcs in magnetic fields. *Proceedings of the IEE - Part C: Monographs, 108*(N14), 463–470.
44. Bauer, A. (1961). Zur Feldbogentheorie bei kalten verdampfenden Kathoden II. *Z fur Physics, 165*(1), 34–46.
45. Bauer, A. (1961). Zur Feldbogentheorie bei kalten verdampfenden Kathoden I. *Z fur Physics, 164*(5), 563–573.
46. Hull, A. W. (1962). Cathode spot. *Physical Review, 126*(5), 1603–1610.
47. Gallagher, C. J. (1950). The retrograde motion of the arc cathode spot. *Journal of Applied Physics, 21*(8), 768–771.
48. Hermoch, V., & Teichmann, J. (1966). Cathode jets and the retrograde motion of arcs in magnetic fields. *Zeitschrift f Physics, 195*, 125–145.
49. Hermoch, V. (1973). On the retrograde motion of arcs in magnetic field by heating cathode. *IEEE Transactions of Plasma Science, PS–1*(N3), 62–64.
50. Murphree, R. P., & Carter, D. L. (1969). Low-pressure arc discharge motion between concentric cylindrical electrodes in a transverse magnetic field. *AIAA Journal (American Institute of Aeronautics & Astronautics), 7*(8), 1430–1437.
51. Seidel, S., & Stefaniak, K. (1972). Retrograde motion of the electric arc in vacuum and its mechanism on the solid electrodes. In *Proceedings of the V International Symposium on Discharge Electrical Insulation, Vacuum* (p. 207). Poznan: Politechnika Poznanska.
52. Sherman, J. C., Webster, R., Jenkins, J. E., & Holmes, R. (1975). Cathode spot motion in high-current vacuum arcs on copper electrodes. *Journal of Physics. D. Applied Physics, 8*, 696–702.
53. Agarwal, M. S., & Holmes, R. (1984). Cathode spot motion in high-current vacuum arcs under self-generated azimuthal and applied axial magnetic fields. *Journal of Physics. D. Applied Physics, 17*(4), 743–756.
54. Carslaw, H. S., & Jaeger, J. C. (1959). *Conduction of heat in solids* (Chap. 10), 2nd ed. Oxford: Clarendon.
55. Djakov, B. E., & Holmes, R. (1970). Cathode spot motion in a vacuum arc under the influence of the inherent magnetic field. In *1st International Conference on Gas Discharges*, London (pp. 468–472).
56. Djakov, B. E., & Holmes, R. (1972). Retrograde motion of a cathode spot and conduction of heat in the cathode. In *2nd International Conference on Gas Discharges*, London (pp. 183–185).
57. Djakov, B. E. (1988). Cathode spot motion and conduction of heat in the cathode of high-current vacuum arc. In *Proceedings of the XIII International Symposium Discharges Electrical Insulation, Vacuum*, France, Paris (pp. 169–171).
58. Moizhes, B. Ya., & Nemchisky, V. (1979). Retrograde motion of a vacuum arc in a magnetic field. *Soviet Technical Physics Letters, 5*, 78–80.
59. Nemchisky, V. (1983). Causes of cathode spot motion in a vacuum arc and an estimation of retrograde velocity in a magnetic field. *Soviet Technical Physics Letters, 28*(2), 150–155.
60. Juttner, B. (1979). Erosion craters and arc cathode spots in vacuum. *Beitrage Plasma Physics, 19*(1), 25–48.
61. Hantzsche, E. (1977). A new model of crater formation by arc Spots. *Beitrage Plasma Physics, 17*, 65–74.

62. Nemchisky, V. (1988). Theory of the retrograde motion of a vacuum arc on mercury cathode. *Soviet Technical Physics Letters, 33*(2), 166–170.
63. Moizhes, B. Ya., & Nemchisky, V. (1991). On the theory of the retrograde motion of a vacuum arc. *Journal of Physics. D. Applied Physics, 24*, 2014–2019.
64. Emtage, P. R. et al. (1977). The interaction of vacuum arc with transverse magnetic fields. In *Proceedings of the XIII International Conference on Phen in Ionized Gases*, Part2, Berlin (pp. 673–674).
65. Daybelge, U. (1985). Theory of the self-induced electric field at a cathodic arc spot. *Physics of Fluids, 28*(1), 312–320.
66. Drouet, M. G. (1981). The physics of the retrograde motion of the electric arc. *Japanese Journal of Applied Physics, 20*(N6), 1027–1036.
67. Drouet, M. G. (1985). The physics of the retograde motion of the electric arc. *IEEE Transactions on Plasma Science, PS-13*(N5), 235–241.
68. Emtage, P. R., Kimblin, C. W., Gorman, J. G., Holmes, F. A., Heberlein, J. V. R., Voshall, R. E., & and Slade, P. G. (1980). Interaction between vacuum arcs and transverse magnetic fields with application to current limitation. *IEEE Transactions on Plasma Science, PS-8*(N4), 314–319.
69. Meunier, J.-L., & Drouet, M. G. (1983). Bouncing expansion of the arc cathode–plasma in vacuum along the transverse applied B field. *IEEE Transactions on Plasma Science, PS-11*(N3), 165–168.
70. Drouet, M. G., & Meunier, J.-L. (1985). Influence of the background gas pressure on the expansion of the arc cathode–plasma. *IEEE Transactions on Plasma Science, PS-13*(N5), 285–287.
71. Leycuras, A. (1978). Observation of high-temperature electron injection in dense argon and mercury: Application to cathode spot. *Journal of Physics. D. Applied Physics, 11*, 2249–2256.
72. Beilis, I. I. (1974). Analysis of the cathode in a vacuum arc. *Soviet Physics Technical Physics, 19*(2), 251–256.
73. Fabre, E., Stenz, C., & Colburn, S. (1973, April). Experimental study of the interaction with a magnetic field of a plasma produced by laser irradiation of solids. *Journal of Physics France, 34*, 323–331.
74. Beilis, I. I. (1986). Cathode arc plasma flow in a Knudsen layer. *High Temperature, 24*(3), 319–325.
75. Beilis, I. I. (2007). Laser plasma generation and plasma interaction with ablative target. *Laser and Particle Beams, 25*(1), 53–63.
76. Fang, D. Y. (1982). Cathode spot velocity of vacuum arcs. *Journal of Physics. D. Applied Physics, 15*(4), 833–844.
77. Fang, D. Y. (1983, September). Temperature dependence of retrograde velocity of vacuum arcs in magnetic fields. *IEEE Transactions on Plasma Science, PS-11*(110–114), N3.
78. Harris, L. P. (1983). Transverse forces and motions at cathode spots in vacuum arcs. *IEEE Transactions on Plasma Science, PS-11*(N3), 94–102.
79. Sanochkin, Yu V. (1984). Balancing of the Amperian force in a current-carrying thermocapillary cell in a transverse magnetic field. *Soviet Physics Technical Physics, 29*(9), 1003–1006.
80. Sanochkin, Yu V. (1985). Velocity of retrograde cathode spot motion and thermocapillarity convection during local heating of liquid surface. *Soviet Technical Physics Letters, 30*, 1052–1056.
81. Schrade, H. O. (1966). On arc pumping and motion of electric arc in a transverse magnetic field. In *Proceedings of the VII International Conference on Phenomena in Ionized gases* (Vol. 1, pp. 740–744).
82. Schrade, H. O. (1973). Stable configuration of electric arc in transverse magnetic fields. *IEEE Transactions on Plasma Science, PS-1*(N3), 47–51.
83. Auweter-Kurtz, M., & Schrade, H. O. (1980). Exploration of arc spot motion in the presence of magnetic fields. *Journal of Nuclear Materials, 93 & 94*, 799–805.

84. Schrade, H. O., Auweter-Kurtz, M., & Kurtz, H. L. (1983). Analysis of the cathode spot of metal vapor arcs. *IEEE Transactions on Plasma Science, PS-11*(N3), 103–110.
85. Schrade, H. O., Auweter-Kurtz, M., & Kurtz, H. L. (1988). Cathode phenomena in plasma thruster. *Journal of British Interplanet Society, 41*, 215–222.
86. Schrade, H. O. (1989). Arc cathode spots: Their mechanism and motion. *IEEE Transactions on Plasma Science, 17*(5), 635–637.
87. Tseskis, A. L. (1978). Theory of the cathode spot in an external magnetic field. *Soviet Physics Technical Physics, 23*(6), 602–604.
88. Schrade, H. O., Auweter-Kurtz, M., & Kurtz, H. L. (1987). Cathode phenomena in plasma thrusters. In *AIAA-87-1096 presented at the AIAA/DGLR/JSASS 19th International Electric Propulsion Conference*, May 11–13, 1–8, Colorado Springs, Colorado.
89. Bobrov, Yu K, Bystrov, V. P., & Rukhadze, A. A. (2006). Physical model of the cathode spot retrograde motion. *Technical Physics, 51*(5), 567–573.
90. Kleberg, I. (2000). Die dynamik von kathodischen brennecken im externen magnetfeld, Dissertation, Humboldt Universoty, Berlin.
91. Jüttner, B., & Kleberg, I. (2000). The retrograde motion of arc cathode spots in vacuum. *Journal of Physics. D. Applied Physics, 33*, 2025–2036.
92. Juttner, B. (2001). Cathode spots of electric arcs. *Journal of Physics. D. Applied Physics, 34*(17), R103–R123.
93. Kadomtzev, V. V. (1970). *Collective phenomena in plasma*. Moscow: Nauka.
94. Mikhailovsky, A. B. (1978). *Instability of a plasma in a magnetic traps*. Moscow: Atomizdat.
95. Lee, K. C. (2007). Gyrocenter shift of low-temperature plasmas and the retrograde motion of cathode spots in arc discharges. *Physical Review Letters, 99*, 065003.
96. Lee, K. C. (2006). Radial electric field formation by the gyrocenter shifts of the charge exchange reactions at the boundary of fusion device. *Physics of Plasmas, 13*, 062505.
97. Pang, X., Wang, T., Xiu, S., Yang, J., & Jing, H. (2018). Investigation of cathode spot characteristics in vacuum under transverse magnetic field (TMF). *Contacts Plasma Science and Technology, 20*, 085502.
98. Beilis, I. I. (2001). State of the theory of vacuum arcs. *IEEE Transactions on Plasma Science, 29*(5), 657–670.
99. Beilis, I. I. (2002). Vacuum arc cathode spot grouping and motion in magnetic fields. *IEEE Transactions on Plasma Science, 30*(6), 2124–2132.
100. Djakov, B. E., & Holmes, R. (1974). Cathode spot structure and dynamics in low-current vacuum arcs. *Journal of Physics. D. Applied Physics, 7*, 569–580.
101. Beilis, I. I. (1995). Theoretical modelling of cathode spot phenomena. In R. L. Boxman, P. J. Martin, & D. M. Sanders (Eds.), *Handbook of vacuum arc science and technology* (pp. 208–256). Park Ridge, NJ: Noyes Publications.
102. Zykova, N. M., Kantsel, V. V., Rakhovsky, V. I., Seliverstova, I. F., & Ustimets, A. P. (1971). The Dynamics of the Development of Cathode and Anode Regions of Electric Arcs I. *Soviet Physics-Technical Physics, 15*(11), 1844–1849.
103. Rakhovsky, V. I. (1976). Experimental study of the dynamics of cathode spots development. *IEEE Transactions on Plasma Science, 4*, 87–102.
104. Anders, A., & Juttner, B. (1991). Influence of residual gases on cathode spot behavior. *IEEE Transactions on Plasma Science, 19*(5), 705–712.
105. Drouet, M. G., & Meunier, J. L. (1985). Influence of the background gas pressure on the expansion of the arc plasma. *IEEE Transactions on Plasma Science, PS-13*(N5), 285–287.
106. Landau, L. D., & Lifshitz, E. M. (1984). *Electrodynamics of continuous media*. Oxford: Pergamon Press.
107. Beilis, I. I. (2002). Mechanism for cathode spot grouping in vacuum arcs. *Applied Physics Letters, 81*(21), 3936–3938.
108. Beilis, I. I., Djakov, B. E., Juttner, B., & Pursch, H. (1997). Structure and dynamics of high-current arc cathode spots in vacuum. *Journal of Physics D: Applied Physics, 30*, 119–130.
109. Dukhopel'nikov, D. V., Zhukov, A. V., Kirillov, D. V., & Marakhtanov, M. K. (2005). Structure and features motion of cathode spot on a continuous titanium cathode. *Measurement Techniques, 48*(N10), 995–999.

110. Perskii, N. E., Sysun, V. I., & Khromoy, Yu D. (1989). Dynamics of vacuum-discharge cathode spots. *High Temperature, 27*(6), 832–839.
111. Miyano, R., Fujimura, Y., Takikawa, H., & Sakakibara, T. (2001). Cathode spot motion in vacuum arc of Zinc cathode under oxygen gas flow. *IEEE Transactions on Plasma Science, 29*(5), 713–717.
112. Beilis, I. I., & Zektser, M. P. (1991). Calculation of the parameters of the cathode stream an arc discharge. *High Temperature, 29*(4), 501–504.
113. Beilis, I. I., Keidar, M., Boxman, R. L., & Goldsmith, S. (1998). Theoretical study of plasma expansion in a magnetic field in a disk anode vacuum arc. *Journal of Applied Physics, 83*(2), 709–717.
114. Beilis, I. I., & Keidar, M. (2000). Theoretical study of plasma expansion and electrical characteristics in the high-current vacuum arc. In *Proceedings of XIXth ISDEIV*, Xi'an, China, September 18–22 (pp. 206–209).
115. Robson, A. E. (1978). The motion of low-pressure arc in a strong magnetic field. *Journal of Physics. D. Applied Physics, 11*, 1917–1923.
116. Robson, A. E. (1960). In *1959 Proceedings of 4th International Conference on Ionization Phenomena in Gases*, Uppsala, Sweden (Vol. 1, p. 346) North-Holland: Amsterdam.
117. Beilis, I. I. (2016). Vacuum arc cathode spot motion in oblique magnetic fields: An interpretation of the Robson experiment. *Physics of Plasmas, 23*, 093501.
118. Litvinov, E. A., Mesyats, G. A., & Sadovskaya, E Yu. (1990). Effect of an external magnetic field on the motion of a cathode spot of a vacuum arc. *Soviet Technical Physics Letters, 16*(9), 722–724.
119. Barengoltzs, S. A., Litvinov, E. A., Sadovskaya, E. Yu., & Shmelev, D. L. (1998). Movement of cathode spot in vacuum arc in an external magnetic field. *Soviet Physics Technical Physics, 68*(N6), 60–64 (in Russian).
120. Garner, A. L. (2008). Cathode spot motion in an oblique magnetic field. *Applied Physics Letters, 92*, 011505.
121. Kesaev, I. G., & Pashkova, V. V. (1959). The electromagnetic anchoring of the cathode spot. *Soviet Physics - Technical Physics, 4*(3), 254–264.
122. Keidar, M., & Beilis, I. I. (2018). *Plasma Engineering*. London, NY: Academic Press, Elsevier.
123. Beilis, I. I. (2013). Cathode spot development on a bulk cathode in a vacuum arc. *IEEE Transactions on Plasma Science, 41*(N8), Part II, 1979–1986.
124. Smith, C. G. (1946). Cathode dark space and negative glow of a mercury arc. *Physical Review, 69*(3–4), 96–100.
125. Coombe, R. A. (1964). *Magnetohydrodinamic generation of electrical power*. London: Chapman and Hall.
126. Beilis, I. I. (2015). Physics of cathode phenomena in a vacuum arc with respect to a plasma thruster application. *IEEE Transactions on Plasma Science, 43*(1), 165–172.
127. Beilis, I. I. (2015). Mechanism of cathode spot splitting in vacuum arcs in an oblique magnetic field. *Physics of Plasmas, 22*, 103510.
128. Zabello, K. K., Barinov, Yu A, Chaly, A. M., Logatchev, A. A., & Shkol'nik, S. M. (2005). Experimental study of cathode spot motion and burning voltage of low-current vacuum arc in magnetic field. *IEEE Transactions on Plasma Science, 33*(5), 1553–1559.
129. Chaly, A. M., Logatchev, A. A., Zabello, K. K., & Shkol'nik, S. M. (2007). Effect of amplitude and inclination of magnetic field on low-current vacuum arc. *IEEE Transactions on Plasma Science, 35*(4), 946–952.
130. Chaly, A. M., & Shkol'nik, S. M. (2011). Low-Current vacuum arcs with short arc length in magnetic fields of different orientation: A review. *IEEE Transactions on Plasma Science, 39*(6), 1311–1318.
131. Heberlein, J., & Porto, D. (1983). The interaction of vacuum arc ion currents with axial magnetic fields. *IEEE Transactions on Plasma Science, PS-11* (N3), 152–159.
132. Smith, C. G. (1948). Arc motion reversal in transverse magnetic field by heating cathode. *Physical Review 3*(N5), 543.

Chapter 20
Theoretical Study of Anode Spot. Evolution of the Anode Region Theory

Limited number of publications was presented concerning to the theoretical study of the anode region of vacuum arcs up to 70–80 years of the twentieth century. The most of arcs were operated with relatively small arc current, and therefore, without or at small contraction of the electrical current at the anode surface. This provokes for long time an opinion that the anode plays only a passive role as electron collector, while the cathode mechanism was traditionally studied as a complicated object supporting a source of electrons. The further arc applications were connected with development plasma accelerators, high-current electro-commutation apparatus, plasma jet generators, and welding plasma devices [1, 2]. It has been an increasing interest in graphite vacuum arc owing to its use in metallurgy and as plasma source for carbon films deposition [3]. The new technologies stimulated progress in experimental research (Chap. 14) and, as consequence, growth of the interest to the fundamental investigation of anode phenomena. Therefore, the experiments were extended and they showed different types of anode plasma region and anode spot appearing depending on discharge conditions [4–6], which caused an attempt to more detailed understanding of the physics of anode processes. Moreover, these attempts show that anode spot mechanism is no less complicated and interesting in comparison with that for cathode spot. Below a state of the anode spot modeling will be considered in the period of the twentieth century and up to present time.

20.1 Review of the Anode Region Theory

Let us consider the published approaches close to those presented by the authors to describe the originality of the results in order to better understand the progress and weakness of the studies.

20.1.1 Modeling of the Anode Spot at Early Period for Arcs at Atmosphere Pressure

In the beginning of the twentieth century, the investigations of arc phenomena were conducted mostly for arcs in air at atmosphere pressure. The results were reviewed by Ecker [7]. A constricted plasma region was detected at the anode, for which mainly the electrical characteristics were studied theoretically and experimentally. An early typical such research was presented by Gunter Schulze in [8].

Before analysis of the explanation, let as consider his observed data. Schulze measured the arc voltage including the anode u_a and cathode u_c potential drops for a number of metallic and metal oxides anodes with carbon cathode. A carbon probe was introduced into the arc gap in order to determine the potential drop at different arc lengths. A precision voltmeter was used and u_a and u_c were measured with the help of a switch immediately one after the other. The arc voltage u was determined as a sum of u_a and u_c. Table 20.1 presents the results corresponding to the group from Mendeleev system for arc current of 4 A. The data for $l = 0$ indicate the measurements for very small arc lengths, and another measurement shows the data for arc lengths of l, most of 8 mm. It was indicated also that for very small arc lengths, the arc voltage is independent of the arc current. The results were explained taking into account the anode evaporation and possibly the vapor deposition on the opposite carbon cathode.

The observation of the arc shape shows that both electrodes emit vapor of their substance and presence of a brightness small arc point at the electrodes (in future in other publications it named as spots). For example, the size of 1 mm was observed for small arc point for iron ($I = 6$ A), while the largest diameter of 5 mm ($l = 7$ mm) was for the discharge column. Furthermore, due to its high brightness, the arc point is so sharply delimited from the surroundings and sends such amounts of steam into the arc that the boiling point of the relevant electrode substance was assumed.

Schulze explained his voltage experiments by discussion an energy balance assuming at the arc spot of the electrodes constant temperature. Let us describe his discussion of the electrode potential measurements. In case of constant temperature, the arc spot must be supplied with as much warmth as is withdrawn from it at every moment. The input heat could be compensated by radiation, by dissipation into the electrodes, and by forming the required amount of steam. In case of the current constant and neglecting the warmth, which arises from the burning of the electrodes (it could at most be considered at carbon), the voltage reduction is caused by the heat loss at the arc spots is proportional to the heat loss. It can be assumed that this loss of power and heat replacement takes place mainly through resistance heating, in that the current meets a high contact resistance until the boiling temperature is reached and the sufficient number of vaporizing molecules allows it to escape. In the case of small arc lengths, the heat loss through radiation must be lower than in the case of larger ones, since in the former a large part of the heat radiated by one electrode meets and heats the other. This mutual irradiation decreases with increasing arc length, so the heat loss and therefore u_a and u_c must increase with the arc length.

20.1 Review of the Anode Region Theory

Table 20.1 Arc voltage and anode potential drop measured by Schulze [8]

Mendeleev group	Element	Arc voltage (V), $l=0$	Arc length, l, mm	Arc voltage (V), at l	Anode voltage (V), at l	Temperature, °C Melting	Temperature, °C Boiling
I, Alkali	Li$_2$O	11.1	8	12.9	7.0	180	–
	Na$_2$O	9.8	6	11.8	6.8	97	900
	K$_2$O	9.8	8	11.7	4.9	60	700
II, Earth-Alkali	MgO	15.4	8	22.4	13.0	750	1100
	CaO	13.0	8	19.1	11.3	red heat	–
	SrO	11.9	8	17.7	9.9	–	–
	BaO	11.2	8	14.9	7.0	1500	–
I	CuO	21.0	5	29.4	19.1	1100	–
	Ag	17.0	–	–	–	957	–
II	ZnO	17.0	8	26.9	17.6	412	950
	CdO	14.4	8	22.4	13.3	318	
IV	C	39.7	6	47.0	36.3	–	–
	Sn	13.9	8	20.6	10.9	233	1530
	Pb	11.0	8	17.2	8.7	334	1520
V	Sb	15.0	8	22.0	13.4	430	1400
	Bi	11.5	8	15.5	8.3	267	1400
VI–VIII	Cr	15.4	8	20.6	10.8	>2000	
	Mn	11.0	8	20.3	11.1	1900	
	Co	13.9	–	–			
VI–VIII	Cr$_2$O$_3$	20.1	8	26.8–	17.0	>2000	
	Cr$_3$O$_4$	20.1	8	32.6	22.5	1580	
	NiO	17.5	8	31.9	22.7	1450	

For an arc with vertically oriented electrodes, the upper electrode is heated by the ascending arc vapor, so it needs less heat itself. Thus, the values of u_a and u_c as function on the arc length must be lower if they belong to the upper electrode. The heat loss suffered by the arc spot is greater and the higher is the boiling point. The area of the arc spot increases with the current and at the larger spot, the lower influence of the heat. Since assumption that the temperature of the point of the arc remains constant and equals to the boiling point [8], the heat dissipation into the electrodes decreased only as the spot of the arc increases, and therefore, the potential drop must decrease with increasing current. This decreases greater in the case of copper than in iron, because in copper the loss of heat through conduction into the electrodes accounts for a greater part of the total voltage loss than for iron.

The most of theoretical study of anode phenomena in the first half of the twentieth century is provided for carbon and mercury arcs in air at atmosphere pressure. This was due to significant application of the discharge with these electrodes for light sources. The potential drop, the heat flux to the anode, anode temperature, and the production of electrode vapor jets mainly were studied experimentally and theoretically developing a thermal theory for plasma region constricted near the anode [9, 10]. The power consumed by the anode was determined by measuring the temperature increase of the cooling water. This temperature was determined with a thermocouple and the result compared with the dissipated electrical power in the arc and the anode region [11]. Most of the research was conducted by Finkelburg, Busz–Peuckert, and Maecker, and it is summarized in the review [12].

According to Finkelburg [13], the experimental results show differences between high-current and low-current arcs. In contrast to the falling volt–current characteristic of the low-current arc, the braking voltage of the high-current carbon arc always increases more or less strongly with the current. In the case of the low-current arc, the ionized vapor radiation occurs completely in the direction of the continuous radiation of the glowing positive crater. A peculiar anode flame is absent.

In contrast, in the high-current arc mostly a very pronounced anode flame always outweighs the column radiation in front of the anode with air hot temperatures up to 4700 and 5800 K, as well appears as the peculiar crater radiation. According to the experiments, all characteristics of the high-current carbon sheets depend exclusively on the material and diameter of the anode carbon and are practically independent of material and dimension of the cathode carbon. The strong increase of the burning voltage with the current strength does not depend on the arc length, but is located in front of the anode [9]. The rising characteristic of the high-current carbon arc is thus conditioned by an anode region, which is strongly current dependent in contrast to the low-current arc.

It was indicated that a good insight to describe the conditions of the high-current carbon arc can be found considering the energy balance of the anode fall (u_a) area, which differs significantly from that of the low-current arc. Assuming that the loss of energy through thermal conduction, convection, and chemical reactions is negligible compared with the emission of the anode crater, the balance was used in form

$$I(u_a + \varphi) = \pi r_a^2 \varepsilon_T \sigma_{sB} T_a^4 + Q_s \qquad (20.1)$$

where φ is the anode material work function, σ_{sB} is the constant of the Stefan–Boltzmann law, ε_T is the absorptivity of the anode surface, T_a is the anode temperature in the crater, and Q_s is the energy with evaporation of the anode material. Using a measured carbon erosion rate of 2.5×10^{-4} g/C for an 80 A arc and assuming a homogeneity, the number of atoms evaporating per second a contribution of anode fall was calculated of about 12 V only due to anode evaporation. For the other, more unlikely, extreme case, which take into account that all carbon atoms recombine to C_2 molecules, releasing the dissociation energy of 8.6 eV volts, the above contribution was obtained as 8.4 V. Thus, it is shown that in the highly loaded homogeneous carbon arc, the anode evaporation causes the anode case to increase by 10 V.

In the next work, Finkelburg [14] investigated the properties of carbon anode flame (anode jet). He considered a mechanism of the flame generation and developed an anode fall theory. Testing of the entire high-current arc with a potential probe yielded

20.1 Review of the Anode Region Theory

expected potential jumps hard before the cathode and the anode (cathode and anode case) and a low potential drop in the arc column from the anode to the cathode (column gradient), however, no potential differences within the very extended anode flame. The results show that (i) the extent and intensity of the anode flame are completely independent of the material of the negative carbon electrode and determined only by that of the positive carbon, and (ii) the length of the anode flame depends on the evaporation rate of the anode material and grows with it.

The adjacent anode drop region was described defining the energy supplied to the anode, as it was used in the previous work through anode potential drop u_a (also named as anode fall), in form

$$W = I_e(u_a + \varphi) \tag{20.2}$$

where I_e is the electron current strength. As a result, this energy, as in the case of the low-current arc, the heat loss of the anode and its directly adjacent anode fall area due to heat conduction, convection, and radiation must be included. Also, the model must consider the total energy to be expended for continuous generation of the anode flame, namely the evaporation heat of the anode material and that for heating the generated steam under its expansion. This anodic flame energy reappears to a considerable extent as radiant energy. The other part was dissipated by convection, and especially by heat conduction in a broader sense, including recombination of electrons and atoms into atoms, and of atoms into molecules as heat to the environment. If we denote by $A(I)$ the current-dependent evaporating anode material quantity in g/s and $Q(I)$ the energy required for vaporization and heating of 1 g of anode material to the current-dependent anode flame temperature, then, neglecting former processes, the energy balance of the anode area is

$$I_e(u_a + \varphi) = A(I)Q(I) \tag{20.3}$$

and the expression of anode potential drop is

$$u_a = \frac{A(I)Q(I)}{I_e} - \varphi \tag{20.4}$$

Depending on the initial velocity of the anode vapor jet $V(I)$ and the anode diameter D, which depends on I again, it is possible to write with the density $\rho = 10^{-5}$ g/cm^3 [14]

$$u_a = \frac{10^{-5} \pi D^2 V(I) Q(I)}{2 I_e} - \varphi \tag{20.5}$$

An anode drop u_a grows with the current if numerator in (20.4) and (20.5) increases the current more strongly than linearly. In uniform carbon arc, the anode temperature T and thus Q very little depend of I, while according to the measurements $A(I) \sim$ const $\times I^2$ [13]. The numerator in (20.4) thus increases approximately quadratically and calculating u_a increases approximately linearly with I in the homogeneous carbon arc. On the other hand, in the Beck arc [13], T and Q, depends significantly stronger with I, while according to the measurements $A(I) \sim \text{const} \times I^3$ [13]. The numerator in (20.4) thus grows stronger than I^3, and the anode drop u_a is stronger than square with the current strength I. Experimentally [9] in the homogeneous carbon arc, this dependence is weaker than linear, but the Beck arc is a powerful, more measured as a linear increase in the characteristic. The striking difference between the measured characteristics of the homogenous carbon arc and the Beck arc is therefore correctly represented qualitatively by the theory, but not the degree of increase.

Finkelburg indicated that according to the observations [14], the cause for this consists in the fact that the measured loss of anode material is not due, exclusively, to evaporation, but partly also due to an explosive spewing of small coal particles. Since the latter effect was just at larger current, the stronger increase of $A(I)$ was substituted with I, to consider a pure evaporation. In view of this fact, we can say that the agreement between theory and measurement is satisfactory. The total arc energy was used in the energy balance to analyze the anode flame formation. Thus (20.4) and (20.5) also show the experimentally established relationship between the value of u_a and the values $A(I)$ and $V(I)$ which are decisive for the extent and intensity of the anode flame. It was discussed the necessity is to consider an influence of chemical reactions in the presence of a surrounding gas on the arc mechanism for experiments with different gases at various pressures.

Thus, the probe measurements also allow, together with measurements of the anode burn, to confirm the theory of the anode [14], wherein it was shown that the anode decomposition actually takes place by evaporation, and only a small part of the burnup chemical degradation is conditional.

Later Finkelburg in [15] further improved his analysis with revised expression for anode energy balance

$$W_A = I_e(u_a + \varphi) = W_N + (A - A_0)Q(T) \qquad (20.6)$$

where W_N is the part of energy delivered to the anode, which expended by heat conduction, convection, and radiation. Another part $W_A - W_N$ should be expended to increase the temperature $T(I)$ under expansion in order to evaporate the anode material and heat the steam under expansion, thus serving to produce the anode flame. $Q(T)$ denotes the energy required to evaporate and heat 1 g to the temperature T. An estimation of different energy components in (20.6) was provided using measured data for carbon anodes of Beck arc, $\varphi = 4.5$ eV and assuming $I_e = I$, neglecting the small value of ion current. The estimations show (Table 20.2) that the anode potential drop (fall) increased from about 10–40 V when the arc current increased from 30 to 100 A and the measured arc voltage from 23.5 to 55 V.

As result of above-reported experimental studies and estimations, the model of formation of the anode potential drop region and mechanism of anodic processes were discussed for relatively low arc current. It was indicated that the anode's role is generally to absorb the electron stream transported by the column. At the same time, the positive ions which are required to compensate for the charge of the space

20.1 Review of the Anode Region Theory

Table 20.2 Calculated anode fall and energy components of anode energy balance for carbon Beck arc according to Finkelburg [15]

Arc current A	Arc voltage V	Anode fall V	W_A Watt	$W_A - W_N$ Watt
30	23.5	11	465	
40	28	14.5	750	
50	33	18.5	1150	480
60	40	24	1710	1010
70	47	29.5	2380	1680
80	50	33	3000	2300
90	53	35	3550	2850
100	55	37.5	4200	3500

in the column and slowly migrate to the cathode in the field have to be supplied by the anode or its gas layer. Since only a very small proportion of ions originate from the anode itself, an ionization of the anode vapor occurs, which in according to the spectroscopic findings takes place. The ionization takes place in front of the anode, and a steep potential gradient forms there integrated over the distance to the beginning of the column in the anode drop region.

In case of the high-current carbon arc, the energy supplied by the incoming electrons to the anode can no longer be dissipated by conduction and radiation, as in the case of the low-current arc, and therefore, a vigorous evaporation occurred. The steam formed in accordance with [14] flows at a speed of the order of 10^3 cm/s from the anode end face. This vapor layer at the anode should be heated by the electrons flowing in from the arc, so that it delivers the ions needed to maintain the column. In this case, the energy required arises by automatically increasing the anode drop and thus correspondingly greater acceleration of the electrons is delivered in front of the anode. By this mechanism of continuous regeneration and efflux of highly heated steam, Finkelburg explained the anode flame of the high-current carbon arc in agreement with the experimental findings [14].

Finkelburg [16] summarized the properties (definitions) of high-current carbon arc and the difference from low-current discharge. It was pointed out the role of high-current carbon arc as the most powerful and one of the most important radiation sources. The rising voltage characteristic, the anodic vapor stream which causes its excellent radiation properties, and the contracted arc stream distinguish the high-current carbon arc from the well-known normal low-current carbon arc. Two observed important distinctions of the high-current carbon arc from the normal low-current carbon arc were indicated: (a) the deeply molded positive crater is filled with a streaming, brilliant, high-temperature vapor ejected from the base of the crater which accounts for the high crater brightness, and (b) for currents above 100 amperes the arc stream appears to be contracted into a slender flame-like column with a current density more than ten times that of the low-current arc stream.

The term *high-current arc* has been introduced to distinguish this general type of discharge from others that included special compounds in the core of the positive

Fig. 20.1 Schematic sketch of the high-current carbon arc with the different regions of theoretical importance

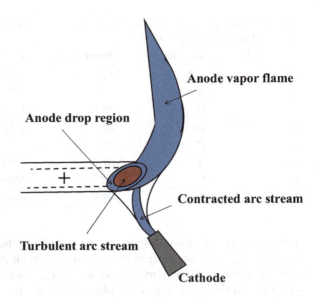

carbon. The vapor eruption from the positive carbon influences the arc mechanism. The general shape of the arc, especially the direction of the anodic vapor flame and the arc stream as shown in Fig. 20.1, is independent of its orientation in space and does not result from the convection of hot vapors or gases. It is, rather, a general property of all high-current arcs of more than 100 A that their specific shape is greatly influenced by their self-magnetic fields. This arc called as field stabilized in contrast to the wall-, electrode-, or convection-stabilized arcs.

The high-current carbon arc is operated with anodic current densities between 100 and 400 A/cm^2 compared with approximately 40 A/cm^2 for the low-current arc. Evidence for a fundamental difference in the arc mechanism is the rising voltage characteristic of the high-current arc, which is caused by an increase of the anode voltage drop with increased current. This rapid evaporation of the positive carbon affects the core more than the harder carbon shell and thus causes the deeply molded crater. As a consequence of this evaporation and an additional heating effect, the positive crater of the high-current carbon arc in contrast to the low-current arc is filled with brilliantly radiating vapors. These vapors, ejected from the carbon, form the anodic vapor "flame" mentioned above and indicated in Fig. 20.1.

The theory of the contracted arc stream has been reported according to which the electric field strength E adjusts itself to such a value that the power input per cm arc length IE is compensated by the radial heat conduction and by radiation. Also, the transition from low- to high-current arc stream was explained. With increased current, the arc plasma stream can either increase its diameter at the constant temperature of 6800 K, corresponding to the minimum of the total heat conductivity, or it can remain at constant diameter with increased temperature and consequently with increased electric conductivity. In the above currents of approximately 100 A, the second

20.1 Review of the Anode Region Theory

possibility was more favorable for the energy balance because of the exponential increase of the electric conductivity with temperature.

Immediately in front of the positive electrode, the total current then is a pure electron current. The resulting negative space charge (anode drop region) causes a potential drop immediately in front of the positive electrode surface. By it, the electrons are accelerated and, by ionizing collisions with atmospheric gas molecules, produce the positive ions which diffuse in the electric field toward the cathode and which are necessary in order to compensate the negative space charge in the quasineutral arc stream. It was indicated that obtaining the anode drop of the low-current carbon arc using an ion-emitting, metal-cored positive carbon was only about 10 V, whereas the anode drop for a pure homogeneous positive carbon without ion emission is approximately 30 V.

The anodic mechanism of the high-current carbon arc is based on one premise: The anodic vapor stream blows away the positive ions from the anode and thus, by an increased negative space charge, causes the increase of the voltage drop in front of the anode with all its consequences, and it is these consequences which are characteristic of the high-current carbon arc. The transition from the low-current arc to the high-current carbon arc was explained using this anodic mechanism. By increasing the anodic current density, more energy is transferred to the anode by the electrons accelerated in the normal anode drop than can be dissipated by heat radiation and conduction. Consequently, the anode surface begins to evaporate more, a vapor is ejected as from a nozzle in the anode surface, and the evaporation results in a vapor stream. The velocity of this stream is given by the amount of electrode material due to anode heating to the temperature of 6000–8000 K [16].

The vapor stream causes an increase of the anodic voltage drop and thus of the electron energy transferred to the anode surface, so that in turn the evaporation rate increases until a state of equilibrium is reached. The energy transferred to the anode surface by the electrons is dissipated by heat conduction and radiation from the anode plus the amount consumed for the production of the high-temperature vapor stream. The rising voltage characteristic of the high-current arc was explained by above-described mechanism. It pointed out the important role which the magnetic field of the arc current (>100 A) plays in the stabilization of the high-current arcs.

Blevin [17] studied the behavior of arcs at the anode by photographing the luminous discharge and by examination of signs of melting and tarnishing left on the electrode by the discharge (see Chap. 14). Considerable radial contraction of the discharge is seen to have occurred near the anode, and a highly luminous and circular "anode spot" is apparent. A Kerr cell shutter has been used to show that such a spot was established within the first few microseconds of the life of the arc. The main goal of the work was calculation of the anode spot temperature.

The temperature was calculated in the central region of the active anode surface area using observation of the extent of melting of the electrode of the transient arc. Usually, the depth of melting is small compared with the anode thickness and the theory of melting in a semi-infinite solid can be applied to the physical problem with reasonable accuracy. It was considered a case of arc time t when the anode can be represent as a thin metallic slab of thickness d_a to stude it thermal regime. If the

Table 20.3 Anode spot temperature according to Blevin's calculated model [17]

Anode metal	Arc duration, μs	$\lambda l c T_{mel}$	$(at)^{0.5}/d_{a0}$	T_a/T_{mel}	Spot temperature, T_a °C
Tin	27	1.06	0.22	7.5	1600
	50	1.06	0.24	6.8	1460
	250	1.06	0.25	6.5	1400
Aluminum	250	0.45	0.63	1.9	1250
Nickel	250	0.43	0.83	1.5	2300

diameter of the discharge greatly exceeds d_a, there is little radial conduction of heat in the metal from the center of the active area, and heat entering the anode in that region is conducted linearly through the slab. It was assumed that for the duration of the arc, the active anode surface is held at a constant temperature T_a (the initial temperature of the anode being zero), and that no heat is lost from the back of the anode slab.

The solution was obtained for parameter $\lambda l c T_{mel}$ (λ is the thermal conductivity, c is the specific heat and melting temperature T_{mel}) as dependence on dimensionless parameter of $(at)^{0.5}/d_{a0}$ (a—thermal diffusivity) for given different values of T_a/T_{mel}. d_{a0} is the critical thickness, at which melting just extends to anode back surface given from experiment. From this solution, the central anode spot temperature T_a was determined. The results are presented in Table 20.3. It was concluded that for each metal, T_a is well below the boiling point. However, no any measurements and some comparison with the calculation were presented and discussed. The calculated average surface flux of heat is consistent with the value derived from the anode drop with given values from 2 to 9 V assuming that all the energy of the anode fall enters the anode.

In general, the theoretical study of the anode region in vacuum arcs is presented by limited number of publications. Cobine and Burger [18] extended Finkelburg's theoretical approach [10] (see above) by studying the anode energy balance. They proposed to evaluate the relative importance of various physical processes occurring at the anodes of high-current arcs by examining the contribution of each process acting alone. This involved some reason of the assumptions appraised by examining their consequences. The anode spot receives thermal and radiant energy from the column immediately in front of it. Part of the thermal energy is carried by the electrons and can be added to the potential component as a term u_T. The power delivered to the anode spot by conversion of the kinetic energy of neutral atoms W_m, and that due to radiant energy from the column W_r, will represent the entire energy received by the anode

$$W_A = I_e(u_a + \varphi + u_T) + W_m + W_r \quad (\text{Watts/cm}^2) \qquad (20.7)$$

Here the energy received from excited atoms and from the arrival of negative ions was omitted, assumed as negligible. Energy balance was analyzed at anode spot for constant surface temperature and for time of 1/120 s, which included thermal

20.1 Review of the Anode Region Theory

energy stored in metal, including melted area, energy carried off by evaporation, heat of fusion of volume melted, and radiation energy. The amount of heat that flow into the metal was estimated from heat conduction equation in linear approximation, which determined an approximate upper limit for penetration of heat in the time under consideration. The use of this equation for the flow of energy into a metal from an anode spot is justified that the depth of penetration of heat in the time of 1/120 s is relatively small compared to the diameter of the spot. The temperature distribution was calculated for given surface temperature of 4000°, 3500°, 3000°, 2500°, and 2000 K for an elapsed time of 1/120 s. The total heat stored in the metal down to 3000 K, as well as the heat stored in metal above the melting point, both obtained by integrating the temperature–distance dependences, as a function of surface temperature. So, for a likely constant value of 3000 K, the temperature is above the melting point for a distance of only 0.115 cm. Table 20.4 demonstrated the results of calculations of the anode spot (minima and maxima) temperatures for, respectively, minimal and maximal parameters of measured anode potential drop and current densities [18]. For comparison, the corresponding experimental anode spot temperatures were indicated also that values are close to the maxima temperatures.

Thus, the main goal of the above-considered works is to study the anode energy balance and anode temperature taking into account the energy loss due to formation of the intense vapor stream of the anode material. The main anode region problem describes transition between cold anode and relatively high-temperature adjacent to plasma. As the electrical current in the anode region was supported mostly by the electron flux, therefore near the anode surface a negative space charge sheath

Table 20.4 Anode spot temperature calculated using anode energy balance according to Cobine and Burger [18]

Material	Boiling point, K	Anode spot temperature, K		Experimental temperature, K
		Minima	Maxima	
Zn	1180	1370	1630	1580
Ag	2485	2390	2980	2880
Cu	2868	2490	3040	2920
Al	2600	2640	3320	3270
Ni	3110	3040	3650	–
Fe	3008	3070	3760	3750
Sn	3000	3760	5760	5930
Ti	3550	3870	5040	5050
Zr	3850	4430	5500	–
C	4640	4600	5330	4780
Mo	5077	5380	6810	6580
W	5950	6700	8300	7470

Fig. 20.2 Schematic presentation of the plasma zones in the near-anode region modeled by Bez and Hocker [19]

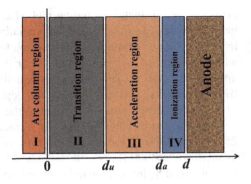

was produced. This space charge forms an anode potential drop. So, further understanding the mechanism of anode region requests an analysis of electron generation and electron flux formation in the near-anode plasma. Such an attempt to study the anode plasma processes was primarily modeled by Bez and Hocker in [19].

Let us consider briefly their theory. The anode region was modeled by four zones presented in Fig. 20.2. The first zone is bound with the arc column plasma in which the electrical field increases from low value which determined the electron drift to a value in the second transition plasma zone, in which formation of free motion of the electrons is occurred. In the third acceleration zone, the electrons acquire the energy enough for atom ionization (named as field ionization) that occurs in the fourth ionization zone. The produced ions collide with the atoms in transition zone, and their distribution is thermalized to temperature corresponded to that in the first zone.

Bez and Hocker calculated the anode potential drop assuming that the ionization probability is largely proportional to the energy of the electrons in the following form:

$$q^* = qC(u(x) - u_i) = qf(u) \qquad (20.8)$$

q is the geometric atomic and molecular cross section, $u(x)$ is the potential drop from the plasma column $f(u)$ yield function, u_i is the potential ionization, and C depends on the nature of each gas and its density. Using the measured current density at the anode and relation (20.8), the solution of Poisson equation for potential u distribution was obtained as follows:

$$x = d_u + a \frac{4T}{\alpha^2} \left[\frac{1}{3} (\alpha u^{0.5} - C_1)^{3/2} + C_1 (\alpha u^{0.5} - C_1)^{0.5} \right] - C_2;$$

$$\alpha = 8\pi j_e \sqrt{\frac{2m}{e}}; \quad j_e = en_e \sqrt{\frac{2eu}{m}} \qquad (20.9)$$

20.1 Review of the Anode Region Theory

There j_e is the electron current density, two integration constants C_1 and C_2 are specified form that the potential curve can be connected to the potential curve in the transition region in a continuous and differentiable manner.

The potential drop was determined for an arc with carbon anode in air using ionization potential 14.3 eV of oxygen and nitrogen. According to (20.9) this ionization potential is reached at $x = d_u = 2.99 \times 10^{-3}$ cm, at which the acceleration zone was ended and the ion current disappears at $x = d_u + 0.45 \times 10^{-3}$ cm. So, the extent of the ionization area is in comparison with the acceleration area. The thickness of the entire anode drop region was calculated as 3.44×10^{-3} cm. For comparison, the electron mean free path is $L_e = 5.5 \times 10^{-3}$ cm. Thus, condition $d \sim L_e$ which is necessary for the application of the condition of free-electron motion is well fulfilled. The calculated anode potential drop becomes $u_{an} = 18.7$ V. The authors indicated that this value good agrees with the experimental value of Finkelnburg and Segal of 20 ± 1 V [20].

The conditions under which field ionization in the anode case of an arc can occur were discussed by Bez and Hocker in their next work [21]. It was shown that the mechanism of field ionization is only possible if the total trap energy is greater than the ionization energy and two additional conditions must be fulfilled: (i) The corresponding potential must be formed over a distance of an electron path length, and (ii) the generated ions must be delivered to the column with a disordered movement, i.e., the length of mean free path for the ions should be lower than the electrons. As the relation between these lengths depends on the temperature and at some critical temperature, the field ionization was impossible. The analysis indicated that these conditions were satisfied for a low-current carbon arc in air under normal conditions with the anode drop mechanism of field ionization, while under the conditions of high-pressure arcs and large temperature, the plasma density increases and the abovementioned conditions are violated.

The plasma was heated with the increase of the current density, and the mechanism of charge particle generation was changed to the thermal ionization. The process of transition from field to thermal ionization was studied taking into account the current, pressure, and gas type of the discharges. The observed anode potential drop in frame of thermal ionization mechanism was discussed. It was taken into account the potential profile in front of the anode of a low-current carbon arc in air, the onset of thermal ionization and appearance of a contraction of the arc column in front of the anode as well as arise of the microspots on the anode [22].

Considering the arcs with different anode materials, Rieder [23] and then Cobine [24] continue to use the anode energy balance like used by Cobine and Burger [18]. Zingerman and Kaplan [25] investigated the anode erosion due to spot action as dependence on short gap distance (Chap. 14). Zolotykh [26] studied the crater diameter (produced by the spot) on Cu anode as function on interelectrode distance for different materials (see details in Chap. 8). Sugawara in [27] suggested an extended analysis of Cobine and Burger approach [18] to lower values of the input power and to propose a useful method of determining the anode fall for argon at atmospheric pressure. The heating of the anode is due to electrons accelerated by the anode fall, the energy of the electrons, and the work function of the anode material were considered. The anode is shaped as a straight wire of small cross section so that the heat flow is one dimensional along its length l without heat loss from the surface. The non-stationary one-dimensional heat conduction was solved. Delay time before the anode surface begins to melt, and length melted during a time t after onset of the arc was determined assuming the melting point T_m at the surface. The comparison of the measured and calculated dependencies of anode length on arc time allows to determine the value of anode potential drop as 4.22 V for arc current of 10 A. Noted that this value is approximately consistent with 7.3 V obtained by Busz–Peuckert and Finkelnburg [11], bearing in mind that this experiment will give little larger fall because the anode was directly water cooled.

Anode energy balance and equations of heat conductions were used in order to study the thermal processes in the anode at different stages of a pulse discharge by Beilis and Zykova [28] as well as by Goloveiko [29] (see Chap. 14). Konovalov [30] and Caap [31] used 1D heat conduction with expression for the heat flux to the anode in the form of Cobine and Burger (see Chap. 10).

Bacon and Watts in [32] discussed the spectroscope measured data using a collisional–radiative model to show how excitation and ionization of Al^{2+} occur in a small volume of plasma within about 0.3 mm near the surface of the anode of the arc discharge. A one-dimensional model assumed to describe the plasma within a few tenths of a millimeter from the anode surface with a constant electron temperature and Maxwellian distribution of electron velocities. The results were concluded, indicating that the theoretical and experimental data are consistent if $kT_e/e = 30$ V and $N_e = 5 \times 10^{16}$ cm^{-3} at the anode surface. At distances greater than 0.3 mm from the anode surface, $kT_e/e < 15$ V. The discontinuity in T_e at distance of 0.3 mm was explained by the presence of a potential hump in the plasma near the anode that is separated from the plasma column by a double sheath. The excited states of Al^{2+} were not in partial LTE. A Boltzmann plot of the theoretical excited-state densities yields a distribution temperature, T_D of $kT_D/e = 2$ V that is practically independent of the T_e. The possible non-LTE conditions are discussed as a result, of the low electron density and the transient nature of the ionization processes.

20.1.2 Anode Spot Formation in Vacuum Arcs

While the anode spot in the arcs at atmosphere pressure appeared at relatively low current (even < 100 A), the spot in vacuum is initiated in high-current arcs. It is occurring due to electrical energy dissipation and gas ionization in high-pressure arcs, while in a vacuum at moderate currents the anode collects electron current from the cathode plasma uniformly over anode surface (passive operation). Therefore, the important questions are (i) what is the condition for active anode operation and (ii) what is the mechanism of anode spot formation. The first theory on anode spot formation was published by Lafferty [33] based on a thermal runaway effect at the anode surface associated with a discharge contraction and local vapor emission from the anode surface.

20.1 Review of the Anode Region Theory

The model is qualitative and takes into account that normally the anode is not a positive ion source and the electric field in the positive column tends to drive the positive ions away from the anode. This causes an electron space charge sheath to build up around the anode with an accompanying anode drop of potential through which all the electrons are accelerated striking the anode. These electrons bombard the anode with energy corresponding to the anode fall of potential, the average electron energy, and the electron heat of condensation, thus subjecting it to a much higher total power input [18]. At a sufficiently high-current density, some local area of the anode where the heat conductance is poor will be heated hot enough to release metal vapor. The vapor is immediately ionized by the incident high-energy electron flux. According to Lafferty [33], the ions neutralize the space charge in the volume adjacent to the anode area emitting the metal vapor and produce a sharp reduction in the anode drop of the potential. The lower-voltage drop causes more current to flow into the anode in this area. This increase in current flow is also aided by an increase in the random electron current density in the plasma resulting from the increased metal vapor density. The local increase in current flow to the anode heats it even more causing additional metal vapor to be emitted and ionized. This leads to a runaway effect causing a constriction in the arc at the anode with the production of an anode spot.

It should be noted that a reduction of the anode potential drop cannot cause more current to flow into the anode, but can reduce the electron energy and further anode heating. As for our opinion, after vapor ionization the electrical field causes ion drift in the direction to the cathode, while the produced electrons flow to the anode (as an additional flux) increasing the electron current and energy flux to the anode. As a result, an instability occurs associated with infinity increase of the local vapor emission from the anode surface, their ionization, and finally a contraction of the anode region.

Kimblin [34] showed that according to the experimental data and theoretical analysis, the spot formation is associated with a large temperature rise at the anode surface. He noted that assuming that spot formation is associated with a temperature rise from about ~1000 °C, anode temperature can possibly influence the formation mechanism. Kimblin mentioned the temperature-dependent mechanism of Lafferty [33] and has suggested that this mechanism is consistent with his concept of vapor starvation. He also concluded that anode spot formation is associated with increase arc stability due to reductions in the magnitude of the high-voltage oscillations and, in the case of the smaller anode, marked reductions in the mean arc voltage, presumably due to vapor emission from the anode into a previously vapor-starved region.

Mitchell [35] formulated a model of anode spot formation. He reported that at low currents (<1000 A), the vacuum arc was diffused with an arc voltage of 20 V. But as the current in the copper arc is increased to several thousand amperes, the collisions between electrons and neutral vapor increase a voltage drop in the interelectrode plasma. The larger voltage reduces the velocity of ions ejected from the cathode spot region toward the anode. Thus, as the current in a vacuum arc is increased, the total arcing voltage increases until the plasma voltage drop reaches a value, which corresponds to the ion kinetic energy. If the arc current is marginally increased beyond this value, or if the contact separation is enlarged at this current, charge neutrality may not be maintained close to the anode by ions emitted from the cathode spot region, and the voltage across this thin region immediately in front of the anode will rise to the enormous values. Thus, once this critical current has been reached, the total arcing voltage may be expected to rise discontinuously to the value necessary to produce ionization at the required rate in the region close to the anode. The current at which ion starvation reached, Mitchell, is termed as the starvation current. At a higher current, as the anode spot formed, the voltage again decreased and became less noisy.

It should be pointed out that Mitchell's assumption about increasing a voltage drop due to collisions between electrons and neutral vapor in the interelectrode plasma can be understood taking into account that presence of large amount of neutral vapor reduces the plasma electrical conductivity. The electron-neutral collisions determine the conductivity in case of very low degree of ionization. However, it is doubtful that this situation can be for cathode plasma [36]. For the case of highly ionized vapor, the plasma conductivity is determined by the electron temperature, which can increase with the arc current. Therefore, the ion starvation can occur by other effects, for example, due to reduce of the plasma density with interelectrode distance or by use of the small active anode surface area.

Ecker [37, 38] studied the instability of the anode region in case when the current in a "vacuum arc" surpasses a certain critical value and an anode spot formation can occur. A comprehensive quantitative analysis of the arc phenomena was conducted including the cathode and anode regions as well the gap plasma (Fig. 20.3). The following assumptions were used: (i) The electrode separation is small enough so that the one-dimensional approach can be treated. The electrodes are planar and infinitely extended without lateral loss; (ii) vaporization at the cathode must be sufficient so that the "collision dominated uniform phase" was reached at a time to short after ignition. A displaced Maxwell distribution throughout the gas except for the sheath regions was considered; (iii) the plasma body is *weakly ionized* and quasineutral. Heat

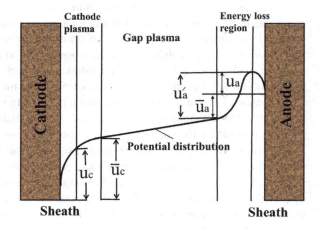

Fig. 20.3 Schematic presentation of the potential distribution and directions of the particle current densities in the vacuum arc according to Ecker anode region model [37]

20.1 Review of the Anode Region Theory

production in the plasma volume goes mainly into kinetic, excitation, and ionization energy of the neutral gas. The voltage applied to the electrodes is a known function of time. In some cases, the current form instead of the voltage form may be prescribed.

The anode sheath with potential drop u_a accelerates the electrons toward the anode. Immediately in front of the anode—in the sheath—there were negative space charge due to the current supported through the diffusion and mobility transport from the arc plasma. When the current density is large, the negative space sheath is dominated by positive charge, which keeps some of the plasma electrons back and allows also partial compensation of the electron Langmuir current through an ion saturation current.

A general system of transient differential equations includes equations of conservation of particle number, their momentum, and energy taking into account the excited and unexcited neutrals. Description of the electron and ion components was presented as balance equations. To calculate the electron temperatures, the energy balances were used in the energy loss region near the electrode. The regions gain energy from incoming particle beams and from the electric field which is balanced by a net loss due to elastic and inelastic collisions. The neutral particle number density n and the neutral particle temperature T are determined by the boundary conditions.

Using the condition of quasineutrality $n_i = n_e$, heat conduction equation, random electron flux to the anode, the Saha equation, the plasma Ohm's law $j_i - j_e = \sigma_{el} E$ and anode energy balance can be written as

$$j_a E_a = f(T_{ea}) \frac{3k T_{ea} n_e}{2\tau_e} \tag{20.10}$$

where $f(T_{ea})$ is related to the average energy loss per collision in the anodic energy loss region. T_{ea} is the electron temperature, and E_a is the electrical field. The energy loss to the anode consists of several contributions: the loss of kinetic energy, the loss of excitation energy, and the loss of ionization energy. Remembering that there is no energy dissipation within the sheaths, the energy loss in the anode region was characterized by an effective potential drop u'_a which was defined through the relation

$$j_a u'_a = l_a f(T_{ea}) \frac{3k T_{ea} n_e}{2\tau_e} + \frac{3k T_a n_a v_{Ta}}{2} \tag{20.11}$$

In essence, this is the potential required across the anode loss region to compensate for the loss of kinetic, excitation, and ionization energy to the anode surface (Fig. 20.3). The average transport velocity of the particles within the anode loss region was determined by the average thermal velocity v_{Ta}. Under these circumstances, the extension of the anode loss region characterized by $l_a = \tau_{de} \lambda/\tau$, where τ_{de} is the lifetime of deexcitation, τ is the lifetime for momentum exchange at excited collision, and λ is the mean free path for momentum exchange.

The modeling regions—electrodes, sheaths, and plasma body—are matched through the requirement of particle number and energy flux continuity at their boundaries taking into account the ratio of spot area to the total electrode area F_s/F. The current pulse was given in the form $I(t) = I_p g(t)$ where I_p is the peak current and $g(t)$ is a function on pulse time. The current density was determined as $j = I/F$ and the ion current fraction as $f_i = u_c/2(u_c + u_i)$. To calculate u_a, the returned ion current density in the anode sheath (dependent on u_a) was used matching it with ion current density in the cathode region. Matching the various regions results in an equation for an area of the appeared anodic spot which was a result of an anode constriction instability. The spot area at the anode surface was calculated as a function of time. This function allows finding a marginal instability criterion for the onset of any instability which may exist. An expression for a critical current density j_{cr} for the instability was obtained solving the differential equation for time-dependent neutral density dn_a/dt. The value of j_{cr} was determined also by condition when the anode evaporation rate dominated term in this equation.

The solution of the differential equation for $n_a(t)$ provides the time dependence of the anode spot area. First, it was solved for the "non-evaporating phase" and then for the "evaporating phase". The solution for the first case showed that practically all of the energy influx goes into heat conduction if the temperature is small in comparison with the boiling point temperature. So, there is no indication for a constriction instability. From this, it was concluded that instability can occur only if the "evaporating phase" is reached.

For the "evaporating phase," a critical current density was calculated. To this end, the characteristic quantities "$T_e, f, n_a, n_e, T_a, u_a, u_a$" as functions of the current density j were calculated using the system of equations. Then these quantities were introduced into the expression for j_{cr} yielding $j_{cr}(j)$ determining marginal instability. Finally, j_{cr} was then found assuming that $j_{cr}(j) = j$. The dependence of the critical current density j_{cr} as a function of j evaluated for typical pulse duration t_0. As a result j_{cr} was obtained about $10^4, 6 \times 10^3, 10^3$ A/cm^2 for $t_0 = 1$, 10 and 100 µs, respectively. This result was compared with experimental data for the instability onset observed by Kimblin [34] 1.5×10^2 A/cm^2, Mitchell [39] 10^4 A/cm^2, and Rich et al. [40] 4×10^2 A/cm^2.

The calculation agrees well with Mitchel data. From the difference between Kimblin's and Rich's experiment, Ecker [38] explained by assumption that the experiments are not really "small gap devices". He indicated, with small electrodes and large electrodes spacing the anode drop u_a experimentally and theoretically is expected to increase. This will also result in an increased energy influx to the anode surface which according to the theory will shift the critical current density to lower values. A similar effect can be expected if the assumption of lateral homogeneity is violated due to the structure of particle beams reaching the anode. Finally, another cause for a decrease of the critical current density may lie in the presence of local surface irregularities, which can produce current densities locally much higher than those calculated from assuming uniform distribution of the total current at the electrode surface.

Generally, it should be noted a very detailed and useful analysis of the gap plasma and near-electrode processes is presented by Ecker also considering the particle elementary collisions in a vacuum arc. Nevertheless, the present mathematical description consists of numerous formulas, and, moreover, the relations between

20.1 Review of the Anode Region Theory

number of them are difficult to understand not only for wide readers, but also for those close to specific researches of the arc. Questions always appear reading the large number of expressions and symbols that sometimes were confused. It is first of all because the assumptions and statements are arbitrary, as well due to the presentation of some not understandable matching conditions. As an example, it can be indicated the use of the condition of equality between plasma pressures of heavy particles in the gap is produced only by heavy particle flow from the cathode and anode sides. However, the electron pressures and the mass velocity of the plasma flow due the electrode evaporation were not considered, although that the convective terms were accounted by formulation of the general system of equations.

The final expression of instability criterion that was derived is not obvious. The model was rather considered for an idealization case that also was indicated by the author when he discussed the comparison with the experiments. A small electrode separation with homogeneous plasma parameters was modeled, while a high-current vacuum arc consists of large numbers of cathode spots with individual plasma jets even when the electrode diameter is significantly larger than the gap size. The plasma in the gap was assumed *weakly ionized*, while the electron temperature is not small. Also, the difference of the Ecker's calculations from Kimblin's data for copper could be understood taking in account the specifics of the experiment. It was conducted for different electrode sizes and for a much long period of arcing time of the order of a half-second, for which an anode spot formed at about 400 A with electrode diameter of 1.25 cm.

Rich et al. [40] observed a strong correlation between the threshold current I_{th} for anode spot formation and the thermal characteristic $T_m(\lambda_T \rho c)^{1/2}$ of the electrode material (see Chap. 14). To understand this correlation, a temperature rise at the anode surface as a function of time was studied considering transient one-dimensional heat conduction equation for a semi-infinite anode, characterized by thermal diffusivity a and conductivity λ_T, respectively:

$$a \frac{\partial^2 T}{\partial x^2} - \frac{\partial T}{\partial t} = 0 \qquad (20.12)$$

Boundary condition:

$$-\lambda_T \frac{\partial T_a}{\partial x}(x=0) = j_{a0} u_{\text{aef}} \sin \omega t \qquad (20.13)$$

Here j_{a0} is the amplitude of the current density at the anode surface, and u_{aef} is the effective anode potential drop, which includes, in addition to the intrinsic anode drop, the anode work function, and the thermal electron energy according to the analysis of Cobine and Burger [18]. Taking into account $a = (\lambda_T/c\rho)$ a solution was obtained in form

$$T_a = \frac{j_{a0} u_{\text{aef}} \sqrt{\frac{2}{\omega}} \Phi(t)}{(\lambda_T c \rho)^{0.5}} \qquad (20.14)$$

Taking $T_a = T_m$ and $j_{a0} = I_{th}/F_a$, the observed dependence of the threshold current I_{th} for anode spot formation and the thermal characteristic $T_m(\lambda_T \rho c)^{1/2}$ of the electrode material were obtained. This dependence, with $\Phi(t_m) = 0.95$, was discussed in frame of experiments conducted by the authors and also using published, respectively, other results. It was indicated that a difference between calculations and the measurements can be explained by quite likely taking into account the variations of value u_a and dependence of the thermal characteristics on the temperature for different electrode materials, which were considered as constant in the present approach.

According to Kimblin [41], the vapor pressure of copper at the melting temperature of 1350 °K is only of the order of 5×10^{-4} Torr. Since this pressure is even below the vapor pressure in the anode region due to the cathode evaporation, it is somewhat surprising that spot formation should be triggered by the onset of anode melting. Kimblin suggested that the onset of localized melting results in enhanced local heating with a resulting rapid vapor pressure increase. A further temperature increase to 1650 °K, for example, would raise the local vapor pressure to 0.1 Torr and the particle density to 5×10^{14} cm^{-3}. At this vapor density, the mean free path of the electrons is low, and ionization can be expected in the anode region.

Further Kimblin discussed the observed increase in the anode voltage drop with increasing arc current. He noted that it can be also attributed to redistribution of the cathode spots, which were mutually repelled to the edge of the cathode, and this reduced the contribution to the anode plasma density due to declined cathode plasma jets. This is again consisting with the concept of vapor and ion starvation in the anode region. According to Kimblin's model prior to the spot formation, the anode voltage drop can exceed the ionization potential and increase the energy flux to anode. When metal vapor is released from the heated anode surface, however, the electrons experience ionizing collisions, the ion density adjacent to the anode increases, and the overall anode voltage drop reduces. The arc voltage not only increases from 20 V (cathode region) but exhibits high-frequency and high-voltage fluctuations. With anode spot formation, the arc voltage is stabilized.

We should note the obvious and known fact that the amount of the cathode plasma reached the anode decreases with decrease of the anode surface area and with electrode distance increase, independently, the reached quasineutral plasma is collisional or collisionless. The observed increase of the anode voltage could be considered as a mechanism (larger resistance) to support an increase of the electron current to the anode due to plasma starvation near the anode. The simultaneous increase of anode heating and its surface temperature leads to increase of intensity of the anode evaporation rate. As the vapor density grows, the probability of the vapor ionization also increased and, as result, decrease of the plasma starvation in the anode region. When this process develops unlimitedly, an anode spot can be formed and preferably local at some inhomogeneity on the anode surface.

Kamakshiah and Rau [42, 43] showed that value of the threshold current depends not only on arcing time but also upon the anode material, surface conditions, contact separation, and electrode diameter, which agree with the previous conclusions. Lateral inhomogeneity in the gap reduces the threshold value by promoting early formation of anode spots. The authors detected that the threshold values obtained

20.1 Review of the Anode Region Theory

for copper and aluminum electrodes with 1.65 μs arcing time are, respectively, 1.75 and 1.85 times the values at 8 μs arcing time. They explained that the spot formation one after another may be due to the fluctuations in metal vapor and plasma density near the anode region due to variations in the volume and direction of vapor emitted from the cathode spots.

Miller [44] reviewed the theoretical approaches described the transition of the arc into the anode spot mode in vacuum at time before 1977. Some point of view on these publications can be found in [45]. Miller in his comprehensive review indicated that a deficiency in the runaway theory [33] is reflected by the observed fact that the anode spot transition seems to occur for anode temperature less than the melting point (for most materials), which implies a vapor pressure near the anode surface far below that necessary to initiate the runaway effect (see above Kimblin's remark [41]). This observation has led several researches to prefer a magnetic constriction explanation. As a result, either magnetic constriction in the gap plasma or gross anode melting can trigger the transition, indeed a combination of the two is a common cause of anode spot formation.

Based on the previous works, Miller suggested his own opinion for a reasonable phenomenological description of the arc transition to an anode spot mode. It is consists of six stages including (i) at low current, cathode flux impinges upon the anode surface to carry the current required by the external circuit, so there is no need for an anode potential drop; (ii) as the arc current slightly increased, the situation changes only slightly due to increase in the electron collection surface by lateral anode sides and the cathode ions still have sufficient energy to reach the anode; (iii) as the current continues to increase, the plasma flow to the anode no longer is enough to sustain the external current, so a small anode drop develops to pull in additional electrons; (iv) further, the anode potential approaches the mean cathode ion potential, so no longer can enough cathode ions reach the region near the anode to compensate for the local electron flux, which is at the ion starvation current.

In order to supply positive ions in this region, an appreciable voltage drop develops across the region. In addition, magnetic constriction could occur before the ion starvation current was reached; (v) the increase in ionization produces a decrease in anode drop, which causes the current to increase over a broad area. The arc then constricts, and the anode power density increases sharply; (vi) the constricted arc is resulting in a large local power density increase rapidly increasing the local anode surface temperature. This leads to intense ionization of the generated vapor near the anode, relieving the ion starvation. The final equilibrium high-current vacuum arc with a fully developed anode spot has a hump potential distribution in the vicinity of the anode surface.

As for the further publications, the series works of Boxman and co-authors can be considered. Boxman [46] developed a model of the anode spot formation considering a fluid approximation for the interelectrode plasma flow from the cathode to the anode and generated by a multiplicity of cathode spots in a vacuum arc. A vacuum arc sustained between 25 mm diameter electrodes separated by 9 mm with a current of 3 kA by a 9 μs duration half-sinusoidal current pulse.

The steady-state mass flow in the interelectrode gap was analyzed by applying the MHD equations considering the conservation of mass (without particle source), momentum (with self-magnetic force), and energy equations (with Joule energy dissipation and radiation) assuming radial symmetry and constant temperature. The mass density, current density, and plasma velocity were given from the experiment at the cathode side as boundary conditions. The mass density reduction inversely as the square of distance from the cathode and the radial profile of the jet was determined by the magnetic force along the distance. In essence, the used approach is one dimensional with given radial profile of the jet opening with distance. Two types of calculations were conducted for (i) mass flow (assuming a constant current flow) and for (ii) current flow (assuming a constant mass flow) in the interelectrode region. The calculations show that both flows have a tendency to constrict in the vicinity of the anode. A constriction in the current flow is also calculated, caused primarily by the Hall current. It was indicated that, when constriction does occur in the interelectrode plasma prior to the release of significant quantities of anode vapor, the constriction may play a significant role in anode spot formation. It was noted that this conclusion is approved by a number of experimental observations, which support a constriction caused by anode spot formation process.

In another work, Boxman et al. in [47] discussed the threshold current transition from the low-current to the high-current vacuum arc mode. They indicated that this current is significantly higher at higher frequencies than at lower frequencies, allowing the application of vacuum switches in certain high-frequency circuits without the formation of deleterious anode spots. Several possible explanations of this phenomenon are suggested and cause of the increase in threshold current with increased frequency was analyzed. The most obvious possibility to that transition is basically a thermal instability at the anode surface, which studied by Ecker [38].

Boxman and Goldsmith [48] developed an anode region model assuming the presence near the anode surface of an uniform plasma, produced by mixing of a number of cathode jet, having a very low electric field and existed up to the boundary of anode collisionless sheath region. The anode surface will be modeled as a perfectly absorbing surface. A simple calculation indicated that the potential difference between the anode and the edge of the plasma region must be negative. The particle flux in the direction of the anode results from thermal motion of the electrons. In order to insure current continuity, the negative sheath potential should arise, which repel most of the incident random electron flux. Thus, the anode must insure that the net current incident to the surface is equal to the circuit current. In order to calculate the particle density, their fluxes, energy fluxes, the potential drop, and electric field in the sheath, the electron distribution function was approximated by a truncated Maxwell–Boltzmann distribution. The ion distribution is represented as a monoenergetic beam. Further generation of additional neutrals in the anode region was accounted by two ways, sputtering and anode evaporation using the Langmuir model. It was pointed out that the sputtered neutrals and the evaporated neutrals were rapidly ionized. Thus, the model of the anode region was taking this additional generation of the charge particle density into account.

20.1 Review of the Anode Region Theory

The overall analysis shows that both components (ions and electrons) have the same order of magnitude, with the ion heat flux dominating at lower values of T_e, while the electron component dominates at higher values of T_e. The overall energy flux is given by ju_a where j is the current density and u_a is an effective anode drop whose value in a Cu arc ranges from 13 V for $T_e = 1$ eV to 33 V for $T_e = 9$ eV. Neutral atoms are evaporated from the anode surface when the anode has been heated sufficiently. The sputtering is significant in the case of the Cu arc, producing a value of the neutral density of about 37% of the ion density. In general, the approach developed for a uniform discharge, with the bulk plasma. Nevertheless, the authors noted that the model may be easily extended by modifying some of the input parameters to describe other cases including non-uniform discharges. It should be indicated that the proposed model allows calculating heat flux, either directly for the case of a uniform discharge as well with modifications, which describe the early stages of anode spot formation.

Later the anode surface melting prior the anode spot onset was studied by Goldsmith et al. [49] using a thermal model [48]. The threshold peak current I_{pt} was determined for the onset of anode surface melting. The melting process of the anode obtained by balancing the energy influx to the anode evaluated according to the anode region model with the energy outflux through heat conduction. The value of I_{pt} was determined both by the thermophysical properties of the anode and by the physical properties of cathode produced plasma. The calculated I_{pt} was inversely proportional to the anode potential drop and directly proportional to parameter $T_m(\lambda_T \rho c)^{1/2}$ [40]. It has been showed that the anodic evaporation occurred at the onset of the arc, which may be increased by at least one order of magnitude when melting of the anode was observed.

Izraeli et al. investigated the influence of the self-magnetic field [50] as well as both the external axial and self-magnetic field [51] on the current distribution in the interelectrode plasma by solving a two-dimensional model using equations of magnetic field transport and arc current conservation. Assuming a uniform conducting medium and a uniform current density at the cathode, they showed that the electrical current was constricted near a boundary layer adjacent to the anode, while application of an external axial magnetic field reduces the degree of the constriction.

Beilis et al. [52] developed a two-dimensional model of plasma jet expansion in the vacuum arc with the free lateral boundary. The plasma expansion was modeled using the steady-state gasdynamic equations and equations for electrical current, which were solved self-consistently with the expanding free lateral boundary of the plasma jet taking into account the influence of the self-magnetic and external axial magnetic fields. The free boundary surface was characterized by the jet angle α, which depends on axial distance. In contrast to previous models, the present model takes into account the plasma jet angle α_0 at the starting plane. The current–voltage characteristic of the arc and its dependence on the magnetic field were also determined. The anode sheath potential distribution was calculated, using the theoretical plasma density, velocity, and current distributions in the near-anode region.

The calculations were based on the experiment, which show that the arc current distributed at the cathode surface as separate groups of spots even in high-current arcs is of few or few tens of kA [1]. As result, the expanding plasma cannot be uniform that already discussed by Schellekens [53] and shown later appeared, especially, in the presence of an axial magnetic field [54]. Therefore, the used spot current was varied in the range of 100–500 A and the boundary conditions at the cathode side used the data of cathode plasma jet study [55]. This approach, in essence, allows to obtain the plasma parameters of the jet expanding from cathode far from the surface in comparison with simple assumption of diffuse plasma near the anode with a total arc current. The present analysis shows that the critical current for the transition from the negative to the positive sheath potential depends on the initial jet angle. The initial plasma jet angle α_0, which has a strong influence on the plasma flow, depends on the processes occurring in the cathode spot and on the plasma expansion in the dense part of the cathodic plasma jet. The self-magnetic field does not substantially influence either the plasma jet shape, density, velocity nor the current density distribution for arc currents $I < 200$ A, while the plasma jet angle at the starting plane and the radial plasma density gradient force in the expansion region strongly influence the plasma and current flow. The plasma jet expands radially rather than constricting even in the $\alpha_0 = 0$ case. The two-dimensional calculations showed that the plasma density decreases in the axial direction if $I < 200$ A and $\alpha_0 > 15°$. However, at the higher arc current ($I = 500$ A), both the centerline electrical current flux and the plasma density increase in the axial direction, despite the overall expansion of the jet boundary.

The most significant result of the described work [52] is the prediction of a critical current to reach $u_a = 0$, which matches previous experiments at an appropriate choice of the free parameter α_0. The cause for the decrease in the magnitude of the negative anode potential with respect to the adjacent plasma is a decrease fraction of the plasma intercepted by the anode when the plasma jet radius at the starting plane increases with current. This effect alone would lead to a much lower value for the critical current. However, magnetic constriction in the plasma flow also increases with current and partially offsets the above effect. As noted by Djuzhev et al. [56, 57] and Kutzner [58], the appearance of a positive sheath potential is closely related to anode spot formation and signifies the inability of the plasma to supply sufficient electron current at the anode sheath to match the arc current. A possible mechanism of anode spot initiation proposes an increase of heat flux associated with the positive sheath, which leads to evaporation and ionization of material from the anode, and this additional plasma source supplies the electron current to the anode.

Harris [59] analyzed the constriction mechanism developed by Boxman and Ecker as an instability effect. He explained the results of the plasma density measurements considering the possible constriction, which caused a localized concentration of the heat flux to the anode, taking into account the cathode plasma jet properties and the characteristic parameters of particle collisions in the plasma. He assumed that at the constriction in the anode surface temperature rises rapidly with increase of the heat flux, and relatively cool anode vapor was emitted. This momentarily cools the anode plasma and increases the recombination rate, and thus causes the sudden decrease in electron density near the anode, which agreed with the experiment. Using this analysis, Harris concluded that the exact nature of the constriction mechanism remains unknown although the preceding discussion suggests that a plasma energy

20.1 Review of the Anode Region Theory

flux constriction, rather than an anode vapor-triggered phenomenon, was a more likely mechanism for the transition to the anode spot mode. Harris noted that no constriction in the electron density profile was observed by the experimental study, but this does not exclude from consideration several proposed magnetic constriction mechanisms [39, 46] since (i) the constriction may occur in the directed velocity profiles rather than the density profiles, (ii) the constriction may be so localized as to be undetectable with a 3 mm diameter interferometer beam, or (iii) the ion and/or neutral mass flows may constrict independently of the electrons.

Schuocker [60] indicated that according to the evaporation instability model in high-current vacuum arcs, a critical vapor density at the anode calculated by Ecker [38] and needed for anode spot formation was by two orders of magnitude larger than that measured by Boxman [61]. However, it should be mentioned that Ecker's mechanism needs a critical gas density immediately in front of the anode and Boxman measured his density right in the middle between the electrodes. Nevertheless, Schuocker developed a model to explain this difference, which is important to describe in order to understand its weakness. The model is based on anode vapor calculation by Herz–Knudson approximation, the electron temperature was taken in form deduced by Ecker [38], and electron density was determined by Saha equation. The anode temperature was determined from anode energy balance including energy with heat conduction loss, but neglecting the electron potential energy (work function). The electric field in front of the anode was obtained from Ohm's law using Spitzer's plasma conductivity. Finally, the equilibrium between magnetic pressure and thermodynamic pressure was taken into account.

The numerical analysis has been carried out in the following sequence. First, the anode surface temperature was calculated for given values of current and column radius as the simultaneous solution. Second, the vapor density was calculated. Third, the electron temperature T_e is varied step by step in order to find the proper value that fits the Ecker's model and then the electron density from Saha was calculated. For these results, the ratio of magnetic pressure to thermodynamic pressure was used, and finally, the electric field is calculated. This procedure was conducted only for one special value of current at stepwise reduced anodic discharge radius. It has been assumed that the cathode and the plasma column, in the middle of the gap and at the anode have the same radius 3 cm. Further, it has been assumed that the initial diameter of the column very near the anode before magnetic constriction is the same as in the middle of the gap. The calculation indicated that for currents lower than 900 A, the initial magnetic pressure is lower than the thermodynamic pressure, and no magnetic constriction takes place near the anode. For current higher than 900 A, an equilibrium radius was reached. The temperature corresponding to that equilibrium radius rises with rising current and approaches the boiling point for very high currents.

Thus, the mathematical analysis of Schuocker's model shows that the predicted constriction near the anode is possible. The vapor density obtained at the anode surface can reach a value by more than two orders of magnitude higher than in the plasma column and the absolute value is high enough to start the anode spot instability due to evaporation of the anode. Nevertheless, the author indicated that neither a pure magnetic constriction model nor a pure anode evaporation model can account for the effects observed. But the self-magnetic field enhances evaporation instability model for anode spot formation in high-current vacuum arcs, and therefore, both effects contribute considerably to the phenomenon of anode spot formation in high-current vacuum arcs.

Commenting this work, it should be pointed out that the presented results are obtained with a number of arbitrary assumptions, and some of these assumptions can be indicated here. The electron temperature T_e is given as input parameter of a certain value referred to the Ecker's work, while T_e depends on the discharge conditions and could be varied. The anode temperature was obtained from an energy balance, in which the anode evaporation rate was assumed in vacuum that not takes place. The anode heat conduction loss was determined using arbitrary value of the heat penetration depth. The anode region parameters cannot be determined from priory given condition indicating the equilibrium between magnetic and kinetic pressures and then conclude about the effect of magnetic constriction.

Drouet [62] showed that the current threshold values associated with the onset of anode spots in a vacuum arc were correlated with the solid angle of the expanding jet emitted from the cathode center and subtended by the anode and by the angular distribution of the current collected at the anode. A study of the mentioned solid angle was presented by Kutzner [63] as well by Kutzner and Zalucky [64] showing that it depends on the electrode gap determining the plasma amount (measuring the ion current) reached the anode surface.

Jolly [65] studied the anode spot formation by developing a simplified model to calculate the temperature rise of the anode surface in vacuum arcs. The known anode energy balance was used including the input energy flux by the electrons and neutrals, while the energy losses were by anode heat conduction (described by the one-dimensional equation), its evaporation, and radiation. Each term of the balance was analyzed including the effective anode fall and time-dependent arc current in order to determine the functional dependence of current at the melting point taking into account experimental variables. The threshold current was found solving the thermal anode problem by setting the anode temperature equal to the melting temperature T_m. The model predicts the threshold current for anode melting as a function of the properties of the arc and electrode materials (W, Mo, Cu, Al, Ag). The calculated and measured currents agree well for considered anode materials. The author noted that his modeling results were consistent with the suggestion of Lafferty that localized heating of the anode trigger anode spot formation by the anode vaporization. It found out a magneto-hydrodynamic pumping of the local liquid metal producing droplets, which were quickly vaporized in the arc column. The current necessary to significantly deform the surface of a melted region was estimated using a force balance between the electromagnetic stresses and the surface tension. The value of such current was estimated as about 1 kA. Again as in other models assumed input parameters determined the results.

Dyuzhev et al. [66] presented a critical analysis of the runway and magnetic constriction models of an anode spot formation in a high-current vacuum arc. Instead a qualitative condition of an anode spot formation in a vacuum arc was discussed. This condition was based on the experimental results obtained by studying anode phenomena in a low-pressure arc [56, 57]. The authors concluded that a change of the sign of the anode voltage drop (from negative to positive) is the key to an anode spot formation (see Chap. 14).

Nemchisky [67] studied the effect of magnetic constriction in vacuum arc simultaneously for both plasma and current constriction suggested by Boxman [46]. The plasma jet and the current in the gap were described by a system of magnetohydrodynamic equations for plane-parallel electrode geometry with radius R and gap width d. This system was simplified to describe the trajectory of the ions by an equation for collisionless motion of an ion acted by Lorentz force. It was indicated that when the arc current increases, ion trajectory passed close to the anode, eventually reaching at the critical current, at which the ion location z_c corresponds to the gap width. The critical current I_{cr} was calculated as function on R/d for given cathode jet velocity, ratio of the ion current to the arc current, and for an effective ion charge state and for different metals. The calculated values of I_{cr} were discussed by comparing with those presented by the experiments. The comparison shows that differences between the calculated and measured results are less than a factor two. The cause of the difference was discussed in frame of uncertainty of used input parameters.

In the further work, Nemchisky [68] studied the plasma compression effect on the anode spot formation due to the presence of an axial magnetic field. The equation of motion was analyzed, in which the force due to plasma acceleration was balanced by the force $\mathbf{j} \times \mathbf{B}$, while neglecting the force due to plasma pressure gradient. An expression for a critical current was obtained using a number of assumption and arbitrary input parameters. The dependence of the I_{cr} on the magnetic field was calculated. The satisfied agreement between calculated and measured data was obtained by selection of a certain ratio of the ion current to the arc current as 0.6. Here, it should be noted that the used above approaches are very simplified ignoring the influence of magnetic field on the plasma flow, plasma pressure gradient, and the current density distribution in the expanding plasma. Such influence was studied by the model developed in the later work [69].

20.1.3 State of Developed Models of Anode Spot in Low Pressure and Vacuum Arcs

In the last few decades, an attention to understand the mechanism of anode spot functioning was further enhanced. Some progress of a number of characteristic works is considered below.

Lyubimov et al. [70] used transient equations of heat conductions in cylindrical symmetry, relations Saha, traditional expression for energy flux to the anode, and random electron flux to the anode in order to calculate the plasma and spot parameters for aluminum anode. The zero anode fall and waveform of discharge pulse current of about 1 μs and time-dependent measured anode spot radius $r_a(t)$ were given as input parameters. The pulse current form $I(t)$ was used to determine the time-dependent energy flux to the anode as boundary conditions for heat conduction equation neglecting terms determined by anode erosion rate. The current density j, plasma density n_e, and electron temperature T_e were calculated as functions on time.

The results show strong increase of the calculated parameters, which then sharply decrease in the initial stage less near 100 μs and after which the parameters attain significantly lower stationary values. Due to the presence of difference for T_e in steady state 0.3 eV and measured 0.76 eV, the results were averaged to value of 0.52 eV. The calculated parameters were compared with experiment [70] (see Chap. 14). It should be noted that the analysis of the obtained results indicated significant difference between the measured current density and that presented after calculations in the work [70]. One of the reasons can be the problematic way for determination of the electron temperature using Saha equation, and, besides, it is difficult to provide correct accuracy of such calculation in case of the highly ionized plasma. The data of anode temperature and the neutral atom density at the anode were not reported. The presented model not considered the plasma energy balance, which is an important issue to describe the anode spot.

Nemchinsky in [71] studied the electron and ion motion in an anode layer solving Poisson equation instead of the usual expression for electron flux from plasma to the anode. He assumed that the ion density determined by atom ionization and electron density was derived from the relation between their thermal velocity and the current density. The solution was obtained by simplifications using a number of assumptions that needed to be based. The anode voltage was found as sum of the potential drop in the space charge sheath and a potential drop in the near-anode quasineutral plasma. The first part of anode potential drop was determined by "3/2" law, in which the thickness of sheath h_a was determined by a distance, at which the most of evaporated atoms were ionized. The potential drop in the quasineutral plasma was determined by plasma electrical conductivity from Laplace equation. The electron temperature T_e of 2.5 eV was given as free input parameter. A "minimal principle" was used. The calculated parameters for Cu anode and current of 400 A were $j = 3.6 \times 10^4$ A/cm^2, $T_a = 2700$ K, and $u_a = 8.6$ V.

In essence, it was assumed that the anode sheath equal to the ion mean free path length l_i was determined by the atom ionization. This value strongly depends on the ionization cross section and for calculated work u_a significantly exceeds ($l_i > 10^{-3}$cm) the ion mean free path length determined by charge-exchange collisions (10^{-4} cm). In addition, at $T_e > 1$ eV the degree of ionizations increases that lead to increase in the role of charge particle collisions, which was not taken into account. Therefore, at given work T_e the anode vapor was fully ionized and the mean free path of charge particles is ~10^{-6} cm. So, the sheath (>10^{-3} cm) cannot be assumed as collisionless. Besides, according to the calculated plasma pressure of 0.4 atm, the plasma density is 10^{18} cm^{-3}, which was at fully ionization state. Taking this into account and that the electron thermal velocity is v_{eT} ~ 10^8 cm/s, the resulting current density determined by the electron flux significantly exceeds the value obtained from the model [71] using the minimal principle. Thus, the assumptions of this model contradict to the calculated results.

20.1 Review of the Anode Region Theory

Lefort and Andonson [72] calculated the anode (Cu and Ag) parameters using anode energy balance, equation of total current by given arc current and condition that gas kinetic and self-magnetic pressures are equal to one to other. The anode current was supported by the electron current from the cathode and by electron current emitted from the heated anode. The space charge sheath was assumed with negative anode potential drop, and the anode erosion rate was determined by Langmuir–Dushman formula (vaporization with a sonic speed) at a temperature of the anode [73]. The current density j, anode temperature T_a, and anode spot radius calculated by the given arc current I as parameter varied up to 10 kA. The linear growth T_a with I and sharp rise of the j were obtained at values of $I < 1$ kA. The last dependence can be caused not only by lower input of the energy due to the lower arc current, but also due to assumption of equality between kinetic and self-magnetic pressures. Here it should be noted that the model did not consider the anode plasma, the plasma parameters were not calculated, and the plasma pressure was determined only by the neutral atoms of the anode vapor. Also, the application of the Langmuir–Dushman formula as well as the assumption that the vapor flows from the surface with sonic speed is correct only for vaporization into vacuum [74].

The model described in [72] was used by Lefort and Parazet [75] to calculate the spot parameters for graphite anode for high current up to 10 kA and also by Salihou et al. [76] for arc study at low current in the range 2–6 A. The model of the last work was modified using calorimetric measurement of the electrical power to the anode for an arc burning in argon. The arc was drawn at fixed current with an electrode spacing of 0.4–1 mm. The temperatures were recorded at different points by thermocouples distributed along the anode axis until the steady state was reached. These values were then approximated to use the temperature as a function of thermocouple position, with which an expression for power dissipated into the anode was obtained. Assuming that the power from this expression is approximately equal to that transferred through the anode spot area into semi-infinity anode, the spot radius r_a was calculated. Using this r_a, the corresponding current density j_a was calculated for circular-shaped anode spot. These parameters determined in such a way were defined as experimental data. This means that the value of r_a was not determined by some observation of the spot geometry, but, in essence, calculated using the experimental data as an input parameter. On the other hand, the values of r_a and j_a were determined by numerical solution of the system of equations repeated from the work [72] (with limitations mentioned by the comments above), but defined by the authors as theoretical data. The calculations were conducted for Cu, Ag, and W anodes, and the anode potential drop was varied in th range from negative -2 V up to positive value of 7 V.

The results show that the power lost by conduction P_{cd} into the anode linearly increases with arc current. Moreover, both results theoretically and defined as experimentally show that P_{cd} is higher for tungsten than for copper and silver anode, respectively. These results are consistent with the thermal properties of these materials such as thermal conductivities, which are higher for silver than for copper or tungsten. The current densities j_a are of the order of $(1–13) \times 10^6$ A/cm^2 and decrease with increasing arc current. The higher values of j_a were for silver than for copper and for tungsten, respectively. In both experimentally and theoretically, the anode spot radii linearly depend versus the arc current. Experimental data were between 2 and 7 μm, while the theoretical results lie between 3 and 18 μm. In both cases, the lower limit of the spot radius was for a silver anode and the upper limit corresponds to tungsten. Finally, it should be noted that here the presented results characterized as a comparison between experiment and theory, apparently, are not correct because both used different input experimental data.

Abbaoui and Lefort [77] studied the energy transmitted by the arc spots to the electrode material. Two simulation approaches 1D and symmetrical 2D are used, and both the results are compared between one with other and together with these found by other authors. To study the anode material phase and temperature evolutions in each point, heat conduction equation was solved in the form of the enthalpy problem as follows:

$$\frac{\partial H}{\partial T} = \text{div}(\lambda_T \nabla T) \tag{20.15}$$

H, k, and T represent, respectively, the material enthalpy, the thermal conductivity, and the temperature. A finite element method is used with moving boundaries and ablation to simulate the time evolution of the solid liquid limit and of the liquid vapor limit. When a node enthalpy is higher than the vaporization enthalpy, this node is removed and the energy flux is transmitted to the next liquid (or solid) node. To formulate the boundary conditions, the power density on the anode surface was analyzed according to that reported by Cobine and Burger [18]. The same approximation was made for the power density, supposed that it is transmitted in the beginning to the anode material with the anode spot carrying a 1000 A current. It finds the radius of anode spot $r_a = 0.5$ mm, the transmitted energy flux to the surface $W = 5 \times 10^5$ W/cm^{-2}, and the arc duration $t_a = 1$ μs.

The calculated results of phase state of the material were presented for a copper anode as the temperature distribution for liquid and vaporization phases along the anode (z-axis) with arc time (0.2–1 μs) as parameter. The temperatures decreased from about 3000 K to the room value at radius of about 0.08 cm for 1D while at 0.06 cm for 2D approach. Also the liquid and vapor volumes were presented as dependences on time used for calculation of the anode erosion rates. It was obtained that at $t_a = 1$ μs the liquid and vapor phase radii are, respectively, equal to 520 and 465 μm. The vaporized volume in 2D approach gives the erosion evaporation rate of 181 μg/C. The results were discussed in frame of comparison with the early published investigations.

Ulyanov [78] considered the anode region assuming a negative anode potential drop taking into account that an effective directed velocity of electrons v_0 from the plasma to the anode should be lower than their thermal speed v_T. A kinetic mathematical model of Langmuir layer was developed, which consists in applying the Maxwell function distribution that is shifted by the value of the directed velocity for electrons in plasma. An equation is derived for determination of the negative anode drop as a function of the ratio v_0/v_T. It is shown that, in the case of small

20.1 Review of the Anode Region Theory

values ($v_0/v_T \ll 1$), the derived expression asymptotically reduces to the Langmuir formula and obviously that the dependence of electron concentration on the potential differs from the Boltzmann law.

As examples, the current density distribution and the anode drop distribution over the anode surface were calculated in case of the high-current vacuum arc in an axial magnetic field. The study was conducted under strong current contraction taken into consideration a dimensionless parameter that is equal to the ratio of the axial magnetic field to the field created by the current of the arc itself. A limit of the Langmuir formula was utilized as a boundary condition with a strong contraction of electron current including the presence of the saturation current was discussed. The calculating results are presented for the low-pressure discharge. It obtained the potential drops across the region of the inhomogeneous near-anode plasma. The space charge layer was obtained using previously published characteristics of the positive column of the electric low-pressure discharge in the axial magnetic field.

A series of work of high-current vacuum-arc simulation using magneto-hydrodynamic (MHD) approach was published in last decade by researches from Xi'an Jiaotong University [79–83]. The anode phenomena were taken into account as a region contacted with the diffuse plasma column. Some last results are discussed below. Wang et al. [84] presented a two-dimensional deflected anode erosion model in vacuum switch devices, where a generated transverse magnetic field deflects the plasma column from the electrode center. The published data for energy flux per unit area from anode plasma to the anode in the anode spot region were used. The anode erosion under different deflection distance was simulated and analyzed, as well the anode surface temperature, anode melting process, and evaporation flux from anode surface were calculated and discussed.

In their work, Wang et al. [85] studied the two fluid three-dimensional (3D) magneto-hydrodynamic (MHD) model of high current of tens kA vacuum arc. In this work, it was accounted that the strong anode evaporation drives the evaporated atoms from the anode surface to the arc column and then to interact with the cathode plasma considering the inelastic particle collisions. According to an analysis of previous works, the temperature of the anode surface assumed to be distributed with a maximum of 2100 and 2500 K in the center, which then exponentially decreases along the radial direction. In the considered simulations, the anode temperature is fixed. The simulation results show that when the anode vapor enters into the region of the arc column, the ionization effect occurs, and the ion density increases rapidly, while the anode vapor density decreases rapidly. Anode vapor has the same temperature with the anode surface, which is much lower than the temperature of arc column plasma, so it has a significant cooling effect on the plasma. When the anode temperature is low, the anode vapor pressure is less than arc column plasma pressure, so the anode vapor cannot enter the arc column, and the anode receives plasma from the arc column region passively.

Tian et al. [86] developed a model of fluid flow and heat transfer to the anode combined with a magneto-hydrodynamics plasma model of the high-current vacuum arc. The numerical simulations for an arcing time of 10 μs (half-cycle of 50 Hz power frequency) are considered using open-source computational fluid dynamic software.

The anode material is pure copper. The diameter of the anode and the contact gap are 40 mm and 10 mm, respectively. An axial magnetic field of 10 mT/kA is applied. Crater formation is observed in simulation for a peak arcing current higher than 15 kA which can be assumed as a possibility formation of an anode spot. The anode temperature distribution and characteristics of liquids flow were studied. However, the energy loss due to anode evaporation was calculated taking into account the evaporation rate in a vacuum which is not correct [74]. It is because the vapor pressure can be significantly large at the calculated temperature level (2800 K at 25 kA), and, therefore, the results presented by the authors should be corrected.

Zenhing Wang et al. [87] simulated the anode phenomena under high-current vacuum arcs of vacuum interrupters considering the anode sheath effects on regulating the energy and momentum transfer from the arc column to the anode surface in vacuum arcs. A model for the anode sheath using particle-in-cell method was developed. The arc currents were in the range from 5 to 15 kA and are subjected to an axial magnetic field of 10 mT/kA, assuming the vacuum arc to be in diffuse mode. The anode was considered as an inactive collector for electrons and ions. The required input parameters are obtained from a magneto-hydrodynamic model for the arc column and for a collisionless sheath near the anode with a negative voltage drop. The ion temperature is set to 3 eV, and the current density is assumed to be uniformly distributed depending on the ratio of the current amplitudes and the contact areas. Two different input anode temperatures, 1600 and 2000 K, were tested. At the boundary between the sheath and plasma column, it was assumed that the electron density is $n_{e0} = Zn_{i0}$, where Z is the average charge of the ion and was taken as 1.85. However, this value cannot be taken into account at this boundary, because it was measured in the vacuum cathode jet and not related to the subject.

The results indicated distributions of the ion temperature, axial ion and electron drift velocities, ion pressure, and axial current density distributions along the radial direction at the anode with arc current as parameter. These data show relatively low variation of the mentioned parameters inside the arc column, while these parameters strongly change at the periphery of the column. The sheath acceleration effect has a significant impact on the kinetic energy, and the random kinetic energy also accounts for a large part that cannot be ignored. The input energy and momentum upon the anode surface are not obviously affected by the evaporated atoms under the surface temperatures of 1600 and 2000 K. The authors concluded that the arc pressure on the anode surface is mainly caused by ion impact, due to the ion accelerating in the anode sheath. It should be noted that this conclusion is not understandable because the electromagnetic momentum was not discussed.

20.1.4 Summary of the Previous Anode Spot Models

In general, the above-considered hypotheses and extending models of anode region for vacuum arcs represent different mechanisms in order to describe the anode phenomena detected experimentally. The first analysis is based on the study of the

20.1 Review of the Anode Region Theory

anode energy balance in order to estimate the anode potential drop in the anode space charge layer or to determine the anode temperature. The most important contribution in the anode region theory was related to understanding the mechanism of anode spot initiation considering the anode thermal mechanism with vaporization runaway (in different approaches) or to a magnetic constriction in the plasma from the diffusion discharge mode before the anode spot onset. While the first mechanism seems to be consisted with cathode large energy fluxes at high-current arcs, the second (magnetic) mechanism meets some difficulties with the cathode current distribution in the form of multi-cathode plasma jet, especially, in the presence of an axial magnetic field. At the same time, experiment showed that the threshold current associated with the onset of anode spots correlated with the solid angle subtended by the anode of the cathode and the angular distribution of the current collected at the anode. This current is also compared with the erosion flux emitted from the cathode indicating importance of the cathode plasma jet to control the onset of anode spots.

As the investigated anode region is enough complicated multi-parametric (similar to the cathode phenomena) subject, depended on the arc conditions, the problem was not completely understood by the existing solutions. The complication of different anode processes can be clear from the very detailed listing by Ecker [37, 38]. The weakness of the approaches is first of all due to disability in modeling the anode region using closed mathematical approaches. Most of the mathematical descriptions used a standard system of equations included the anode energy balance, equation for the total current, electron flux to the anode, and some relation for anode vapor pressure and its evaporation rate. Because the system of equation is not closed, the different authors used some free input parameters such as anode radius, electron temperature, or arbitrary conditions such as Steenbeck's minimum principle, equality of self-magnetic, and the vapor pressure in the spot.

In some early theories, the erosion processes were not considered. When these processes were considered, the plasma velocity in spot was arbitrarily assumed to be equal to the sonic speed. In this case, the erosion rate was calculated using the Dushman relationship. However, the application of the Dushman formula and the assumption that the vapor flows with sonic speed are not correct, while in the arc the vapor pressure in the anode region is very high. Most of the theories did not consider the plasma processes and the plasma energy balance for the anode spot besides Ecker's work [38]. However, his plasma energy balance was determined only by the equality between the energy gains from an electric field in the plasma (with doubtful expression for the thickness) and corrected by a f-factor accounted non-elastic collisions. In additional it was assumed low degree of anode vapor ionization, while the electron temperature was proposed and calculated using Saha equation. Most of the mathematical descriptions remain at a state of mathematical formulation without any calculations. The data about electron temperature were absent, and therefore, the degree of ionization cannot be obtained. Therefore, the Ecker's way to study the plasma parameters not only considerable complicate, but also represents a confusing approach. Thus, the main question of the anode region theory related to the mechanism of the plasma electron heating and information of the degree of anode vapor ionization inside of the anode spot remains unsolved and requests further study.

A fundamental issue in the study of anode spot in vacuum arcs is the relationship between arc current and plasma parameters (density, temperature) and anode erosion rate. The understanding of the anode spot phenomena related to the study of the mechanism of plasma generation should be based on a description of the kinetics of the anode vaporization and following atom ionization. In contrast to the cathode, for which the energy and the region of its produce arise by the presence of accelerated beam of the electron emission, in the anode region such a possibility is absent. This feature does not allow correctly to formulate the energy balance in the anode region, which is necessary to calculate the electron temperature. An indirect analysis shows that it is impossible to obtain degree of vapor ionization with enough accuracy due to strong dependence of the ionization rate on T_e. Analogical problem arises using experimental value of T_e. However, the necessity of correct data for plasma ionization is an important issue. The estimation shows that even for constant current density, the velocity of the plasma flow and, consequently, the rate of anode erosion strongly depend on the relation between neutral and charged component in the plasma. Therefore, the further development of the anode processes model should be considered taking into account a relation between the anode erosion and the spot parameters. An understanding of coupling the different anode processes was studied below in frame of kinetic model allowed to calculate the jump of gasdynamic parameters in the non-equilibrium region at the anode vaporization and to obtain their relations with the vapor velocity, vapor ionization, and the anode spot parameters. This kinetic approach and the mentioned relations are described below based on the theory developed by Beilis [88, 89].

20.2 Anode Region Modeling. Kinetics of Anode Vaporization and Plasma Flow

Anode region is described firstly considering the spot and plasma region as a common subject. The mechanism of anode vaporization with non-equilibrium (Knudsen) layer is presented using kinetic approach of velocity particle function distribution development. A physics of the transition between kinetic and gasdynamic regions is described. The mathematical model of plasma parameters in gasdynamic region is formulated, and a solution for the spot parameters is obtained and discussed.

20.3 Overall Characterization of the Anode Spot and Anode Plasma Region

An analysis of the experimental results shows that the behavior of the constricted anode region even before the spot appearance depends on the anode material and, therefore, the anode spot initiated in the metal vapor. This followed from the spectral

20.3 Overall Characterization of the Anode Spot and Anode ...

study of the anode region and from observation of the spot dynamics, the time of the spot initiation, which strongly depends on constant, characterized the pressure vapor of the anode material. Basically, the anode spots have low mobility. The observed spot has correctly circular form with a radius of 160 μm for luminous region and 250 μm for the crater observed at Al anode. Thus, it may consider the anode spot as an object operated in the vapor of the anode material.

Let us consider the plasma parameters on example of Cu anode spot at spot current of 400 A. We consider a case when all the arc current is equal to the spot current. In this case, the current density is about 10^4 A/cm^2, anode erosion rate is about 10 μg/C, and the arc burning occurred at a positive anode potential drop. Therefore, the electron flux to the anode can be determined by the random particle flux to a wall and, consequently, the electron density is about 10^{16} cm^{-3}. Assuming that the electron temperature for Cu is close to that measured for Al anode of about 1 eV, then the equilibrium degree of the vapor ionization can be reached about $\alpha \approx$ 0.1–0.01. In this case, the characteristic length of mean free path for electrons is $l_e \sim 10^{-3}$ cm and for ions $l_{ia} \sim 10^{-3}$ to 10^{-4} cm. The length of negative space charge sheath determined by the "3/2" law is $h_{an} \sim$ 0.1 μm. Taking into account that energy flux Q_a to the anode surface is determined at least by the work function, this value is already enough to anode heating to a temperature $T_a = Q_a/2\pi r_a \approx (2–3) \times 10^3$ K, at which an intense vaporization takes place.

An important parameter characterizing the mass flow in the anode region is the ratio between direct mass velocity v of the heavy particles to their thermal velocity v_T. Let us define this ration v/v_T using the experimental data of anode erosion rate G_a by following expression $10^5 \alpha G_a (T_e/T)^{0.5}$. The estimation shows $v/v_T \sim 10^{-2}$ to 10^{-1}. The vapor ionization in anode region occurs preferably by thermal mechanism. An estimation indicates relatively low electric field in the quasineutrality plasma (Chap. 16). In order to estimate the ionization length l_i, let us use the expression for the diffusion particle fluxes (see Chap. 4). After simple computation, the following expression can be derived as

$$D_a \frac{\partial^2 n_e}{\partial x^2} = \beta n_0^2 n_e \left(1 - \frac{n_e^2}{n_0^2}\right) \quad (20.16)$$

According to (20.16), the value of l_i is determined by coefficient of ion ambipolar diffusion D_a, equilibrium density n_0 of the charge particles, and coefficient recombination β in form

$$l_i = \sqrt{\frac{D_a}{\beta n_0^2}} \quad (20.17)$$

In the considered conditions, the neutral density $n_a \sim 10^{18}$ cm^{-3}, $\beta \sim 10^{-26}$ cm^{-6} s^{-1}, and therefore $l_i \sim 10^{-3}$ cm, i.e., of order of l_e. This estimation shows that the ionization processes occur basically at distance significantly larger than the length of space charge region, and that the diffusion phenomena in the anode region can be

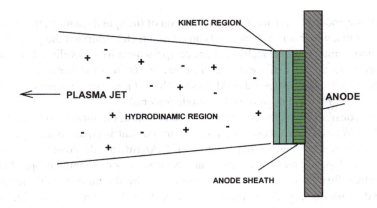

Fig. 20.4 Schematic presentation of the anode physical model including expanding plasma jet

negligible. The electron flux to the anode can be determined by the plasma parameters close to the anode surface.

Consequently, at the anode surface, similarly as at the cathode, the locally heated anode produces enough dense and relatively highly ionized cloud of the vapor, which is separated from the surface by collisionless sheath and with an equilibrium heavy particle velocity function distribution. The overall anode region and expanding anode plasma jet are schematically presented in Fig. 20.4. This means that the description of the anode vaporization can be conducted like that for cathode vaporization taking into account the specifics of plasma heating and plasma generation characterized the anode region.

During the anode current carried plasma expansion, it can be considered that the plasma electrons are heated by the Joule energy dissipation as well by energy of accelerated cathode jet in which particles can scatter in the anode plasma. First of all, let us estimate the characteristic length l_J of energy dissipation assuming that the Joule energy heats the electrons to temperature of ~1 eV. Such estimation shows that $l_J = 2T_e \sigma_{el}/j \sim 6 \times 10^{-3}$ cm, i.e., at distance of order of anode region and corresponds to the hydrodynamic plasma flow. The estimations show that the lengths, l_e, l_i, and l_J, are significantly lower than the anode radius r_a, and therefore, the plasma parameters near the anode can be the treatment in one-dimensional approximation. The problem consists in quantitative analysis of the gasdynamic plasma parameters at the beginning of the anode jet and in a determination a degree of their declination at the anode surface from the equilibrium values. These data allow to find the relation between direct plasma velocity of the anode erosion flux and the anode temperature as well the anode current density.

20.3.1 Kinetic Model of the Anode Region in a Vacuum Arc

The model is based on a kinetic treatment of atom evaporation [36, 90, 91, Chap. 17], as well as on plasma production and plasma energy dissipation processes at the anode. The anode surface is assumed to be plane. The plasma consists of ionized electrode vapor taking into account the experimental results indicated that in spot mode, the anode potential is positive with respect to anode plasma [56, 57]. The region near the anode, at the spot location, consists of (1) a space sheath, (2) non-equilibrium plasma layer, and (3) a plasma acceleration region. Four characteristic boundaries can thus be distinguished in the near-anode region, defined by a coordinate system whose origin is at the spot center, where an x-axis is set normal to the electrode surface (Fig. 20.5). The x-axis coincides with the direction of vapor flow and with generated anode plasma jet. The four boundaries are parallel to the electrode surface. Boundary 1 is set at the coordinate origin, boundary 2 is set at edge of the near-anode space charge zone, boundary 3 is set at the edge of the non-equilibrium (Knudsen) layer, whose length is of the order of several heavy particle mean free path. Boundary 4 is set at the location where the plasma is enough heated by Joule energy dissipation in order to provide significant atom ionization and after which the plasma accelerated to its velocity equals the anode jet velocity.

Due to the different particle fluxes to the anode from the first flow region, a strong jump in the plasma parameters takes place between boundaries 1 and 3. This analysis is made here by use of the bimodal velocity distribution function method that was used to study laser–target evaporation [92] and cathode spot physics (Chap. 17). The gasdynamic parameters (density, velocity, and temperature) at the boundaries are denoted as $n_{\alpha j}$, $v_{\alpha j}$, $T_{\alpha j}$, where indices α—e, i, a are the electrons, ions, and atoms, respectively, and $j = 1, 2, 3, 4$ indicates the boundary number, and n_{e0a} and n_{0a} are the equilibrium electron and heavy particle densities determined by the anode surface temperature T_{0a}. The velocity function distribution (VDF) is half-Maxwellian (positive velocity $v_{\alpha x}$) with a temperature at the first boundary in accordance with particle emission from the anode determined by evaporation for atoms [93] and for electrons [94]. The evaporated atoms are significantly ionized within one mean free path [95], therefore the reverse flux consists of atoms and ions, whose velocity

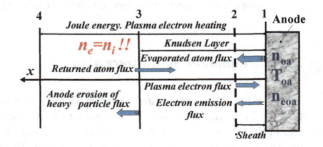

Fig. 20.5 Schematic presentation of the anode kinetic processes at characteristic boundaries

distribution (negative velocity v_x) at boundary 1 can be approximated as shifted half-Maxwellian ones by the plasma parameters determined at this boundary from the problem solution [92].

At the external boundary 3 of the Knudsen layer, the heavy particles are in equilibrium. Therefore, the heavy particle distribution function is a shifted full Maxwellian with parameters $n_{\alpha 3}$, $u_{\alpha 3}$, and $T_{\alpha 3}$. For electrons, the VDF consists of a shifted full Maxwellian for the plasma electrons moved to the anode surface and a shifted half-Maxwellian for the electron emission beam. At boundary 4, the particles are in equilibrium, and therefore, there the distribution functions are shifted full Maxwellians with parameters $n_{\alpha 4}$, $u_{\alpha 4}$, and $T_{\alpha 4}$. Between the boundaries, 1 and 2 are the space charge sheath. At boundary 2, the beam of electron emission and the returned ions to the anode are considered, while the electrons from the plasma are accelerated to the anode with energy eu_{an}, where u_{an} is the anode potential drop.

The velocity distribution function of the particles for $v_{\alpha x} < 0$ is denoted by $f^-_{\alpha j}$ and for $v_{\alpha x} > 0$ is by $f^+_{\alpha j}$ where

$$f^{\pm}_{\alpha j} = n_{\alpha j}\left(\frac{m_\alpha h_{\alpha j}}{\pi}\right)^{3/2} \exp\{-m_\alpha h_{\alpha j}[(v_{\alpha x} - v_{\alpha j})^2 + v_{\alpha y}^2 + v_{\alpha z}^2]\} \quad (20.18)$$

and where $h_{\alpha j} = (2kT_{\alpha j})^{-1}$, $v_{\alpha y}$, $v_{\alpha z}$ are the random particle velocity components in the y- and z-directions, m_α is the particle mass and k the Boltzmann constant.

Next, the conditions at the boundaries are written.

At boundary 1: $\quad f_{\alpha 1} = f^+_{o\alpha 1} + \xi^-_{\alpha 1} f^-_{\alpha 1}, \quad f^+_{oi1} = 0 \quad (20.19)$

$f^+_{o\alpha}1$ is the half-Maxwellian VDF of particles emitted from the anode at anode temperature $T_a = T_0$ with neutral density $n_a = n_0$ and electron density $n_e = n_{e0}$.

At boundaries 2 and 3: $\quad f_{\alpha j} = \xi^+ f^+_{\alpha j} + \xi^-_{\alpha j} f^-_{\alpha j} \quad j = 2, 3 \quad (20.20)$

At boundary d 4 $\quad f_{\alpha 4} = f^+_{\alpha 4} + f^-_{\alpha 4} \quad (20.21)$

where $f^+_{o\alpha}1 = 0$ for $\alpha = i$ and $\xi_{\alpha j}$ is the correction factor to the Maxwellian function distributions. The correction factors are about of unit [92].

20.3.2 Equations of Conservation

Two atomic fluxes exist near the anode surface, consisting of evaporated atoms, flowing from the electrode, and a return flux of atoms that formed in the non-equilibrium plasma layer (before boundary 3). The difference between these fluxes determines the anode mass erosion in vapor phase. The return flux of ions is small due to positive anode potential with respect to the adjusted plasma. To determine the

20.3 Overall Characterization of the Anode Spot and Anode ...

spot parameters, it is only necessary to know the parameters on the abovementioned boundaries, and not their variation between the boundaries. Hence, the problem is reduced to the integration of the conservation equations, for each particle between these boundaries. The general form of these equations is

$$\int v_x f(v)dv = C_1; \quad \int v_x^2 f(v)dv; = C_2 \quad \int v_x v^2 f(v)dv = C_3; \quad (20.22)$$

where $f(v)$ is the velocity distribution function, v is the velocity vector, vx is the velocity component along the x-axis, and C_1, C_2, and C_3 are integration constants. Equation (20.22) at the above boundaries are expressed in following form [94, 96]:

$$n_\alpha(x) = \int f(x, v_{0\alpha})dv_\alpha; \quad \int v_{\alpha x} f(x, v_\alpha)dv_\alpha = n_\alpha(x)v_{0\alpha}(x);$$

$$\frac{m}{2}\int \left[(v_{\alpha x} - v_{0\alpha})^2 + v_y^2 + v_z^2\right]f(x, v_\alpha)dv_\alpha = \frac{3n_\alpha(x)kT_\alpha(x)}{2}; \quad (20.23)$$

where $f(x, v_\alpha)$ is the velocity distribution function at the control boundaries, v_α is the velocity vector, $v_{\alpha x}$ is the particle velocity component in the x-direction, $v_{0\alpha}$ is the shifting velocity of particle flow, and v_α is the velocity of the ionized vapor.

The integration is done between boundaries 1 and 3. These integrals enable us to calculate the differences and the relations between the various particle fluxes, and differences between the equilibrium and non-equilibrium values (at the boundary 3) of the parameters: particle densities, particle temperatures, and particle velocities. The plasma region after boundary 3 is much larger than the heavy particles mean free path. The plasma is quasineutral, and its flow can be described using a hydrodynamic approximation. Thus, plasma flow occurs in two different flow regions: a non-equilibrium, kinetic flow before boundary 3 and a hydrodynamic flow after boundary 3.

20.3.3 System of Kinetic Equations for Anode Plasma Flow

Using the definitions and (20.18)–(20.23), the procedure of integrating brings following system of equations to describe the heavy particles and electrons flow after their vaporization from the anode surface and taking into account specifics of atom ionization and plasma generation during the flow of the particles.

Equation of heavy particles conservation:

$$n_1 = \frac{n_0}{2} + \frac{n_{a1}}{2}[1 - \Phi(b_{a1})] + \frac{n_{i1}}{2}[1 - \Phi(b_{i1})] \quad (20.24)$$

Equations of heavy particles continuity:

$$n_1 v_1 = n_j v_j \quad j = 2, 3; \quad n_j = n_{aj} + n_{ij} \qquad (20.25\text{–}20.26)$$

$$n_2 v_2 = \frac{n_{a0}}{2\sqrt{\pi d_{a0}}} - \frac{n_{a1}}{2\sqrt{\pi d_{a1}}}\left[\exp(-b_{a1}^2) - b_{a1}\sqrt{\pi}[1 - \Phi(b_{a1})]\right] - A_{i1} \quad (20.27)$$

$$A_{i1} = \frac{n_{i1}}{2\sqrt{\pi d_{i1}}}\left[\exp(-b_{i1}^2) - b_{i1}\sqrt{\pi}[1 - \Phi(b_{i1})]\right]$$
$$= \frac{n_{i2}}{2\sqrt{\pi d_{i2}}}\left[\exp(-b_{i2an}^2) - b_{i2an}\sqrt{\pi}[1 - \Phi(b_{i2an})]\right] \qquad (20.28)$$

Equations of heavy particles momentum flux:

$$\frac{n_0}{4d_0} - \frac{n_{a1}}{2d_{a1}}\left[(0.5 + b_{a1}^2)[1 - \Phi(b_{a1})] - \frac{b_{a1}}{\sqrt{\pi}}\exp(-b_{a1}^2)\right]$$
$$+ \frac{n_{i1}}{2d_{i1}}\left[(0.5 + b_{i1}^2)[1 - \Phi(b_{a1})] - \frac{b_{i1}}{\sqrt{\pi}}\exp(-b_{i1}^2)\right] + B_{e1}$$
$$= B_{a2} + B_{i2} + B_{e2} + F(E_{12}) \qquad (20.29)$$

$$B_{e1} = \frac{n_{e0}}{2d_{e0}}\left[\exp(-b_{ean}^2)\frac{b_{ean}}{\sqrt{\pi}} + 0.5[1 - \Phi(b_{t1})]\right]$$
$$= \frac{n_{e1}}{2d_{e1}}\left[\exp(-b_{e1an}^2)\frac{b_{e1an}}{\sqrt{\pi}} + (0.5 + b_{e1an}^2)[1 + \Phi(b_{e1an})]\right]$$

$$B_{i2} = \frac{n_{i2}}{2d_{i2}}\left\{\begin{array}{l}(0.5 + b_{i2}^2)[2 + \Phi(^-b_{i2an}) - \Phi(^+b_{i2an}) + 2\Phi(b_{i2})] \\ + \dfrac{b_{i2}}{\sqrt{\pi}}[2\exp(-b_{i2}^2) - \exp(-^+b_{i2an}^2) - \exp(-^-b_{i2an}^2)]\end{array}\right\}$$

$$B_{\alpha j} = \frac{n_{\alpha j}}{2d_{\alpha j}}(1 + 2b_{\alpha j}^2); \quad j = 2, \alpha = e, a; \quad j = 3; \quad \alpha = e, i, a;$$

$$F(E_{i2}) = F_e(E_{i2}) - n_i k T_i\left[1 - \exp\left(-\frac{eu_{an}}{kT_i}\right)\right]$$

$$F_e(E_{i2}) = j_e\left(\frac{2eu_{an}}{m_e}\right)^{0.5} - n_{e0}kT_0\left[1 - \exp\left(-\frac{eu_{an}}{kT_0}\right)\right]$$

Equations of heavy particles energy flux

$$\frac{n_0}{\sqrt{\pi}d_0^{3/2}} + \frac{n_{a1}}{\sqrt{\pi}d_{a1}^{3/2}}\left[(2.5 + b_{a1}^2)[1 - \Phi(b_{a1})]b_{a1}\sqrt{\pi} - (2 + b_{a1}^2)\exp(-b_{a1}^2)\right]$$
$$+ \frac{n_{i1}}{\sqrt{\pi}d_{i1}^{3/2}}\left[(2.5 + b_{i1}^2)[1 - \Phi(b_{i1})]b_{i1}\sqrt{\pi} - (2 + b_{i1}^2)\exp(-b_{i1}^2)\right] = W_{a2} + W_{i2}$$

$$(20.30)$$

20.3 Overall Characterization of the Anode Spot and Anode ...

$$W_{\alpha j} = \frac{n_{\alpha j}}{\sqrt{b_{\alpha j}^2/d_{\alpha j}^3}}(2.5 + b_j^2) \quad j = 2, \alpha = e, a; \quad j = 3, \alpha = e, i, a$$

$$W_{i2} = \frac{n_{i2}}{2\sqrt{\pi} d_{i2}^{3/2}} \left\{ \begin{array}{l} [(2.5 + b_{i2}^2)[2 + 2\Phi(b_{i2}) + \Phi(^-b_{i2an}) - \Phi(^+b_{i2an})]b_{e2}\sqrt{\pi} -] \\ -3\sqrt{\frac{d_{i2}}{d_{ian}}}\exp(-^-b_{i2an}^2)b_{i2}b_{ian} + 2(2 + b_{i2}^2)\exp(-b_{i2}^2) \\ -(2 +^- b_{i2an}^2)\exp(-^-b_{i2an}^2) - (2 +^+ b_{i2an}^2)\exp(-^+b_{i2an}^2) \end{array} \right\}$$

$$\sum_{\alpha} B_{a2} = \sum_{\alpha} B_{a3} + F(E_{23}) \quad \alpha = a, i, e \quad (20.31)$$

$$\sum_{\alpha} W_{a2} = \sum_{\alpha} W_{a3} - q_{23}^{el} \quad \alpha = a, i \quad (20.32)$$

$$n_{e1} = \frac{n_{e0}}{2}[1 - \Phi(b_{ean})] + \frac{n_{e1}}{2}[1 + \Phi(b_{e1an})] \quad (20.33)$$

Equations of electron flux continuity:

$$A_{e1} - n_{e1}v_{e1} = \frac{n_{e0}}{2\sqrt{\pi} d_0} \exp\left(-\frac{eu_{an}}{kT_0}\right) \quad (20.34)$$

$$A_{e1} - n_{e1}v_{e1} = -n_{ej}v_{ej} \quad j = 2, 3 \quad (20.35\text{--}20.36)$$

$$A_{e1} = \frac{n_{e1}}{2\sqrt{\pi} d_{e1}}\left[\exp(-b_{e1an}^2) + b_{e1an}\sqrt{\pi}[1 + \Phi(b_{e1an})]\right]$$

$$= \frac{n_{e2}}{2\sqrt{\pi} d_{e2}}\left[\exp(-b_{e2}^2) - b_{e2}\sqrt{\pi}[1 - \Phi(b_{e2})]\right] \quad (20.37)$$

Equation of electron momentum flux:

$$B_{ej} = B_{ej+1} + F_e(E_{jj+1}) \quad j = 1, 2 \quad (20.38\text{--}20.39)$$

Equations of electron energy flux:

$$W_{e2} = W_{e3} + q_{23}^{el} + q_{23}^{n-el} - q_{23}^{g}; \quad q_{23}^{n-el} = (n_3v_3 - n_2v_2)u_i \quad (20.40)$$

$$W_{e2} = \frac{n_{e0}}{2\sqrt{\pi} d_{e0}^{3/2}}\exp(-b_{ean}^2)(2.5 + b_{ean}^2) - en_{e1}v_{e1}u_{an}$$

$$+ \frac{n_{e1}}{2\sqrt{\pi} d_{e1}^{3/2}}\left[(2.5 + b_{e1an}^2)(1 - \Phi(b_{e1an}))b_{e1an}\sqrt{\pi} - (2 + b_{e1an}^2)\exp(-b_{e1an}^2)\right]$$

$$(20.41)$$

$$n_{ej} = n_{ij} \quad j = 2, 3 \quad (20.42\text{--}20.43)$$

Definition of the following expressions:

$$h_{\alpha j} = (2kT_{\alpha j})^{-1}; \ m_\alpha h_{\alpha j} = d_{\alpha j}; \ m_\alpha h_{\alpha 0} = d_{\alpha 0}; \ m_\alpha h_{ean} = d_{\alpha an}; \ \alpha = e, i$$

α—e, i, and a are the electron, ion, and neutral atom, m_α—mass of the particles

$$\pm b^2_{\alpha j an} = (v_{\alpha an} \pm v_{\alpha j})^2 d_{\alpha j}; \ b^2_{\alpha j} = v^2_{\alpha j} d_{\alpha j}; \ b^2_{\alpha an} = v^2_{\alpha an} d_{\alpha 0}; \ v^2_{\alpha an} = \frac{2eu_{an}}{m_\alpha}$$

q^g_{jj+1} is the input energy, in regions 1–2 from the electric field E and determined by u_{an}. q^{el}_{jj+1} and q^{in}_{jj+1} are the energy losses by plasma electrons in elastic and inelastic collisions with heavy particles. $\Phi(b_{\alpha j})$ is the probability integral function, j_{e0} is the emission current density, and u_i is the ionization potential. $F(E_{jj+1})$ is the input momentum from the electric field in the corresponding regions. Estimations using calculated data from gasdynamic model show that these terms are important in regions 1–2 where they are determined by u_{an}. The term q^{el}_{jj+1} is small in comparison with the convective term which follows from ratio $jm/(ev_e m_e n_e x)$. This ratio reached about 1 at $x = 1$ cm, i.e., at distance significantly larger than the spot size r_a. The above system of equations was obtained assuming that the spot current density before boundary 3 is constant, and the emitted electron beam from the anode is scattered after the boundary 2. Description of the charge particles in the space charge sheath takes into account the corresponding velocity shift and partial plasma ion reflection at the boundary 2 due to the presence of the potential barrier. In order to describe the parameter between boundary 3 and conditional boundary 4, it is accounted for the plasma expansion, which lies to increase of the cross-sectional area of the jet between considered boundaries. The corresponding equations are derived in form:

$$n_3 v_3 F_3 = n_4 v_4 F_4 \quad F_3 = I/j \tag{20.44}$$

$$n_{e3} v_{e3} F_3 = n_{e4} v_{e4} F_4 \tag{20.45}$$

$$\sum_\alpha B_{a3} F_3 = (mn_4 v_4^2 + m_e n_{e4} v_{e4}^2) F_4 \quad \alpha = a, i, e \tag{20.46}$$

$$B_{e3} F_3 = n_{e4} v_{e4} F_4 + F_e(E_{34}) \tag{20.47}$$

$$\sum_\alpha W_{a2} F_3 = v_{e4}(mn_4 v_4^2 + m_e n_{e4} v_{e4}^2) F_4/2 - q^g_{34} + q^{n-el}_{34} \quad \alpha = a, i \tag{20.48}$$

$$W_{e3} F_3 = m_e n_{e4} v_{e4}^2 F_4/2 + q^{el}_{34} + q^{n-el}_{34} - q^g_{34} \quad q^{n-el}_{34} = (n_4 v_4 - n_3 v_3) u_i \tag{20.49}$$

At four boundaries, for three types of particles (a, i, e) and three gasdynamic parameters (T, n, v), the total unknown consists of 36. This number can be reduced to 27 assuming equality of the parameters for the heavy particles $T_{ij} = T_{aj}$, $n_{ij} =$

n_{aj}, $v_{ij} = v_{aj}$. The system of (20.24)–(20.49) consists also of 27 equations including additionally the Saha equation in the plasma equilibrium region, while the spot current I and the anode positive potential drop are given data. The above system should be added by equations for electron emission, anode energy balance, and pressure of vaporized atoms (described in Chap. 16) in order to determine j_{e0}, n_{e0}, n_{a0}, and T_a. Besides, in general, the parameters of cathode plasma jet can be used to describe the current continuity of the arc operated in an electrical circuit. A solution of such complicated general system is very complex. The estimations show that it is because of a strong sensitivity of the solution and its dependence on the emitted electron current density j_{eo}, due to very small value of j_{eo}.

Taking this into account, the mathematical approach could be simplified in order to study the physics of the kinetic and gasdynamic processes. Therefore, to demonstrate the jump parameters in the Knudsen layer, and to determine the parameters of the anode region in the spot mode, a simplified approach is illustrated below, considering the equations for kinetics of heavy particles describing the anode plasma flow and a resulting anode vaporization flux.

20.4 System of Gasdynamic Equation for an Anode Spot in a Vacuum Arc

The analysis of the gasdynamic flow region is provided using the boundary condition from solution of kinetic model for plasma parameters on boundary 3. In the gasdynamic region, the atoms are ionized by plasma electrons. The plasma electrons are heated by the Joule energy dissipation which produced in current carried expanding plasma. It is assumed that the heavy particles temperature T in the gasdynamic region equals T_3, at boundary 3. Electron temperature T_e differs from the temperature T, and both are assumed constant in the dense plasma region. These assumptions are based on the calculations of plasma particle temperature distribution in cathode jet, where the mentioned temperatures are not changed substantially (Chap. 16). The heavy particle density n_T and velocity v vary from their values n_3 and v_3, respectively, at the boundary 3 to their values in plasma jet during the plasma acceleration in the gasdynamic region at boundary 4. In this region, the plasma is quasineutral, having a considerably weak plasma electrical field E_{pl}, [88].

For the plasma in the gasdynamic region as well for the anode body, the system of equations is presented in the integral form. The processes of energy and mass transfer are described.

1. *Equation continuity of anode spot current density.*

$$j = j_e - j_{eo} - j_i \qquad (20.50)$$

where j is the net anode spot current density, j_e is the current density of the electrons moving from anode plasma to the anode surface, j_{eo} is the anode electron emission current density, and j_i is the current density of the ions returned in the anode sheath from the anode plasma determined by Boltzmann relation [88]. The electron flux to the anode Γ_{eD} is determined by the equation of mass transfer including the fluxes due to plasma particle gradient and electric field mechanism (Chap. 16). In the case when the plasma particle gradient near the anode wall appeared at distance of about mean free path, the flux Γ_{eD} is approximately equal to random electron flux Γ_{eT} [95] from the anode plasma at the boundary 3. The plasma density is determined by the assumption that the electron density inside the gasdynamic region is equal to the ionization equilibrium value n_e. This assumption is based on the fact that at boundary 3 the velocity distribution function of the particles is in equilibrium [74] and the plasma density is relatively large ($>10^{16}$ cm^{-3}). The value of n_e is obtained from Saha equation taking into account the heavy particle density n_3 at the boundary 3, when the electron temperature T_e is known.

2. *Equation of the electron plasma current:*

$$j_e = e\Gamma_{eT} \quad \Gamma_{eT} = \frac{n_e v_{eT}}{4} \tag{20.51}$$

where v_{eT} is electron thermal velocity and e is the electron charge.

3. *Equation of the electron emission:*

$$j_{e0} = \frac{4e\pi m_e k T_a^2}{h^3} Exp(-\frac{\varphi + eu_a}{kT_a}) \tag{20.52}$$

Equation (20.52) corresponds to electron current density j_{e0} by thermionic emission [88] from the hot anode in the spot. φ is the work function, h is the Planck's constant, and u_a is the positive anode potential drop in the anode sheath.

4. *Equation of total spot current I with spot area F_a:*

$$I = jF_a, \quad F_a = pr_a^2 \tag{20.53}$$

5. *Plasma energy balance.* The plasma electron temperature T_e is determined by the balance between the energy carried by the Joule energy in the anode plasma jet, as well by the cathode jet energy, and the energy dissipation caused by ionization of atoms, energy convection brought about by the electric current and energy, required for plasma acceleration in the plasma anode jet

$$Iu_p + Q_{ce} + Q_{ci} = Iu_i\alpha\varsigma + 2kT_e\frac{I}{e}(1+\alpha\varsigma) + \frac{GV_{ja}^2}{2} + Q_{el} + Q_r \tag{20.54}$$

$$Q_{el} = I \int_0^\infty \frac{3m_e n_i}{jm\tau} k(T_e - T)dx \quad \varsigma = \frac{eG}{mI}$$

20.4 System of Gasdynamic Equation for an Anode Spot in a Vacuum Arc

Here G(g/s) is the anode erosion rate and u_p is the anode jet voltage, which is determined by Ohm's law in the anode plasma jet. Q_{ce} and Q_{ci} are the cathode jet energy by the electrons and ions, respectively, Q_{el} is the energy losses by elastic collisions, τ is the effective collisions time for elastic collisions between electrons and heavy particles, Q_r is the energy plasma loss by radiation, and V_{ja} is the anode plasma jet velocity.

6. *Anode plasma jet Ohm law:*

$$u_p = \left(\frac{Ij}{4\pi\sigma_{el}^2}\right)^{1/2} \tag{20.55}$$

where σ_{el} is the plasma electrical conductivity

7. *Equation of plasma state:*

$$p = n_T T + n_e T_e, \quad n_T = n_a + n_i \quad p_a(T_a) = n_T k T_a \tag{20.56}$$

where $p_a(T_a)$ and n_a are equilibrium pressure and heavy particle density as function of anode temperature T_a, respectively [88], p is the plasma pressure, and n_T is the plasma heavy particle density. The relationship between the values p and p_a as well between T and T_a was obtained from the analysis of equations for the kinetic layer (see above and in [88]).

8. *Equation of plasma flow momentum:*

$$GV_{ja} = pF_a \tag{20.57}$$

9. *Anode energy balance* [88]:

$$j(u_a + \varphi + 2kT_e) + q_J = q_T + (\lambda_s + \frac{2kT_a}{m})\frac{G}{F_a} + \sigma_{sB}T_a^4 \, ; \quad q_J = 0.32 r_a I j \rho_{el} \tag{20.58}$$

where λ_s is the specific heat of evaporation of the anode material, ρ_{el} is the anode-specific electrical resistance, q_T is the heat flux into the anode due to heat conduction, and σ_{sB} is the Stefan–Boltzmann constant. The value of q_T is obtained from the solution of the non-stationary heat conduction equation for the anode bulk (Chaps. 3 and 16). It is assumed that the temperature within the spot is equal to T_a, the calculated temperature in the spot center. It is further assumed that the time dependence of spot parameters is determined only by the time dependence of the anode heat conduction (Chap. 16). Anode heating takes place in electron bombardment of the anode, and by Joule heating of the anode body q_j according to Rich [97] and used also by Ecker [98, 99]. The terms of the left part of (20.58) correspond to these processes. Anode cooling results from thermal conductivity, electron emission, evaporation of the electrode (erosion rate G in g/s), and radiation from the spot according to Stephan–Boltzmann law.

10. *Equation of mass flux conservation*

$$G = mn_T v_a F_a \qquad (20.59)$$

The system of 10 (20.50)–(20.59) together with Saha equation allows the determination the following 11 anode spot parameters: $j, j_{e0}, j_e, T_a, r_a, n_a, n_i, v_a, T_e, u_p, V_{ja}$, as functions of the erosion rate G. Also the anode potential drop is a given parameter. The relations between n_3 and n_{a0} as well between T_a and T_3 are obtained from analyzing the kinetic approach considering equations for the heavy particle by conditions $v_a = v_3$ and $n_T = n_3$. This system of equations describes in a self-consistent manner the mutual processes in the near-anode plasma as well as in the anode bulk and on the anode surface. The relation between the net anode mass flux (anode erosion rate) and the evaporated anode material flux determined by Langmuir–Dushman formula at the anode temperature, denoted as K_{er}, will be also calculated below.

20.5 Numerical Investigation of Anode Spot Parameters

The dependence of anode spot current density, anode surface temperature, plasma density and plasma temperature, and anode plasma jet velocity on electrode erosion rate for given spot current and spot lifetime is also investigated. In the case of the vacuum arc with planar copper electrodes (anode and cathode), the anode spot appeared when the threshold current reached 400 A reported by Kimblin [34]. However, the anode spot threshold current can decrease in the case of planar anode and rod cathode in the presence of an ambient gas at relatively low pressure [28]. In this case, the experiments [28, 44] showed that when ambient gas pressure was in the range of 0.1 to 100 mTorr, an anode spot was formed on different anode materials at currents that could be low as 5–50 A. Therefore, below parameters of the anode region are calculated for spot current in the range of 10–400 A and for initial room temperature. The anode potential drop in the anode sheath is relatively small, taken as 2 V [72]. The influence of the cathodic plasma jet was not taken into account considering appropriate electrode configuration. This is the case for a relatively large interelectrode distance, when only very small fraction of the cathode plasma jet is in contact with anode spot area.

The influence of the initial temperature and spot lifetime was calculated in the previous work for an arc with graphite anode [89]. The calculations showed that the effect of initial anode temperature in the range 300–1500 K on main of spot parameters is not large, while the degree of ionization and spot current density strongly depends on spot lifetime. As an example, the spot current density increases with decreasing spot lifetime from steady-state operation to 0.1 μs by an order of magnitude. The degree of ionization decreases with decreasing spot lifetime.

20.5 Numerical Investigation of Anode Spot Parameters

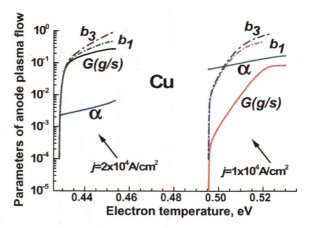

Fig. 20.6 Anode erosion rate G (in g/s for show in the scale), degree of atom ionization α, dimensionless velocity at the anode surface boundary b_1, and at the external boundary of Knudsen layer b_3 as function on electron temperature with current density as parameter

20.5.1 Preliminary Analysis Based on Simple Approach. Copper Anode

A preliminary study of the solution shows that some parameters are very sensitive to variation of some others. Therefore, the equations of kinetic layer for heavy particles were solved with the gasdynamic system without (20.54) and (20.57) in order investigate the influence of varied T_e for given j. The results of calculations for Cu anode are presented in Fig. 20.6 and illustrated by the values in Table 20.5. It can be seen that plasma parameters (G, b_1, b_3) sharply rise at some critical T_e at which the solutions exist, and then the rate of increase of their values is reduced. Especially, it is related to the plasma velocity at external boundary of Knudsen layer v_3, characterized by dimensionless parameter b_3. This parameter is the ratio of the plasma velocity on the boundary 3 to the atom thermal velocity. Therefore, the increase of G approaches the same dependence as it is determined by the velocity v_3. The degree of ionization also significantly increases with weakly change of T_e.

The anode erosion rate sharply increases with electron temperature T_e and then saturated, and degree of ionization increases with T_e more than order of magnitude larger for lower value of j because of lower density of the heavy particle density (Table 20.5). The absolute values and the range of T_e are very small for fixed j, but larger T_e was obtained for $j = 2 \times 10^4$ A/cm^2 than for $j = 1 \times 10^4$ A/cm^2. So, the solution is significantly sensitive to change the T_e.

20.5.2 Extended Approach. Copper Anode

The extended study of the anode region was conducted by considering the equations of kinetic layer for heavy particles which were solved with the gasdynamic system including (20.54) and (20.57) in order investigate dependences on the anode erosion

Table 20.5 Anode temperature, heavy particle density, erosion rate G, and plasma and gasdynamic parameters in Knudsen layer as function on the plasma electron temperature

Current density A/cm^2	Electron temperature T_e eV	Anode temperature T_a K	Density $n_0 10^{18}$ cm^{-3}	Velocity b_1	Velocity b_3	Density Ratio n_{30}	Velocity v_3 cm/s	Erosion rate G, g/s	Erosion rate G, g/C	Degree ionization α
2 × 10^4	0.42853	~2985	6.04	1 × 10^{-4}	1.1 × 10^{-4}	0.8394	10.5	1.1 × 10^{-4}	2.75 × 10^{-7}	0.0224
	0.4535	3003	6.5	0.461	0.95	0.32	5.8 × 10^4	0.278	6.95 × 10^{-4}	0.00667
1 × 10^4	0.49546	2233	0.098	0.013	0.014	0.823	1.18 × 10^3	4 × 10^{-4}	1 × 10^{-6}	0.065
	0.5233	2247	0.11	0.4734	0.87813	0.33	5.5 × 10^4	0.082	2.05 × 10^{-4}	0.149

20.5 Numerical Investigation of Anode Spot Parameters

rate for anode spot current varied in the range of 10–400 A for usual spot lifetime of 1 μs. According to the calculations (Fig. 20.7), the anode temperature ($T_a = T_0$) in the spot region increases with anode erosion rate $G(\mu g/C)$, but decreases with spot current I.

Similar dependences are shown in Fig. 20.8 for heavy particle density n_0 calculated by the saturated anode vapor pressure at the respective anode temperature. As it is expected, the value of n_0 increases in accordance with increase of the anode temperature. Figure 20.9 shows the dimensionless plasma density n_{30} at the boundary 3, which is the ratio n_3 to n_0, as function on erosion rate with arc current as parameter. It can be seen that the value of n_{30} decreases in the non-equilibrium (Knudsen) layer. The density n_3 relatively weakly different from n_0 indicating on large returned particle flux to the anode and that the value of n_{30} decreases with G but increases with I. As it follows from this calculation returned to the anode, surface heavy particles flux at

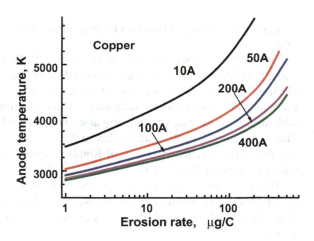

Fig. 20.7 Anode temperature in the spot as function on erosion rate with arc current as parameter

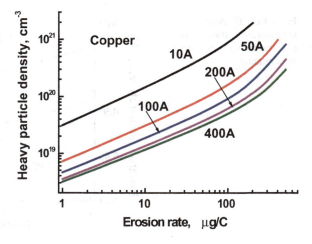

Fig. 20.8 Heavy particle density n_0 as function on erosion rate with arc current as parameter

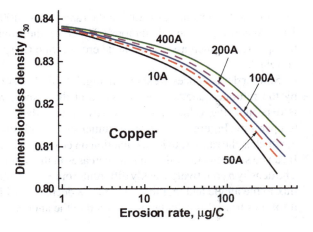

Fig. 20.9 Dimensionless plasma density n_{30} at the external boundary of Knudsen layer as function on erosion rate with arc current as parameter

the external boundary of Knudsen layer decreases with increase of the anode erosion rate.

The electron temperature weakly decreases with G at fixed I but it also decreases with I (Fig. 20.10). Taking into account the dependence of the heavy particle density, electron temperature, and decrease of the degree of ionization with G (Fig. 20.11), it can understand the calculated relatively weakly increase of the spot current density with anode erosion rate (particularly at <200 μg/C), which is shown in Fig. 20.12. The large returned heavy particle flux to the anode provokes relatively low plasma velocity at the external boundary of Knudsen layer b_3 as dependences on G and I, which is shown in Fig. 20.13.

This dependence indicated the presence of subsonic plasma flow in the anode region formed in the non-equilibrium layer and relatively low anode erosion fraction (Fig. 20.14) indicating significant difference from the anode evaporation rate in vacuum.

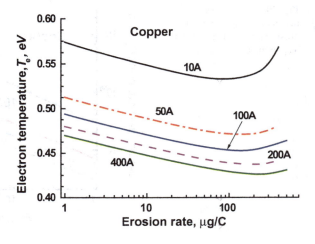

Fig. 20.10 Electron temperature as function on erosion rate with arc current as parameter

20.5 Numerical Investigation of Anode Spot Parameters

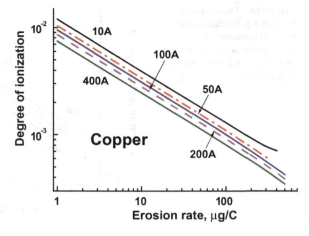

Fig. 20.11 Degree of anode vapor ionization as function on erosion rate with arc current as parameter

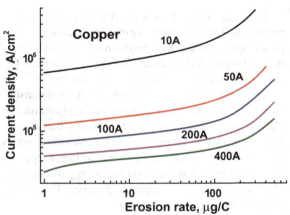

Fig. 20.12 Spot current density as function on erosion rate with arc current as parameter

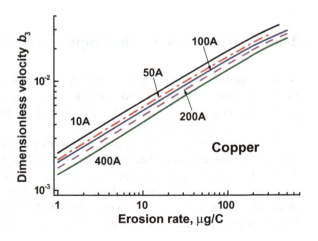

Fig. 20.13 Dimensionless plasma velocity b_3 at the external boundary of Knudsen layer as function on erosion rate with arc current as parameter

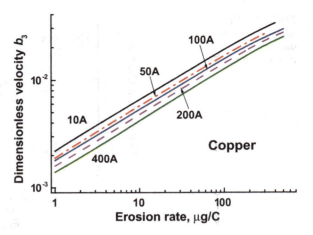

Fig. 20.14 Anode erosion fraction K_{er} formed at the external boundary of Knudsen layer as function on erosion rate with arc current as parameter

Thus, the plasma flow is not free (impeded) as that for plasma flow in the cathode region. In fact, the flow regime is determined by the plasma contraction in the anode spot, formed by condition of self-sustained discharge. The heat source in the spot supported necessary vaporization regime for the spot operation depends on the formed plasma flux at the anode surface.

This follows from the fact that an electron density flux (determined the heat flux) from plasma to the anode depends on the plasma density, which in turn depends on the electron temperature, degree of ionization determined on the velocity of plasma flow. In other words, the regime of plasma motion is determined by relationship between rate of anode evaporation dependent on the energy flux density to the anode, degree of atom ionization dependent on the energy dissipation in the plasma flow, and degree of plasma contraction (current density). Therefore, the regime of plasma flow is different from the sonic regime.

20.6 Extended Approach. Graphite Anode

As mentioned above (Sect. 20.1), the electrical arc with graphite electrode was investigated for many years [12] due to its use in welding and light devices. An increasing interest is in graphite vacuum arc owing to its use in metallurgy and as plasma source for carbon films deposition [3].

The graphite anode is interesting because its electrical conductivity is by three orders of magnitude lower than good electrical conductivity of the copper. To study the graphite anode region, the calculations were conducted for steady-state case at wide range of parameters indicated above. The solutions of the equation system are obtained for spot currents in the range of 10–400 A, fork function of 4.34 eV, specific heat of evaporation 6.25×10^4 W/g, and for graphite specific resistance 10^{-3} Ω cm.

20.6 Extended Approach. Graphite Anode

The calculations indicate that some anode spot parameters vary weakly before about 100 μg/C. Figure 20.15 shows that the anode temperature T_a weakly increases up to 60 μg/C in the range of 3000–4000 K, and then the increase occurs sharply, while T_a decreases with spot current. As a result, the heavy particle density n_0 determined by the saturated anode vapor pressure demonstrates this same dependence (Fig. 20.16). However, the dimensionless plasma density n_{30} at the external boundary of Knudsen layer initially decreased passing a minimum value and then also sharply increased (at 100 μg/C) with erosion rate (Fig. 20.17). The evaporated atoms are ionized due to thermal mechanism with the plasma temperature that weakly changed with G up to 100 μg/C and then sharply increased. However, the range of T_e change is relatively small (Fig. 20.18).

The calculated degree of atom ionization as shown in Fig. 20.19 monotonically decreases with anode erosion rate and increases with increasing spot current. For

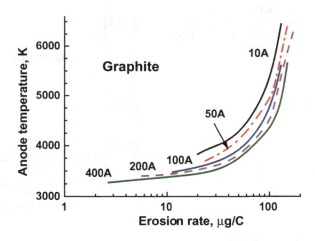

Fig. 20.15 Anode temperature in the spot as function on erosion rate with arc current as parameter

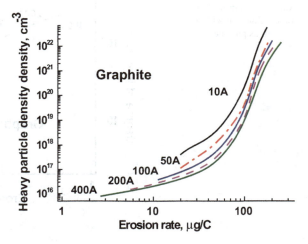

Fig. 20.16 Heavy particle density n_0 as function on erosion rate with arc current as parameter

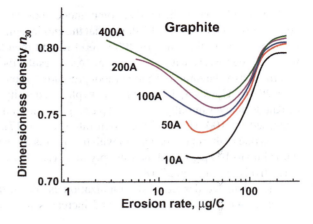

Fig. 20.17 Dimensionless plasma density n_{30} at the external boundary of Knudsen layer as function on erosion rate with arc current as parameter

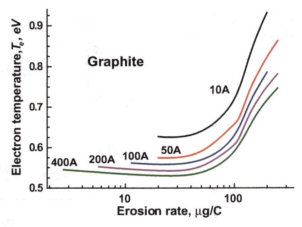

Fig. 20.18 Electron temperature as function on erosion rate with arc current as parameter

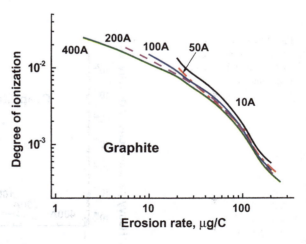

Fig. 20.19 Degree of anode vapor ionization as function on erosion rate with arc current as parameter

20.6 Extended Approach. Graphite Anode

the considered region of anode erosion rate and spot current, the calculated value of degree ionization is relatively small and is in region 3×10^{-4} to 2×10^{-2}. Figure 20.20 presents the dependence of anode spot current density on erosion rate, when spot current is a parameter. Again, the current density weakly increases with increasing erosion rate and decreases with increasing spot current. Mainly, the value of j increases weakly by increase of G in the range of 3–60 µg/C and then significantly increases with G up to 300 µg/C.

Figure 20.21 shows dependence on G and I of the dimensionless plasma velocity b_3 at the external boundary of Knudsen layer, which is significantly small due to large returned heavy particle flux. As a result, the small anode erosion fraction K_{er} formed at the external boundary of Knudsen layer, shown in Fig. 20.22, indicates significant difference of the metal vaporization in the anode region in comparison with that into vacuum determined by the Langmuir–Dushman formula. It can be seen that K_{er} initially increases passing through a maximum and then decreases with G.

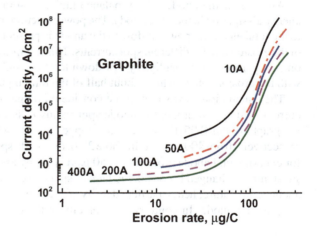

Fig. 20.20 Spot current density as function on erosion rate with arc current as parameter

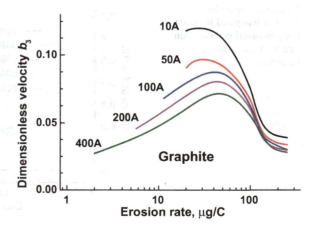

Fig. 20.21 Dimensionless plasma velocity b_3 at the external boundary of Knudsen layer as function on erosion rate with arc current as parameter

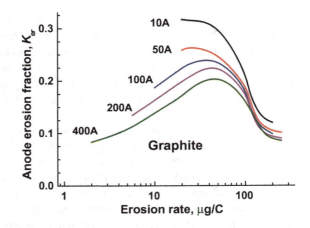

Fig. 20.22 Anode erosion fraction K_{er} formed at the external boundary of Knudsen layer as function on erosion rate with arc current as parameter

Analysis of the anode energy balance indicates some contribution of the Joule energy dissipation in the anode body. The power fraction of resistive heating relatively to the total electrical power inflow to the anode is presented in Fig. 20.23 as function of the erosion rate for different spot currents. This fraction practically does not depend on spot current and is negligibly low down to 100 μg/C and then sharply increases with the erosion rate reaching about half of the total input power.

The calculation shows that in the considered ranges of anode erosion rate, spot current, and spot lifetime, the anode spot radius r_a is in the range 0.38–0.012 cm for graphite and 0.05–0.0015 cm for copper anodes when the current decreases, respectively, from 400 to 10 A. In Table 20.6, the anode spot parameters are presented for experimental anode erosion $G = 50$ μg/C [75] by assumption that this value not substantially changed is considered here as spot current range. It can be seen that the anode temperature, heavy particle density, and current density are larger for Cu than for graphite anode. In contrary, the velocity at the external boundary of Knudsen

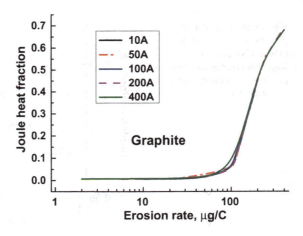

Fig. 20.23 Joule heat fraction dissipated in anode body as function on erosion rate with arc current as parameter

20.6 Extended Approach. Graphite Anode

Table 20.6 Anode spot parameters calculated for anode erosion rate of 50 μg/C

I, A	T_a, K	n_0, 10^{17} cm^{-3}	$j \times 10^3$ A/cm^2	$v_3 \times 10^3$ cm/s	α	K_{er}, cm/s	T_e, eV	n_{30}	b_3
10	4156.6	25.16	21.34	29.62	0.0055	0.311	0.629	0.719	0.118
50	3895.8	6.026	4.35	24.67	0.0048	0.274	0.578	0.735	0.101
100	3807	3.536	2.361	22.59	0.0045	0.256	0.57	0.742	0.0913
200	3735.5	2.26	1.381	20.41	0.0041	0.2364	0.545	0.751	0.085
400	3682.5	1.6046	0.884	18.2	0.0037	0.2148	0.532	0.759	0.0761
10	4710.7	4648.4	1359.7	1670.2	0.0018	0.044	0.534	0.824	0.0139
50	3853.3	896.5	209.7	1332.9	0.0015	0.0389	0.475	0.826	0.0123
100	3640.3	524.0	110.9	1204.4	0.0014	0.0362	0.457	0.827	0.0114
200	3509.1	363.9	68.79	1074.6	0.0013	0.033	0.443	0.828	0.0104
400	3448.3	304.2	50.50	942.34	0.0011	0.0292	0.434	0.830	0.0092

layer v_3 is larger for graphite anode than for Cu. Therefore, the erosion fraction K_{er} is also larger for graphite anode. The calculation also shows that for graphite anode, the potential drop u_p in the plasma jet varied in the range of 5–8 V and the plasma jet velocity in the range of 1–1.6 cm/s. The obtained values are correlated with the experimental data for low-voltage vacuum arc at relatively low values of G = 10–100 μg/C. These parameters calculated for copper anode are significantly larger for graphite anode. However, they can be reduced by improving the calculated model taking into account that not all plasma enthalpy in the spot is realized during the plasma jet expansion already shown for cathode region [36].

20.6.1 Concluding Remarks

Anode spot parameters including anode temperature, current density, plasma temperature, density, and the return atom ion fluxes to the anode are determined self-consistently by considering the processes in the kinetic layer and in the plasma acceleration region. The calculation shows that the degree of atom ionization in anode plasma is small (~0.001) in comparison with highly ionized cathode spot plasma (Chap. 16). This result can be understood taking into account that the anode current density was supported by the electrons whose velocity is significantly larger in comparison with the ion velocity supporting the cathode current density. As a result, a lower electron density in the anode plasma in comparison with that density in the cathode spot is necessary for reaching this same current density.

Almost all above calculated anode spot parameters sharply increase by some critical anode erosion value, while all these parameters weakly depend before this critical value. This value occurs due to large energy losses in the plasma energy and anode energy balances by some value of anode erosion rate. It should be noted that the anode erosion in the present model is considered in vapor phase. An experiment [100] indicates that even the graphite erosion can be also in macroparticle form and the MP fraction strongly depends on the graphite material processing. Our calculations, however, indicate that in general the influence of MP contribution on the spot parameters is through the anode energy balance and plasma parameters obtained by taking into account the MP fraction in this case are not substantially changed due to their relatively small velocity and therefore low kinetic energy.

The calculated results show that, in general, the anode spot parameters are in agreement with the existing experimental data [37, 38, 75]. However, only small numbers of measurements (no complete anode experimental data existed) occur for vacuum-arc copper and graphite anode that can be compared with the theoretical results. According the experimental data [75], the graphite anode erosion rate is 50 μg/C measured for arc current 70 A. The measurement graphite anode spot temperature is about 4000 K [12] and 4780 K [18, 75], while the present calculation for graphite is in the range 3500–4900 K. The experiment for copper anode indicated some range of data (see Sect. 14.4.1), while it is comparable with present calculation (Fig. 20.7).

The plasma density and electron temperature were measured for Al anode (see above). Therefore, it can be compared by some trends of these parameters predicted by the theory. The calculated electron density, in general, is agreed with the density measured in anode spot for Al ($\sim 10^{17}$ cm^{-3}) [9] and in near-anode spot region for Ni ($\sim 10^{15}$ cm^{-3}) (Chap. 14, [101]) and illustrates the lower anode plasma density in comparison with the plasma density in cathode spot ($\sim 10^{19}$ cm^{-3}) (Chaps. 16 and 17). This same conclusion is related to electron temperature, which also sufficiently is lower in anode spot (measured value for Al is 0.67–0.47 eV (see above) and present calculation is about 0.5–0.8 eV) in comparison with electron temperature in cathode spot (1–2 eV). The order of magnitude of plasma jet velocity (10^6 cm/s) is also comparable with results discussed by Ecker [7]. The high velocity anode plasma jets were observed by Hermoch [102] for a number of materials.

The net anode mass flux can be relatively small due to large returned atom flux from the plasma to the anode. The anode erosion fraction for graphite varies in wide range (8–30%,) and it is larger in comparison with cathode spot erosion fraction in vacuum arc (3–5% [103]). The calculated, here, anode erosion fraction for copper 0.2–10% indicated lower difference from that calculated for cathode of about 4% at spot lifetime larger than 100 ns. This fact can be explained by smaller anode spot plasma pressure in comparison with plasma pressure in the cathode spot for graphite material, while these pressures for copper electrodes differ less.

References

1. Boxman, R. L., Martin, P. J., & Sanders, D. M. (eds.) (1995). *Handbook of vacuum arc science and technology*. Park Ridge, NJ: Noyes Publications.

References

2. Anders, A. (2008). *Cathodic arcs: From fractal spots to energetic condensation.* Berlin: Springer.
3. Chhowalla, M., Davis, C., Weiler, M., Klelnsorge, & Armaratunga, G. (1996). Stationary carbon cathodic arc: Plasma and film. *Journal of Applied Physics, 79*(N5), 2237–2244.
4. Craig Miller, H. (1977). Vacuum arc anode phenomena. *IEEE Transactions on Plasma Science, 5*(N3), 181–196.
5. Craig Miller, H. (1983). Vacuum arc anode phenomena. *IEEE Transactions on Plasma Science, PS-11*(N2), 76–89.
6. Miller, H. C. (1995). Anode phenomena. In R. L. Boxman, P. J. Martin, & D. M. Sanders (Eds.), *Handbook of vacuum arc science and technology* (pp. 308–364). Park Ridge, NJ: Noyes Publications.
7. Ecker, G. (1961). Electrode components of the arc discharge. *Ergeb. Der Exakten Naturwissenschaften, 33,* 1–104.
8. Schulze, G. (1903). Über den Spannungsverlust im elektrischen Lichtbogen. *Annalen der Physik, Vierte Folge, 12,* 828–841.
9. Finkelnburg, W. (1939). Untersuchungen Uber Hochstromkohlebogen. I. Die Strom-Spannungs charakteristiken verschiedener Hochstromkohleb6gen. *Zs. f Physics, 112*(N5-6), 305–325.
10. Finkelburg, W. (1948). A theory of the production of electrode vapor jets by sparks and arcs. *Physical Review, 74*(10), 1475–1477.
11. Busz-Peuckert, G., & Finkelburg, W. (1956). Zum Anodenmechanismus des thermischen Argonbogens. *Zeitschrift fur Physik, 144*(1–3), 244–251.
12. Finkelnburg, W., & Maecker, H. (1956). Electrische Bogen and thermishes plasma. In S. Flugge (Ed.), *Handbuch der Physik* (Bd. 22, pp. 254–444). Berlin: Springer.
13. Finkelnburg, W. (1939). Leuchtdichte, Gesamtstrahlungsdichte und schwarze Temperatur von Hochstromkohlebogen (Untersuchungen fiber HochstromkohlebSgen. II.). *Zs. f Physics, 113*(N9–10), 562–581.
14. Fihkelnburg, W. (1939). Zum Mechanismus des Hochstromkohlebogens (Untersuchungen fiber Hochstromkohlebogen. III.). *Zs. f Physics, 114*(N11–12), 734–746.
15. Finkelnburg, W. (1940). Anodenfall, Anodenabbau und Theorie des Hochstromkohlebogens. (Untersuchungen uber Hochstromkohlebogen IV.). *Zs. f Physics, 116*(N3–4), 214–233.
16. Finkelburg, W. (1949). The high current carbon arc and its mechanism. *Journal of Applied Physics, 20*(5), 468–474.
17. Blevin, W. R. (1953). Further studies of electrode phenomena in transient arcs. *Australian Journal of Physics, 6*(N2), 203–208.
18. Cobine, J. D., & Burger, E. E. (1955). Analysis of electrode phenomena in the high current arc. *Journal of Applied Physics, 26*(7), 895–900.
19. Bez, W., & Hocker, K. H. (1954). Theorie des Anodenfalls I. *Zeitschrift fur Naturforschung, A9,* 72–81.
20. Finkelburg, W., & Segal, S. M. (1951). The potential field in and around a gas discharge, and its influence on the discharge mechanism. *Physical Review, 83*(3), 582–585.
21. Hocker, K. H., & Bez, W. (1955). Theorie des Anodenfalls II. *Zeitschrift fur Naturforschung, A10,* 706–714.
22. Bez, W., & Hocker, K. H. (1955). Theorie des Anodenfalls III. *Zeitschrift fur Naturforschung, A10,* 714–717.
23. Rieder, W. (1956). Leistungsbilanz der Elektroden und Charakteristiken frei brennender Niederstrombogen. *Zeitschrift fiir Physik, 146,* 629–643.
24. Cobine, J. D. (1980). Vacuum arc anode phenomena. In J. M. Lafferty (Ed.), *Vacuum arcs* (Chap. 5). New York: Wiley.
25. Zingerman, A. S., & Kaplan, D. A. (1958). Electric erosion of an anode as a function of interelectrode distance. *Soviet Physics - Technical Physics, 3*(2), 361–367.
26. Zolotykh, B. N. (1960). The mechanism of electrical erosion of metals in liquid dielectric media. *Soviet Physics - Technical Physics, 5*(12), 1370–1373.

27. Sugawara, M. (1967). Anode melting caused by a d.c. arc discharge and its application to the determination of the anode fall. *British Journal of Applied Physics, 18*(314), 1777–1781.
28. Beilis, I. I., & Zykova, N. M. (1968). Dynamics of the anode spot in low current arc. *Soviet Physics-Technical Physics, 38*(2), 323–325.
29. Goloveiko, A. G. (1968). Thermophysical processes at the anode at conditions of high power pulse discharge. *Journal of Engineering Physics, 14*, N6.
30. Konovalov, A. E. (1969). Heat fluxes to the electrodes in condition of high current discharge. *High Temperature, 7*(1), 91–96.
31. Caap, B. (1972). The power balance in electrode-dominated arcs with a tungsten anode and a cadmium or zinc cathode in nitrogen. *Journal of Physics. D. Applied Physics, 5*, 2170–2178.
32. Bacon, F. M., & Watts, H. A. (1975). Vacuum arc anode plasma. II. Collisional-radiative model and comparison with experiment. *Journal of Applied Physics, 46*(N11), 4758–4766.
33. Lafferty, M. (1966). Triggered vacuum gaps. *Proceedings of the IEEE, 54*(1), 23–32.
34. Kimblin, W. (1969). Anode voltage drop and anode spot formation in DC vacuum arcs. *Journal of Applied Physics, 40*, 1744–752.
35. Mitchell, G. R. (1970). High-current vacuum arcs Part 2—Theoretical outline. *Proceedings of the Institution of Electrical Engineers, 117*(N12), 2327–2332.
36. Beilis, I. I. (1995). Theoretical modeling of the cathode spot phenomena. In R. L. Boxman, P. J. Martin, & D. M. Sanders (Eds.), *Handbook of vacuum arc science and technology* (pp. 208–256). Park Ridge, NJ: Noyes Publications.
37. Ecker, G. (1970). *Theory of the vacuum arc. I. Anode constriction instability, General Electric report N70-C-23* (25 p). Schenectady, NY: RD Center.
38. Ecker, G. (1974). Anode spot instability. *IEEE Transactions on Plasma Science, PS-2*(N3), 130–146.
39. Mitchell, G. R. (1970). High current vacuum arcs. Part 1—An experimental study. *Proceedings of the Institution of Electrical Engineers, 117*(N12), 2315.
40. Rich, J. A., Prescott, L. E., & Cobine, J. D. (1971). Anode phenomena in metal-vapor arcs at high currents. *Journal of Applied Physics, 42*(2), 587–601.
41. Kimblin, C. W. (1974). Anode phenomena in vacuum and atmospheric pressure arcs. *IEEE Transactions on Plasma Science, PS-2*(N6), 310–319.
42. Kamakshiah, S., & Rau, R. S. N. (1977). Anode phenomena in triggered vacuum gaps. *IEEE Transaction on Plasma Science, PS-5*(N1), 1–6.
43. Kamakshiah, S., & Rau, R. S. N. (1977). Vacuum arc phenomena in triggered vacuum gaps. In *Proceedings of the 3rd International Conference on Switch Arc Phenomena*, Poland, Lodz, Part 1 (pp. 206–209).
44. Miller, H. C. (1977). Vacuum arc anode phenomena. *IEEE Transactions on Plasma Science, 5*(3), 181–196.
45. Malkin, P. (1989). The vacuum arc and vacuum interruption. *Journal of Physics. D. Applied Physics, 22*, 1005–1019.
46. Boxman, R. L. (1977). Magnetic constriction effects in high-current vacuum arcs prior to the release of anode vapor. *Journal of Applied Physics, 48*(6), 2338–2345.
47. Boxman, R. L., Harris, J. H., & Bless, A. (1978). Time dependence of anode spot formation threshold current in vacuum arcs. *IEEE Transaction on Plasma Science, PS-6*, 233–237.
48. Boxman, R. L., & Goldsmith, S. (1983). A model of the anode region in a uniform multi cathode spot vacuum arc. *Journal of Applied Physics, 54*(2), 592–602.
49. Goldsmith, S., Shalev, S., &Boxman, R. L. (1983). Anode melting in a multi cathode spot vacuum arc. *IEEE Transactions on Plasma Science PS-11*, (N3), 127–132.
50. Izraeli, I., Goldsmith, S., & Boxman, R. L. (1983). The influence of the self-magnetic field on the steady state current distribution in an axially flowing conducting medium. *IEEE Transactions on Plasma Science, PS-11*(N3), 160–164.
51. Izraeli, I., Boxman, R. L., & Goldsmith, S. (1987). The current distribution and the magnetic pressure profile in a vacuum arc subject to an axial magnetic field. *IEEE Transactions on Plasma Science, PS-15*, 502–505.

References

52. Beilis, I. I., Keidar, M., Boxman, R. L., & Goldsmith, S. (1998). Theoretical study of plasma expansion in a magnetic field in a disk anode vacuum arc. *Journal of Applied Physics, 83*(2), 709–717.
53. Schellekens, H. (1985). The diffuse vacuum arc: A definition. *IEEE Transactions on Plasma Science, PS-13*(N5), 291–295.
54. Keidar, M., & Shulman, M. B. (2001). Modeling the effect of an axial magnetic field on the vacuum arc. *IEEE Transactions on Plasma Science, 29*(5), 684–689.
55. Beilis, I. I., & Zektser, M. P. (1991). Calculation of the parameters of the cathode stream an arc discharge. *High Temperature, 29*(4), 501–504.
56. Dyuzhev, G. A., Shkol'nik, S. M., & Yur'ev, V. G. (1978). Anode phenomena in the high current density arc. *Soviet Physics Technical Physics, 23*, 667–671.
57. Dyuzhev, G. A., Shkol'nik, S. M., & Yur'ev, V. G. (1978). Anode phenomena in the high current density arc. *Soviet Physics Technical Physics, 23*, 672–677.
58. Kutzner, J. (1981). Voltage-current characteristics of diffusion vacuum arc. *Physica C, 104*, 116–123.
59. Harris, J. (1979). Electron density measurements in vacuum arcs at anode spot formation threshold. *Journal of Applied Physics, 50*(2), 753–757.
60. Schuocker, D. (1979). Improved model for anode spot formation in vacuum arcs. *IEEE Transactions on Plasma Science, PS-7*(N4), 209–216.
61. Boxman, R. L. (1974). Interferometric measurement of electron and vapor densities in high-current vacuum arc. *Journal of Applied Physics, 45*(11), 4835–4846.
62. Drouet, M. G. (1981). Anode spots and cathodic plasma flow in a vacuum arc. *Journal of Physics. D. Applied Physics, 14*(12), L211–214.
63. Kutzner, J. (1978). Angular distribution of ion current in dc copper vacuum arc. In *Proceedings of the 8th International Symposium on Discharges Electrical Insulation in Vacuum* (Albuquerque, NM) (p. A1).
64. Kutzner, J., & Zalucki, Z. (1977). Ion current emitted from the cathode region of a DC copper vacuum arc. In *Proceedings of the 3rd International Conference on Switch Arc Phenomenon*, Poland, Lodz, Part 1 (pp. 210–216).
65. Jolly, D. C. (1982). Anode surface temperature and spot formation model for the vacuum arc. *Journal of Applied Physics, 53*, 6121–6126.
66. Dyuzhev, G. A., Lyubimov, G. A., Shkol'nik, S. M. (1983). Conditions of the Anode Spot Formation in a vacuum arc. *IEEE Transaction on Plasma Science, PS-11*(N1), 36–44.
67. Nemchisky, V. A. (1983). Formation of an anode spot in a vacuum arc. *Soviet Physics-Technical Physics, 28*(2), 146–149.
68. Nemchisky, V. A. (1989). Calculation of the influence of an axial magnetic field on the formation of the anode spots in a vacuum arc. *Soviet Physics-Technical Physics, 34*(9), 1014–1017.
69. Beilis, I. I., Keidar, M., Boxman, R. L., & Goldsmith, S. (1996). 2-D expansion of the low density interelectrode vacuum arc plasma in an axial magnetic field. *Journal of Applied Physics, 29*(7), 1973–1983.
70. Lyubimov, G. A., Rakhovsky, V. I., Seliverstova, I. F., & Zektser, M. P. (1980). A study of parameters of the anode spot on aluminum. *Journal of Physics. D. Applied Physics, 13*, 1655–1664.
71. Nemchinskii, V. A. (1982). Anode spot in a high-current vacuum arc. *Soviet Physics-Technical Physics, 27*(1), 20–25.
72. Lefort, A., & Andanson, P. (1985). Characteristic parameters of the anode spot. *IEEE Transaction on Plasma Science, P-13*(N5), 296–299.
73. Dushman, S. (1962). *Scientific foundation of vacuum technique*. In J. M. Lafferty (Eds.) (p. 742) NY: Wiley.
74. Beilis, I. I., & Lyubimov, G. A. (1975). Parameters of cathode region of vacuum arcs. *High Temperature, 13*(6), 1057–1064.
75. Lefort, A., Parizet, M. J., El-Fassi, S. E., & Abbaoui, M. (1993). Erosion of graphite cathodes". *Journal of Physics. D. Applied Physics, 26*(8), 1239–1243.

76. Salihou, H., Guillot, J. P., Abbaoui, M., & Lefort, A. (1996). Anode parameters of short arcs at low current. *Journal of Physics. D. Applied Physics, 29*, 2915–2921.
77. Abbaoui, M., & Lefort, A. (2009). Arc root interaction with the electrode: a comparative study of 1D-2D axisymmetric simulations. *The European Physical Journal of Applied Physics, 48*, 11001.
78. Londer, Ya. I., & Ul'yanov, K. N. (2013). Theory of the negative anode drop in low pressure discharges. *High Temperature, 51*(N1), 7–16.
79. Wang, L., Jia, S., Zhang, L., Yang, D., Shi, Z., Gentils, F., & Jusselin, B. (2008). Current constriction of high-current vacuum arc in vacuum interrupters. *Journal of Applied Physics, 103*, 063301.
80. Wang, L., Jia, S., Yang, D., Liu, K., Su, G., & Shi, Z. (2009). Modelling and simulation of anode activity in high-current vacuum arc. *Journal of Physics. D. Applied Physics, 42*, 145203.
81. Wang, L., Jia, S., Liu, Y., Chen, B., Yang, D., & Shi, Z. (2010). Modeling and simulation of anode melting pool flow under the action of high-current vacuum arc. *Journal of Applied Physics, 107*, 113306.
82. Huang, X., Wang, L., Deng, J., Jia, S., Qin, K., & Shi, Z. (2016). Modeling of the anode surface deformation in high-current vacuum arcs with AMF contacts. *Journal of Physics. D. Applied Physics, 49*(N7), 075202.
83. Wang, L., Huang, X., Jia, S., Deng, J., Qian, Z., Shi, Z., et al. (2015). 3D numerical simulation of high current vacuum arc in realistic magnetic fields considering anode evaporation. *Journal of Applied Physics, 117*, 243301.
84. Wang, L., Huang, X., Jia, S., Zhou, X., & Shi, Z. (2014). Modeling and simulation of deflected anode erosion in vacuum arcs. *Plasma Science Technology, 16*(N3), 226–231.
85. Wang, L., Huang, X., Zhang, X., & Jia, S. (2017). Modeling and simulation of high-current vacuum arc considering the micro process of anode vapor. *Journal of Physics. D. Applied Physics, 50*(N9), 095203.
86. Tian, Y., Wang, Z., Jiang, Y., Ma, H., Liu, Z., Geng et al. (2016). Modelling of crater formation on anode surface by high-current vacuum arcs. *Journal of Applied Physics, 120*(N18), 183302.
87. Wang, Z., Zhou, Z., Tian, Y., Wang, H., Wang, J., Y. Geng, & Liu, Z.(2017). Effects of an anode sheath on energy and momentum transfer in vacuum arcs. *Journal of Physics. D. Applied Physics, 50*(N29), 295203.
88. Beilis, I. I. (1985). The anode region of electrical arc in the electrode material vapor. *Izvestiya Siberian Akad. Nauk, Ser. Tekhn. Nauk, 16*(N3), 77–82 (in Russian).
89. Beilis, I. I. (2000). Anode spot vacuum arc model. *Graphite Anode, IEEE Transaction on Components and Packaging Technologies CPMT, 23*(2), 334–340.
90. Beilis, I. I. (1982). On the theory of erosion processes in the cathode region of an arc discharge. *Soviet Physics Doklady, 27*, 150–152.
91. Beilis, I. I. (1985). Parameters of kinetic layer of arc discharge cathode region. *IEEE Transactions on Plasma Science, PS-13*(5), 288–290.
92. Anisimov, S. I., Imas, Y., Romanov, G., & Khodyko, Y. (1971). *Action of high-power radiation on metals*. Springfield, Virginia: National Technical Information Service.
93. Knake, O., & Stranski, I. N. (1956). Mechanism of vaporization. *Progress in Metal Physics, 6*, 181–235.
94. Dobretzov, L. N., & Homoyunova, M. V. (1966). *Emission electronics*. Moscow: Nauka.
95. Beilis, I. I. (1974). Analysis of the cathode spots in a vacuum arc. *Soviet Physics - Technical Physics, 19*(2), 251–260.
96. Anisimov, S. I. (1968). Vaporization of metal absorbing laser radiation. *JETP, 37*(1), 182–183.
97. Rich, I. A. (1961). Resistance heating in the arc cathode spot zone. *Journal of Applied Physics, 32*(6), 1023–1031.
98. Ecker, G. (1971). The existence diagram a useful theoretical tool in applied physics. *Z. fur Naturforschg, 26a*, 935–939.
99. Ecker, G. (1971). Zur theorie des vakuumbogens. *Beitrage aus der Plasmaphysik., 11*(5), 405–415.

100. Kandah, M., & Meunier, J. (1995). Study of micro droplet generation from vacuum arcs on graphite cathodes. *Journal of Vacuum Science and Technology, A13*(5), 2444–2450.
101. Boxman, R. L. (1974). Interferometric measurement of electron and vapor densities in high-current vacuum arc. *Journal of Applied Physics, 45,* N114835–4846.
102. Hermoch, V. (1959). The vapor jets of electrode material of a short-time high intensity electrical discharge. *Czechoslovak Journal of Physics, 9,* 221–228.
103. Beilis, I. I. (1999). Application of vacuum arc cathode spot model to graphite cathode. *IEEE Transactions on Plasma Science, 27*(4), 821–826.

Part IV
Applications

Chapter 21
Unipolar Arcs. Experimental and Theoretical Study

Igor Tamm and Andrei Sakharov in 1950 are the first physicists who proposed an idea to use toroidal camera for high-temperature plasma confinement named then as Tokamak. The first tokamak, the T-1, began operation in 1958. Also in 1958, the model of the Harwell controlled Zero Energy Thermonuclear Assembly (ZETA) was demonstrated in London. High-temperature plasma exposes a high heat load the facing a divertor material and first wall in fusion reactors. Different materials were used for tokamak construction. One of these materials, tungsten, was used as a metal with high durability against the high heat load and low sputtering yield. When the sprayed tungsten was exposed to the helium plasma, the surface was covered with arborescent nanostructured tungsten containing many helium bubbles inside the structure.

The below description reports first the experimental investigation and the main observations of the arcing in the fusion devices. The next part of the chapter presents results of theoretical study of unipolar arc occurred at the walls in fusion devices including Tokamak.

21.1 Experimental Study

Although that the first tokamak was constructed in 1954, the unipolar arc appearance was already observed in period of few years after. The principal problems of such discharges are the contamination of the plasma by material eroded from the walls. Some presentation about first investigations of arcing phenomena in fusion devises can be found considering the materials of workshop, Tennessee [1]. In this section, the arcing experiments are described including the primary and developed observations of the arcing on the metals as well on the nanostructured surface presented in the last decades.

21.1.1 Unipolar Arcs on the Metal Elements in Fusion Devices

Robson and Hancox in [2] informed that the metals are superior to insulators in thermal properties and ease of fabrication, but tend to form "unipolar" arcs when exposed to plasma. The occurrence of unipolar arcs is a general disadvantage of metals exposed to plasma. Considering the presentations at previous on International Conferences [3], they discussed results of experimentally detected unipolar arcs both in transient form on isolated metal plates exposed to pulse discharges, and on the walls of Zero Energy Thermonuclear Assembly (ZETA). Noted, that in a thermonuclear reactor it would be essential to eliminate unipolar arcs completely, since a single arc carrying only a few amperes will introduce more impurities than can be tolerated. The authors analyzed the problems of the torus, the plasma contaminations, and the fundamental characteristics, which affect the choice of material and method of construction.

A few conditions at the surface of the metal were studied in order to prevent the arc initiation. One approach to the elimination of unipolar arcs relies on careful choice and preparation of the material of the torus wall. The results of such experimental data determined by the arc formation conditions are given. A second method of eliminating unipolar arcs relies on the empirical observation that a minimum current is required to maintain an arc on a metal cathode. Known, that the current in a unipolar arc is proportional to the exposed area of the metal surface. When this area is restricted so that the surface cannot collect the minimum current required for a cathode spot, the unipolar arc should not form. A third possible method of eliminating arcs is to coat the metal with a thin layer of an insulating material. As result, a critical review is given of a number of possible torus designs.

The effect of inclusions on the arcing behavior was studied by Pfeil and Griffiths in [4] and then by Hancox [5] on initiation of the arc using metal specimens known to contain microscopic alumina inclusions. Hancox suggested that the dielectric charging and subsequent breakdown of these inclusions initiated arcing process. The experiment shows that the time lag before arcing occurred was inversely proportional to the ion current using alumina inclusions of 1 μm size. The time lag was independent of voltage for any given ion current. The arc initiation was also independent of the nature and pressure of the gas. Pfeil and Griffiths microscopically examination of the surface after experiments with unipolar arcing revealed that surface damage of the specimens had occurred around certain types of inclusion in the metals while other part of the surface being unaffected. A correlation has been observed between the resistivity of inclusions and their influence as arc-initiators.

Maskray and Dugdale continue investigation the importance of contamination in arc initiation on molybdenum [6] and on stainless steel in [7] exposed a toroidal hydrogen discharge, considering the effects of oxygen, nitrogen, and neon. The results demonstrate that using Mo the arcing occurs through the agency of oxide and nitride particles formed by impurities present in the molybdenum. In the case of the nitride-forming impurities, the concentration near the metal surface can be

21.1 Experimental Study

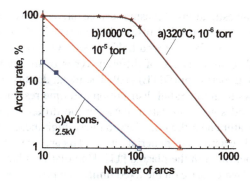

Fig. 21.1 Arcing rate ~1 s number of discharges for fingered specimens of stainless steel at 1 kV, 8 A/cm^2 after **a** degreasing and vacuum baking at 320 °C, 10^{-6} torr; **b** degreasing and vacuum baking at 1000 °C, 10^{-5} torr; **c** bombardment with argon ions at 2.5 kV, 1 mA/cm^2 for 2 h [7]

reduced by repeated testing to a stage at which arcing becomes improbable with nitrogen present. It was explained by the evaporation of the precipitated particles by action of the arcs. The addition of oxygen or nitrogen gas to the torus has weak effect while the arc initiation is entirely due to both organic and inorganic contaminants on the cathode surface. The involved inclusions in the steel were ineffective arc initiators because the contamination has been removed by arcing, by sputtering with argon ions, or vacuum heat treatment.

The arcing rate measured during the experiments was defined as a number of arcs, which occurred in twenty-five discharges. This parameter was measured as a function of the number of discharges and as a function of temperature. Figure 21.1 shows these dependences for different conditions. It can be seen that increase of the wall temperature and cleaning by argon ion bombardment decrease the probability of arc initiation. The results show a statistical variation in the arcing rate which explained as a variations in the ion current density from one discharge to another.

Panayotou et al. [8] studied the plasma–wall materials interacted with a dense, hot deuterium plasma, which approximates the energy and particle fluence expected to bombard the first wall in a CTR tokamak power reactor. The wall was made of 1.27 cm thick OFHC copper, nickel, stainless steel and others. Some of the damage was found to penetrate into the bulk of the material. One of the damage areas, the unevenness of topology and the appearance of ruffled edges, is indicative of areas that may have experienced localized melting and resolidification. The average equivalent diameter of these melted areas is about 5 µm.

Miley [9] reported that compared to vacuum breakdown, arcing between plasma and a conducting surface introduces entirely a new phenomenon. They discussed the importance of both vacuum breakdown and unipolar arcs occurrence in CTR devices. The experiment demonstrated that arc spot superimposed on a blister on a 6 mm niobium wire. Such arcing shown by microprobe detection of traces of cathode material.

A wide kind of investigations of plasma surface interactions, of origin of impurity, of unipolar arcs, etc., in controlled fusion devices provided by McCracken and co-authors in a series works from second half of 1970. One of the first, the impurity concentration studied in a variety of different types of discharge in the Divertor Injection Tokamak Experiment (DITE) [10]. The metal density was detected with a probe, which allows both the radial distribution and time resolution measurements. The results indicated some correlation between the different metal densities and the electron temperature near the boundary, namely, high these temperatures lead to high metal influx. The amount of material removed by arcing was correlated with the density of metal measured in the plasma [11]. Unipolar arcs have been observed in the DITE tokamak, which arise on the fixed limiter, on probes inserted in the plasma and on parts of the torus structure. The technique of thin layer activation used for erosion measurement of the metallic elements. The arc tracks were observed. The tracks were approximately 10 μm wide and varied from continuous overlapping of melted regions to a discontinuous series of craters. In all cases, a retrograde direction of motion was observed [12]. The observations of arc tracks in the torus and results of measurements reported by the authors show that the impurity levels erosion rates were consistent with the arcing phenomena being the dominant mechanism for metallic impurity production.

An attention to the arcing caused also by observation of metal film deposited on collectors during tokamak discharges. This metal deposition was responsible by arcs, which leave arc tracks on limiters and probes observed directly [12]. Observations with the scanning electron microscope (SEM) show that the linear tracks are typically of 10 μm wide. They are normally discontinuous with small craters at irregular intervals and also consist of a series of overlapping melted regions.

The results of published researches were reviewed by McCracken [13] to analyze the role of arcing and sputtering in experiments of tokamaks. Different aspects of the investigations were considered including characteristics of unipolar arcs and their initiation, time dependence of arcing, the ways to reduce arcing phenomena, evidence for sputtering such as sputtering due to the plasma ions and by impurity ion.

The review shows significant progress presented in the last few years in the study of both arcing and sputtering in tokamaks. The arcing has been observed in most tokamaks, and in most cases, it occurred during the initial rise of the discharge current. Noted, about the presence of a critical density for arcing and it occurred when the local density increases for short periods during the instability. The evidence from DITE indicated the instabilities occurring at some rational heat flux at the limiter.

The most evidence of arcing is the observation by the tracks, which the arc left on a surface after exposure to a plasma [12]. The tracks are often visible even by eye as separate lines when the surface has to be well polished for the arc tracks to be easily seen (Fig. 21.2). On surfaces that parallel to the toroidal field, the arc spot runs along the $\boldsymbol{B} \times \boldsymbol{j}$ direction (where the current j was electrons emitted from the surface). Observation with an optical or scanning electron microscope shows the track to be made up of a series of small craters, typically -10 diameter (Fig. 21.3).

Mioduszewski et al. [14] studied the arcing phenomena in tokamak by using the electrical and optical measurements. Arc currents on the samples varied in range

21.1 Experimental Study

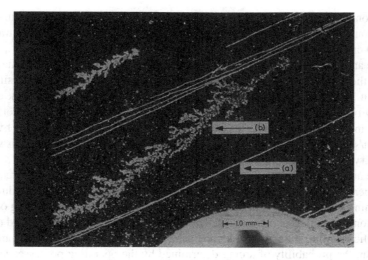

Fig. 21.2 Two types of arc track observed on a molybdenum probe exposed for many discharges in the DITE tokamak. **a** The "linear" arc; **b** the "fern" arc. Photographed using an optical microscope. Figure taken from [13]. Used with permission

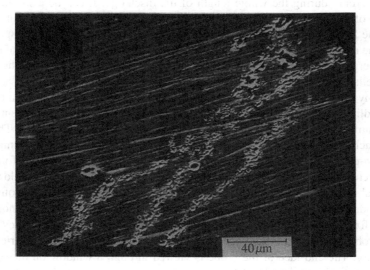

Fig. 21.3 Scanning electron micrograph of a linear arc track on a molybdenum surface exposed in the DITE tokamak. Figure taken from [13]. Used with permission

between 10 and 50 A, but mainly the current amplitude was around 20 A. The arcing signals were related to pin-diode measurements detecting the soft x-ray fluctuations, which were a measure for MHD activity. Each disruption of the plasma was indicated by a breakdown in the pin-diode signal. The arc tracks in the ISX tokamak were observed on the limiter, in the inner-wall surface and in samples. The tracks were found on various materials such as stainless steel at the limiters, walls, samples, as carbon at the limiter as aluminum and gold at samples. The many of these tracks are several cm long and up to 50 μm wide.

The arcing signal shows several fast spikes at the initial breakdown of the plasma and in the quenching phase at the end of the discharge. The duration of a single arc is about 50 μs. A comparison with a SEM picture showed that the track consists usually of a number of single cathode spots. Using typical data of 1 μs total pulse length and a 10 mm total path length, the resulting estimation indicates spot velocity as 10 m/s, in a magnetic field of typically several tenths of a tesla. Noted, this velocity may vary a lot with the surface conditions.

The authors indicated that arcing occurred usually during the initial breakdown, i.e. at period when plasma is not confined. Also, the arcing was observed during the quenching phase at the end of the discharge. At the stable discharge, arcing occurred only along with plasma disruptions. They noted that described results valid only for a specific clean-up state of the tokamak, which conditioned with respect to arcing. In addition, the probability of arcing determined by the operating processes appeared in the tokamak.

The sputtering phenomena also occurred in different tokamaks, but mainly during the constant current phase. This was because the impurity influx is not only local during arcing during the whole length of the discharge. It can be also due to the energy of the ions reaching the limiter as the sheath potential is considerably higher than the local electron temperature. Although that not clear whether the sputtering was due mainly to plasma ions or to the light impurities ions, the last data of the flux and energy of ions arriving at the limiter indicate that the sputtering of the limiter may be larger than previously indicated.

Many objectives in tokamak plasma require some control. One of them consists in handling the MHD instabilities and improving the plasma confinement, while maintaining some current, temperature, and pressure density profiles [15]. This goal was reached determining some specific nonuniform profiles of the 1D plasma safety factor q-profile, an important parameter for both plasma stability and performance. This factor can be characterized the ratio between the magnetic toroidal field and the poloidal one, and it is approximated by inversion of the plasma current profile [15].

Goodall [16] detected the arcs using time-resolved probe at the same minor radius as the fixed limiter. The arc detection probe consists of two concentric cylinders. The probe placed in a top vertical port of the torus and connected electrically by a resistor. The end face of the probe is viewed through a window on the opposite bottom vertical port. Arcs are detected by measuring the current at the inner electrode while simultaneously observing the emission of a neutral spectral line of the probe material. A good correlation between the arc current and optical signals found. This agreement makes it possible to use both signal as an indication of arcing and the optical signals. For example, use data of Langmuir probes, and MHD activity, i.e., signals with plasma different variations simultaneously.

The measurements show that arcing in DITE is characterized by prominent current pulses during the initial stage of the discharge. The correlation between the arcing periods, the loop voltage pulses, and rational safety factors q shows that the prominent arcs in the initial phase of the discharge are caused by plasma instabilities. For most discharges, the arc pulses with amplitude of 5–20 A were observed in the first 20 μs of the discharge during the plasma current rise time. This has been observed in hydrogen, deuterium, and helium using probes from molybdenum and titanium. The strong arcing in disruptive discharges shows that conclusions drawn from the visual observation of arc tracks could be misleading, since a few disruptive discharges could well dominate the normal arc. Also, arcing signals often observed during neutral injection and at small amplitude current was registered.

To obtain independent observations of the occurrence of arcing in DITE, high-speed time films were also taken of the region near an inner fixed limiter. The films show evidence of events occurring with lifetimes <1 μs, consistent with arcing. In addition, however, much longer events were observed, some lasting more than 10 μs, in which glowing particles entered the plasma from the wall with a velocity of a few m/s. At later stages of the discharge, the arcing contributes significant impurity levels. At this stage, the arcs in DITE appeared as a lower intensity and two types of arc tracks have been observed for the same discharge, which indicate substantially different amount of erosion. This result was compared with experiment reported by Zykova et al. [17]. They have also identified two types of arcs on molybdenum probes exposed in DITE and have estimated the arc current from the dimensions of craters in the tracks. Thus, the arcs have currents in the range 1–3 A with velocities of 10^4 cm/s for one group and 5–10 A with velocities from 3×10^3 to 10^4 cm/s for the second group. McCracken (1980) also discussed this experiment [17].

The Nedospasov [18] group continued the study of arcing phenomena in tokamaks. Bogomolov et al. [19] reported about experimental investigations provided to clear up the nature of the emf causing arcing in the course of minor-disruption MHD instability.

The experiments [19] were conducted in a hydrogen torus with plasma density ~10^{13} cm^{-3}, toroidal magnetic field of 1.4 T, plasma current of 13–15 kA, discharge duration 7–8.5 μs. Segmented movable arc test probes of various configurations were used as plasma limiters. Movable rail limiters were inserted through the horizontal ports on the outer side of the torus. Local magnetic probes registered the MHD activity. Periodic oscillations of the rate of magnetic field increase having frequency of 160 kHz were observed at safety factor of $q_{pr} > 3$. Minor disruptions occurred recurrently when at safety factor of $q_{pr} < 2.5$. During the disruptions, the plasma current at first increased by 100–300 A and later decreased by 300–800 A. The most intensive arcing is initiated between the limiter and the liner or through the limiter. The current through the limiter from the anode to the cathode flows along a helical magnetic line of force. If the current from the limiter (anode) flows along the helical line to the liner (cathode), a cathode spot was formed where the magnetic line of force touches the wall. Indicated, that the emf required for arcing initiation was of inductive nature. Therefore, the major part of arc currents was observed during the final stage of the disruption when plasma current decreased.

Nedospasov et al. [20] presented a review considering the main characteristics of the unipolar arcs occurred in different tokamaks. Noted, that the traces of the cathode spots on the surfaces parallel to the magnetic field are linear, while the tracks were branched and consist of melted craters up to 100 μm in diameter on the surfaces perpendicular to the magnetic field. The correlation between the arc appearance and the development of MHD instability was discussed. In the case of a stable discharge, arcing occurred during the initial and final stages of the discharge. The current of the arc in different tokamaks varied from 4 A to 1.6 kA. The arc duration varied from 10 μs to several hundred microseconds. The maximum value of the arc current 1.6 kA in tokamak T-3 M was 5–7% of the main current. Arcing probability on the limiter in in tokamak TV-1 exponentially decreases with increasing distance from the limiter edge facing the plasma. The mean current densities on the anode and cathode on the limiter surfaces attain 5×10^2 A/cm^2. The authors concluded that the existence of the unipolar arcs in the future tokamaks is dependent to a considerable extent on the plasma temperature near the wall. Noted, the experiments showed that when the plasma temperature near the surface decreased to few eVs by using a divertor then no arcing was observed. If the temperature near the divertor plate in the tokamak is about 100–200 eV, then hundreds of unipolar arcs can occur.

Stampa and Kruger [21] modeled the unipolar arcs ignition without applying an external voltage using a plasma produced in a pulsed high-frequency discharge with a frequency of 1 MHz and duration of 1 μs. The produced plasma flux was of density 10^{13} cm^{-3} and a temperature of 5–25 eV in point located at distance of 50 cm outside the discharge. The investigations were provided with wall probes from Cu, Cd, stainless steel, and brass of areas in range of 2–50 cm^2. The smoothed probes were polished and finally cleaned chemically. A homogeneous magnetic field <0.1 T was applied. The arc current was measured with a circular copper disk of radius 15 mm, and a central stainless steel pin, radius 3 mm.

The results showed that mostly the arcs have a current 9–15 A with a most probable value 11 A. No arc with a current lower than 4.5 A was observed. The direction of propagation of the spot in a static magnetic field parallel to the surface of the wall is retrograde, i.e., in the anti-Amperian direction. The arc velocities were in range of (3–30) $\times 10^3$ cm/s. The erosion was concentrated on a linear trace 0.3–1 mm wide and 5–30 mm long. In some cases, it was observed a splitting into two or three branches, which made it possible to determine the direction of propagation of the spot and to investigate the starting point of the arc. Mostly, fast arcs caused the observed erosion phenomena, while the overlapping craters at the end of a trace indicate a transition to low moving arcs. The spot formation typical was predominated for contaminated surfaces at the beginning of a trace, while the spot formation was typical for clean surfaces at end of the trace. Therefore, it was suggested that the arc arises at a more contaminated part of the surface and is extinguished in a relatively clean region.

Herrmann et al. [22] focused their investigation of the arc characteristics at the inner divertor with arc tracks analysis using modern techniques. The results show that the track width as well as the effective depth decrease with increasing distance from the edge of the tile. The large width (150 μm) at the tile edge is partly due to an overlap of arc tracks. The crater depth is in average 5.5 μm at the edge of the tile. A broad of length distribution of the arc track was observed between 0.5 and 3 mm with a pronounced maximum at 1.5 mm. The lifetime of an arc was estimated in the range of 5–37 μs using the averaged spot velocity of 290 m/s from measurements at 1 μm thick tungsten coatings and that measured at massive tungsten (40 m/s for 10 A).

The arc track density as derived from a scanning region of 11 × 12 mm^2 at the edge was 12%. The arc did not evaporate a significant fraction of the tungsten material and splashed of as liquid metal, and redeposited as droplet. Using SEM and profilometry technique, such droplets were detected not only in the crater bottom, but also at other surface areas. For 10% of the material eroded as droplets, 10^8 droplets (5 μm diameter) were ejected. The region of the inner divertor where the arcs are observed shows a local enhancement of plasma density, $\sim 10^{14}$ cm^3, and pressure of the neutrals as 0.1 mbar, increasing probability of the arc ignition. A high sheath potential at the beginning of an ELM is a trigger of the arc.

Rohde et al. [23] studied the arcs induced by isolating layers. For this investigation, two similar tiles were installed at the baffle of the inner divertor. One tile was coated with a tungsten layer, and the other one with an additional isolating TiO$_2$ layer on top. Arcing was studied by probe current and by arc light emission. In addition, photographic techniques, profilometry, and scanning electron microscopy were developed to investigate damage by the arc. A fast camera was used to monitor the inner divertor baffle with a high spatial and temporal resolution. It was detected that the typical dimensions of arcs are about 10 μm. The W layer at the TiO$_2$-coated tile is by a factor of 5 more eroded than at the pure tungsten tile, showing significant erosion at deposition-dominated areas. Video observations with high temporal and spatial resolution showed correlation of the onset of arcing with the appearance of ELMs at the inner divertor baffle. At the inner divertor baffle region, erosion by arcing dominates in comparison with the physical sputtering by a factor of 7. The number of arc traces only depends on the position of the tile. The effective erosion strongly depends on the surface resistivity.

21.1.2 Unipolar Arcs on a Nanostructured Surfaces in Tokamaks

The exposed to energetic helium bombardment as in the plasma-facing components of fusion devices, tungsten will grow a mat of "fuzz" and form narrow grass-like mats of tendrils. The fuzz is typically a few tens of nanometers in diameter and up to several microns in length. Melting traces were found on the surface after the laser pulses irradiated the surface even though the pulse energy was lower than that for melting bulk tungsten. A numerical temperature calculation of the sample suggested that the effective thermal conductivity near the surface dramatically decreased by several orders of magnitude due to the formation of nanostructured tungsten. It was shown that tendrils grown under lower energy helium plasmas bombardment are both finer and smoother than those grown under bombardment by high energy plasma. Additionally, the both families of tendrils are nanocrystalline with fine-scale cavities, believed to be helium bubbles, just a few nanometers in size.

21.1.2.1 Surface State Under He Plasma–Wall Interaction in Fusion Devices

Doerner et al. [24] studied an equilibrium conditions between a fuzz development in a steady-state plasma (surface temperature and energetic helium flux) and also the fuzz erosion with a rate that will determine the equilibrium thickness of the surface fuzz layer. The parameters necessary for fuzz growth is an energetic helium bombardment of a tungsten surface at a certain surface temperature (in the range 1000–2000 K). However, the plasma can remove material from a surface in one of two ways. Firstly, due to overheating of the surface, when the temperature excursion either results in sublimation from the surface or melting of the surface followed by splashing of the melt layer. It was also assumed that at the used temperature the tungsten fuzz would not chemically erode. The second way to remove material from the surface is due to sputtering, which depends on the incident energy, charge state, and composition of the plasma ions. It was indicated that both the fuzz growth rate and the sputtering yield decrease with the thickness of the fuzz layer. So, as the two competing processes will balance each other, an equilibrium layer thickness should be established in steady-state plasma.

The effects of a transient heat load on the damaged material were investigated in the divertor simulator NAGDIS-II using ruby laser pulses with a pulse duration of ~0.6 μs [25]. When the ruby laser pulses irradiated the bulk tungsten surface, the sub-micrometer holes disappeared at the pulse energy higher than 58 J/cm^2. According to the numerical calculation, the surface temperature became close to the melting point, and consequently, the recrystallization process alleviated the surface damage during heating. It was indicated that the heat load used by ruby laser irradiation corresponds to that of typical in fusion device. Therefore, this result suggests that the transient heat load may alleviate helium holes/bubbles even if helium holes/bubbles are temporarily formed on the surface by helium plasma irradiation.

A similar formation of micron-sized structures on tungsten surface due to helium ion irradiation has been investigated in the linear divertor plasma simulator NAGDIS-II [26]. The surface variation during the helium plasma exposure was determined by studying the optical reflectivity measured by using a He–Ne laser and a photodiode. It was detected that a great amount of bubbles and holes were formed by the exposure with the ion fluence of 1.8×10^{23} cm^{-2} and for helium ion energy of 15 eV. At the incident ion energy of 30–50 eV, the surface is covered with fiber of nanostructured tungsten. When the laser pulse energy exceeded 120 J/cm^2 with duration of 0.6 μs, the surface reflectivity recovers to the initial value.

Kajita et al. [27] investigated the formation of the nanostructured tungsten. The nanostructured samples processed by the focused ion beam (FIB) milling was observed by transmission electron microscope (TEM) during the helium plasma

21.1 Experimental Study

irradiation and the nanostructure formation. The experiments were performed in the linear plasma device NAGDIS (Nagoya Divertor Simulator)-I and II. In the both devices, the electron density was on the order of 10^{12} to 10^{13} cm^{-3}, and the temperature was 5 eV.

The result indicated that the growth of helium bubbles under the nanostructured layer could lead to the formation of tungsten dusts, which can influence to the tokamak operation at certain conditions. Figure 21.4 shows the cross-sectional TEM micrographs of W with helium fluence of 2.4×10^{21} cm^{-2}. The thickness of the nanostructured layer is approximately 450 nm in average. Figure 21.4a is the bottom part of the fiber form structure. There is a layer packed with nanometer-sized bubbles on the bottom part. The thickness is in the range of 10–100 nm. Figure 21.4b shows a plate-like structure in the part A, and ring-like structure is shown in the part B. Figure 21.4c shows the cross-sectional TEM micrographs of W-5, the helium fluence of which is 5.5×10^{21} cm^{-2}. As increasing the helium fluence, the structure grows and becomes more complicated.

The above-described tungsten surface state determined a phenomenon of unipolar arc, which appeared by helium plasma interaction with the wall of fusion devices. Let us consider the experiments in context of a cathode spot arise.

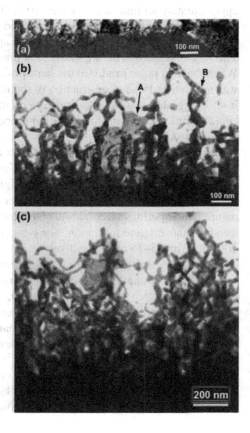

Fig. 21.4 Cross-sectional TEM micrographs of helium irradiated tungsten with the helium fluence of **a** and **b** 2.4×10^{21} cm^{-2}, and **c** 5.5×10^{21} cm^{-2}. Figure taken from [27]. Used with permission

21.1.2.2 Arcing on a Nanostructured Surfaces

A series of experimental study [28–30] showed some damage of a tungsten coating on graphite. Kajita et al. [31] demonstrated unipolar arc experiments using irradiated by laser tungsten in a helium as an arcing electrode in a linear plasma device. They observed that a unipolar arc initiated and continued for a much longer time than the laser pulse of 0.6 μs duration and power of ~500 J/cm^2. Before a laser pulse ignited an arc, the plate, immersed in He plasma, was biased to -75 V, because of which the surface was heated to 1900 K and a net-like nanostructure was formed on the plate surface. The action of laser radiation modeled type-I ELM led to the result that the electrode potential increased and arc spots were generated. The helium fluence was of the order of 10^{22} cm^{-2}. The laser pulse width and power were similar to the heat load accompanied by edge localized modes in ITER resulting in the electrode potential increase, and arc spots were generated. The plasma density and the electron temperature of NAGDIS-II were 2×10^{13} cm^{-3} and ~6 eV, respectively. The moving unipolar arc was detected by a fast camera and by traces on the surface.

At arc ignition, the potential increased by 20–30 V from the floating potential, which was approximately of 60 V and continues for ~2.8 μs far beyond the pulse width of the laser. In response to increase of the potential, the current was increased approximately tenfold and exceeded 10 A. The observed luminous size of the spot did not significantly change during exposed time in range from 0.352 to 0.592 μs. However, it was also observed that the arc spot was amendable to a fragmentation. This fact indicates that the arc spot is sometimes composed of a group of sub-arc spots. When the laser pulse irradiated the same position more than once, the unipolar arcing was not initiated due to destruction of the nanostructure on the surface. Therefore, it was suggested that the nanostructure on the tungsten surface as a fusion product could significantly change the ignition property of arcing, and become a trigger of unipolar arcing.

Kajita et al. [32, 33] studied the unipolar arc motion initiated on a nanostructure tungsten a helium plasma using the laser, fast framing camera and other diagnostic technique described by Kajita et al. [31]. Ignition conditions of arcing were investigated systematically by changing the laser power, plasma conditions, and surface nature. The experimental condition of the investigations satisfies the necessary condition for nanostructure formation. The typical electron density and temperature were 10^{13} cm^{-3} and 5–15 eV, respectively.

It was observed that the arc trace depend significantly on the applied external magnetic field strength.. The width of the track (from SEM micrographs) varies and can be even greater than 50 μm. However, the track has an inner structure, which consists of sub-arc tracks of ~10 μm. Character of these tracks is the same as that of a single arc spot randomly moving on the nanostructured surface. This result shows that several arc spots form a group and move together. Kajita et al. [34] demonstrated the group spot formation.

It was revealed that the retrograde speed of arc spot linearly depends on transverse magnetic field as it was observed early for bipolar arcs. The measured velocity was about $(0.3–1.5) \times 10^4$ cm/s, and the coefficient of proportionality was estimated as $(3-15) \times 10^4$ cm/s/T, which is larger than the previous reported values for Cu (6×10^3 cm/s/T). It was concluded that the obtained large value of this coefficient determined by the nanostructured tungsten surface.

The arc spot, appearing at two different thicknesses of the W nanostructured layer, half of which has a thin and the other half has a thick nanostructured layer, has been investigated [35]. The bright arc spot was observed with a fast framing camera, and the velocity of the arc spot was evaluated. The cross-sections of nanostructured layers were observed by scanning electron microscope (SEM), and the thickness of the eroded layer caused by arcing was obtained. At arc initiation, the target potential increased from -83 to -38 V, and the current increased from 0.27 to 4.2 A. The camera images showed that the width of the single arc trace was 5 μm in the narrow trail, and 10 μm in the wide trace. It was detected that the trail width changed sharply as the arc spot moves across the boundary of the two different parts of the nanostructured surface. An analysis of the surface of the specimen showed that the averaged trace width on the thick layer was 3.3 times wider than that on the thin layer. It was thought that the difference in the velocities resulted from the formation of the group spots.

It was also observed that the velocity of the arc spots significantly varied when the thickness of the nanostructured layer varied. The velocity of the arc spot on the thin layer was larger by factor of 6.7 than that on the thick layer. The thin nanostructured layer was estimated to be ~1 μm for the thin layer, and 5 μm for the thick layer. The erosion rate was measured as 2 mg/C for the thin layer, while 10 mg/C was for the thick layer. It can be noted that the observed difference of the velocities and the trace width for thin and thick nanostructures can be understood in frame of film cathode theory [36]. According to this theory (for Cu film deposited on a glass substrate), the spot velocity is larger for thinner film due to lower time of the evaporation of the film under the moving spots (Chap. 17).

A recent review of arcing on fiber form nanostructured W produced by the interaction between He plasma and W plate under fusion conditions was presented by Kajita et al. in [37]. He described experiments on tungsten substrates conducted in NAGDIS-II device. A ruby laser with the pulse of ~0.6 μs and of 60 J/cm^2 was used to ignite arcing on the surface biased at 90 V before arc ignition, and the current was limited by power supply in range of 5–20 A. Figure 21.5 shows framing camera observation at different time indicated in the frames.

At 0.325 s, the brightness of laser irradiation can be seen before development of the further spot dynamics. Figure 21.6 shows SEM micrographs of an arc trace formed on nanostructured tungsten surface. The arc spot (although random local motion) moves in average in retrograde direction due to influence of magnetic field. It can be seen that the traces detected under fusion conditions are similar to that observed in vacuum arc.

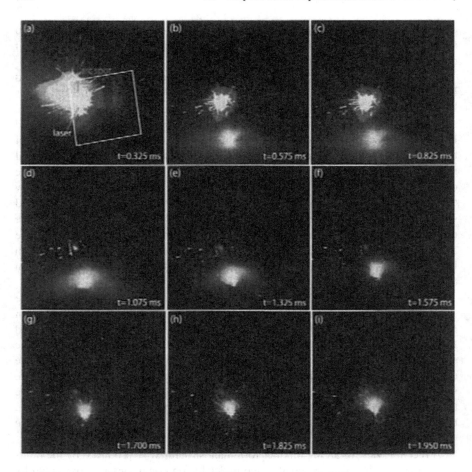

Fig. 21.5 Images of arc spots in an arc ignited from laser irradiation [37]

21.1.3 Film Cathode Spot in a Fusion Devices

As example of spot appearing on the film cathode can be considered the arcs observed in fusion devices by Laux et al. [38–40]. The measurements were provided on tokamak tiles (ASDEX Upgrate, IPP Garching) where some of graphite tiles have been coated by a W-film using plasma arc deposition and on laboratory system. The experiments have been carried out in vacuum to study arc behavior arise on graphite cathode covered by a tungsten film. It was originated from the usage of similar materials at first wall in fusion devices where a large number of arcs appeared at the contact with plasma. The standard measurements of current and voltage, optical observations of the arc spot in transverse magnetic field to determine the spot velocity were conducted.

Fig. 21.6 Spot trails on nanostructured W surface by different magnification obtained from micrographs of scanning electron microscope [37]

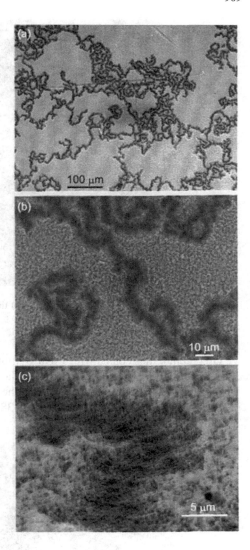

21.1.3.1 Tokamak Experiments

The experiments conducted at 0.3 and at 1 μm thin films [39, 40]. Numerous arc tracks have been identified at several locations (Fig. 21.7). One of the favored places are plates of the divertor baffle. Generally, they have a linear configuration due to the retrograde motion in the strong parallel to the surface magnetic field (~2.5 T). For both layer thicknesses (0.3 and 1 μm), the tracks consist of perforations of the layer. Arc traces on the 0.3 μm films consist of multiple chains with numerous divisions and dead ends, and they do not exhibit typical molten forms. These traces appeared as mosaic of adjoining polygonal holes.

Fig. 21.7 Arc tracks on a lower W-coated (1 lm) tile of the divertor baffle of ASDEX Upgrade. Figure taken from [40]. Used with permission

On graphite tiles coated with 1 μm W-film, the arc tracks consist of broader (10–30 μm) continues furrows intercepting the layer and their rims clearly exhibit molten structures. The traces also extend though the whole film until the graphite substrate (Fig. 21.8). The number of tracks found [38] on pure graphite was about 10–15%

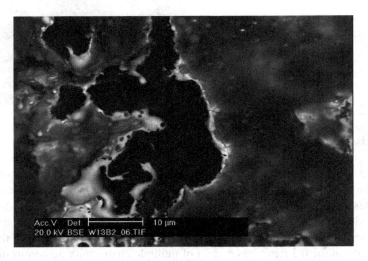

Fig. 21.8 Eroded region from an arc track at a baffle tile with 0.3 μm films demonstrated that the layer is removed completely and the graphite substrate is exposed. Figure taken from [40]. Used with permission

lower than on the tungsten coated tiles. The distribution of the trace lengths follows an exponential law, thereby defining a characteristic length that was 12 and 6.3 mm for the two different W-covered tiles, respectively, and 8.1 and 5.6 mm for the pure graphite tiles.

21.1.3.2 Laboratory Experiments on W Layer

The laboratory arcs were ignited between a Mo wire anode and a sample as cathode in a vessel with ultra-high vacuum. An external magnetic field of about 0.4 T was applied parallel to the sample surface. A capacitor bank with a time constant of about 300 μs allowed provide arc currents up to 15 A was used. An IMACON 486 camera was used to study the behavior of the cathode spot. Electron microscopy and surface analysis techniques have been used to specify the geometric structure of the tracks.

The behavior of arcs on 1 μm W-coated graphite vacuum arcs was obtained from a number of frame-series of a high-speed camera. The traces of arcing show a bundle of tracks starting at each ignition location, which extended into the retrograde direction orientated by the external field until the edge of the sample. The arc velocity of the retrograde motions of 2.9×10^4 cm/s was estimated. Using an electron microscope, numerous scattered holes observed, which perforating the W layer locally up to clean graphite [38]. The degree of perforation of the layer was found to vary along a bundle of laboratory arcs in the same manner as the arc current did, although the variation of the current was rather poor. The holes differ significantly from usual arc craters on bulk metals, although their edges show faint traces of melting. The diameters of the holes are typically 10–20 μm, and they are larger than those for craters on bulk W by about a factor of 2. The fraction of the area perforated was about 8% at a current of about 10 A. Through the perforation, substrate-carbon was available to all erosion processes. The holes differ significantly from usual arc crater on bulk metals, although their edges show faint traces of melting. The diameter of the holes were typically of about 10–20 μm.

21.2 Theoretical Study of Arcing Phenomena in a Fusion Devices

The specifics of an arc occurred in high electron temperature plasma–wall layer are studied. The description begins from first principle ideas of the arc processes, and then continued by involving the publications, where these ideas were developed. Finally, the last models including some mathematical approaches were discussed in order understand the plasma parameters in the vicinity of the spot and in not disturbed plasma of a fusion device.

21.2.1 First Ideas of Unipolar Current Continuity in an Arc

Robson and Thonemann in [41] were one of first to show that the floating potential of the electrical layer produced at the plasma–wall contact can be enough large (at large electron temperature) to support significant energy flux to the wall. The zero total current of electron and ion fluxes to the wall at high value of the electron temperature supported a large floating potential u_f. As result, a local hot area can be appeared from which a thermionic electron emission arises. The strong local emission of electrons from it will reduce the potential difference between the wall and plasma, i.e., from u_f to some potential u_c, which is the cathode potential drop when the current increased to a value for an arc development.

As noted by Robson and Thonemann, with increase of the electron emission more electrons can reach the wall against the retarding potential and, except in the immediate vicinity of the cathode spot, net electron current flows from the plasma to the wall. This current returns to the plasma through the arc spot, thereby satisfying the condition that the total current to the wall should be zero. Two equilibrium potentials are therefore possible. In the first case, the wall is at floating potential u_f with respect to the plasma ($u_f < 0$) and receives a uniform and equal current density of particles of each sign, and in the second case, the wall was at u_c ($u_c < 0$) and there is a circulating current through a cathode spot.

$$I_c = Aen_e \left(\frac{kT_e}{\sqrt{2\pi m_e}}\right)^{1/2} \left\{\exp\left(-\frac{eu_c}{kT_e}\right) - \exp\left(-\frac{eu_f}{kT_e}\right)\right\} \quad (21.1)$$

This arc is maintained by the thermal energy of the plasma electrons as a source, which converts the thermal energy to the electricity. As this arc requires only one electrode, it is the reason to call it a "unipolar" arc. This model assumed a constant electron density and a constant potential of the sheath over the whole plasma–wall boundary, contributing over a large area of the wall.

The further theoretical investigation of the unipolar arc mainly related to that arc occurred in fusion devices including Tokamak. As indicated by Robson and Thonemann, the main concept of the models consists in presence of relatively high electron temperature of fusion plasma bordered with the wall. As the emitted electrons influence not only on the electrical sheath, but also on the plasma temperature and density the further investigation considered different approaches to describe the plasma parameters in the unipolar arcs. The above-reviewed publications demonstrate the unipolar arcs with different form of spot.

Miley [9] used the above model [41] to describe their experiments of vacuum breakdown and unipolar arcs appearance in CTR devices. The model assumed that the plasma takes on a positive potential relative to a surrounding plate or electrode. Unipolar arcing then occurs due to surface imperfections; small hot spots occur on the wall. Pfeil and Griffiths discussed this fact [4], which has been established a correlation between the resistivity of inclusions in the metal and their effectiveness as unipolar arc-initiators. After unipolar arcing, the spots vigorously emit electrons

21.2 Theoretical Study of Arcing Phenomena in a Fusion Devices

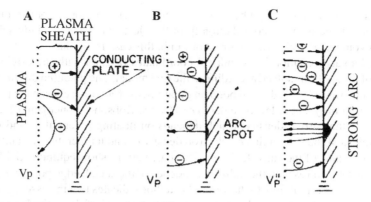

Fig. 21.9 **a** Plasma potential V_p at equalize ion and electron currents to the wall, **b** plasma potential V'_p at electron emission from the hot spot developing, **c** the potential V''_p has decreased to a point where the increased electron current over the broad surface area sustains a very intense "unipolar arc" [9]

that flow back into the plasma. As an equivalent to an outward flow of ions, this electron current allows the plasma potential to fall below its original value, and this in turn allows larger electron leakage currents from the plasma electrical load.

Figure 21.9 shows different stages of the potential development. Initially (A), a positive plasma potential V_p develops to equalize ion and electron currents. The floating potential $V_p = u_f$. When the hot spot developed, the balance of the currents violated due to presence of electron emission leading to plasma potential $V'_p < V_p$ shown in (B). Finally, in (C) the potential has decreased to a point $V''_p < V'_p$ (potential $V''_p = u_c$) at which the increased electron current over the broad surface area sustains the unipolar arc.

Thus, a balance is established where the excess plasma electron current just matches the return current from the spot. In essence, electrons are collected over a broad area of the wall, but they are returned from a small area corresponding to the arc spot. Considering for a copper arc the current densities of the order of 10^5–10^7 A/cm^2 and rate of mass loss 10^{-5} to 10^{-4} g/C, the author [9] showed that the unipolar arc formation at a plasma–wall interface lead to erosion rates of several orders of magnitude larger than for normal ion sputtering.

Wieckert [42] studied the unipolar arc by considering the mechanism of electromotive forces together with the arc spot mechanism at the contacting walls and the current flow. Two models considered for the external plasma, which exemplify with respect to the induced electromotive forces and with respect to the current flow. The typical spot parameters calculated previously by Ecker [43, 44] are accounted: a minimum current $I_0 = 13$ A that cannot exceed the value of $2I_0$, current density of $j_0 = 5 \times 10^5$ A/cm^2, and the cathode potential drop $u_c = 15$ V. These data are taken as prescribed values. The first model considers a situation, where the electromotive force is solely present in the sheath, and the current flow is unipolar. The second model studies the situation, where in addition to an electromotive force induced in the sheath, there is also a contribution from the electromotive force induced in the plasma volume. The electrical current origin in this case is bipolar.

The Laplace equation in cylindrical coordinates (r, z, φ) with the origin in the center of the spot was solved according to the first model. The potential distribution in the plasma was calculated using boundary conditions, for which chosen the potential as zero at infinity, but at left boundary $(r = 0)$, the Robson–Thonemann (21.1) was chosen for the current density, which depends on floating potential u_f and u_c. For given values of the spot radius r_s, electron density n_e and temperature T_e and for the prescribed value of the cathode drop u_c, the calculations show equipotential lines in a hydrogen plasma as well the radial dependence of the sheath edge potential and the corresponding current flow to the wall. The author indicates that the essential part of the potential drop occurs in the immediate neighborhood of the cathode spot. This solution of Laplace equation cannot be considered as surprise due to given small spot radius r_s.

The second approach modeled a bipolar current flow in a spherical geometry. Assumed that plasma is weakly ionized with a constant neutral particle density occurred between two half-spheres of the radii r_s and r, which connected by an ideal conductor. Charged particles are generated by electron impact proportional to n_e. The flux of the particles assumed move under the influence of mobility and diffusion. The calculated result shows that the plasma volume may contribute a significant part to the electromotive force, which even can dominate the floating potential.

Wieckert [42] concluded from his study that in any plasma contact with conducting walls, a large variety of arc discharges are possible, which can be supported by electromotive forces induced in the sheaths or in the volume or—for the mixed types—by a combination of external electromotive forces and induced ones. They also can show different current flow pattern in the sense that the anode flow returns to the same electrode (unipolar flow), respectively, goes to another one (bipolar flow) or to many other ones (multipolar flow).

21.2.2 Developed Mechanisms of Unipolar Arc Initiation

Hantzsche [45] developed an extended model that include the superposition of two completely different plasmas. The first is the tokamak plasma, and second is the cathode spot plasma of the arc. The wall layer potential u_c was determined taking in account the ion and electron currents to the wall as well the secondary electrons, which liberated from the wall in the case when an ion hits the wall. This potential differs from the floating potential u_f due to equality plasma ion and electron currents to the wall. The existence conditions of the unipolar arc were studied. As first condition, the relation between u_c and u_f should satisfied according to model [41]. The second decisive condition for the unipolar arc existence—apart from the potential condition, namely the entire current flowing through the cathode spot must return in the anode zone. To this end, the current distribution at the anode site was studied

21.2 Theoretical Study of Arcing Phenomena in a Fusion Devices

taking into account that the integral of such distribution should be equal to the spot current at the cathode site. Different mechanisms of electron emission from the wall were considered. A third condition arises from the requirement that the arc current in the plasma must be transported by electrical conduction and by plasma convection.

The model allows derivation of the radial dependencies of the current densities and voltages produced between plasma and wall and of the plasma parameters. The result indicated that the anode region collected the arc current is a ring-shaped area, which situated at distances from the arc cathode spot between 0.3 and 1 cm. The plasma density decreased with radius r according to r^{-2} to r^{-3} that followed from the current balance condition. The ignition problem and the effects of magnetic fields were discussed.

In another work, Hantzsche [46] developed a simple model of the electrical current rise in tokamaks when a plasma flux tube intersected by limiters. He showed that the limiter space charge sheaths act as additional resistances. Therefore, a part of voltage was dropped at this resistance limiting the current. Additional diminishing the current density was limited both by the plasma conductivity and by the ion saturation current of the sheaths. The analytical analysis show that the induced secondary electron emission has no influence at all on the obtained results.

Schwirzke and Taylor [47] used laser-produced plasma of short duration to model the onset and development of arcing on a stainless steel surface. It was considered a laser-produced Fe–Cr plasma with electron temperature of 100 eV, which expand rapidly from the focal spot on the target surface in the normal and in radial directions. Although no external voltage was applied, about 20,000 unipolar arc craters were observable on the stainless steel surface, which was exposed to the radially expanding plasma for the short time of a few hundred nanoseconds. The observed crater sizes were explained by estimations the rate of Fe ionization, the value of produced floating potential and an electric field at the surface in the electrical layer assuming equal to Debye length. As the provided analysis consist of number arbitrary assumptions and arbitrary parameters, the calculations indicate a qualitative result.

Schwirzke [48, 49], one of the first, modeled an onset of the unipolar arc considering the increase of plasma density due to ionization of neutrals produced by desorption of a gas and evaporation of metal atoms. The further ion bombardment leads to a locally increased surface temperature, which in turn furthermore increases the plasma density. It was assumed that the local increase of the plasma pressure provoked an electric field, which drives the arc current and promoted the electron return current flow to the surface. The increased electric field strength on surface protrusions will also increase the field emission and the ion flux from the plasma to these spots. Similar scenario of cathode spot development (increase of plasma density produced by cathode evaporation and ionization in time) was previously also considered for a conventional vacuum arc (for example [50]).

However, the model [48] does not enough justified mathematically. The model remains mainly qualitative due to arbitrary assumption at estimations of an additional neutral density and their rate of ionization. For example, the atom density calculated of about 2×10^{18} cm^{-3} in a Debye layer could not be obtained at Fe melting temperature of 1526 °C (see [48]). At such low temperature, the atom density is only about 10^{14} cm^{-3} at Fe saturated vapor pressure. The estimated additional ion density is lower than ion density in the background plasma of the fusion device.

Moreover, not clear what is the mechanism of the current continuity in the unipolar arc at the device wall. The mechanism was explained using the ambipolar electric field (gradient of plasma density) and, associated with this field, a decrease of the plasma potential in a ring-like area surrounding the higher plasma pressure above the arc spot in radial direction. Then, it is assumed that in a ring-like area, the potential was reduced and a more electrons than ions will return to the cathode surface, which closes the electron current loop of the unipolar arc. However, at the spot periphery, the current structure only controlled by the background plasma conditions producing the floating potential independent from the spot plasma. In addition, it should be taken in account that the ambipolar electric field, which is the gradient of electron pressure in direction perpendicular to the surface, is reduced due to plasma acceleration, which not analyzed (Chap. 18).

Considering the relationship between current density due to radial gradient of electron pressure and current density limited in the space charge layer, Schwirzke [49] concluded that the surface heating by ions at high current density of the unipolar arc could provide the "explosive" formation of cathode spot plasma. To explain the explosive process, the authors assumed that the ion energy is deposited only within a few atomic layers at a time instead of an entire whisker volume. Since the neutral contaminants are only loosely bound to the surface, the onset of breakdown by this mechanism requires much less current than the Joule heating mechanism. However, no any explosion mechanism due to action of surface heat source by ion flux has been reported, which more preferable than due to evaporation or sputtering. In addition, no numerical data of ion current density was discussed to analyze this model for fusion device.

Hothker et al. [51] demonstrated experimentally that for metal–plasma contact there a possibility of purely plasma-induced unipolar arcs, and the existence of thresholds for the electron temperature and for the arc current which sustain plasma-induced unipolar arcs. The results are consistent with a Robson and Thonemann theoretical model calculating the requested floating potential using minimum electron temperatures for unipolar arcing from the minimum voltage for bipolar arcs [52, 53]. These results indicate that unipolar arcs are maintained by the potential drop of the sheath, which is determined by the electron temperature, and that the cathode processes for unipolar arcs and vacuum arcs are similar. This conclusion can be seen also considering the crater from photograph in Fig. 21.10 [51], which is similar to that obtained in case of vacuum arcs [53].

Different models of unipolar arcs as an impurity source in tokamaks were widely studied by Nedospasov and his group. As in the previously mentioned papers, these models also take in account the presence of relatively high electron temperature of the plasma bordered with the wall. In order to understand the phenomena in fusion devises the presence of a magnetic field was taken in account.

Some conditions required for arcing were considered by Nedospasov and Petrov [54] by modeling the wall as a smooth surface, which is parallel to magnetic field

21.2 Theoretical Study of Arcing Phenomena in a Fusion Devices

Fig. 21.10 Erosion crater on a stainless steel. Typical sizes is in range of 3 to 15 μm. Figure was corrected from [51]. Used with permission

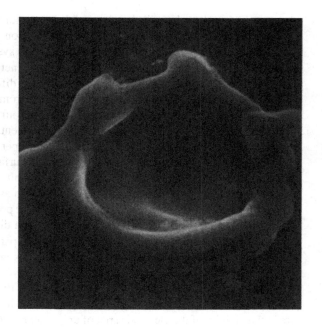

lines. In this case, at large distance from the wall, the plasma diffusion is ambipolar and the electric field vanishes. However, at a distance of about an ion mean Larmor radius, the fastest ions begin to reach the wall and a plasma potential will build up. At small distances from the wall, less than the electron Larmor radius, the electric field will retard electrons and so it directed toward the wall. At this distance, quasineutrality no longer exists. Under these conditions, the cold electrons emitted from the arc spot have the possibility of intersecting the magnetic field lines and mixing with the hot ones coming from the plasma bulk. Thus, arcing is possible even under ideal coincidence of the wall surface with a magnetic surface. In modern tokamaks, magnetic field lines periodically hit the limiters. The more rapid leakage of electrons along magnetic field lines results in a positive potential of the plasma relative to the surrounding walls and so the possibility of arc formation emerges. The model allows calculation of the ratio J/eI as dependence of $T_0/\Delta\varphi$, where $\Delta\varphi$ is the potential drop of an arc (20–40 V), J is the arc current, I is the ambipolar leakage of the ions and electrons to the limiters. The calculation show that the ratio J/eI increased sharply up to about 8 at $T_0/\Delta\varphi = 16$ V and then passed to saturation value of 12 at $T_0/\Delta\varphi = 40$ V. So, the arc current is larger than the ion leakage. An additional estimate shows that plasma pollution due to arcing can be a great danger to tokamak operation.

The further their theoretical model of unipolar arc was developed in [55] to study the discharge at the wall as an impurity source in tokamaks. The plasma parameters in the limiter shadow was used as $T_e = T_i = 10\text{--}100$ eV, the plasma density $n = 10^{11}\text{--}10^{13}$ cm^{-3}, which are approximately constant along the magnetic lines. The plasma density is assumed to be exponentially decreasing with the radius r. The hot plasma entering the tube with low potential supports a stationary arc current. The ion influx to the tube from the central tokamak region was due to the diffusivity, which determined by the Bohm diffusivity. The used system of equations included the balance equations for the particle fluxes, electron energy, and ion energy balance, provided an equality between diffusivity and thermal diffusivity and Ohm law.

The calculations show that the ratio J/eI varied from 1 to 3.5 when the plasma temperature characterized plasma entering the tube from the central tokamak region increased from 25 to 150 eV. The model shows agreement with the observed impurity quantities. The effective erosion can be 0.2–0.3 atoms per ion. Commonly, 10^{13}–10^{14} impurity atoms per 1 cm^2 are deposited on the wall surface during a discharge. The arc erosion also occurred due to sputtering by the limiter material ions arrived into plasma from the arc spot. They gain energy from the plasma and then hit the limiter. The ions gain the additional energy passing through the potential drop of the electrical layer at the limiter. The arcs at the initial period of a discharge supply the plasma with the primary impurity ions that subsequently take part in self-sputtering. The arc becomes impossible if the radiation due to impurities and recycling strongly cools the plasma in the limiter shadow.

The plasma energy balance taking into account the plasma conductivity and plasma heat conduction was analyzed to calculate distribution of the plasma parameters in the vicinity of the electrode spot of unipolar arc [56]. The Ohm law accounted the gradient of electrical potential and gradient of the plasma pressure. Calculated results indicate dependencies of the minimal plasma temperature and minimal plasma density, at which the given arc current can be supported. Considering the plasma energy balance of the unipolar arc, it was shown grows of the current density with distance from the arc spot in contrast to that result obtained by Wieckert [42]. Another result indicated that the plasma potential far from the spot should be larger than the potential drop in the spot in order to eject the electrons from the cold plasma with impurity near the spot.

The unipolar arc arises due to interaction of a hot plasma with a metal surface when an intense evaporation initiated at some heated local area. The local thermal mechanism of arc spot initiation in vacuum was developed early by Lafferty [57] for anode and then for cathode by Beilis and Lyubimov [50]. This thermal mechanism was considered also to study a discharge contraction at divertor plate of tokamaks [58]. A potential drop in a space charge layer at the metal–plasma boundary was calculated from balance between ion and electron currents taking into account secondary and thermionic electron emission. The plate energy balance was used to calculate plate temperature T. The balance includes the ion energy flux to the plate due to ion energy acquire in the layer and its temperature. The energy loss was due to heat-surface exchange by Newton law and by Stephan–Boltzmann radiation. The critical heat flux as dependence on plate temperature and critical plasma density as dependence on electron temperature were calculated for different arbitrary combination between spattering coefficients and given electron temperatures. These critical values used in discussion to interpreting how the discharge contraction can develop from slow increase of average heat flux and electron temperature. As it can be understood, the developed model of discharge contraction mainly characterized by the undisturbed plasma and plate geometry in tokamaks. The arc initiation due to local grows of the plate vaporization was not considered.

Tokar [59] studied the mechanism of the discharge transition to the detached plasma state based on the edge plasma cooling instability. The edge plasma in tokamaks was detached from a limiter due to change of its parameters. In the state of "detached plasma" (DP), the discharge power was transferred to the wall in a form of the neutral and radiation of light impurity The charge exchange of the light impurities on the hydrogen atoms arising as a result of the plasma recombination on the linear surface take in account. The study was based on an analysis of particle flux and energy balances.

Tokar et al. [60] further developed the mechanism of enhancement of the heat flux from the plasma to the wall with the increase in the thermionic emission of electrons (see [58]), by investigating an instability of small perturbations of homogeneous stationary states. According to the observation, the hot spots were generally localized on those areas of the tokamak walls, which are oriented perpendicularly to the magnetic field **B**. Taking into account the fact, an analytical approach was developed considering the energy transfer. It was studied the heat transport in the SOL plasma (see Rozhansky et al. [61]) the non-stationary equation governing the heat conduction in a wall bounding a plasma and the stationary heat regime along the y-axis, which is orthogonal to the xz plane (the z-axis is directed along the magnetic field vector **B**) of the wall–plasma system. The dependences of the plasma temperature and the wall surface temperature on the heat flux density in stationary were calculated. From the analysis of the homogeneous plasma, the authors concluded that states might develop, which are non-homogeneous along the y-direction. They can be both non-stationary processes and as steady–state auto-solitons. The stability of plasma–wall system was studied by standard method, using linearization of equations for heat transfer in the plasma and in the wall for small perturbations of the variables near their stationary values.

This relatively complex auto-soliton theory takes in account the characteristic parameters of hot spots and a comparison of its analytical results with the measured data from JET and TFTR. The authors concluded that the formation of hot spots and the development of carbon blooms in tokamaks could be explained by contraction of the thermal contact area between the plasma and wall surfaces, which is due to intensification of the thermionic emission of electrons. It should be noted that this conclusion follows from assumption that the plasma potential with respect to the wall determined by the condition of zero electric current onto its surface. However, the assumption of zero current can be problematic at the case of spot ignition and development.

Philips et al. [62] also showed that a unipolar arc formation with hot spot is due to thermal electron emission. This evidence was provided developing a similar calculating model for carbon limiters. The observed development of the hot spot was explained by an increase of the local power loading due to the breakdown of the sheath potential by thermal emission of electrons from the carbon surface.

The sharply increase of the carbon release from the surface associated with the appearance of the hot spot. This increase explained by carbon thermal sublimation. The intense sublimation leads to the surface cooling resulting in establishment of the quasiequilibrium temperature at about 2700 °C.

21.2.3 The Role of Adjacent Plasma and Surface Relief in Spot Development of a Unipolar Arc

The distribution function for ions near the plates is necessary for determining heat and particle fluxes in the designed fusion reactors. The scale length of transition zone between a quasineutral plasma and the Debye space charge zone near the plates is an important issue. To study the ion distribution function, Igitkhanov [63] developed one-dimensional kinetic model, in which a plasma flux was given, which value formed from infinity (at the symmetry plane) to a divertor/limiter plate. The plate was assumed as completely absorbing. A collisional term is represented by the BGK approximation. Igitkhanov considered two cases. In first, a kinetic equation considered for the ion distribution function with a collision term represented a low-energy ion acceleration under electrostatic potential. Assuming that the electron current very small in comparison with random electron motion and that the electron density distributed in form of Boltzmann's distribution with constant electron temperature T_e, potential distribution was obtained using the Poisson equation. The second case considers the ion velocity significantly larger than their thermal velocity. In this case, a change in the ion density under electric field is mainly related to acceleration of the ions. A change in the density of such ions was estimated in the hydrodynamic approximation. The ion density, their velocity, and mean energy were found from the solutions of the integral equations at a given potential.

The numerical solution obtained in the constant relaxation time approximation indicated a deviation from the Maxwellian distribution when approaching the plate. Near the plate, the distribution becomes almost one-velocity, with a minimal velocity corresponding to a potential difference in the layer. The amount of such ions somewhat exceeds an equilibrium value, meanwhile the amount of higher-velocity ions becomes smaller. A noticeable deviation from Maxwellian distribution was observed only for the particle with velocities exceeding 2–3 of thermal velocities.

Igitkhanov [64] analyzed the existing models of the unipolar arcs. He concluded that the published mechanisms cannot completely explain the observed experimentally dependence of arc ignition and stationary burning from the treatment of surface. The author reveals another possible mechanism for a transversal current diffusion related with electron scattering on the surface roughness. His model assumed that under radial electric field formed by the beam of emitted electrons from the spot, the electrons drift along the electrode surface. Colliding with the surface roughness, they lose an initial drift velocity, and, as a result of each collision, shift along the plate surface by the cycloid height, in the direction of a radial electric field. The

21.2 Theoretical Study of Arcing Phenomena in a Fusion Devices

current provided by collisions of the electrons with the surface roughness flows in a thin, near-wall layer, an electron Larmor radius thick. This current determined by the so-called near-wall conductivity. These electrons enter a weak radial electric field region, reach the adjacent magnetic field lines, and reduce the potential on them. A magnitude of the surface current depends on the nature of electron scattering by the surface roughness, and therefore, a quantitatively uncertainty of such scattering in the general case makes difficulties to obtain an expression for this current. Some estimations show that the considered mechanism takes place, when the roughness size h > 5 μm. For dimensions of a microrelief at a level 1 μm and less, the roughness size is lower than the Debye radius. Therefore, the Debye layer screens the roughness and the surface not support the above mechanism. The author noted that the developed mechanism of spot initiation predict the way of avoiding the impurity generation due to thermal unipolar arcs by careful treatment of the wall surface, to prevent the second type (see Chap. 7) of spot occurred.

Igitkhanov and Bazylev [65] developed model of unipolar arc initiation using an analysis the specifics of the electric field and the role of the corrugated surface. The modeled relief initially consists of the typical diameter of brushes 0.5–1.0 cm; depth of the gaps between the brushes of 1 cm, width of the gap, 0.5–1 mm. In order to avoid sharp corners, each brush element was rounded with a radius R that varies from 0.5 to 1 mm. The surface roughness is much less than 1 mm. However, an evolution of the corrugated surface was observed under recurrent impact of a heat flux density of 1.6 MJ/m^2 with pulse exposition time of 0.5 μs. After exposure to number of pulses, the process of surface heating, melting, and molten material resolidifies in between pulsed eventually leads to the appearance of W-wedge-like shape of a brush elements. The electric field profile in the vicinity of corrugated W-wedges was determined by solving the 2D Poisson's equation. The numerical analysis show that the electric field of ~3.5 × 10^7 V/cm at the tips vicinity is sufficient for triggering the field emission current on the level of ~1 A/cm^2.

Furthermore, authors modeled similar to that for vacuum spot initiation. Namely, the emitted electrons accelerate within the sheath potential and acquire a kinetic energy of ~100 eV. At that energy, the tungsten atoms are ionized and the ions accelerate toward the tips of the wedge. The tungsten ion bombardment leads to heating of some area, augmenting the electron thermal emission and vaporization. The initial electron field emission breaks the sheath potential and eventually drops itself. The high temperature at the spot allows the arc current sustained by increased thermal electron emission and evaporation of tungsten atoms from the hot spot. Some estimations allow the authors to conclude that calculated current density exceeds the minimum arc current (determined in [66]) needed for sustaining an unipolar arc and the associated surface heating by ions can result in the formation of hot spots and a strong tungsten impurity ejection.

It should be noted that the electrode processes are mainly studied qualitative. For some calculations, arbitrary input parameters were used, which can be significantly varied in the considered conditions and that can be obtained by self-consistent study.

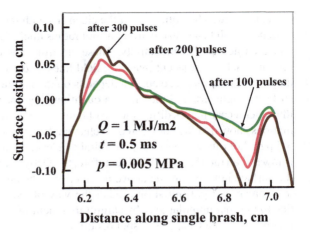

Fig. 21.11 View of a single W-brush after melting and displacement of the molten layer for various numbers of pulses [65]

Another problem is the calculated very large electrical field of ~3.5 × 10^7 V/cm. The value cause big doubt taking into account presented sizes of a single W-brush after melting and displacement of the molten layer due to a number of exposure pulses [67] (see Fig. 21.11).

Chodura [68] discussed basic problems of modeling the particle and energy flux flowing from a magnetically confined plasma core from the plasma boundary to the wall. An analysis of previous publications was provided, which considered the stationary, kinetic solutions under the assumptions of a perpendicular magnetic field and without collisions among plasma particles. The modeling of the ion transport including collisions also was analyzed. This analysis stimulated the authors to consider another phenomenon, which arise when a magnetic field B strikes the target under an oblique angle of incidence. In this case, ion flux does not directly parallel to the magnetic field in front of the target. As result, the ion space charge produces a new layer length scale into the potential profile near the wall. The following model was discussed. The electric field, perpendicular to the target, tries to pull the ions away from their path along a magnetic field line. On the other hand, the electrons motion is along magnetic field due to their small mass and gyro-radii. Thus, a double structure of the electric sheath ahead of the target arises. One of them is a magnetic sheath, which thickness determined by the combination action of electric and Lorentz forces on ions. The other is the electrostatic sheath of Debye thickness. A mathematical model was developed. The calculated fluxes of the particle mass and energy in the layer were discussed using analytically derived expressions.

Gielen and Schram [69] developed a three-dimensional approach modeling of an axisymmetric arc, in which both sheath and plasma phenomena were taken in account. An isothermal plasma was considered assuming that both the background and cathode plasma have the same electron temperature of 3 eV. The sheath thickness assumed of the order of the Debye length is small compared to the size of the cathode spot. Therefore, the description of the plasma above the sheath, the electric potential of the sheath was used as boundary conditions. The sheath potential was

21.2 Theoretical Study of Arcing Phenomena in a Fusion Devices

calculated using (21.1) for given perpendicular component of the current density. This potential was kept constant in further calculations taking into account that the sheath potential drop in vacuum arc for Cu cathode is of 15 V (measured data for vacuum arc). The plasma considered as only singly ionized atoms assuming the high degree of ionization to neglect the electron-neutral friction in the electron momentum equation. This equation was presented as the Ohm law, which determined the current density j by different forces, including the electron pressure gradient, the electric field, and plasma resistivity and by the magnetic field. The electrons carry the current almost completely. The magnetic field was as a sum of external and a self-generated one due to electric currents in the plasma components. The electron density and temperature profiles were assumed known and constant during iteration procedure. The calculation described a quasistationary expansion of a plasma from the cathode spot on an electrode surface. For this quasistationary expansion, the current density, electric field and magnetic field distributions were calculated. The most important force term in the electron momentum equation is the pressure gradient term. The obtained result indicated that the electrical current distributed over one electrode, which serves both as cathode and as anode approving the discharge as a unipolar arc.

Rozhanskij et al. [70] noted that unipolar arcs can exist only when there is a return current flowing from the plasma to the surface close an electrical current circuit. This current produced by the difference between the electron and ion fluxes to the surface due to intense electron emission from a hot spot. The authors concerned with understand of a physical mechanism responsible for the transport of plasma polarization across a magnetic field over relatively large distances from the surface. To this end, they studied the potential distribution near the emitting surface for a small emission current to determine the area collecting the return current. This case is related to the situation before arc formation, when the plasma density in front of the emitting surface is almost unperturbed. The model considered nonuniform fully ionized plasma with a density $n(z)$ restricted by a conductive surface. Electrons and ions are created in pairs by the source $I(z)$ and are driven toward the wall by the pressure gradient. Electron and ion temperatures are considered as constant values. The mathematical formulation consists of continuity equations of particle fluxes, momentum equations for ions and for electrons including terms of the electron pressure gradient, the electric field, plasma resistivity, and by the magnetic field. As for boundary condition, the potential inside the plasma at the sheath edge was determined by the current, flowing from the plasma to the wall. The potential profile was determined by the classical conductivity along the magnetic field and by the perpendicular ion viscosity across the field. The potential distribution was calculated both analytically and numerically. The main advantage of the work is to establish a criterion of unipolar arc formation. According to the criterion—when the reduced potential in front of the emitting surface is larger than the cathode voltage drop of the arc and the return current exceeds the minimal current typical for the arc, the ignition of the arc is possible. It was shown that the perpendicular ion viscosity determines the characteristic transverse scale of the return current region. This scale is much larger than the emitting channel width, while the longitudinal scale depends on the classical plasma conductivity.

21.2.4 Explosive Models of Unipolar Arcs in Fusion Devices

Mesyats in [71] proposed another application of the local explosive model and phenomena of explosive electron emission. He indicated that these processes must play a part in unipolar arcs. This claim based on the observations showed identical behavior of the cathode spots in vacuum discharges and in unipolar arcs: the presence of microcraters on the surface, the retrograde motion of the spots in a magnetic field, the readiness of their formation on surface impurities, etc. It was stated that when a cathode spot appears under the hot plasma, then during operation of it cell, conditions arise for the formation of new dense near-cathode plasma, which interacts with the surface irregularities resulting in initiation of new emission center. Thus, in contrary to EC model (Chap. 15), for which the EC functioning is initiated by explosion of the protrusion, the new interpretation of EC considers the thermal interaction of a protrusion with a hot plasma resulting in following its explosion during the thermal process.

The "explosive" formation of cathode spot plasma of the unipolar arc was further developed in a series works based on the model of EC for spot initiation under hot plasma of fusion devices development up to explosion of the EC. Barengolts et al. [72] developed a calculating schema based on significantly modifying Uimanov's model [73] (Chap. 15). They calculated the increase of protrusion temperature and current distribution as dependencies with time for a wide range of plasma densities (10^{14}–10^{20} cm^{-3}) and electron temperatures (0.1–10^4 eV), which were free varied as parameters. The thermal regime of a microprotrusion contacting with plasma was studied taking into account the energy carried by plasma ions and electrons, as well as Ohmic heating and heat removal due to heat conduction.

The calculations were conducted for a protrusion of 0.8 μm base diameter placed at center of flat surface of 4 μm size. The results show the temperature T growth and increase of the current density j with time in nanoscale. The final stage of heating and increase of T and j was assumed at a moment, when the values of temperature and current density reached 10^4 K and 10^8 A/cm^2, respectively, after which should be a subsequent "explosion" of the microprotrusion. It was found that such an explosive type of heating of the microprotrusion is determined by flux of energy transferred to the surface at certain critical parameters of the plasma. Such threshold value of energy flux was calculated as 200 MW/cm^2. This energy density corresponds to Ohmic overheating of the microprotrusion heated by a plasma with a density exceeding (3–5) × 10^{18} cm^{-3} during approximately 10 ns. The presence of the protrusion on the surface ensures the necessary density of released energy in a microvolume.

The EC explosion model [74] was considered to describe the experimental results of Kajita et al. [31, 34] in which an arc was ignited by a laser pulse radiation of the nanostructured W plate immersed in He plasma. This structure consisted of nanosize protrusions. Their thermal regime and overheating conditions of such protrusions were analyzed. The laser radiation was absorbed in the nanostructured W layer leading to it intense vaporization.

It was assumed that ionization of the evaporated atoms gives rise to W plasma of density ~10^{19} cm^{-3} with T_e ~ 6 eV. For these parameter range, a large electric field in the Debye layer was estimated providing current density greater than 10^8 A/cm^2 through the protrusion and as result it overheating and explosion. Note, that the laser plasma velocity is also significantly high (~10^6 cm/s) and therefore the produced plasma will be shifted to 10–100 μm already during 1–10 ns. As result, after finishing of the laser pulse the plasma density can be significantly decreased at the immediate vicinity of the surface. Thus, when the spot will be initiated during the laser pulse, then it could develop without any explosion phenomena.

Similar study of explosion phenomenon under interaction of a W-fuzz with relatively hot plasma was conducted in work of [75]. The heating of W-fuzz nanowires, the sheath–wall interaction, plasma–wall interaction, and the emitted electron beam interaction with plasma were studied as function on heat flux density varied as input parameter in range of 10^4–10^8 W/cm^2. In essence again, the vaporization model was used considering Mackeown and electron emission equations, etc. At some certain input parameters of the plasma, the electron emission beam heated the W plasma enhancing both the electric field and emission current density up to about tens of megavolts per centimeter and an emission current density to j_e ~ 100 MA/cm^2.

As result, at such large value of j_e, the volume was strong heated by Joule energy and that leads to electric explosion of the W-fuzz nanowire layers. The calculated critical condition for the explosion is similarly to that obtained previously in [71]. It should be noted that, as the hot plasma interacted with very small size of a subject like W-fuzz nanowires, it is obvious that at some critical point, a strong overheating can be induced, but the question is how the W-fuzz nanowire is destroyed.

Note, the used above approach considered the thermal regime of a microprotrusion contacting with plasma which studied taking into account the energy carried by plasma ions and electrons, as well as Ohmic heating and heat removal due to heat conduction is similar to the GDM model described in Chap. 16 and applied to study a protrusion cathode in Chap. 17. However as the used model is not closed the obtained unlimited increase of the temperature and current density finally was interpreted as explosion of the protrusion.

21.2.5 Briefly Description of the Mechanism for Arcing at Film Cathode

According to the observations (Sect. 21.1.3 film spot), the spot after initiation rapidly move out the initial point. In essence, the spot move on the tungsten film (also between arising holes). Taking into account that the electrical conductivity and heat conductivity of the graphite substrate are significantly decreased with the temperature, the energy loss to this substrate can be neglected. Therefore, the mechanism of spot motion and its operation will be determined by the tungsten film evaporation. Using the Beilis and Lyubimov model [36] for film cathode, the erosion rate can be determined from equation $G = \gamma df\ \delta vs$ (Chap. 17). Taking the track width of $\delta = 1$ μm (the hole size is much larger because the spot was fixed), the erosion rate $G \approx 6 \times 10^{-3}$ g/s for above-mentioned spot velocity (2.9×10^4 cm/s) with tungsten thickness (1 μm) in laboratory experiment (Sect. 21.1.3.2). According to the experiment, [35] the erosion rate was measured as 2 mg/C for the thin layer, which is comparable with above calculated value at spot current about 3 A. Therefore, in order to detail description of such spot behavior and approval the model, it is necessary additionally to measure the current through the tracks and its width, spot velocity and the erosion rate for given film thickness, like to that measurements conducted by Kesaev [76].

21.3 Summary

In general, at the contact wall–plasma, the charged negative dies to faster electron motion. A usual electrical sheath arises due to ambipolar plasma particle diffusion when the particle energy is relatively low. Another case is such contact with hot plasma in fusion devices. The hot electrons bring significant energy flux to the wall casing an electron emission into the plasma from strong hot spot appeared on irregularities at the not ideally smooth surface. This fact reduces the negative charge of the wall. As an equivalent to an outward flow of ions, this electron current allows the plasma potential to fall below its original value, and this in turn allows larger electron leakage currents from the plasma. There the current flow is in the sheath region named as unipolar and a circulating current through a hot spot described by Robson and Thonenmann [41]. This is not the case when an external electrical load determined the current between two electrodes.

According to Wieckert's study in any plasma contact with conducting walls, a large variety of arc discharges are possible, which can be supported by electromotive forces induced not only in the sheaths but also in the plasma volume. The plasma volume resistivity can play as an internal electrical load. The current flow in this case is bipolar. The discharge in fusion devices can be considered similar to arcs in appeared at the wall with a plasma flow across a magnetic field in channels. In essence, it can be discharge between wall as one electrode and the plasma as the second electrode. In tokamaks is the case when the potential drop between wall and plasma deviated from floating value. So, the current can be not closed at the same wall as an electrode.

Considering the research results of the discharge appeared in the fusion devices, it is obvious that there are two main issues: (1) understanding the mechanism of arc initiation under specifics of hot plasma with energetic electrons and (2) explaining the spot parameters arising in such conditions. As for first point, the above-reviewed theories extensive studied the unipolar arc initiation taking in account different device conditions showing the plasma parameters determined the arc development. A progress of such study was demonstrated by Hantzsche, Schwirzke's group and by Nedospasov's theoretical group. The second issue concerned to the spot operation is

21.3 Summary

like to the mechanism of usual spot behavior in vacuum arcs. It is follows from the observations of the spot dynamics in the developed unipolar arcs.

As for vacuum arcs, the experiment show that the unipolar arc appearance significantly depends on the surface state, its geometry and on the applied external magnetic field strength. The SEM micrograph showed relatively wide arc tracks even higher than 50 μm, but having an inner structure, which consist of sub-arc tracks of ~10 μm. Thus, a group spot formation was demonstrated. A several arc spots form a group and move together. The arc spot randomly moves on the nanostructured surface; however, in the presence of transverse magnetic field, the spot move in retrograde direction. The velocity can reached large values 30–150 m/s, when this motion occurred on the nanostructured tungsten surface or on deposited thin films.

Therefore, the parameters of a formed cathode spot in a unipolar arc can be studied using the theory developed for a cathode spot operated in a vacuum arc considering the specifics of surface devices, such as material, films, nanostructured objects, and others. Namely, the similar behavior of the both cathode spots in vacuum and in fusion devices, stimulate the application of explosion approach (EC) to the unipolar arc study. The example of application of film cathode theory is demonstrated above. However, the complete descriptions of the spot behavior from it initiation and spot development can be provided using the gasdynamic (Chap. 16) and kinetic (Chap. 17) theories. As for the initial conditions, the resulting plasma parameters can be served obtained from the researches of the unipolar arc initiation. Some such attempt was conducted relatively recently by Igitkhanov and Bazylev, by Gielen and Schram as well by Rozhanskij (see above).

In contrast to vacuum arc cathode spot, description of the unipolar spot behavior can meet distinctive features such as spot sudden extinction and new initiation during stepwise spot motion and especially under transverse magnetic field in the fusion devices. These specifics of unipolar arc should be taken in account in some cases by studying the craters and tracks.

Another important question is how the unipolar arc can be suppression. Levchenko et al. [77] studied some possible mechanism of arcing suppression. It was shown that the external high-frequency electrical field could effectively suppress the unipolar arcing. The sharp cutoff was found independently for both breakdown rate and the eroded mass. The quantity of unipolar arcs per second (throughout referred to as breakdown rate) influences directly the electrode surface erosion. The dependence of this value for two different pressures is shown in Fig. 21.12. It can be seen that the breakdown rate decreases considerably with the frequency approaching 10 MHz of high-frequency generator (HFG), and approaches zero with the HFG frequency approaching 15 MHz. This phenomenon, which is very important for practical applications, was called "HF cutoff."

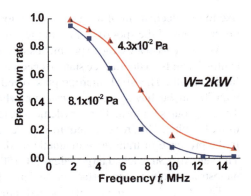

Fig. 21.12 Breakdown rate as a function of the HFG frequency f with background pressure as a parameter [77]

References

1. Arcing phenomena in fusion devises. In *Proceedings of Workshop*, Langley (Ed.), US Department of Energy, Knoxville, Tennessee, April 5–6 (1979).
2. Robson, A. E., & Hancox, R. (1959). Choice of materials and problems of design of heavy-current toroidal discharge tubes. *Proceedings of the IEE—Part A: Power Engineering, 106*(2S), 47–55.
3. Craston, J. L., Hancox, R., Robson, A. E., Kaufmann, S., Miles, A. T., Ware, A. A., & Wesson, J. A. (1958, September). The role of materials m controlled thermonuclear research (F,S). In *Proceedings of the Second United Nations International Conference on the peaceful uses of atomic energy*, Geneva, P/34 (Vol. 32, p. 414).
4. Pfeil, P. C. L., & Griffiths, L. R. (1959). The effect of inclusions on the arcing behaviour of metals. *Journal of Nuclear Materials, 1*(3), 244–248.
5. Hancox, R. (1960). Importance of insulating inclusions in arc initiation. *British Journal of Applied Physics, 11*(10), 468–471.
6. Maskrey, J. T., & Dugdale, R. A. (1962). Arc initiation on heated molybdenum exposed to a toroidal hydrogen discharge contaminated with impurity gases. *Journal of Nuclear Materials, 7*(2), 197–204.
7. Maskrey, J. T., & Dugdale, R. A. (1963). The importance of contamination in arc initiation on stainless steel exposed to a toroidal discharge. *Journal of Nuclear Materials, 10*(3), 233–242.
8. Panayotou, N. F., Tien, J. K., & Gross, R. A. (1976). Damage of a candidate CTR material in a high energy fluence deuterium plasma. *Journal of Nuclear Materials, 63*, 137–150.
9. Miley, G. H. (1976). Surface effects related to voltage breakdown in CTR devices. *Journal of Nuclear Materials, 63*, 331–336.
10. McCracken, G. M., Dearnaley, G., Gill, R. D., Hugill, J., Paul, J. W. M., Powell, B. A., et al. (1978). Time resolved metal impurity concentrations in the dite tokamak using RBS analysis. *Journal of Nuclear Materials, 76*(77), 431–436.
11. Goodall, D. H. J., Conlon, T. W., Sofield, C., & McCracken, G. M. (1978). Investigations of arcing in the DITE tokamak. *Journal of Nuclear Materials, 76*(77), 492–498.
12. McCracken, G. M., & Goodall, D. H. J. (1978). The role of arcing in producing metal impurities in tokamaks. *Nuclear Fusion, 18*(4), 537–543.
13. McCracken, G. M. (1980). A review of the experimental evidence for arcing and sputtering in tokamaks. *Journal of Nuclear Materials, 93–94*, Part 1, 3–16.
14. Mioduszewski, P., Clausing, R. E., & Heatherly, L. (1979). Observations of arcing in the ISX tokamak. *Journal of Nuclear Materials, 85*(86), 963–966.
15. Wu, T., Nouailletas, R., & Lefèvre, L. P. (2016). Plasma q-profile control in tokamaks using a damping assignment passivity based approach. *Control Engineering Practice, 54*, 34–45.

16. Goodall, D. H. J. (1980). Arcing studies in the DITE tokamak using a time resolved arc detector. *Journal of Nuclear Materials, 93–94*, 154–160.
17. Zykova, N. M., Beilis, I. I., & Kurakina, T. C., Private communication.
18. Zykova, N. M., Nedospasov, A. V., & Petrov, V. G. (1983). Unipolar arcs. *Teplofiz. Vysokikh Temp., 21*(4), 778–787.
19. Bogomolov, L. M., Zykova, N. M., & Kabanov, V. N. (1983). Electric arc discharge in the Tokamak TV-1. *Journal of Nuclear Materials, 162–164*, 443–447.
20. Nedospasov, A. V., Petrov, V. G., & Zykova, N. M. (1985). Unipolar arcs. *IEEE Transactions on Plasma Science*, PS-13 N5, 253–256.
21. Stampa, A., & Kruger, H. (1983). Simulation experiments on unipolar arcs. *Journal of Physics. D. Applied Physics, 16*, 2135–2144.
22. Herrmann, A., Balden, M., Laux, M., Krieger, K., Müller, H. W., Pugno, R., et al. (2009). Arcing in ASDEX Upgrade with a tungsten first wall. *Journal of Nuclear Materials, 390–391*, 747–750.
23. Rohde, V., Endstrasser, N., Toussaint, U. V., Balden, M., Lunt, T., Neu, R., et al. (2011). Tungsten erosion by arcs in ASDEX upgrade. *Journal of Nuclear Materials, 415*, S46–S50.
24. Doerner, R. P., Baldwin, M. J., & Stangeby, P. C. (2011). An equilibrium model for tungsten fuzz in an eroding plasma environment. *Nuclear Fusion, 51*, 043001.
25. Kajita, S., Takamura, S., Ohno, N., Nishijima, D., Iwakiri, H., & Yoshida, N. (2007). Sub-ms laser pulse irradiation on tungsten target damaged by exposure to helium plasma. *Nuclear Fusion, 47*(9), 1358–1366.
26. Sakaguchi, W., Kajita, S., Ohno, N., & Takagi, M. (2009). In situ reflectivity of tungsten mirrors under helium plasma exposure. *Journal of Nuclear Materials, 390–391*, 1149–1152.
27. Kajita, S., Yoshida, N., Yoshihara, R., Ohno, N., & Yamagiwa, M. (2011). TEM observation of the growth process of helium nanobubbles on tungsten: Nanostructure formation mechanism. *Journal of Nuclear Materials, 418*, 152–158.
28. Behrisch, R. (1979). Surface erosion from plasma materials interaction. *Journal of Nuclear Materials, 85–86*, 1047–1061.
29. Ye, M. Y., Ohno, N., & Takamura, S. (1997). Study of hot tungsten emissive plate in high heat flux plasma on NAGDIS-I. *Journal of Nuclear Materials, 241–243*, 12431247.
30. Yang, Q., You, Y.-W., Liu, L., Fan, H., Ni, W., Liu, D. et al. (2015). Nanostructured fuzz growth on tungsten under low-energy and high-flux He irradiation-Scientific Reports. *Five*, N10959.
31. Kajita, S., Takakura, S., & Ohio, N. (2009). Prompt ignition of a unipolar arc on helium irradiated tungsten. *Nuclear Fusion, 49*(N3), 032002.
32. Kajita, S., Takamura, S., & Ohno, N. (2011). Motion of unipolar arc spots ignited on a nanostructured tungsten surface. *Plasma Physics and Controlled Fusion, 53*, 074002.
33. Kajita, S., Ohno, N., Yoshida, N., Yoshihara, R., & Takamura, S. (2012). Arcing on tungsten subjected to helium and transients: ignition conditions and erosion rates. *Plasma Physics and Controlled Fusion, 54*, 035009.
34. Kajita, S., Ohno, N., Takamura, S., & Yo, T. (2009). Direct observation of cathode spot grouping using nanostructured electrode. *Physics Letters, A 373*, 4273–4277.
35. Hwangbo, D., Kajita, S., Barengolts, S. A., Tsventoukh, M. M., & Ohno, N. (2014). Transition in velocity and grouping of arc spot on different nanostructured tungsten electrodes. *Results in Physics, 4*, 33–39.
36. Beilis, I. I., & Lyubimov, G. A. (1976). Theory of the arc spot on a film cathode. *Soviet Physics—Technical Physics, 21*(6), 698–703.
37. Kajita, S. (2018). Ignition and behavior of arc spots under fusion relevant condition. In *Proceedings of 28th International Symposium on Discharges and Electrical Insulation in Vacuum, Germany*, Greifswald, September (pp. 1–6).
38. Laux, M., Schneider, W., Juttner, B., Linding, S., Mayer, M., Balden, M., et al. (2004). Modification of tungsten layers by arcing. *PSI (Plasma Surface Interaction)-16, Germany*, P3-28.
39. Laux, M., Schneider, W., Juttner, B., Balden, M., Linding, S., Beilis, I. I., & Jakov, B. (2005). Ignition and burning of vacuum arcs on tungsten layer. *IEEE Transactions on Plasma Science, 33*(N5), 1470–1475.

40. Laux, M., Schneider, W., Jüttner, B., Lindig, S., Mayer, M., Balden, M., et al. (2005). Modification of tungsten layers by arcing. *Journal of Nuclear Materials, 337–339*, 1019–1023.
41. Robson, A. E., & Thonemann, P. C. (1959). An Arc maintained on an Isolated Metal Plate exposed to a Plasma. *Proceedings of the Physical Society, 73*(3), 508–512.
42. Wieckert, C. (1978). Plasma induced arcs. *Journal of Nuclear Materials, 76*(77), 499–503.
43. Ecker, G. (1971). Zur theorie des vakuumbogens. *Beitrage aus der Plasmaphysik, 11*(5), 405–415.
44. Ecker, G. (1976). The vacuum arc cathode. A phenomenon of many aspects. *IEEE Transactions on Plasma Science, PS-4*(N4), 218–227.
45. Hantzsche, E. (1980). Unipolarbogen. *Beitrage Plasmaphysics, 20*(5), 329–342.
46. Hantzsche, E. (1988). Currents in intersected tokamak flux tubes. *Contributions to Plasma Physics, 28*(4–5), 411–416.
47. Schwirzke, F., & Taylor, R. J. (1980). Surface damage by sheath effects and unipolar arcs. *Journal of Nuclear Materials, 93–94*, Part 2, 780–784.
48. Schwirzke, F. (1984). Unipolar arc model. *Journal of Nuclear Materials, 128 &129*, 609–612.
49. Schwirzke, F., Hallal, M. P., Jr., & Maruyama, X. K. (1993). Onset of breakdown and formation of cathode spots. *IEEE Transactions on Plasma Science, 21*(5), 410–415.
50. Beilis, I. I., & Lyubimov, G. A. (1976). Signature" determination of current density at the cathode spot in an arc. *Soviet Physics—Technical Physics, 21*, 1280–1282.
51. Hothker, K., Bieger, W., Hartwig, H., Hintz, E., & Koizlik, K. (1980). Plasma-induced arcs in an RE-discharge. *Journal of Nuclear Materials, 93 & 94*, 785–790.
52. Reece, M. P. (1963). The vacuum switch. *Proceedings of IEE, 110*, 793–811.
53. Kesaev, I. G. (1964). Laws governing the cathode drop and threshold currents in an arc discharge on pure metals. *Soviet Physics. Technical Physics, 9*, 1146–1154.
54. Nedospasov, A. V., & Petrov, V. G. (1978). Model of the unipolar arc on a tokamak wall. *Journal of Nuclear Materials, 76 & 77*, 490–491.
55. Nedospasov, A. V., & Petrov, V. G. (1980). Unipolar arcs as impurity source in Tokamaks. *Journal of Nuclear Materials, 93 & 94*, 775–779.
56. Petrov, V. G. (1982). Gometry of current close in an unipolar arc considering plasma energy balance. *High Temperature, 20*(2), 220–224.
57. Lafferty, M. (1966). Triggered vacuum gaps. *Proceedings of the IEEE, 54*(1), 23–32.
58. Nedospasov, A. V., & Petrov, V. G. (1983). Thermal contraction during heat exchange between a hot plasma and metal surface. *Soviet Physics. Doklady, 28*(N3), 293–295.
59. Tokar, M. Z. (1988). Tokamak edge plasma transition to the state with detachment from limiter. *Contributions Plasma Physics, 28*(4–5), 355–358.
60. Tokar, M. Z., Nedospasov, A. V., & Yarochkin, A. V. (1992). *Nuclear Fusion, 32*(1), 15–23.
61. Rozhansky, V., Kaveeva, E., Senichenkov, I., & Vekshina, E. (2018). Structure of the classical scrape-off layer (SOL) of a tokamak. *Plasma Physics. Control Fusion, 60*(N3), 035001.
62. Philips, V., Summ, U., Tokar, M. Z., Unterberg, B., Pospieszczyk, A., & Schweer, B. (1993). Evidence of hot spot formation on carbon limiters due to thermal electron emission. *Nuclear Fusion, 33*(6), 953–961.
63. Igitkhanov, Yu L. (1988). Calculation nonequilibrium distribution function ions. *Contributions Plasma Physics, 28*(4–5), 333–339.
64. Igitkhanov, Yu L. (1988). On the mechanism of stationary burn of unipolar micro arcs in the Scrape-Off tokamak plasma. *Contributions Plasma Physics, 28*(4–5), 421–425.
65. Igitkhanov, Yu L, & Bazylev, B. (2011). Electric field and hot spots formation on divertor plates. *Journal of Modern Physics Open A, 2*(3), 131–135.
66. Granovski, V. (1971). *The electric current in a gases*. Moscow: Nauka. (in Russian).
67. Bazylev, B., Janeschitz, G., Landman, I., & Pestchanyi, S. (2005). Erosion of tungsten armor after multiple intense transient events in ITER. *Journal of Nuclear Materials, 337–339*, 766–770.
68. Chodura, R. (1988). Basic problems in edge plasma modelling. *Contributions Plasma Physics, 28*(4–5), 303–312.

References

69. Gielen, H. J. G., & Schram, D. C. (1990). Unipolar arc model. *IEEE Transactions on Plasma Science, 18*(1), 127–133.
70. Rozhanskij, V. A., Ushakov, A. A., & Voskobojnikov, S. P. (1996). Electric field near an emitting surface and unipolar arc formation. *Nuclear Fusion, 36*(2), 191–198.
71. Mesyats, G. A. (1984). Microexplosion on a cathode aroused by plasma-metal interaction. *Journal of Nuclear Materials, 128&129,* 618–621.
72. Barengolts, S. A., Mesyats, G. A., & Tsventoukh, M. M. (2008). Initiation of ecton processes by interaction of a plasma with a microprotrusion on a metal surface. *Soviet. Physics JETP, 107*(6), 1039–1048.
73. Uimanov, I. V. (2003). A two-dimensional nonstationary model of the initiation of an explosive center beneath the plasma of a vacuum arc cathode spot. *IEEE Transactions on Plasma Science, 31*(5), 822–826.
74. Barengolts, S. A., Mesyats, G. A., & Tsventoukh, M. M. (2010). The ecton mechanism of unipolar arcing in magnetic confinement fusion devices. *Nuclear Fusion, 50,* 125004.
75. Barengolts, S. A., Mesyats, G. A., & Tsventoukh, M. M. (2011). Explosive electron emission ignition at the "W-Fuzz" surface under plasma Power Load. *IEEE Transactions on Plasma Science, 39*(9), 1900–1904.
76. Kesaev, I. G. (1964). *Cathode processes in the mercury arc.* NY: Consultants Bureau.
77. Levchenko, I. G., Voloshko, A. U., Keidar, M., & Beilis, I. I. (2003). Unipolar arc behavior in high frequency fields. *IEEE Transactions on Plasma Science, 31*(1), 137–141.

Chapter 22
Vacuum Arc Plasma Sources. Thin Film Deposition

In the last decades metal plasmas deposition techniques have progressed and have been utilized in different applications. Ion bombardment is an important tool in producing new materials and structures. Thin metallic films are extensively used in microelectronics. Among the important applications is the formation of diffusion barriers and conducting metal layers in high aspect ratio vias and trenches in integrated circuits [1, 2].

Several metallization techniques using ion bombardment are used in industry and described in the literature, including ionized physical vapor deposition (IPVD), chemical vapor deposition (CVD), electron beam (EB) evaporation, DC magnetron sputtering and others. Vacuum arc deposition (VAD) is an IPVD technique, which is extensively used in the tool industry. Vacuum arc-generated plasma has different important technological applications. The vacuum arc is a well-known plasma source for producing metallic coatings, which has a long history [3–5]. The most common application is the deposition of metallic and ceramic thin films, utilizing the cathodic plasma jet.

Some of applications require high-quality metallic plasma, free from liquid droplets or solid particles (collectively known as macroparticles—MPs) which are also generated from cathode spots of the vacuum arc [6]. Such plasma can be produced using magnetic filtered cathodic vacuum arc plasma jets, by anode evaporation in hot anodic vacuum arcs [7]. During the past decade, two VAD variants, the hot refractory anode vacuum arc (HRAVA) [8] and vacuum arc with black body assembly (VABBA), have been investigated at Tel Aviv University (TAU). In this chapter, different techniques including HRAVA and VABBA plasma source measurements are characterized. The plasma sources using vacuum arcs with refractory anodes and producing by the source metallic film deposition characteristics are compared with other deposition techniques and the results are summarized.

22.1 Brief Overview of the Deposition Techniques

In this section, we consider the typical systems in order to overview main feature and the deposition rate possible to compare of different deposition systems.

22.1.1 Advanced Techniques Used Extensive for Thin Film Deposition

PVD techniques include evaporation, magnetron sputtering (MS) (including DC, r.f., and pulsed), ion beam sputtering, and different types of arc plasma for coating applications in the metal, biomedical, optical and electrical industries. Various ionized PVD (IPVD) sputtering techniques (DC and impulse) have appeared that can achieve a high degree of ionization of the sputtered atoms and recently reviewed by Helmersson et al. [9]. A novel biasing technique for magnetron sputtering that controls the ion flux and energy to the insulated samples was considered by Barnet [10].

Self-sputtering magnetron deposition (SSMD) was discussed for DC [11, 12] and pulsed magnetrons [13]. This technique can work without gas and can reach the deposition rate of about 2 μm/min at a distance of 10 cm with 80 W/cm^2 for Cu film. It has been demonstrated that self-sputtering in vacuum can deliver extraordinarily high metal–ion current [14]. Richter et al. [15] used a circular magnetron source with graphite target placed 50 mm above the substrate. For the typical discharge power of 900 W and argon pressure 0.4 Pa the deposition rate was obtained as 0.7 nm/s. Horwat and Anders [16] showed that the dense *High power impulse magnetron sputtering (HIPIMS)* plasma occupies a near-target zone, and therefore, the sputtered atoms have possibility before they arrive to the substrate, to be an appreciable probability of ionization. Copper plasma with significant directed velocity 8.8 eV and low temperature 0.6 eV has been obtained using self-sputtering far above the runaway threshold. In other work Anders et al. [17] considered the HIPIMPS studying instabilities that are essential in providing the physical, non-classical mechanism for electron transport across magnetic field lines. The specific feature of this system that the target supplies a large fraction of the to-be-ionized neutrals. It was established that the plasma is concentrated in ionization zones that travel azimuthally in the same direction as the electrons drift.

Wang et al. [18] studied a performance of Zr–Ti–Ni thin film deposited with or without nitrogen atmosphere by magnetron sputtering as a diffusion barrier between silicon and copper layers. Compared to the sample without nitrogen purging, nitrogen atoms in the deposit were found to increase the failure temperature from 700 to 800 °C by retarding crystallization of the deposit and diffusion of copper. The failure mechanism of the barrier was also investigated.

Recently, Wahl et al. [19] studied the widely used thermal evaporation of Al, the industrially important high deposition rate processes sputtering and electron

22.1 Brief Overview of the Deposition Techniques

beam evaporation for aluminum electrodes. The authors examined the influence of different deposition methods on the solar cell performance. The results show that industrial deposition techniques as sputtering or e-beam evaporation are possible alternatives for applying front contact layers for solar cells and modules. Also studied a combination of these Al deposition methods in order to further enhancements of the layers.

Electron Beam (EB) evaporation is extensively used to produce thin films [20] and for the deposition of metals may reach deposition rate of about 1000 μm/min and for compounds of about 60 μm/min [21]. Illés et al. [22] used evaporator with the electron beam—physical vapor deposition (EB-PVD) method to the Sn thin film deposition onto 1.5-mm-thick Cu and ceramic substrates. The applied cathode heating current was 100 mA with 7 kV acceleration voltage. A high vacuum (10^{-3} Pa) was used, and the evaporation time was 25 min. The spontaneous whisker formation studied, which can cause reliability issues in microelectronic appliances. The longest filament of such whiskers reached 300 μm in length. Hershcovitch et al. [23, 24] developed devices and techniques for coating of various surface contours or very long small aperture pipes by technique of physical vapor deposition.

CVD techniques are important for formation of thin films on complex structures and selective deposition used in mechanical, optical, electronic and other fields [25–27]. Typically, the deposition rate is low—about hundreds Å/min at optimal temperatures of 200–300 C [28]. Plasma-assisted CVD (PA CVD) is used when low deposition temperatures (<200 °C) are required [29]. Babayan et al. [30] reported using plasma-enhanced CVD about deposition rates of 0.1 μm/min.

Electroplating is suitable for filling complexes structures. *Electroless* is a method of deposition from solution where the electrochemical reaction consists of two reactions in which one generates electrons, while the second neutralizes the metal ions with the electrons from the first reaction [31, 32]. The electroless copper deposition is a low-cost, highly selective process and widely used for depositing Cu films with low resistivity ~2 μΩ cm and deposition rate of about 0.1 μm/min. Richter et al. [15] using ion plating technique working with pressure of about 0.2 Pa indicate deposition of hydrocarbon films with rate in range of 0.5–1 nm/s.

22.1.2 Vacuum Arc Deposition (VAD)

VAD utilizes plasma jets with a very high degree of ionization of atoms evaporated from the electrodes and with a high velocity (and kinetic energy) [6, 33]. Let us summarize the main vacuum arc characteristics from above chapters to represent this arc as a plasma source for metallic deposition. The arc plasma is attached to the cathode by visible "cathode spots" [6, 34, 35]. The spots are very small (10–100 μm) and move randomly with velocities of $10-10^4$ cm/s and current density of 10^5-10^7 A/cm^2. The ions in the energetic plasma jet have a multiplicity of charge states [36,

37]. The ion current fraction in the jet is 0.07–0.1, and the heavy particles in the jet are mostly fully ionized. Different types of cathode spot and of mechanism operation on different cathode metals [38] were detected including a transition spot [39], spot grouping and motion (including in transverse magnetic fields) [40] and plasma jet generation [41, 42].

While the cathode spot generates energetic plasma, the arc also produces MPs [43], which must be separated from the plasma jet to deposit high quality films. Different methods to reduce or eliminate the MPs have been proposed in the last decades [6, 44–46].

Rother et al. [47] reported about DC cathodic arc evaporation process for graphite. A computer controlling of the electric potential of a cathode shield and the external magnetic fields distribution to fix the arc spot position were utilized. This allowed reaching a uniformity of the eroded cathode surface. The arc ignited at pressure 10^{-3} Pa with arc current 125 A. The distance between the cathode and the substrate was 30 cm. The film was formed by the carbon ions at *bias voltage* of about −600 V. The deposition rate was ~0.1 μm/min, but MPs were not prevented.

Filtered vacuum arc deposition. The first magnetically collimated arc plasma transport in a straight duct was described by Aksenov et al. [48] who obtained an output ion current of 17 A with an arc current of 190 A. The overarching idea is to remove large macroparticles (MP) from the plasma jet. Aksenov et al. [49] were also first to use a quarter-torus filter to separate the metallic plasma from MPs generated by cathode spots. Various types of MP filters have been proposed, and their advantages and disadvantages were reviewed [6, 7, 50, 51]. Falabella [52] indicated that his knee sources deliver over 3 A of ions and a maximum deposition rate of 0,8 μm/min of TiN was obtained for an arc current of 100 A. Other filtering systems include the in-plane 'S' and twisted 'S' filters [53].The ion transmission of in twisted 'S' systems was only 6%. Trench filling by Ta and Cu using a magnetic filtering system with a pulsed [54] and plasma gun was demonstrated [55]. It was theoretically predicted [56] and then was experimentally observed [57] the plasma remains consisting with a MPs that reached the coated substrate [58, 59].

A filter using simple shielding of the substrate by an intermediate deflection plate [60] filtered the droplets diameter over 5 μm and reduce the number of large sizes of MPs (1–5 μm) by an order of magnitude, while the deposition rate is low, ~1 μm/h of TiN. Pulsed high-current (~1 kA) vacuum arc and obstacle plates preventing the MPs from reaching the substrate were described by Ryabchikov et al. [61]. A rectangular dual-filtered-arc source deposited uniform films with a deposition rate of 1–2 μm/h [62, 63]. A large area coating system using a rectangular duct with two 45 bends in opposite directions produced an in-line deposition rate of 100 nm/min [64, 65]. Therefore, alternative to conventional filtering techniques was proposed and considered in the next sections.

Siemroth and Schuke [54] reviewed the problems of metallization in microelectronics and the needs for a higher activated PVD and CVD techniques. Filtering vacuum arc evaporation was described as a process having the potential, to solve some actual deposition problems, because the droplets can be eliminated. Noted that disadvantage of filtering is the reduction of the deposition rate due to plasma

22.1 Brief Overview of the Deposition Techniques

losses in the filter. Schultrich [66] developed modified vacuum arc using pulsed arcs to control the spot motion in a suitable manner instead of modifying the spots themselves working with pulse currents in the kiloampere range.

The laser arc method [67] is restricted the propagation of the cathode spots by short-pulse times below 100 μs in the immediate neighborhood of the ignition point. This method allows displacing the area of arc-induced target evaporation in a very controlled manner. On the other hand, the high-current arc method can be used longer pulse times of about 1 ms. Thus, with the high-current arc and the related laser arc method, effective and intense ion sources are available. They yield ion currents up to 100 A and more and thus meet the demands of activated high-rate coating. Pulse frequencies up to 300 Hz have been successfully tested corresponding to deposition rates up to 300 nm/s.

A practically achievable deposition rate for a 300 mm wafer is in the order of 60 nm/min. Anders et al. [68] used a dual-cathode arc plasma filtered source combined with a computer-controlled bias. The method has been applied to the synthesis of metal-doped diamond-like carbon films, where the bias was applied and adjusted when the carbon plasma was condensing and the substrate was at ground when the metal was incorporated. This species-selective bias method could be extended to multiple material plasma sources and complex materials

High-current arc (HCA) deposition current pulses with peak of 4–5 kA and filtered high-current pulsed arc deposition (f-HCA) were described by Siemroth et al. [69], where deposited copper, pure aluminum and an aluminum alloy. The deposition rate filling SiO was reported of 10 nm/s (0.06 μm/min) for Cu and Al and ~2 μm for ultra thin carbon coating [70].

Hot cathode vacuum arc (HCVA) is a discharge with a hot volatile cathode in vacuum [71] in which the cathode is heated by the arc current itself. Kajioka [72] used an electron beam (EB) to heat the cathode and termed this arc as an "arc-like discharge" to deposit titanium nitride films. Goedicke et al. considered spotless arc deposition (SAD) that was a cathodic arc with EB heating of a rod [73, 74]. The deposition rate was 50–1000 nm/s (0.05–1 μm/s) using an arc with EB evaporator of 300 kW. Chayahara et al. [75] demonstrated a metal plasma source using an "arc-like discharge" in order to develop the MP-free plasma-based ion implantation (PBII) without a magnetic filter.

A different vacuum arc plasma source is based on the *hot anode vacuum arc (HAVA)* [76, 77] where metallic plasma is produced by the evaporation of anode material. In this arc mode, the arc current heats the anode until its temperature reaches sufficiently high values so that the anode surface becomes an intensive source of vapor. The HAVA plasma is diffusely attached to the hot electrode surface and metallic plasma has not contaminated the droplets as in conventional cathode spot or anode spot vacuum arcs. Therefore, cleaner coatings could be obtained. The HAVA occurs when the anode is more volatile and smaller than the cathode [78–80]. The arc with 20–40 A and Al, Ti, and Cr anodes initially operates by cathode evaporation, but in steady state mostly in anode vapor reported by Erich and co-authors [81–84]. The ion energy is relatively low 5 eV. The deposition rate was in range 1–10 nm/s depending on the arc current and anode material. The degree of atom ionization was 5-10%.

22.2 Arc Mode with Refractory Anode. Physical Phenomena

The cathode material of HAVA plasma sources is usually selected to be less volatile than that of the anode. In contrast, in considered below arcs the anode should of refractory material.

22.2.1 Hot Refractory Anode Vacuum Arc (HRAVA)

HRAVA. This arc mode has been first studied [85, 86] using copper cathode graphite anode and then developed as a plasma source for metallic film deposition [87]. The HRAVA used cylindrical electrodes with planar faced onto other surfaces (Fig. 22.1). In this discharge, material evaporated from the cathode is transported to the refractory anode.

As the arc developed, two phenomena were observed: (i) anode heating, (ii) condensation of the cathode material at initial stage when the anode was cold and the arc operated in multi cathode spot mode, and (iii) re-evaporation of the cathode material deposited on the hot anode from on the developed stage of the arc (Fig. 22.2). As the HRAVA evolves, the anode significantly heated (~2000 K [85]), and an intensely radiating plasma plume is created at the anode surface, expanding with time in the axial and radial direction. Spectroscopically, the plume radiates lines of the cathode material [86].

The HRAVA plasma is sufficiently hot and dense to result in the evaporation of the macroparticles produced by the cathode spots during their passage through the interelectrode gap. Thus, the radially expanding HRAVA plasma (shown schematically in Fig. 22.3 and during the arc burning in Fig. 22.4) has the potential to be used as macroparticle-free plasma source in technological applications, and in particular in thin film deposition [88]. To optimize the usability of such source evolution of the HRAVA plasma parameters, i.e., electron temperature, particle density, particle fluxes, and heat flux for different arc currents was determined [89].

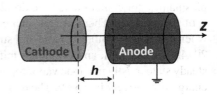

Fig. 22.1 Electrode configuration in a HRAVA systems

22.2 Arc Mode with Refractory Anode. Physical Phenomena

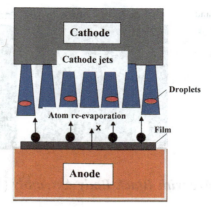

Fig. 22.2 Mechanism of condensation and re-evaporation of cathode material from hot refractory anode

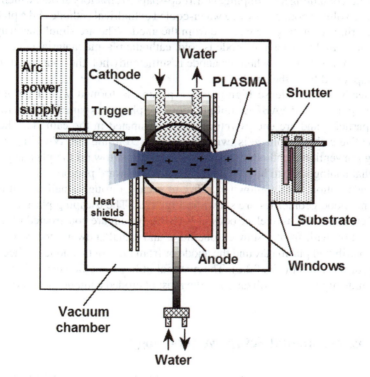

Fig. 22.3 Schematic presentation of the radially expanding HRAVA plasma, which has the potential to be used as macroparticle-free plasma source

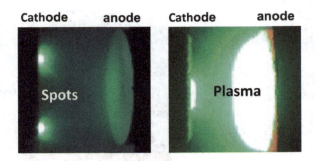

Fig. 22.4 Cathode spots and plasma jets in the initial stage and anode plasma plume at the development HRAVA stage

22.2.2 Vacuum Arc with Black Body Assembly (VABBA)

VABBA. In this arc mode [90], the material eroded from the cathode spots as plasma jets containing MPs impinges on a cup-shaped refractory anode, which almost encloses a volume bounded by it, a water-cooled cylindrical cathode and a planar BN insulator ring separating the cathode from the anode. The arc simultaneously heats the anode. Initially, when the anode is cold, cathodic plasma material condenses on the anode. With arc time, when the anode is sufficiently hot, the condensed material is re-evaporated from the anode.

In steady state, a dense high-pressure plasma is formed within the enclosed volume. In the closed VABBA configuration, the plasma expansion is not free. Part of the particles and arc energy returned and dissipated in the cathode. Therefore, the effective cathode voltage is twice that measured in the HRAVA [91] or in free burning conventional cathodic arcs. The VABBA acts toward the plasma and MPs somewhat analogously to how a "black body" acts toward photons, i.e., not permitting condensation, while allowing a flux to escape through small apertures [90]. Different anode geometries are shown in Fig. 22.5. The escaping plasma forms an expanding flow. The used electrodes in VABBA system are constructed of different configuration with frontal shower, one-hole, and radial shower anodes in order to obtain distributed plasma around the anode or from the frontal anode surface as well in form of plasma jet (Fig. 22.6). The material utilization efficiency using VABBA can be quite high, due to utilization of the most of eroded cathode included the MPs.

22.3 Experimental Setup. Methodology

Experimental apparatus and electrodes assembly include two cylindrical stainless steel chambers: (a) 400 mm length, 160 mm diameter (Fig. 22.7) and (b) 530 mm length and 400 mm diameter (Fig. 22.8). Both chambers were water-cooled using U-shaped channels welded onto the chamber walls. The chamber was pumped by diffusion pump to $(0.6–2) \times 10^{-5}$ Torr. During the arc, the pressure in the chamber

22.3 Experimental Setup. Methodology

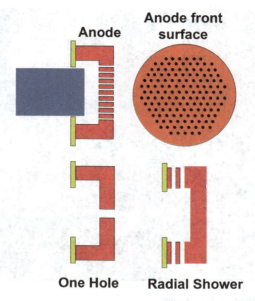

Fig. 22.5 Electrode configuration in a VABBA systems with frontal shower, one-hole, and radial shower anodes

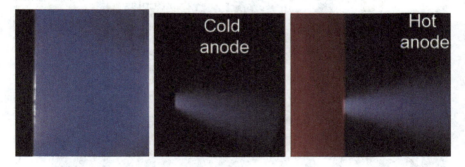

Fig. 22.6 Plasma plume generated in VABBA from frontal shower, one-hole cold and hot anode

was about (0.05–1) mTorr. The discharge was ignited between a water-cooled cathode and anodes of different geometry.

Two cylindrical Mo radiation shields, 60 and 70 mm diameter, surrounded the anode to reduce radiative heat losses. The anode was supported by a water-cooled thin tungsten rod, which also connected it to the electrical circuit. A shutter was used to determine the deposition time window. A substrate holder kept the substrate in the mid-plane of the arc, facing normally the plasma flux. The anode was produced from different refractory metals. Arc current was supplied by a welding power supply (Miller XMT-400 CC/CV). The anode was grounded, and the chamber was floating. The arc currents were $I_{arc} = $ 150–340 A and the arc duration up to 200 s. The most

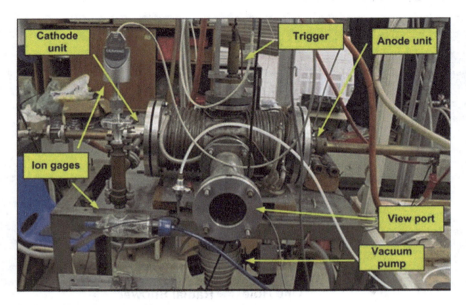

Fig. 22.7 Vacuum arc deposition system

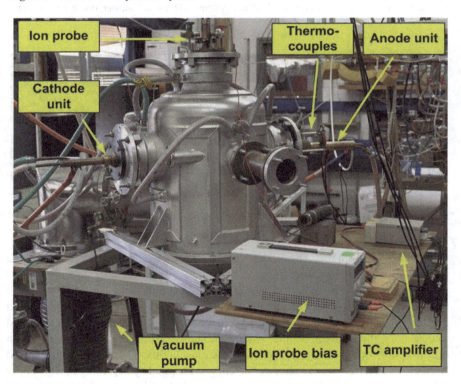

Fig. 22.8 Vacuum arc deposition system

22.3 Experimental Setup. Methodology

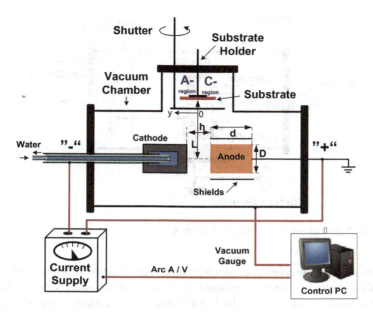

Fig. 22.9 Schematic of the HRAVA configuration

used interelectrode gap h was 10 mm. Some specific parameters will be indicated in the results of the experiments described below [85–87].

A few causes for the cooling system using (i) the measurements conducted in the vacuum chambers with the power of arc operation about 3–4 kW and (ii) the heat flux by the plasma and by radiation heated the camber walls causing intensive outgassing and corrupting the vacuum. The used a cooling system was supported with refrigerator and closing water circulation that allow firstly, to save the total amount of the water in each experiment and secondly, to thermostating the electrode and chamber walls and to choosing the necessary electrode heat regime. The electrode configuration with electrical supply is shown schematically for HRAVA in Fig. 22.9 and for VABBA in Fig. 22.10.

Substrate. Stainless steel and glass sheet substrates were used to collect depositions from the arc [87]. The substrates were mounted at distances L from the electrode axis on a holder, so that they faced the plasma flux emanating from the interelectrode gap. The holder was placed in a substrate chamber, which was separated from the arc chamber by a shutter. The substrate was placed at two positions: (i) along the mid-plane of the electrodes normally facing to the plasma flux and (ii) shifted relative to the mid-plane of the electrodes on the holder plane, in direction parallel to the electrode axis toward cathode. The cathode sometimes was surrounded by a thin shield (Fig. 22.9) protruding 2 mm below the cathode surface, to block any MP flux originating at cathode spots on the side of the cathode from reaching the substrate. The shutter shown in Fig. 22.1 was used to determine the exposition time window. The substrates were cleaned with cotton wool dampened with alcohol.

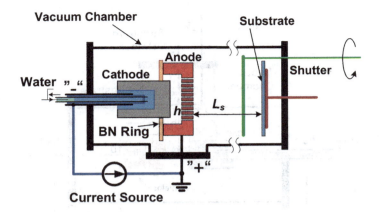

Fig. 22.10 Schematic of the VABBA configuration

Methodology for coating characterization. As observed previously [8, 87, 92], the cathode shield position determined the location of a boundary between regions on the substrate with low and high MP contamination, designated as the anodic (A) and cathodic (C) regions, respectively (Fig. 22.9). Figure 22.11 illustrated the detail configuration of the different deposited regions together with electrode-shields geometry. When the cathode shield was not used, these regions were depended by the cathode size and the gap. The C-region faced the cathode and was exposed to MPs emitted from the cathode spots. The A-region faced the anode and collected anode plasma that was almost free of MPs.

The MPs and deposited films in the A-region, in a strip 2 mm adjacent to the A–C boundary at distance between 3 and 5 mm, were characterized. The MPs were counted using an optical microscope equipped with a digital camera and a video display [90]. The MPs which could be visually resolved (larger than approximately 1–3 μm) were manually counted on the video display in regions with a width of 100 μm (measured with a transparent scale placed on the substrate), and a length of

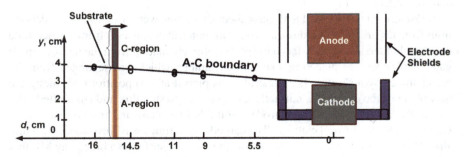

Fig. 22.11 Diagram of the substrate placement along y-axis and the definition of the A and C regions (HRAVA system). The boundary line indicates the placement of A and C regions at different distance from the electrode axis

10 mm, which was determined by the field of view of the microscope/camera/display combination. The substrate was then moved to examine an adjacent region using a micrometer-controlled microscope stage translator, until the coated width of the substrate (10 mm) was examined. The number of MPs in each 100 μm × 10 mm region was summed and the sum divided by the total area of the all examined number of strips, and reported as the MP density. The anode surface contamination was also observed visually, and the microscope studied the substrate deposit after different arc time duration beginning from 10 s.

Cathode erosion rate. The cathode erosion rate was determined by weighing the cathode before and after arcing with a known arc current and arc duration.

Deposition rate. Coatings were deposited on substrates with various arc currents. Using the shutter, the substrates were exposed to the arc during a controlled time window. The coating thickness was measured by profilometry. The deposition rate (V_{dep}) was determined by dividing the film thickness by the exposure time. Sometimes the deposition rate was obtained by weighing the substrate (with an accuracy 10^{-5} g) before and after deposition.

The above methodology mainly was used to obtain the described below experimental results. Some additional details will be explained using different conditions of film deposition. However, before let us consider the theoretical approach and calculated results in order to understand the mechanism of vacuum arc operation with refractory anode as a new mode of arc discharge with relatively high current.

22.4 Theory and Mechanism of an Arc with Refractory Anode. Mathematical Description

The HRAVA plasma phenomena are very complex, and a model of the mutual processes was extensively described by Beilis et al. [93]. The theoretical description includes non-steady and nonlinear plasma-anode heat transfer, evaporation of deposited cathode material from the anode surface, interaction of the evaporated atoms with the cathode spot plasma jet, as the formation of the *anodic plasma*, i.e., the plasma obtained by the ionization of the evaporated material, and its radial expansion beyond the interelectrode gap. The mathematical formulation permits for given arc current, electrode geometry and materials calculate the anode temperature, its spatial and temporal distribution, anode heat balance, anode-effective voltage, plasma electron temperature and density, and plasma particle and energy fluxes to the anode. The physics of particle condensation, the radial plasma flow in the gap, the anode plasma generation, density, plasma energy balance, and others were studied. Below we describe the main equations and the results of the model [92] to understand the evolution of the anode heating

22.4.1 Mathematical Formulation of the Thermal Model

The one-dimensional thermal model for a solid cylindrical anode with radius R and length L used to determine the anode temperature T is a function of the arc time t and the distance x from the surface in contact with the plasma, $x = 0$. The thermophysical coefficients of the anode depend on the temperature, which initially is at room temperature and then increases with time to a steady-state value. The temperature distribution in the anode cylinder was calculated from the solution of the following time-dependent and nonlinear heat conduction equation:

$$\rho(T)c(T)\frac{dT}{dt} = \frac{d}{dx}\left(\lambda(T)\frac{dT}{dx}\right) + \rho_e(T)j^2 - \frac{2}{R}\varepsilon_{\text{eff}}(x)\sigma_{sB}T^4 \quad (22.1)$$

The boundary and initial conditions are:

$$-\lambda(T)\frac{dT}{dx} = q_{\text{in}}(T) - q_{r0}(T) - q_{ev}(T) \quad \text{at } x = 0 \quad (22.2)$$

$$-\lambda(T)\frac{dT}{dx} = q_{rL}(T) \quad \text{at } x = L \quad (22.3)$$

$$q_{r0}(T) = \varepsilon_g \sigma_{sB} T^4(x = 0) \quad (22.4)$$

$$q_{r0}(T) = \varepsilon_g \sigma_{sB} T^4(x = L) \quad (22.5)$$

$$T(t = 0, x) = T_0 \quad (22.6)$$

where $\rho(T)$, $c(T)$, $\lambda(T)$, $\rho_e(T)$, and ε_g are the mass density, heat capacity, thermal conductivity, electrical conductivity, and emissivity of the anode material, respectively; $q_{\text{in}}(T)$, $q_{r0}(T)$, $q_{rL}(T)$, $q_{ev}(T)$ are the temperature-dependent incoming heat flux, the flux radiated by the upper surface, the flux radiated by the lower surface, and the evaporation heat flux, respectively. The radiation energy flux from side surfaces is taken into account in this model by the last term of (22.1), using the Stefan–Boltzmann law with emissivity correction. The parameter $\varepsilon_{\text{eff}}(x)$ is the radiation flux that escapes from the sides of the anode via the gap between the anode and radiation shields [8, 85]. The evaporation heat flux $q_{ev}(T)$ was calculated using published data for the vapor pressure and the heat of vaporization of Cu. The heat flux to the anode, q_{in}, is calculated below, using the above described plasma model.

22.4.2 Incoming Heat Flux and Plasma Parameters

The random electron current to the anode surface is much larger than the arc current (100–300 A), and therefore, an anode sheath with a negative potential drop u_a is formed near the anode surface. As the electron temperature in the cathode spots is about 1–2 eV and the density is high, the plasma jet emitted from the cathode in the beginning of the HRAVA discharge is highly ionized. The anode is heated by the energy fluxes of the cathode electrons (q_{eb}) and ions (q_{ib}), and the corresponding interelectrode plasma fluxes (q_{ep}) and (q_{ip}). Hence, the incoming heat flux to the anode surface is:

$$q_{in} = q_{eb} + q_{ib} + q_{ep} + q_{ip} \tag{22.7}$$

The electron energy flux of the cathode plasma is:

$$q_{eb} = (I/A_a/e)(1+f)(2T_{ek} + \varphi)\text{Exp}(-h/L_{ef}) \tag{22.8}$$

where φ is the work function of the anode material.

The ion beam energy flux is:

$$q_{ib} = (I/A_a/e) f \sum \text{Exp}(-h/L_{if}) f_{iz}(u_{iz} + T_{ib})$$
$$+ \sum E_{kin} - \sum z f_{iz} \varphi + \sum z f_{iz} e u_a \tag{22.9}$$

where, f_{iz} is the ion current fraction of ions of ionicity z [94] and u_{iz} is the ionization energy of ions with charge state z. The ion beam energy flux includes the ionization energy of the ions and their kinetic energy [93], which is determined by the jet velocity and acceleration in the anode sheath. The anode plasma electron energy flux to the anode surface is:

$$q_{ep} = \Gamma_{ep}(2T_{ep} + \varphi) \tag{22.10}$$

$$\Gamma_{ep} = \Gamma_{rd}\text{Exp}(-u_a/T_{ep}) \tag{22.11}$$

$$u_a = T_{ep}\text{Ln}(\Gamma e_p/I) \tag{22.12}$$

$$\Gamma_p = \Gamma_{rd} - \Gamma_{ip} + \Gamma_{eb}(0,t) - \Gamma_{ib}(0,t) \tag{22.13}$$

where Γ_{rd} and Γ_{ip} are the random electron and ion particle fluxes from the plasma to the anode surface, respectively. Equations (22.9)–(22.14) facilitate the calculation of q_{in} by (22.8). The ion energy flux from the anode plasma toward the anode surface is:

$$q_{ip} = q_{ipp} + q_{ipk} = \Gamma_{ip}(2T_{ip} + u_{i1} - \varphi + eu_a) \tag{22.14}$$

The radial plasma energy losses q_r can be determined as the sum

$$q_r = q_{ep} + q_{ir} \tag{22.15}$$

$$q_{er} = \Gamma_r(2T_{ep} + u_{i1}) \quad \Gamma_r = (IG[1 - k_c c(T)] - G_c)/(mA_r)$$

$$q_{ir} = \Gamma_r(2T_{ip} + E_{rkin}), \quad A_r = 2\pi Rh,$$

where E_{rkin} is the loss of kinetic energy due to radial plasma expansion.

The anodic plasma ion energy balance can be written in the form:

$$q_{ic}(0) = q_{ir} + q_{ipk} + q_{els} \tag{22.16}$$

where q_{ipk} is the kinetic energy flux due to ion flux Γ_{ip}, q_{ir} is the energy fluxes transported by the particle fluxes Γ_{ir}, and q_{els} is the electron energy flux lost by elastic collisions between ions and electrons. Using (22.16), the electron energy balance of the plasma could be written in form:

$$q_{ec}(0) + f(I/A_a/e)\sum\left[1 - \mathrm{Exp}(-h/L_{ef})\right]f_{iz}u_{iz} + (I - I_e(x))eu_p/A_a$$
$$= \Gamma_{ep}(2T_{ep} + eu_a) + q_{er} + 3\varepsilon(T_{ep} - T_{ip})hn\nu_{ei} \tag{22.17}$$

where ν_{ei} is the electron–ion collision frequency, $\varepsilon = m_e/m$. The plasma ion energy balance is:

$$q_{ic}(0) + 3\varepsilon(T_{ep} - T_{ip})hn\nu_{ei} = q_{ir} + (\Gamma_{ip} + \Gamma_{ic})2T_{ip} \tag{22.18}$$

where $\Gamma_{ic} = G_c/A_c/m$. The last term in (22.18) and (22.19) describes the energy transfer by elastic collisions between electrons and heavy particles. Taking into account that the anode vapor is highly ionized, the anode plasma density is determined by:

$$n(T) = \frac{IG(1 - k_c(T)) - G_c}{2\pi Rmhv_R} \tag{22.19}$$

The arc voltage u_{ca} is calculated from the power, Iu_{ca}, which is dissipated in the electrodes and in the interelectrode plasma, and by use of the expressions for anode and radial energy losses:

$$u_{ca} = u_{\text{ef-c}} + u_{\text{ef-a}} + u_r + u_p - u_{ms} \tag{22.20}$$

where, $u_{ef\text{-}c}$ is the effective cathode voltage. This voltage was measured (Chap. 8) and calculated previously using cathode spot theory [94], $u_{ef\text{-}a} = q_{in}A_a/I$ is the effective anode voltage, $u_r = A_r/I$ is the effective voltage associated with the radial energy losses, $u_p = Ih/(\pi R^2 \sigma_{el})$ is the ohmic voltage in the interelectrode plasma and σ_{el} is the electrical conductivity of the plasma. The value of E_{rkin} (the average ion energy in the radial flow) was measured up to 20 eV [95]. The value of the effective voltage associated with the radial energy loss in the stage of conventional vacuum arc was also measured (Chap. 8) to be $u_{ms} = 3$ V.

22.4.3 Method of Solution and Results of Calculation

The radial expansion of the plasma from the gap assumed as the main loss mechanism, and that the ion flux to the cathode is negligible compared to the radial flux. The plasma ion temperature was assumed equal to the anode temperature and the plasma velocity at the exit of the gap v_R is equal to the sound velocity v_{sn}. There are ten variables—$T(t, x)$, q_{ro}, q_{rL}, q_{in}, q_{ep}, q_r, T_{ep}, n, u_a and u_{ca}, that should be determined from the ten equations—(22.1), (22.4), (22.5), (22.7), (22.10), (22.12), (22.15), (22.17), (22.19), and (22.20). This equation system was solved for a discharge with a copper cathode and a graphite (DFP-1 Poco Graphite Inc.) anode with length $L = 30$ mm and diameter $2R = 32$ mm, $\varepsilon_g = 0.84$, and arc currents $I = 340$ and 175 A. The electron temperature of the cathode plasma jet is 1–2 eV [6]. The above-coupled nonlinear system of (22.1)–(22.7) was solved by finite difference technique.

The calculated time-dependent temperatures of the upper anode surface $T(0)$, of the anode body 5 mm from the surface $T(5)$ and on the lower surface $T(L)$ were compared with the respectively experimental dependences of $T_{exp}(0)$, $T_{exp}(5)$, and $T_{exp}(L)$ [8, 85]. The results are presented in Fig. 22.12a, for $I = 175$ A, and in

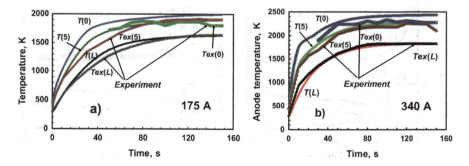

Fig. 22.12 Time dependence of anode temperature. Solid curves $T(0)$, $T(5)$, and $T(L)$ are the calculated anode temperatures at $x = 0.5$ mm and $L = 2.5$ cm, respectively. $T_{exp}(0)$, $T_{exp}(5)$, and $T_{exp}(L)$ are measured anode temperatures at the same locations [8, 75]: **a** $I = 175$ A and **b** $I = 340$ A

Fig. 22.12b for $I = 340$ A. As can be seen, the calculated temperature–time dependencies agree well with the experimental data [8, 85]. The calculated plasma electron temperature T_{ep} shown in Fig. 22.13a (175 A) and Fig. 22.13b (340 A) decreases with time from the initial value of (1.5, 1.8) eV to a steady-state value (0.9, 1.05) eV within ~(80, 30) s for 175, 340 A arcs, respectively. The measured electron temperature [89] agrees well with that calculated.

The calculated time dependence of the effective anode voltage for the total incoming heat flux, several of its components, and effective voltages for different components of the anode energy losses (from upper and down surfaces and lateral side), are shown in Fig. 22.14a (175 A) and Fig. 22.14b (340 A). The effective anode voltage $u_{ef\text{-}a}$ is initially 11 V for 175 A, and 12 V for 340 A, and both decrease with time to the steady-state values of ~6.6 V after ~70 s for $I = 175$ A and after ~20 s for $I = 340$ A. These steady-state values are slightly larger than the steady-state

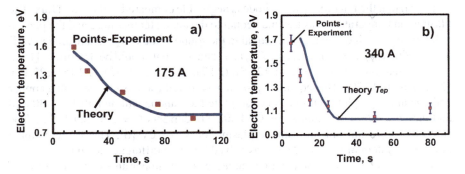

Fig. 22.13 Time evolution of electron temperature. T_{ep}-theory—calculated, $T_{ep\text{-}exper\text{-}measured}$ [89]: **a** $I = 175$ A and **b** $I = 340$ A

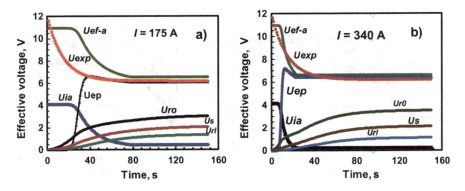

Fig. 22.14 Time evolution of effective anode voltage $u_{ef\text{-}a}$, effective voltages corresponded to plasma electron heat flux, u_{ep}, and to energy flux of the ions of the cathode jet u_{ia}, effective voltage of the fluxes radiated by upper surface of the anode u_{r0}, by the lower surface u_{rL} and by the lateral anode surface us and measured effective anode voltage rex [8, 75]: **a** $I = 175$ A and **b** $I = 340$ A

22.4 Theory and Mechanism of an Arc with Refractory Anode ...

values of plasma electron heat flux, $u_{ep} = q_{ep}/I = 6.1$ V (175 A) and 6.3 V (340 A). The difference between u_{ep} and $u_{\text{ef-a}}$ is due to energy flux of the ions of the cathode jet, which is larger for the 175 A arc (u_{ia} ~0.5 V) than for 340 A arc (~0.3 V). The effective anode voltage agrees well with u_{ex}, the measured effective anode voltage, which are also shown in Fig. 22.14a, b. The agreement in the transitory period is better for $I = 340$ A than for $I = 175$ A. It should be noted that the contribution of the anode plasma ion energy flux to the anode is smaller than that associated with the anode plasma electrons.

Regarding the energy losses from the anode, in the steady state approximately half of the energy flux into the anode is radiated by upper surface of the anode (u_{r0}) and the remainder is divided between radiation by the side surface (u_s) and the lower surface (u_{iL}). The calculated arc voltage u_{ca} in initial stage of HRAVA obtained about 21 V, and it increases with time to the steady-state values of ~23 V (175 A) and 24.5 V (340 A), which is in agreement with the measured values [6, 36].

The calculated and measured plasma densities (n) anode sheath potential drop (u_a) reach a steady-state value after ~(80, 25) s for $I = (175, 340)$ A, respectively. The calculated steady-state values are $n = (2-3) \times 10^{14}$ cm^{-3} and $u_a = 2.5-3$ V. The measured plasma density is in region $(3-10) \times 10^{14}$ cm^{-3} and exceeds the calculated values by factor 2–3 [89]. The discrepancy between the calculated and measured n can exceed an order of magnitude in the first few seconds, while the general tendency of the density evolution with time (initially a sharp grow and then saturation) agrees with observation.

The observed characteristic time to steady state of the luminous plasma expanding from the anode toward the cathode is smaller than all other characteristic times, and this difference is larger for a 175 A arc (Table 22.1). The characteristic time to steady state for the plasma density is shorter than that of the anode temperature. The calculated characteristic times of the arc voltage, electron temperature, effective anode voltage and anode sheath potential difference are approximately equal. The agreement between the calculated results and measurements is generally reasonable.

Table 22.1 Experimental and calculated characteristic times (s) to reach steady state for different arc parameters [93]

Characteristic time to reach steady state (s)				
Arc parameters	$I = 175$ A		$I = 340$ A	
	Experiment	Theory	Experiment	Theory
Luminous plume	40–50	–	20–30	–
Anode temperature	100	100–110	50–60	50–60
Arc voltage	120	80	40–50	30–40
Electron temperature	80	80	30–40	30
Plasma density	80–100	80	30–40	25
Effective anode voltage	–	70–80	–	25–30
Anode potential drop	–	70	–	20–25

22.5 Application of Arcs with Refractory Anode for Thin Films Coatings

This section demonstrates the possibility of thin metallic film producing by both electrode configurations (HRAVA and VABBA) with strongly reduced MPs contamination. The wide thermophysical properties of cathode materials were tested including volatile, intermediate, and also refractory metals. The typical dependences on arc time, arc current and other parameters are considered below to show the effect of clean deposition due to MPs re-evaporation from the hot refractory anode. All investigations conducted in discharge laboratory of Tel Aviv University and the data presented for volatile in works [97–99], intermediate in works [87, 90, 100–104] and for refractory materials in works [105]. The results of these researches are reviewed and analyzed below.

22.5.1 Deposition of Volatile Materials

Different techniques were used for thin film deposition of volatile materials, the physical vapor deposition (PVD) such as evaporation source [106, 107], magnetron sputtering [108–110], the pulse laser deposition [111], the electroplating [112].

Sn thin films have several applications. Bimetallic Cu–Sn thin films are used as bonding components in many electrical and electronic devices to join Cu conductors [113]. Bimetallic Cu-Sn thin films were prepared by a consecutively depositing 560 nm of Cu and either 200 or 500 nm of Sn by e-beam vaporization onto 2.54 cm diameter-fused quartz disks at a rate of about 0.5 nm/s [1]. Sn coated Cu plates were used to make rotating capacitors [114]. Inaba et al. [115] used electroplating to deposit thin Sn films on copper substrates. Sn thin films are used to produce an anode layer in thin film Li-ion batteries due to its high Li storage capacity in comparison to graphite anodes. Nimisha et al. [116] fabricated Sn thin films by radio-frequency discharge sputtering. The substrate to target distance was 5 cm with a deposition rate of 1 nm/s. Thus, the mentioned methods had relatively low deposition rates. Traditional deposition techniques, such as physical vapor deposition (PVD), chemical vapor deposition (CVD), or pulsed laser deposition (PLD) were extendable used. The specifics of these techniques were reviewed in [8]. The goal of the present section is to measure and understand deposition rate, MP contamination, and cathode erosion rate during Sn HRAVA film deposition for different arc parameters. In contrast to previously used techniques the relatively larger deposition rate using arcs with hot anodes is demonstrated.

22.5.1.1 Al and Zn Film Deposition Using a HRAVA Plasma Source

A diffusion pump pumped the chamber of cylindrical vacuum system (**b**) to pressure of 1.3×10^{-2} Pa before arc ignition. During the arc, the pressure in the chamber was 5.3×10^{-2} Pa. The arc was sustained between a water-cooled mainly Al, Zn source cathodes and refractory (graphite, Mo, and W) anodes, for times up to 150 s, and current (I) of 100–225 A. Cathode and anode material and geometry is presented in Table 22.2. Partially Sn cathode was tested for comparison, but detailed investigation for Sn was provided separately and the results presented in the next section.

A stainless steel cylindrical radiation shield of 70 mm diameter surrounded the anode to reduce radiative heat losses during Al and Zn deposition. The cathode was surrounded by a boron nitride square box-shaped shield with side 65 mm and positioned at different axial locations relative to the cathode surface. The shield blocked MP flux originating in cathode spots located on the side of the cathode from reaching the substrate. The shield prevented also the MP flux originating on the cathode face from reaching a portion of the substrate. The W anode used with Sn cathode was relatively thin, and the radiation shield was not required. In addition, no shield was placed around the Sn-filled Cu cup. The interelectrode gap was 10 mm.

Substrate preparation and mounting. The substrates were 76×26 mm glass microscope slides. The substrates were pre-cleaned by liquid soap and water and were then soaked in alcohol. The substrate mounted on a holder which was movable in the radial direction was positioned at distances of 80–165 mm from the electrode axis, facing the plasma flux emanating from the interelectrode gap. The holder was separated from the arc chamber by a shutter (Figs. 22.9 and 22.10) which controlled the deposition onset and exposure duration of 15 s. The MP and the deposited films were characterized in the A-region (about 3–5 mm from the boundary of A–C regions). The cathode was surrounded by a square boron nitride box-shaped shield with side 65 mm. This shield is positioned at different axial locations relative to the cathode surface for three cases (Fig. 22.15): (i) cathode surface recessed 3 mm, (ii) flush, and (iii) protruding 3 mm with respect to the shield edge. The shield blocked MP flux originating at the cathode from reaching a portion of substrate and designated the A-region.

Results of arc visualization. Photographs of 175 A arcs with Al, Zn, and Sn cathodes (without shields) 30 s after arc ignition are shown in Fig. 22.16. The Al arc emission color is light green while Sn is green color and the Zn is blue color.

Table 22.2 Electrode characteristics

Cathode			Anode		
Material	Length (mm)	Diameter (mm)	Material	Length (mm)	Diameter (mm)
Al	40	30	Graphite	30	32
	25			9	32
Zn		30	Mo	30	32
Sn	10	60 (in Cu cup)	W	10	60

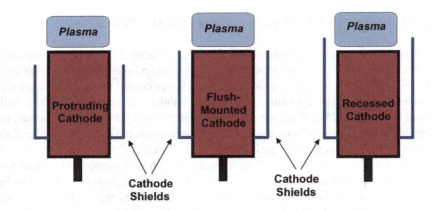

Fig. 22.15 Schematic diagram of the cathode surface location with respect to the cathode shield: (left to right) cathode surface protruded 3 mm, flush-mounted, and recessed 3 mm

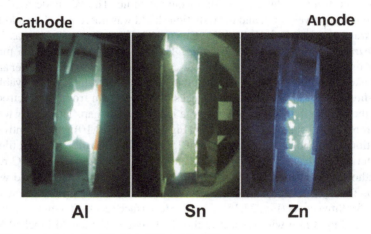

Fig. 22.16 Arc photographs with Al and Zn and Sn cathodes

Cathode spot appeared on the cathode surface, while the anodes were covered by a diffuse discharge. Later ($t > 30$–60 s) an anodic plasma plume filled most of the electrode gap and expanded in the radial direction.

Film morphology. Two characteristic regions with sharply different MP densities were observed on the substrate surface. Typical micrographs of Al films from the two regions are shown in Fig. 22.17: the "A-region" facing the anode and mostly without MPs and the "C-region" facing the cathode with many MPs.

Deposition Rate. *Aluminum* time-dependent Al deposition rates V_{dep} for a graphite anode are presented in Fig. 22.18 for flush-mounted and in Fig. 22.19 with cathode protruding 3 mm with respect to the anode shield edge (Fig. 22.20).

It can be seen that V_{dep} increased with time to a peak and then decreased to a steady-state value. The peak value weakly depended on the current and the peak

22.5 Application of Arcs with Refractory Anode for Thin Films Coatings

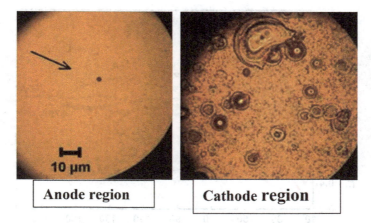

Fig. 22.17 Al film clean deposition in the anode region and with MPs contamination in the cathode regions

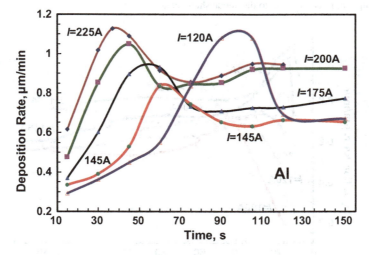

Fig. 22.18 Time-dependent rate of Al film deposition, $L = 110$ mm and 30 mm graphite anode for flush-mounted cathode

time decreased with I. The maximal deposition rate increased with arc current, for both the cathode surface was flush-mounted or protruded 3 mm above the end of the shield. The dependence of maximal deposition rate on the cathode position with respect to the cathode shield was $V_{dep} \sim 1.1$ μm/min when cathode surface was flush-mounted with the shield edge and ~1.6 μm/min when the cathode surface protruded above the shield (for $I = 225$ A and $L = 110$ mm). $V_{dep} \sim 0.8$ μm/min when cathode surface was recessed mounted with the shield edge (not shown).

The deposition rate with the shorter (9 mm) anode and for an Al cathode surface flush-mounted with the shield edge is shown in Fig. 22.20. The peaks occurred much

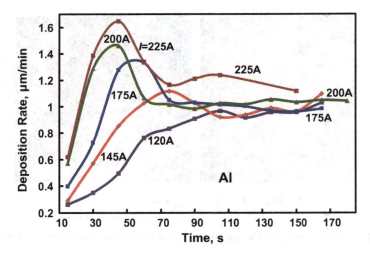

Fig. 22.19 Time-dependent rate of Al film deposition, $L = 110$ mm and 30 mm graphite anode for protruding 3 mm with respect to the anode shield edge

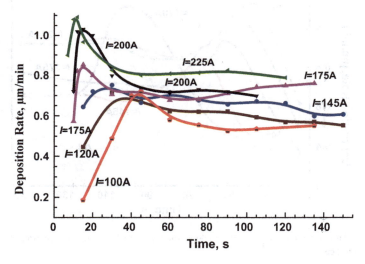

Fig. 22.20 Time-dependent rate of Al film deposition, $L = 110$ mm and 9 mm graphite anode for flush-mounted cathode

earlier and the peak is narrower than with the 30 mm anode. V_{dep} at the peak and at steady state were about the same as with the 30 mm anode.

The Al film thickness distribution on the substrate was measured in y-direction (i.e., perpendicular to the electrode axis) from A–C boundary (Fig. 22.11) in the A-region for currents 175, 200 and 225 A, with distance L from the electrode axis to the substrate as a parameter. The result indicated that the film distribution close

22.5 Application of Arcs with Refractory Anode for Thin Films Coatings

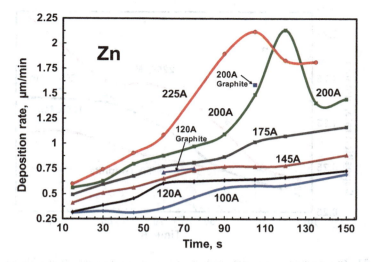

Fig. 22.21 Zn deposition rate versus time with arc current as a parameter, 30 mm Mo anode, $L = 110$ mm. The measured deposition rate with a graphite anode and recessed cathode for $I = 120$ A and 200 A is presented at three times

to uniform for $L \geq 110$ mm at y up to 5 cm, while for lower L the thickness is larger [96, 97].

Zink. The deposition rate with a Zn cathode recessed 3 mm below the shield edge and 30 mm height anode is shown in Fig. 22.21. The rate increased with current in all of the arc stages. V_{dep} increased with time during arcing of about 120 s from about 0.3 to 0.7 μm/min for 100 A, and from 0.5 to 1.2 μm/min for 175 A. At higher currents (200 and 225 A) a peak of 2.1 μm/min appeared after ~2 min of arcing, and then V_{dep} decreased to 1.4 and 1.8 μm/min, respectively. The deposition rate with a Zn cathode recessed 3 mm below the shield edge and 10 mm height anode is shown in Fig. 22.22. The deposition rate (short anode) increases with arc time and arc current monotonically (without peak, except 225 A). For comparison, the Zn deposition rate for short graphite anode (175 A) is also presented indicating lower V_{dep} than for Mo anode for $t > 45$ s. Also the Zn film thickness distribution on the substrate in y-direction, co-axial with electrode axis in the A-region (from the boundary of A-C regions) for 200 and 225 A is relatively uniform for $L \geq 80$ mm.

The preliminary study of Sn deposition rate [96] showed that initially V_{dep} was almost constant, but then increased with time after some critical time which decreased with I. The arc current affects significantly on V_{dep}. It can be seen that V_{dep} for $I = 175$A exceed the V_{dep} for 120 A by factor 5 at 2 min of arc duration. Let us consider the extended study of the Sn deposition [98].

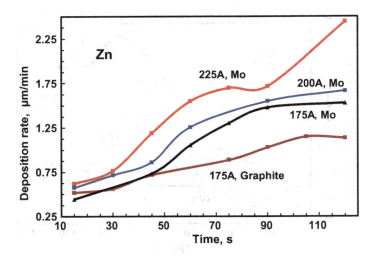

Fig. 22.22 Zn deposition rate vs time with arc current as a parameter, 10 mm Mo anode, $L = 110$ mm. The measured deposition rate with 9 mm graphite anode and recessed cathode for $I = 175$ A is presented also

22.5.1.2 Sn Thin Film Deposition Using a HRAVA Plasma Source

The vacuum system (a) with water-cooled cylindrical Cu cathode modified in form of a cup to fill with Sn shown schematically in Fig. 22.23 [99]. The chamber was evacuated by an oil diffusion pump to 1.3×10^{-2} Pa. During the arc, the pressure in the chamber was 5.3×10^{-2} Pa. The arc was sustained between the cathode and a non-consumable cylindrical anode, for times up to 180 s, operating with a current $I = 60$–175 A. The outside diameter of the Cu cup was $D = 30$ or 60 mm and the Sn filling had a thickness of $d = 10$ mm (Fig. 22.24). The 30-mm-diameter cathode was tested with two graphite anodes having $d = 9$ and 15 mm thickness. Both graphite anodes were $D = 32$ mm, and the gap between the cathode and the anode was $h = 10$ mm. The cathode with $D = 60$ mm was used with a W anode with $d = 10$ mm thickness and $D = 60$ mm diameter and gaps of $h = 10$ and 15 mm.

A cylindrical stainless steel radiation shield of 70 mm diameter surrounded the graphite anodes to reduce radiative heat losses. The W anode was used without a radiation shield. The $D = 30$-mm cathode was surrounded by a square boron nitride box-shaped shield with 65-mm sides, and the cathode was recessed 3 mm behind the shield. The $D = 60$ mm cathode was used with a flat boron nitride plate shield positioned between the cathode and the substrate. The cathode was recessed 7 or 10 mm behind the shield for the case with $h = 15$ mm, while with $h = 10$ mm no cathode shield was used. The purpose of above shields was to block MPs emitted from the cathode from reaching a portion of substrate, designated the A-region (Fig. 22.11). The cathode erosion rate was measured by weighing the cathode before and after arcing.

22.5 Application of Arcs with Refractory Anode for Thin Films Coatings

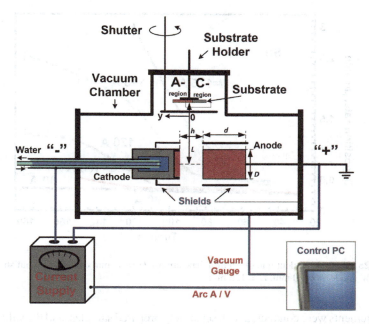

Fig. 22.23 Schematic diagram of the chamber, cathode–anode assembly, Cu cup filled with Sn and the substrate position, illustrating the A and C regions

Fig. 22.24 Cylindrical Cu cathode modified in form of a cup to fill with Sn

All substrates were 75 × 25 mm glass microscope slides. The pre-cleaned substrates were mounted on a holder which was movable in the radial direction and was positioned at distances of $L = 110$ or 125 mm from the electrode axis (95 mm from cathode edge), facing the plasma flux emanating from the interelectrode gap. The holder was separated from the arc plasma by a shutter, which controlled the deposition onset and duration of 15 s in all of the experiments. At least three or four

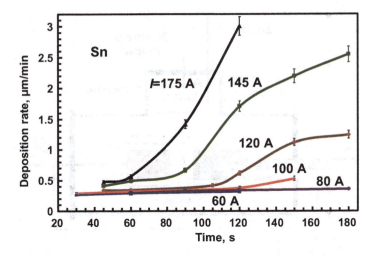

Fig. 22.25 Time-dependent rate of Sn film deposition—$D = 60$ mm cathode without shield, $d = 10$ mm W anode, $h = 10$ mm, $L = 125$ mm

measurements were conducted with separately prepared samples, and the values were averaged for each graph point in the results reported below.

Deposition Rate. Figure 22.25 presents the measured Sn deposition rate dependence on time beginning from arc ignition, with I as a parameter, using a $D = 60$ mm cathode, W anode, $L = 125$ mm, electrode gap $h = 10$ mm, and without a cathode shield. It can be seen that V_{dep} increased with time with all arc currents studied. For $I < 100$ A, the V_{dep} was mostly not changed in comparison with that for larger arc currents. For $I > 100$ A, the V_{dep} changed very significantly, i.e., V_{dep} increased from 0.5 to 3 µm/min when the arc time increased from 45 to 105 s ($I = 175$ A).

The measurements showed that for this cathode with the BN cathode shield protruding for 7 mm above the cathode surface and $h = 15$ mm V_{dep} increased from 0.3 to 2 µm/min ($I = 175$ A). When the cathode shield is protruding 10 mm, V_{dep} is slightly lower. V_{dep} is 0.58 µm/min in comparison to 0.62 µm/min measured for 7 mm cathode protruding above the shield, at arc time of 150 s.

The time-dependent deposition rates for $D = 30$ mm cathode and $d = 9$ and 15 mm graphite anodes are shown in Figs. 22.26 and 22.27, respectively. The BN cathode shield protruded 3 mm above the cathode and $h = 10$ mm. It can be seen that V_{dep} increased with time to a peak (for $I \geq 80$ A), and then decreased to a steady-state value.

The peak value depended on the current (with $d = 9$ mm, 0.74 and 0.84 µm/min for $I = 120$ and 175 A, respectively) and the peak time decreased with I (from 60 to 30 s for $I = 120$ and 175 A respectively, $d = 9$ mm). At steady state with $d = 9$ mm, V_{dep} increased from 0.3 to 0.7 µm/min when the arc current was increased from 60 to 175 A. The deposition rate for the thicker anode ($d = 15$ mm) also had peaks for same arc currents (Fig. 22.27), but the peaks occurred much later (90 and 60 s for $I = 120$ and 175 A, respectively) and were wider than with the $d = 9$ mm anode. In addition,

22.5 Application of Arcs with Refractory Anode for Thin Films Coatings

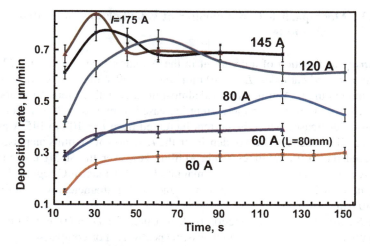

Fig. 22.26 Time-dependent rate of Sn film deposition—$D = 30$ mm cathode recessed 3 mm behind box shield, $d = 9$ mm graphite anode, $h = 10$ mm, $L = 110$ mm. (The case for $L = 80$ mm is also shown for $I = 60$ A)

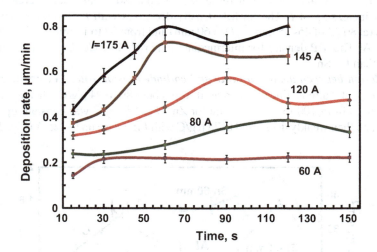

Fig. 22.27 Time-dependent rate of Sn film deposition—$D = 30$ mm cathode recessed 3 mm behind box shield, $d = 15$ mm graphite anode, $h = 10$ mm, $L = 110$ mm

V_{dep} at the peak for $d = 15$ mm was lower than with the $d = 9$ mm anode (0.57 and 0.80 μm/min for $I = 120$ and 175 A, respectively). The measurements shown that the film thickness decreased with distance in y-direction for the considered range of I, and it reached about half of the maximal value at $y = 2$–3 cm, depending on I.

22.5.1.3 Macroparticle Contamination at Volatile Films and Cathode Erosion Rate

MP density as function of arc current in the A-region is shown in Fig. 22.28, for different recessed cathodes, $L = 110$ mm, and $t = 60$ s after arc ignition. The MP contamination in the Al film increased almost linearly with current, from about 5 to 32 particles per mm^{-2} when the current increased from 100 to 250 A. For Zn, the MP density decreased when the current increased from $I = 100$ to 145 A and then increased till $I = 225$ A. The MP density in the A-region of Sn films increased with I and exceeded by a several times that in Al films. However, the Sn MP contamination in the A-region was a few orders of magnitude lower than in C-region. In the all C-region, the MP density was ~10^3 mm^{-2} and the MP diameters were in the range of 15-40 μm.

The Sn MP density as a function of I measured for $D = 60$ and 30 mm cathodes with 10 and 9 mm anode thickness, respectively. For comparison, previously measured MP density for Al, Zn, Cu, and Cr cathodes [7, 8] is also presented. The substrates were exposed for 15 s after 60 s of arcing, $L = 110$ mm, $h = 10$ mm. The MP density for Sn is larger than for other relatively low melting temperature materials like Al and Zn, and increases with I up to 80 A, and then decreased with further increasing of I. The Sn MP density decreases with I from their maximum values (at 80 A) of 45–35 mm^{-2} ($D = 60$ mm) and from 37 to 17 mm^{-2} ($D = 30$ mm) at 175 A. The MP density for intermediate materials like Cu, Cr significantly was lower than for Sn.

Relation between deposition rate and cathode erosion rate. Al cathode erosion rates G (μg/C) almost linearly increased from about 80 to 94 μg/C when arc current I increased from 100 to 225 A, while for Zn, $G = 85$ μg/C was constant up to $I = 150$ A and then weakly decreased to 78 μg/C with I increase up to 225 A (Fig. 22.29).

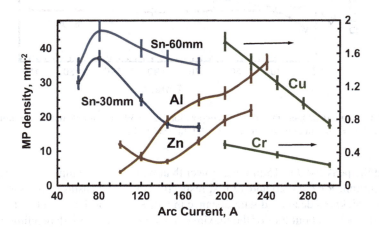

Fig. 22.28 Dependence of MP density on arc current for Sn, Zn, and Al (left axis) and for Cu, Cr (right axis), recessed cathodes. The substrate was exposed after 60 s for 15 s, $L = 110$ mm

22.5 Application of Arcs with Refractory Anode for Thin Films Coatings

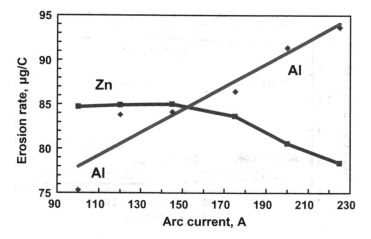

Fig. 22.29 Erosion rate of Al and Zn cathodes as function on arc current

It should be noted that the erosion rates are lower by factors 1.5 and 4 for Al and Zn, respectively, than those measured by [6] in conventional cathodic arcs. The cathode erosion rates G (μg/C) for Sn with $D = 30$ mm as functions of arc current I. The Sn erosion rate decreased from ~500 to ~320 μg/C for $d = 9$ mm when I was increased from 60 to 140 A and then increased to ~550 μg/C for $I = 175$ A. For $d = 15$ mm, the erosion rate decreased from 470 to 380 μg/C when I was increased from 60 to 130 A and then increased to 410 μg/C for $I = 175$ A.

For $d = 15$ mm the erosion rate decreased from 470 to 380 μg/C when I was increased from 60 to 130 A and then increased to 410 μg/C for $I = 175$ A. The Sn cathode erosion rate measured here exceeds that of Cu by a factor of four, which in general agrees with the previous data for Sn cathodic arcs with $I = 100$ A [VAST] (~300 μg/C—total and ~100 μg/C—ion erosion rate).

To compare the deposition rates of different cathode materials let us introduce parameter

$$F = \frac{V_{1dep}\rho_1 G_2}{V_{2dep}\rho_2 G_1},$$

where the indices 1,2 designate different cathode materials. F is the ratio of $S = V_{dep}\rho/GI$ for two materials and characterizes the ratio of the portion of the mass flux of two materials condensing at a given location in the A-region near the A–C boundary. F for different material pairs as a function of I is presented in Fig. 22.30 by taking into account the peak of V_{dep} for Al and Zn cathodes and $G = 110$–115 g/C for Cu [8] (recessed 30 mm cathodes). Figure 22.30 shows that the parameter F varied weakly with I. $F(\text{Cu/Al}) > 1$ and slightly increases with I. $F(\text{Cu/Al})$ for steady state significantly exceeds $S(\text{Cu/Al})$ at the deposition rate peak. In contrast,

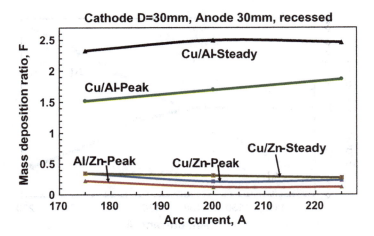

Fig. 22.30 Mass deposition ratios F for different Cu, Al, and Zn cathode pairs as function of arc current

the difference in $F(Cu/Zn)$ between peak and steady state was small. It also can be seen that $F(Cu/Zn) < 1$, while $F(Al/Zn) < 1$, both at peak and steady state.

Thus, the above calculations indicate that the Cu mass deposition rate was larger than that for Al cathode and lower for Zn cathode, whereas Al mass deposition rate is lower than that for Zn cathode. This result can be understood taking into account that the film thickness is determined by the large difference of metal densities ($\rho(Cu)/\rho(Al) = 3.3$) when the cathode boiling temperatures are identical such as Cu-Al pair. On contrary, the film thickness is determined by the low difference temperatures T_m and T_b (more MPs evaporation and their conversion the plasma due to lower T_b) when the metal densities are identical ($\rho(Cu)/\rho(Zn) = 1.25$) such as Cu–Zn pair.

22.5.1.4 Remarks to the Mechanism of Deposition Rates and MPs Generation

The appearance of a peak in the **deposition rate** versus time graphs can be understood taking into account that the MPs accumulated early during the arc on the cold anode surface, and evaporated from the anode during a relatively brief interval when the anode was heated to an appropriate temperature. At the steady state, the MPs previously accumulated at the cold stage significantly disappeared. The deposition rate then declined toward its steady-state value when a balance was achieved between material impingement and evaporation at the hot anode surface. The deposition rate peak reached significant level for low melting temperature metals such as Al and Zn cathodes, and negligible with Cu, Cr, and Ti cathodes (see below). The peak in the time dependence of deposition rate occurred earlier when using a short anode than a long anode, because the shorter anode was heated faster since it had smaller mass and reached a higher steady-state temperature due to its radiative heat loss rate was less because of its smaller surface area.

The deposition rate depended on the cathode surface position with respect to the cathode shield. The rate V_{dep} was lower when cathode surface was recessed or flush-mounted with respect to the shield comparing to the cathode surface protruded above the shield. This is because the material ejected from the cathode remained on the internal walls of the shield. Therefore, the amount of the cathode material reached the anode surface depends on cathode position, and in turn, different amount of condensed material was re-evaporated from the anode. As result, different intensity of anode plasma plume is radially expanded to the substrate.

Sn. According to the present measurements, the deposition rate behaved differently with time, depending on anode geometry and arc current. A V_{dep} peak for Sn was observed with the smaller ($D = 30$ mm) anode, for $I \geq 80$ A. In this case, V_{dep} at the peak increased with I and the time of the peak appearance increased with d. A V_{dep} peak was not observed with $D = 60$ mm in the measured range of arc current. These effects can be understood, firstly, taking into account that in the HRAVA the thermal mass of the anode increases with d, and power delivered to the anode increases with I, and thus the heating rate decreases with d and increases with I [117]. Secondly, the relatively large MP production from the Sn cathode spots should be noted.

The rate of rise of the anode temperature influences the rate of rise of the evaporation rate of the anode material, and this effect can explain the different time-dependent V_{dep} for different electrode geometries. For relatively large I and small D, the anode was quickly heated [116]. Therefore, the rate of vaporization of the previously condensed cathode material (including MPs) on the anode also quickly rose, increasing the anode plasma density during the transient period, and leading to the V_{dep} peak appearing before the steady-state level of V_{dep} was reached. In contrast, for small I and large D the anode heating rate is lower due to the larger anode mass. In this case, it can be assumed that the deposited cathode plasma and the deposited MPs were re-evaporated gradually from the anode surface, while the anode temperature T increased with time t. This explains the observed monotonic increase of V_{dep} for $I < 80$ A with $D = 30$ mm and for considered range of I with W anodes having $D = 60$ mm.

In addition, the W anode reached a higher temperature and had a higher rate of rise of the temperature than the graphite anode, and that also explains the monotonic $V_{dep}(t)$ dependence. As the heating time [116] was larger with thicker anodes, the V_{dep} peak for $d = 15$ mm appeared later than for $d = 9$ mm. Similar results were observed with Zn cathodes. When the $D = 60$ mm cathode was not shielded, all of the generated plasma flux was extracted from the gap and a higher V_{dep} (3 μm/min) was obtained. When the cathode surface was recessed with respect to the shield, the plasma flux to the substrate was partially shielded and therefore V_{dep} was reduced.

Another experimental result showed that V_{dep} with $D = 60$ mm electrode pairs exceeded that for $D = 30$ mm electrode pairs by a factor of 2.5–3.0 ($I = 175$ A, time >80–90 s, Figs. 22.25, 22.26 and 22.27). As the erosion rate is independent of the cathode diameter, the above-mentioned difference of V_{dep} can be understood because of a significant accumulation of the cathode material from plasma jets on the large

Table 22.3 Melting and boiling temperatures

Material	T_m (°C)	T_b (°C)	$(T_b-T_m)/T_m$
Cu	1083	2595	1.40
Al	660	2447	2.71
Zn	420	907	1.16
Cr	1903	2642	0.28
Ti	1668	3280	0.966
Sn	232	2687	0.086

anode. The larger amount of the deposited material (including MPs) on the anode surface is vaporized with gradually heating in time leading to monotonic increase V_{dep} (Fig. 22.25). Therefore, the larger V_{dep} for $D = 60$ mm electrode pair was also reached at time 150–180 s.

In general, the **MP density** in the deposited films depends on the relation between the melting point T_m and boiling point T_b. Table 22.3 illustrates the boiling and melting temperatures and difference between T_b and T_m, characterized by parameter $(T_b-T_m)/T_m$ which is large for Al and very low for Sn. The boiling temperatures of Al and Sn are close to that for Cu and the generated from the cathodes MP will be evaporated with close rate. However, T_m is lower for Al, Zn, and much lower for Sn than for Cu, and therefore, the rate of MP generation for Al, Zn, and Sn occurs with larger rates than for Cu. The difference of T_m and T_b for Zn is lower than for Al, and therefore, the MP density for Zn was observed less than for Al. The difference of T_m and T_b for Sn is larger than for Al, and therefore, the MP density for Sn was observed larger than for Al. The observed MP density for Al, Zn, and Sn cathodes was significantly exceeded this density for cathodes from Cu and Cr. The MP density for Al, Zn was comparable with that for Cu films deposited only when $I < 140$ A.

Sn. Microscopic observation of the present substrates also suggests that the cathode mass loss occurs largely in droplet form (Fig. 22.17). Due to large difference between the very low Sn melting temperature ($T_m = 232$ C) and large Sn boiling point ($T_b = 2687$ °C), the rate of MPs generation is relatively large, the in-flight evaporation of the MPs is relatively weak, and MPs account for a large fraction of the Sn cathode erosion. The relatively large spread between T_m and T_b explains also the large MP contamination detected in Sn films (Fig. 22.17), compared to Cu, Ti, and to Al, Zn ($I < 150$ A) for which the difference between T_m and T_b is significantly less (Table 22.3). The Sn MP contamination in the deposited films was reduced with arc current I due to MP interaction with the anode plasma, which density increased with I (see Fig. 22.28).

22.5.2 Deposition of Intermediate Materials

In this section, the time dependence of Cu film deposition from a plasma generated in a HRAVA and VABBA configurations having either a single-hole and shower-head anode was studied.

22.5.2.1 HRAVA Copper Thin Film Deposition

The first measurements of the HRAVA Cu deposition film were conducted in 2000 [87] with graphite anode in the stainless steel chamber (Fig. 22.7) of above described system (**a**). Coatings were deposited on 20 × 20 × 0.5 mm ground stainless steel sheet substrates with various arc currents (175–340 A). Using the shutter, the substrates were exposed to the arc during a controlled time window. The deposition rate was obtained by weighing the substrate (with an accuracy 10^{-5} g) before and after deposition. The Cu films were deposited on stainless steel and glass substrates displaced 15 mm away from the mid-plane parallel to the electrode axis and toward the cathode. Thus, these substrates were shielded from a direct line of sight from the cathode by the cathode shield in order minimize the MPs in the A-region of the deposited substrate. Coatings were deposited on substrates with area A_s for different arc currents and exposure time Δt_s. The deposition rate V_{dep} was calculated as the ratio

$$V_{dep} = \Delta m_s / (A_s \Delta t_s r)$$

where Δm_s is the substrate mass gain. It was assumed that the deposited film density ρ is that of solid copper, 8.9 g cm^{-3}. The measured mass deposit Δm_s as well as the calculated rate V_{dep} for different arc currents and exposure time are presented in Table 22.4. The deposition rate was relatively large, about 2 μm/min.

The next study was conducted with a tungsten anode [101], which reached relatively high surface temperatures of ~2500 K at an arc current of 250 A in comparison to the graphite temperature of about 2000 K at 340 A. The higher anode temperature significantly reduced the MP contamination in the deposit. The arc was contained in the stainless steel chamber (Fig. 22.8) of above described system (**b**). A diffusion pump, before arc ignition, pumped the chamber down to a pressure of 0.67 mPa.

Table 22.4 Measurement data of the deposition rate on stainless steel substrates

I (A)	Δm_s (mg)	Δt_s (s), shutter closed/open	V_{dep} (μm/min)
175	7.5	0/60	2.11
175	1.6	60/15	1.8
175	3.5	60/30	1.97
250	5.3	0/80	2.3
340	2.2	60/15	2.5

During the arc, the pressure in the chamber increased to about 13 mPa. The arcs were operated at currents of $I = 150–300$ A for periods up to 190 s. The cathode–anode assembly is shown schematically in Fig. 22.9. The arc was sustained between a 30-mm-diameter cylindrical, water-cooled, copper cathode and a 32-mm-diameter, 30-mm-length tungsten anode. Two Mo cylindrical radiation shields, 60 and 70 mm diameter, surrounded the anode to reduce radiative heat losses. The cathode was surrounded by Mo shield with diameter of 50 mm (Fig. 22.9), whose position determines location of the boundary between regions on the substrate with small ("A") and high ("C") MP contamination. The interelectrode gap h was set to 5 or 10 mm.

The substrates for deposition were 75×26 mm² glass microscope slides. The slides were mounted on a water-cooled holder, located at distances of $L = 80–165$ mm from the electrode axis. A Mo shutter was used to control the exposure time of the substrate to the plasma. The film thickness (H) distribution was measured in the A-region by profilometry along the y-axis (in the substrate plane, perpendicular to the C–A boundary, for each value of L as shown in Fig. 22.11). Film thickness distributions $H(y)$ were determined for $L = 90, 110, 140$, and 165 mm, $h = 10$ mm, $I = 200$ A for different L. It was observed that (1) the thickness was maximal at $y = 0$, (2) the film thickness decreased with L, and (3) thickness distribution was more uniform with larger L. The film uniformity was characterized by the non-uniformity coefficient k_{n-u} representing an average rate of thickness variation with y: $k_{n-u} = <\Delta H/(H \Delta y)> \times 100\%$/cm. The estimations showed that k_{n-u} decreased from ~50%/cm at $L = 90$ mm to more uniform distribution after 140 mm reaching about 10%/cm at $L = 165$ mm.

Deposition rate. The temporal evolution of the deposition rate at $y = 0$ (see Fig. 22.11) is presented in Fig. 22.31 for $h = 10$ mm, $L = 80$ mm, $I = 200$ A. The deposition rate V_{dep} was determined from the film thickness measured after 15, 45, 60, 90, and 120 s deposition times beginning from arc ignition. V_{dep} was determined as the ratio between the incremental film thickness growth ΔH in an incremental deposition time divided by differences of deposition times Δt, i.e., $V_{dep} = \Delta H/\Delta t$. It

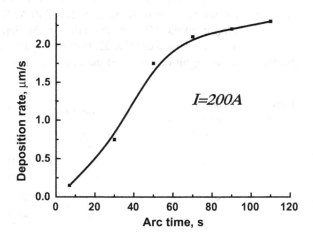

Fig. 22.31 Deposition rate versus of arc time ($L = 80$ mm, $I = 200$ A, $h = 10$ mm, $y = 0$)

22.5 Application of Arcs with Refractory Anode for Thin Films Coatings

can be seen that the deposition rate started from ~0.5 μm/min in the initial stage of the arc and reached a steady state of ~2 μm/min in the developed HRAVA (≥70–80 s). About the same value (~2 μm/min) was found previously with a Mo anode [99]. V_{dep} in steady state was also determined in experiments, where substrates were coated for 15–25 s beginning 60 s after arc initiation, and the subsequent film thickness was divided by the exposure time. Steady-state V_{dep} increased almost proportionally to the arc current, from 2 μm/min at $I = 200$ A to 3.6 μm/min at $I = 300$ A ($h = 10$ mm, $L = 80$ mm and $y = 0$). With L increasing from 80 to 110 mm, the steady-state V_{dep} decreased from ~3.6 to ~2.0 μm/min for $I = 300$ A.

MP contaminations were determined by two characteristics: (1) the MP size distribution function (number of MPs per μm diameter (D) and per Coulomb charge transfer in the arc per mm cylindrical height like to MP size distribution function presented by Daalder [43]), and (2) the total MP flux density (the average number of MPs per mm^2 per minute). The MP size distribution functions for a Mo anode [90] and for the W anode [101] were obtained previously and are presented in Ch. 9. The MP number decreased with the MP diameter and was slightly lower with the W anode than with the Mo anode. MPs with diameters more than 20 μm were not observed. It was observed that MP flux density in the A-region decreased approximately linearly with the arc current and reached ~3 mm^{-2} min^{-1} for $I = 300$ A.

Cathode erosion rate. The cathode mass loss was determined in two experimental series. In first series, the cathode mass loss, measured after running a 175 A arc for 390 s and a 250 A arc for 360 s, was $\Delta m_c = 17.5$ g. In second series, the cathode mass loss, measured after running a 175 A arc for 1055 s, was $\Delta m_c = 18.7$ g. The cathode erosion rate, defined as the cathode mass loss Δm_c divided by the charge transfer, was thus $G = 110$ μg/C in first and 101 μg/C in the second series.

22.5.2.2 VABBA Copper Thin Film Deposition

Experiments were conducted in stainless steel vacuum chamber (Fig. 22.8) of above described system (**b**) that was diffusion pumped to a vacuum of about 0.67 mPa (5 μTorr) before arc ignition and about 13 mPa (0.1 mTorr) due to outgassing pressure rise during arcing. The arcs were operated at a current of $I = 200$ A for a **one hole** (OH) and for **frontal shower** (FS) anodes and $I = 150–275$ A for a **radial shower** (RS) anode and these anode configurations described above.

The VABBA configuration, with a cylindrical shower-head anode, is shown schematically in Fig. 22.10. The arc was sustained between the 30-mm-diameter water-cooled Cu cathode and a 50-mm-diameter refractory W or Ta anode. The gap between the front cathode surface and inner anode surface was about $h = 10$ mm. The arc was triggered by explosion a Cu wire placed between the cathode and anode. In the FS configuration, plasma was ejected through an array of 250 holes of 0.6 mm diameter for Ta and 1 mm diameter for W anodes. In OH anode, the diameter of the single aperture in the center was 4 mm. The RS anode had 230 holes of 0.8 mm diameter arranged on the lateral side of a W anode. The close of the interelectrode volume was completed with a boron nitride insulating ring. Figure 22.32 shows a

Fig. 22.32 Photograph of the Ta shower-head cap-shaped anode with 0.6 mm diameters of 250 holes

photograph of the Ta FS anode.

The substrates were 75 × 25 × 1 mm glass microscope slides. In addition, a 185-mm-diameter glass disk substrate was used to deposition on large distances. The substrate was shielded from the depositing flux by a shutter, which was opened for 15 s (exposure time) at all different times after arc ignition. For RS anode, the substrates were located on cylindrical holder with radius of 190 mm. The distance between anode lateral surface with perforation and glass substrate was $L = 70$ mm.

The deposited during 15 s films were observed by an optical microscope, and XRD analysis was conducted. SEM images were obtained using Quanta 200 FEG Environmental Scanning Electron microscopy. The images of the plasma plume and the hot anode were recorded with a digital camera. Cathode erosion was determined by weighting it before and after several arcs with known duration and arc current.

Results. The anode temperature increasing was observed by changing the anode color from black to red and then to white. Partially it is demonstrated in Fig. 22.33 for VABBA with one-hole Ta anode and with W shower anode. When the shower anode

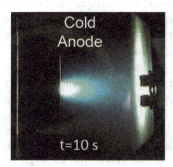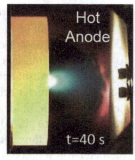

Fig. 22.33 Photographs of a Cu plasma jet ejected from a one-hole Ta anode, cold (left) and hot (middle), and from a W frontal shower anode (right), at indicated times after arc ignition

22.5 Application of Arcs with Refractory Anode for Thin Films Coatings

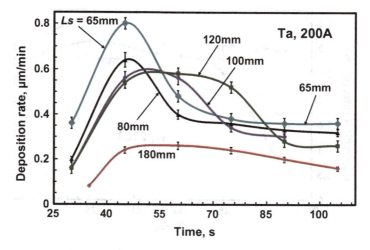

Fig. 22.34 Frontal shower anode deposition rate with distance L as a parameter Ta-06mm holes diameter, $I = 200$ A)

reached white color, the blue plasma already difficult seen due to large brightness area for shower anode. Thus, when the arc was ignited, cathode plasma heated the anode and it relatively high temperature was reached with arc time [118]. In the case of Cu films, a few MPs can be observed when the arc time was lower than 40–60 s while at larger arc time >60 s, the deposited films were practically without MPs. Below the maximal thicknesses (in the substrate center) was used for rate of deposition calculation.

VABBA with FS anode. The time-dependent Cu deposition rate using the Ta FS is shown in Fig. 22.34. The measurements were conducted at arc current $I = 200$ A with anode-to-substrate distances of $L = 65, 80, 100, 120,$ and 180 mm. It can be seen that V_{dep} increased with time to a peak, and then decreased to a mostly steady state. The time width of the peaks was larger for larger L and was weak for $L \geq 120$ mm. At relatively large distance ($L = 180$ mm) V_{dep} increased from about 0.08 μm/min at 35 s to 0.24 μm/min at 45-60 s, and further decreased to 0.16 μm/min at 105 s.

At $L = 80, 100, 120$ mm the peak of V_{dep} was about 0.65–0.55 μm/min (but with different duration), and at $L = 65$ mm was 0.8 μm/min. V_{dep} decreased to its steady state of about 0.28, 0.3, 0.32, 0.36 μm/min at L of 120, 100, 80 and 65 mm respectively. Figure 22.35 presents the deposition rate with the W FS anode as a function of time from arc ignition before plasma exposure. These experiments were conducted at $I = 200$ A with the anode-to-substrate distance $L = 80$ and 120 mm. When the substrate was placed at distance from the anode frontal surface $L = 120$ mm, the deposition rate increased with time after arc ignition from 0.4 μm/min at $t = 30$ s to maximal $V_{dep} = 3.12$ μm/min and at $L_s = 80$ mm increased from 1.48 μm/min at $t = 30$ s to maximal to $V_{dep} = 3.6$ μm/min. The maximal deposition rate occurred at 75 s for both substrate distances.

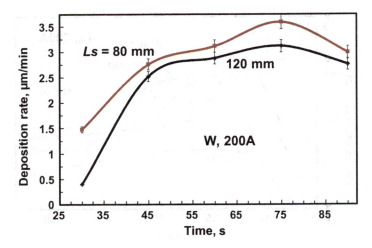

Fig. 22.35 Frontal shower anode deposition rate with distance L as parameter (W—1 mm holes, $I = 200$ A)

The next experiment with FS W (1 mm) anode shows the VABBA thin film deposition on relatively large substrates and at large distances. The deposition with $I = 200$ A was done at $L_s = 280$ mm with 15 s substrate exposition and at 70 s after arc ignition.

The deposition rate profile on substrate of 185 mm diameter glass disk is presented in Fig. 22.36. The maximal $V_{dep} = 0.47$ μm/min was at the substrate center and then decreased to about 0.3 μm/min at 30-40 mm and then weakly changed up to a radius of 80 mm of the disk substrate.

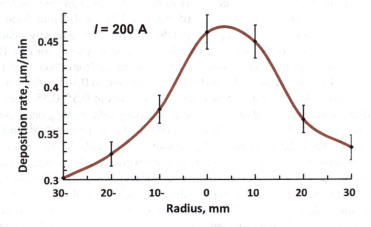

Fig. 22.36 Deposition rate for large distance and substrate area, frontal shower anode (W—1 mm holes, $I = 200$ A, $Ls = 280$ mm, 70 s after arc ignition)

22.5 Application of Arcs with Refractory Anode for Thin Films Coatings

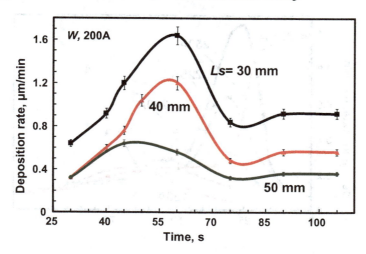

Fig. 22.37 One-hole anode deposition rate with distance L as a parameter (W—4 mm holes, $I = 200$ A)

VABBA with OH anode. The deposition rate from OH (W-4 mm) anode is shown in Fig. 22.37 for $I = 200$ A and $L_s = 30$, 40, and 50 mm. The deposition rate had peaks of about 0.6, 1.2, and 1.7 μm/min for $L_s = 50$, 40, and 30 mm, respectively. After about 90 s of arcing, V_{dep} decreased to its steady state value of 0.4, 0.6, and 0.9 μm/min for $L_s = 50$, 40, and 30 mm, respectively.

VABBA with RS anode. Experiments using the RS anode were conducted to determine the possibility of its application for deposition on the inside surface of hollow objects (e.g., inner surface of tubes). Arc current was $I = 175$ A, and the film was deposited at $L = 70$ mm without any shutter for 90 s. The deposition thickness was mostly equally relatively to different substrate locations, with maximum of about 0.22 μm at the center of the substrates (corresponds to the central circumference of perforated lateral anode surface) and decreasing to about 0.07 μm at the substrate (specified as 70 mm) periphery. Figure 22.38 presents the measured time dependence of RS deposition rate with arc current as a parameter. For all arc currents considered, there were peaks in the deposition rate with values of about 0.2, 0.3, 0.47, 0.64, 0.8, and 1 μm/min for $I = 150$, 175, 200, 225, 250, and 275 A, respectively. The peak time decreased from about 70 s for $I = 150$ A to 25 s for $I = 275$ A. After the peak, the deposition rate declined to mostly a steady-state values depended on arc current in range 200–275 A. For 175 and 150 A arcs, the deposition rates continue weakly decrease from their relatively low peaks.

The XRD pattern is obtained for Cu thin film deposited with the FS Ta anode, $I = 200$ A, $L = 100$ mm, and with an exposure time of 20 s beginning 60 s after arc initiation. Only polycrystalline Cu was detected, and the grain size was 30 nm. Figure 22.39 shows an SEM image of this Cu film. Optical microscopy of the deposited film shows that rarefied MPs were deposited for times 30–60 s, while

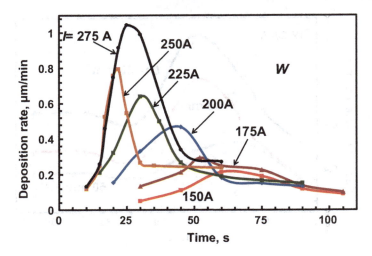

Fig. 22.38 Radial shower anode deposition rate with arc current as a parameter (W—0.8 mm, L_s = 70 mm)

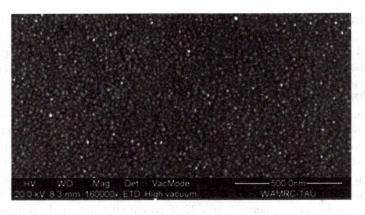

Fig. 22.39 SEM image of a Cu thin film, Ta (0.6 mm) anode, $I = 200$ A, $L_s = 100$ mm, 60 s shutter was closed and deposition time was 20 s (shutter was opened)

at more than 60 s after arc ignition, no MPs were detected anywhere on the substrate area (25×75 mm^2).

Cathode erosion was found to be about 26, 43, and 55 µg/C for the FS, RS, and OH anodes respectively. This is lower than in a normal cathodic arc configuration having a freely expanding arc plasma jet (~100 µg/C) and in HRAVA electrode configuration [119]. The relatively low cathode erosion rate was possible due to large plasma back flux to the cathode reflected from the hot anode in the closed electrode assembly. This result indicates larger VABBA deposition efficiency using the cathode erosion material comparing to the erosion rate for vacuum arc with open electrode configuration.

22.5 Application of Arcs with Refractory Anode for Thin Films Coatings

The peak appearance in the V_{dep} dependence on arc time was possibly due to MP's contamination in the cathode plasma flux, which was deposited on the cold anode at the initial stage of VABBA deposition re-evaporated when the anode was sufficiently hot producing the additional plasma contribution. Larger W anode aperture diameter significantly increased the film deposition rate that was demonstrated by the result obtained from W (1 mm) shower anode in comparison with also refractory Ta (0.6 mm) cathode reaching relatively large rate of 3.6 µm/min at $L = 80$ mm. This conclusion based on the assumption that the temperature reached sufficiently high for both refractory anodes (W and Ta) to completely cathode material re-evaporation from the anodes.

The VABBA demonstrated that the plasma plume expansion from radial electrode configuration can be used for thin film deposition on an internal tube surfaces with a large deposition rate (about 1 µm/min with $I \geq 275$ A). Also at large substrate distances from the frontal surface and at large substrate (~160 mm diameter), the metallization can be produced with rate of ~0.3 µm/min at relatively large distance ~280 mm. The deposition uniformity can be improved using rotating tube.

22.5.2.3 Cu Filling Trenches in a Silicon Wafer

Thin metallic films are extensively used in microelectronics. One of the important steps in integrated circuit (IC) manufacturing is the production of metal connections between different parts of the circuit [120]. For example, submicron high conductivity interconnects are a limiting factor in further miniaturization of ultra-large-scale integrated circuits (ULSI). These connections are in the form of trenches or vias etched in dielectric material. It is required to fill them with metal completely. Usually, Al was commonly used as an interconnect material. Recently, a preferred material for ULSI metallization is copper due to its high conductivity and better electro migration reliability [121].

Several processes have been demonstrated to deposit Cu into vias and trenches. CVD [122], with a typical deposition rate less than 100 nm/min, electrochemical plating [123] and electroless plating methods [124, 125] with typical deposition rates of about 150 nm/min and physical vapor deposition (PVD), e.g., based on thermal evaporation and sputtering, which produce mostly neutral particles [119]. Another approach, ionized physical vapor deposition (I-PVD) [126] techniques, was developed from conventional sputtering and has a typical deposition rate of about 200 nm/min [9]. The main disadvantages of the PVD processes are relatively low deposition rate, and high sidewall and low bottom deposition, creating a void. The main disadvantage of the electrochemical and electroless processes is that plating baths must be disposed, posing both practical and environmental concerns.

The main disadvantage of using cathodic vacuum arc plasma is the presence of droplets, known as macroparticles (MPs), which also generated from the cathode spot. Several techniques for eliminating or reducing the MP density in the coating are used [6, 55, 127], but generally have poor material utilization or pulse vacuum arc. The radially expanding anodic plasma in a HRAVA can be used to produce mostly MP-free metallic coatings with deposition rates of up to 2 μm/min [89, 90]. In this section, the trench filling using the HRAVA is demonstrated.

Cathode–anode assembly The arcs were conducted in a stainless steel chamber (Fig. 22.8) of system (b). A diffusion pump before arc initiation pumped the chamber down to a pressure of 0.67 mPa. The experimental system is shown schematically in (Fig. 22.10). During the arc, the chamber pressure increased to approximately 13 mPa. An arc current of 200 A was applied for a period of 180 s between a 30-mm-diameter cylindrical, water-cooled, Cu cathode and a W anode, 32 mm in diameter and 30 mm in length. The arc was initiated by momentarily touching the cathode with a trigger electrode, electrically attached to the anode through a current-limiting resistor. The anode was supported on a thin W rod that also connected it to the electrical circuit. A molybdenum shield surrounded the cathode, 40 mm in diameter, placed flush with the cathode surface. The interelectrode gap was approximately 10 mm.

The substrate was a silicon wafer with a top layer of 1.2 μm thick SiO_2, in which trenches of 300 nm depth and widths down to 100 nm were etched. The wafer was cut into samples with areas of approximately 10×10 mm^2. The substrates were mounted on a holder, which allowed orienting the substrate perpendicularly face the plasma flux emanating from the interelectrode gap in the anode region [90]. The substrates were located at a distance L from the electrode axis, which was varied from 74 to 125 mm.

The substrate consisted of two main layers, a top SiO_2 layer facing the plasma flux and a bulk silicon layer to which bias voltage was connected. As the top SiO_2 layer was an insulator, applying a DC bias via the substrate does not affect the potential of the surface exposed to the plasma, and hence does not influence the energy of the incoming ions. Barnat et al. [10, 128] studied the influence of pulsed bias to eliminate surface charging, in order to deliver an approximately mono-energetic ion flux while varying the pulse width. It was shown that for bias a voltage of -100 V, pulse frequencies between 0.1 and 1000 kHz and a duty cycle of 70–90%, the average surface potential of the insulating substrate was minimized to within about 1–5% in a frequency range of 50–90 kHz.

In the considered work, Cu deposition with a HRAVA system using conditions close to the optimum found by Barnat et al. [10, 127] was studied. Specifically, in this case a bias voltage of -100 V a duty cycle of 80% and pulse frequency of 60 kHz were used. In order to best utilize the HRAVA properties, the substrate was isolated from the particle flux by a shutter during the first 60 s after arc initiation in order to allow the HRAVA mode to be established [90], and thus producing MP-free plasma and the highest deposition rate.

The deposited films were examined [129] using a high-resolution scanning electron microscope (HRSEM). The cross section of the deposited substrates was examined after cleaving the substrate, to determine fill quality, film thickness, and deposition rate. The sheet resistance was measured on the top surface of the coating using a Signatone S-301-6 manual four-point probe, with a 62085TRS head. The resistivity was calculated based on the film thickness measured with HRSEM in flat regions, i.e., without trenches. An X-ray diffractometer equipped with Cu–K_α radiation (λ

22.5 Application of Arcs with Refractory Anode for Thin Films Coatings

= 1.5406 Å) in Bragg geometry was used to analyze the structure of the deposited films.

Results. The HRAVA-deposited films indicate a Cu-deposited substrate with a row of successfully filled trenches. A bias amplitude of 100 V, 60 kHz bias frequency, duty cycles of 80 at a distance of about 110 mm from the electrode axis was applied for aspect ratios of 0.5, 1, 1.5, and 3 trenches. The exposure time to the plasma of 2 min. The narrowest trenches were 100 nm wide and 300 nm deep (aspect ratio of 3). Figure 22.40 shows the cross-sectional microstructure of deposition.

The deposition rate dependence on the distance L of the substrate from the arc axis is presented in Fig. 22.41. It can be seen that the deposition rate decreased with distance. The film thickness was in range from 350 to 850 nm. The maximum deposition rate was 425 nm/min. Figure 22.42 shows the dependence of average resistivity of the deposited films versus distance from the electrode axis. The films were deposited using a bias voltage amplitude of −100 V, duty cycle of 80%, and frequency of 60 kHz. The resistivity decreased with decreasing distance. At the distance of 55 mm, the resistivity decreased to a value of 1.72 $\mu\Omega$ cm, while the intrinsic resistivity of copper is 1.67 $\mu\Omega$ cm.

The average resistivity increased with distance from the electrode gap axis. This may be caused by lower plasma density and lower heat flux density at the substrate with the lower deposition rate at the increased distance, due to radial plasma expansion [11]. With increased distance, a smaller part of the emitting flux will reach the substrate, thus reducing the plasma density and the substrate temperature, and as consequence the defect density and microstructure of the film are affected [130].

It was found that the trenches were always completely filled using 80% duty cycle. With duty cycles about 50%, some of the trenches were filled and some were not.

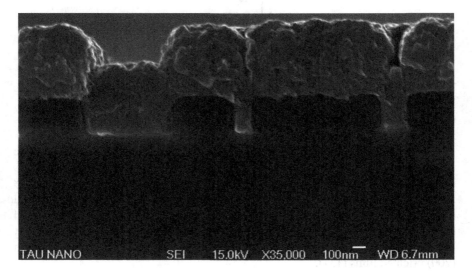

Fig. 22.40 Trench depth 300 nm, widths 600, 100, and 200 nm. Deposition time—2 min, arc current 200 A, $L = 110$ mm, pulsed bias of −100 V, 60 kHz and 80% duty cycle

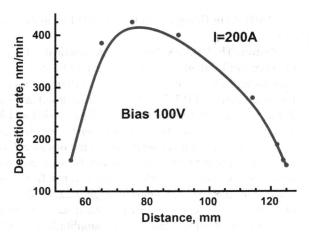

Fig. 22.41 Deposition rate versus distance from the electrode axis, bias voltage: −100 V, 80% duty cycle, 60 kHz. The dashed curve is a fourth-order fit to the experimental data

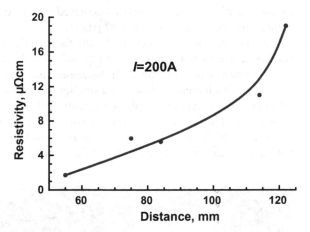

Fig. 22.42 Average resistivity and deposition rate of the deposited films, using a pulsed bias voltage of 100 V, with a frequency of 60 kHz and a duty cycle of 80%

The observed films on the overlying flat surfaces and partially filled wide trenches had the same average grain size, 45 nm. Figure 22.43 shows a typical XRD pattern, showing strong (111) texture—the intensity ratio between the (111) and (200) planes was 3.4.

The deposited film texture depends on the deposition processes. Typically, the Cu film texture is (111) and (200) [121]. In our case, (111) and (200) texturing was observed in the deposited Cu films, and the diffraction peaks at $2\alpha = 43.3°$ and $50.4°$ were assigned to the (111) and (200) reflection of Cu. The ratio between (111) and (200) is more than twice stronger than that ratio (about 1.5) according to the standard random sample (JCPDS card no. 04-0836). This is important and beneficial for circuit lifetime because (111)-oriented Cu films have better electromigration reliability than (200) oriented films [131].

22.5 Application of Arcs with Refractory Anode for Thin Films Coatings

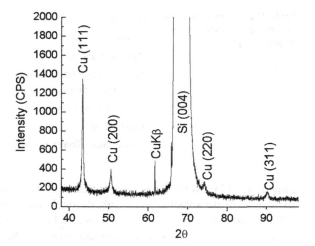

Fig. 22.43 Typical XRD pattern of the deposited film (figure taken from [129]. **Used with permission**)

The obtained result confirmed the validity of our initial selection of the bias parameters. As discussed by Barnat et al. [126], the charging effect depends on the bias voltage, the structure of the substrate and on the plasma parameters—especially the plasma density. The values of first two parameters were close in our experiment and in Barnat's. The plasma density near the substrate in our case was about 10^{11} cm^{-3}, and it is in the 10^{10}–10^{12} cm^{-3} range in Barnat's work [127]. Therefore, despite the difference in the systems, the experimental results show that in both cases, pulsed bias was effective for biasing the surface of insulating SiO$_2$ top layers facilitating metal deposition. The much higher deposition rates in the HRAVA system are because in the case of magnetron sputtering, the plasma consisted of Ar ions, which do not condense and form a metal film, while the HRAVA plasma was composed of Cu, which condenses on the substrate and forms a metal film.

22.5.2.4 Chrome and Titanium Thin Films

Vacuum Chambers and Electrodes. Cr and Ti HRAVAs were studied in separate stainless steel chambers of systems: (a) 400 mm length, 160 mm diameter and (b) 530 mm length and 400 mm diameter, respectively. The cathode–anode assembly was mounted in the chambers as shown schematically in Fig. 22.9. The chambers were diffusion pumped down to pressures of (a) 2.6 mPa and (b) 0.67 mPa before arc ignition. During the arc, the pressure in the chamber increased to about (a) 26 mPa or (b) 13 mPa. The arcs were operated at currents of $I = 200$–300 A for periods up to 120 s. The anode was grounded, and the chamber was floating. Both the water cooled cathode and the anode were surrounded by Mo shields. The anode shield reduced radiative heat loss from the side of the anode. The electrode and shield dimensions, as well as the chamber differed according to the cathode material, as detailed in Table 22.5 [103]. The interelectrode gap was $h = 10$ mm in all experiments. The

Table 22.5 Systems type and dimensions of electrodes and shields [103]

Cathode	Cr	Ti
Systems	a	b
Cathode diameter (mm)	30	60
Cathode height (mm)	20	25
Cathode shield diameter (mm)	50	70
Anode diameter (mm)	32	60
Anode height (mm)	30	1.5
Anode shield diameters (mm)	60, 70	None

cathode shield prevented the MPs emitted from the cathode spots from reaching the "A region" of the substrate. The cathode shield position determined the location of the boundary, designated as $y = 0$ (Fig. 22.11), between regions on the substrate with low and high MP contamination, designated as the anodic (A) and cathodic (C) regions, respectively.

The substrates were 75×26 mm² glass microscope slides. The slides were mounted on a water-cooled holder, located at distances of $L = 80$–145 mm from the electrode axis. MP flux was determined using an optical microscope equipped with a digital camera. MPs with diameters $> 3\mu m$ were counted at $y = \sim 5$ mm in the A-region on substrates exposed to the plasma for a 30 s deposition time starting 60 s after arc ignition. The film thickness (H) was measured by profilometry.

The thickness growth (ΔH) was determined between times 0–15, 15–30, 30–45, 45–60, 60–5, 75–90, 90–105, and 105–120 s from arc ignition. The arc current was varied as $I = 200, 250$ and 300 A. Film thickness growth was measured as a function of distance from the electrode axis for the Ti and Cr cathodes for $I = 200$ A and $L = 80, 100, 115, 130$ and 145 mm.

Results *Deposition rate*. The temporal evolution of the deposition rate V_{dep} is presented in Fig. 22.44. For Cr with $L = 80$ mm, and in Fig. 22.45 for Ti with L

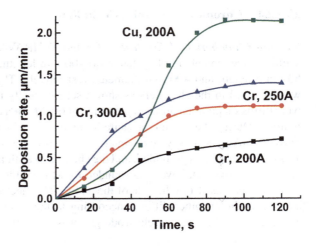

Fig. 22.44 Temporal evolution of the deposition rate at $L = 80$ mm for Cr ($I = 200, 250$ and 300 A). The curve for Cu ($I = 200$ A) from [102] is presented to comparison

22.5 Application of Arcs with Refractory Anode for Thin Films Coatings

= 100 mm, with I = 200, 250, and 300 A as a parameter. It can be seen that the deposition rate for Cr with I = 300 A was ~0.4 μm/min at ~15 s after arc ignition, and reached a steady-state level of ~1.4 μm/min in the developed HRAVA, approximately after 60–70 s from arc ignition. It was observed that the deposition rate increased with arc current. For comparison, the temporal evolution of the deposition rate for Cu (I = 200 A) was also presented in Fig. 22.44. It can be seen that V_{dep} for Cu significantly exceeds the deposition rate for Cr after 45 s of arc operation.

For Ti with I = 300 A (Fig. 22.45), the deposition rate was ~0.54 μm/min ~20 s after arc ignition and reached a steady state of ~1.8 μm/min in the developed HRAVA stage. The dependence of steady-state deposition rate versus distance for Ti and Cr cathodes with I = 200 A is shown in Fig. 22.46. The deposition rate decreased with

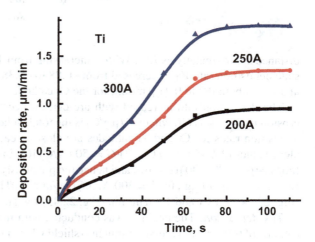

Fig. 22.45 Temporal evolution of the deposition rate for Ti (L = 100 mm, I = 200, 250, and 300 A)

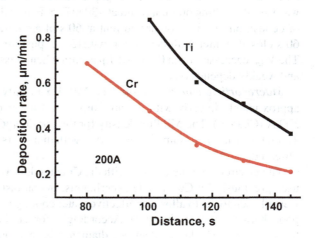

Fig. 22.46 Dependence of the steady-state deposition rate versus distance for Ti and Cr cathodes (I = 200 A)

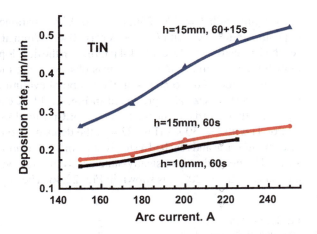

Fig. 22.47 HRAVA titanium nitride deposition as dependence on arc current

distance approximately as L^{-2}. With L increasing from 100 to 145 mm, the steady-state deposition rate V_{dep} decreased from ~0.88 to ~0.38 μm/min for the Ti cathode and from about 0.7 to 0.21 μm/min for the Cr cathode.

The deposition rate increased with arc current because the total cathode mass erosion increases as GI, where G (μg/C) is the total cathode erosion coefficient. The deposition rates for Cr and Ti cathodes are close to each other because they have close values of G (Cr-40 μg/C, Ti $G = 50$ μg/C) [6]. In contrast, G for Cu is 2.5–3 times larger, 100–130 μg/C [6], and hence the previously determined deposition rate for Cu was larger (e.g., for $I = 300$ A, 3.6 μm/min [101]), compared to 1.4 μm/min (Fig. 22.44.) for Cr and 1.8 μm/min (Fig. 22.45) for Ti.

TiN deposition. The position was conducted with tungsten anode of 60 mm. Ti cathode of 60 mm was recessed 1 mm into shield when it was used. The distance from electrode axis was $L = 100$ mm and gap distance h 10 and 15 mm. The pressure of N_2 was varied reaching optimal result at ~30 mTorr. Figure 22.47 shows the dependence of deposition rate V_{dep} on arc current at 60 s of arc duration at open shutter and at 60 s closed shutter with deposition with 15 s exposition for N_2 pressure 30 mTorr. The V_{dep} decreased from 0.55 to 0.1 μm/min when pressure increased to 200 mTorr and weakly depends on h.

Macroparticle contamination. The MP flux density in the A-region decreased approximately linearly with the arc current and reached ~1 mm^{-2} min^{-1} for $I = 300$ A (Chap. 9). The MP flux density for the considered arc currents was larger for Ti, on the order of 10 mm^{-2} min^{-1}. MPs with diameters more than 20 μm were not observed.

The electrode geometry used with the Cr cathode was chosen to be this same that used previously for Cu cathode experiments. In contrast, larger electrode diameters were chosen for Ti to allow for effective water cooling of the cathode, because of the poor thermal conductivity of Ti. Accordingly, for Ti, the minimum L was selected to be 100 mm. The larger electrode diameters in this case produced a lower plasma density and a lower anode temperature for a given arc current, and therefore, MP flux density was larger for Ti than for Cr. The Ti cathode will be investigated with $I > 350$ A in the future.

22.5.3 Advances Deposition of Refractory Materials with Vacuum Arcs Refractory Anode

In general, the cathode spot is particularly suitable for generating plasma comprised of refractory metallic elements (i.e., with high melting temperatures and low equilibrium vapor pressure), such as W, Nb, C, Ti, Zr, and Ta. In some applications, such metals requested to be deposited as thin films and coatings. Refractory films have been used for investigating arcing in tokamaks (on tungsten coated graphite) [132], reaction of Ti, Nb, W, Zr, and Mo films (deposited by rf sputtering) with Al_2O_3 (motivated by microelectronics applications) [133], and properties of metal containing diamond-like carbon coatings [(Ti, Nb, W)-DLC] deposited by magnetron sputtering [134]. Hoffman and Thornton [135] presented data for up to 0.3 μm thick sputtered onto glass substrates at a nominal deposition rate of 1 nm/s over the pressure range 0.067–4.0 Pa of argon. Lia et al. [136] reported on CVD carbon growth. Girginov et al. [137] reported electrodeposition of refractory metals (Ti, Zr, Nb, Ta). Pogrebnyak et al. [138] fabricated new nanocomposite (Ti Zr Hf V Nb)N coatings using vacuum arc deposition. Thus, while refractory metal films were deposited by several methods with low rate of deposition, it is important investigate a possibility of the deposition method from a vacuum arc with a refractory anode. The present section shows the deposition rate of W, Nb and Mo films with HRAVA and VABBA electrode configurations [104].

The VABBA electrode configuration is shown schematically in Fig. 22.9. The cathode–anode distance was $h = 10$ mm. The aperture diameters were 1 mm (250 holes) and 1.5 mm (113 holes) for W and Nb anodes, respectively, and one hole of 4 mm diameter for Ta anode. The front wall thickness (with aperture) of the anodes was 8 mm. The HRAVA electrode configuration with $h = 7$ mm, $L = 110$ mm (except Zr cathode, $h = 10$ mm, $L = 125$ mm) is shown schematically in Fig. 22.10. The cathodes were mainly water cooled. In the HRAVA configuration, the anode was surrounded by Mo shields, which reduced radiative heat loss from the side of the anode. A long-duration arc was ignited between the cathode and anode, which were both fabricated from refractory materials. The characteristics of electrode pair configurations are summarized in Table 22.6.

The substrates were $75 \times 25 \times 1$ mm glass microscope slides. Also a 185-mm-diameter glass disk substrate was used for deposition at large distances from the electrodes. They were pre-cleaned with detergent and alcohol and placed at different distances L from the VABBA anode surface (Fig. 22.10) and at $L = 110$ mm from the arc axis in the HRAVA configuration (Fig. 22.9). The substrate was shielded from the depositing plasma flux by a shutter, which was opened for an exposure time of 15 s at various times after arc ignition. The deposition rate V_{dep} was determined by dividing

Table 22.6 Electrode materials and pair electrode configurations [105]

Number	Electrode configurations				
	Cathode	Anode	Cathode	Anode	
(1)	**(I) HRAVA**, cathode not water cooled		**(II) HRAVA**, cathode not water cooled		
	Zr 60 mm diam	W 60 mm diam	W 32 mm diam	W 32 mm diam	
	Cylindrical electrodes. The pressure was 26 mPa (0.2 mTorr) during arcing.		Cylindrical electrodes. The pressure was 78–104 mPa (0.6–0.8 mTorr) during arcing. $h = 6$ mm, $L = 110$ mm		(5)
	(I) HRAVA, water-cooled cathode.		**(II) VABBA**, water-cooled cathode		
(2)	Mo 32 mm diam	Mo 32 mm diam	Nb, W 32 mm diam	Ta, W 50 mm diam	
			The pressure—0.013 mPa and 0.01 mTorr during arcing		
	Cylindrical electrodes. The pressure was 78–104 mPa and 0.6–0.8 mTorr during arcing.		W shower-head anode, 250 holes, 0.85 mm diameter with W cathode		(6)
(3)	Nb 32 mm diam	W 32 mm diam	W shower-head anode, 250 holes, 1 mm diameter with Nb cathode		(7)
	Cylindrical electrodes. The pressure was 26 mPa (0.2 mTorr) during arcing		Ta one-hole 4-mm-diam cup anode		(8)
(4)	W 32 mm diam	W 32 mm diam	Nb 32 mm diam	Nb 32 mm diam	
	Cylindrical electrodes. The pressure was 26 mPa (0.2 mTorr) during arcing		Nb shower-head anode with 113 holes of 1.5 mm diameter		(9)

the film thickness by the exposure time. HRSEM images were obtained using JEOL JSM-6700F high-resolution scanning electron microscopy.

Results. A photograph of Mo and Nb films obtained using Mo and W anodes, respectively, deposited with configuration (**I**) with a 200 A arc and $L = 110$ mm is shown in Fig. 22.48. As can be seen, both films had a mirror-like finish, as seen by the reflection of a fluorescent lamp. It should be noted that the anodes worked with different refractory cathodes were clean after 20–30 s of arcing.

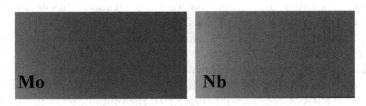

Fig. 22.48 Photograph of Mo and Nb films obtained using configuration a with a 200 A arc, $L = 110$ mm

22.5 Application of Arcs with Refractory Anode for Thin Films Coatings

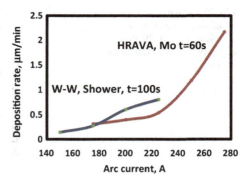

Fig. 22.49 Water-cooled Mo cathode-Mo anode. Deposition rate as function of current, at distance $L = 100$ cm, for arc time before deposition 60 s, with $h = 7$ mm (configuration 2 in Table 22.6)

Below, the maximal deposition rate is presented for the VABBA, in the center of the substrate, and for the HRAVA, at side deposited from the anode plasma. For the VABBA, the thickness was parabolic distributed like shown for Cu in [103]. In the HRAVA, the thickness was slightly decreased from anode to the cathode side. In a previous Cu deposition experiment, a sharp boundary between an anode and a cathode deposition region (Fig. 22.11, A and C regions) was observed for the HRAVA [8].

In contrast, however, the present refractory films had no such sharp boundary (see Fig. 22.48). Here the thickness distribution along the substrate was not measured. Such similar distributions can be observed considering the results for Cu by VABBA [103] and for Sn by HRAVA in [98] depositions.

Tungsten film deposition rate as function of arc current with a W shower-head anode, 0.85 mm hole diameter, for arc time before deposition 100 s, (configuration 6 from Table 22.6). Figure 22.49 shows molybdenum and tungsten deposition rate at 200 A. The measured V_{dep} was 0.8 μm/min. A water-cooled Mo cathode-Mo anode was used to deposit a Mo film using system (**I**)-HRAVA, $h = 7$ mm (configuration (2) in Table 22.6). The deposition rate sharply increased when the arc current exceeds 225 A and reached about 2.2 μm/min at 275 A and $L = 100$ mm. V_{dep} increased up to 0.7 μm/min after 60 s of arcing. Tungsten film deposition with a W head shower anode (with 0.85 mm holes) as a function of I at $L = 60$ mm and 100 s is presented for configuration (6) from Table 22.6. In this case, V_{dep} reached 0.6 μm/min.

The Zr film produced using system (**I**), configuration (1) in Table 22.6, was deposited at $L = 110$ mm and 90 s after arc ignition with 15 s exposure time, arc current 200 A and $h = 10$ mm. The measured V_{dep} was 0.8 μm/min. The time dependences of Nb and W film deposition rates with HRAVA configurations (3) and (4), respectively, from Table 22.6 are presented in Fig. 22.50. V_{dep} sharply increased after 1 min for W up to ~1 μm/min. The dependence of V_{dep} on time after arc ignition for W films deposited using configuration (5) from Table 22.6 with two cylindrical W electrodes, $L = 60$ mm, and $h = 8$ mm is shown in Fig. 22.51.

Figure 22.52 presents the VABBA Nb film deposition rate at 30 s after arc ignition as a function of arc current using anode configurations (7)–(9) of Table 22.6. V_{dep} increased with arc current for all anode types. With the W shower-head anode, when

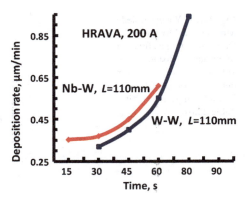

Fig. 22.50 Nb (configuration 3) and W (configuration 4) HRAVA deposition rates as functions of time

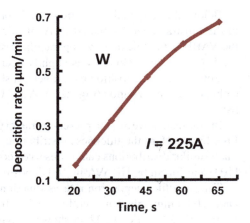

Fig. 22.51 W deposition rate as a function of time after arc ignition by both planar W cathode and W anode configuration (5), $L = 110$ mm, $h = 8$ mm

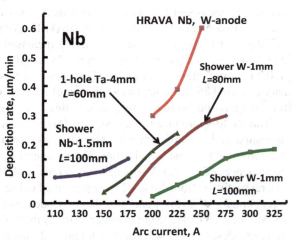

Fig. 22.52 Nb (cathode) deposition rate as function on arc current with a single-hole Ta anode, a shower-head W anode, and a shower-head Nb anode in the VABBA configuration, and with a Nb cathode and W anode in the HRAVA (configuration 7, 8, and 9 from Table 22.6)

22.5 Application of Arcs with Refractory Anode for Thin Films Coatings

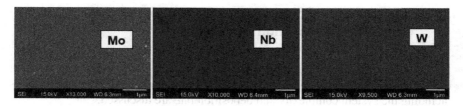

Fig. 22.53 HRSEM images of the films deposited with Mo (HRAVA, (2), 200 A), Nb (VABBA, (9), 325 A), and W (HRAVA, (5), 225 A)

arc current increased from 200 to 275 A V_{dep} increased from 0.13 to 0.3 μm/min. When the distance to the substrate decreased from 100 to 80 mm, V_{dep} increased from 0.15 to 0.3 μm/min. When arc current increased from 150 to 225 A for one-hole Ta anode V_{dep} increased from 0.05 to 0,25 μm/min. The HRAVA Nb film deposition rate as a function of arc current was also presented for comparison. It was significantly larger than the VABBA deposition rate with all other conditions the same. Generally, no MPs were observed in the VABBA-deposited films. Figure 22.53 shows HRSEM images of Mo, Nb, and W films. Only a few MPs were observed in the HRAVA-deposited films on the anode side.

In general, the results show that the refractory material (W, Nb, Mo) film deposition rates were lower than previously obtained for Cu, Cr, or Al deposition [8]. This is due to the relatively low rate of evaporation from the refractory cathode. In addition, the overall feature of the deposition indicates a relatively good re-evaporation from the hot anode of the previously condensed cathode material. It might be explained by very low adhesion on the hot surface, which increases during the anode heating with arcing time [139]. The relatively lower MP contamination in the deposited films was due to low MP generation show for high melting temperature and low vapor pressure cathode materials [9].

The deposition rate strongly depends on the electrode configuration and distance to the substrate due to strong expansion of the ejected plasma. V_{dep} reached much higher values with HRAVA (Figs. 22.49 and 22.50) than with VABBA (Figs. 22.51 and 22.52). This is because the plasma expansion was relatively free in the planar disk configuration comparing to the plasma ejection from the small apertures of mostly closed discharge volume. However, the VABBA technique allows obtain mostly total MPs conversion to the plasma state significantly increasing the performance of the cathode erosion products. The obtained results demonstrated that mainly the deposition rate from VABBA was influenced by the aperture sizes and distance to the substrate (Nb deposition, Fig. 22.52).

22.6 Comparison of Vacuum Arc Deposition System with Other Deposition Systems

In this section, the main features of various deposition techniques are compared with deposition using vacuum arc plasma source with refractory anode. Different aspects including the system complexity and deposition rate are discussed.

22.6.1 Vacuum Arc Deposition System Compared to Non-arc Deposition Systems

Conventional PVD includes evaporation and sputtering. In evaporation systems, the source material is held in a crucible or hearth, which is heated resistively, inductively, or by an electron beam (EB). The deposition rate is limited by the evaporation rate of the source material. Usually, the source material is molten, requiring upward orientation of the source. Large power supplies are required maintain refractory deposition materials at sufficiently high temperatures to obtain a useful deposition rate. The deposition is effective when the sample surface is normally oriented to the vapor stream, which propagates in a straight line from the evaporation source. Consequently, the coverage on vertical walls is poor. It is generally difficult to evaporate alloys from a single source, if its various components have different vapor pressures. The energy of the evaporated atoms is determined by the source surface temperature and is relatively low. As a result, the coating density and adhesion are also low.

EB evaporation can have deposition rates in range of 0.1–100 μm/min on relatively cool substrates with very high material utilization efficiency. However, the hardware is complex as radiation damage, and X-ray and secondary electron emission must be addressed.

Magnetron sputtering (MS) can be conducted at lower substrate temperatures comparing to other systems, but it leaves residual compressive stress in the film. In MS systems, the ions from a low-pressure glow discharge bombard the target and eject mostly neutral atoms. Conventional MS cannot fill vias and trenches—often voids are formed due to the broad angular distribution of the sputtered atoms especially for trenches with high aspect ratio [1, 9]. The deposition rate and deposition quality for samples having surface structures were significantly improved by relatively complex plasma-assisted IPVD or SSMD techniques [1, 30].

CVD, electroless and electroplating techniques have some advantages over PVD technology. Highly conformal films can be deposited on very complex shaped substrates. The gas composition or the substrate temperature can control CVD film composition. However, in CVD, the film growth takes place at high temperatures, and corrosive gaseous products may be formed and incorporated into the film. The energy of the depositing atoms is low in chemical techniques. Film morphology, texture, and other properties depend strongly on the deposition parameters. The Cu film structure in electroplating has more randomly oriented grains than by MS PVD

[54]. Electroplating baths must be disposed, posing both practical and environmental concerns.

In comparison, the HRAVA deposition system is relatively simple working without magnetic systems, plating baths, and complex power sources. A plasma source using HRAVA or VABBA configurations generates highly energetic (~20 eV) and highly ionized (~0.6) metallic plasma. The expanding plasma has potential to metallize different substrate shapes including plane, bands, rings, and internal surfaces. High deposition rates can be achieved, e.g., ~3.6 μm/min for Cu films. The deposition rate obtained with various technologies is summarized in Table 22.7. It can be seen that only by EB-PVD exceeds the HRAVA deposition rate. However, the apparatus is complex, and the electrical power required to reach the cited deposition rate is much larger (~100 kW) [73] than needed for HRAVA deposition (~6 kW). According to Schultrich et al. [66], this advantage of EB-PVD is restricted by the negligible vapor activation (low degree of ionization) in the conventional electron beam techniques caused by the low interaction of high-energy electrons with the vapor.

22.6.2 Comparison with Other Vacuum Arc-Based Deposition Systems

Two vacuum arc deposition systems using different techniques, filtered vacuum arc deposition (FVAD) with a quarter-torus MP filter, and HRAVA, were directly compared experimentally using $I = 200$ A, deposition time 120 s, and a Cu cathode in both systems [101]. The deposited area, cathode utilization efficiency, thickness deposition rate, mass deposition rate, ion flux fraction in the total depositing flux, and MP contamination were compared. The HRAVA films were deposited on a cylindrical region co-axial with the electrode axis with a characteristic area of $S = 2\pi R \Delta h = 100$ cm^2, where $\Delta h = 2$ cm is the characteristic half-width of the thickness distribution in the axial direction. The mass deposition rate Δm_{tot} was calculated by integrating the measured film thickness on the whole deposited area. The mass deposition in the FVAD system was calculated taking into account its axially symmetric thickness distribution. The average cathode mass utilization efficiency f_m was found as the ratio between the total mass of the whole deposited film and the total mass eroded from the cathode [101].

The measured HRAVA mass deposition rate (400 mg/min) was about 40 times higher than for the FVAD system (9.5 mg/min), and the cathode mass utilization efficiency f_m for the HRAVA system was about 36% in comparison with 0.8% for FVAD. The HRAVA coated a much larger substrate area than the FVAD system (100 cm^2 compared with 30 cm^2), azimuthally uniform around the electrode axis [101]. V_{dep} was an order of magnitude greater by HRAVA deposition than by FVAD, 2.3 and 0.25 μm/min, respectively. In the FVAD system, the substrate is placed at a relatively large distance from the cathode to reduce MP flux, resulting in considerable plasma losses to the filter walls, as well as a bulky system. In the HRAVA system,

Table 22.7 Characteristics of film and method deposition

Method	Material	Deposition rate (μm/min)	Substrate temperature (°C)	References
PVD (magnetron sputtering) Ionized physical vapor deposition High-power impulse magnetron sputtering	Ti	0.017	25	Rossnagel [1]
	Al on SiO2	0.01	25	Barnet [10]
	Cu and Ti films	By factor of 2 lower	–	Helmersson [9]
	AlOx from an Al	By factor of 3–4 lower than with DC sputtering	–	Helmersson [9]
Self-sputtering magnetron deposition	Cu	2	25	Radsimski [11]
	Cu on Si wafer	0.14	25	Posadowski [13]
	Reactive: Al2O3	0.02–0.4	25	Posadowski [13]
Electron beam PVD	Metals and alloys	30–50	–	Movchan [20]
	Oxides, carbides, and borides	15–20	–	Movchan [20]
CVD	Cu on TiN sample	0.2–0.05	250–200	Cho [28]
	Cu on Si wafer	0.05	190–140	Bollmann [27]
	Cu on TiN	0.05–0.01	140–110	Bollmann [27]
Plasma-assisted CVD	Si2O2 on Si(100) substrate	0.1	115–350	Babayan [30]
	Cu on SiO2, Al, Ta, and TiN samples	0.02–0.004	240–160	Lakshmanan [29]
Electroless deposition	Cu on Si wafer	0.1–0.16	70	Hasegawa [32]
	Cu on Si wafers	0.1–0.12	55–90	Shacham [31]
VAD with straight duct	C on Si wafer	0.1	–	Rother [47]
	Cu and Al	0.18	–	Boxman [80]
Shield filter	TiN	0.016	–	Miernik [60]
Filtered high-current arc	Cu on Si wafer	Few	–	Simroth [54]
Hot anode vacuum arc	Al, Ti, and Cr	0.06–0.6	<70	Erich Hasse [79]
	Al	0.48		Erich [82]
Arc-like (EB gun) or spotless arc deposition-SAD	Ti- stainless plate	0.132	220	Kajioka [72]
	Ti and TiN	3–60		Goedicke [73]
	FeCrNi on steel	0.1–2		Scheffel [74]

(continued)

22.6 Comparison of Vacuum Arc Deposition System ...

Table 22.7 (continued)

Method	Material	Deposition rate (μm/min)	Substrate temperature (°C)	References
Hot refractory vacuum arc (on glass)	Cu	2.3 (200 A), 3.6(300 A)	25	[102]
	Ti; Cr	1.8; 1.4 (300 A)		[103]
	Al; Zn; Sn	1.6(22 5A); 2(225); 3(175 A)		[97–99]
	Mo; W;	2.2(275 A); 1(200 A)		[105]
	Nb; Zr	0.6(200 A); 0.8(200 A)		[105]
Vacuum arc black body assembly	Cu	~3(200 A)	25	[104]
	W; Nb	0.7(225 A); 0.3(275 A)		[105]

almost MP-free plasma flux is obtained directly from the anode plasma plume; a large source–substrate separation was not required. The plasma jet in the FVAD system was sometimes not stable due to the presence of the magnetic field near the cathode surface. [140]. In contrast, in the HRAVA system, a magnetic field was not required.

The mass utilization rates obtained with other FVAD systems reported in the literature were also examined. Their cathode mass utilization was calculated from the reported ion current assuming that the depositing flux is fully ionized from:

$$f_m = \frac{I_{\text{ion}}}{I_{\text{arc}}} \frac{m}{ZeG} \quad (22.21)$$

where G (g/C) and Z are the erosion rate and ion charge [6, 36]. Table 22.8 shows that $f_m = 36\%$ for HRAVA exceeds the highest value for a quarter-torus FVAD system [141, 142] by three times. The ratio of ion current at the filter exit to I_{arc} is less than 2.5% for most FVAD systems [33], corresponding to $f_m = 8.8\%$. It was also presented the data of steady state [143] and pulsed [144] FVAD sources for comparison in Table 22.8. Small MPs have been observed to be transmitted with plasma flow even in curved filter ducts [57]; this result explained theoretically by MP reflection from the duct walls [56]. In the HRAVA, most of the MPs are evaporated in the dense plasma and on the hot anode surface.

The maximum deposition rate was 425 nm/min, and the resistivity was 1.7 μΩ·cm obtained for Cu film filling trenches with aspect ratio of 3 by HRAVA source. The resistivity was higher than in bulk Cu. It should be noted that thin Cu films with low electrical resistivity (decreased from 3.6 to 1.8 μΩ cm when film thickness increased form to 25 to 135 nm) have been deposited on glass substrate with DC arc by FCVA technique [147, 148]. Trench and vias filling using a pulsed vacuum arc technique was demonstrated by Witke and Siemroth [149] (deposition rate of 10 nm/s with 4 kA arc current) and Montiero [55] (the deposition rate was not reported).

Table 22.8 Comparison of the cathode utilization efficiency for the HRAVA system and several FVAD systems

Apparatus	Reference	Operation type	Cathode	Iarc (A)	Iion (A)	Cathode utilization efficiency (%)
HRAVA	Shasuurin et al. [101]	Steady state	Cu	200	–	36
Quarter torus	Shasuurin et al. [101]	Steady state	Cu	200	0.37	0.8
	Davis [142]	Steady state	Cu	110	0.25	0.8
	Zhitomirsky et al. [140]	Steady state	Ti	320	0.9	1.3
	Aksenov et al. [49]	Steady state wall bias + 20 V	Ti	100	0.85	4
	Anders et al. [145]	Pulsed	Ti	150	3.8	12
	Bilek et al. [146]	Pulsed	Ti	300	4.5	7

Table 22.7 also presents the deposition rate using straight ducts, systems using filtering shields, and evaporators using hot cathode and hot anode vacuum arcs. A relatively high deposition rate is observed with Ψ-HCA and spotless hot cathode arc, comparable with HRAVA deposition. However, very high EB power (up to 300 kW) is required for intense cathode evaporation, and hence, the apparatus is relatively complex. The plasma flux generated in the low-current hot anode source is weakly ionized.

22.7 Summary

Vacuum arc with refractory anodes is a discharge that serves as a long time highly ionized energetic plasma source produced from the cathode material. This relatively new type of investigated in last two decades as a source for thin metallic film deposition. The deposition rate increases with time because the anode temperature increases with time [93, 101, 150]. Initially, the cold anode collects almost all of the cathode plasma and MPs incident on it, and thus, the deposition rate of the substrate, which is shielded from the cathode surface, is nearly zero. Only when the anode heats up and finally reaches a temperature (~2000 K), at which cathodic plasma will not condense and MPs (if they should reach the anode surface) are quickly evaporated, the deposition rate in the A-region (faced the anode in HRAVA) will become significant. The small area at the C-region become directly the cathode plasma (including MPs) which cannot be used and could be excluding from the process by respectively

22.7 Summary

shifting of the substrate. Furthermore, with formation of the anode plasma plume, the interelectrode plasma density increases, increasing the rate at which MPs evaporate during their flight through the A-region, and this evaporated material adds to the depositing flux.

It should be noted that due to the MP evaporation in the anode plasma plume, the MP sizes were significantly lower (<20 μm [101]) when deposited in the A-region of the HRAVA than when deposited with conventional cathodic arcs (for Cu, up 60–70 μm [90]). This is one of the advantages of using the HRAVA for deposition of metallic films. Cu deposition characteristics were improved by increasing the arc current and using tungsten anode. In a 300 A HRAVA, the deposition rate was very high, e.g., 2.0 and 3.6 μm/min at $L = 110$ and 80 mm, respectively. The MP contamination was very low, e.g., ~3 mm^{-2} min^{-1} at $L = 110$ mm. Trenches with an aspect ratio of 3 were filled with Cu using the HRAVA plasma source and pulsed biasing A deposition rate of 425 nm/min (0.42 μm/min) was reached. The minimum film resistivity was 1.7 μΩ cm.

Cu film was successfully deposited on glass substrates using a vacuum arc with VABBA closed cathode–anode configuration and with one-hole, frontal, or radial shower anodes. The rate of deposition was relatively high with frontal shower W (1 mm holes) anode. The cathode erosion rate in the VABBA was about by a factor of 2–3 lower than in the conventional cathodic arc, presumably due to the large plasma back flux to the cathode from the hot anode. One experiment was demonstrated VABBA wafer (160 mm diam) deposition rate of 0.47 μm/min on the substrate center at distance 280 mm (200 A), which then decreased to about 0.3 μm/min at 30–40 mm and then weakly changed up to a radius of 80 mm.

Thin films of low melting temperature cathode materials (Al, Zn and Sn) were obtained by HRAVA deposition, where the MPs are significantly converted to the plasma state at the non-consumable hot refractory anode surface and in dense anode plasma in the electrode gap.

Important issue is deposition refractory metal thin films for the first time using a refractory anode vacuum arc. The deposition rates using the HRAVA mode were lower for refractory materials, (up to about 1 μm/min, except Mo at $I > 250$ A) than for intermediate materials (Cu, about 2–3 μm/min) under similar conditions. With the HRAVA cathode, for Zr and Nb cathodes with a W anode, V_{dep} was 0.8 and 0.6 μm/min, respectively, and for a pair of cylindrical W electrodes, V_{dep} was ~1 μm/min. For the VABBA configuration, Nb deposition with one-hole Ta anode reached 0.25 μm/min at $I = 225$ A. The rate of deposition of 0.3 μm/min was measured using a Nb cathode and a W shower-head anode, while with both electrodes fabricated from W, V_{dep} reached 0.6 μm/min. W deposition rate as a function of time after arc ignition by both W planar cathode–anode configuration (5), $L = 110$ mm, $h = 8$ mm.

A comparison of deposition techniques including PVD, CVD, and their plasma enhanced variants shows that they mostly produce film with relatively low deposition rates. The exception is technique using EB heating; these, however, require high power, and these systems are complex. In comparison, HRAVA apparatus is simple, and the main disadvantage of cathodic vacuum arc deposition, MP contamination, is avoided. In the HRAVA, the MPs are converted to plasma by evaporation at the hot refractory anode surface as well as in the dense gap plasma, and subsequent ionization of the neutrals. This increased the extracted ion current and the deposition rate in comparison to those using cathodic arcs. HRAVA apparatus requires no additional ducts or magnets, the plasma losses are minimal, and the cathode mass utilization efficiency is relatively high. The high degree of ionization permits effective substrate biasing to increase film adhesion. The radial plasma flow deposits a film over a relatively large cylindrical area placed around the source. MP contamination can be almost eliminated by mounting components only in the A-region. Rapid Cu trench filling without MP contamination was demonstrated.

Acknowledgements The authors gratefully acknowledge S.Goldsmith, R. L. Boxman, H. Rosenthal, M. Keidar, J. Heberlein, E. Pfender, V. Paperny A. Shashurin, D. Arbilly, A. Nemirovsky, A. Snaiderman and D. Grach, Y. Yankelevich for their contributions as co-authors at different stages of HRAVA and VABBA investigation.

References

1. Rossnagel, S. M. (1998). Directional and ionized physical vapor deposition for microelectronics applications. *Journal of Vacuum Science and Technology, B16*(5), 2585–2609.
2. Hopwood, J. A. (2000). Ionized physical vapor deposition (Vol. 27). Academic Press, S Diego, N.Y.
3. Boxman, R. L. (2001). Early history of vacuum arc deposition. *IEEE Transactions on Plasma Science, 29*(5), 759–761.
4. Boxman, R. L. (2001). Recent developments in vacuum arc deposition. *IEEE Transactions on Plasma Science, 29*(5), 762–767.
5. Anders, A. (2008). Cathodic arcs: From fractal spots to energetic condensation, Springer.
6. Boxman, R. L., Martin, P. J., & Sanders, D. M. (Eds.) (1995). Handbook of vacuum arc science and technology. Park Ridge, N.J.: Noyes Publ.
7. Sanders, D. M., & Anders, A. (2000). Review of cathodic arc deposition technology at the start of the new millennium. *Surface and Coat. Technol., 133–134*(1–3), 78–90.
8. Beilis, I. I., & Boxman, R. L. (2009). Metallic film deposition using a vacuum arc plasma source with a refractory anode. *Surface & Coatings Technology, 204*, 865–871.
9. Helmersson, U., Lattemann, M., Bohlmark, J., Ehiasarian, A. P., & Gudmundsson, J. T. (2006). Ionized physical vapor deposition (IPVD): A review of technology and applications. Thin Solid Films, 513, 1–2.
10. Barnat, E., & Lu, T. (1999). Pulsed bias magnetron sputtering of thin films on insulators. *Journal of Vacuum Science and Technology, A17*(6), 3322–3326.
11. Radsimski, Z. J., Posadowski, W. M., Rossnagel, S. M., & Shingubara, S. (1998). *Journal of Vacuum Science and Technology, B16*(3), 1102–1108.
12. Posadowski, W. M., Wiatrowski, A., Dora, J., & Radzimski, Z. J. (2008). Magnetron sputtering process control by medium-frequency power supply parameter. *Thin Solid Films, 516*, 4478–4482.
13. Posadowski, W. M. (1999). Pulsed magnetron sputtering of reactive compounds. *Thin Solid Films, 343–344*, 85–89.
14. Andersson, J., & Anders, A. (2009). Self-sputtering far above the runaway threshold: An extraordinary metal-ion generator. *Physical Review Letters, 102*, 045003.

References

15. Richter, F., Bewilogua, K., Kupfer, H., Muhling, I., Rau, B., Rother, B., Schumaher, D. (1991 August). Preparation and properties of amorphous carbon and hydrocarbon films. In Y. Tzeng, M. Yoshikawa, M. Murakawa, & A. Feldman (Eds.), Proceedings of the First International Conference on the Applications of Diamond Films and Related Materials. Alabama, USA: Elsevier Science Publisher.
16. Horwat, D., & Anders, A. (2010). Ion acceleration and cooling in gasless self-sputtering. *Applied Physics Letters, 97,* 221501.
17. Anders, A., Ni, P., & Rauch, A. (2012). Drifting localization of ionization runaway: Unraveling the nature of anomalous transport in high power impulse magnetron sputtering. *Journal of Applied Physics, 111,* 053304.
18. Wang, C.-W., Yiu, P., Chu, J. P., Shek, C.-H., & Hsueh, C. H. (2015). Zr–Ti–Ni thin film metallic glass as a diffusion barrier between copper and silicon. Journal of Material Science, 50, 2085–2092.
19. Wahl, T., Hanisch, J., & Ahlswede, E. (2018). Comparison of the Al back contact deposited by sputtering, e-beam, or thermal evaporation for inverted perovskite solar cells. *Journal of Physics. D. Applied Physics, 51,* 135502.
20. Movchan, B. A. (2006). Inorganic materials and coatings produced by EBPVD. *Surface Engineering, 22*(1), 35–46.
21. Schiller, S., Goedicke, K., & Metzner, C. (1993). Plasma-activated high-rate electron-beam evaporation for coating metal strips. *Materials Science and Engineering A, 163*(2), 149–156.
22. Illés, B., Skwarek, A., Bátorfi, R., Ratajczak, J., Czerwinski, A., Krammer, O., et al. (2017). Whisker growth from vacuum evaporated submicron Sn thin films. *Surface & Coatings Technology, 311,* 216–222.
23. Hershcovitch, A., Blaskiewicz, M., Brennan, J. M., Fischer, W., Liaw, C.-J., Meng, W., & Todd, R. (2015). Novel techniques and devices for in-situ film coatings of long, small diameter tubes or elliptical and other surface contours. Journal of Vacuum Science and Technology, B33, 052601.
24. Hershcovitch, A., Blaskiewicz, M., Brennan, J. M., Custer, A., Dingus, A., Erickson, M., et al. (2015). Plasma sputtering robotic device for in-situ thick coatings of long, small diameter vacuum tubes. *Physics of Plasmas, 22,* 057101.
25. Zhang, M., Gu, B., Wang, L., & Xia, Y. (2005). Preparation and characterization of (100)-textured diamond films obtained by hot-filament CVD. *Vacuum, 79,* 84–89.
26. Zhang, S., Sun, D., Fu, Y., & Du, H. (2005). Toughening of hard nanostructural thin films: a critical review. Surface & Coatings Technology, 198 2–8.
27. Bollmann, D., Merkel, R., & Klumpp, A. (1997). Conformal copper deposition in deep trenches. *Microelectronic Engineering, 37*(38), 105–110.
28. Cho, N.-I. (1997). Microstructures of copper thin films prepared by chemical vapor deposition. Thin Solid Films, 308–309 465–469.
29. Lakshmanan, S. K., & Gill, W. N. (1998). Experiments on the plasma assisted chemical vapor deposition of copper. *Journal of Vacuum Science and Technology, A16*(4), 2187–2198.
30. Babayan, S. E., Jeong, J. Y., Tu, V. J., Park, J., Selwyn, G. S., & Hicks, R. F. (1998). Deposition of silicon dioxide films with an atmospheric-pressure plasma jet. *Plasma Sources Science and Technology, 7,* 286–288.
31. Shacham-Diamand, Y., & Lopatin, S. (1997). High aspect ratio quarter-micron electroless copper integrated technology. *Microelectronic Engineering, 37*(38), 77–88.
32. Hasegawa, M., Okinaka, Y., Shacham-Diamand, Y., & Osaka, T. (2006). Void-free trench-filling by electroless copper deposition using the combination of accelerating and inhibiting additives. *Electrochemical and Solid-State Letters, 9*(8), C138–C140.
33. Martin, P. J., & Bendavid, A. (2001). Review of the filtered vacuum arc process and materials deposition. *Thin Solid Films, 394,* 1–15.
34. Juttner, B. (2001). Cathode spots of electric arcs. *Journal of Physics. D. Applied Physics, 34,* R103–123.
35. Beilis, I. I. (2001). State of the theory of vacuum arcs. *IEEE Transactions on Plasma Science, 29*(5), 657–670.

36. Davis, W. D., & Miller, H. C. (1969). Analysis of the electrode products emitted by dc arcs in a vacuum ambient. *Journal of Applied Physics, 40*(5), 2212–2220.
37. Ecker, G. (1961). Electrode components of the arc discharge. *Erg. exakt. Naturwiss., 33*, 1–104.
38. Beilis, I. I. (1995). Theoretical modelling of cathode spot phenomena. In R. L. Boxman, P. J. Martin, & D. M. Sanders (Eds.), *Handbook of vacuum arc science and technology* (pp. 208–256). Park Ridge, NJ: Noyes Publications.
39. Beilis, I. I. (2007). Transient cathode spot operation at a microprotrusion in a vacuum arc. IEEE Transactions on Plasma Sciences, 35 N4, Part 2, 966–977.
40. Beilis, I. I. (2002). Vacuum arc cathode spot grouping and motion in magnetic fields. IEEE Transactions on Plasma Sciences, 30 N6, 2124–2132.
41. Beilis, I. I. (2003). The vacuum arc cathode spot and plasma jet: Physical model and mathematical description. *Contributed Plasma Physics, 43*(3–4), 224–236.
42. Beilis, I. I., Keidar, M., Boxman, R. L., & Goldsmith, S. (1998). Theoretical study of plasma expansion in a magnetic field in a disk anode vacuum arc. *Journal of Applied Physics, 83*(2), 709–717.
43. Daalder, J. E. (1976). Components of cathode erosion in vacuum arcs. *Journal of Physics. D. Applied Physics, 9*, 2379–2395.
44. Sanders, D. M., Boercker, D. B., & Falabella, S. (1990). Coating technology based on the vacuum arc-a review. *IEEE Transactions on Plasma Science, 18*(6), 883–894.
45. Boxman, R. L., Beilis, I. I., Gidalevich, E., & Zhitomirsky, V. N. (2005). Magnetic control in vacuum arc deposition: A review. *IEEE Transactions on Plasma Science, 33*(5), 1618–1624.
46. Boxman, R. L., & Zhitomirsky, V. N. (2006). Vacuum arc deposition devices. Review of Scientific Instruments, 77 N2, 021101.
47. Rother, B., Siegel, J., & Vetter, J. (1990). Cathodic arc evaporation of graphite with controlled cathode spot position. *Thin Solid Films, 188*(2), 293–300.
48. Aksenov, I. I., Padalka, V. G., Tolok, V. T., & Khoroshikh, V. M. (1980). Motion of plasma strems from a vacuum arc a long, straight plasma-optic system. Soviet Journal of Plasma Physics, 6 N4, 504–507.
49. Aksenov, I. I., Belous, V. A., Padalka, V. G., & Khoroshikh, V. M. (1978). Transport of plasma streams in a curvilinear plasma-optics system. *Soviet Journal of Plasma Physics, 4*(4), 425–428.
50. Karpov, D. A. (1997). Cathodic arc sources and macroparticle filtering. *Surface & Coatings Technology, 96*, 22–33.
51. Anders, A. (2007). Metal plasmas for the fabrication of Nanostructures. *Journal of Physics. D. Applied Physics, 40*, 2272–2284.
52. Falabella, S., & Sunders, D. A. (1992). Comparising of two filtering cathodic arc sources. *J. Vac. Sci. Techhol., A10*(2), 394–397.
53. Anders, S., Anders, A., Dickenson, M. R., MacGill, R., & Brown, I. G. (1997). S-shaped magnetic macroparticle filter for cathodic arc deposition. *IEEE Transactions on Plasma Science, 25*(4), 670–674.
54. Siemroth, P., & Schuke, T. (2000). Copper metallization in microelectronics using filtered vacuum arc deposition—Principles and technological development. *Surface and Coating Technology, 133–134*, 106–113.
55. Monteiro, R. (1999). Novel metallization technique for filling 100-nm-wide trenches and vias with very high aspect ratio. *Journal of Vacuum Science and Technology, B17*(3), 1094–1097.
56. Keidar, M., Beilis, I. I., Boxman, R. L., & Goldsmith, S. (1996). Transport of macroparticles in magnetized plasma ducts. *IEEE Transactions on Plasma Science, 24*(1), 226–234.
57. Keidar, M., Beilis, I. I., Aharonov, B. R., Boxman, R. L., & Goldsmith, S. (1997). Macroparticle distribution in a quater-torus plasma duct of a filtered vacuum arc deposition system. *Journal of Physics. D. Applied Physics, 30*, 2972–2978.
58. Monteiro, R., & Anders, A. (1999). Vacuum-arc-generated macroparticles in the nanometer range. *IEEE Transactions on Plasma Science, 27*(4), 1030–1033.

59. Bilek, M. M. M., Monteiro, O. R., & Brown, I. G. (1999). Optimization of film thickness profiles using a magnetic cusp homogenizer. *Plasma Sources Science and Technology, 8*, 88–93.
60. Miernik, K., Walkowicz, J., & Bujak, J. (2000). Design and performance of the microdroplet filtering system used in cathodic arc coating deposition. *Plasmas & Ions, 3*, 41–51.
61. Ryabchikov, A. I., Stepanov, I. B., Dektjarev, S. V., & Sergeev, O. V. (1998). Vacuum arc ion and plasma source Raduga 5 for materials treatment. *Review of Scientific Instruments, 69*(2), 893–895.
62. Gorokhovsky, V. I., Bhattacharaya, R., & Bhat, D. G. (2001). Characterization of large area filtered arc deposition technology: part I—Plasma processing parameters. *Surface Coating and Technology, 140*(2), 82–92.
63. Gorokhovsky, V. I., Bhat, D. G., Shivpuric, R., Kulkarnic, K., Bhattacharyad, R., & Rai, A. K. (2001). Characterization of large area filtered arc deposition technology: part II—Coating properties and applications. *Surface Coating and Technology, 140*, 215–234.
64. Zhitomirsky, V. N., Boxman, R. L., & Goldsmith, S. (2004). Plasma distribution and SnO2 coating deposition using a rectangular filtered vacuum arc plasma source. *Surface Coating and Technology, 185*(1), 1–11.
65. Zhitomirsky, V. N., Çetinörgü, E., Boxman, R. L., & Goldsmith, S. (2008). Properties of SnO2 films fabricated using a rectangular filtered vacuum arc plasma source. *Thin Solid Films, 516*, 5079–5086.
66. Schultrich, B., Siemroth, P., & Scheibe, H.-J. (1997). High rate deposition by vacuum arc methods. *Surface Coating and Technology, 93*, 64–68.
67. Scheibe, H.-J. (1994). Schultrich, DLC-Elm deposition by laser-arc and properties study. *Thin Solid Films, 246* N1–2, 92–102.
68. Anders, A., Pasaja, N., & Sansongsiri, S. (2007). Filtered cathodic arc deposition with ion-species-selective bias. *Review of Scientific Instruments, 78*, 063901.
69. Witke, T., Schuelke, T., Schultrich, B., Siemroth, P., & Vetter, J. (2000). Comparison of filtered high-current pulsed arc deposition (φ-HCA) with conventional vacuum arc methods. *Surface Coating and Technology, 126*, 81–88.
70. Petereit, B., Siemroth, P., Schneider, H. H., & Hilgers, H. (2003). High current filtered arc deposition for ultra thin carbon overcoats on magnetic hard disks and read-write heads. *Surface Coating and Technology, 174–175*, 648–650.
71. Vasin, A. I., Dorodnov, A. M., & Petrsov, V. A. (1979). Vacuum arc with distributed discharge on an expendable cathode. *Soviet Technical Physics Letters, 5*, 634–636.
72. Kajioka, H. (1997). Characterization of arc like Ti vapor plasma on the high-voltage electron-beam evaporator. *Journal of Vacuum Science and Technology, A15*, 2728–2739.
73. Goedicke, K., Sheffel, B., & Shiller, S. (1994). Plasma-activated high rate electron beam evaporation using a spotless cathodic arc. *Surface & Coatings Technology, 68*(69), 799–803.
74. Sheffel, B., Metzner, C., Goedicke, K., Heiness, J.-P., & Zywitzki, O. (1999). Rod cathode arc-activated deposition (RAD)—A new plasma-activated electron beam PVD process. *Surface & Coatings Technology, 120–121*, 718–722.
75. Chayahara, A., Mokuno, Y., Kinomura, A., Tsubouchi, N., Heck, C., & Horino, Y. (2004). Metal plasma source for PBII using arc-like discharge with hot cathode. *Surface & Coatings Technology, 186*(1–2), 157–160.
76. Miller, H. C. (1985). A review of anode phenomena in vacuum. *IEEE Transactions on Plasma Science, 13*(5), 242–252.
77. Miller, H. C. (1995). Chapter 5 in Handbook of vacuum arc science and technology, R. L. Boxman, P. Martin, & D. Sanders (Eds.). Ridge Park NJ: Noyes Publishing.
78. Dorodnov, A. M., Kuznetsov, A. N., & Petrsov, V. A. (1979). New anode vapor vacuum arc with a permanent hollow cathode. *Soviet Technical Physics Letters, 5*(8), 418–419.
79. Ehrich, H., Hasse, B., Mausbach, M., & Muller, K. G. (1990). The anodic vacuum arc and its application to coating. *Journal of Vacuum Science and Technology, A8*(3), 2160–2164.
80. Boxman, R. L., & Goldsmith, S. (1990). Momentum interchange between cathode spot plasma jets and background gases in vacuum arc anode spot development. *IEEE Transactions on Plasma Science, 18*(2), 231–236.

81. Ehrich, H., Hasse, B., Muller, K. G., & Schimidt, R. (1988). The anodic vacuum arc. II. Experimental study of arc plasma. Journal of Vacuum Science and Technology, A6 N4, 2499–2503.
82. Ehrich, H. (1988). The anodic vacuum arc. Basic construction and phenomenology. Journal of Vacuum Science and Technology, A6 N1, 134–138.
83. Katsch, H. M., Mausbach, M., & Muller, K. G. (1990). Investigation of the expanding plasma of an anodic vacuum arc. Journal of Applied Physics, 67(8), 3625–3629.
84. Meassik, S., Chan, C., & Allen, R. (1992). Thin film deposition techniques utilizing the anodic vacuum arc. Surface Coatings and Technology, 54/55, N1, 343–348.
85. Rosenthal, H., Beilis, I., Goldsmith, S., & Boxman, R. L. (1995). Heat fluxes during the development of hot anode vacuum arc. Journal of Physics D: Applied Physics, 28, N1, 353–383.
86. Rosenthal, H., Beilis, I., Goldsmith, S., & Boxman, R. L. (1996). Spectroscopic investigation of the development of hot anode vacuum arc. Journal of Physics D: Applied Physics, 29, 1245–1259.
87. Beilis, I. I., Boxman, R. L., & Goldsmith, S. (2000). The hot refractory anode vacuum arc: A new plasma source for metallic film deposition. Surface & Coatings Technology, 133–134(1–3), 91–95.
88. Beilis, I. I., Shashurin, A., Nemirovsky, A., Boxman, R. L., & Goldsmith, S. (2005). Imaging of the anode plasma plume development in a hot refractory anode vacuum arc. IEEE Transactions on Plasma Science, 33(2), 408–409.
89. Beilis, I. I., Keidar, M., Boxman, R. L., & Goldsmith, S. (2000). Interelectrode plasma parameters and plasma deposition in a hot refractory anode vacuum arc. Physics of Plasmas, 7(7), 3068–3076.
90. Beilis, I. I., Shashurin, A., & Boxman, R. L. (2008). Anode plasma plume development in a vacuum arc with a "black-body" anode-cathode assembly. IEEE Transactions on Plasma Sciences, 36, N4, Part 1, 1030–1031.
91. Beilis, I. I., Koulik, Y., & Boxman, R. L. (2013). Effective cathode voltage in a vacuum arc with a black body electrode configuration. IEEE Transactions on Plasma Sciences, 41 N8, Part II, 1992–1995.
92. Beilis, I. I., Shashurin, A., Arbilly, D., Goldsmith, S., & Boxman, R. L. (2004). Copper film deposition by a hot refractory anode vacuum arc. Surface & Coatings Technology, 177–178(1–3), 233–237.
93. Beilis, I. I., Boxman, R. L., & Goldsmith, S. (2002). Interelectrode plasma evolution in a hot refractory anode vacuum arc: Theory and comparison with experiment. Physics of Plasmas, 9(7), 3159–3170.
94. Kutzner, J., Miller, H. C., & Trans, I. E. E. E. (1989). Plasma Science, 17, 688.
95. Beilis, I. I. (1977). Cathode spots on metallic electrode of a vacuum arc discharge. High Temperature, 15(5), 818–824.
96. Beilis, I. I., Boxman, R. L., Goldsmith, S., & Paperny, V. L. (2000). Radially expanding plasma parameters in a hot refractory anode arc. Journal of Applied Physics, 88(11), 6224–6231.
97. Beilis, I. I., Koulik, Y., Boxman, R. L., & Arbilly, D. (2010). Thin film deposition using a plasma source with a refractory anode vacuum arc. Journal Materials Science, 45(23), 6325–6331.
98. Beilis, I. I., Koulik, Y., Boxman, R. L., & Arbilly, D. (2010). Al and Zn film deposition using a vacuum arc plasma source with a refractory anode. Surface and Coatings Techn, 205, 2369–2374.
99. Beilis, I. I., Koulik, Y., & Boxman, R. L. (2013). Sn thin film deposition using a vacuum arc plasma source with a refractory anode. Surface & Coatings Technology, 232, 936–940.
100. Beilis, I. I., Shashurin, A., Boxman, R. L., & Goldsmith, S. (2005). Influence of background gas pressure on copper film deposition and ion current in a hot refractory anode vacuum arc. Surface & Coatings Technology, 200, 1395–1400.
101. Shashurin, A., Beilis, I. I., Sivan, Y., Goldsmith, S., & Boxman, R. L. (2006). Copper film deposition rates by a hot refractory anode vacuum arc and magnetically filtered vacuum arc. Surface & Coatings Technology, 201(7), 4145–4151.

102. Beilis, I. I., Snaiderman, A., Shashurin, A., Boxman, R. L., & Goldsmith, S. (2007). Copper film deposition and anode temperature measurements in a vacuum arc with tungsten anode. *Surface & Coatings Technology, 202*(4–7), 925–930.
103. Beilis, I. I., Snaiderman, A., & Boxman, R. L. (2008). Chromium and titanium film deposition using a hot refractory anode vacuum arc plasma source. *Surface & Coatings Technology, 203*(5–7), 501–504.
104. Beilis, I. I., Koulik, Y., & Boxman, R. L. (2014). Cu film deposition using a vacuum arc with a black-body electrode assembly. *Surface & Coatings Technology, 258,* 908–912.
105. Beilis, I. I., Y. Koulik, Y. Yankelevich, D. Arbilly, & Boxman, R. L. (2015). Thin-film deposition with refractory materials using a vacuum arc. IEEE *Transactions* on Plasma Sciences, 43, N8, Part I, 2323–2328.
106. Bandyopadhyay, S. K., & Pal, A. K. (1979). Grain-boundary scattering in aluminium films deposited on to calcite substrate. *Journal Materials Science, 14*(6), 1321–1325.
107. Park, H. W., & Danyluk, S. (1991). Residual stress measurement in filament-evaporated aluminium films on single crystal silicon wafers. *Journal Materials Science, 26*(1), 23–27.
108. Yoshii, K., Inoue, S., Inami, S., & Kawabe, H. (1989). Microstructural changes of Al/amorphous SiC layered films subjected to heating. *Journal Materials Science, 24*(9), 3096–3100.
109. Wuhrer, R., & Yeung, W. Y. (2002). A study on the microstructure and property development of d.c. magnetron co-sputtered ternary titanium aluminium nitride coatings, Part II Effect ofMagnetron discharge power. Journal of Material Science, 37, 3477–3482.
110. Liu, Z. W., Yeo, S. W., & Ong, C. K. (2007). An alternative approach to in situ synthesize single crystalline ZnO nanowires by oxidizing granular zinc film. Journal of Material Science, 42 6489–6493.
111. Fan, X. M., Lian, J. S., Guo, Z. X., Zhao, L., & Jiang, Q. (2006). Influence of the annealing temperature on violet emission of ZnO films obtained by oxidation of Zn film on quartz glass. *Journal Materials Science, 41,* 2237–2241.
112. Chandrasekar, M. S., Shanmugasigamani, S., & Pushpavanam, M. (2010). Structural and textural study of electrodeposited zinc from alkaline non-cyanide electrolyte. *Journal Materials Science, 45,* 1160–1169.
113. Tu, K. N., & Thompson, R. D. (1982). Kinetics of interfacial reaction in bimetallic Cu-Sn thin films. *Acta Metallurgica, 30*(5), 947–952.
114. Tu, K. N. (1996). Mater, Cu/Sn interfacial reactions: thin-film case versus bulk case1. *Chemistry & Physics, 46*(2–3), 217–223.
115. Inaba, M., Uno, T., & Tasaka, A. (2005). Irreversible capacity of electrodeposited Sn thin film anode. Journal of Power Sources, 146 N1–2, 473–477.
116. Nimisha, C. S., Venkatesh, G., Yellareswara Rao, K., Mohan Rao, G., & Munichandraiah, N. (2012). Morphology dependent electrochemical performance of sputter deposited Sn thin films. Materials Research Bulletin, 47, 1950–1953.
117. Beilis, I. I., Koulik, Y., & Boxman, R. L. (2011). Temperature distribution dependence on refractory anode thickness in a vacuum arc: Experiment. IEEE *Transactions* on Plasma Sciences, 39, N6, Part 1 1303–1306.
118. Beilis, I. I., Koulik, Y., & Boxman, R. L. (2017). Anode temperature evolution in a vacuum arc with a black body electrode configuration. IEEE *Transactions* on Plasma Sciences, 44, N8, Part II, 2115–2118.
119. Shashurin, A., Beilis, I. I., & Boxman, R. L. (2008). Angular distribution of ion current in a vacuum arc with a refractory anode. Plasma Sources Science and. Technology, 17 015016.
120. Muraka, S. P. (1993). *Metallization: Theory and practice For VLSI and ULSI.* Boston: Butterworth-Heineman.
121. Du, M., Opila, R. L., & Case, C. (1998). Interface formation between metals (Cu, Ti) and low dielectric constant organic polymer (FLARE™ 1.0). Journal of Vacuum Science and Technology, A16 N1, 155–162.
122. Morand, Y. (2000). Copper metallization for advanced IC: Requirements and technological solutions. *Microelectronic Engineering, 50*(1–4), 391–401.

123. Hafezi, H., Huang, Y.-C., Zhang, A., Singh, S., & Ngai, C. (2005). Enhanced copper ECP for 45 nm devices. *Semiconductor International, 11*, 55–60.
124. Lin, J.-H., Hsieh, W.-J., Hsu, J.-W., Liu, X.-W., Chen, U.-S., & Shih, H. C. (2002). Gap-filling capability and adhesion strength of the electroless-plated copper for submicron interconnect metallization. *Journal of Vacuum Science and Technology, B20*(2), 561–565.
125. Hasegawa, M., Okinakab, Y., Shacham-Diamand, Y., & Osaka, T. (2006). Void-free trench-filling by electroless copper deposition using the combination of accelerating and inhibiting additives. Electrochemical and Solid-State Letters, 9 N8, C138–C140.
126. Hopwood, J. A., (Ed.) (2000). Thin films advances in research & development, Academic Press, V27.
127. Siemroth, P., Wenzel, Ch., Klimes, W., Schultrich, B., & Schülke, T. (1997). Metallization of sub-micron trenches and vias with high aspect ratio. *Thin Solid Films, 308–309*, 455–459.
128. Barnat, E., & Lu, T.-M. (2001). Transient charging effects on insulating surfaces exposed to a plasma during pulse biased dc magnetron sputtering. *Journal of Applied Physics, 90*(12), 5898–5903.
129. Beilis, I. I., Grach, D., Shashurin, A., & Boxman, R. L. (2008). Filling trenches on a SiO2 substrate with Cu using a hot refractory anode vacuum arc. Microelectronic Engineering, 85, 1713–1716.
130. William, D., & Callister, Jr. (2003). Materials science and engineering an introduction. Wiley.
131. Ryu, C., Kwon, K. W., Loke, A. L. S., Lee, H., Nogami, T., Dubin, V. M., et al. (1999). *Electron Devices, 46*, 1113–1120.
132. Laux, M., Schneider, W., Jüttner, B., Balden, M., Lindig, S., & Beilis, I. I. (2005). Ignition and burning of vacuum arcs on tungsten layers. *IEEE Transactions on Plasma Science, 33*, 1470.
133. Zhao, X.-A., Kolava, E., & Nicolet, M. A. (1986). Reaction of thin metal films with crystalline and amorphous Al2O3. *Journal of Vacuum Science and Technology, A4*(6), 3139–3141.
134. Bewilogua, K., Wittorf, R., Thomsen, H., & Weber, M. (2004). DLC based coatings prepared by reactive d.c. magnetron sputtering. Thin Solid Films, 447/448, 142–147.
135. Hoffman, D. W., & Thornton, J. A. (1977). The compressive stress transition Al, V, Zr, Nb and W metal films sputtered at low working pressures. *Thin Solid Films, 45*, 387–396.
136. Lia, W.-L., Dinga, Y. S., Suiba, S. L., DiCarloc, J. F., & Galasso, F. S. (2005). Controlling the growth of CVD carbon from methane on transition metal substrates. Surface & Coatings Technology, 190, 366–371.
137. Girginov, A., Tzvetkoff, T. Z., & Bojinov, M. (1995). Electrodeposition of refractory metals (Ti, Zr, Nb, Ta) from molten salt electrolytes. *Journal of Applied Electrochemistry, 25*, 993–1003.
138. Pogrebniak, A. D., Beresnev, V. M., Kolesnikov, D. A., Bondara, O. V., & Takeda, Y. (2013). Multicomponent Nanostructure Coatings. *Acta Physica Polonica A, 123*(5), 816–818.
139. Beilis, I. I. (1982). On vapor flow from cathode region of a vacuum arc. Izvestiya Siberian branch of Academy Nauk SSSR, Seriya Thekhicheskih Nauk, 3 N1, 69–77 (In Russian).
140. Zhitomirsky, V. N., Boxman, R. L., & Goldsmith, S. (1995). Unstable arc operation and cathode spot motion in a magnetically filtered vacuum-arc deposition system. *Journal of Vacuum Science and Technology, A13*(4), 2233–2240.
141. Anders, S., Anders, A., & Brown, I. (1993). Macroparticle-free thin films produced by an efficient vacuum arc deposition technique. Journal of Applied Physics, 74 N6, 4239.
142. Anders, S., Anders, A., & Brown, I. (1994). Focused injection of vacuum arc plasmas into curved magnetic filters. Journal of Applied Physics, 75 N10, 4895.
143. Davis, C. A. (1993). Ph.D. Thesis, University of Sydney, Australia.
144. Bilek, M. M. M., & Anders, A. (1999). Designing advanced filters for macroparticle removal from cathodic arc plasmas. *Plasma Sources Science and Technology, 8*, 488–493.
145. Anders, S., Anders, A., & Brown, I. (1993). Macroparticle-free thin films produced by an efficient vacuum arc deposition technique. *Journal of Applied Physics, 74*(6), 4239–4241.
146. Bilek, M. M. M., & Anders, A. (1999). Designing advanced filters for macroparticle removal from cathodic arc plasmas. *Plasma Sources Science and Technology, 8*(3), 488–493.

147. Lau, S. P., Cheng, Y. H., Shi, J. R., Cao, P., Tay, B. K., & Shi, X. (2001). Filtered cathodic vacuum arc deposition of thin film copper. *Thin Solid Films, 398,* 539–543.
148. Shi, J. R., Lau, S. P., Sun, Z., Shi, X., Tay, B. K., & Tan, H. S. (2001). Structural and electrical properties of copper thin films prepared by filtered cathodic vacuum arc technique. Surface & Coating Technology, 138, 250–255.
149. Witke, T., & Siemroth, P. (1999). Deposition of droplet-free films by vacuum arc evaporation-results and applications. *IEEE Transactions on Plasma Science, 27*(4), 1039–1044.
150. Beilis, I. I., Koulik, Y., & Boxman, R. L. (2011). Temperature distribution dependence on refractory anode thickness in a vacuum arc: Experiment. IEEE Transactions on *Plasma Science,* 39, N6, Part 1, 1303–1306.

Chapter 23
Vacuum-Arc Modeling with Respect to a Space Microthruster Application

The vacuum arc has generated a great interest as very promising mechanism for micropropulsion is capable of providing orbit maneuvers of a spacecraft weighing on the order of few 10 kg or smaller.

23.1 General Problem and Main Characteristics of the Thruster Efficiency

The vacuum arc is a unique source of supersonic, highly ionized, high energetic, and quas-neutral metallic plasma jet [1–4]. The metallic plasma accelerators based on vacuum arc and their applications in industrial devices were developed in the early reported works [5–8]. It has high pulse stability and lower energy consumption per ionized metal mass due to the relatively low ionization potential of atoms and operation with higher repetition rates since the metal melting temperature can be relatively higher than that, for example, of polymer propellants in another type of a thruster. The advantageous and the relatively simple design of the arc plasma source has potential to use metallic plasma as a propellant for an arc thruster in order to control the microsatellite orbit [9, 10]. Therefore, the vacuum-arc thrusters produce efficiently highly ionized supersonic and directional plasma at very low average power. The characteristics of the vacuum arc thrusters were widely reviewed and discussed [11–16] recently by Kolbeck et al. [17], where various aspects of electric propulsion technology based on vacuum arcs are considered and emphasized the role of vacuum-arc thrusters in space missions.

The vacuum-arc thrusters are characterized by a few important parameters [13, 17]. One of them is the specific impulse that describes the thrust efficiency of a propulsion system and can be defined as follows

$$I_{sp} = \frac{V}{g} \quad (23.1)$$

where V is the effective exhaust velocity of the work medium and g is the gravitational acceleration. The unit of the I_{sp} is seconds. The specific impulse (23.1) is used to calculate the thrust T in form

$$T = I_{sp}\frac{dm}{dt}g = V\frac{dm}{dt} \qquad (23.2)$$

where dm/dt is the propellant mass flow. Electric propulsion systems have very high specific impulses in the range of thousands of seconds and thrust levels ranging from micronewtons (vacuum-arc thrusters) and other devices such as electro-spray thrusters to a few newton's [18].

The thruster performance (efficiency) is defined as

$$\eta = \frac{P_{jet}}{P_{in}} \; or \; \eta = \frac{mV^2}{2Iu_{arc}} \qquad (23.3)$$

where P_{jet} is the plasma (or ion beam) jet power, P_{in} is the total electrical power supplied to the thruster, I is the arc currentand u_{arc} is the arc voltage.

One of first in 1966 Gilmor [19, 20] suggested the application of the cathode jet of a vacuum arc to space propulsion. He demonstrated experimental model using the vacuum arc–coaxial arc diode configuration showing thrust with an efficiency of 30% for an OFHC copper cathode and a magnetic field of 500 gauss. The thrust was computed to be over 10 millipounds, and the specific impulse was over 1200 s. Efficiencies over 20% were obtained with type-304 stainless steel and titanium for a variety of magnetic field and power levels. Gilmor et al. [21] presented results of a program of research and development for thrusters having total impulse capabilities from 300 to 2000 Ns, impulse per pulse capabilities from specific impulse of 1000 s, and pulse repetition frequencies from less than one pulse per second to several tens of pulses per second. Experimental results presented include thruster characteristics as functions of voltage, capacitance, time, and magnetic field.

Dethlefsen [22] measured the thrust impulse per pulse, specific impulse of different cathode, and performance for two electrode configurations of vacuum-arc thrusters using high-current pulse with magnesium and copper electrodes. The test thrusters were provided with pendulum, so that the integrated impulse of about 1000 shots was determined by the change in pendulum deflection. The high velocity of the plasma jets was determined by a time-of-flight technique on thruster with anode and cathode erosion. Two Faraday cups were placed at 65 cm to measure the signal of the ion current. A velocity of 4.2×10^6 cm/s corresponding to a specific impulse of 4300 s for the ions was obtained for a magnesium plasma. This value indicates a rough agreement with the specific impulse determined from thrust measurements on the cathode jet. The thruster performance of magnesium was reported as about 12%. For copper electrode, the specific impulse was measured as 1500–1900 s and performance as 7.7–9.6 depending on the applied voltage.

The presence of droplets by arcing tends to reduce the thrust efficiency because they are not accelerated to the high velocities as the ions [23]. However, the droplets

23.1 General Problem and Main Characteristics ...

were minimized through the action of thermal and mechanical inertia if a current pulse of sufficiently short duration on the order of 10 μs is used in a pulsed vacuum-arc thruster. Use of the pulsed operation was demonstrated, and the thrust efficiency was found to be dependent on the electrode material, the amplitude, and duration of the current pulse.

In general, a vacuum arc is established by using a coaxial geometry with a relatively small diameter cathode, which is surrounded by an insulator and an anode. The breakdown mechanism in the microcathode arc thruster has been studied [24]. The different developed designs and the advantages of the vacuum-arc application were widely discussed [25, 26] comparing to other microthrusters used a gas as propellant [27]. It employed the vacuum-arc source as a plasma source for the ion thruster [9]. For this case, the measured energy efficiency was about 80% that is comparable to that of the xenon ion thruster. An important characteristic is the performance of the vacuum-arc thruster that is determined by the plasma jet parameters. Polk et al. [18] showed that the performance could be improved when the arc plasma was subsequently accelerated electrostatically. It was found also that the application of a magnetic field leads to ion acceleration by a factor of ~2 in the free expanding plasma plume [28].

23.2 Vacuum-Arc Plasma Characteristics as a Thrust Source

The measurements of the force originated at the cathode of electrical arc and produced highly energetic plasma jet (Fig. 23.1) were analyzed in Chaps. 11 and 12, and the nature of cathode jet rise is considered in Chap. 18. Some progress in understanding of the phenomena is demonstrated. Let us briefly summarize the data of cathode force measurements in order to determine the thrust characteristics. Tanberg [29] measured for a Cu cathode with arc current in the range of 11–32 A, and the vacuum chamber pressure was ~10^{-4} torr.

Fig. 23.1 Cathode plasma jet configuration

Fig. 23.2 Schematic diagram of the spot, plasma jet, and the cathode reaction due to plasma pressure

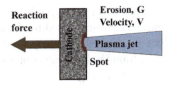

The average cathode jet velocity of the order of 1.6×10^6 cm/s and the average reaction force from the freely swinging cathode of ~17 dyn/A were obtained. Robertson [30] measured the reaction force of a copper cathode for a range of pressures from 1 to 10 torr. The largest force measured was ~15 dyn/A at a pressure of 1 torr with arc currents of 7–20 A. Kobel [31] measured the force on a mercury cathode indicating the force in range of 19–53 dyn/A for currents of 30–37 A and pressures of 1–0.5 torr. Chabrerie et al. [32] measured for Ag cathode 15.33 dyne/A, and the maximal force for Cu cathode was about 14.7 dyne/A at ~1.5 kA. Our measurements [33] using pendulum deflection due to cathode plasma jet action showed about 40 dyn/A for Cu and about 20 dyn/A for Al cathodes.

According to the theory (Chap. 18), the cathode force, in general, can arise due to electron pressure gradient [34], self-magnetic axial action [33, 35], additional to self-magnetic axial action by the sputtering, neutralization, and reflection of the accelerated ions, as well by the emitted electrons accelerated in the space charge, and deliver their momentum through the plasma acceleration [36, 37].

The theoretical analysis show that the main mechanism generated the force in the cathode region, and the plasma jet acceleration is the gradient of plasma pressure adjacent to the cathode surface region (Chap. 18). The force acted on the cathode is the reaction of the plasma expansion that is schematically shown in Fig. 23.2. The force at the cathode surface was determined by the erosion mass flow and the jet velocity as it was used for the measurements [29, 38]. According to the measurements, relation between the force F(dyn/A), the cathode erosion rate G(g/C), and plasma velocity V (cm/s) was determined as

$$F = GV \qquad (23.4)$$

The force (23.4) is in essence of the thrust (23.2) defined to characterize the thruster systems. The plasma velocity V and the erosion rate G measurements indicated a relatively narrow range of V from 6×10^5 to 3×10^6 cm/s and (20–60) μg/C, respectively, for different cathode materials [1, 18]. The data of V ~1×10^6 cm/s and G ~35 μg/C for copper cathode correspond to the force, which agrees with that measured by Marks et al. [33]. The measured arc parameters G, V, and F represented averaged values over some time of arc duration. At present, the time-dependent data of the mentioned quantities as well as of the thrust performance were absent for transient arc operation. An important question is: *how the arc plasma parameters in time during the cathode spot development and how the conditions of spot initiation, pulse duration, cathode material etc. influence the trust performance?*

To understand the mechanism of time-dependent cathode mass loss and plasma jet formation, the above physical model for transient spot is used. The time-dependent cathode phenomena in a vacuum arc produced the force by the cathode spot development on a bulk cathode which is considered. The kinetics of cathode vaporization into the adjacent to the surface plasma generated from the cathode erosion material is taken in account considering also the plasma flow in hydrodynamic acceleration region. The analysis is performed for microscale electrode configuration as well for configuration of free jet expanding by plasma spot at bulk cathode.

23.3 Microplasma Generation in a Microscale Short Vacuum Arc

A short microscale vacuum arc is investigated in order to understand what arc and plasma parameter can be used as propellant for microscale vacuum-arc thruster.

23.3.1 Phenomena in Arcs with Small Electrode Gaps

The calculation using gasdynamic model of the cathode plasma expansion showed that the plasma is accelerated to the supersonic velocity at the distance about 2–3 of cathode spot radii Chap. [18, 39], and therefore, the vacuum-arc source can be used in devices of very small sizes for micropropulsion applications. This calculated result is important because the velocity measurements cannot be conducted at such small distance from the cathode surface. Usually, the plasma jet was produced in the cathode region using conventional (large gap) vacuum arcs [1]. The plasma flow in the expanding part of the jet was calculated using free boundary expansion model [40]. It was found that the plasma jet had a conical shape, and for axial distances relatively far from the cathode surface, the radial velocity becomes comparable with the axial velocity if no magnetic field is imposed. However, in microscale devices, the plasma is generated in a short gap of the vacuum arc where the phenomena in the cathode and anode regions should be taken in account. Therefore, the short vacuum arc (SVA) is an important scientific subject to study the microplasma generation.

The SVA with gap of 1–10 μm and low current (<10 A) pulse (0.3–3 μs) was investigated for contact systems [41–43] and in electrical discharge machining [44]. The plasma pressures 100–400 atm, and the spot current density about 10^6 A/cm^2 was determined by the electrode craters examination. A low-current short vacuum arc with gap about 200 μm was used for investigation of soft X-ray generation, when negative high-frequency voltage was applied to the cathode after arc extinction [45]. In accordance with this work, the absent of knowledge in the short vacuum arc about plasma parameters (density, temperature), which are determined the sheath phenomena and X-ray generation after the arc extinction, is most important problem.

Similar conclusion regarding the necessity of the plasma parameters knowledge in a short gap vacuum arc to understand the phenomena in vacuum switchgears and nature of SVA maintenance was indicated also by Dong et al. [46]. The short vacuum arc was considered experimentally [47] and theoretically [48] in case of high-current commercial interrupters. No diffuse mode was observed for gap distance less than 4 mm. Strong coupling of the anode, cathode, and the discharge plasmas was indicated.

Meng et al. [49] indicated that electrical breakdown in atmospheric air across micrometer gaps is critically important for the insulation design of micro- and nano-electronic devices, and planar aluminum electrodes with gaps ranging from 2 to 40 μm were fabricated by microelectromechanical system technology. The influence factors including gap width and surface dielectric states were experimentally investigated. Radmilovic–Radjenovic and Radjenovic [50] studied the phenomenon of field emission role in the deviation of the breakdown voltage from that predicted by Paschen's law within the range of high electric fields. It was shown that high fields obtained in small gaps may enhance the secondary electron emission and such enhancement could lead to a lowering of the breakdown voltage and a deviation from the result indicated by Paschen curve. Kolachinski [51] developed a unified mathematical description of re-ignition phenomena in a short arc with analysis of the cathode and anode layers. The author concluded that the influence of the thermionic emission of non-refractory cathode on arc re-ignition is negligible, while the volume and surface Townsend processes of charge particle generation are mainly responsible for the re-ignition formation

Thus, while the conventional arc (large gap) was relatively studied, the SVA remains mostly unknown subject. The main questions are: *(i) how the near-cathode and near-anode phenomena influence on the current continuity in a SVA? (ii) How the anode mass losses influence the plasma jet formation? (iii) What is the nature of the force generated in the microscale gaps?* A physical model is formulated in the present section, and calculation of the plasma and electrode characteristics are studied to understand the principal mechanism of the microscale SVA operation.

23.3.2 Short Vacuum Arc Model

A calculation using the cathode spot model showed that the cathode plasma density in a microscale vacuum arc (MVA) gap near the anode surface was sufficiently large, and therefore, the anode spot was created (Chap. 14). As a result, the common plasma in the MVA was produced from the cathode and anode spots. To obtain the plasma jet, a configuration with planar anode and ring cathode was considered (Fig. 23.3). Circular cathode and anode spots are assumed. In general, the anode and cathode spots can have different radius. Kinetic models developed for cathode (Chap. 17) and anode [52] spots were used.

According to these models, the ionized electrode vapor structure consists of the several partially overlapping regions, which include a ballistic zone comprising a

23.3 Microplasma Generation in a Microscale Short Vacuum Arc

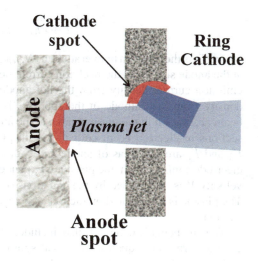

Fig. 23.3 Schematic presentation of the plan anode and ring cathode of the cathode and anode regions and plasma jet

space charge sheath at the surface, a non-equilibrium (Knudsen) plasma layer, and an electron emission relaxation zone. Two heavy particle fluxes (evaporated and returned) are formed in the Knudsen layer. The difference of these fluxes determines the plasma velocity v and the net rate of mass evaporation. In the cathode relaxation region, the atoms are ionized by electrons emitted from the cathode as well as by plasma electrons.

The ionization in the common plasma and in the anode region is by the plasma electrons. The system of Saha equations determines plasma multi-charge state. Two erosion fluxes are produced: from the cathode $G_c(g/C) = m_c n_c v_c / I$ and from the anode $G_a(g/C) = m_a n_a v_a / I$, where I is the current, $m_{c,a}$, $n_{c,a}$, and $v_{c,a}$ are atom mass, heavy particle density, and velocity for cathode and anode, respectively. The resulting erosion mass flow $G(g/C)$ from both electrodes is $G = G_c + G_a$.

Both the electrode bodies are heated ohmically and in addition (i) the cathode by the incident ions accelerated in the ballistic region (sheath) with cathode potential drop u_c and by returned electron flux; (ii) the anode by electrons and returned ions from the adjacent plasma. The anode is also heated by a part of electron beam emitted from cathode that is not relaxed in the MVA gap. An anode's negative potential drop is produced when the flux of thermal electrons exceeds the arc current I. The returned ion and electron fluxes (toward the cathode) and ions (toward the anode) are formed in the ballistic and in Knudsen layers. The electrode is cooled due to body heat conduction, electron emission, and electrode erosion. The plasma electrons are heated by part of the electron beam emitted from the cathode that relaxed in the MVA gap as well as ohmically, which is determined by Iu_p, where $u_p = jh/\sigma_{pl}$, is voltage drop in the gap with distance h, j is the current density, and σ_{pl} is the electrical conductivity of the highly ionized plasma. The negative potential drop u_a formed near the anode surface determined by the electrical current conservation as:

$$u_a = T_e \ln\left[\frac{j_{\text{etha}} F_a}{I + (j_{\text{ia}} + j_{\text{ema}})F_a - j_{\text{eb}} K_h F_c}\right] \quad (23.5)$$

where T_e is the electron temperature in eV, j_{etha} is the thermal electron current density at the anode side, j_{ia} is the back ion current density to the anode, j_{ema} is the electron emission current density from the hot anode, j_{eb} is the electron emission current density from the cathode in the spot, K_h is the coefficient of cathode's electron beam absorption in the plasma gap h at the anode side which depended on the cross sections of electron beam collisions with plasma atoms, ions, and electrons, and F_a and F_c are the areas of anode and cathode spots, respectively. For considered electrode configuration, the plasma is expanded into vacuum as a jet, and the plasma velocity V is determined by the equations of momentum and energy conservation. The plasma is accelerated by the ion and electron pressures and by the electron–ion friction.

The mathematical model also includes the equations of electron emission in general form, and equations for total spot current, electric field E at the cathode surface, kinetics of direct and back heavy particle fluxes, fluxes of the charge particles to the electrodes, electrode energy balances, and plasma energy balance (see Chap. 16, [52]). The calculated parameters are T_e, u_a, E, heavy particle density n, degree of ionization α, cathode T_c and anode T_a temperatures, erosion rate G, total current density j, ion current density j_i in the cathode spot, plasma jet velocity V for given u_c, spot current I, and anode spot radius r_a. The calculated parameter is also the net cathode K_{er} and anode K_{era} evaporation fractions, which are the ratio of the net of atom evaporation into the ambient dense plasma to the Langmuir evaporation rate (i.e., into vacuum). Below, the calculation for copper electrodes and $u_c = 19$ V is presented in order to characterize the parameters in a MVA as dependence on gap distance and cathode potential drop.

23.3.3 Calculation Results. Dependence on Gap Distance

The calculation were conducted according to above-described model for $I = 25$, 50 A, arc duration $t = 1$ ms, and gap $h = 2$–100 μm. The results presented in Figs. 23.4, 23.5 and 23.6 were calculated for $r_a = 2r_c$. According to the solution, a dense plasma $\sim 10^{20}$ cm^{-3} was obtained with mainly single charged ions and degree of ionization ~ 0.1, $T_e \sim 1$ eV, plasma jet velocity few of 10^5 cm/s, $u_a = -(5 \div 6)$ V, and $E \sim 10^7$ V/cm.

The cathode spot current density j decreases from 2 to 0.2 MA/cm^2 when h increases from 2 to 100 μm and j exceeds j_i by factor of about three. The electrode temperatures in the spots decrease with h (Fig. 23.4), and T_a in the anode spot significantly exceeds T_c in the cathode spot and is weakly dependent on arc current. The anode evaporation fraction K_{era} is significantly larger than K_{er}, which increases with h (Fig. 23.5) indicating that the degree of anode plasma non-equilibrium in the anode Knudsen layer is larger than that of the cathode plasma. The dependence

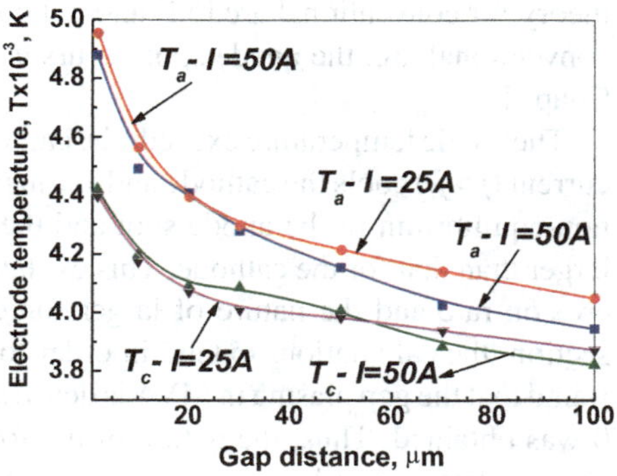

23.3 Microplasma Generation in a Microscale Short Vacuum Arc

Fig. 23.4 Cathode T_c and anode T_a temperatures in spots as dependence on interelectrode distance

Fig. 23.5 Net of cathode K_{er} and anode K_{era} evaporation fractions as dependence on interelectrode distance

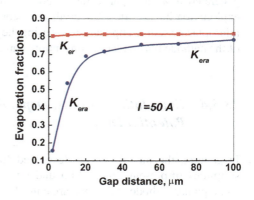

Fig. 23.6 Cathode G_c and anode G_a erosion rates in (g/C) as function on interelectrode distance

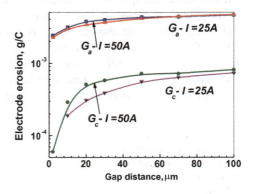

of cathode G_c and anode G_a erosion rates in (g/C) as function on h is presented in Fig. 23.6. It can be seen that G_c increases with h, and G_c is significantly lower than G_a, which weakly depends on h. The calculated force per Coulomb, which is equal GV/I and generated in the anode region, increases from 3.4 to 4.9 mN/A while this force for conventional cathodic arc is much lower, about 0.2 mN/A [29]. The obtained large anode evaporation fraction in the short arc agrees with the anode spot theory for conventional arc [52] and not contradicts the measured result [53], but in conventional arc, the anode spot occurs usually for very large arc current (few kA, Chap. 14).

The anode temperature exceeds the cathode temperature because the large electron current ($j > j_i$) cools the cathode and warms the anode. As result, the degree of plasma non-equilibrium in the anode spot and the mass flux evaporated from the anode is larger than that for the cathode. This explains calculation results for the larger anode erosion rate and the nature of larger force generated in the anode region. In next section, the calculations extend in order to study the influence of ratio r_a/r_c. It was found that the gap plasma in MVA is dense and highly ionized, and therefore, the $u_{an} < 0$ was obtained. Thus, the nature of the force and mass flux in microscale vacuum arc are determined by the anode phenomena, and they exceed the force and flux generated in a large gap cathodic arc. Therefore, the MVA can be suggested for exploitation in exceptional microthruster concept for spacecraft propulsion and for microplasma source applications by using the jet from MVA common plasma instead of the cathode plasma jet in conventional arc.

23.3.4 Calculation Results. Dependence on Cathode Potential Drop

In general, the model can be developed for different anode configurations. For simplicity of presentation and for understanding the plasma phenomena in a short arc, a planar cathode and anode are considered. The arc is sufficiently short that the cathode and anode spots produce a common plasma column. A circular configuration of the spots is assumed. In general, the anode and cathode spots can be of different radii. Therefore, the column geometry is assumed as a truncated cone (Fig. 23.7). As above, the kinetic models for cathode and anode spots were used. Below, an illustration of simple example of calculation for arc with copper electrodes, $I = 50$ A, and gap $h = 2$ μm is presented in order to characterize the wide parameters in a short vacuum arc as dependences on cathode potential drop. The calculations were

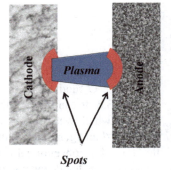

Fig. 23.7 Schematic presentation of the cathode and anode regions and the truncated cone plasma column

23.3 Microplasma Generation in a Microscale Short Vacuum Arc

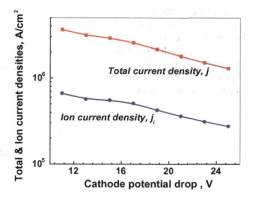

Fig. 23.8 Total and ion current density in the cathode spot as dependence on cathode potential drop

performed according to above-described model varying u_c and spot lifetime t. The results presented in Figs. 23.8, 23.9, 23.10, 23.11, 23.12 and 23.13 were calculated for $r_a = 2r_s$.

It is obtained that the relatively dense plasma is with mainly single charged ions and neutrals with heavy particle density ~10^{20} cm^{-3}, degree of ionization ~0.1,

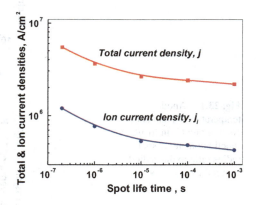

Fig. 23.9 Total and ion current densities in the cathode spot as dependence on spot lifetime

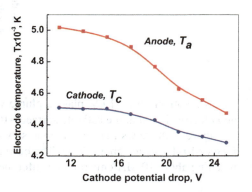

Fig. 23.10 Cathode T_c and anode T_a temperatures in spots as dependence on cathode potential drop

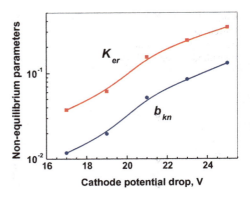

Fig. 23.11 Cathode evaporation fraction K_{er} and normalized velocity b_{kn} of the heavy particle flux at the external boundary of Knudsen layer as function on cathode potential drop

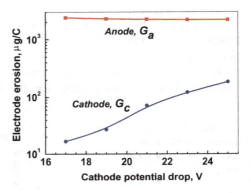

Fig. 23.12 Cathode G_c and anode G_a erosion rates in (μg/C) as function on cathode potential drop

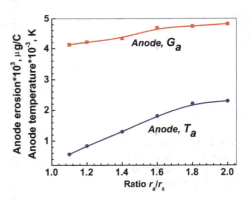

Fig. 23.13 Anode temperature T_a and anode erosion rate G_a in (μg/C) as function on anode spot radius ratio to the cathode spot radius ($u_c = 19$ V)

electron temperature ~1 eV, plasma plume velocity few of 10^5 cm/s, $u_a = -5$–6 V, and electrical field at the cathode surface ~10^7 V/cm.

Figure 23.8 shows that the cathode spot current density decreases with u_c, and j exceed j_i by factor about three. The current density decreases with spot lifetime reaching about 10 MA/cm² for short lifetime of about 100 ns (Fig. 23.9).

23.3 Microplasma Generation in a Microscale Short Vacuum Arc   1015

The electrode temperatures in the spots decrease with u_c (Fig. 23.10), and T_a in the anode spot significantly exceeds T_c in the cathode spot. Figure 23.11 shows that at low u_c, the cathode evaporation fraction K_{er} as well as the normalized (by $(2T/m)^{0.5}$) velocity b_{kn} of the heavy particle flux at the external boundary of Knudsen layer is relatively small. These parameters are increased with cathode potential drop (the cathode spot ionized vapor is strongly non-equilibrium). The calculation indicates that the degree of non-equilibrium of the anode plasma in the anode Knudsen layer is larger than for the cathode plasma. The dependence of cathode G_c and anode G_a erosion rates in (μg/C) as function on cathode potential drop is presented in Fig. 23.12. It can be seen that G_c increases with considered range of u_c by the order of magnitude, and G_c for the cathode is significantly lower than G_a for the anode, which weakly depends on u_c. The anode erosion rate and the anode temperature in the spot increase with ratio r_a/r_s (Fig. 23.13). The increase is more for T_a than for G_a.

According to the calculation, the electron current is relatively large (the fraction is about 0.7) in a short arc. The emission electrons cool the cathode and warm the anode. Therefore, the anode temperature exceeds the cathode temperature. As a result, the plasma non-equilibrium in the anode spot as well as the mass flux evaporated from the anode is larger than that for the cathode. This fact explains the calculated larger anode erosion rate. The relation between cathode and anode erosion rates is determined by cathode and anode energy balances as well as by plasma energy balance, which are depended on the ratio r_a/r_s. The expanding plasma plume velocity is determined by the energy dissipated in the dense gap plasma and by amount of the expanding mass flux.

The smaller plasma plume velocity in comparison with cathode jet velocity for conventional arc is due to larger total electrode erosion rate in the short arc. However, the force per Coulomb, which is equal GV and generated in the short vacuum arc, exceeds such force generated in the conventional cathodic arc for all considered ranges of u_c. This result is important for a microthruster designing using a short microscale vacuum arc.

23.4 Cathodic Vacuum Arc Study with Respect to a Plasma Thruster Application

In this section, time dependent of the plasma jet parameters and thruster performance were studied for vacuum arc with free expansion of the cathode plasma jet for intermediate and refractory cathode material thermophysical properties.

23.4.1 Model and Assumptions

The theoretical study describes the time-dependent cathode phenomena in a vacuum arc producing the force during the development of the cathode spot in time on a bulk cathode. It is taken into account the kinetic model of cathode vaporization into the adjacent to the surface plasma generated from the eroded material from the cathode and by considering the hydrodynamics of plasma flow (Chap. 17). According to the kinetic model, the difference between evaporated and returned heavy particle fluxes produces the net cathode mass loss flux at the external boundary of Knudsen layer in form

$$G(g/s) = m(n_{a3} + n_{i3})v_3, \qquad (23.6)$$

Equation (23.6) describes the cathode erosion rate $G(g/C) = G(g/s)/I$, where $m = m_i = m_a$ and I is the electrical current. The relation of $b_3 = v_3/v_{th}$ defined the velocity fraction relatively to the thermal velocity $v_{th} = (kT_3/m)^{0.5}$ at the external boundary of the Knudsen layer.

Previously, the distribution of Cu cathode plasma flow parameters along the jet was calculated using the equations of momentum and energy conservation in differential form [54], (Chap. 18). The result obtained from equations in form of differential approximation was in good agreement with the measurements [1, 2] and showed that the plasma expands with a mostly frozen electron temperature of about 1–2 eV for 100 A spot and that sometime only part of internal momentum was converted to the direct jet velocity. It is also important when the jet velocity is calculated from a simpler model using the momentum equation in an integral approximation (Chap. 18). In this case, the calculated velocity using the complete spot plasma pressure at the cathode site and zero pressure at the vacuum condition can exceed sometime the measured jet velocity [1]. This is a critical point for cathode force calculation, for which requested a high accuracy of the velocity estimation, for example, to compare with experimental value of the force.

Therefore, as will be shown below, results were obtained for transient spot on bulk cathode according to the model presented in Chap. 17 for the parameters u_c, electron temperature T_e, heavy particle density n, degree of ionization, cathode temperature T, erosion rate G, cathode electric field E_c, current density j, electron current fraction s, and the fraction of cathode material evaporation K_{er}. However, the jet velocity is calculated using the experimental value of the cathode force. The given parameters are the cathode material properties and a characteristic small lifetime τ, during which the system of equation was solved to determine the primary plasma parameters that initiates the spot. The calculations were conducted for cathode materials of Cu, Ag, and Al, which have intermediate thermophysical properties, and for W, a refractory material. The behavior of mentioned above parameters was already presented in Chap. 17, and the plasma jet velocity was determined from simple integral equation in the expanding region using the value of measured force [55]. The obtained jet velocity was used to calculate the thrust efficiency η.

23.4.2 Simulation and Results

As it was mentioned above, the character of cathode plasma jet generation and the cathode plasma force were determined by the time-dependent energy dissipation in the cathode body and in the cathode spot dense plasma and by returned plasma flux. As a result, the time-dependent u_c, electron temperatures, cathode temperature, current density, electron current fraction, etc., were produced. For Cu, Al, and Ag, the time of spot initiation was chosen in range $\tau = 6\text{–}100$ ns, and the spot current was about 10 A. The calculation shows (Chap. 17) that u_c dependence on time was very sensitive near the value of τ. The larger u_c was at first step with $\tau = 10$ ns. For Al cathode, the u_c was relatively lower and increases from 20 to 50 V when τ decreased from 100 to 6 ns. The cathode evaporation fraction K_{er} decreases from about 0.25 to about 0.04 with the spot time in range shown for Cu and Ag, while for Al lower $K_{er} = 0.2\text{–}0.05$ was calculated. The characteristic parameters for Al were: $G = 7.4\text{–}12$ µg/C (28 µg/C-experiment [1]), $V \sim 2 \times 10^6$ cm/s (1.5×10^6 cm/s-experiment [1, 2]), $s = 0.75\text{–}0.65$, $j \sim (4\text{–}2) \times 10^6$ A/cm^2; $T_e = 6\text{–}1.5$ eV.

The jet velocity dependencies are obtained using the calculated parameters of G(g/C) for Cu (Fig. 23.14), for Ag (Fig. 23.15), and for Al (Fig. 23.16). It used the

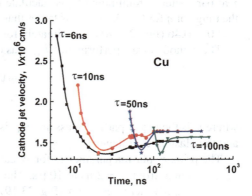

Fig. 23.14 Jet velocity as dependence on spot time t with τ as parameter for Cu cathode

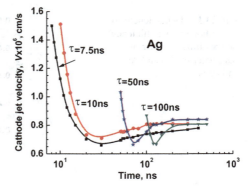

Fig. 23.15 Plasma jet velocity as dependence on spot time t with τ as parameter for Ag cathode

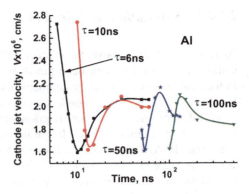

Fig. 23.16 Plasma jet velocity as dependence on spot time t with τ as parameter for Al cathode

measured force for Cu cathode about 40 din/A (also assumed this value for Ag and W) and 20 din/A for Al [33]. As shown in figures, the jet velocity depends on primary time τ of spot ignition and sharply decreased with spot time near τ. While the value of V is around the measured 10^6 cm/s, the velocity for Cu is comparable with that for Al and larger than for Ag by factor two.

Figure 23.17 illustrates the time dependencies of η for Cu and Ag cathodes with τ as parameter. A minimum of η is calculated in the dependencies of Fig. 23.17, and the range of η for Cu (0.25–0.5) is larger than that for Ag (0.1–0.22). For Al cathode, $\eta = 0.2$–0.46 (Fig. 23.18) approximately for all values of τ.

The cathode thrust performance K_{en} was defined as

$$\eta = \frac{mV^2}{2Iu_{\text{arc}}} \tag{23.7}$$

where m is the heavy particle mass and u_{arc} is the arc voltage, which is the sum of u_c and the plasma jet voltage u_p.

The primary plasma parameters for W cathode was calculated varying the initial times in range from $\tau = 2$ ns to 10 µs. The spot current was 10 A. The plasma jet velocity slightly increases with τ (Fig. 23.19), and it is around 1.6×10^6 cm/s (1.4 ×

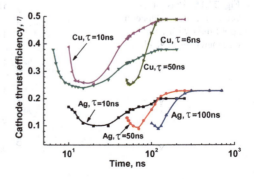

Fig. 23.17 Dependencies of the cathode thrust efficiency on spot time t, with τ as parameter for Cu and Ag cathodes

23.4 Cathodic Vacuum Arc Study with Respect to a Plasma ...

Fig. 23.18 Cathode thrust efficiency η as dependencies on spot time t with τ as parameter for Al cathode

Fig. 23.19 Cathode thrust efficiency η and plasma jet velocity as dependencies on spot initial lifetime τ for W

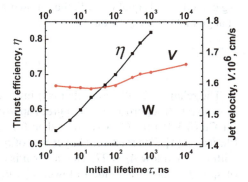

10^6 cm/s -experiment [2]). Figure 23.19 illustrates the dependence of cathode plasma thrust efficiency η. It can be seen that η is relatively large and increases from 0.5 to 0.85 with τ while calculation indicated relatively small values of b_3 (0.02) and K_{er} (0.08) slightly changed with τ (not shown in the figure).

In the calculated range of τ, the plasma heavy particle density significantly decreases (from 7×10^{20} to 1×10^{19} cm^{-3}), the cathode erosion rate is mostly constant (25–24 μg/C), and electron current fraction slightly increases from 0.7 to 0.8 when τ increase in the above indicated range. Also, the plasma in the spot is fully ionized and mainly consists of ions with charge fraction of 1$^+$ (0.09), 2$^+$ (0.40), 3$^+$ (0.44), and 4$^+$ (0.06) for $\tau = 2$ ns and 1$^+$ (0.03) 2$^+$ (0.39), 3$^+$ (0.56), and 4$^+$ (0.02) for $\tau = 10$ μs.

Explanation. The difference between direct and back heavy particle fluxes in the Knudsen layer produces the erosion rate that according to the calculations changed approximately by factor two during the considered ranges of spot time. This result explains the change of plasma jet velocity by using the measured cathode plasma force 40 dyn/A. The difference between the calculate velocities for Cu and Ag cathodes is due to difference of G(g/C) for these cathodes. The small minimum in the dependencies of V on time t for both cathodes is caused by those dependencies of G(g/C) on t, which is determined by a time-dependent combination of heavy particle and plasma velocity at the external boundary of Knudsen layer b_3.

The minimum in V(t) dependence produces similar minimum in the time dependence of cathode thrust performance, which is adequately different for Cu and Ag cathodes and reaches about 0.5 for Cu and 0.25 for Ag. It should be noted that η is lower for Ag although its larger atom mass ($A_i = 108$) in comparison with Cu ($A_i = 64$). It can be understood considering the expression for jet kinetic energy and force F as

$$K_{ev} \sim mV^2 = m\left(\frac{F}{G}\right)^2 = m\left(\frac{F}{mn_3v_3}\right)^2 \quad (23.8)$$

Taking into account that the calculated heavy particle density n_3 and plasma velocity v_3 at the external boundary of the Knudsen layer are close for both Cu and Ag cathodes and equal force F, the following dependence can be derived (with weak change of F and Iu_{arc}) from (23.8) as:

$$K_{ev} \sim \frac{1}{m} \quad (23.9)$$

The relation (23.9) shows that is η inversely proportional to the atom mass. The erosion rate (proportional to m) agrees approximately with that measured and which for Cu (0.3–0.4 μg/C) was lower than for Ag (0.8 μg/C) and comparable with Al (28 μg/C) [1]. The relatively larger part of η for Al was determined by reducing u_{arc} with time and low $A_i = 27$. The larger η is obtained for W cathode at primary time of spot development τ due to lower u_{arc}, b_3, and lower rate of calculated cathode erosion that is determined by lower heavy particle density (low evaporation rate) for W with respect to the density for Cu and Ag cathodes.

In frame of proposed transient model, the calculation shows that G and V are changed during the spot development. This change is relatively weak (factor two after time 10 ns) and indicates the resulting relatively weak change of cathode force with spot time (even for without use of the measured force). It should be noted that the force and the jet velocity could be obtained correctly independently using the more complicated approach for jet expansion in differential form. The present simpler approach illustrates the mechanism of the force formation and the importance of cathode plasma flow in the kinetic region, which studied self-consistently with the plasma flow in hydrodynamic regions during the spot initiation and development. This result allows to understand and to determine the contribution of cathode mass loss and plasma jet acceleration in the reactive force formation at the cathode of a vacuum arc.

23.5 Summary

A mechanism describing the momentum generation in the vacuum arc has been proposed in this Chapter. The plasma plume generation and plasma jet formation in a microscale vacuum arc (MVA) is unique subject of the force generated by a common

23.5 Summary

plasma plume extracted from cathode and anode spots in MVA. A calculating model for microplasma origin was developed taking into account the atom ionization and the phenomena in the non-equilibrium layers near evaporated surface in the cathode and anode regions. The interelectrode plasma in the short microscale vacuum arc is relatively dense $\sim 10^{20}$ cm^{-3}. The anode temperature in the anode spot is larger than that parameter in the cathode spot. It was found that relatively large anode erosion rate and force at the anode surface in MVA in comparison with the force usually occur at the cathode in arcs with large gap. This result of MVA is important to use in an exceptional microthruster concept for spacecraft propulsion and for anode microplasma source.

The calculations for free expanding plasma show that the time-dependent range of G agrees well with the measured data, and the jet velocity V calculated from the experimental data of cathode force is also close to that velocity obtained in the literature. The trust performance was obtained larger for W cathode than that for Cu and Ag cathodes. When the spot time increased from few ns to ~1 μs, the cathode evaporation fraction K_{er} decreases from about 0.25 to about 0.04–0.07. The plasma flow and the cathode vaporization rate in a vacuum arc are far from that into vacuum due to relatively large density of the cathode plasma that supported the request current density. The observed cathode force was determined by the kinetics of cathode mass loss flow and plasma jet expansion. The mechanism of the force formation was due to mutual influence of the cathode and plasma phenomena during the spot initiation and its development.

One particular recent implementation of vacuum arc technology for spacecraft propulsion is the microcathode arc thruster (μCAT) [12]. The technology is scalable and could be used for spacecraft of up to 50 kg in mass, which would roughly represent a 50 U CubeSat. The μCAT is an electric propulsion system that is based on the well-researched (see this Chapter) vacuum-arc or "cathodic arc" process. This physical phenomenon is known to erode the negative electrode (cathode) with every discharge. For space propulsion, this is highly desirable as the cathode material is the thruster's propellant. Therefore, during each discharge, a small amount of metallic propellant is eroded, ionized, and accelerated. The efficiency is enhanced by a magnetic field. The arc current generates the magnetic field as it travels through a magnetic coil prior to arcing between the electrodes. Summary of some μCAT parameters and its comparison with other thruster technologies are shown in Table 23.1.

In general, number of space applications can be considered in which electric micropropulsion such as μCAT can be beneficial. This includes station keeping, altitude control that generally require small thrust of about 10 μN and high efficiency of propellant utilization. More significant space maneuvers such as orbit transfer might require of mN thrust level and in this case more powerful μCAT thrusts or array of multiple thrusters can be considered.

The system uses a solid metal as a propellant and therefore, does not require any pressurized tanks or other components that may be required when dealing with gaseous propellants such as xenon. This is advantageous because it greatly reduces the system's complexity and risk involved. Additionally, only an electrical connection

Table 23.1 Comparison of micropropulsion technologies [12]

	μCAT (GWU)	PPT (Clyde space) [56, 57]	PPT (Busek Co) [58–60]	Electrospray (MIT) [61–63]	Electrospray (Busek Co) [58, 59]	VAT (Alameda) [64]
System mass (g)	200	160	550	45	1150	600
System volume (cm^3)	200	200	500	300	500	200
Propellant	Metal	Teflon	Teflon	Liquid	Liquid	Metal
I_{sp} (s)	3000	590	700	3000	800	150
Propellant mass (g)	40	10	36	20	75	40
Delta-V (for 4 kg satellite) (m s^{-1})	300	15	63	150	151	151
Efficiency (%)	15	4.7	16	71	31	9.4
Thrust-to-mass ratio (μN g^{-1})	0.63	0.03	0.18	0.5	0.65	0.22
Ionization degree	High	Low	Low	High	High	High
Cost	Low	Low	Low	High	High	High
Technical readiness level (TRL)	6	7	7	2–3	5	4

is required to operate the thruster, since the propellant and all necessary components are integrated within the thruster's structure.

The μCAT system has been flown on BRICSat-P, which is a 1.5 U CubeSat mission led by the United States Naval Academy (USNA) (Fig. 23.20) [65]. Preliminary reports show that the propulsion system was capable of de-tumbling the satellite to a rate of less than 1 degree per second on all axes within 48 h of deployment. Another recent mission was Canyval-X, which is a joint project between NASA Goddard Space Flight Center, the Korean space agency KARI and the Yonsei University in Korea. Canyval-x was launched on Jan 12, 2018.

23.5 Summary

Fig. 23.20 μCAT thrusters installed on the NASA/KARI CANYVALx spacecraft. Courtesy of Dr. Michael Keidar

References

1. Boxman, R. L., Martin, P. J., & Sanders, D. M. (Eds.) (1995). *Handbook of vacuum arc science and technology*. Park Ridge, N.J.: Noyes Publ.
2. Davis, W. D., & Miller, H. C. (1969). Analysis of the electrode products emitted by dc arcs in a vacuum ambient. *Journal of Applied Physics, 40*, 2212–2221.
3. Anders, A. (2008). *Cathodic arcs: From fractal spots to energetic condensation*. Springer.
4. Beilis, I. I. (2018). Vacuum arc cathode spot theory: history and evolution of the mechanisms. *IEEE Transactions on Plasma Sciences, 47*(N8), 3412–3433.
5. Dorodnov, A. M. (1974). Some applications of plasma accelerators in a technology. In A. I. Morosov (Ed.), *Physics and applications of plasma accelerators*. Minsk: Nauka & Technika Publisher. (In Russian)
6. Gilmour, A. S., & Lockwood, D. (1972). Pulsed metallic-plasma generators. *Proceedings of the IEEE, 60*, 977–991.
7. Dorodnov, A. M. (1978). Technical applications of plasma accelerators. *Soviet Physics. Technical Physics, 23*, 1058–1065.
8. Dorodnov, A. M., & Petrosov, B. A. (1981). Physical principals and types of technological vacuum plasma devices. *Soviet Physics. Technical Physics, 26*, 304–315.
9. Qi, N., Gensler, S., Prasad, R. R., Krishnan, M., Visir, A., & Brown, I. G. (1998). A pulsed vacuum arc ion thruster for distributed small satellite systems. In *34th AIAA/ASME/SAE/ASEE Joint Propulsion Conference and Exhibit, AIAA* (Vol. 98, p. 3663). Cleveland, OH.
10. Schein, J., Qi, N., Binder, R., Krishnan, M., Ziemer, J. K., Polk, J. E., & Anders, A. (2002). Inductive energy storage driven vacuum arc thruster. Review of Scientific Instruments, *73*(N2), 925–927.
11. Zhuang, T., Shashurin, A., & Keidar, M. (2011). Microcathode thruster μCT plume characterization. *IEEE Transactions on Plasma Sciences, 39*, 2936–2937.
12. Keidar, M., Zhuang, T., Shashurin, A., Teel, G., Chiu, D., & Lukas, J. (2015). Electric propulsion for small satellites. *Plasma Physics and Controlled Fusion, 57*(N1), 014005.
13. Wright, W. P., & Ferrer, P. (2015). Electric micropropulsion systems. *Progress in Aerospace Science, 74*, 48–61.

14. Keidar, M., & Beilis, I. I. *Plasma engineering: Applications from aerospace to bio and nanotechnology*, (London-NY: Academic Press, Elsevier, 2013). *Plasma Engineering*, 2nd edn. (London-NY: Academic Press, Elsevier, 2018).
15. Levchenko, I., Bazaka, K., Ding, Y., Raitses, Y., Mazouffre, S., Henning, T., et al. (2018). Space micropropulsion systems for cubesats and small satellites: From proximate targets to furthermost frontiers. *Applied Physics Review, 5*, 011104.
16. Kronhaus, I., Laterza, M., & Linossier, A. (2018). Experimental characterization of the inline-screw-feeding vacuum-arc-thruster operation. *IEEE Transactions on Plasma Sciences, 46*(N2), 283–288.
17. Kolbeck, J., Anders, A., Beilis, I. I., & Keidar, M. (2019). Micro-propulsion based on vacuum arcs. *Journal of Applied Physics, 125*, 220902.
18. Polk, J. E., Sekerak, M. J., Ziemer, J. K., Schein, J., Qi, N., & Anders, A. (2008). A theoretical analysis of vacuum arc thruster and vacuum arc ion thruster performance. *IEEE Transactions on Plasma Sciences, 36*(N5), 2167–2179.
19. Gilmour, A. S. (1966). Concerning the feasibility of a vacuum arc thruster. In *5th Electric Propulsion Conference*, San Diego, CA.
20. Gilmour, A. S. Concerning the feasibility of a vacuum arc thruster, Paper 66–202, March 1966, AIAA.
21. Gilmour, A. S., Clark, J. R. R., & Veron, H. (1967). Pulsed vacuum-arc microthrustors. In *6th Electric Propulsion and Plasmadynamics Conference*, American Institute of Aeronautics and Astronautics.
22. Dethlefsen, R. (1968). Performance measurements on a pulsed vacuum arc thruster. *AIAA, 6*(N6), 1197–1199.
23. Dethlefsen, R. (1967 March). Cathode spot phenomena of pulsed vacuum arc utilized for micro thrusters. In *Proceedings of the 8th Symposium on the Engineering Aspects of MHD*, Stanford, California.
24. Teel, G., Shashurin, A., Fang, X., & Keidar, M. (2017). Discharge ignition in the micro-cathode arc thruster. *Journal of Applied Physics, 121*(N2), 023303.
25. Keidar, M., Schein, J., Wilson, K., Gerhan, A., Au, M., Tang, B., et al. (2005). Magnetically enhanced vacuum arc thruster. *Plasma Sources in Science and Technology, 14*(N4), 661–669.
26. Kemp, M. A., & Kovaleski, S. D. (2008). Thrust measurements of the ferroelectric plasma thruster. *Plasma Sources in Science and Technology, 36*(N2), 356–362.
27. Wirz, R., Sullivan, R., Przybylowski, J., & Silva, M. (2008). Hollow cathode and low-thrust extraction grid analysis for a miniature ion thruster. *International Journal of Plasma Science and Engineering, 1*(N1), Article ID 693825.
28. Zhuang, T., Shashurin, A., Beilis, I. I., & Keidar, M. (2012). Ion velocities in a micro-cathode arc thruster. *Physics of Plasmas, 19*(N6), 063501.
29. Tanberg, R. (1930). On the cathode of an arc drawn in a vacuum. *Physical Review, 35*(N9), 1080–1090.
30. Robertson, R. M. (1938). The force on the cathode of a copper arc. *Physical Review, 53*(N7), 578–582.
31. Kobel, E. (1930). Pressure and high velocity vapor jets at cathodes of a mercury vacuum arc. *Physical Review, 36*(N11), 1636–1638.
32. Chabrerie, J. P., Devautour, J., Gouega, A. M., & Teste, Ph. (1995). A sensitive device for the measurement of the force exerted by the arc on the electrodes. *IEEE Trans on Components, Packaging and Manufacturing Technology, Part A, 18*(N2), 322–328.
33. Marks, H. S., Beilis, I. I., & Boxman, R. L. (2009). Measurement of the vacuum arc plasma force. *IEEE Transactions on Plasma Sciences, 37*(N7), 1332–1337.
34. Tonks, L. (1934). The pressure of plasma electrons and the force on the cathode of an arc. *Physical Review, 46*(N9), 278–279.
35. Maecker, H. (1955). Plasmastromungen in lichtbogen infolge eigenmagnetischer kompression. *Z. f Physik, Bd 141*, S198–S216.
36. Ecker, G. (1961). Electrode components of the arc discharge. *Ergeb. Exakten Naturwiss, 33*, 1–104.

37. Lyubimov, G. A. (1977). Dynamics of cathode vapor jets. *Soviet Physics. Technical Physics*, 22(N2), 173–177.
38. Beilis, I. I. (1995). Theoretical modeling of cathode spot phenomena. In R. L. Boxman, P. J. Martin, & D. M. Sanders (Eds.), *Handbook of vacuum arc science and technology* (pp. 208–256). Park Ridge, N.J.: Noyes Publ.
39. Keidar, M., Beilis, I. I., Boxman, R. L., & Goldsmith, S. (1997). Voltage of the vacuum arc with a ring anode in an axial magnetic field. *IEEE Transactions on Plasma Sciences*, 25(N4), 580–585.
40. Germer, L. H., & Haworth, L. H. (1949). Erosion of electrical contacts on make. *Journal of Applied Physics*, 20, 1085–1109.
41. Kisluk, P. (1954). Arcing an electrical contacts on closure. The cathode mechanism of extremely short arcs. *Journal of Applied Physics*, 25(N7), 897–900.
42. Boyle, W. S., & Germer, L. H. (1955). Arcing an electrical contacts on closure. The anode mechanism of extremely short arcs. *Journal of Applied Physics*, 26(N5), 571–574.
43. Abbas, N. M., Solomon, D. G., & Bahari, Md. F. (2007). A review on current research trends in electrical discharge machining (EDM). *International Journal of Machine Tools & Manufacture*, 47, 1214–1228.
44. Yanagidaira, T., Nose, D., Abiko, H., Miura, H., & Tsuruta, K. (2004). Soft x-ray generation after arc extinction of a short vacuum gap operated with low-current, repetitive discharge. *Plasma Source Science & Technology*, 13, 116–120.
45. Dong, H., Zou, J., Liao, M., Wu, Y., & Zhao, S. (2005). Experimental investigation of vacuum arc characteristics in short gap. In *Proceedings of the XIVth International Symposium on High Voltage Engineering*, Tsinghua University, Beijing, China, August 25–29.
46. Ecker, G., & Paulus, I. (1988). The short vacuum arc-Part I: Experimental Investigations. *IEEE Transactions on Plasma Science*, 16(N3), 342–347.
47. Ecker, G., & Paulus, I. (1988). The short vacuum arc-part II: Model & theoretical aspect. *IEEE Transactions on Plasma Science*, 16(N3), 348–351.
48. Meng, G., Cheng, Y., Dong, Ch., & Wu, K. (2014). Experimental study on electrical breakdown for devices with micrometer gaps. *Plasma Science and Technology*, 16(N12), 1083–1089.
49. Radmilovic-Radjenovic, M., & Radjenovic, B. (2008). Theoretical study of the electron field emission phenomena in the generation of a micrometer scale discharge. *Plasma Source Science & Technology*, 17, 024005.
50. Kolacinski, Z. (1993). Modelling of short arc re-ignition. *Journal of Physics D: Applied Physics*, 26, 1941–1947.
51. Beilis, I. I. (2000). Anode spot vacuum arc model. Graphite Anode *IEEE Transaction on Components and Packaging Technologies*, 23(N2), 334–340.
52. Beilis, I. I., & Zektser, M. P. (1991). Calculation of the parameters of the cathode stream an arc discharge. *High Temperature*, 29(N4), 501–504.
53. www.aasc.net/micropropulsion.
54. Beilis, I. I. (2015). Physics of cathode phenomena in a vacuum arc with respect to a plasma thruster application. *IEEE Transactions on Plasma Sciences*, 43(N1), Part I, 165–172.
55. Guarducci, M., & Gabriel, S. B. (2011). Design and testing of a micro pulsed plasma thruster for Cubesat application. In *32nd International Electric Propulsion Conference (Wiesbaden, Germany, 11–15 Sept. 2011)* IEPC-2011–239.
56. www.clyde-space.com/cubesatshop/propulsion/303cubesat-pulse-plasma-thruster.
57. Hruby, V. (2012 May 29–30). High is cubesat propulsion. In *1st Interplanetary CubeSat Workshop (Massachusetts Institute of Technology, Cambridge, MA)*.
58. www.busek.com/.
59. Mueller, J., Ziemer, J., Hofer, R., Wirz, R., & O'Donnell, T. (2008 April 9–11). A survey of micro-thrust propulsion options for microspacecraft and formation flying missions. In *Cube Sat—5th Annual Developers Workshop (California Polytechnic State University, San Luis Obispo, CA)*.
60. Martel, F., & Lozano, P. (2012 May 29–30). Ion electrospray thruster assemblies for CubeSats. *iCubeSat Workshop (Cambridge, MA)*.

61. Gassend, B. L. P. (2007). Fully microfabricated two-dimensional electrospray array with applications to space propulsion. *Ph.D. Thesis Massachusetts Institute for Technology.*
62. Vel´asquez-Garc´ıa, L. F., Akinwande, A. I., & Mart´ınez-S´anchez, M. (2006). Planar array of micro-fabricated electrospray emitters for thruster applications. *Journal of Micromechanical Systems, 15*, 1272.
63. Plyutto, A., Ryzhkov, V., & Kapin, A. (1965). High speed plasma streams in vacuum arcs. *Soviet Physics. JETP, 20*, 328–337.
64. Kimblin, C. W. (1974). Anode phenomena in vacuum and atmospheric pressure arcs. *IEEE Transactions on Plasma Sciences, PS-2*(N6), 310–319.
65. Hurley, S., Teel, G., Lukas, J., Haque, S., Keidar, M., Dinelli, C., & Kang, J. (2016). Thruster subsystem for the United States naval academy's (USNA) ballistically reinforced communication satellite (BRICSat-P). Transactions of JSASS, Aerospace Technology Japan, 14(N30), Pb157–Pb163.

Chapter 24
Application of Cathode Spot Theory to Laser Metal Interaction and Laser Plasma Generation

Numerous applications indicate that laser–matter interaction is technologically important [1]. In particular, the interaction of intense laser beams with plasmas is highly relevant to laser fusion [2], material science and medicine [3], and nanoparticles generation [4]. The interferometric diagnostics was established as a technique to probe dense plasmas using optical wavelength laser beams [5]. A large electric field was generated in plasma by high-power lasers in order to accelerate electrons [6–8]. The laser plasmas are used for target surface treatment [9, 10], for thin film deposition [11–13], for high-energy atomic beam generation [14], laser welding [15], for laser beam machining (see review Dubey et al. [16]), and for ion beam generation [17].

24.1 Physics of Laser Plasma Generation

One of the important phenomena is plasma generation near a target surface by the impingement of laser radiation of moderate power density ($10^6 - 10^{10}$ W/cm^2) focused onto a very small area. Several phenomena occur during laser–target interaction (**LTI**). When the laser pulse is absorbed by the target, energy is converted into thermal and mechanical energy, ablating the target. When the laser pulse is absorbed by the vapor, energy is converted into thermal energy of the vapor particles and then into plasma formation. Then, the laser radiation will interact with the generated plasma. Understanding the physics of energy conversion is useful for choosing the laser and target characteristics suitable for different applications, taking into account ablation. This understanding can be achieved by theoretically modeling laser plasma generation and plasma expansion. Below the developed models are based on experimental studies of the laser interaction phenomena. Therefore, first, the state of plasma measurements is discussed to formulate the main problem of moderate power laser interactions with metallic targets, and then, results are used for modeling of the laser–target interaction and laser plasma generation.

24.2 Review of Experimental Results

A laser pulsed beam with 6 ns pulse durations and 3.2 mJ energy impinges on an Al target and modified it. Using scanning electron microscopy (SEM), Henc-Bartolic et al. [18] observed that molten droplets with radii between 0.1 and 0.5 µm, Al vapor, and surface craters of ~30 µm depth were formed after 50 laser pulses. The electron temperature of the Al plasma was estimated as 1.4 eV, using spectroscopic data and that the laser energy was mostly absorbed by the plasma particles. The plasma produced by a pulsed nitrogen laser beam (6 ns, 9 mJ) impinging on a Ti target was studied spectroscopically by Henc-Bartolic et al. [19]. It was found that the electron density was $n_e \sim (1.5 - 3) \times 10^{18}$ cm^{-3} and that Stark broadening is the dominant broadening mechanism. The electron temperature T_e measured from the relative intensities of two Ti spectral lines in vacuum was approximately 2.7 eV.

Hermann et al. [20] recorded time- and space-resolved spectra to investigate the plasma formed from 20 ns pulsed excimer laser irradiation of Ti targets. The laser intensity was varied from the vaporization threshold at 25 MW/cm^2 up to 500 MW/cm^2, which generated plasma. Electron temperature T_e at a distance of 2 mm from the target decreased with time, from 2 to 3 eV at 150 ns after the laser pulse, to 1.5 eV at 300 ns. At fixed time of 100 ns after the laser pulse, density n_e decreased with distance, from 10^{18} cm^3 at 0.5 mm to 10^{17} cm^3 at 2 mm, as measured using the Stark broadening of Ti atom and ion lines. When the power density increased from 25 to 500 MW/cm^2, the directed ion velocity increased from 2.10^5 to 2.10^6 cm/s.

Chang et al. [21] used a copper vapor laser (CVL) and varied the intensity from 0.1 to 10 GW/cm^2 to investigate the parameters of plasma generated from aluminum and carbon steel targets. Schlieren images were captured by a charge-coupled device (CCD). A narrowband filter centered at the probe laser wavelength was used to block the plume radiation and scattered CVL light. The measured speed $(3 - 20) \times 10^5$ cm/s was used to determine vapor density, temperature, and pressure during the end of the CVL pulse through the hydrodynamic relations describing adiabatic shock expansion and Knudsen layer jump conditions. It was found that T_e was 1–2 eV and $n_e = 10^{20} - 10^{21}$ cm^{-3}. These data correlated with published measurements even though the reverse approach was uncertain.

The energy distribution of ions ejected from solid and liquid Si and Ge and from solid Cu targets by a high-fluence (1–8) J/cm^2 excimer laser was determined by analyzing the time of flight (TOF), by measuring the current pulse delay of ions overcoming an adjustable electrostatic potential barrier applied before a detector [22]. For both solid and liquid targets, the ion energies were found of the order of 100 eV per ion charge and depended little on the laser fluence.

Abdellatif and Imam [23] determined the spatial distribution of T_e using Boltzmann plots of Al II lines and the spatial profile of n_e using Stark broadening, produced by a 7 ns pulse laser with wavelengths 1064, 532, and 355 nm. When the laser wavelength was 1064 nm, $T_e \sim 1.17$ eV at the target surface, and it increased gradually to 4.2 eV at a distance of 500 mm from the target surface and then decreased with

24.2 Review of Experimental Results

distance to about 1 eV. When the laser wavelength was 355 nm, maximum T_e was about 5.7 eV and $n_e = (0.85 - 0.35) \times 10^{18}$ cm^{-3}. Rieger et al. [24] investigated the emission Si and Al plasmas produced by a KrF laser with 10 ps pulse and power densities (0.05–50) GW/cm^2 and also with 50 ps pulse and power densities (10–10^4) GW/cm^2. It was shown that 50 ps, 100 GW/cm^2 pulses generated Al plasma with $n_e \sim 10^{22}$ cm^{-3}, $T_e \sim 14$ eV, while 10 ns, 0.5 GW/cm^2 pulses produced 10^{20} cm^{-3} and ~2.4 eV.

A 7 ns pulsed Nd:YAG laser producing a 52 mJ, 335 nm, 10^{10} W/cm^{-2} beam was irradiated on a Cu target in vacuum and argon [25]. T_e was determined by a Boltzmann plot and n_e by Stark broadening. Also, a single Langmuir probe was located 3.5 mm from the target. In vacuum, T_e increased from 1.6 eV at the surface to 3.8 eV at a distance of 2 mm from the surface, and then decreased, reaching 0.6 eV at 10 mm. The plasma velocity was found as 5.3×10^5 cm/s. The electron density was $n_e \sim 1.4 \times 10^{16}$ cm^{-3} at the surface and decreased by more than one order of magnitude at a distance of 10 mm from the target.

Plasmas produced by a 200 mJ, 8 ns pulse in air on a target from a ~60% Cu, ~40% Zn brass alloy, were studied spectroscopically [26]. It was reported that n_e was $10^{16} - 10^{18}$ cm^3 and T_e $0.7 - 1.2$ eV. Aguilera et al. [27] studied plasma induced by a (3.5–4.5) GW/cm^2, 4.5 ns, laser beam in atmospheric pressure air and argon and focused onto a Fe target and determined the temporal and spatial distributions of T_e and n_e and neutral atom density. In air, the Fe emissivity and atom density were maximized when the focal plane of the lens was placed 10 and 12 mm, respectively, below the sample surface, whereas higher temperature (1 eV) was obtained when the focus was 5 mm below the surface and decreased for deeper focusing positions. This behavior was explained by plasma shielding at high laser irradiance. The emission intensity in Ar was about twice of the intensity in air due to higher temperature in Ar. Also, a maximum of the emission intensity was found for both gases at a focus distance of ~10 mm.

Laser-generated metallic plasma was widely investigated by Torrisi and co-authors, and the details of their methods were described by Margorone et al. [28] and Torrisi et al. [29]. They used a Nd:Yag laser with a 3–9 ns pulse and fluence of (0.1–9.3) J/cm^2. Electrostatic ion energy analyzers and the TOF technique were used to measure the ion energy and charge-state distributions. The mass loss per pulse was obtained by measuring the crater depth profile. The radiation emission spectrum, T_e and n_e, was investigated by optical spectroscopy using a CCD camera. They found, for metal targets with a wide range of thermophysical properties, that the ion distribution function was a shifted Maxwellian distribution, which depended on laser fluence, and that the energy shift increased linearly with the ion charge. For carbon it was ~65 eV/charge [30] and for Si ~300 eV/charge, with ion charge from +1 to +4 at 150 mJ laser energy [31]. Using the maximal energy ion distribution shift, the estimated singly charged ion velocity for carbon was 2.8.10^6 cm/s and for Si was 4.5.10^6 cm/s. The energy shift increase with the ion charge was explained assuming that an electrostatic field is present within a Debye length similarly with that electric field occurred for high intensity fs and ps laser pulse irradiation.

For Ta target, the ion energy was higher, ~200 eV/charge at 30 J/cm² and ~600 eV/charge state at 100 J/cm², [32]. The average width of the distribution indicated a relatively large ion temperature, $(7 - 26) \times 10^5$ K, i.e., ~60–200 eV, while the average plasma velocity was reported as $(0.7 - 1.4) \times 10^6$ cm/s. Later, for Ag plasma $T_e \sim 5$ eV and $n_e \sim 2.10^{16}$ cm^{-3} were obtained spectroscopically [28]. Similarly shifted Maxwellian ion distributions were observed by Bleiner et al. [33]. The average width of the distribution indicated ion temperature depended on the target material: 18, 27, 30, and 45 eV for Al, Fe, Sn, and Zn, respectively. Torrisi et al. [34, 35] measured the target mass loss from the craters by weighing the targets before and after irradiation by 1000 pulses of 9 ns and 875 mJ and found 0.25, 0.51, 0.55, and 10.5 μg/pulse for Cu, Al, Ni, and Pb, respectively. Another group of materials, Au, Ti, W, and Ta, had higher mass loss, 0.8–1 μg/pulse. A linear dependence of mass loss on fluency in range of 1–6 J/cm² was measured for Al, Cu, and Ta targets by Caridi et al. [36]. Kasperczuk et al. [37] showed that the craters have a semi-toroidal shape when the laser was focused inside of the target, and when the laser was focused at the front position, they resemble a hemisphere.

Zeng et al. [38] spectroscopically measured T_e and n_e in plasma produced by nanosecond laser impinging on Si target with a flat surface or cavity. T_e was 3 eV and ~4 eV, and n_e was ~5×10^{18} cm^{-3} and $(1 - 3) \times 10^{19}$ cm^{-3} with flat surface and cavity, respectively. T_e and n_e were uniform up to 1.5 mm from the target surface. The plasma expanded at a velocity of 5×10^5 cm/s.

The spatial evolution of T_e and n_e in plasma generated on SiC samples by a Nd:YAG laser beam at wavelength of 1064 nm, pulse width of 10 ns, and power of 0.2–2 GW was spectroscopically studied by Chen et al. [39]. They found that n_e decreased from 1.8 to 0.73×10^{17} cm^{-3} and T_e from $1.4 - 1.7$ to 1 eV with distance from the target up to 9 mm. In the region between 3 and 9 mm, n_e fluctuated and T_e decreased slowly. Further investigations of these authors determined the T_e and n_e dependence on the time after the pulse [40–42]. It was shown that T_e substantially and n_e moderately decreased with time in range of 50–600 ns. For example, at 1 mm from the target surface, T_e decreased from 3.3 to 1.75 eV and n_e decreased from 2.6 $\times 10^{17}$ to 2.25×10^{17} cm^{-3} at 50 and 400 ns, respectively, after the laser pulse. T_e and n_e had maxima at ~2 mm from the target surface.

Cristoforetti et al. [43] spectroscopically observed that n_e increased from 0.7×10^{16} to 2×10^{17} cm^{-3} and T_e from 0.85 to 1.15 eV when the irradiance of a Nd:YAG laser beam of Al target increased from 3×10^8 to 5×10^9 Wcm^{-2}. Ablated plasma flow from a liquid Ga-In target was studied with laser power density 1 GW/cm² and pulse duration 2.7 ns [44]. The average ion energy and the directed ion velocity were measured using a mass-energy and time-of-flight methods, respectively. They found an energy of ~400 eV, indicating that there was an electrostatic field in the plasma plume. However, a relatively low velocity of about 3.10^5 cm/s was measured. Measured characteristic plasma parameters published in the last decades are summarized in Table 24.1.

Concluding remarks: When the laser power density exceeds 10^8 W/cm², the plasma was produced with broad ion kinetic energy distributions. Furthermore, when multiply charged ions are detected, it was found that their most probable kinetic

24.2 Review of Experimental Results

Table 24.1 Methods and measured characteristic plasma parameters (ion velocity and energy, electron temperature and density) depending on laser type, power, pulse duration, and spot area in experiments with laser irradiation of different target materials [45]

Target	Laser	Pulse (ns)	Energy, fluence	Spot diameter, area	Method	V (cm/s)	Ne, cm^{-3} Craters	T_e (eV)	Shift, V/charge	References
Ti	Excimer laser	20	25–500 MW/cm^2		TOF, stark broadening	$(0.2-1) \times 10^6$	$10^{18} - 10^{17}$	3–1.5	–	Hermann et al. [20]
Si, Ge, Cu	Excimer laser	25	(1–8) J/cm^2		Electrostatic analyzer TOF	$(1.8-3.1) \times 10^6$	10^9 Ions/pulse	2	100	Franghiadakis et al. [22]
Al and carbon steel	Cu laser		0.1–10 GW/cm^2		CCD camera	$(3-20) \times 10^5$	$10^{20} - 10^{21}$	1–2	–	Chang et al. [21]
Ti	N$_2$ laser	6	9 mJ	–	Stark broadening	–	$(1.5 - 3) \times 10^{18}$	2.7	–	Henc-Bartolic et al. [19]
Al	N$_2$ laser	6	3.2 mJ	–	SEM	–	Droplets 0.1–0.5 μm Craters 30 μm	1.4 (estimated)	–	Henc-Bartolic et al. [19]
Ta	Nd:YAG Wavelengths 1064	9	1–900 mJ	0.5 mm^2	Electrostatic analyzer TOF	1.6×10^6	–	–	200–600	Torrisi et al. [32]
Al	Nd:YAG Wavelengths 1064, 532, and 355 nm	7	500, 100, and 60 mJ	300 μm	Boltzmann plot, stark broadening	–	$(0.85 - 0.35) \times 10^{18}$	1.17–6.3	–	Abdellatif and Imam [23]
Al, Si	KrF laser	50 ps – 10 ns	(0.05–50) and 10,000 GW/cm^2	5 μm	Spectra Pro 500,	–	$10^{20} - 10^{22}$	2.4 14	–	Rieger et al. [24]

(continued)

Table 24.1 (continued)

Target	Laser	Pulse (ns)	Energy, fluence	Spot diameter, area	Method	V (cm/s)	Ne, cm^{-3} Craters	T_e (eV)	Shift, V/charge	References
Cu	Nd:YAG Wavelengths 335 nm	7	52 mJ	10^{10} W/cm^{-2}	Spectroscopy CCD camera, langmuir single probe	5.3×10^5	1.4×10^{16}	1.6 – 0.6 at 10 mm $T_{emax} =$ 3.8 eV at 2 mm	–	Hafez et al. [25]
Fe	Nd:YAG 1064 nm	4.5	100 mJ	(3.5–4.5) GW/cm^2	Spectrometer CCD	–	2×10^{16}	1.0	–	Aguilera et al. [27]
60% Cu, ~40% Zn brass	Nd:YAG 1.06 μm	8	200 mJ	–	Echelle-type spectrometer CCD camera	–	$10^{16} - 10^{18}$	0.7–1.2	–	Corsi et al. [26]
Pyrolytic graphite	Nd:Yag 532 nm	3	20–170 mJ	Area 3 mm^2	Quadrupole spectrometer	2.8×10^6 estimated	6.7×10^{17}	–	65	Torrisi et al. [30]
Al, Cu, and Ta	Nd:Yag 532 nm	3	1.7–170 mJ	Area 0.8 mm^2	Quadrupole spectrometer	–	–	–	83 and 112	Caridi et al., [36]
Si	Nd:Yag 266 nm	3	40 J/cm^2	40 μm	Spectroscopic measurements	2×10^6	$(0.5 - 3) \times 10^{19}$	~5	–	Zeng et al. [38]
Al, Fe, Sn, Zn	Eximer 248 mm	23	1.3 GW/cm^2	Area 1 mm^2	Faraday cup TOF	$(2 - 4) \times 10^6$	–	Width: 18, 27, 0.45 eV	–	Bleiner et al. [33]
Si	Nd:Yag 532 nm	3	25–150 mJ	Area 0.5 mm^2	Quadrupole spectrometer TOF	4×10^6 estimated	–	6–9 (estimated)	259–306	Torrisi et al. [31]

(continued)

24.2 Review of Experimental Results

Table 24.1 (continued)

Target	Laser	Pulse (ns)	Energy, fluence	Spot diameter, area	Method	V (cm/s)	Ne, cm^{-3} Craters	T_e (eV)	Shift, V/charge	References
Ag	Nd:Yag 1064 nm	3–9	(0.3–9.3) J/cm^2	Area 3 mm^2	Spectroscopy TOF energy analyzer	10^6	2 × 10^{16}	5	280	Margorone et al. [28]
Al, Zn, Pb, Ta	Nd:Yag 532 nm	3	50–150 mJ	Area 2 mm^2	Electrostatic mass spectrometer	–	–	–	80,90,95 115—Pb, Zn, Al Ta	Torrisi et al.[29]
SiC	Nd:YAG 1064 nm	10	1.9–11 mJ	1 mm	Spectroscopic Boltzmann plot, CCD	–	(1.8 – 0.73) × 10^{17}	1.1–1.6	–	Chen et al. [39]
SiC Time delay 50–400 ns	Nd:YAG 1064 nm	10	130	1 mm	Spectroscopic Boltzmann plot, CCD	–	(3 – 1) × 10^{17}	3.5–1	–	Chen et al. [40]
KTiOAsO$_4$ at distance 1–19 mm Time delay 50–600 ns	Nd:YAG 1064 nm 532 nm	10	1.9–11 mJ	1 mm	Spectroscopic Boltzmann plot, stark CCD	–	(2.9 – 1.55) × 10^{17} (3.5 – 0.7) × 10^{17}	3–0.7 2.2–1.5	–	Chen et al. [41], Sun et al. [42]
Al	Nd:YAG 1.06 μm	32	2.6–38 mJ	(0.3–5) × 10^9 Wcm^{-2}	Echelle-type spectrometer CCD camera	–	(0.07 – 2) × 10^{17}	0.85–1.15	–	Cristoforetti et al. [43]
Ga-In	Nd:YAG 1064 nm	2.7	1 GW/cm^2	100 μm	Mass-energy analyzer, TOF	3 × 10^6	–	–	Width 400 eV	Popov et al. [44]

energy increases linearly with the ion charge. This energy reaches few hundred eV per ion charge and was identified with a correspondingly large potential drop in the plasma, which was not always correlated with the moderate plasma velocities $(5 - 20) \times 10^5$ cm/s that were observed by TOF. Also, the electron temperature was usually low, 1–2 eV, and the direct spectroscopic measurements never exceeded 5–6 eV, while the measured average width of the ion velocity distribution function corresponded to relatively large plasma temperatures, reaching few tens eV depending on target material and laser fluence.

While these measured plasma parameters were briefly explained by the discussions in the corresponding publications, they considered only separate phenomena and were limited usually by comparison with other observations. The mechanism of ion acceleration developed for high-power ($>10^{14}$ W/cm^2) laser impingement was used to explain the ion energy measured by the moderate laser power ($<10^9$ W/cm^2) irradiation, which is not correct due to different physical phenomena and plasma parameters. The measured relatively large ion energies were not correlated with the low electron temperatures measured spectroscopically, and this fact was not explained. Let us consider the theoretical works and the calculated results for moderate power density.

24.3 Overview of Theoretical Approaches of Laser–Target Interaction

Material ablated from the target from the impingement of a laser beam expands away from the surface in form of a plasma jet. Jet expansion into vacuum is described by the hydrodynamic equations, expressing the conservation of mass, momentum, and energy. Plasma formation by the impingement of nanosecond (3–7) J/cm^2 pulsed laser beams on Cu, Ba, and Y targets was modeled by Singh and Harayan [46]. After interaction of the laser beam with the target the plasma formation and heating were considered. The plasma expansion was studied by two approaches, as initial 3D isothermal expansion and then the adiabatic expansion and thin film deposition were investigated. The dynamics of thin film laser deposition was obtained using the time evolution of the calculated plasma velocity and measured T_e.

Gamaly et al. [47] described the process of vaporization of a target material; the properties of the vapor and plasma flow and film deposition on a substrate for high repetition rate laser evaporation were studied. For this purpose, the evaporation atom flux was previously described for a single pulse and for different conditions of absorbed laser radiation ($<10^{10}$ W/cm^2), solving the target one-dimensional heat conduction equation. In case of absorption at the vapor–solid interface, all parameters at the interface between solid and vapor are related to the absorbed laser intensity via conservation laws for mass, momentum, and energy. The vapor was considered as an ideal gas with an adiabatic exponent and sound velocity. It was indicated that

24.3 Overview of Theoretical Approaches of Laser–Target Interaction

the proposed theoretical method allowed defining the optimal conditions for efficient evaporation of a target, with given thermodynamic properties.

An important part of LTI study is to correctly define the temperature-dependent boundary conditions at the target–plasma interface. Itina et al. [48] proposed a combined model to describe the laser-induced plasma plume expansion in vacuum or into a background gas. The model takes into account the mass diffusion and the energy exchange between the ablated and background species, as well as the collective motion of the ablated species and the background gas particles. The system of gasdynamical equations in the divergent form for non-equilibrium (i.e., $T_e \neq T$) plasma expansion into surrounding gas was simulated. The thermal evaporation model is used, so that the vapor pressure at the surface is obtained from the Clasius–Clapeyron equation. The flow parameters are calculated using Anisimov's [49] jump conditions at the Knudsen layer boundary at sound plasma velocity, which derived, however, for expansion in a vacuum and not in a gas pressure. Only a singly charged ion was assumed, and no information about the data of electron temperature as a boundary condition was presented. The developed approach bridges the gap between the Monte Carlo modeling of the plume expansion in a low-pressure regime and hydrodynamical models in the high-pressure regime. The authors indicated, "Despite a number of simplifications, the modeling has provided a physical picture of the complicated phenomenon of laser plume expansion into a gas. This picture has yielded the explanations of a number of experimental results obtained in the presence of both inert and reactive background gases."

A three-dimensional computer code was used by Wang and Chen [50] to solve the hydrodynamic governing equations, to study the plasma plume characteristics in laser welding for iron vapor in an ambient gas. The temperature, vapor, and gas velocity as boundary conditions at the target were used as parameters.

A model of laser ablation in an ambient gas taking into account the phenomena of target–vapor interaction was proposed by Gusarov et al. [51, 52]. A weakly ionized vapor was considered, and the given plasma pressure at the target–plasma kinetic boundary was used as a parameter, assuming plasma equilibrium (i.e., $T_e = T$). Gusarov et al. [52] considered the thermal emission of electrons from the Cu and Al targets, kinetics of ions and neutrals, and an electrostatic sheath that was formed at the surface. The potential drop in the electrostatic sheath was determined taking into account the electron fluxes from the target and plasma. The problem was solved for given plasma pressure at the external boundary of the Knudsen layer as a parameter assuming temperature equilibrium for the plasma particles.

A theoretical modeling of expanding plasma plume induced during welding of iron sheets with CO_2 laser in a shielding gas (argon, helium) was developed, a 2D approach by Moscicki et al. [15]. The set of equations consists of the equations of conservation of mass, energy, momentum, and the diffusion equation was solved for a given vapor velocity and temperature, as well as given gasdynamic parameters of the flow shielding gas. The main goal of this work was to study the interaction of vapor–gas taking into account two plasmas: shielding gas plasma and metal plasma.

An overview of different modeling approaches was reported by Bogaerts et al. [53] for laser interaction with a matter operating in different regimes of wavelength

(UV, Vis, IR), laser irradiance ($10^4 - 10^{10}$ W/cm^2), and pulse length (fs, ps, ns). It was concluded that the entire process of laser ablation and the subsequent behavior of the ablated material cannot be described with one single model, and therefore, it should be described step by step taking into account target heating, melting and vaporization, plume expansion in vacuum, and plasma shielding of the incoming laser light. Such a numerical model was developed by Bogaerts et al. [53] and by Chen and Bogaerts [54] for Cu target. The equation for target heat conduction was solved in a one-dimensional approximation. The plasma formation was considered near the target surface, where the ions and electrons emission from the heated surface are described by the Langmuir–Saha equation and in the vapor volume by Cu atom ionization to Cu$^+$ and Cu^{++} ions by the Saha–Eggert equations, assuming common temperature for the electrons, ions, and neutrals. This temperature was determined by the laser beam absorption in the plasma, which was accelerated during its expansion. The plasma expansion was calculated using an equilibrium condition for the pressure and vapor density and thermal velocity for evaporated atoms.

Thus, different approaches were considered previously assuming an equilibrium boundary condition for plasma expansion, thermal equilibrium in the plasma, or weakly ionized vapor, neglecting collisions between charged particles. A local speed of the vapor flow in the Knudsen layer was used as sound, which is not correct at large laser power density (≥ 10 MW/cm^2). The calculated results for moderate power density were obtained using some arbitrary parameters (plasma equilibrium (i.e., $T_e = T$), given plasma pressure at kinetic boundary, temperature, vapor velocity, and low ionization assumption). The used Langmuir–Saha equation to describe the surface ionization mechanism is not available for materials with the potential of ionization of the atom larger than the target work function. The multi-charged (>2) ion generation and the ion energy dependence on charge in the expanding plasma were not considered.

At the same time, the previous calculation indicated a relatively large particle temperature and a large degree of ionization of the vapor. Therefore, a space charge region with high potential drop and large electrical field at the surface and as a result large ion and electron emission currents at the target is expected and should be investigated. An understanding of the measured results requests to consider the LTI as mutually dependent phenomena. In this case, there were the following important questions: *What is the plasma structure in the near-target region by plasma–target interaction? How does the ion heat flux influence the target heat balance? How does the energy of emitted electrons influence the plasma energy balance?*

As the main point of the present chapter is to study the physics of plasma interaction with a target, the laser power absorption in the expanding plasma which is taken into account by using an effective coefficients K_L from published results is described as a well-known mechanism of such absorption. The tractability of moderate energy LTI measurements will be clarified considering the coupled physical phenomena which determined the produced laser plasma parameters as ion energy, plasma temperature, etc. To this end, the physics of laser ablation and plasma generation will be considered, taking into account the kinetics near the target-plasma interface, the hydrodynamics of plasma flow based on the existing mechanisms of

24.3 Overview of Theoretical Approaches of Laser–Target Interaction

plasma acceleration, and using a previously developed self-consistent approach to LTI. Therefore, the phenomena description will be detailed separately to base the developed plasma model for LTI. Thus, the LTI is studied by analyzing vaporization and electron emission, considering vapor breakdown mechanisms, sheath formation, electron emission and mechanism of plasma heating, the plasma expansion by analysis of the existing mechanisms of plasma acceleration.

24.4 Near-Target Phenomena by Moderate Power Laser Irradiation

The pulsed laser ablation and plasma formation are complex, and various phenomena occur simultaneously (Fig. 24.1). First, these phenomena are characterized separately and the present state of art is analyzed. The obtained results are involved for modeling of self-consistent processes by laser interaction with a target. The laser power density q_L was absorbed in the plasma by value $K_L q_L$ and so, reaching the value at the target by $(1 - K_L)q_L$.

24.4.1 Target Vaporization

During the laser irradiation, the target evaporates, and evaporating layer determines the heavy particle flow. The kinetics of vaporization described in Chap. 2. Aden et al. [55] studied laser-induced vaporization of Al and iron targets in air at atmospheric pressure. It was assumed that $u = u_{sn}$ is an upper bound for the vapor velocity, and for an absorbed intensity, the vaporization is independent of the ambient gas. However, in general, the vaporization depends on the near-target pressure, which determines

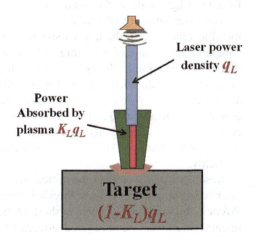

Fig. 24.1 Laser–target interaction. Power density q_L distribution between a plasma and irradiated target

the flow properties in the Knudsen layer. The definition of plasma flow as a *non-free* (*impeded*) when $M < 1$ and as *free* into vacuum when $M = 1$ was introduced in Chap. 17. A strongly impeded mass flow was characterized the cathode evaporation in a vacuum arc due to energy dissipation in the near-surface dense plasma. The developed kinetic approach in Chap. 17 could be used to study the vaporization target material and determine the structure of the Knudsen layer in case of zero electrical current of laser irradiation. In this case, the parameters at the external boundary of the Knudsen layer determine the boundary condition for the hydrodynamic flow along a path from the surface.

24.4.2 Breakdown of Neutral Vapor

When the laser power density is relatively low (<1 MW/cm^2), the evaporating flux consists of neutral atoms. At significant laser power density, the laser irradiation accompanied by an electric field provoking a vapor breakdown and plasma generation. Gas breakdown by laser irradiation was discovered in 1962–1963 and found to be similar to spark discharges [56]. The electrical breakdown occurs when an electric field of $3 \times 10^6 - 10^7$ V/cm, depending on the pressure and gas, was produced by the light wave, which can be reached in laser radiation with power density of $10^{10} - 10^{11}$ W/cm^2 [57]. The atoms can be directly ionized by the quantum, or by the multi-quantum photo effect and due to excited atom ionization by energetic electrons. Atoms can be ionized by a mechanism, similar to the tunneling effect, i.e., an electron emitted from the atom by a static electric field. The breakdown develops by electron avalanches created by the primary electrons. The gas breaks down when the laser power density exceeds some threshold value. Gili et al. [58] measured the threshold breakdown electric field for a 50 ns, 30 MW laser beam as a function of Ar, He, and N pressure. Minimum breakdown fields have been found taking into account the electron impact ionization and electron heating through energy transfer from laser light wave to the electrons.

The laser intensity of breakdown threshold is lower for plasma generation in metallic vapor due to the lower ionization and excitation energies of metal atoms, compared to gas atoms. This threshold intensity is determined by a mathematical model as a function of the laser wavelength [59]. Mathematical modeling with the given shape and duration of the pulse showed that the threshold intensity of the radiation was 2×10^7 W/cm^2, and the optical breakdown time did not exceed 30 ns. This time reduced to about 10 ns for ~10^9 W/cm^2. The phenomenon was studied considering charged particle generation kinetics and taking into account different atom ionization mechanisms. However, target heating and evaporation were not considered explicitly.

Rosen et al. [60] studied the optical breakdown in aluminum vapor by excimer laser radiation where the laser threshold intensity for a 0.5 μs pulse was 5×10^7 W/cm^2. The vapor breaks down at shorter times with laser power density. The experimental result was found in good agreement with predicted by a model for plasma initiation included evaporation, photoionization, atom excitation, and others. Thus, the plasma was initiated during the time comparable with the regular small laser pulse time for laser power significantly larger than the threshold laser intensity.

24.4.3 Near-Target Electrical Sheath

A general comprehensive review by Eliezer and Hora [61] reviewed the problems related to laser produced plasma. They showed that in laser-generated plasma, the target–plasma transition comprises a sheath formed by plasma floating potential with respect to a negatively charged wall. The sheath should be taken into account considering the LTI [52]. Considering a self-consistent study of sheath-surface transition, Beilis [62] showed that the ion and electron fluxes across the sheath with potential drop u_{sh} determined a new phenomenon, namely the back energy flux to the target previously absorbed in the plasma by laser irradiation (see below).

24.4.4 Electron Emission from Hot Area of the Target

During LTI, targets are irradiated by highly energetic photons, which transport substantial energy flux which heated the target. At another hand, an electric field is generated in the sheath at the target–plasma interface. Thus, the electrons can be emitted from the target by several mechanisms including photoemission, thermionic (T) emission, field emission by electron tunneling (F), and combined thermal and field (T-F) emission. The photo effect and electron emission current were generated by UV laser action on Zn target [63]. Relatively, low laser power density ~1 MW/cm^2 was studied when mainly photoelectric mechanism determined the electron emission. Dolan and Dyke [64] calculated the energy distribution of emitted electrons for different temperatures T and electric fields E. Beilis [65] developed this method showing the transition from T- to F-emission and the intermediate case. According to these calculations, electron emission by T- and T-F emission mechanisms can be important for laser power density $>10^7$ W/cm^2.

24.4.5 Plasma Heating. Electron Temperature

An important question is how the plasma can be heated and what is the mechanism of energy transfer to the electrons by laser irradiation? According to Raizer [56], plasma electrons can acquire energy in a laser radiation field. Energy absorption from the field is quantum-like. Electrons can acquire instantaneously a large amount of energy from the field from collisions of many photons [57]. Energy absorption is frequency-dependent process. Photon energy is absorbed by free electrons in the inverse Bremsstrahlung process, producing higher energy free electrons. Another mechanism is due to acceleration of the electron emission in the near-target sheath. The energy acquired by the electron beam heated the plasma electrons in an electron beam relaxation region [66]. The electron energy, absorbed from the field electron beam, is lost via atom excitation, ionization, and plasma outflow. Thus, a balance between energy gain and energy loss determines the electron temperature T_e. This balance will be studied by LTI mathematical formulation.

24.4.6 Plasma Acceleration Mechanism

Let us consider the possible mechanisms for accelerating fully ionized plasma having conductivity σ. The equation of ion motion in general form is [67]:

$$m\frac{d\vec{V}_i}{dt} = -\frac{\nabla P_i}{n_i} - \frac{ej}{\sigma} + e\left(E + \frac{1}{c}\left[\vec{V}_i \vec{H}\right]\right); \tag{24.1}$$

where P_i is the ion pressure, V_i is the ion velocity, j is the electrical current density, and $\vec{H}$ is the magnetic field vector. The electric field E is limited by high electron mobility in a quasineutral plasma. In order to study the electrical field generation mechanisms while maintaining quasineutrality, the equation of electron motion with velocity V_e and pressure P_e may be considered in the following form:

$$e\left(E - \frac{j}{\sigma}\right) = -\frac{\nabla P_e}{n_e} - \frac{m_e d\vec{V}_e}{dt} + \frac{e}{c}\left[\vec{V}_e \vec{H}\right]; \tag{24.2}$$

The sum of (24.1) and (24.2) in simple case where $H = 0$ and $j = 0$ for moderate laser power is:

$$m\frac{d\vec{V}_i}{dt} = -\frac{\nabla(P_i + P_e)}{n_i} - m_e\frac{d\vec{V}_e}{dt}; \tag{24.3}$$

Considering (24.3), the following forces accelerating the plasma can be noted:

(1) Forces due to gradient of charged particle density and by particle heating in the flow when the inertial term of electrons can be neglected. This is the **gasdynamic** or heat mechanism (see below).
(2) When $E \gg j/\sigma$, electron inertia from the last term in (24.2) gives:

$$E \approx \frac{m_e}{e}\frac{d\vec{V}_e}{dt}, \tag{24.4}$$

24.4 Near-Target Phenomena by Moderate Power ...

This mechanism contributes with strong V_e variation in a small time. For example, when an electron velocity increases to 5×10^7 cm/s in 0.1 ns, the electrical field of about 280 V/cm can be induced.

(3) Force in the rarefied part of the expanding plasma is caused by an ***electric field at the plasma front*** due to significantly fast motion of the hot electrons in comparison with the ions, resulting in a strong violation of quasineutrality. The ions stream into vacuum with electrons is preceded by a characteristic length at which a space charge can be induced. This space charge at the boundary of the expanding plasma produces the accelerating electrostatic field at the ion front. The most energetic electrons extend into vacuum, maintaining an accelerating field determined by the electron temperature.

Eliezer and Ludmirsky [68] estimated electric fields between ~5×10^5 and 5×10^6 V/cm at widths from 10 to 100 Debye lengths for a sheath produced by Nd:YAG laser intensities between 10^{12} and 10^{15} W/cm^2. The ion acceleration with the electrostatic field E caused by the hot electrons in the freely expanding plasma model was considered by Denavit [69]. The expanding ions were described by the ion continuity and motion equations taking into account the electron–ion collisions. The electrons were assumed in isothermal equilibrium, and their distribution was given by the Boltzmann relation taking into account the quasineutrality of the plasma plume. A self-similar solution gives $E = T_e/L$; $L = v_{sn} t$, where v_{sn} is the ion sound speed and t is the expanding time. Thus, in essence, gasdynamic acceleration of the ions due to the electron temperature gradient was considered. The problem of ion acceleration in a time-dependent ambipolar field was developed for the case when the electron energy strongly depends on time during a powerful sub-picosecond laser pulse by Gamaly [70]. Wickens et al. [71] observed a significant difference with the well-known case of isothermal expansion.

The shifted Maxwellian ion energy distribution was measured for relatively moderate laser power density ($\leq 10^{10}$ W/cm^2). The shifting of the ion energy Torrisi and Gammino [72] proposed to explain by an electrostatic acceleration of the ions. The corresponding electric field was estimated as the measured equivalent voltage ratio to the Debye length. It was indicated that this electric field is in accordance with that obtained for significantly larger power density $(1 - 5) \times 10^{19}$ W/cm^2 for fs and ps pulse experiments [73–75]. The electrostatic field model for ion acceleration was developed by Hatchett et al. [76] with scale length L given by the Debye length.

However, using one- and two-dimensional numerical simulations, Liseikina et al. [77] indicated that in contrast to the electrostatic (and similar TNSA) mechanism used to explain most of the experiments with ion acceleration from solid targets, laser radiation pressure acceleration (RPA), predicted from theoretical studies, becomes dominant at intensities exceeding 10^{23} W/cm^2 and produces highly collimated ions with energies approaching GeV values. Previously, Attwood et al. [5] presented interferometric data that confirmed the significant role of radiation pressure during laser–plasma interaction (10^{23} W/cm^2). The electrostatic acceleration mechanism or TNSA and RPA models are mainly applicable for extremely high-intensity LTI (see discussion below).

24.5 Self-consistent Model and System of Equations of Laser Irradiation

The plasma produced by laser irradiation near the target [47, 53, 78] and the plasma generated in the cathode region of a vacuum arc [79] have similar characteristics. Both surfaces interacted with large power density, and the plasmas are very dense and appear in a minute region, known as laser spot and cathode spot, respectively. It is expected that the character of plasma interaction with the matter in both cases is also similar. Cathode spot theory was modified for current-less laser-generated plasma [66]. The developed approach based on a previously presented kinetic model of the cathode was spot by Beilis [79], (see Chap. 17) and modify it for the particular features of a laser-generated plasma taking into account that the net electrical current is zero in the laser spot, in contrast to the current carrying vacuum arc cathode spot. The laser spot phenomena also described non-equilibrium plasma flow and a jump of plasma parameters in the Knudsen layer at the solid surface–plasma interface. The mechanism of near-target laser plasma generation and flow using the kinetics of target vaporization, plasma heating and atom ionization was considered self-consistently in a mathematical model without arbitrary given parameters [62].

The laser spot located on a smooth surface is assumed as a circular with radius R_s. The near-target vapor is a partially ionized vapor that separated from the surface by an electrical sheath across which there is a potential drop u_{sh}. The sheath voltage accelerates the charged particles. Similar to the cathode spot, here are two groups of electrons: (i) accelerated electrons emitted from the heated target and (ii) slow plasma electrons. The energy of emitted electrons dissipates in a zone of the electron beam relaxation. Since the energy of accelerated electron beam exceeds the slow electron energy by an order of magnitude, then the beam relaxation zone is much larger than the plasma ion or electron mean free path. Thus, inside the relatively large electron beam relaxation zone, a collision-dominated plasma is formed, which is heated by the accelerated electron beam.

As the ionized vapor near the heated surface is not in equilibrium, a kinetic (*Knudsen*) layer of several ion mean free path lengths is originated. Recall that at Knudsen layer external boundary, the velocity function distribution of the particles approaches equilibrium. This layer is considered in the hydrodynamic treatment as a discontinuity region [49], where two heavy particle fluxes are produced. The first is a flux of atoms evaporated from the target (*Langmuir flux*) at a rate of evaporation $W(T_0)$, where T_0 is the surface temperature. The second flux is a returned atoms and ions from the plasma toward the target surface. The difference of these fluxes determines the mass velocity flow and the net target evaporation rate G. The ratio of net evaporation rate to the total evaporated mass flux determines the target mass loss fraction $K_{er} = G/W(T_0)$.

As in the cathode spot kinetic theory (Chap. 17) in the laser spot model, four characteristic boundaries are distinguished in the near-cathode region indicated by $j = 1, 2, 3,$ and 4 (Fig. 24.2). The origin of the coordinate system is defined at the target surface, and the x-direction coincides with the direction of the vapor flow.

24.5 Self-consistent Model and System of Equations of Laser Irradiation

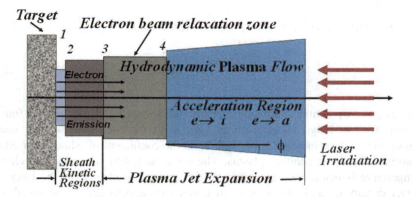

Fig. 24.2 Schematic presentation of the kinetic regions and hydrodynamic laser plasma expansion

Also, four regions are the space charge sheath or the ballistic zone, Knudsen layer, electron beam relaxation zone bounded by the boundaries 2, 3, and 4, respectively. After boundary 4 the expanding region is arise, in which plasma accelerated. The gas parameters at these boundaries denoted by density $n_{\alpha j}$, velocity $v_{\alpha j}$, and temperature $T_{\alpha j}$, where indices α—e, i, a are the electrons, ions, and atoms, respectively, and $j = 1, 2, 3, 4$ indicates the boundary number. The equilibrium electron n_{e0} and heavy particle n_0 densities are determined by the target surface temperature T_0 from equations for electron emission current density j_{em} and saturated heavy particle flux from the target.

Anisimov [49] proposed the distribution function within the non-equilibrium layer to approximate by the sum of two known terms before and after the discontinuity, by considering the laser evaporation of metals in form of neutral atoms. The model was extended by Beilis [80] (Chap. 17) for electron emitted metals in form of charge particle evaporation and plasma flow in the Knudsen layer.

The evaporated atoms have a half-Maxwellian VDF, and this DF was served as a condition at the evaporated surface $x = 0$ with the surface temperature and a given density. In the region between boundaries 1 and 2, the plasma electrons are returned, while the ions and emitted electrons are accelerated with energy eu_{sh}. At boundary 2, the ion velocity is determined by the condition at the sheath–plasma interface, and then, the ions are accelerated toward the target (to boundary 1) in the space charge sheath. At boundary 3, the heavy particles VFD are in equilibrium shifted by mass flow velocity. At boundary 4, the electron beam disappears. It is assumed that $T_{ij} = T_{aj}$ ($j = 1, 2, 3$) and $v_{ij} = v_{aj}$ ($j = 3$). In general, the ion flux Γiw from the electron beam relaxation zone to the cathode is determined by a diffusion mechanism with Saha's equations for different ions at boundary 4 (see Chaps. 16 and 17). The ion current to the cathode consists of an ion current with different ion charge $z = 1, 2, 3, 4$ and $j_i = \sum_1^z j_{iz}$. If the ionization–recombination length and plasma velocity are small in comparison with the length of the relaxation zone and to the thermal velocity, respectively, then Saha's equations for different ions are fulfilled in the volume of this zone. The electron energy balance in the relaxation zone is in form:

$$K_e K_L q_L + j_{em}(U_{sh} + 2T_s/e) = j_{et}(2T_e/e + U_{sh}) + j_i \sum_1^z \left(\frac{f_z}{ez} \sum_1^z u_{iz}\right)$$
$$+ \sum_1^z \left(n_{zi4} v_4 \sum_1^z u_{iz}\right) + 2T_e \Gamma_e \qquad (24.5)$$

where u_{iz} is the potential of ionization of ions with ionicity z, f_z is ion current fraction of ions with ionicity z determined as j_{iz}/j_i, K_e is the coefficient of laser energy absorption in the relaxation zone, and K_L is the coefficient of whole laser energy absorption in the expanding plasma. The equation (24.5) show that the electron temperature is determined by inflow of the energy absorbed from the laser with power density q_L and energy of emitted electrons accelerated in the sheath with potential drop u_{sh}. The energy losses are due to dissipation by atom ionization, convective transport of the electrons $\Gamma_e = n_{e4} v_4$, and ions $n_{i4} v_4$ outflow through external boundary of relaxation zone. Plasma quasineutrality requires that $n_e = \sum z n_{iz}$, where n_{iz} is the density of ions with ionicity $z = 1, 2, 3, 4$, and n_e is the electron density. The plasma pressure $P_j = \sum n_{aj} T_{aj}$ ($j = 3, 4$). The target is heated by the incident ions with current density j_i (bringing energy flux $j_i U_{sh}$) and by reverse electron with current density j_{et} from the adjacent to the target–plasma and cooled by electron emission and due to body heat conduction.

$$(1 - K_L) q_L + j_i \left(u_{sh} + \sum_1^z \left(f_z \sum_1^z u_{iz} - f_z z\varphi\right)\right)$$
$$+ j_{et} 2T_e/e = j_{em}(\varphi + 2T_s/e) + q_G + q_T \qquad (24.6)$$

$$q_G = \left(\lambda_s + \frac{2kT_s}{m}\right)\frac{G}{S};$$
$$q_T = 2\frac{\lambda_T}{\sqrt{S}}(T_s - 300)\left(\frac{2}{\pi} \arctg \sqrt{\frac{4\pi a t}{S}}\right)$$

where q_G is the energy loss due to target ablation and q_T is the energy loss in the target body due to its heating which is obtained by the solution of the three-dimensional heat conduction equation. The zero electrical total current at the target–plasma interface is also considered, and therefore, the following equation can be used:

$$j_e + j_i = 0, \quad j_e = j_{em} - j_{et} \qquad (24.7)$$

The expression for potential drop in the sheath can be obtained using (24.7) in form:

$$u_{sh} = T_e Ln \left(0.6 \sqrt{\frac{2\pi m_e}{m}} + \frac{j_{em}}{en_{es}} \sqrt{\frac{2\pi m_e}{T_e}}\right)^{-1} \qquad (24.8)$$

24.5 Self-consistent Model and System of Equations of Laser Irradiation

where m_e is the electron mass and m is the atom mass. The rate of target ablation which is a mass flow of the ionized vapor G is obtained by

$$G(g/s) = S m n_3 v_3, \quad m = m_i = m_a, \quad n_3 = n_{a3} + n_{i3}. \tag{24.9}$$

Electron emission current density is determined in the form (See Chap. 2)

$$j_{em} = \frac{4 e m_e k T_s}{h^3} \int_{-\infty}^{\infty} \frac{\mathrm{Ln}\left[1 + \mathrm{Exp}(-\frac{\varepsilon}{kT_s})\right] d\varepsilon}{\mathrm{Exp}\left[\frac{6.85 \times 10^7 (\varphi - \varepsilon)^{1.5} \theta(\tilde{y})}{E}\right]} - j_{eT} \tag{24.10}$$

$$\theta(\tilde{y}) = 1 - \tilde{y}\left[1 + 0.85 \sin\left(\frac{1 - \tilde{y}}{2}\right)\right];$$

$$\tilde{y} = \frac{\sqrt{e^3 E}}{|\varepsilon|}; \quad j_e = j_{em} - j_{eT}; \quad s = j_e/j$$

The equation for electric field at the surface target was obtained solving the Poisson equation, assuming that positive ion flux and negative beam electron emission and back electrons from the plasma to the target form the volume charge (See Chap. 2):

$$E^2 - E_{pl}^2 = 16 \varepsilon_0^{-1} \sqrt{u_{sh} \frac{m_e}{e}} \left\{ j_i \left(\frac{m}{m_e}\right)^{0.5} \exp(0.5) \left[\sum_1^z \left(\left(1 + \frac{kT_e}{2ze u_{sh}}\right)^{0.5}\right.\right.\right.$$

$$\left.\left.\left. + \left(\frac{kT_e}{2ze u_{sh}}\right)^{0.5} - \sqrt{\frac{kT_e}{2e u_{sh}}}\left(1 - \exp(-\frac{e u_{sh}}{kT_e})\right)\right)\right] - j_{em}\right\} \tag{24.11}$$

The plasma parameters at boundary 3 serve as boundary conditions for hydrodynamic equations of mass, momentum, and energy in the expanding plasma jet. In general, the plasma jet parameters can be described by a set of 2D hydrodynamic equations, which was developed by Keidar et al. [81, 82]. Assuming plasma expansion in the geometrical form of a truncated cone (Fig. 24.2) with cross section depended on distance from the target, the mathematical model is developed by Beilis [83]. With increasing distance from the target, the plasma expands due to large plasma pressure, and the velocity V of the plasma increases in the jet. The plasma is accelerated by the thermal mechanism in which the enthalpy of the particles is transformed into kinetic energy. To demonstrate the influence of laser power density, for simplicity, the resulting jet velocity was calculated using the hydrodynamic equations in the following integral form.

The integral form of jet momentum equation:

$$GV = p_3 S, \quad S = \pi R_s^2, \tag{24.12}$$

Jet energy equation:

$$[(1 - K_e)K_L q_L + 2T_e \Gamma_e]S = (V^2 - V_0^2)G/2 + Q_r \quad (24.13)$$

Equation (24.13) is determined by the ion velocity V of the accelerated jet by the plasma pressure p_3 at the boundary 3 and spot area S [83]. Here, Γ_e is the electron flux density flowing from the electron beam relaxation zone into the plasma jet expansion region, and Q_r is the energy loss by the radiation [53]. In the case when the characteristic gasdynamic time is much shorter than a time for solid thermal diffusion, the plasma parameters in the transition period will be determined by the target heating time [78]. The above mentioned system of equations, taking into account the heavy particle flow in the kinetic layer, was solved to study the target ablation. In the above formulation, the spot radius R_s and laser power density q_L were the given parameters. The plasma density n, T_e, T_s, j_i, j_{em}, j_{et}, α_{iz}, u_{sh}, E, V, K_{er} and other spot parameters were calculated. An example of the calculation is presented below, showing the contribution of plasma heat flux in target heating, rate of target ablation, and plasma density.

24.6 Calculations of Plasma and Target Parameters

The above system of equations was solved numerically by iteration method for different target materials in accordance with kinetic model of vaporization phenomena. The results present the dependence of the target and plasma parameters on laser power density q_L and time of laser irradiation. The given parameters were spot radius $R_s = 100$ μm, the coefficient characterized the power absorption in plasma of expanding jet K_L, and the coefficient which characterized the radiation absorption in plasma of electron beam relaxation zone K_e. For simplicity, in most of the calculations, K_L was chosen constant as 0.4 and K_e was chosen as 10% of K_L according to the numerical analysis [53, 54]. The values of these coefficients also varied in order to study their influence on the calculated result while q_L was held constant.

24.6.1 Results of Calculations for Copper Target

As an example, the calculations were provided for the typical copper target. Pulse duration t was varied as 10 ms, 1 μs, 10, 1 ns. The laser radiation intensity density q_L was assumed constant during the laser pulse and varied in a wide range ($10^{-3} - 1$) GW/cm^2. It was considered also q_L close to minimal intensity for which the solution of the system of equations can be found, i.e., when the laser plasma can be created.

It can be seen that the potential drop in the sheath u_{sh} decreased with q_L (Fig. 24.3),

24.6 Calculations of Plasma and Target Parameters

Fig. 24.3 Potential drop in the near surface sheath as a function on laser power density

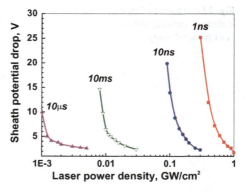

and it is larger for lower pulse duration varying from 2 to 25 V. The electrical field at the target surface E is also larger for lower pulse duration (Fig. 24.4), but the dependence on laser power density is non-monotonic and a maximum of E is calculated for some certain q_L determined by t. This electric field can be very large, more than 10 MV/cm. The target temperature increases with q_L (Fig. 24.5) and

Fig. 24.4 Electric field at the target surface as a function on laser power density

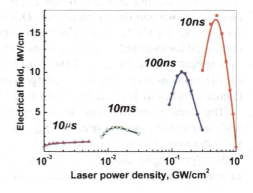

Fig. 24.5 Electron temperature T_e and surface target temperature T_0 versus laser power density

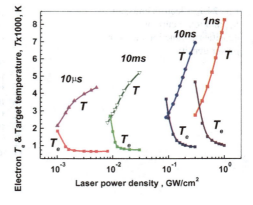

Fig. 24.6 Degree of atom ionization dependence on laser power density

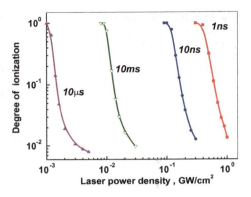

decreases with pulse duration when a lower range of q_L was used. For shorter pulse duration, the rate of temperature increase is larger.

Figure 24.5 shows that the electron temperature sharply decreases to a relatively small certain value and then weakly changes with q_L. T_e can reach up to about 5 eV for short laser pulse duration (~1 ns). Similarly, the degree of plasma ionization depends on q_L (Fig. 24.6), and plasma is passed from weakly ionized ($\alpha \sim 0, 01$) to fully ionized state increases to few eV. In the fully ionized state, the plasma also consists of two and three charge ions. Different fractions of high charge ions (α_{i1} for Cu^+, α_{i2} for Cu^{++}, α_{i3} for Cu^{+++}) as dependence on laser power density were calculated for laser pulse duration 10 ns (not shown in the figure). The high charge ions appear mainly for $q_L = (0.09 - 0.12)$ GW/cm^2, and for $q_L > 0.12$ GW/cm^2, the plasma consists mostly of one charge ions. The ion fractions α_{i2} and α_{i3} sharply decreased when q_L increased from 0.1 to 0.15 GW/cm^2.

The electron emission current density dependence (Fig. 24.7) is in accordance with temperature dependence, i.e., j_{em} also increases with q_L and decreases with pulse duration for a lower range of q_L. For each pulse duration, j_{em} sharply increases with q_L (growing by about $10^5 - 10^6$ times) in the beginning, and then, the rate of this increase significantly decreases with q_L. At the same time, the ion current density to

Fig. 24.7 Electron emission j_{em} and ion j_i current densities as function on laser power density

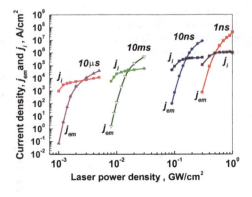

24.6 Calculations of Plasma and Target Parameters

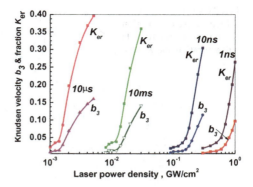

Fig. 24.8 Normalized velocity at the Knudsen layer $b_3 = v_3(m/kT_3)^{0.5}$ and evaporation fraction K_{er} dependences on laser power density

the target (Fig. 24.7) grows significantly smaller with q_L for each pulse duration (by about 10–15) times and saturated to some certain value with laser power increasing.

Figure 24.8 shows that the plasma flow is sub-sonic near the target surface in the dense part of the plasma jet. This fact is characterized by the dependence of normalized velocity at the external boundary of the Knudsen layer $b_3 = v_3(m/kT_3)^{0.5}$ on the laser power. Parameter b_3 increases with q_L indicating that plasma flows with velocity smaller than the ion thermal velocity by a factor less than 10^{-2}. As a consequence, the ratio of net evaporation rate to the total evaporated mass flux—the parameter K_{er}—also increases with q_L (Fig. 24.8).

The heavy particle density increases by a factor of about 10^4 with q_L for each laser pulse duration (Fig. 24.9). The similar dependence is calculated for target ablation rate in g/s, which increases according to the heavy particle density growth. The degree of vapor non-equilibrium near the target surface is characterized by normalized heavy particle density at the external boundary of the Knudsen layer $n_{30} = n_3/n_0$. The calculated dependence of n_{30} on laser power density shows that this parameter is in range of 0.8–0.7, indicating the degree of vapor non-equilibrium with q_L.

The calculation shows that the jet velocity decreases with q_L for constant K_L because of the increase of the target ablation rate (see Fig. 24.14, $K_L = 0.4$). In

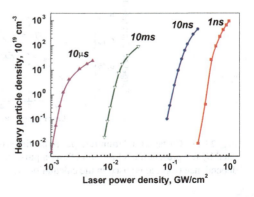

Fig. 24.9 Heavy particle density as a function on laser power density

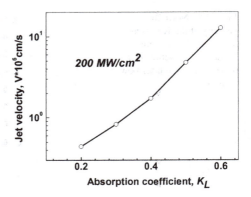

Fig. 24.10 Plasma jet velocity versus absorption coefficient K_L for 10 ns laser pulse duration and $K_e = 0.1$

the case of $K_L \neq$ const, this velocity depends on the energy absorbed in the plasma expanding region and characterized by K_L. Therefore, the jet velocity was calculated as a function on laser power density with given K_L as parameter.

The solution presented in Fig. 24.10 indicated that the velocity increases with K_L. In order to compare the measured rate of target ablation with an experiment, the calculation was also taken into account with the change of K_L. The dependencies of target ablation rate per pulse vs. full pulse energy density (Fig. 24.11) demonstrate good agreement between the measured by Caridi et al. [36] and calculated results when K_L changed with pulse energy in accordance with dependence shown in Fig. 24.11 ($K_e = 0.05$).

To understand the plasma energy contribution by its interaction with the target, the calculation was also provided considering only laser irradiation, i.e., neglecting the plasma energy fluxes. This result was compared with that considering the plasma contribution. The calculations indicate significant difference of all parameters in both mentioned approaches.

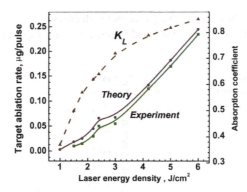

Fig. 24.11 Cu target ablation rate per pulse time measured by Carridi et al. [36] and that calculated by the present model as a function on pulse energy density (left axis) and corresponding K_L dependence (right axis) used in the calculation ($K_e = 0.05$)

24.6 Calculations of Plasma and Target Parameters

To demonstrate the plasma contribution in the target heat regime, the target temperature T_q was calculated considering only laser irradiation, i.e., neglecting the heat flux from the plasma by the ions and energy flux to the plasma by the electron emission. Figure 24.12 shows the calculation of T_q and T_0 as a function on pulse energy density for condition presented in Fig. 24.11. The difference between T_0 and T_q is in region of 300–3000 °C. The calculations show that this difference substantially depends on coefficient of energy absorption in the electron beam relaxation region K_e.

Figure 24.13 shows the temperatures T_q and T_0 as a function on coefficient of energy absorption K_e with q_L as a parameter. It can be seen that the difference between T_q and T_0 significantly grows as K_e and q_L increase. For $q_L = 0.2 \, \text{GW/cm}^2$,

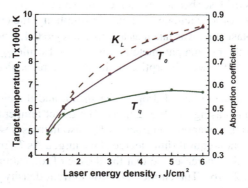

Fig. 24.12 Target surface temperature T_0 calculated taking into account the plasma energy flux for condition presented in Fig. 24.11 ($K_e = 0.05$) and target temperature T_q calculated without plasma energy inflow as a function on pulse energy density (left axis) and corresponding K_L dependence (right axis) used in the calculation

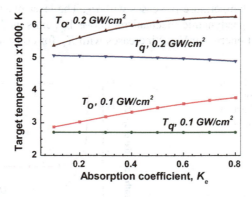

Fig. 24.13 Target surface temperature T_0 calculated taking into account the plasma energy flux and target temperature T_q calculated without plasma energy inflow as a function for the coefficient of energy absorption K_e in the electron beam relaxation region with q_L as a parameter

the difference between T_0 and T_q reaches more than 1000 °C. As a result, the rate of target ablation is also larger in comparison with that calculated without energy flux from the plasma, which originated due to laser radiation absorption. This is an important result showed that the ion energy flux contributes significantly to the target heating and reduces the critical laser power radiation when the laser plasma appears.

24.6.2 Results of Calculations for Silver Target

The plasma parameters were calculated for the conditions corresponded to conditions of the experiment by Margarone et al. [28]. In this experiment a 10 ns, power density $q_L = 10$–100 MW/cm^2 laser beam was incident on a 3 mm^2 spot of an Ag target. The dependencies on q_L were calculated, when, for simplicity, the coefficient of power absorption in the plasma jet was taken as a constant $K_L = 0.4$. And, to determine the influence of K_L on the plasma parameters, it was varied while q_L was held constant [53]. The coefficient of power absorption in the electron relaxation region was taken as $K_e = 0.1$.

It was found that the potential drop in the sheath and the electron temperature T_e decrease with increasing q_L (Fig. 24.14). The target temperature T_0 increases, and therefore, the target evaporation fraction K_{er} also increases with q_L (Fig. 24.15) indicating the vapor non-equilibrium degree at the target. As T_0 increases, the heavy particle density n_0 also increases, so that the degree of ionization α decreases with increasing q_L (Fig. 24.16). The emitted electron current density j_e increases by six orders of magnitude, while ion current density j_i increases by one order of magnitude with increasing q_L [45].

The calculated electrical field at the target surface increases with q_L to a maximum of 4 MV/cm at $q_L = 15$ MW/cm^2 and then decreases with further increase of q_L, while the target ablation rate G increases by six orders of magnitude with q_L. G increases because the target temperature increases with q_L. The laser plasma is significantly accelerated by the gasdynamic mechanism, above 10^6 cm/s, and the velocity V as well the kinetic jet energy ($mV^2/2$) increases with K_L for constant q_L (Fig. 24.17).

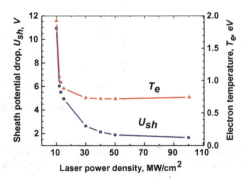

Fig. 24.14 Potential drop in the space charge region near the target surface and plasma electron temperature as a function on laser power density, 10 ns pulse

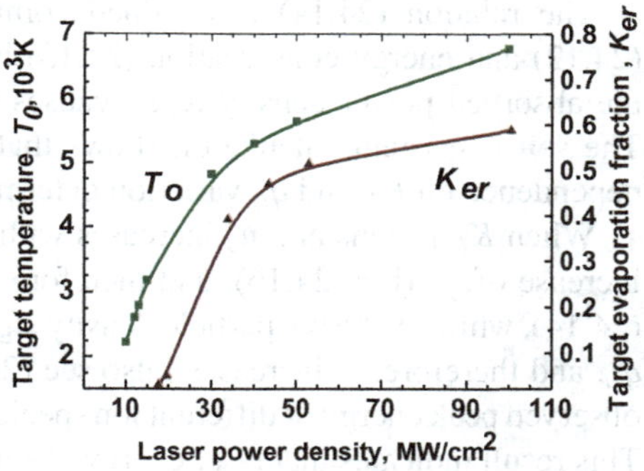

24.6 Calculations of Plasma and Target Parameters

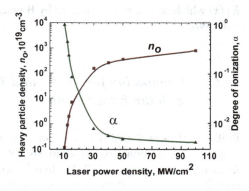

Fig. 24.15 Surface target temperature T_0 and target evaporation fraction K_{er} dependencies on laser power density, 10 ns pulse

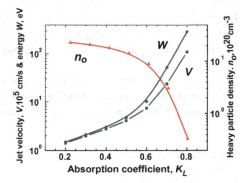

Fig. 24.16 Dependencies of heavy particle density n_0 and degree of ionization α in the beam relaxation region as function of laser power density, 10 ns pulse

Fig. 24.17 Jet velocity V, jet energy W, and heavy particle density n_o as function of coefficient of energy absorption αL in the plasma jet calculated for Ag target with laser pulse 9 ns, $q_L = 33.333$ MW/cm^2

However, when K_L is constant, the velocity V decreases with q_L.

These different dependencies can be understood from relation between K_L, q_L, and n_0 in form:

$$V \approx \frac{K_L}{n_0 kT} \frac{q_L}{(1 + \alpha T_e/T)} \qquad (24.14)$$

The relation (24.14) is obtained combining the equations of jet momentum (24.12) and energy conservation (24.13) in integral form, taking into account that the absorbed power density $K_L q_L$ causes the plasma heating and its acceleration. The self-consistent calculation shows that the character of heavy particle density dependence on K_L and q_L variation determines the dependence of V.

When K_L is constant, n_0 increases with q_L which is significantly larger than the increase of q_L (Fig. 24.16), and therefore, V decreases with q_L, as it follows from (24.14), while the heavy particle density significantly decreases with α_{IB} for constant q_L, and therefore, V increases [also see (24.14)]. The experiments showed that the observed peak energy of different ion species increased with laser energy [29, 32, 36]. This result indicates that laser energy absorption is nonlinearly function on distance in the expanding plasma jet, and the plasma velocity is determined by a dependence $K_L(x)$ which, as example, was used by Bogaerts et al. [53] for Cu and power density $>10^8$ W/cm^2.

24.6.3 Calculation for Al, Ni, and Ti Targets and Comparison with the Experiment

Beilis [84] was found the solution for laser irradiation of Al, Ni, and Ti targets in conditions of experiments [85]: spot area $S = 0.7$ mm^2, laser fluence in the range of 3-21 J/cm^2, and pulse duration of 3 ns. Parameter $K_L = 0.97 - 0.98$ characterized the target reflection and therefore was taken as the coefficient of laser power density absorption [85] in the plasma. Figure 24.18 shows the ablation yield for Al, Ni, and Ti measured (solid lines) and calculated (dash lines) as functions of laser fluence. It can be seen that the theoretical results fit sufficiently well the experimental linear dependence ($K_e = 0.01$).

The number of heavy particles N_t per laser pulse for Ni target measured by Torrisi et al. [86] is presented in Fig. 24.19 (solid line) as a function of laser fluence (in range of 5–80 J/cm^2). A good agreement with data calculated by the theoretical

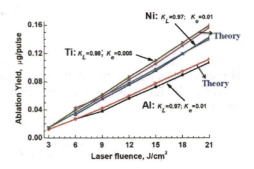

Fig. 24.18 Target ablation yield per pulse as function on laser fluence for Al, Ni, and Ti, shown for the experimental and calculation data with $K_e = 0.01$

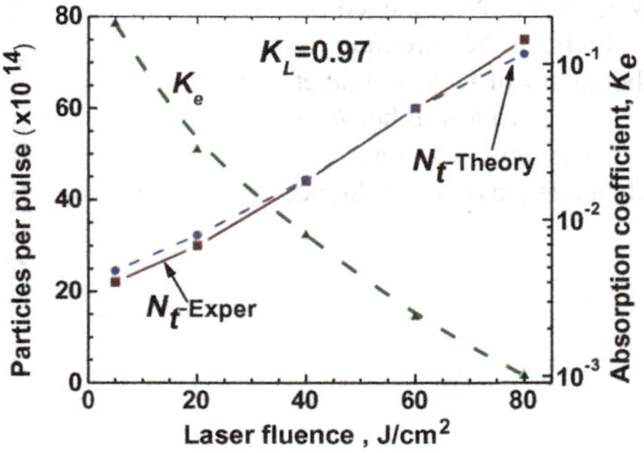

24.6 Calculations of Plasma and Target Parameters

Fig. 24.19 Number of heavy particles per pulse Nt as function on laser fluence for Ni with fitted coefficient of K_e: experiment (solid lines) and calculation (dash lines)

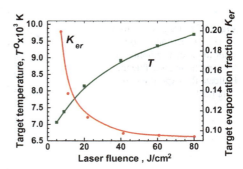

Fig. 24.20 Target temperature T and target evaporation fraction K_{er} as dependence on laser fluence

model (obtained using heavy particle density n_h) was reached when the coefficient K_e decreased with laser fluence according to the dependence illustrated in Fig. 24.19.

Let us consider the plasma and target parameters calculated, as an example, for Ni with $K_e = 0.01$. The dependencies of electrical field and sheath potential drop on the laser fluence are presented in Fig. 24.20. These parameters decrease with laser fluence, but the electric field remains sufficiently large (1–10 MV/cm). While the target temperature T increases with laser fluence and slightly exceeds 9000 K, the target evaporation fraction K_{er} significantly decreases as it is shown in Fig. 24.20.

The change of parameters n_{30} and b_3 on the external boundary 3 of the Knudsen layer as functions of the laser fluence for Ni is illustrated in Fig. 24.21. When the ratio n_{30} increases, the dimensionless velocity b_3 decreases with the laser fluence. Figure 24.22 indicates that the calculated degree of atom ionization near the target decreases with the laser fluence. The measured data [86] at laser fluence of 10 J/cm^2 is in good agreement with the calculated value at this fluence. The electron current inessentially changes at the level of $(2–3) \times 10^7$ A/cm^2, but the ion current decreases from 4×10^6 to 2×10^5 A/cm^2 with laser fluence for Ni. Velocity of the plasma jet increases in range of $(1.1–3) \times 10^6$ cm/s, and the electron temperature decreases from 1.3 to 0.9 eV with the laser fluence. These values are in region of that data calculated above for other target metals.

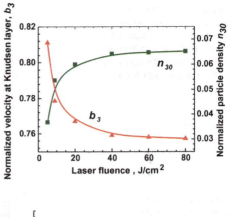

Fig. 24.21 Normalized velocity at the external boundary of Knudsen layer b_3 and normalized heavy particle density n_{30} as dependence on laser fluence

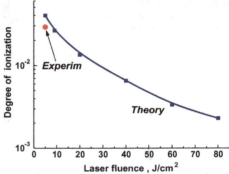

Fig. 24.22 Degree of atom ionization as function on laser fluence. The points at low and large laser fluence indicate the experimental data [85]

24.7 Feature of Laser Irradiation Converting into the Plasma Energy and Target Shielding

According to the calculation, the most important phenomena occurred at the plasma–target transition, which is realized by an electrical sheath. The electric field at the target surface depends mainly on two parameters u_{sh} and j_i. u_{sh} decreases while j_i increases with q_L in region of relatively low q_L, and j_i saturates with q_L for relatively large q_L. Therefore, a maximum of electric field in the dependence on q_L is calculated. The plasma for sufficiently high q_L mostly consists of one charge ions. However, the ion charge state dramatically changed for the low region of q_L (at the minimal q_L), and two and even three charged ions occurred due to a sharp increase of the plasma temperature.

When the laser power is down, this potential drop increases in order to support the further plasma generation by increase of the plasma electron temperature and as a consequence by the degree of atom ionization growth. According to the calculation, the surface temperature decreased, and therefore, the vapor density decreased substantially (exponentially with T_0) when the laser radiation intensity and pulse duration decreased. As a result, the vapor approaches to fully ionization, and α

increases to 1. The high rate of atom ionization supports the high level of j_i, and therefore, the large heat flux to the target surface increases the plasma energy contribution.

The calculation shows the presence of minimal and maximal values of q_L, at which the solution is obtained. Let us discuss the dependencies of plasma parameters on q_L and time t for a given absorption fraction of the incident laser power. (1) The ion current and electron emission decrease considerably when q_L decreases to the minimal value of q_L and t. Below some critical laser irradiation q_L, plasma cannot be generated because the incident laser power is sufficient only for target vaporization and the laser irradiation interacts the matter as an independent heat source without plasma contribution. Above this lower critical radiation, the plasma generates additional heat flux, and the surface temperature is increased. As a result, the rate of target ablation increased in comparison with that calculated without energy flux from the plasma, which originated due to laser radiation absorption. The lower critical power was determined as a minimal laser irradiation, for which the self-consistent solution of the mathematical model is absent. Consequently, the sum of low plasma heat flux and laser irradiation to the target *lower* than the critical power cannot reproduce the plasma density, even in case of fully ionized vapor. (2) On the other hand with further increase of q_L, the heavy particle density and electron emission current density substantially increase due to increase of the target temperature. Therefore, the degree of ionization is dropped with heavy particle density n_o. The electron temperature also dropped due to relatively large energy losses by the returned electron current j_{et}, which increases because u_{sh} decreased with q_L. As a result, for some large critical q_L, the plasma cannot be reproduced due to very low degree of ionization (density n_o) of the vapor in the electron beam relaxation zone. The larger critical power was determined as a maximal laser irradiation, for which the self-consistent solution of the mathematical model is absent.

In both cases, the plasma significantly shields the target, and this phenomenon can be considered as a mechanism of instability by laser–target interaction during the radiation pulse. According to the calculation, a relatively large difference between plasma parameters, as well as between target temperatures, is obtained using the present model and model where the heat flux from the plasma by the ions and energy flux to the plasma by the electron emission were neglected. The obtained result indicated that the plasma energy flux contributes significantly to the target heating. This means that the laser-generated plasma not only passively shields the target from the laser radiation but also converts the absorbed laser energy to kinetic and potential energy of plasma particles, which were transported not only in the expanding plasma jet, as it was considered previously, but also to the target surface.

The solution indicates that the plasma flow in the Knudsen layer is sub-sonic, and the low values of normalized plasma velocity b_3 and the heavy particle evaporation fraction K_{er} are much less than a unit. This means that returned flux of the heavy particle is relatively large and that the net of target evaporating flux significantly differs from the Langmuir rate of target evaporation into the vacuum. The relatively large returned flux from the high density plasma is formed near the target surface by high-intense laser radiation, and therefore, the vaporization is not free. As b_3

and n_0 increase with q_L, then the rate of target mass loss G increases by the target evaporation. In the case of constant K_L, the jet velocity decreases because larger ablative mass was accelerated with q_L due to the increase of G. On other hand, the velocity increases with K_L. The last result agrees with results calculated by Bogaerts et al. [53] and with measured plasma energy by Caridi et al. [36] indicating the important role of laser absorption efficiency in comparison with increase of G with q_L in real time-dependent process. The validation of this conclusion follows also from the comparison of calculated dependence of the rate of target ablation on a pulse energy density with such dependence measured by Caridi et al. [36].

24.8 Feature of Expanding Laser Plasma Flow and Jet Acceleration

The experiments and theoretical study show that the laser–target interaction (LTI) consists of number of complex phenomena. The observed and calculated behaviors of the plasma parameters, plasma expansion, and ion acceleration are result of number physical processes, which appear during laser plasma formation and its expansion. The kinetics of vaporization in the region adjacent to the target surface with further hydrodynamic plasma flow considered together with target heating determined the phenomena of laser irradiation. The phenomena are developed with time at different stages.

The first stage of LTI is the intense target ablation when mainly neutral atoms are produced and the breakdown of neutral vapor occurred in the second stage with the laser plasma plume formation. The next stage occurs when the near-target–plasma has been generated and the plasma plume expands simultaneously with target heating and ablation, and the part of laser energy dissipated in the plasma. At this stage, the important issue is to determine the plasma parameters, taking into account the mutual phenomena solving the mathematical problem self-consistently.

In general, the above calculations show that the target vaporizes in a dense near-target–plasma that is separated from the solid surface by an electrical sheath with a relatively large potential drop and by a Knudsen layer across which there is a jump of plasma parameters and a mass flux returned toward the target. The electron beam emitted from the hot target is accelerated in the sheath, and its energy is dissipated in the electron relaxation region whose length is significantly larger than the length of the Knudsen layer. The dense plasma and the beam energy dissipation impede the plasma flow near the target surface. *This is very important point in formation of the target ablation rate.*

The **laser plasma acceleration** depends on the laser energy absorbed in the target and in the generated plasma. The important issue is to understand the mechanism producing the observed ion energy distribution, and the shifted equivalent voltage per charge state studied the LTI with intermediate values of power density $10^8 - 10^{10}$ W/cm^2. One of the approaches proposed in these published works uses Target

24.8 Feature of Expanding Laser Plasma Flow and Jet Acceleration

Normal Sheath Acceleration (TNSA) mechanism developed by works in [87–89]. The TNSA theory (named also electrostatic) considered laser interaction with a target assuming that at the first stage, the laser energy is converted into hot electrons usually having energy of several 100 keVs. When the hot electrons get out through the rear surface of the target, they create a large electric field. This field ionizes the atoms, and then, the ions are pulled and accelerated by the electric field.

Although this mechanism was widely used, the sheath structure was not considered and, especially, for moderate laser power density when the electron temperature is relatively small, the ion acceleration in the expanding electrical sheath is not detailed. The authors assumed that the plasma quasineutrality is violated, and a space charge region with a potential drop u_f (whose value depends on T_e) can appear. In most cases, the characteristic length of the space charge region was chosen to be the Debye length. The electric field E was calculated using the measured shifted equivalent voltage as u_f and Debye length L_D, i.e., $E = u_f/L_D$. In this case, the acceleration mechanism remains unknown because namely u_f should be determined. Although the particle acceleration determined by the electrical field, the particle energy is determined by potential drop in the space charge region. Note also that Debye length is a parameter that characterizes the scale of E where the quasineutrality violated and the potential of charged particle interaction $\sim T_e/e$. The electrostatic TNSA mechanism cannot explain the observed relatively large ion energy shift per charge also because the spectroscopic direct measurements showed that the expanding plasma has relatively low T_e for intermediate laser power density.

The effect of quasineutrality violation takes place also at the expanding plasma front, but the characteristic length depends on the whether the plasma is expanding into vacuum, into a gas, or impinges onto a solid wall. In this case, the particle energy acquired by electrostatic acceleration in the space charge region cannot exceed the electron temperature although there is a relatively large electric field in a small region (whose size is the Debye or other length). Wilks et al. [90] concluded that the electrostatic acceleration mechanism (i.e., TNSA) is applicable only for ultra-intense ($10^{17} - 10^{20}$ W/cm^2), short-pulse (fs and ps) lasers when $T_e \sim$ MeV and $L \sim 10$ μm. Therefore, the ion acceleration by the electrostatic mechanism described well the experiment with ultra-intense laser irradiation when T_e is about MeV.

The ion acceleration can be described taking into account that there a strong gradient of plasma pressure in the expanding plasma [91]. Thus, the laser plasma acceleration is determined by the energy absorbed in the relaxation region that produces a plasma gradient in the expanding plasma. As a result, the observed plasma jet velocity is determined by the gasdynamic mechanism of plasma acceleration. In the gasdynamic region, the plasma is quasineutral, and therefore, the electric field is relatively small. The ion energy W due to the pressure gradients can be considered by analysis of (24.3) for multi-charged ion density n_{iz} with charge z, taking into account the mass density $\rho = m \Sigma n_{iz}$, ion pressure $p_i = T \Sigma n_{iz}$, and electron pressure $p_e = T_e \Sigma z n_{iz}$.

$$\frac{dW_i}{dx} = \frac{d(mV^2/2)}{dx} = -\frac{1}{\sum_z n_{iz}} \left[\sum_z n_{iz} \frac{dT}{dx} + T \frac{d\sum_z n_{iz}}{dx} + \sum_z z n_{iz} \frac{dT_e}{dx} + T_e \frac{d\sum_z z n_{iz}}{dx} \right]$$
(24.15)

Integrating (24.3) across some characteristic length in the plasma expansion region between distances x_0 and x, the ion energy increase $\Delta W = W(x) - W(x_0)$ can be obtained in form:

$$\frac{\Delta W}{z_{ef}} = \left[\int_{T(x)}^{T(x_0)} dT + \int_{f(x)}^{f(x_0)} T d\left(\text{Ln}\left(\sum_z n_{iz}\right)\right) + \right] \frac{1}{z_{ef}}$$
$$+ \int_{T_e(x)}^{T_e(x_0)} dT_e + \int_{f(x)}^{f(x_0)} T_e d\left(\text{Ln}\left(\sum_z z n_{iz}\right)\right) \quad (24.16)$$

And in simple form, this dependence is:

$$\frac{\Delta W}{z_{ef}} = \int_{f(x)}^{f(x_0)} \left[dT_e + T_e d\left(\text{Ln}\left(\sum_z z n_{iz}\right)\right) \right] + F(T, n_{iz}, z) \quad (24.17)$$

where $z_{ef} = \frac{\sum_z z n_{iz}}{\sum_z n_{iz}}$, $W = mV^2/2$ is the ion kinetic energy, $f(x)$ are the limits of integral function with boundary conditions at x_0, and $F(T, n_{iz}, z)$ is a function expressing the sum of the first terms in square brackets on the right side of (24.16) which describes the influence of the pressure gradient by ion temperature. Taking into account that the ion pressure gradient is lower than electron pressure gradient, the (24.17) shows that the ion energy linearly depends on the effective ion charge state, as observed experimentally. The ion energy also depends on a jump of plasma parameters. A strong jump of T_e and n_e can appear over a region of few mean free paths at the plasma front [91, 92]. Eliezer and Hora [61] indicated the simultaneous existence of "hot" electrons and "cold" electrons, which induce rarefaction shock waves with discontinuities in the plasma potential. The jump characteristic should be studied taking into account the plasma front interaction with the incident laser pulse.

Another result that should be understood is the relatively wide dispersion observed in the ion energy distribution that increases for ions with larger charge. This dispersion was interpreted as a very large ion temperature, up to about 10^5 K. However, the measurements and calculations show that T_e is low in the dense plasma region, as mentioned above (see Table 24.1). The plasma is in non-equilibrium, i.e., $T_e \neq T$, and the efficiency of energy transfer from the electrons to the heavy particle is too weak to reach equilibrium. One possible interpretation considers that the effect of the wide ion energy distribution is similar to that measured in plasma jets of a vacuum arc [93]. The ion energy distribution in the vacuum arc was explained by superposition of the jets generated by multiple cathode spots [94]. It can be taken into account also that the ion energy substantially increases when the erosion rate decreases in arcs with large rate of current or power rise during the discharge pulse [95].

Thus, similar to the vacuum arc phenomena, the energy dispersion of laser produced ions can be explained taking into account different plasma jet parameters generated, for example, in experiments with multi-pulse laser irradiation of the target. In this case, the plasma parameters measured by a laser pulse irradiation depend on the target surface features produced by the previous pulse. Indeed, Schwarz-Selinger [96] showed roughening of Si surfaces treated by nanosecond laser pulses. Consequently, the plasma parameters are determined by the vaporization rate of surface irregularities, and a target erosion rate will not be uniform. Surface non-uniformity causes the observed ion energy variation because the ion acceleration depends on the target local erosion rate, which depends on the local target surface morphology. The morphology varies from pulse to pulse. The origin of ion energy variation can be explained taking into account the large incoming laser power intensity in short laser pulses and the mechanism of ion acceleration depending on target ablation rate by analogy to the cathode erosion rate in arcs [95]. It should be noted that plasma structure formation due to non-uniformity of power density distribution within the laser spot in single laser pulse also was indicated by Hora [97].

24.9 Summary

A physical model of laser plasma generation taking into account the electron emission relaxation region with self-consistent mass and energy exchange at the solid-plasma interface was described in this Chapter. The model includes the plasma phenomena in a space charge sheath near the target surface, electron emission, kinetics of target ablation, vapor heating, atom ionization, and plasma jet expansion. The specifics of the theory consist in finding additional to the laser radiation energy two types of energy fluxes. One heats the near-target–plasma by energy flux from the emitted electrons accelerated in the sheath. As result, the vapor is highly ionized. For Cu target, the ions with charge up to Cu^{+3} were calculated for laser power density <1 GW/cm^2.

The second is due to incoming energy flux to the target from the adjacent plasma by the accelerated ion and by electron flux returned from the plasma toward the target. The large ion current to the target was supported by a relatively large electric field at the target surface in the space charge sheath formed near the target. It was shown that the absorbed laser power in the plasma was converted into kinetic and potential energy of the plasma particles and was returned to the target. The target surface temperature calculated, taking into account the plasma energy flux, was sufficiently larger than that without this flux. These energy fluxes in the electrical sheath, which arise at the target surface, was not considered previously.

The minimal and maximal laser power densities were detected that determined as values when near-target–plasma can be reproduced by an additional plasma energy flux to the target depending on the vapor ionization state. It was indicated that for power density q_L lower than the maximal value, the plasma produced self-consistently in the relaxation region near the target, while for larger q_L, the plasma can be produced by vapor interaction with laser radiation in the expanding vapor jet.

The calculation for Al and Cu targets shows:

(i) the dependence of target rate ablation on pulse energy density well agrees with experiments;
(ii) the energy flux from the adjacent plasma significantly contributes to heating the target;
(iii) the calculated electron temperature of 1–5 eV agrees well with other calculated results and is the range measured by direct spectroscopic methods (see Table 24.1).

References

1. Hoffmann, D. H. H., Blazevic, A. P., Rosmej Ni, O., Roth, M., Tahir, N. A., Tauschwitz, A., et al. (2005). Present and future prospectives for high energy density physics with intense heavy ion and laser beams. *Laser and Particle Beams, 23*, 47–53.
2. Miller, G. H., Moses, E. I., & Wuest, C. R. (2004). The national ignition facility: enabling fusion ignition for the 21st century. *Nuclear Fusion, 44*(N12), S228–S238.
3. Gamaly, E. G., Rode, A. V., Luther-Davies, B., & Tikhonchuk, V. T. (2002). Ablation of solids by femtosecond lasers: Ablation mechanism and ablation thresholds for metals and dielectrics. *Physics of Plasmas, 9*(N3), 949–957.
4. Eliezer, S., Eliaz, N., Grossman, E., Fisher, D., Gouzman, I., Henis, Z., Pecker, S., Horovitz, Y., Fraenkel, M., Maman, S., Ezersky, V., & Eliezer, D. (2005). Nanoparticles and nanotubes induced by femtosecond lasers. *Laser and Particle Beams, 23*, 15–19.
5. Attwood, D. T., Sweeney, D. W., Auerbach, J. M., & Lee, P. H. Y. (1978). Interferometric confirmation of radiation-pressure effects in laser-plasma interactions. *Physical Review Letters, 40*(N3), 184–187.
6. Tajima, T, & Dawson, J. M. (1979). Laser electron accelerator. *Physical Review Letters, 43*(N4), 267–270.
7. Steinke, S., Henig, A., et al. (2010). Efficient ion acceleration by collective laser-driven electron dynamics with ultra-thin foil targets. *Laser and Particle Beams, 28*(1), 215–221.
8. Shoucri, M., Lavocat-Dubuis, X. et al. (2011). Numerical study of ion acceleration and plasma jet formation in the interaction of an intense laser beam normally incident on an overdense plasma. *Laser and Particle Beams, 29*(3), 315–332.
9. Tsui, Y. Y., & Redman, D. G. (2000). A laser ablation technique for improving the adhesion of laser-deposited diamond-like carbon coatings to metal substrates. *Surface and Coatings Technology, 126*(N2–3), 96–101.
10. Fernandez, J. C., Hegelich, B. M., Cobble, J. A., Flippo, K. A., Samulel, A. et al. (2005). Laser-ablation treatment of short-pulse laser targets: Toward an experimental program on energetic-ion interactions with dense plasmas. *Laser and Particle Beams, 23*(3), 267–273.
11. Anisimov, S. I., Bauerly, D., & Luk'yanchuk, B. S. (1993). Gas dynamics and film profiles in pulsed laser deposition of materials. *Physical Review, B48*(N16), 12076–12081.
12. Gamaly, E. G., Luther-Davies, B., Kolev, V. Z., Madsen, N. R., Duering, M., & Rode, A. V. (2005). Ablation of metals with picosecond laser pulses: Evidence of long-lived nonequilibrium surface states. *Laser and Particle Beams, 23*, 167–176.

13. Shukla, G., & Khare, A. (2010). Spectroscopic studies of laser ablated ZnO plasma and correlation with pulsed laser deposited ZnO thin film properties. *Laser and Particle Beams, 28*(N1), 149–155.
14. Zheng, J. P., Shaw, Z. Q., Kowk, D. T., & Huang, H. C. (1989). Generation high-energy atomic beams laser. *Applied Physics Letters, 54*(N3), 280–282.
15. Moscicki, T., Hoffman, J., & Szymanski, Z. (2006). Modelling of plasma plume induced during laser welding. *Journal of Physics D: Applied Physics, 39,* 685–692.
16. Dubey, A. K., & Yadava, V. (2008). Laser beam machining—A review. *The International Journal of Machine Tools and Manufacture, 48,* 609–628.
17. Schaumann, G., Schollmeier, M. S., Rodriguez-Prieto, G., Blazevic, A., Brambrink, E., Geissel, M., et al. (2005). High energy heavy ion jets emerging from laser plasma generated by long pulse laser beams from the NHELIX laser system at GSI. *Laser & Particle Beams, 23,* 503–512.
18. Henc-Bartolic, V., Boncina, T., Jakovljevic, S., Pipic, D., & Zupanic, F. (2008). The action of a laser on an aluminium target. *Material & Technology, 42*(3), 111–115.
19. Henc-Bartolic, V., Andreic, Z., & Kunze, H. J. (1994). Titanium plasma produced by a nitrogen laser. *Physica Scripta, 50,* 368–370.
20. Hermann, J., Thomann, A. L., Boulm, Leborgne C., & Dubreuil, B. (1995). *Journal of Applied Physics, 77*(N7), 2928–2936.
21. Chang, J. J., & Warner, B. E. (1996). Laser-plasma interaction during visible-laser ablation of methods. *Applied Physics Letters, 69*(4), 473–475.
22. Franghiadakis, Y., Fotakis, C., & Tzanetakis, P. (1999). Energy distribution of ions produced by excimer-laser ablation of solid and molten targets. *Applied Physics A. Materials Science & Processing, A68,* 391–397.
23. Abdellatif, G., & Imam, H. (2002). A study of the laser plasma parameters at different laser wavelengths. *Spectrochimica Acta, B57,* 1155–1165.
24. Rieger, G. W., Taschuk, M., Tsui, Y. Y., & Fedosejevs, R. (2003). Comparative study of laser-induced plasma emission from microjoule picosecond and nanosecond KrF-laser pulses. *Spectrochimica Acta B, 58,* 497–510.
25. Hafez, M. A., Khedr, M. A., Elaksher, F. F., & Gamal, Y. E. (2003). Characteristics of Cu plasma produced by a laser interaction with a solid target. *Plasma Sources Science and Technology, 12,* 185–198.
26. Corsi, M., Cristoforetti, G., Giuffrida, M., Hidalgo, M., Legnaioli, S., Palleschi, V., et al. (2004). Three-dimensional analysis of laser induced plasmas in single and double pulse configuration. *Spectrochimica Acta, Part B, 59,* 723–735.
27. Aguilera, J. A., Bengoechea, J., & Arago, C. (2004). Spatial characterization of laser induced plasmas obtained in air and argon with different laser focusing distances. *Spectrochimica Acta, B59,* 461–469.
28. Margarone, D., Torrisi, L., Borrielli, A., & Caridi, F. (2008). Silver plasma by pulsed laser ablation. *Plasma Sources Science and Technology, 17,* 035019.
29. Torrisi, L., Caridi, F., Margarone, D., & Borrielli, A. (2008). Plasma–laser characterization by electrostatic mass quadrupole analyzer. *Nuclear Instruments and Methods in Physics Research, B266,* 308–315.
30. Torrisi, L., Caridi, F., Margarone, D., Picciotto, A., Mangione, A., & Beltrano, J. J. (2006). Carbon-plasma produced in vacuum by 532 nm–3 ns laser pulses ablation. *Applied Surface Science, 252,* 6383–6389.
31. Torrisi, L., Caridi, F., Margarone, D., & Borrielli, A. (2008). Characterization of laser-generated silicon plasma. *Applied Surface Science, 254,* 2090–2095.
32. Torrisi, L., Gammino, S., Ando, L., & Laska, L. (2002). Tantalum ions produced by 1064 nm pulsed laser irradiation. *Journal of Applied Physics, 91*(7), 4685–4692.
33. Bleiner, D., Bogaerts, A., Beiloni, F., & Nassisi, V. (2007). Laser-induced plasmas from the ablation of metallic targets: The problem of the onset temperature, and insights on the expansion dynamics. *Journal of Applied Physics, 101,* 083301.
34. Torrisi, L., Ciavola, G., Gammino, S., Ando, L., & Barna, A. (2000). Metallic etching by high power Nd:yttrium–aluminum–garnet pulsed laser irradiation. *Review of Scientific Instruments, 71*(N11), 4330–4334.

35. Torrisi, L., Ando, L., Ciavola, G., Gammino, S., & Barna, A. (2001). Angular distribution of ejected atoms from Nd:YAG laser irradiating metals. *Review of Scientific Instruments, 72*(N1), 68–72.
36. Caridi, F., Torrisi, L., Margarone, D., Picciotto, M., Gammino, A. A., & Mezzasalma, S. (2006). Energy distributions of particles ejected from laser generated pulsed plasmas. *Czechoslovak Journal of Physics, 56,* B449–B456.
37. Kasperczuk, A., Pisarczyk, T., Kalal, M., Martinkova, M., Ullschmied, J., Krousky, E., et al. (2008). PALS laser energy transfer into solid targets and its dependence on the lens focal point position with respect to the target surface. *Laser and Particle Beams, 26*(2), 189–196.
38. Zeng, X., Mao, X., Mao, S. S., Sy-Bor, Wen, Greif, R., & Russo, R. E. (2006). Laser-induced shockwave propagation from ablation in a cavity. *Applied Physics Letters, 88,* 061502.
39. Chen, M., Liu, X., Yanga, X., Zhao, M., Sun, Y., Qi, H., et al. (2008). The dimension of the core and the tail of the plasma produced by laser ablating SiC targets. *Applied Physics Letters, 372,* 5891–5895.
40. Chen, M., Liu, X., Zhao, M., Chen, C., & Man, B. (2009). Temporal and spatial evolution of Si atoms in plasmas produced by a nanosecond laser ablating silicon carbide crystals. *Physical Review E, 80,* 016405.
41. Chen, M., Liu, X., Zhao, M., & Sun, Y. (2009). Early-stage evolution of the plasma over KTiOPO4 samples generated by high-intensity laser radiations. *Optics Letters, 34*(17), 2682–2684.
42. Sun, Y., Chen, M., Li, Y., Qi, H., Zhao, M., & Liu, X. (2008). Analysis of plasma profile over KTiOAsO4 surface produced by 532 and 1064 nm laser radiations. *Journal of Applied Physics, 104,* 123303.
43. Cristoforetti, G., Lorenzetti, G., Benedetti, P. A., Tognoni, E., Legnaioli, S., & Palleschi, V. (2009). Effect of laser parameters on plasma shielding in single and double pulse configurations during the ablation of an aluminium target. *Journal of Physics D, 42,* 225207.
44. Popov, S., Panchenko, A., Batrakov, A., Ljubchenko, F., & Mataibaev, V. (2011). Experimental study of the laser ablation plasma flow from the liquid Ga-In target. *The IEEE Transactions on Plasma Science, 39*(N6), 1412–1417.
45. Beilis, I. I. (2012). Modeling of the plasma produced by moderate energy laser beam interaction with metallic targets: Physics of the phenomena. *Laser & Particle Beams, 30*(3), 53–63.
46. Singh, R. K., & Narayan, J. (1990). Pulsed-laser evaporation technique for deposition of thin films: Physics and theoretical model. *Physical Review B, 41*(13), 8843–8859.
47. Gamaly, E. G., Rodea, A. V., & Luther-Davies, B. (1999). Ultrafast ablation with high-pulse-rate lasers. Part I: Theoretical considerations. *Journal of Applied Physics, 85,* 4213–4221.
48. Itina, T. E., Hermann, J. Delaporte, & Sentis, M. (2002). Laser-generated plasma plume expansion: Combined continuous-microscopic modeling. *Physical Review E, 66,* 066406.
49. Anisimov, S. I. (1968). Vaporization of metal absorbing laser radiation. *Journal of Experimental and Theoretical Physics, 37*(1), 182–183.
50. Wang, H.-X., & Chen, X. (2003). Three-dimensional modelling of the laser-induced plasma plume characteristics in laser welding. *Journal of Physics D-Applied Physics, 36*(6), 628–639.
51. Gusarov, A. V., Gnedovets, A. G., & Smurov, I. (2000). Gas dynamics of laser ablation: Influence of ambient atmosphere. *Journal of Applied Physics, 88*(7), 4352–4364.
52. Gusarov, V., & Aoki, K. (2005). Ionization degree for strong evaporation of metals. *Physics of Plasmas, 12,* 083503.
53. Bogaerts, A., Chen, Z., Gijbels, R., & Vertes, A. (2003). Laser ablation for analytical sampling: what can we learn from modeling? *Spectrochimica Acta, B58,* 1867–1893.
54. Chen, Z., & Bogaerts, A. (2005). Laser ablation of Cu and plume expansion into 1 atm ambient gas. *Journal of Applied Physics, 97,* 063305.
55. Aden, M., Beyer, E., & Herziger, G. (1990). Laser-induced vaporization of metal as a Riemann problem. *Journal of Physics D, 23,* 655–661.
56. Raizer, Yu P. (1965). Breakdown and heating of gases under the influence of a laser beam. *Sovereign Physics Uspekhy, 8*(5), 650–673.
57. Raizer, Yu P. (1980). Optical discharges. *Sovereign Physics Uspekhy, 23*(N11), 789–806.

References

58. Gili, D. H., & Dougal, A. A. (1965). Breakdown minima due to electron-impact ionization in super-high-pressure gases irradiated by a focused giant-pulse laser. *Physical Review Letters, 15*(N22), 845–847.
59. Mazhukin, V. I., Nossov, V. V., Nickiforov, M. G., & Smurov, I. (2003). Optical breakdown in aluminum vapor induced by ultraviolet laser radiation. *Journal of Applied Physics, 93*(1), 56–66.
60. Rosen, D. I., Mitteldorf, J., Kothandaraman, G., Pirri, A. N., & Pugh, E. R. (1982). Coupling of pulsed 0.35-μm laser radiation to aluminum alloys. *Journal of Applied Physics, 53*(N4), 3190–3200.
61. Eliezer, S., & Hora, H. (1988). Double layers in laser produced plasma. *Physical Reports, 172*, 339–406.
62. Beilis, I. I. (2007). Laser plasma generation and plasma interaction with ablative target. *Laser & Particle Beams, 25*, 53–63.
63. Caretto, G., Doria, D., Nassisi, V., & Siciliano, M. V. (2007). Photoemission studies from metal by UV lasers. *Journal of Applied Physics, 101*, 073109.
64. Dolan, W. W., & Dyke, W. P. (1954). Temperature and field emission of electrons from metals. *Physical Review, 95*(2), 327–332.
65. Beilis, I. I. (1974). Emission processes at the cathode of an electric arc. *Soviet Physics Technical Physics, 19*(2), 257–260.
66. Beilis, I. I. (2006). Mechanism of laser plasma production and of plasma interaction with a target. *Applied Physics Letters, 89*, 091503.
67. Morozov, A. I. (1978). *Physical foundations of space electric propulsion engines. Elements of the dynamics of flows in the electric propulsion.* In M. Atomizdat.
68. Eliezer, S., & Ludmirsky, A. (1983). Double layer formation in laser produced plasma. *Laser Part Beams, 1*, 251–269.
69. Denavit, J. (1979). Collisionless plasma expansion into vacuum. *Physics of Fluids, B22*(7), 1384–1392.
70. Gamaly, E. G. (1993). The interaction of ultrashort, powerful laser pulses solid target: Ion expansion and acceleration with time-dependent ambipolar field. *Physics of Fluids, B5*(N3), 944–949.
71. Wickens, L. M., Allen, J. E., & Rumsby, P. T. (1978). Ion emission from laser-produced plasmas with two electron temperatures. *Physical Review Letters, 41*(N4), 243–246.
72. Torrisi, L., & Gammino, S. (2006). Method for the calculation of electrical field in laser-generated plasma for ion stream production. *Review of Scientific Instruments, 77*, 03B707.
73. Davies, J. R., Bell, A. R., & Tatarakis, M. (1999). Magnetic focusing and trapping of high-intensity laser-generated fast electrons at the rear of solid targets. *Physical Review E, 59*(N5), 6032–6036.
74. Pukhov, A. (2001). Three-dimensional simulations of ion acceleration from a foil irradiated by a short-pulse laser. *Physical Review Letters, 86*(N16), 3562–3565.
75. Hegelich, M., Karsch, S., Pretzler, G., Habs, D., Witte, K., Guenther, W., et al. (2002). MeV ion jets from short-pulse-laser interaction with thin foils. *Physical Review Letters, 89*(N8), 085002.
76. Hatchett, S. P., Brown, C. G., Cowan, T. E., et al. (2000). Electron, photon, and ion beams from the relativistic interaction of Petawatt laser pulses with solid targets. *Physics of Plasmas, 7*(N5), 2076–2082.
77. Liseikina, T. V., Prellino, D., Cornolti, F., & Macchi, A. (2008). Ponderomotive acceleration of ions: circular versus linear polarization. *IEEE Transactions on Electrical Insulation, 36*(4), 1866–1871.
78. Anisimov, S. I., Imas, Y. A., Romanov, G. S., & Khodyko, Y.V. (1971). Action of high-power radiation on metals, National Technical Information Service. Virginia: Springfield.
79. Beilis, I. I. (1995). Theoretical modelling of cathode spot phenomena. In R. L. Boxman, P. J. Martin, & D. M. Sanders (Eds.), *Handbook of vacuum arc science and technology* (pp. 208–256). N.J.: Noyes Publications Park Ridge.

80. Beilis, I. I. (1982). On the theory of erosion processes in the cathode region of an arc discharge. *Soviet Physics-Doklady, 27*, 150–152.
81. Keidar, M., Beilis, I. I., Boxman, R. L., & Goldsmith, S. (1996). 2-D expansion of the low density interelectrode vacuum arc plasma in an axial magnetic field. *Journal of Physics D, 29*(7), 1973–1983.
82. Keidar, M., & Beilis, I. I. (2018). *Plasma Engineering*. Elsevier, London-NY; Academic Press.
83. Beilis, I. I. (2003). The vacuum arc cathode spot and plasma jet: Physical model and mathematical description. *Contributions to Plasma Physics, 43*(N3–4), 224–236.
84. Beilis, I. I. (2016). Metallic targets ablation by laser plasma production in a vacuum. *Journal of Instrumentation (JINST), 11*(3), 03056.
85. Caridi, F., Torrisi, L., Margarone, D., & Borrielli, A. (2008). Laser-generated plasma investigation by electrostatic quadrupole analyzer. *Radiation Effects and Defects in Solids, 163*, 357–363.
86. Torrisi, L., Andò, L., Gammino, S., Kràsa, J., & Làska, L. (2001). Ion and neutral emission from pulsed laser irradiation of metals. *Nuclear Instruments and Methods, B184*, 327–336.
87. Mora, P. (2003). Plasma expansion into avacuum. *Physical Review Letters, 90*, 185002.
88. Mora, P. (2005). Thin-foil expansion into a vacuum. *Physical Review E, 72*, 056401.
89. Nishiuchi, M., Fukumi, A., Daido, H., Li, Z., Sagisaka, A., Ogura, K., et al. (2006). The laser proton acceleration in the strong charge separation regime. *Physical Letters A, 357*(N4–5), 339–344.
90. Wilks, S. C., Langdon, A. B., Cowan, T. E., Roth, M., Singh, M., Hatchett, S., et al. (2001). Energetic proton generation in ultra-intense laser–solid interactions. *Physics and Plasmas, 8*(2), 542–549.
91. Zeldovich, Y. B., & Raizer, Y. P. (1966). *Physics of shock waves and high-temperature hydrodynamic phenomena*. New York: Academic Press.
92. Raizer, Y. P. (1974). *Laser spark and discharge expansion*. Nauka. Moscow (in Russian).
93. Davis, W. D., & Miller, C. H. (1969). Analysis of the electrode products emitted by dc arcs in a vacuum ambient. *Journal of Applied Physics, 40*(5), 2212–2221.
94. Beilis, I. I., Keidar, M., Boxman, R. L., & Goldsmith, S. (1998). Theoretical study of plasma expansion in a magnetic field in a disk anode vacuum arc. *Journal of Applied Physics, 83*(N2), 709–717.
95. Beilis, I. I. (2004). Nature of high-energy ions in the cathode plasma jet of a vacuum arc with high rate of current rise. *Applied Physics Letters, 85*(14), 2739–2740.
96. Schwarz-Selinger, T., Cahill, D. G., Chen, S.-C., Moon, S.-J., & Grigoropoulos, C. P. (2001). Micron-scale modifications of Si surface morphology by pulsed-laser texturing. *Physical Review B, 64*, 155323.
97. Hora, H. (1981). *Physics of laser driven plasmas*. New York: John Wiley & Sons.

Chapter 25
Application of Cathode Spot Theory for Arcs Formed in Technical Devices

One of the important issues is the electrical arc initiation and development in devices with flowing plasmas or in high-pressure systems. Such arcing occurred in generators of low-temperature plasmas, in plasma accelerators, in plasma of products of combustion, and in a rail gun [1–8]. The plasma devices consist of complicated designs and different types of electrode assemblies. In this case, different problems arise by the projection of the assemblies and their optimization. In the presence of hot plasma, a large number of electrode parameters such as design and conditions of cooling to reach electrical characteristics for different work regimes should be taken in account. To understand the approaches for optimization, a modeling of the plasma–electrode interaction is a significant factor. It is obvious that a universal theory that describes any new apparatus and electrode configurations are absent. Therefore, the developed models should consider the typical phenomena in order obtain rules to prevent the destruction of the assembly. Some of the most important processes are following: the influence of the heat flux from the plasma to the surface, large temperature gradients, effects of current constriction, conditions for saving the plasma and surface doping, and the influence of the geometry of the electrode-plasma contact. Two types of current carrying electrodes are typically used: hot ceramic and cold metallic electrodes. Usually, the front surfaces of both types of electrodes faced the high-temperature plasma flow, while the rear surfaces are water cooled. In case of arc at high pressure, the discharge column configuration was formed by surrounding gas stabilization. Let us consider different electrode types and specifics of their thermal and electrical characteristics for discharges in different conditions.

25.1 Electrode Problem at a Wall Under Plasma Flow. Hot Boundary Layer

When a cathode contacts with plasma flow, a boundary layer arises in which the temperature increased from the value at the cathode surface T_w to the value of plasma volume temperature T_{pl}. The cooled rear cathode surface is at value T_0. At relatively low-current density j, the value of T_w is determined by the heat flux from the plasma. With increase j, an addition heating occurs in the cathode volume due to Joule energy dissipation resulting in an increase in the cathode surface temperature. The electrode-plasma characteristics that occurred with these phenomena are described below.

25.2 Hot Ceramic Electrodes. Overheating Instability

In this section, different phenomena are studied and analyzed of a condition for transient thermal regime, unstability and stability state of the current carrying electrodes due to their overheating by Joule energy dissipation in the volume and by heat flux from a hot plasma to the electrode surface. The double layer approximation is described based on the previous research [9, 10].

25.2.1 Transient Process

The problem includes the transient heat conduction equation with nonlinear boundary condition. The mathematical formulation of the process can be presented in form:

$$\frac{\partial T}{\partial t} = a\nabla^2 T + \frac{j^2}{\sigma(T)}; \quad \sigma(T) = A_0 \text{Exp}(B_0 T)$$

$$\text{div } j = 0, \quad j = \sigma(T)E, \quad E = -\frac{\partial \varphi}{\partial x} \qquad (25.1)$$

Boundary conditions:

$$T|_{x=0} = T_0 \quad T(x)|_{t=0} = T_0$$

$$-\lambda_{T2} \frac{\partial T_2}{\partial x}\bigg|_{x=0} = \alpha(T_w - T_1) \qquad (25.2)$$

Dimensionless form

$$h^2 \frac{\partial \theta}{\partial t} - a \frac{\partial^2 \theta}{\partial \xi^2} = A_n e^{-B\theta}, \quad A_n = \frac{jh}{c\gamma\sigma_0(T_1 - T_0)}$$

$$\theta = \frac{T - T_0}{T_1 - T_0}, \quad \theta_w = \theta|_{x=w} \quad B = B_0(T_1 - T_0) \qquad (25.3)$$

Boundary condition

25.2 Hot Ceramic Electrodes. Overheating Instability

$$\theta|_{\xi=0} = 0, \quad \theta|_{\xi=1} = \theta_w$$

$$-\frac{\partial \theta}{\partial \xi}\bigg|_{\xi=1} = Bi(\theta_w - 1); \quad Bi = \frac{\alpha h}{\lambda_T} \quad (25.4)$$

Characteristic time

$$h^2 \frac{\partial \theta}{\partial t} = A_n e^{-B\theta} \quad \text{and after integrating} \quad \theta|_{t=0} = 0, \theta|_t = \theta_w$$

$$\frac{e^{B\theta_w} - 1}{B} = \frac{A_n t}{h^2} \quad (25.5)$$

where $\xi = x/h$, λ_T are the heat conductivity. A_0 and B_0 are the experimental constants. Let us determine $t = \tau$, when $\theta_w = 1$, i.e., the temperature in the electrode volume exceeds the plasma temperature and direction of the heat flux can be reversed. In this case, it is easy to see that $\tau \sim j^{-2}B$. According to characteristic parameters $j \sim 0.5$–1 A/cm^2 and $B \sim 6$ [10] the time of transient regime is a few second. Therefore, for long-time working of the devices, the problem can be considered in steady-state approximation.

25.2.2 Stability of Arcing and Constriction Conditions

The transition from diffuse discharge form to an arc form occurs as a jump process. The arc current can be controlled is by an additional load resistance R_r. In general, for arc stability, the sum of arc voltage u_{arc} and voltage at the load resistance $u_r = IR_r$ should be equal to ponderomotive voltage ε of the source [11, 12]. A *dropping volt–current characteristic* of the arc and a linear volt-load characteristic are presented in Fig. 25.1. When the current varied, the voltage of load characteristic can be expressed as $u_d = \varepsilon - u_r$. Figure 25.1 shows two points (A and B), at which the condition of

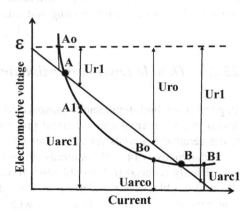

Fig. 25.1 Schematic presentation of arc dropped volt–current and resistance Dropping volt-current characteristic characteristics and arc stability condition at point B

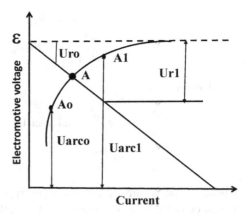

Fig. 25.2 Schematic presentation of arc with growing volt–current and linear load characteristics and the arc stability condition

$u_d = u_{arc}$ is fulfilled, however, only at one point the arc is stable when the current fluctuated. Let us consider the point A. When the current increases to point A_1, the voltage $\varepsilon - u_r - u_{arc} > 0$, i.e., $u_d > u_{arc}$ and therefore the source voltage allows and further increase the current.

When the current decreases to point A_0 the voltage $\varepsilon - u_r - u_{arc} < 0$, i.e., $u_d < u_{arc}$, and, therefore, the current decreases, which then leads to further decrease up to state that the arc extinguish. Thus, the arc at point A is unstable under the current fluctuation. At point B, when the current increases to point B_1, the voltage $\varepsilon - u_r - u_{arc} < 0$, i.e., $u_d < u_{arc}$, and therefore the current decreases returning to point B. When the current fluctuated to point B_0, the voltage $\varepsilon - u_r - u_{arc} > 0$, i.e., $u_d > u_{arc}$, As a result, the current increases passing again to point B and thus the arc at this point is stable.

A case with **growing volt–current characteristic** of the arc and a linear volt-load characteristic are presented in Fig. 25.2. At point A, when the current increases to point A_1, the voltage $\varepsilon - u_r - u_{arc} < 0$, i.e., $u_d < u_{arc}$, and therefore the current decreases returning to point A. When the current fluctuated to point A_0, the voltage $\varepsilon - u_r - u_{arc} > 0$, i.e., $u_d > u_{arc}$. As a result, the current increases passing again to point A. Thus, the arc with growing volt–current characteristic is always stable.

25.2.3 Double Layer Approximation

In general, the high-temperature plasma devices should operate with minimal power losses due to thermal and electrical processes. It can be achieved with hot wall using ceramic materials for electrodes cooled at rear surface. However, the electrical conductivity of such materials strongly depends on the temperature and it is significantly reduced in the cold zone, limiting the circuit current [9]. To prevent this effect, a block of two elements was used. The part faced to the cooling system (current terminal) consists of a ceramic with high thermal and electro-conductivity in

25.2 Hot Ceramic Electrodes. Overheating Instability

comparison with the ceramic part (current collector) faced the hot plasma. Analysis of thermal processes of such current carried electrodes related to study of the overheating problem [13, 14]. The specifics of the problem are present in a volume heat source (Joule energy dissipation) characterized by strong dependence on the temperature. The previous study of this problem allowed to understand the breakdown phenomena in the dielectrics [15] and in the low-temperature plasma [16, 17].

The volt–current characteristic and the temperature distribution were studied for ceramic electrode in case of exponential dependence of the electrical conductivity on temperature and for heat flux according to the Newton law at the front surface. The result shows that the temperature distribution in the body passes through a maximum, which level is determined by intensity of the volume heat source and volt–current characteristic, and then the temperature increased asymptotically to a maximum value. In case of contact two ceramics with different thermal and electrical properties (double layer model), it is important to understand the influence of additional energy dissipation in the different parts of the body on the electrode parameters at different discharge current as well the dependence on the relation between the sizes of the current terminal and current collector.

Let us consider the double layer model in one-dimension approximation assuming that the current density j is constant along the electrode thickness and the value of j and the coefficient of heat exchange α in the Newton law can be given as varied parameters [10]. The simple schema of a double layer electrode with thickness h in a flowed plasma with temperature T_{pl} is presented in Fig. 25.3. Coordinate x begins at the cooled side ($x = 0$) with temperature T_0 and thickness δ of the first layer (current terminal) directed to the front surface with temperature T_w. The temperature of the boundary between two layers is T_δ.

The mathematical formulation of the problem includes the steady-state nonlinear heat conduction equation with nonlinear boundary condition and equations for electrical characteristic.

Nonlinear heat conduction equation:

$$\lambda_{Tj} \frac{\partial^2 T_j}{\partial x^2} + \frac{j^2}{\sigma_j(T_j)} = 0 \tag{25.6}$$

Fig. 25.3 Schematic presentation of double layer electrode model with two different thermal and electrical properties

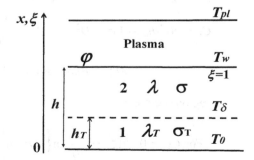

Boundary condition

$$T|_{x=0} = T_0$$

$$T_1|_{x=\delta} = T_2|_{x=\delta}, \quad \lambda_{T1}\frac{\partial T_1}{\partial x}\bigg|_{x=\delta} = \lambda_{T2}\frac{\partial T_2}{\partial x}\bigg|_{x=\delta}$$

$$-\lambda_{T2}\frac{\partial T_2}{\partial x}\bigg|_{x=h} = \alpha(T_2(h) - T_{pl}) \tag{25.7}$$

where λ_{Tj} are the heat conductivity of the layers with index "j" (current removable 1 or current terminal 2, layers respectively),

$$\sigma_j = A_{0j} e^{b_j T_j} \tag{25.8}$$

Dimensionless form

$$\frac{\partial^2 \theta_j}{\partial \xi^2} = -A_j e^{-B_j \theta_j} \tag{25.9}$$

Boundary conditions:

$$\theta_1(0) = 0, \quad \theta_1(s) = \theta_2(s)$$

$$\frac{\partial \theta_1}{\partial \xi}(s) = \lambda_0 \frac{\partial \theta_2}{\partial \xi}(s), \quad \frac{\partial \theta_1}{\partial \xi}(\xi = 1) = Bi(1 - \theta_2(1))$$

Definitions:

$$\theta = \frac{(T_1 - T_0)}{(T_{pl} - T_0)}, \quad Bi = \frac{\alpha h}{\lambda_2}, \quad A_j = \frac{j^2 h^2}{\sigma_0 \lambda_j (T_{pl} - T_0)} \tag{25.10}$$

where $\xi = x/h$, $B_j = b_j(T_{pl} - T_0)$, $\sigma_{0j} = \sigma_j(T_0)$, $s = \delta/h$, $\lambda_0 = \lambda_1/\lambda_2$.

The solution of (25.6)–(25.10) can be represent for different electrode layers taking into account that heat flux can change the direction, characterized by sign of following parameter C_T and C_s:

$$C_T = \lambda^{-2}\left[D^2(M^- - N) + Bi^2(\theta_w - 1)^2\right] - D_T^2 \mathrm{Exp}(-B_T \theta_g)$$

$$C_s = Bi^2(\theta_w - 1)^2 - D^2 N, \quad D_j^2 = \frac{2A_j}{B_j}, \quad R_j = \frac{2A_j}{B_j c_j} = \frac{D_j^2}{C_{sj}} \tag{25.11}$$

$$M^\pm = \mathrm{Exp}(\pm B\theta_g), \quad N = \mathrm{Exp}(-B\theta_w)$$

At region $0 < \xi < s$:

For $C_T < 0$ $\quad \theta(\xi) = \frac{2}{B_T} \mathrm{Ln}\left\{\sqrt{R} \sin\left[\frac{B_T \sqrt{C_T}}{2} s\xi + \arcsin \sqrt{R^{-1}}\right]\right\}$

25.2 Hot Ceramic Electrodes. Overheating Instability

For $C_T = 0$ $\theta(\xi) = \dfrac{2}{B_T} \mathrm{Ln}\left(1 + \dfrac{D_T B_T}{2} s\xi\right)$

For $C_T > 0$ $\theta(\xi) = \dfrac{2}{B_T} \mathrm{Ln}\left\{0.5\left[Exp(B_T\sqrt{C_T}s\xi)\left(1 + \sqrt{R_T + 1}\right)\right.\right.$
$\left.\left. + 1 - \sqrt{R_T + 1}\right]\right\} - \sqrt{C_T}s\xi$ (25.12)

At region $s < \xi < 1$:

For $C_s < 0$ $\theta(\xi) = \dfrac{2}{B}\mathrm{Ln}\left\{R\sin\left[\dfrac{B\sqrt{C_s}}{2}(1-s)\xi + \arcsin\left(\sqrt{R^{-1}M^+}\right)\right]\right\}$

For $C_s = 0$ $\theta(\xi) = \dfrac{2}{B}\mathrm{Ln}\left[\sqrt{M^+} + \dfrac{DB}{2}(1-s)\xi\right]$

For $C_s > 0$ $\theta(\xi) = \dfrac{2}{B}\mathrm{Ln}\left\{0.5\sqrt{M^+}\left[Exp(B\sqrt{C_s}(1-s)\xi)\left(1 + \sqrt{RM^- + 1}\right)\right.\right.$
$\left.\left. + 1 - \sqrt{RM^- + 1}\right]\right\} - \sqrt{C_s}(1-s)\xi$ (25.13)

Electrical characteristics:

$$\mathrm{div}\, j = 0, \quad j = \sigma(T)E, \quad E = -\dfrac{\partial \varphi}{\partial x} \qquad (25.14)$$

The electrode voltage is obtained by the integration of (25.14).

$$u = j\left\{\int_0^{h_T} \dfrac{dx}{\sigma_T[T(x)]} + \int_{h_T}^{h} \dfrac{dx}{\sigma[T(x)]}\right\} \qquad (25.15)$$

Dimensionless parameters:

$\bar{u} = \bar{\lambda}\sqrt{\dfrac{2}{B}\left[\dfrac{D_T^2}{D^2}(1 - \mathrm{Exp}(-B_T\theta_g(s))) + \bar{\lambda}^{-2}\left(\dfrac{2}{B}(M_s^- + N_s)\right)\right.}$
$\left. + \dfrac{Bi^2}{A}(\theta_w - 1)^2\right]^{0.5} - \dfrac{Bi}{A^{0.5}}(\theta_w - 1)$

$\bar{j} = \sqrt{A} = \dfrac{jh}{[\sigma_0 \lambda(T_1 - T_0)]^{0.5}}, \quad \bar{\varphi}^2 = \dfrac{\varphi^2 \sigma_0}{\lambda(T_1 - T_0)},$

$M_s^- = \mathrm{Exp}(-B_s\theta(s)), \quad N_s = (-B_T\theta_w)$ (25.16)

The calculated parameters are electrode temperatures and volt–current characteristics, while the varied parameters are Bio Bi, D_2^2, and B_2, which characterized the convective heat change in plasma boundary layer, current density, and electrical properties of the electrode materials, respectively. It was used the following experimental values [10, 18]: $T_{pl} = 2600$ K; $\lambda = 0.38$ W/m/K, $A_{02} = 6.8 \times 10^{-4}$ O^{-1} m^{-1}, $A_{01} = 3.79$ O^{-1} m^{-1}, $b_1 = 2.4 \times 10^{-3}$, $b_2 = 6 \times 10^{-2}$.

Figure 25.4 shows the temperature distribution in the electrode body (on ξ) for two layers as dependence of dimensionless varied parameters and parameter $s = \delta/h$ indicated ration between the layer sizes. The curves at $s = 0$ and $s = 1$ correspond to calculated distribution at for one layer of current terminal and current collector, respectively. The parameter Bi influence mainly at the electrode front surface. When the thickness of the first layer increases, the temperature gradient near the cooled surface significantly decreases. When the current density relatively low ($D_2^2 \equiv D < 10$) the temperature linearly increases, and the inclination angle changed passing from layer 1 to layer 2 depending on s (Fig. 25.4).

With current growing, firstly the temperature arise as a nonlinear dependence at contact location and then the dependence saturated and passes through a maximum, which location is determined by to Joule energy dissipation in the electrode volume. Decrease of B_2, (below, for simplicity, denoted as B) characterized the rate of conductivity rise with electrode temperature and leads to decrease the influence of layer properties difference on the calculated parameters that depend on discharge current. For $B = 6$, this influence begin at $D = 10^5$, (Fig. 25.5).

Let us consider the electrical electrode parameters. Figure 25.6 presents a dimensionless electrode voltage as dependence on dimensionless current density D at different thicknesses of the current terminal part of the electrode. The curves at s

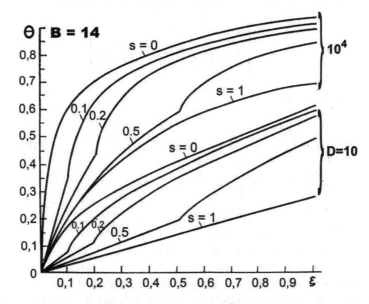

Fig. 25.4 Dimensionless temperature as function distance in two-layer electrode with current density D as parameter, $B = 14$, Bi $= 1$

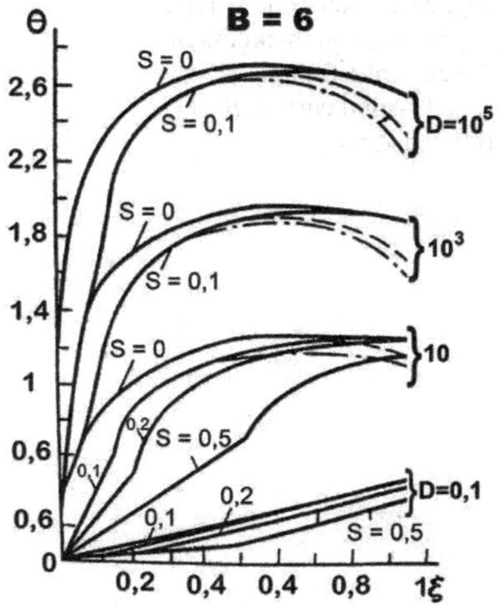

25.2 Hot Ceramic Electrodes. Overheating Instability

Fig. 25.5 Dimensionless temperature as function distance in two-layer electrode with current density D as parameter, $B = 6$, Bi = 1-solid curves, Bi = 5-doted curves, Bi = 10-dash-dot curves

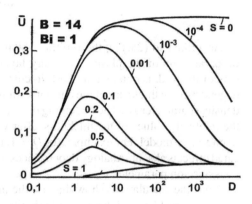

Fig. 25.6 Two-layer electrode voltage as function on dimensionless current density D with thickness fraction s as parameter, Bi = 1, $B = 14$

$= 0$ and $s = 1$ correspond to calculation in one-layer approximation. The mentioned dependences are monotonic functions of discharge current density.

However, presence even very thin layer ($s = 10^{-4}$) with higher conductivity remove the saturated part of the volt–current characteristic and to arise a branch with a dropping dependence. With increasing of parameter s, in the beginning, the dropped fraction increase and then the potential drop decreases at the constant current on the increasing branch of the curve. The obtained characteristic in double layer approximation is due to sharp electrical conductivity increase in the current terminal layer. The higher limit of the electrode voltage with current increasing can be expressed as

$$\bar{u} = \left(\frac{2\sigma_{02}\lambda_1}{B_1\sigma_{01}\lambda_2}\right)^{1/2} \tag{25.17}$$

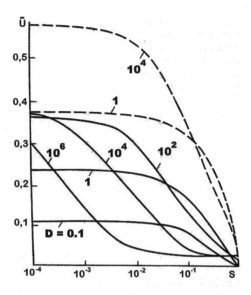

Fig. 25.7 Electrode voltage dependence on thickness of the current terminal, Bi = 1, B = 14-solid curves, B = 6-doted curves

According to (25.17), the limiting voltage of the electrode does not dependent on parameters of the plasma boundary layer. With B_2 decrease, the more expression of the peak is obtained in the volt–current characteristic. The electrode voltage dependence on thickness of the current terminal is shown in Fig. 25.7 at fixed current density (parameter D). The beginning of saturation in the dependences on s indicates the relatively values of the thickness, after which the calculated results correspond to one-layer model. For example, for $D = 1$, the $s = 0.01$, and for $D = 10^6$, the indicated effect is absent. Analogy dependences are calculated for the electrode electrical resistance on parameter s.

Thus, the calculations show that use the materials with higher electrical conductivity in the cold part of the electrode significantly influence on the thermal and electrical electrode parameters. This influence is ambiguously. At one hand, increase of the thickness fraction with large electrical conductivity cause significant decrease of the temperature gradient in the cooled layer of the electrode and, consequently, decrease of the heat losses. When the fraction s increase, the electrode voltage decreases. Therefore, it is preferable to increase contribution of the current terminal layer. At another hand, strong increase of parameter s, the front surface temperature is significantly decreasing and decreases the current density, at which a dropped branch appeared in the volt–current characteristic. This is a difficult case because the role of plasma boundary layer increases by increase of the heat flux to the electrode surface and arise conditions for overheating instability.

Analysis of the current instability at the branch of dropping characteristic at the contact of different materials occurs due to violation of the energy balance at the boundary layers. This current instability interpreted as a condition for a current constriction [14]. This effect can be appeared as heat breakdown similar to that occurred in the dielectrics [15] promoting an electrode destruction. According to the

25.2 Hot Ceramic Electrodes. Overheating Instability

above calculation, the probability of the current instability reducing can be reached by decrease the thickness of the current terminal layer. Herewith, for values $s < 0.5$, the front surface temperature decreases weakly.

The study in 2D approximation [19] shows the influence of two-dimensional effects during the ceramic filling of the combined electrode under current and the role of the Joule energy release in its volume. When calculating the temperature field, the effects of two-dimensionality become especially important near the cold metallic framework of the combined electrode. The latter circumstance, as well as the contribution of the Joule energy dissipation, largely depends on the level of the wall cooling temperature. The temperature distribution in the metallic framework of the combined electrode as well the discharge parameters are determined by presence of a film dopant on the surface condensed from the plasma flow containing a doping [20] (see below).

25.2.4 Thermal and Volt–Current Characteristics for an Electrode-Plasma System

In the above double layer model of the plasma thermal characteristic is approximated by variation criterion Bi in accordance with heat change process at Newton law. To understand the correspondence between electrode and plasma parameters determined by thermal and electrical phenomena, the triple layer approximation should be studied [21]. In general, at the electrode-plasma interface a space charge layer can be considered. However, at the hot electrode, this influence can be assumed significantly lower than the complex heat flux from the hot plasma. The calculated schema is based on the model of turbulent layer, assumed local thermodynamic equilibrium and exponential electro-conductivity dependence on the plasma temperature considered by Khait [13, 14]. In essence, in order to study the conjugate combined electrode-plasma problem, the above double layer mathematical formulation should be added by similar equations described the plasma processes (Fig. 25.3).

For $x_{0p} = x - h$ equation of plasma heat conduction is

$$\frac{\partial}{\partial x_{0pl}}\left[(\lambda_{Tp} + \lambda_{Tr})\frac{\partial T}{\partial x_{0pl}}\right] + \frac{j^2}{\sigma_{pl}(T)} = 0, \quad \sigma_{pl} = A_{0pl}\mathrm{Exp}(B_{0pl}T) \quad (25.18)$$

Using Prandtl hypothesis about turbulent viscosity and universal logarithm profile of velocity in the boundary layer according to Khait [13, 14], for study the stability of electric discharge in a dense plasma (25.18) can be converted to form

$$\frac{\partial}{\partial x_{0pl}}\left(1 + \frac{x_{0pl}}{\delta_0}\right)\frac{\partial T}{\partial x_{0pl}} + \frac{j^2}{\lambda_{pl}\sigma_{pl}(T)} = 0 \quad (25.19)$$

Introduce for plasma region dimensionless coordinate $\xi_{tr} = \text{Ln}(1 + k_p \xi_0)$, and temperature $\theta = \theta + \xi_{tr}/B_{pl}$, where $k_p = \delta/\delta_0$, $\xi_0 = x_{0pl}/\delta$, and δ are the thickness of plasma boundary layer, determined by "1/7" profile at absence of the current. In this case, (25.19) can be presented as

$$\frac{\partial^2 \theta_{tr}}{\partial \xi_{tr}^2} = -A_{pl} e^{-B_{pl}\theta_{tr}}, \quad A_{pl} = \frac{j^2 \delta^2}{\lambda_{pl} \sigma_{0pl}(T_1 - T_0)}, \quad \sigma_{0pl} = \sigma_{pl}(T_0) \quad (25.20)$$

Boundary conditions:

$$\theta|_{\xi_0 = 0} = \theta_w \text{ and } q_{pl} = q_w, \quad \theta|_{\xi_0 = 1} = 1 \quad (25.21)$$

According to (25.20) and (25.21), the plasma dimensionless temperature distribution is:

For $C_{pl} < 0$ $\theta(\xi_0) = \dfrac{2}{B_{pl}} \text{Ln}\left\{ \sqrt{F_{pl}} \sin\left[\dfrac{B_{pl}\sqrt{C_{pl}}}{2} \text{Ln}(k_0) + \arcsin\left(\sqrt{F^{-1} M_{pl}}\right) \right] \right\}$

For $C_{pl} = 0$ $\theta(\xi_0) = \dfrac{2}{B} \text{Ln}\left[k_0^{-0.5}\left(M_{pl}^{0.5} + \dfrac{D_{pl} B_{pl}}{2} \text{Ln}\dfrac{k_0}{2} \right) \right]$

For $C_{pl} > 0$ $\theta(\xi_0) = \dfrac{2}{B_{pl}} \text{Ln}\left\{ 0.5\sqrt{M_{pl}}\left[k_0^{2\Psi_{pl}}(1 + C_{pl}) + (1 - C_{pl}) \right] k_0^{-(1+2\Psi_{pl})/2} \right\}$

(25.22)

where some parameters in (25.22) are defined as:

$$C_{pl} = \left[\frac{\lambda}{\lambda_{pl} s_0} Bi(1 - \theta_w) + B_{pl}^{-1} \right]^2 - D_{pl}^2 \text{Exp}(-B_{pl}\theta_w)$$

$$B_{pl} = B_{0pl}(T_1 - T_0), \quad D_{pl}^2 = \frac{2 A_{pl}}{B_{pl}}, \quad k_0 = 1 + k\xi_0 \cdot \Psi_{pl} = \frac{B_{pl}\sqrt{C_{pl}}}{2}$$

$$F_{pl} = \frac{2 A_{pl}}{B_{pl} C_{pl}} = \frac{D_{pl}^2}{C_{pl}}, \quad s_0 = \frac{h}{\delta_0}, \quad G_{pl} = \left[1 + \frac{D_{pl}^2}{C_{pl}} \text{Exp}(-B_{pl}\theta_w) \right]^{1/2} \quad (25.23)$$

Here, the heat flux to the electrode from the plasma is expressed as above through criterion Bi, which now determined by the common solution of system (25.12), (25.13), (25.22), and (25.23). The equations for electrical parameters are similar to that presented in (25.15), (25.16). The calculations were provided for parameters indicated for double layer calculations and for $s = 0.1$. Dependence $\sigma_{pl}(T)$ was taken in form presented by Khait [13, 14]. Figure 25.8 shows the temperature distribution in triple layer region. It can be seen that significant change of the plasma temperature profile $T(\xi_0)$ occurred at relatively high volume heat source in the electrode body ($D > 10^3$). However, the temperature distribution in the electrode body practically is not changed in comparison with the result obtained in double layer approximation.

25.2 Hot Ceramic Electrodes. Overheating Instability

Fig. 25.8 Temperature distribution of triple layer electrode-plasma system with dimensionless current density as parameter

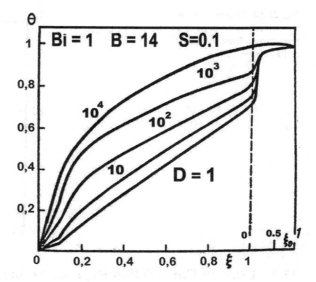

This result indicates that the electrode thermal regime is determined only by its electrical properties. At larger current density ($D \leq 10^4$) corresponding to the measured values, an interesting effect was found, namely, a temperature maximum arises in the plasma layer. The volt–current characteristic of electrode-plasma system is presented in Fig. 25.9, which is similar to that obtained according to the double layer model (doted curve). An interesting dependence of calculated parameter Bi on D is presented in Fig. 25.10. The sharp increase of Bi at $D = 10^3 - 10^4$ indicate on the unlimited rise of the heat flux from the plasma to the electrode surface due to Joule energy dissipation and on possible electrode destruction.

Thus, the developed complex model of electrode-plasma phenomena allows (using tabulated material properties) not only calculate the parameter distribution in the different layers of the system, the intensity of volume heat generation of electrically load electrode, but indicate a local region (current terminal, current collector, or plasma regions) possible current instability development, due to overheating phenomenon.

Fig. 25.9 Volt–current characteristic of electrode-plasma system with plasma processes (solid curve) and without the processes (dotted curve) for $B = 14$, Bio $= 1$, and $s = 0.1$

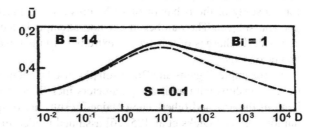

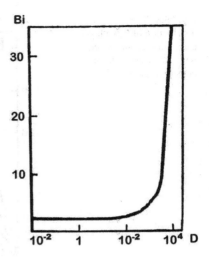

Fig. 25.10 Dependence of parameter Bio on dimension less current density D for $B = 14$ and $s = 0, 1$

25.3 Current Constriction Regime. Arcing at Spot Mode Under Plasma Flow with a Dopant

Relatively new phenomenon—arise of cathode spot at contact of the electrode with a plasma flow contained light-ionization dopant. The specifics of the condition consisted of dopant deposition and thin film formation determined the spot properties. This phenomenon is described in this section based on the developed models [22, 23].

25.3.1 The Subject and Specific Condition of a Discharge

A detailed study of the observed luminescing regions on metallic cathodes of a constricted discharge under plasma flow consisted dopant shows that behavior of these formations (named also cathode spots) differs from analogous formations arising under vacuum or under a noble gas. The difference is related to character of spot motion, of spot current, lifetime and condition of their appearances. Therefore, a study of the behavior of such spots arising under doped plasma is of independent subject. The number of existing attempts in the literature described of the constrict discharges of this type is extremely limited and, as a rule, are incomplete and erroneous.

For example, Agams and Robinson [24] used a not mathematically closed system of relationships with arbitrary parameters including spot size, the angle of plasma cone of the arc, and other. The obtained results are of qualitative character. These authors, as well Dicks et al. [25, 26] assumed model, in which the spot sustains its existence at the expense of vaporization of the material of the electrode, although,

25.3 Current Constriction Regime. Arcing at Spot Mode ...

as it is follows from experiment, such an assertion is unfounded. Below, we describe the cathode spot models developed for discharge with cathode at metallic electrodes taking into account that they arise under dopant plasma flow. It was used the data of actual experimental investigations, reported in references, which can be found in mentioned works [22, 23]. Let us briefly consider the data. The cathode spots on copper in a flow of ionized gas with potassium additive have a velocity about 10 – 10^2 cm/s, radius $r_s \sim 0.01$ cm, a spot current of $I \sim 10$ A, lifetime >0.1 ms and a heat flux is taken off to the electrode with passage of the current $Q_T \sim 20 \times I$ in Watt.

The cathode erosion rate was not measured. However, it was evaluated using an indirect empirical data. The experiment shows that at the moment, when the current is switched on, the original layer of additive on the electrode, with a thickness of $d \approx 0.1$ cm, is rapidly (~1 min) cleaned off by the discharge, which subsequently burn on the "clean" surface. When the current is switched off, the above-mentioned layer of additive redeposits in a few minutes. Here, according to the data, erosion of the copper itself is insignificant and lies within the limits $(1-5) \times 10^{-6}$g/C. Therefore, the spots vaporize mainly the film of previously deposited additive (a potassium compound). In a first approximation, the mass flow rate from the spots can be evaluated from the condition that a volume of additive Sd (S is the area of the surface of the electrode on which the spots burning) with the density ρ is vaporized in time t. From this $G = Sd\rho/Nt$ (N is the number of spots). Taking into consideration that the diameter of the electrode is equal to 40 mm, $d \sim 0.1$ mm, $t \sim 100$ s, $N \sim 10$, the obtained value of $G \sim 3 \times 10^{-3}$g/s. A more exact calculation can be made using the analogy between heat and mass transfer. Such estimation shows $G = (6-30) \times 10^{-4}$g/s depending on number of spots 2–10. This result confirms the mentioned above estimation about considerably greater erosion of the potassium compounds in the spots in comparison with erosion rate of the pure copper.

On the other hand, an evaluation of the temperature arising at the electrode under the action of the heat flux arriving from an arc spot can serve as a confirmation of the smallness of the erosion of copper. Under the assumption that the electrode is a semi-infinite body, the calculated temperature of the spot is ~1000 K, for a moderate heat flux Q_T and for its observed size under fully established conditions. The pressure of copper vapor at such a temperature is negligibly small and it is not surprising that there practically no erosion of copper. Thus, it is found that copper does not participate in the formation of charge carries of the near-cathode plasma of the arc spot and, consequently, it is not possible to explain the experimental data based on Cu vaporization model [26]. In addition, these data are also not explained by the transfer of current at the spots due to the density of potassium ions in the flow, since its value is found to be too low (10^{16} cm^{-3}).

25.3.2 Physical Model of Cathode Spot Arising Under Dopant Plasma Flow

An attempt is made to elucidate the mechanism of the burning of cathode spots taking account of the above analysis of the participation of the additive in the burning of the spots, on the basis of theory describing the corresponding processes at film cathode in a vacuum [27]. In accordance with this model, at a film cathode there can exist a heat spot, whose rate of displacement is determined by the rate of vaporization of the material beneath it. It has been shown that the spot radius is close to the width of the autograph that it leaves. Let us make the evaluations needed to formulate the model assuming that after the removal of the originally deposited thick layer of additive d, on the electrode there remains a thin layer of it, δ, which determines the further existence of the spots. According to model [27], the thickness of this layer is equal to $\delta = G/2r_s v_s \rho$. Substituting into this expression the value of the mentioned above quantities, we obtain a thickness of the thin film on the order of $\delta = 10^{-5} - 10^{-3}$ cm. The relatively high cathode temperature in the region of a spot (>1000 K) leads to a situation in which the potassium is decomposing state, and the transfer of the current takes place due to potassium ions, because the corresponding atoms have the smallest ionization potential.

Following [28] it can be evaluated that the density of potassium ions at an arc spot from the expression for the random component of the ionic current as: $n_i = 3 \times 10^{14} j(1-s) = 3 \times 10^{18}(1-s)$ cm^{-3}, where s is the electron current fraction, j (A/cm^2) is the current density. Thus, with a degree of ionization of the plasma of the spot $\alpha \sim 0.1$ [28], the total density of the heavy particles is found to be on the order of 10^{19} cm^{-3}. The presence of a high density of atoms makes it possible, in what follows, to use an analysis of the equations for the near-cathode plasma at a spot in accordance with those presented by model of Ch16. Here, a satisfaction of the condition of the smallness of the rate of outflow of the cathode vapor, $10^5 \times \alpha G/4A_i j) \ll 1$ (A_i is the atomic weight), is also required. With the conditions under consideration, this ratio is $\sim 10^{-3}$ and, consequently, fulfilled.

Now, the spot model can be formulated as follow. A cathode spot, arising at the electrodes under the condition of MHD ionized doped gas flow, burns because of the vaporization of a thin layer of potassium compounds deposited from the flow. Its displacement to a new location is determined by the time of vaporization of the layer of thin film right down the surface of the solid electrode. The film is heated by the bombardment of ions, which transmit to it their potential and kinetic energy. Cooling of the film takes place mainly by it thermal conductivity, by vaporization of the copper material and by the emission of electrons. An evaluation of the relative value of the temperature drop ε over the thickness of the film for characteristic values of the specific heat flux q, the temperature T of the spot, and the specific thermal conductivity λ_f of the film material shows that ε is not great, of order of 0.1–0.01. The work function is assumed to be work function of the potassium. This assumption is controlled by a change in the value of φ on the results of the solution. The emission electrons, passing through the region of the space charge formed by ions at the cathode, heat up the near-cathode plasma, in which the generation of new ions takes place due to thermal ionization.

25.3.3 Specifics of System of Equations. Calculations

In general, the parameters of the model can be calculated by the system of equation described in Chap. 16, modified in the region of the near-cathode plasma for condition of dopant plasma flow. As a first approximation [22], an analysis of experimental data and corresponding calculation of the arc base parameters was conducted for copper considering the equations for processes for the cathode and the space charge region immediately around it including cathode energy balance and ion flux to the cathode. The arc is maintained as a result of the evaporation of a thin film of impurity deposited at the cathode surface. The results show good agreement between calculated spot radius and that obtained experimentally and presence of relatively high plasma density (10^{19} cm^{-3}) at high degree of ionization. A more complete analysis was provided taking into account the processes occurring in the near-cathode plasma [23]. Corresponding estimation shows that the spot parameters such as the fraction of the electron current and the cathode temperature sensitive to the self-consistent description of the plasma and cathode processes. The system of equation (Chap. 16) was solved for given spot current and measured effective voltage determined the heat flux to the cathode (instead of the cathode potential drop, which here is the calculated parameter).

In using the system of equation for conditions at coated cathode, it is important to correct describe the heat losses by heat conduction in the energy balance for cathode. The specific of these losses depends mainly on the thermophysical properties and thickness of the film, mainly on the thermophysical properties and thickness of the film, and also on the substrate [26]. For electrodes of thermal conductivity comparable with that of the film, it is necessary, in the case of thick films, to take into account the temperature distribution both in the bulk of the cathode and over the film thickness. For considered arc according to the above estimations, the film at the electrode surface is so thin (<1 μm) that the temperature drops over its thickness may be neglect. In this case, the energy balance equation at the cathode may by take analogously to that balance for a bulk cathode (Chap. 16), in which the term taking into account the heat losses in the cathode is determined by the conductivity of the bulk substrate. The film temperature is then equal to the temperature of the electrode surface in the region of the arc spot.

The system of equation (Chap. 16) was solved for experimental region of spot current 5–15 A and varied heat flux in the electrode (effective voltage) [23]. The mean electrode surface temperature T_0 and the time t that the arc remains at some location were taken in account. Steady electrical and hydrodynamic parameters were assumed, and the temperature was calculated from non-steady equation. The results are shown in Table 25.1 and in Fig. 25.11.

Table 25.1 Cathode spotCathode spot parameters under external doped plasma [23]

Electrode material	Q_T, W/A	I, A	T_O, K	u_c, V	α	$W_e = 1.5 kT_e$ eV	s	T_s, K	$j \cdot 10^4$ A/cm^2	$E \times 10^6$, V/cm	(n_a+n_i), 10^{20} cm^{-3}	$n_i \cdot 10^{18}$, cm^{-3}
Steady-state process												
Copper	20	5	300	23	0,08	0,81	0,216	1460	4,1	5,6	1,1	8,8
			1000	22,8	0,013	0,6	0,200	1570	0,98	2,8	1,65	2,1
			1400	22,7	0,001	0,4	0,188	1520	0,044	0,5	1,36	0,1
		15	300	22,8	0,023	0,65	0,206	1540	1,55	3,5	1,5	3,3
			600	22,8	0,012	0,59	0,201	1570	0,96	2,7	1,65	2
			1000	22,8	0,004	0,51	0,196	1560	0,32	1,6	1,6	0,7
		60	300	22,8	0,006	0,52	0,197	1550	0,41	1,8	1,54	0,88
	15	60	300	17,8	0,008	0,56	0,244	1620	0,79	2,2	1,9	1,6
		5	300	13,3	0,27	1,1	0,366	1450	16	8,6	1,04	28
			800	12,9	0,08	0,83	0,342	1580	7,26	5,9	1,7	13
			1200	12,8	0,02	0,66	0,324	1650	2,41	3,4	2,13	4,24
		15	300	13	0,068	0,82	0,34	1580	6,6	5,6	1,7	11,6
			800	12,8	0,024	0,68	0,327	1650	2,9	3,8	2,1	5,1
Stainless steel	20	10	300	22,7	$2,7 \cdot 10^{-4}$	0,356	0,184	1460	0,013	0,32	1,05	0,3
1X18H10T	10	10	300	12,7	$5,1 \cdot 10^{-4}$	0,392	0,296	1620	0,053	0,52	1,93	0
Non-stationary process, $Q_T = 20$ W/A												
t, s												
10^{-3}	Copper	5	300	22	0,09	0,83	0,218	1460	4,5	5,9	1,08	9,8
			1000	22,8	0,016	0,615	0,203	1560	1,16	3	1,56	2,5
10^{-4}		5	300	22,8	0,07	0,544	0,198	1570	0,55	2,1	1,7	1,2

(continued)

25.3 Current Constriction Regime. Arcing at Spot Mode ...

Table 25.1 (continued)

Electrode material	Q_T, W/A	I, A	T_0, K	u_c, V	α	$W_e = 1.5 kT_e$ eV	s	T_s, K	$j \cdot 10^4$ A/cm^2	$E \times 10^6$, V/cm	(n_a+n_i), 10^{20}, cm^{-3}	$n_i \cdot 10^{18}$, cm^{-3}
Stainless steel 1X18H10T	10^{-2}	5	1200	22,8	0,002	0,457	0,192	1550	0,15	1,1	1,5	0,32
			300	22,7	0,001	0,42	0,189	1520	0,067	0,72	1,3	0,15
	10^{-4}		1300	22,7	0,0001	0,33	0,183	1440	0,006	0,21	0,97	0,014
		10	300	22,8	0,007	0,54	0,198	1580	0,54	2	1,7	1,16
			300	22,7	0,0025	0,47	0,193	1560	0,19	1,2	1,6	0,4
	10^{-3}	5	1300	22,7	0,0005	0,383	0,187	1530	0,031	0,5	1,4	0,07

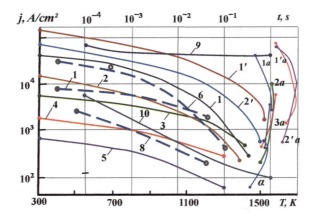

Fig. 25.11 Dependence of current density ($Q_T = 20$ W/A) on mean electrode temperature T0. Calculation (steady process), $I = 5$ A (1, 1′), 15 A (2, 2′). Curves 1′ and 2′ are calculated for $Q_T = 10$ W/A. Steel, $I = 5$ A, $t = 10^{-4}$s (3), 10^{-3} (4), 10^{-2} (5). Experiment [23]: (6), (7), (8). The relations between current density and spot temperature T_s is shown at right side by curves 1a–3a, 5a, 1′a, 2′a at parameters corresponding to curves 1–3, 5, 1′, and 2′. Dependence of current density with arc time t for $T_0 = 300$ K at (9) copper, I–5 A, (10) steel, $I = 10$ A [23]

The analysis of the calculated results reveals that, whereas the current density (curves 1–3, 5, 1′, 2′) and hence the spot are may change significantly with change in T_0, the spot temperature (curves 1a–3a, 5a, 1′a, 2′a) is a weak function of the electrode temperature in the range investigated. It is interesting that the solution can be obtained up to values close the spot temperature T_s. At $T_0 = T_s$, numerical analysis shows that there is no solution of the system and the spot radius $r_s \to \infty$. Physically, this means that only at $T_0 > T_s$ conditions for transition to distributed discharge mode may be realized even at cold metallic cathodes. Evidently, this may serve to explain the presence of arc spots, recorded experimentally in a wide range of electrode background temperature.

According to experiments in the absence of a current, the impurity layer melts at $T_0 \sim 1100$ K and at $T_0 = 1200$ K may be completely absent. In the latter case, the mechanism of an electric discharge on pure electrodes can be differ fundamentally from that outlined above. For example, it should be taken into account that in collisions of potassium atoms (ionization potential 4.3 eV) with the metallic electrodes of work function ≥ 4.5 eV (Cu, Ni, W, etc.) surface ionization may occur. From the experimental and theoretical data [29], it follows that for these metals in the temperature range 1000–2000 K the surface ionization of potassium $n_i/n_a \sim 0.8$, i.e., almost all the atoms are ionized. For the typical potassium density ($n_a \sim 10^{16}$ cm^{-3}), the ionic current density is found to be of order 10^2 A/cm^2, i.e., close to value of j at the electrodes at high temperature.

The calculation of cathode erosion rate shows a dependence on observed spot types. According to the experiment [22], one spot type is observed in incoming-flow region (a type due to the evaporation of the impurity film) and the other in the flow-breakaway region. In terms of external parameters, the second is distinguished from the first only by a slight increase in area and a low displacement velocity (~10 cm/s). Since this type of spot burns at the boundary between the pure surface and surface with a thick layer of impurity (0.1 cm), the mechanism underlying its activity may differ from that described by the faster spot. For example, the role of copper atoms in the near-cathode processes of low moving spot (the lifetime of which at one point is considerable larger than that for faster type) may be significantly larger, the erosion rate would then also increase. This conclusion considers with experimental finding that the main damage (even presence of craters) observed precisely in the flow-breakaway region.

Overall, it should be noted that the main features of the calculated parameter agree with experimental data. In particular, as in the experiment the spot area rises significantly when the mean electrode temperature is increased. Numerical analysis of the effect on the spot parameters of its time of existence, on the assumption that $Q_T = $ const shows that in contrast to copper, the arc discharge on stainless steel is significantly non-steady. The above calculations also demonstrate the effect of Q_T, the experimental value of which has considerable scatter, on the spot parameters. It follows that, generally speaking, the spot parameters are changed when Q_T is halved. The most important change is in the u_c, which also roughly halves. It is noteworthy that all the other parameters have no effect on u_c at all in the present model. Therefore, the accuracy with which u_c is determined in solving the system of equation (Chap. 16) is determined by the accuracy with which Q_T is specified.

25.4 Arc Column at Atmospheric Gas Pressure

Different published approaches described the constriction of arc column was analyzed. A mathematical model allowed to calculate the plasma column without use the "minimal principle" is presented, which take in account the physics of column formation and developed previously [30].

25.4.1 Analysis of the Existing Mathematical Approaches Based on "Channel Model"

In comparison with the near-electrode region, the plasma column in a high-pressure electrical arc was studied more extensively. The plasma column is mainly isothermal and in thermodynamic equilibrium, with Maxwellian distribution of the particle velocities and by absence of layers with sharp gradients of the parameters. Nevertheless, study even this simple subject meets some difficulties. The problem mainly related to the specifics of column origin as a result of discharge plasma constriction.

The main goal of arc plasma study consists in the development of a methodology to determine the radial distribution of the temperature $T(r)$, the electric field $E(r)$, and the radius of column contraction r_0. The existing approach to solve this problem was developed so-called channel model [7, 31, 32]. The mathematical formulation consists of equation of energy balance in differential form (Elenbaas-Heller equation), Ohm's law and Steenbeck's "minimal principle" [33] of the power for three unknown. As the minimal principle is only a condition, the column theory is not closed and therefore it was further developed in order to find some closed approach [34–38].

In general, the problem can be solved if dependence of the electrical conductivity on the temperature of the discharge plasma is known. This point discussed by Raizer [35] considering the channel model (constant conductivity inside of the channel) deriving corresponding additional relationships. Raizer [7, 35, 36] developed a physical model taking into account that the energy dissipated in the discharge draws mainly at the external boundary of the channel by heat conduction, which determined by a small difference of the temperatures at the axis and at the boundary of the channel. In order to determine of the channel radius, an expression was obtained following both from physics of the process and from the integral theorem for the vector of heat flux. Although the expression contains the arbitrary type of electrical conductivity dependence on the temperature $\sigma_{el}(T)$, nevertheless, at its use the condition of σ_{el} = const (channel approximation) was assumed in the region of main dissipation of the electrical energy. Leper [39] used variation method to calculate the characteristics of arc column. He showed that Raizer's approach allow to obtain results indicated on possibility solution the problem of the arc column without use of Steenbeck's minimal principle.

In contrast, Zhukov et al. [40] developed an integral approach by a condition that the total energy dissipated in the discharge is absorbed by the wall. An arbitrary temperature profile was chosen to satisfy the given conditions at the axis and at the wall. A volt–current characteristic of the discharge was determined using Ohm's law and the integral energy balance used the chosen temperature profile. Therefore, the solution significantly depends on type of the temperature profile. It was noted that the calculated results well agree with the experiment for parabolic temperature profile. Benilov [41] analyzed the characteristics of a cylindrical arc by method of matching of asymptotic expansions. He also obtained the solution without use the minimal principle. The mathematical formulation of this method is available in presence of a small parameter, indicated a ratio of twice value of the temperature at the discharge axis to the potential of ionization of the atoms of used gas, which not always satisfied to the requested condition.

The above analysis shows that models were developed primarily in order to solve the problem of arc column without use the minimal principle. Practically absent calculation of the temperature distribution. The considered thermal regime not takes in account a real dependence of the electrical conductivity in the discharge column. In some cases, the temperature is given as an integral profile or assumed as constant (channel model) . When the temperature distribution is calculated, the electrical conductivity is given in form of stepwise approximation [42]. A model described

25.4 Arc Column at Atmospheric Gas Pressure

below was developed without any assumption of channel approximation of the electrical conductivity in the discharge plasma column. Let us consider this model and the calculation approach based on the results of [30].

25.4.2 Mathematical Formulation Using Temperature Dependent Electrical Conductivity in the All Discharge Tube

A cylindrical arc is considered at relatively low arc current I when the energy loss by the radiation can be neglected and electrical conductivity is an exponential dependence on the plasma temperature. A not moving gas in the discharge tube is at atmosphere pressure in thermodynamic equilibrium. The energy balance is described by Elenbaas-Heller equation in following form [32]:

$$\frac{1}{r}\frac{d}{dr}rq + \sigma_{el}(T)E^2 = 0$$

$$q = -\lambda_T(T)\frac{dT}{dr}, \quad \sigma_{el}(T) = A_T \mathrm{Exp}\left(-\frac{u_i}{2T}\right) \tag{25.24}$$

where $\lambda_T(T)$ is the dependent on the temperature, A_T is the conductivity constant, r is the radial coordinate of a cross section of the discharge in the tube with radius R, u_i is the ionization potential and the wall temperature of T_w. As $rotE = 0$, the electrical field E not depends on r.

Boundary conditions:

$$q(0) = 0, \quad T(R) = T_w \tag{25.25}$$

Ohm law:

$$I = 2\pi E \int_0^R \sigma_{el}(r) r\, dr \tag{25.26}$$

Energy dissipation in the arc volume:

$$W = IE \tag{25.27}$$

Let us introduce a potential of heat flux

$$S_p = \int_0^R \lambda_T(T) dT$$

Taking into account function $\sigma(S_p)$ (25.24) is

$$\frac{1}{r}\frac{d}{dr}\left(r\frac{dS_p}{dr}\right) + \sigma_{el}(S_p)E^2 = 0, \quad \varphi(T) = \frac{u_i}{2T}, \quad \sigma_{el}(S_p) = A_T\text{Exp}[\varphi(S_p)] \quad (25.28)$$

The strong dependence of $\sigma_{el}(S_p)$ leads to presence of the main fraction of the current near the discharge axis [7]. Therefore, it is convenient to divide into two regions-internal regions (main part of current) with radius r_0 and external region $r_0 < r < R$, in which the heat flux mainly transferred to the wall. In this case, the problem can be solved in the mentioned regions taking into account matching conditions of continuity of the temperature and heat flux on the boundary r_0 between the regions.

Equation for external region

$$\frac{d}{dr}\left(r\frac{dS_p}{dr}\right) = 0, \quad S_p(R) = S_w \quad (25.29)$$

The solution of (25.29):

$$S_p = S_w + CLn\left(\frac{R}{r}\right) \quad (25.30)$$

C is the constant determined by matching condition. In order to obtain the solution for internal region, let us expand the exponential function of the conductivity on gas temperature in a series and take it first term. Then, (25.28) can be presented in form:

$$\frac{1}{r}\frac{d}{dr}\left(r\frac{dS_p}{dr}\right) + \sigma_{el0}E^2\text{Exp}\left[\left.\frac{d\varphi(S_p)}{dS_p}\right|_{S_p=S_{p0}}(S_p - S_0)\right] = 0, \quad \sigma_{el0} = A_T\text{Exp}[\varphi(S_0)] \quad (25.31)$$

where S_0 is the potential of heat flux, which corresponds to the temperature T_0 at the discharge axis ($r = 0$). Note that the applicability of expansion of the exponential function is limited by condition of small temperature drop ΔT in comparison with the temperature T_0 at the discharge axis.

Now for simplicity find of the solution, let us introduce a function

$$\theta = \left.\frac{d\varphi(S_p)}{dS_p}\right|_{S_p=S_{p0}}(S_p - S_0)$$

And take in account that

$$\left.\frac{d\varphi(S_p)}{dS_p}\right|_{S_p=S_{p0}} = \left.\frac{d\varphi}{dT}\frac{dT}{dS_p}\right|_{S_p=S_{p0}} = \frac{u_i}{2\lambda_{T0}T_0^2}$$

25.4 Arc Column at Atmospheric Gas Pressure

$$\varphi(T) = \frac{u_i}{2T}, \quad \frac{dT}{dS_p} = \left(\frac{dS_p}{dT}\right)^{-1} = \lambda_T^{-1}(T), \quad \lambda_{T0} = \lambda_T(T_0)$$

Then, (25.31) can be written in dimensionless form

$$\frac{1}{\xi}\frac{d}{d\xi}\left(\xi\frac{d\theta}{d\xi}\right) + 8\text{Exp}(\theta) = 0, \quad \xi = \frac{r}{a_{ch}}, \quad a_{ch}^2 = \frac{16\lambda_{T0}T_0^2}{u_i\sigma_0 E^2} \quad (25.32)$$

For boundary condition $d\theta/d\xi = (0)$ at $\xi = 0$ the solution of (25.32) is

$$\theta = \text{Ln}\left(\frac{C_1 r^2}{a^2} + \frac{1}{C_1}\right)^{-2} \quad (25.33)$$

Taking into account that $C_1 = 1$ for $\theta(0) = \frac{u_i(S_p(0)-S_{p0})}{2\lambda_{T0}^2} = 0$ then (25.33) is

$$S_p = S_{p0} - \frac{Q}{4}\text{Ln}\left(1 + \frac{r^2}{a^2}\right), \quad Q = \frac{16\lambda_{T0}T_0^2}{u_i} \quad (25.34)$$

Using condition of equality of heat fluxes at $r = r_0$, for which the corresponding expression can be derived by differentiation of (25.30) and (25.34) for S_p, it can be obtained

$$C = \frac{Q}{2}\left(1 + \frac{r^2}{a^2}\right)^{-1}, \text{ then } S_p = S_{pw} + \frac{Q}{2}\frac{K_{ch}}{1+K_{ch}}\text{Ln}\frac{R}{r}, \quad K_{ch} = \frac{r^2}{a^2} \quad (25.35)$$

According to the condition of equality of the temperatures at $r = r_0$ it can be obtained:

$$S_{p0} = S_{pw} + \frac{Q}{4}\text{Ln}\left[(1+K_{ch})\left(\frac{R}{r}\right)^{\frac{K_{ch}}{1+K_{ch}}}\right] \quad (25.36)$$

The total electrical power loss due to heat conduction from internal region can be expressed as:

$$W_{in} = -2\pi\left(r\frac{dS_p}{dr}\right)_{r=r_0} = \pi Q\frac{K_{ch}}{1+K_{ch}} \quad (25.37)$$

Equations (25.30), (12 25.35)–(25.37) and Ohm's law is fully determine volt–current characteristics and temperature distribution in the discharge tube. The parameter a represents the channel radius, at which the dissipated electrical power transferred by the heat conduction process. This radius should be equal to the radius r_0, when the current totally concentrated in the channel and in this case $K_{ch} = 1$. This conclusion corresponds to the below results of calculations and it is consequence of the mentioned methods [7, 36, 37, 40] of thermal calculation of the arc when the function $\sigma_{el}(T)$, is known. It should be noted that by introducing a channel radius r_0, at which the solutions are matched is a conditioned approximation it is easy to obtain an analytical expressions. Therefore, the solution can be obtained with some accuracy. In reality, the plasma parameters across the discharge in the tube change smoothly up to the values at the wall. Using a numerical approach (parameter r_0 is absent), the accuracy of the solution is determined by an approximation schema of the equations in a differential form.

25.4.3 Results of Calculations

In order to study the applicability of the above analytical approach, the calculations were conducted for discharges in different gases, by given T_w and varying the parameter K_{ch}. For comparison, also (25.24) with boundary conditions (25.25) was solved numerically for all arc zones using modified method of Runge-Kutta. In last case, a Koshi problem was analyzed using experimental value of T_0 and condition of symmetry at the arc axis as boundary conditions [30]. The results are compared with experimental data and with other calculation [42]. The data of dependence $\lambda_T(T)$ [43, 44], dependence of $\sigma_{el}(T)$, and the effective potential ionization u_i [7] were taken from experiment.

Figure 25.12 shows the calculated volt–current characteristic of a discharge with different tube diameter $d_t = 2R$ in air ($d_t = 3$ mm), nitrogen ($d_t = 3$ mm), CO_2 ($d_t = 5$ mm), and Ar ($d_t = 4$ mm). It can be seen that good agreement exists between numerical and analytical results (solid lines) at $K_{ch} = 1$ and also with the experiment [30] (dash lines). Small deviation between calculated and experimental data at large arc current can be due to neglecting the radiation. The calculation shows that the

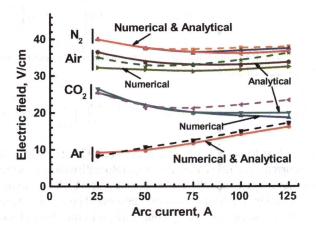

Fig. 25.12 Volt–current characteristics of a discharge in air, nitrogen, CO_2, and Ar. Experiment-dashed lines. Numerical and analytical calculations-solid lines (shown by captions). For nitrogen and argon, the numerical and analytical results are practically coincided one to other

25.4 Arc Column at Atmospheric Gas Pressure

channel model [42] relatively worse agree with experiment than the developed here analytical approach.

The calculated temperature distribution along the tube radius r/R is presented in Figs. 25.13, 25.14 and 25.15. A comparison of the analytical with numerical results is provided for $K_{ch} = 0.5$, 1, and 2. It can be seen that better coinciding between these results is obtained when $K_{ch} = 1$.

This fact confirms the above condition of $a \approx r_0$. An analysis of the temperature profile in the discharge plasma shows that calculation at $K_{ch} > 1$ leads to strong

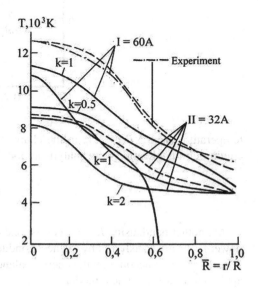

Fig. 25.13 Temperature profile for discharge in air ($d_t = 5$ mm): I–I = 60 A, II–I = 32 A. Solid lines calculation according to present model at $K_{ch} = 0.5$, 1, 2. Dash line-numerical calculation, dash-dot line experiment [30]

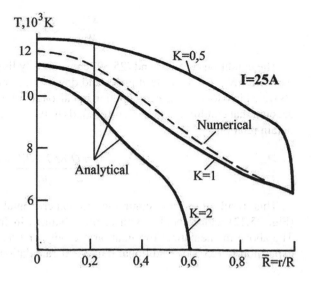

Fig. 25.14 Temperature profile for discharge in CO_2 ($d_t = 4$ mm) for $I = 25$ A. Solid lines calculation according to present model at $K_{ch} = 0.5$, 1, 2. Dash line-numerical calculation

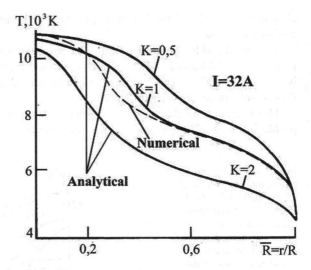

Fig. 25.15 Temperature profile for discharge in N_2 ($d_t = 5$ mm) for $I = 32$ A. Solid lines calculation according to present model at $K_{ch} = 0.5, 1, 2$. Dash line-numerical calculation

temperature drop ΔT or ΔS_p in the channel with r_0 and to violation of the base condition of applicability the analytical model

$$\frac{\Delta S_p}{S_{p0}} < 1 \qquad (25.38)$$

At another hand, using $K_{ch} < 1$ corresponds to process, at which not all dissipated power in the arc transferred by the heat conduction loss, i.e., some of power fraction ΔW not taken in account by the energy balance, i.e., violated the other condition of applicability the analytical model

$$\frac{\Delta W_{in}}{W_{in}} \ll 1 \qquad (25.39)$$

The conditions (25.38) and (25.39) consequently limited the chosen deviation of parameter K_{ch} from unit to it upper and down values at relatively low arc currents, above which not only heat conduction but also radiation will take place. Taking into account that condition $K_{ch} = 1$ corresponds to the experiment, from (25.37), we can obtain relation

$$W_{in} = \frac{\pi Q}{2} \frac{8\pi \lambda_{T0} T_0^2}{u_i}$$

This result does not contradict to experimental volt–current characteristic (Fig. 25.12) and coincides with results obtained in frame of channel model [7]. The above mathematical formulation use only first term of the series of exponent expansion in (25.31). Additional numerical calculations were provided in order to estimate the contribution of the subsequent terms. Such analysis shows neglecting role of these terms [30].

The influence of the wall temperature is conducted by calculation varying T_w, because the corresponding experimental data is limited. For example, for nitrogen at 60 A in discharge tube of $R = 0.25$ cm at $T_w = 0$ the calculated axis temperature is $T_0 = 11700$ K, for $T_w = 5000$ K, the value of $T_0 = 11680$ K, and for $T_w = 6000$ K the value of $T_0 = 11250$ K. These calculating data and analysis of the temperature distribution show that profile $T(r)$ in the discharge tube practically not changed up to $T_w < 6000$ K.

25.5 Discharges with "Anomalous" Electron Emission Current

Two types of the hot-cathode vacuum arc (HCVA) were observed experimentally: those a) with a hot refractory (non-consumable) cathode in a noble gas atmosphere and b) with consumable hot-cathode thermally insulated in a crucible.

In the first case, the specifics of discharge characteristics at refractory cathodes in ambient of inert gases (He, Ne, Ar, Kr, and Xe) for pressure ranges from 1 to 760 torr was investigated by Dorodnov et al. [45] with W rod of 1 mm diameter and by Anikeev [46] for rod (3–10 mm) and flat electrodes with thickness up to 2 mm. The arc duration was up to 60 s. The cathode temperature was measured by optical pyrometer. The main goal of the experiment was measurement of the current density as dependence on the cathode temperature, i.e., $j = f(T_c)$ for different gases and pressures. The cathode potential drop u_c was determined using electrical probe. The results show that the no cathode erosion was observed (spotless regime), dependence $j = f(T_c)$ was not linear, the value of j increases up to 400 A/cm^2 when T_c increases to 2600–3100 K. The cathode was destructed when a large Joule energy dissipation in the volume was reached. The volt–current characteristic weakly depends on the gas pressure and only on gas type. The value of u_c observed lower than ionization potential of gas atom. The main result was related to comparison of the experimental dependence $j = f(T_c)$ with that current density dependence $j_R(T)$ calculated according to thermionic electron emission of Richardson law. It was obtained that the ratio j/j_R is significantly large and varied from 10^4 to 10 depending on cathode temperature for current densities of 100–500 A/cm^2 and in considered gas pressures. The ratio is lower for higher gas pressure. The contradiction between measured arc current and calculations of the current based on known mechanisms of cathodic electron emission, the authors interpreted as effect of "anomalous" electron emission. This conclusion was discussed using measured data for heat conduction q_T, radiation q_r losses, heat flux from the plasma q_{pl} (not explained), and by Joule energy q_J. As a relation between these components is obtained in form of $q_{pl} > q_J > q_T + q_r$. the author claim (without justification of a mechanism) that q_{pl} (of unknown nature) should be compensated by an energy loss due to some current of electron emission [45]. The authors [47] further developed analogical study, calculating the thermal regime of a rod electrode indicating presence of a maximum temperature in the electrode body near the surface faced the arc plasma. An another also limited energy balance [48] was studied using Richardson-Schottky law with Mackeown electric field and natural assumption of presence of ion current from the plasma to the cathode in form

$$I_i(u_c + u_i - \varphi_{ef}) + q_{pl} = I_e \varphi_{ef}$$

where q_{pl} is the radiation flux from the plasma to the cathode, which was not explained how it is determined. Also, the cathode heat conduction loss q_T and the electron current due to ion charge neutralization at the surface in the total current were not considered. As a result, a small value of the ion current fraction for argon was obtained as $f_i = (\varphi_{ef} - q_{pl})/(u_c + u_i) = 0.2$. With this low value of f_i, the measured current density cannot be explained. Therefore, the authors again claimed that this current is result of an "anomalous" electron emission. Also, to explain stability of the discharge, they arbitrary assumed an effect of "electron" cooling of the cathode.

Nevertheless, Nemchinsky [49] and Zektser [50] suggested that current measured by Dorodnov et al. [45] can be maintained at the cathode surface completely by ion flux. Their approach is analogous to the model used in explaining cathode spots on cold cathodes. In the model [49], the cathode plasma is considered as isothermal, without accounting the heat conduction loss in the cathode energy balance for temperature T_c. The gas temperature T_p was determined from a relation described the ion current to the cathode and the plasma energy balance determined by the beam energy of electron emission j_{em} acquired in the cathode sheath u_c. The arc parameters T_c, T_p, and u_c in the cathode region were determined as dependence on current density for values of heavy particles corresponding to given two gas pressures 20 torr and 1 atm. It was concluded that the ion current increases with total current density increasing. However, no data about any contribution or some fraction of the ion current density was presented. In the model [50], the ion diffusion mechanism determines the ion flux to the cathode. The electron temperature T_e was determined according to Kerrebrock formula [51] by the elastic collisions between electron and heavy particles, instead of using the full plasma energy balance taking in account the non-elastic and other losses. The cathode energy balance and effect Schottky in the thermionic emission are not considered. The result presented a total current density as dependence on varied cathode temperature for Ar and Ne arc at different pressures. Also, the contribution of the ion current density in the total that value was not discussed. Therefore, the role of ion flux to the cathode from the plasma remains unclear.

In the second case, diffuse attachment to the cathode is obtained with current density in range of 10–100 A/cm^2 with a hot volatile cathode in vacuum [52] and arc current 40–400 A. For a Cr cathode, the arc voltage was 12–14 V, the cathode temperature was 1900–2200 K, the electron temperature was $T_e \approx 2$–4 eV and the electron density was about 10^{13} cm^{-3} [52]. There is also the main problem is "anomalous electron emission" because the experimental arc current significantly exceeds the calculations of the current based on known mechanisms of cathodic

25.5 Discharges with "Anomalous" Electron Emission Current

electron emission. Thus, the calculated thermionic emission is between factors of 10^2–10^5 lower than the arc current. This means that the electron current fraction is 10^{-2}–10^{-5}. Dorodnov [45, 52, 53] suggests that the discrepancy could be explained if the work function of the cathode material were lowered by approximately 1.5 eV, but offers no explanation about what might cause this. The observed experiments cannot be explained by role of individual electric field of ions [49, 54] and due to presence hot electrons in the metals (Chap. 15).

The main problem, however, is the possibility to explain the measured high electron temperature in case of assumption that the measured current provided by the ion flux. Baksht and Rybakov [55] theoretically analyzed of the energy relationships in the cathode region of a high-current atmospheric arc. They show that the energy required generating ions in an atmospheric arc with planar cathode and other energy losses in the cathode region cannot be made up by the energy supplied from distant regions of the plasma. As the electron beam is low and corresponding heating of the plasma electrons heating can be provided by Joule energy dissipation ($j^2 l_J / \sigma_{el}$) in the plasma volume due to the electrical current (Ohmic plasma). The characteristic length of l_J at which the plasma reaches of $T_e \approx 2$ eV can be determined from the plasma energy balance assuming the small energy losses by the returned plasma electrons to the cathode as well by elastic collisions between electrons and heavy particles (Chap. 16). In this case, the expression for energy balance is

$$j^2 l_J / \sigma_{el} = j_i (u_i + T_i) + 3 j T_e$$

from which, assuming only ion current $f_i = 1$ the length is

$$l_J = (u_i + 3T_e) \sigma_{el} / j.$$

Taking in account the mentioned above data for the arc at Cr cathode ($u_i = 6.8$ eV) the calculated length $l_J = 6$ cm, i.e., relatively large even for $j = 100$ A/cm^2.

To understand the mechanism of arc current supporting by the ion current, let us taken in account that the arc is operated with expendable Cr cathode and the presence of an expanding jet and it is highly ionized state established measuring the rate of cathode erosion by weighing and using the saturation ion current by Faraday cup. Considering this fact, the role of the ion current can be demonstrated in frame of gas dynamic model, considering for simplicity, only part of equations from the total system of equations [56].

The mathematical formulation includes the equation of heat conduction for a disk Cr cathode [see experiment 52], the equation for the cathode screening sheath, like as for Hg [56], cathode energy balance, equation for electron emission, and the equations for plasma momentum and for plasma energy of expanding the cathode jet. The solution was obtained for given current of 40 A, initial temperature of $T_0 = 1500$ K, and erosion rate of $G = 10^{-4}$ g/C as dependence on jet velocity.

The calculation shows that the electron current fraction s very low decreasing from 0.08 to 4×10^{-5} (Fig. 25.16), the arc voltage increases from 11 to 18 V and the cathode temperature decreases from 2800 to 2000 K, and degree of ionization

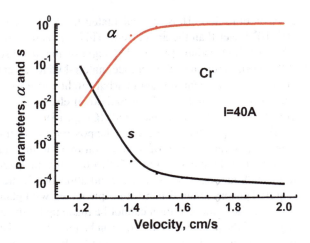

Fig. 25.16 electron current fraction s and degree of atom ionization α versus plasma jet velocity in a hot-cathode vacuum arc (Cr cathode)

increased up to 0.99 when the jet velocity increases from 1.2×10^6 to 2×10^6 cm/s. These calculated values are in good agreement with the experimental data mentioned above and the value of s indicates that cathode current continuity in the HCVA can be supported by ion flux. The predicted values of the ion density $\sim 10^{15}$ cm^{-3}, which necessary to account for current continuity at the cathode surface, are exceed measured the probe value of 10^{13} cm^3 [52]. Also, the calculated electron temperature is near the value of 1 eV, i.e., lower than that measured. The difference can be caused by relatively far probe location from the surface, where the density decreased. Thus, further investigation, in particular, of the plasma density distribution, is necessary for understanding of the mentioned parameters.

25.6 High-Current Arc Moving Between Parallel Electrodes. Rail Gun

One of the promising methods of high-speed acceleration of bodies is the electrodynamic method implemented in magneto-plasma accelerators (MPA) of rail gun type. The acceleration of a dielectric body placed in the MPA channel is carried out when the arc plasma is exposed to it in a vacuum. The efficiency of acceleration is limited not only by the strength of the rail gun and working body, but also depends on the arc plasma properties. Modeling of the arc phenomena appeared in specific conditions of a rail accelerator is the subject of this section based on our work [57].

25.6.1 Physics of High-Current Vacuum Arc in MPA. Plasma Properties

Artsimovich et al. [58] first developed a method (named as electromagnetic) of acceleration of a plasma produced in an arc between parallel electrodes. The mechanism is based on a force due to interaction of the self-magnetic field with the arc current. Brast and Sawle [59] considered such mechanism of the plasma acceleration in a rail gun to accelerating solid-state shells using the arc plasma as a pushing gas. Boxman [60, and this see Ref. 2 (1973)] studied experimentally the motion of the arc column sustained between Cu electrodes in a vacuum when driven with currents of 15–35 kA. The column velocities in the positive Amperian direction of (0.1–3.5) $\times 10^3$ m/s were observed in a self-magnetic field of ~50 G/kA.

An analysis of a high-current arc moving at 1 km/s and more between elongated planar parallel electrodes was conducted in a number of theoretical works [61–67]. They show that at current 10^5–10^6 A, the field strength is about 100 V/cm, plasma length in the propagation direction ~10 cm and transfer dimension about 1 cm. The models assumed that the plasma is single ionized and isothermal, while the gaskinetic pressure and longitudinal dimension were fixed and correspond to experiment [60]. The plasma parameters have been determined from measured data for resistivity using Ohm's law, in which the potential difference across the core of the plasma was taken as a certain part (~1/3) of the potential difference between the electrodes. The plasma temperature was determined of $(3–6) \times 10^4$ K.

Powel and Batteh 1984 [62] considered the stationary case with plasma parameters distributed along the discharge migration direction. The force-balance equation was combined with the energy equation, which incorporated heat transfer due to radiation from the bulk plasma. For the [60] typical conditions, the plasma length and maximum temperature were calculated correspondingly as about 10 cm and about 7×10^4 K on the assumption that the plasma mass was constant and equal to the mass of the metal foil used to initiate the discharge. In [63], the same researchers gave a two-dimensional description, where by comparison with [62] the maximum temperature was calculated of about 4×10^4 K. In works [64, 68], a stationary description was given in the lumped-parameter approximation. The solutions were presented as functions of the plasma mass and discharge current. For conditions close to those considered in [61–63], the temperature was calculated as $(3, 4) \times 10^4$ K.

In all the above studies, the temperature estimates were at the $(3–7) \times 10^4$ K level, and although they agree with the measured temperatures [61, 65], there should be indicate a number of factors to critique of the models [61–64, 67]. Firstly, in all experiments conducted for similar conditions, it has been found that the potential difference between the electrodes is constant independent on the arc current [69], being 150–300 V for 10^5–10^6 A. However, the above calculations give a voltage dependence on the discharge current that increases more rapidly than linearly. Secondly, the models involve the assumption that the plasma mass does not vary as it moves, whereas measurements [70] show that a satisfactory description can be obtained on the assumption that the mass increases.

One needs to mention also the later theoretical study of Zhukov et al. [71], which show that surrounding gas and electrode ablation mass can influence on acceleration of the plasma in a rail gun. However, the rate of electrode ablation was used as an arbitrary input parameter. Finally, none of those studies contains a description of the arc model or incorporates processes governing the current continuity at the plasma-electrode boundary or the arc maintenance mechanism during rapid displacement, or phenomena related to mass and heat transfer near the walls and to formation of the plasma channel. The main purpose of this study is to devise a model for a high-current discharge migrating between long plane-parallel electrodes based on coupling between the processes in the electrodes and in the plasma, as well as phenomena in the layers near the electrodes. The result in a self-consistent treatment enables one to determine the parameters for the moving plasma and for the working surfaces.

25.6.2 Model and System of Equations

Let us consider our developed model of moving arc between parallel electrodes. When the arc moves at a speed over 1 km/s, the time it spends within the characteristic length scale $l \sim 10$ cm is $\sim 10^{-4}$ s (Fig. 25.17). During that time, the contact with the plasma heats an electrode layer having thickness $\leq (at)^{0.5} \sim 10^{-2}$ cm (a is thermal diffusivity for the electrode material, in particular, copper), which is much less than the electrode thickness (~ 1 cm), so one can take the electrodes as infinitely extended in the thermal-wave direction in order to consider the thermal processes there.

An estimation shows that the thickness of the electromagnetic skin layer is $\sim(t/\sigma_{el}\mu_0)^{0.5}$ (σ_{el} is the electrical conductivity), which is much greater than the heated thickness $\sim(at)^{0.5}$. The volume density of the energy dissipated in an electrode because of the current is $q_e \leq \mu_0(I/b)^2/2\pi$ (b is electrode width), which for these conditions is much less than the density of the thermal energy deposited in the thermal skin layer in time t. Therefore, we neglect the bulk heat sources in describing the heat propagation in the electrode. The contact with the arc ($I \sim 10^5$ A) for $t \sim 10^{-4}$ s heats the surface to $T \sim (Iu_{ef}/bl)(at)^{0.5}/\lambda_T \approx 10^3$ K (λ_T is the thermal conductivity of copper and $u_{ef} \approx 10$–20 V is the voltage equivalent of the thermal effect from the arc on the electrode), which can produce melting and evaporation.

With these plasma parameters, kinetic processes occur so quickly (in about 10^{-8} s) that one can use the electrical and gasdynamic equations subject to thermodynamic

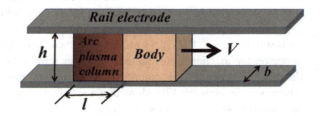

Fig. 25.17 Schematic configuration of arc plasma-body assembly in a magnetoplasma accelerator

25.6 High-Current Arc Moving Between Parallel Electrodes. Rail Gun

equilibrium there, while the processes in the electrode layers are described utilizing the concepts for relatively slowly moving erosion arcs (Chap. 7). The impedance changes as the arc moves, which in general affects the current. To simplify the description and analysis, we consider the current as set and constant.

The heat deposited in the gap is dissipated in melting and evaporating the surface and by conduction into the electrodes, as well as heating and ionizing the electrode material and producing radiation. The electrodes evaporate essentially in the closed arc column volume bounded by the electrodes and the insulating walls, namely by the inertial boundary on one side and the magnetic wall on the other. The balance between the flux of heavy particles from the electrodes and the reverse one to the channel wall from the vapor governs the heavy particle density. The heavy particle density can be determined via the effects of the ponderomotor force on the conducting plasma, although that complicates the treatment; therefore, to identify the major general trends, we neglect those forces. There are considerable differences in thermophysical parameters between the electrodes and the insulating walls, so we neglect the reverse particle fluxes from the latter and assume that the fluxes to the insulating walls correspond to the electrode consumption rate.

The current is maintained during migration by the electron emission from the heated metal; in that model, the task is to determine the discharge parameters such that the electrode heating provides for carrier reproduction in the amount necessary to maintain the set current at a high migration speed. This involves considering the coupled processes in the electrode, the arc column, and the electrode layers. For a given current, the unknown parameters are the cathode temperature T_s, anode temperature T_a, and plasma temperature T, along with the heavy particle density n_h, is the degree of ionization α, the current density j, the fraction of electron component s, the arc longitudinal dimension l, and the electrode erosion rate G in g/C, in conjunction with the potential drop across the plasma column u_{pl}.

The description is based on a system derived for an arc burning in the electrode vapor (Chap. 16, [57]) with the above conditions, particularly, the initial electrode temperature, as the transverse electrode size is finite in relation to the arc longitudinal dimension. The energy balances at the cathode and anode are represented (Chap. 16, [57]) by

Cathode:

$$Iu_{ef}^c + F\left[\sigma_{SB}(T^4 - T_c^4) + \Gamma^c(m\lambda_s + 2kT) - W(T_c)(\lambda_s + \frac{2kT_c}{m})\right] = Q_T^c$$

(25.40)

Anode:

$$Iu_{ef}^a + F\left[\sigma_{SB}(T^4 - T_a^4) + \Gamma^a(m\lambda_s + 2kT) - W(T_a)(\lambda_s + \frac{2kT_a}{m})\right] = Q_T^a$$

(25.41)

Here, m is atomic mass, σ_{sB} is the Stefan's constant, λ_s is the latent heat of evaporation, k is the Boltzmann's constant, Γ is the random heavy particle flux from the plasma to the electrodes, W are the mass fluxes from the electrodes, which are governed by the material properties, F is the total area of the arc contact with the electrode. Index "c" and "a" is related to the cathode and anode, respectively. The effective potential u_{ef} characterize flux to the electrodes from anode bombardment by electroflux and from cathode bombardment by ion flux, being dependent on the cathode potential drop u_c and the anode one u_a, as well as the fraction of the electron component in the current and effects associated with electron transition at the metal-plasma boundary. Q_T is the total heat flux due to conduction into the electrodes, which is determined from the two-dimensional non-stationary treatment of the heat propagation along the x-axis into the electrode from a normally distributed source moving along the surface in the y-direction:

$$\frac{dT}{\partial t} = a\Delta T - v\frac{\partial T}{\partial y} \tag{25.42}$$

Boundary condition:

$$-\lambda_T \frac{\partial T}{\partial x}\bigg|_{x=0} = q(y,l), \quad T|_{t=0} = T_0 \tag{25.43}$$

where $q(y, l)$ is the heat flux per unit area, which is uniformly distributed along the z-axis. Equation (3) has been solved and examined in [72], and we use those results, which were obtained in an analytical form, in examining (25.40) and (25.41).

We represent the plasma energy balance as

$$sj\left(u_c + \frac{2kT_c}{e}\right) + ju_{pl} = \sum_i \Gamma_i u_i + \xi j\frac{kT}{e} + \sigma_{SB}T^4 + q_c(T) + q_{bs}(T) \tag{25.44}$$

where ξ is a coefficient that corrects for the electron energy transport [73], while q_c and q_{bs} represent the convective energy transport by the heavy particles and the backscattered electrons (Chap. 16). The first term on the right in (25.44) describes the total energy loss by ionization, while u_i are the ionization potentials for order i,

$$\Gamma_i = n_i\sqrt{\frac{kT}{2\pi m}} \tag{25.45}$$

being the corresponding ion-flux densities and n_i the concentration of ions with charge i. The expression for Γ_i does not incorporate the diffusion-dependent component of the particle flux because the thickness of the non-equilibrium layer here does not exceed the ion mean free path. We put Γ_{inz} as the random particle flux to the insulating wall in order to write the heavy particle flux balance equation as

25.6 High-Current Arc Moving Between Parallel Electrodes. Rail Gun

$$\frac{1}{m}[W(T_c) + W(T_a)] = \Gamma^c + \Gamma^a + 2\Gamma_{\text{inz}} = n_T \sqrt{\frac{8kT}{\pi m}} \quad (25.46)$$

The total current is

$$I = Fj = \sqrt{\pi} jbl \quad (25.47)$$

Ohm's law gives the potential difference across the arc column:

$$u_{pl} = \frac{jh}{\sigma_{el}(T)} \quad (25.48)$$

where h is the distance between the electrodes. Equations (25.40)–(25.48) may be supplemented with a condition for the arc speed and by Saha's formulas together with an expression for the electro-erosion transport $G = 2\, m\Gamma_{\text{inz}}/j$ and the electron emission equation (Chap. 16) in order to obtain a complete system of equation for the above unknowns. The characteristic arc speeds are taken as the values typical of [61, 67, 69] as well estimated from equation of plasma motion (see below).

25.6.3 Numerical results

A numerical iterative method was used for copper electrodes at 10^4–10^6 A, while the conductivity was calculated from the fully ionized gas model [74]. Table 25.2 gives results for $u_c = 20$ V. The metal vapor is almost completely ionized, and with $T \sim 2$ eV

Table 25.2 Parameters of the arc in a magneto-plasma accelerator for different arc current [57]

Conductivity	$1 \times \sigma_{el}$					$2 \times \sigma_{el}$
Parameters	Electrode length 1 m				2 m	
Arc current, A	10^6	5×10^5	10^5	10^4	5×10^5	5×10^5
T_c K	3180	3185	3310	3495	3230	3410
T_a K	2700	2690	2690	2700	2775	2600
n_h, 10^{18} cm^{-3}	1.4	1.4	1.5	3.6	1.7	2.7
T, eV	1.87	1.89	1.9	2.1	2	2.15
G, mg/C	1.16	1.15	1.16	1.01	1.2	1.1
u_{pl}, V	200	200	214	520	218	174
j 10^4, A/cm^2	1.63	1.65	1.78	5.02	2	3.5
P_r, V	78	78	74	40	86	63.3
$P_{\text{sum-}c}$, V	106	105	102	65	115	92
$P_{\text{sum-}a}$, V	88	87	84	50	97	68
l, cm	34.6	17.1	3.2	0.1	14.1	4.6

consists mainly of singly charged ions. In that current range, the plasma parameters vary comparatively little, particularly, the temperature and current density, so the potential difference is almost independent of the current at about 200 V, which agrees well with measurements. The length of the plasma bunch is proportional to the current as the temperature is relatively constant. Below 10 kA, the energy flux to the electrode decreases substantially. In order to provide the necessary density of the charge particles, the solution of the system is modified by change of current densities and corresponding potential drop in the plasma column, although that variations of these quantities are not large.

The energy reaching the cathode from the plasma as the radiation P_r and the total energy P_{sum-c} together suggest that the contribution from the processes near the electrode cannot be neglected although it is only slightly dependent on the current. Anode processes are less important in the current transport because the anode is cooler. G has a weak dependence on the current in the range used and is about 10^{-3} g/C. The conductivity may be dependent on factors not considered here, so we performed calculations with variable conductivity. For example, when it was doubled relative to the Spitzer values (=$2\sigma_{el}$), the plasma length decreased so greatly that the increase in current density almost balanced the increase in conductivity, so the potential difference altered only slightly and remained in the 200 V range. G was substantially reduced because the eroded area fell as the current density increased.

Considering the measurements [61, 66, 67, 69, 70], it can be concluded that the model gives a satisfactory description of various parameters: arc scale $l \sim 1$–10 cm and current density $j \sim 10$ kA/cm^2, together with the plasma temperature $T \sim 2$ eV and the potential difference, which is only slightly dependent on the current and is about 200 V. This description gives a comparatively low heavy particle concentration, which is due to neglecting the ponderomotor force. Estimates show that the correction by accounting this force not merely increases n_h but also somewhat increases the electrode consumption. Although the formulation is restricted, the model is better than in [62–65, 68] in representing the mechanism that maintains the moving arc, while given satisfactory description of various observed [57, 66, 67, 69, 70] parameters such as u_{pl}, j and T, where in particular it shown that such a discharge can exist at relatively low heavy particle density.

25.6.4 Magneto-Plasma Acceleration of a Body. Equations and Calculation

The physical phenomena in the plasma dynamic discharge, occurred by conversion of the electric energy in a kinetic energy, are determine the acceleration mechanism in rail gun devices. Understanding this mechanism is important issue for obtaining high efficiency of the conversion process and maximal velocity at given power of the source of the magneto-plasma accelerator. The ponderomotor force acts on the arc causing plasma motion and the plasma accelerated the attached body. On one

25.6 High-Current Arc Moving Between Parallel Electrodes. Rail Gun

hand, the acceleration determined by electrical energy dissipated in the plasma. On the other hand, different reasons could limit the velocity of accelerating bodies in the accelerator. The phenomena breaking the acceleration process include the mechanical strength of the device, thermal loads inside it volume, instability of the discharge, the power system forming the current pulse. These reasons were detail discussed by Ostashev et al. [75].

In this section, we develop a calculating model which describe the arc characteristic related to the mechanism of it operation in vacuum, which can influence the acceleration parameters. Namely, the vacuum arc can be ignited and developed when a certain density of electrode vapor can be reached and the arc current can be supported. This means that with increase of the accelerator power and therefore with increase the arc current the plasma density should follow to the current value and to the current distribution. So, the plasma mass in the discharge increase with the current. The self-consistent processes of the plasma-electrode system determine the relation between indicated parameters. The magnitude of the plasma column mass should not be confused with the electrode erosion rate, which can influence in form of an energy loss or developed separately in case, when the erosion amount is not large. Below the physics of plasma-body acceleration is analyzed when the arc current increase with the source power and taking into account the self-consistent change of column mass. The mathematical formulation is reduced to consideration of momentum equations for the plasma and for a body.

Equations of momentum for plasma column:

$$\frac{d(\rho_{pl} v)}{dt} = -\nabla p + jB \rightarrow v\frac{\partial(\rho_{pl})}{\partial t} + \rho_{pl} v \frac{\partial(v)}{\partial x} = -\frac{\partial p}{\partial x} + jB \quad (25.49)$$

Equations of momentum for body

$$m_b \frac{dv}{dt} = pS_b \rightarrow v\frac{\partial \rho}{\partial t} + (m_b + m_{pl})v\frac{\partial v}{\partial x} = pS_b \quad (25.50)$$

Multiply both sides by a volume (hbl_{pl}, Fig. 25.17) of (25.49) is

$$v\frac{\partial(\rho_{pl})}{\partial t} + m_{pl} v \frac{\partial(v)}{\partial x} = hbl_{pl}\left(-\frac{\partial p}{\partial x} + jB\right) = -pS_b + \mu_0 I^2 \quad (25.51)$$

Here is assumed that $l_{pl} dp/dx = \Delta p \approx p$, i.e., the plasma pressure change occurred at the size l_{pl}. Considering steady-state case, after sum of (25.49) and (25.50) it is

$$(m_b + m_{pl})v\frac{\partial v}{\partial x} = \mu_0 I^2 \quad (25.52)$$

Let us determine mass of the plasma column m_{pl} in (25.50) using plasma pressure as $p = (1 + \alpha)n_T kThb$, and then from modified (25.50) can be derived m_{pl}:

Fig. 25.18 Influence of mass grows of the arc plasma with arc current increasing resulting in deviation from linearly dependence of the arc velocity increase with the current

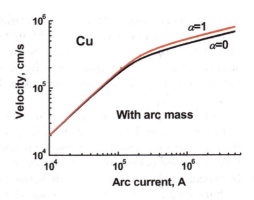

$$m_b \frac{\partial(v^2)}{2\partial x} = (1+\alpha)n_T kThb, \quad \text{as } m_{pl} = n_T h\frac{I}{j}, \quad \text{then } m_{pl} = \frac{m_b}{1+\alpha} \frac{v^2}{2x} \frac{m}{kTb} \frac{I}{j} \quad (25.53)$$

Substitute m_{pl} from (25.53) into (25.52) it can be obtained relation for plasma-body velocity in form:

$$v^4 + A_b v^2 + C = 0$$
$$B_b = \frac{2(1+\alpha)kTbx}{m}\frac{j}{I}, \quad C_b = \frac{4(1+\alpha)kTbj}{m_T m}\mu_0 I x^2 \quad (25.54)$$

The calculation result for body of 1 g and electrode length of $x = 100$ cm by (25.54) is presented in Fig. 25.18 using the j and T calculated in Sect. 25.6.3. It can be seen that in the beginning the plasma-body velocity increases linearly as it is follow from Newton law, i.e., proportional to the body mass. However, at current about 100 kA, the dependence declined from linearly and then it is approached to saturation.

The analysis shows that such dependence is caused by increase of the plasma mass that exceeds the mass of body. The plasma velocity reached value of 8×10^5 cm/s when the arc current increase to 8×10^6 A. In general, the above calculating saturated velocity can be corrected taking into account the complete model (with equations of Sect. 25.6.3 or the condition of some certain experiment. Although of this, it should be noted, that the experimental value of the velocity is reported as $(5–7) \times 10^5$ cm/s and in a separately cases as $(8–11) \times 10^5$ cm/s for body of 1 g [75].

25.7 Summary

Different arc types and plasma and electrode phenomena occurred in the technical devices. The phenomena depend on the device working regimes. In one of most extended type of electrical apparatus, the electrode wall is surrounded by the hot plasma flow. The discharge arises in the boundary layer at the wall containing a hot ceramic or water-cooled metallic electrodes. In order to effective working of the apparatus, a stable regime of the discharges is required for the two types of electrodes.

The overheating phenomena are due to strong dependence of the electrical conductivity on temperature $\sigma(T)$ as described, which triggers unstability of the current in the hot ceramic electrodes. For this case, the main problem occurs in region of the rear water-cooled electrode side due to increase of the resistivity and large Joule energy dissipation. Complex electrodes were constructed with layers using materials of different dependences of $\sigma(T)$ in order to reduce the probability of the overheating problem. The developed calculating approach for triple layer (included double layer electrode and boundary plasma layer) allows study the temperature distribution and the volt–current characteristic of such current carrying system. The numerical analysis of the characteristics is provided in a wide range of input parameters. As a result, the material properties in different electrode layers and the values of discharge current densities were determined as conditions, which can predict the stable state of the system.

The instability of discharge at the water-cooled metallic electrodes appeared in form of an arc with a number of cathode spots. The mechanism of existing of the spots were described in frame of a model taking into account the experimental condition for plasma flow contained a light ionized dopant. The model takes into account the assumption about presence of thin film of dopant material deposited on the cold cupper electrodes and uses the previously developed theory for film cathode (Chap. 16). The numerical analysis allows explaining the experimental data for these spot (size, lifetime, current etc.), which cardinally different from these data observed in a vacuum.

Another phenomenon is related to the century-old problem regarding the mechanism of electron emission from the cathode for different arc discharges. This problem was again discussed due to the experimental data for electrical current and temperature at refractory cathode for atmosphere pressure arcs as well for arcs with hot expendable cathodes at low gas pressure. The significant difference between the experimental dependence of current density $j = f(T_c)$ and that current density dependence $j_R(T)$ calculated according to thermionic electron emission of Richardson law (by factor varied from 10^4 to 10 depending on cathode temperature and current density) the authors interpreted as an effect of "anomalous" electron emission. The existing literature works attempt to explain the mentioned difference taking in account presence of an ion current instead the not understandable "anomalous" process. The analysis of these works show that these attempts contain free input parameters and assumptions indicating that the models are not without lacks. Finally, the authors not presented any data about the ion fraction in the total arc current. The presence of significant contribution of ion current was demonstrated using the gasdynamic model of cathode spot (Chap. 16) taking into account the experimental fact of plasma expanding in arcs burning with hot expendable cathodes at low gas pressure.

The last decades were commemorated by developing approaches described the long-time problem about the nature of plasma column constriction in atmospheric pressure arcs. The proposed approaches attempt to solve the problem without traditionally used Steenbeck's "minimal principle." In the present chapter, the proposed approaches were reviewed, and the weakness of different works was analyzed. Considering the specifics of such arc, we developed an analytical mathematical model including the nonlinear heat conduction equation with a volume source of Joule energy dissipation, Ohm's law for plasma column and relation for total arc power. The solution is searched for two separate regions. The model takes in account that in the internal region of radius r_0, the main fraction of arc current and of dissipated electrical energy are concentrated, in which the electrical conductivity exponentially depends on the temperature, (not constant) while in the external region take place the heat conduction process. The conjugation of the parameters at the boundary between two regions was provided at the radius r_0 determined by equality of all dissipated electrical energy to the energy loss by the heat conduction process to the tube wall in the second region. The temperature distribution and volt–current characteristics were calculated. These results of analytical solution were compared with the exact solution obtained numerically using finite-difference approximation of the equations, as well as with experimental data for arc burning in different gases. Good agreement between the mentioned data indicated that the developed analytical approach could be used to explain the phenomenon of constriction of the plasma column for atmosphere arc without consideration of the "channel" approach and of the "minimal principle."

Finally, a high-speed acceleration of bodies by the electrodynamic method in magneto-plasma accelerators (MPA) of rail gun type is discussed in frame of above-proposed theory in Chap. 16. This method was intensively developed recently due to presence the new tasks of physics, modern engineering, and technology and powerful power supply systems. In the experiments, the arc column scale, current density, the plasma temperature, and the potential difference at the plasma column were determined. At the same time, however, it was experimentally detected a presence some limiting plasma-body velocity in spite of significant increase the power of the supply system. In this chapter, in first step, the theoretical models of cathode and anode spots were used to understand the arc plasma column generation between two parallel rail electrodes. The calculation according to the formulated mathematical system of equations shows the possibility to explain the observed plasma column parameters. At the second step, the arc motion in the rail gun was studied taking into account the dependence of the plasma column mass on the power of the supply, i.e., on the arc current. The calculation shows that dependence of the velocity of the plasma-body assembly on arc current is deviated from linear character at some certain value of arc current, indication the influence of plasma mass increase. This dependence then approaches to a saturation indicating the experimental effect of limiting velocity presence at large arc currents. The analysis shows that the calculated velocity saturation is caused by increase of the plasma mass that is larger than the mass of body.

References

1. Dorodnov, A. M. (1974). Some applications of plasma accelerators in a technology. In A. I. Morosov (Ed.), *Physics and applications of plasma accelerators* (pp. 330–365). Minsk: Nauka & Technika Publisher.
2. Dorodnov, A. M., & Petrosov, B. A. (1981). Physical principals and types of technological vacuum plasma devices. *Soviet Physics - Technical Physics, 26,* 304–315.
3. Coombe, R. A. (1964). *Magnetohydrodinamic generation of electrical power.* London: Chapman and Hall.
4. Heywood, J. B., & Womack, G. J. (1969). *Open-cycle MHD power generation results of research carried out by members of the British MHD collaborative committee.* London: Pergamon Press.
5. Zhukov, M. F. (Ed.). (1993). *Low temperature plasma* (Vol. 11). Novosibirsk: Nauka.
6. Waymouth, I. J. (1971). *Electrical discharge lamps.* London: MIT Press.
7. Raizer, Yu P. (1991). *Gas discharge physics.* Berlin: Springer-Verlag.
8. Benilov, M. S. (2008). Understanding and modelling plasma–electrode interaction in high-pressure arc discharges: a review. *Journal of Physics. D Applied Physics, 41,* 144001.
9. Beilis, I. I., & Kirillov, V. V. (1986). The heat regime of ceramic elements of electrically loaded electrodes. *High Temperature, 24*(4), 616–621.
10. Beilis, I. I., & Kirillov, V. V. (1988). Investigation of the heat and electrical characteristic of two-layer ceramics electrodes. *High Temperature, 26*(2), 287–293.
11. Kaufman, W. (1900). Elektrodynamiche eigentumlichkeiten leitender gase. *Annalen der Physik, 2*(4), 158–178.
12. Kapzov N.A. *Electrical phenomena in gases and in vacuum.* Gostekhisdat. M. (1947), p. 500 (in Russian).
13. Khait, V. D. (1980). Stability of electric discharge in a dense plasma. *High Temperature, 18*(3), 386–392.
14. Khait, V. D. (1986). Theory of thermal contraction of the current at the anode in a turbulent flow of thermal plasma. *High Temperature, 24*(2), 149–157.
15. Skanavi, G. I. (1958). *Physics of dielectrics.* Moscow: Gosfizmatizdat.
16. Nedocpasov, A. V., & Khait, V. D. (1991). *Basic of physic processes in devices with low temperature plasma.* Moscow: Energoatomisdat. In Russian.
17. Sinkevich, O. A., & Stakhanov, I. P. (1991). *Physics of plasma.* Moscow: High School.
18. Zelikson, Yu M, Ivanov, A. B., Kirillov, V. V., Reshetov, E. P., & Flid, B. D. (1971). Electrode potential falls and resistance of zirconium oxide—base electrodes in a flow of ionized gas. *High Temperature, 9*(3), 425–429.
19. Beilis, I. I., Zalkind, V. I., Kirillov, V. V., & Zshigel, S. S. (1990). 2D analysis of Joule energy dissipation in a combined electrodes of MHDG. *High Temperature, 28*(6), 938–944.
20. Beilis, I. I. (1986). Thermal conditions for the metal framework of a combined MHDG electrode in a contracted discharge. *High Temperature, 24*(6), 881–889.
21. Beilis, I. I. (1991). Thermal and electrical characteristic of electrode-plasma system in an channel MHD generator. *High Temperature, 29*(2), 281–284.
22. Beilis, I. I., Zalkind, V. I., & Tikhotsky, A. S. (1977). Cathode spot on metallic electrodes under the conditions of the channel of an MHDG. *High Temperature, 15*(1), 131–135.
23. Beilis, I. I. (1977). The near cathode region of contracted discharge at MHDG metallic electrodes. *High Temperature, 15*(6), 1088–1094.
24. Adams, R. C., & Robinson, E. (1968). Processes on electrodes of a MHDG. *Proceedings of the IEEE, 56,* 1519–1535.
25. Dicks J.B., Wu, J.C.L., Crawoford, J., Chang, P., Stephens, J.W. ()1970. The performance of a family of diagonal conducting wall MHD open-cycle generators. In *Proceedings of XI Symposium on Engineering Aspects of MHD,* pp. 16–28.
26. Muchlhauser, J.W., Dicks, J.B. (1974). Arc spot and voltage losses in a hall generator. In: *Proceedings of XIV Symposium on Engineering Aspects of MHD,* pp. VIII.2.1–VIII.2.8.

27. Beilis, I. I., & Lyubimov, G. A. (1976). Theory of the arc spot on a film cathode. *Soviet Physics—Technical Physics, 21*(6), 698–703.
28. Beilis, I. I. (1974). Analysis of the cathode in a vacuum arc. *Soviet Physics—Technical Physics, 19*(2), 251–256.
29. McDaniel, E. W. (1964). *Collision phenomena in ionized gases.* London: Wiley.
30. Beilis, I. I., & Sevalnikov, A Yu. (1991). Column of an electrical arc of atmospheric pressure. *High Temperature, 29*(5), 669–675.
31. Engel, A., & Steenbeck, M. (1934). *Electrische gasentladungen ihre physic und technik.* Zweiter band: Berlin, Verlag von Julius Springer.
32. Finkelnburg, W., & Maecker, H. (1956). Electrische Bogen and thermishes plasma. In S. Flugge (Ed.), *Handbuch der, Physik* (Vol. 22, pp. 254–444). Berlin: Springer.
33. Steenbeck, M. (1932). Energetik der Gasentladungen. *Zeitschrift f. Physik, 33*(21), 809–815.
34. Peter, Th. (1956). Uber den Zusammenhang des Steenbeckschen minimumprinzips mit dem thermodynamischen prinzip der minimalen Entropieerzeugung. *Zeitschrift f. Physik, 144*, 612–631.
35. Raizer, Yu P. (1972). About the missing channel equation an arc model that replaces the minimum condition voltage. *High Temperature, 10*(6), 1152–1155.
36. Raizer, Yu P. (1979). On the channel arc model. *High Temperature, 17*(5), 1096–1098.
37. Khait, V. D. (1979). About the channel model of electric arcs and the principle of minimum power. *High Temperature, 17*(5), 1094–1095.
38. Rutkevich, I. M., & Sinkevich, O. A. (1980). Properties of nonstationary modes of Joule heating of a low temperature plasma. *High Temperature, 18*(1), 24–36.
39. Leper, D. N. (1973). Variation principle and Steenbeck's principle for a theory of cylindrical arc. *Soviet Physics—Technical Physics, 43*(7), 1501–1506.
40. Zhukov, M. F., Koroteev, A. S., & Uryukov, B. A. (1975). *Applied dynamics of a thermal plasma.* Novosibirsk: Nauka.
41. Benilov, M. S. (1986). Asymptotic calculation of the characteristics of a cylindrical arc. *High Temperature, 24*(41:N1), 45–51.
42. Zarudi, M. E. (1968). Methods for calculating the arc characteristics in a channel. *High Temperature, 6*(1), 35–43.
43. Asinovskii, E. I., & Kirillin, A. V. (1965). Experimental definition thermal conductivity of argon plasma. *High Temperature, 3*(5), 677–685.
44. Asinovskii, E. I., Drokhanova, E. V., Kirillin, V. A., & Lagar'kov, A. N. (1967). Experimental and theoretical study of the heat conductivity coefficient and total radiation in a nitrogen plasma. *High Temperature, 5*(5), 739–750.
45. Dorodnov, A. M., Kozlov, H. P., & Pomelov, Ya. A. (1971). Anomalously high emission of a thermionic cathode in an arc discharge in inert-gas media. *High Temperature, 9*(3), 442–445.
46. Anikeev, V. N. (1981). Investigation of thermal cathodes of an electrical arc in a low pressure inert gases. *Bulletin of the Academy of the Science Siberian branch, 3*(N4), 60–67.
47. Dorodnov, A. M., Kozlov, H. P., & Pomelov, Ya. A. (1973). on the effect of "electron" cooling of a thermionic arc cathode. *High Temperature, 11*(4), 724–727.
48. Dorodnov, A. M., Kozlov, H. P., & Pomelov, Ya. A. (1974). Arc regimes of a thermionic cathode with nomalously high current density. *High Temperature, 12*(1), 10–16.
49. Nemchinsky, V. (1974). On the problem of anomalous high emission ability of a non-vaporizing cathode in an arc discharge. *Zhurnal Tekhnicheskoi Fiziki, 44*(12), 2548–2550.
50. Zektser, M. P. (1975). On the problem of anomalous high emission ability of a thermo-cathode in an inert-gas arc discharge. *Teplofizika Vysokikh Temperatur, 13*(3), 491–494.
51. Kerrebrock, J. L. (1964). Nonequilibrium ionization due to electron heating. *AIAA Journal, 2*(N6), 1072–1080.
52. Vasin, A. I., Dorodnov, A. M., & Petrsov, V. A. (1979). Vacuum arc with distributed discharge on an expendable cathode. *Technical Physics Letters, 5*, 634–636.
53. Porotnikov, A. A., Petrosov, V. A., & Octrezov, I. A. (1974). Near electrode processes. In A. I. Morosov (Ed.), *Physics and applications of plasma accelerators* (pp. 239–261). Minsk: Nauka & Technika Publisher.

54. Dorodnov, A. M., Davydov, V. B., Kozyrev, A. V., & Pomelov, Ya. A. (1974). Effect of ion microfields on emission of an arc cathode. *Soviet Physics —Technical Physics, 19*(3), 390–393.
55. Baksht, F. G. (1994). Rybakov, The fraction of ion current at the cathode of an arc discharge. *Journal of Technical Physics, 39*(8), 769–772.
56. Beilis, I. I. (1990). The nature of an arc discharge with a mercury cathode in vacuum. *Journal of Technical Physics Letters, 16*(5), 390–391.
57. Beilis, I. I., & Ostashev, E. (1989). Model for a high current dischargemoving between parallel electrodes. *High Temperature, 27*(6), 817–821.
58. Artsimovich, L. A., Lukyanov, S Yu., Podgorny, I. M., & Chuvatin, S. A. (1957). Electrodynamic acceleration of plasma blob. *JETP, 33*(1), 3–8.
59. Brast, D. E., & Sawle, D. R. (1964). Study of a rail-type MHD hypervelocity projectile accelerator. *Proceedings VII Hypervelocity Impact Symposium, 1,* 187.
60. Boxman, R. L. (1977). High current vacuum arc column motion on rail electrodes. *Journal of Applied Physics, 48*(5), 1885–1889.
61. Roshleigh, S. C., & Marshall, M. A. (1978). Electromagnetic acceleration of macroparticles to high velocities. *Journal of Applied Physics, 49*(4), 2540–2542.
62. McNab, J. R. (1980). Electromagnetic macroparticle acceleration by a high pressure plasma. *Journal of Applied Physics, 51*(5), 2549–2551.
63. Powel, J. D., & Batteh, J. H. (1982). Arc dynamics in the rail gun. *IEEE Trans. on magn., 18*(1), 7–10.
64. Powel, J. D., & Batteh, J. H. (1983). Two-dimensional plasma model for the arc-driven rail gun. *Journal of Applied Physics, 54*(5), 2242–2254.
65. Powel, J. D., & Batteh, J. H. (1984). Analysis of plasma arcs in arc-driven rail guns. *IEEE Transactions on Magnetics, 20*(N2), 336–339.
66. Marshall, M. A. (1986). Structure of plasma armature of a railgun. *IEEE Transactions on Magnetics, 22*(N6), 1609–1612.
67. Kondratenko, M. M., Lebedev, E. F., Ostashev, B. E., Safonov, V. I., Fortov, B. I., & Ul'yanov, A. V. (1988). Experimental investigation of magnetoplasma acceleration of dielectric projectiles in a rail gun. *High Temperature, 26*(N1), 139–144.
68. D'yakov BB, B. B., & Reznikov, B. I. (1987). Computer-model of an electromagnetic accelerator. *High Temperature, 25*(N1), 128–136.
69. Clark, G. A., & Bedford, A. J. (1984). Performance results of a small-calibre electromagnetic launcher. *IEEE Transactions on Magnetics, 20*(2), 276–279.
70. Parker, J.V., Parsons,W.M., Cummings, C.E., Fox, W. (1985). Performance loss due to wall ablation in plasma armature railguns. In Proceedings of AIAA 18th Fluid dynamics and plasmadynamics and laser conference Cincinnati, Ohio, July, pp 1575–1584.
71. Zhukov, M. F., Reznikov, B. I., Kurakin, P. O., & Rosov, S. I. (2007). About influence of the gas density on motion of free plasma piston in channel of a rail gun. *Soviet Physics—Technical Physics, 77*(7), 43–49.
72. Beilis, I. I. (1979). Normal distributed heat source moving on lateral side of a thin semi-infinite plate. *Physics Chemistry of a Material Treatment, 13*(N4), 32–36.
73. Beilis, I. I., Lubimov, G. A., & Rakhovskii, V. I. (1972). Diffusion model of the near cathode region of a high current arc discharge. *Soviet Physics Doklady, 17*(1), 225–228.
74. Spitzer, L. (1962). *Physics of fully ionized gases.* NY-London: Willey.
75. Ostashev, V. E., Lebedev, E. F., & Fortov, V. E. (1993). Reasons for limiting the acceleration speed of macrobodies in a magnetoplasma accelerator. *High Temperature, 31*(2), 274–281.

Conclusion

The importance of the study of vacuum arcs is motivated by multiple of their applications, some of which are reported in this book. Numerous experimental and theoretical publications on the topic of cathode spots (the key phenomenon that supports existence of the arc) demonstrate the complicated nature of the problem. The evolution of the investigations described in this book illustrates the progress over the years in obtaining improved experimental data and theoretical understanding. Starting with idealized hypotheses that considered isolated processes in the cathode spot, further studies have evolved into considering important stages leading to the more consistent recent works. While the primary hypotheses suggested in the early days to understand the mechanisms of electron emission or the cathode thermal regime are still used, the later attempts are based on systems of equations describing a number of coupled cathode processes to explain the observed spot parameters (current density, minimal arc current and others). Nevertheless, the main problem of current continuity in the cathode region, which is the relation between electron and ion currents, remained unsolved for a long time. New experimental spot data stimulated further theoretical analyses and treatments.

At the same time, further theoretical approaches continued to use non-closed systems of equations, i.e. studying the coupled cathode phenomena, as a rule, the models attended to some selective effects, while being based on inadequate assumptions regarding to others. As a result, the mathematical descriptions employed some arbitrary parameters. Namely, models consisted of a system of equations with parameters given arbitrarily, such as ion current density, electric field, surface temperature, and/or electron temperature.

Another problematic point is use of initial geometric factors (spike, protrusions, and initial craters, determining initially given current density) in models considering development of the cathode spot or thermal mechanism of crater formation. It is obvious that the obtained solution depends on given geometric factors. As an example, the employment of a very small initial crater size (~0.1 μm) as an initial spot radius (extremal value of current density and Joule heat source presence) brings a violation of the cathode energy balance due to radius increasing at the transient thermal regime of the cathode and to cessation of the electron emission at time in nanoseconds range. This time was interpreted as a time of an elementary act, so named, Emission Center (EC), which further used to explain the cathode spot mechanism. Note that this result is changed dramatically according to spot model involved, which instead of mentioning some problematic assumptions used other assumptions or arbitrary parameters.

Sometimes, the microscopic geometric factors of the emission area at the cathode surface are also assumed for studying the cathode spot numerically using standard commercial software requiring these initial factors as input parameters. The problem of such approach is due to necessity to use a number of arbitrary given multiple parameters characterizing different sides of spike and its variation on time at cathode surface. As result, the significance of these calculations and conclusion based upon them is limited, because they depend on input conditions, which are sensitive (especially on the given initial characteristic time) and can be different from the initial conditions in a real situation. Moreover, results obtained by such models cannot be compared with results of other models that used different sorts of assumption and, in particular, that did not use protrusion sizes as initial geometric condition.

Further advances in development of the cathode spot theory occur due to development of the modern plasma science in low- and high temperatures and astrophysical plasma. In particular, the physics of plasma-wall transition, mass and energy transfer in multicomponent plasma, the particle elementary processes in the plasma including electron atom and charge particle collision, the physics in partially- and highly-ionized gases as well the plasma particle interaction with the metallic wall and energy flux formation to the surface.

This book summarized our new approaches taking into account different scientific disciplines considering plurality of physical processes occurring in the cathode and adjacent plasma. Such approaches are based on the modern plasma physics. The most important detailed analysis of the cathode plasma illustrated the importance of coupling between electron-atom, atom-ion and charge particles between one another while considering the modern state of characteristics of elementary collisions in the atomic physics. The analysis indicated the important role collisions of *resonance charge exchange and relationship between the mean free path of the plasma ions and electrons to the electron beam relaxation length*. The corresponding mathematical treatment allowed description of the origin of mass and energy fluxes in hydrodynamic approximation considering the diffusion phenomena in the cathode plasma based on the developed theory of multicomponent plasma physics. The derived cathode plasma equations together with equations for cathode body and cathode surface make a mathematically closed self-consistent system of equations in accordance with gasdynamic model of the cathode spot. This system enabled understanding of the nature of the long standing problem of electron current fraction in the cathode spot. It also determined directly its time dependent value, largely unknown, in addition to analyzing the main parameters of cathode body and cathode plasma.

The next development of the spot theory takes into account the electrode vaporization into the adjacent dense plasma separated by an electric sheath from the surface. Two Knudsen layers for heavy particles and emitted electron beam were studied

Conclusion

considering the kinetics of cathode evaporation and plasma generation. The model describes the returned fluxes of heavy particles and determines the net erosion of cathode material and potential barrier controlling the flux of the returned plasma electron, finally determining the cathode potential drop. Thus, this advanced approach constitutes the physically closed mathematical system of equations in the frame of the Kinetic Model of the cathode spot. The closed self-consistent model was defined as an approach, which allows obtaining the solution without parameters given arbitrarily or taken from an experiment. The spot mechanism and spot parameter were investigated together with gasdynamic plasma flow in the dense plasma region and in the expanding plasma jet. The nature of spot motion in transverse and oblique magnetic fields was understood using the kinetic model taking in account the relations between pressures of self-magnetic and external magnetic fields. The nature of different spot types and the character of spot splitting were considered taking in account the cathode surface irregularities, the metal melting and self-organized energy due to the plasma-cathode interaction limit, the spot lifetime, and its parameters. These parameters of course can be different from that in the steady state and spot lifetime can be shorter than the time to reach the steady state. The theory can show some dependence on time of the ion current fractions and erosion rate per coulomb unit. This result can explain their changes for different types of cathode spots. The corresponded experiments indicate mostly constant their values because the measurements were obtained as integral characteristics as averaged on large number of arcs and time.

The main difference between the various models is that in some, the cathode potential drop u_c was assumed, whereas in the kinetic model u_c was calculated as part of a self-consistent set of equations. This last model newly examines the role of the arc voltage at the moment of arc initiation and spot development. With u_c self-consistently calculated, the spot temperature and the current density do not unlimitedly increase with time (interpreted as unstable state leading to explosion process) in case of planar cathode or it uncertain determination in case of given the plasma (crater) spatial and temporal distribution, as occurred when a constant u_c or arc voltage was assumed. In additional, the kinetic model allows description of the super-fast spot type (of very small current) in contrast to models with given constant u_c as an input parameter. The developed cathode spot theory allows consideration of the role of the arc discharge in broad types of applications explaining the electrode phenomena in systems for metallic thin film deposition and in vacuum arc micro thruster. Understand the mechanism of body acceleration in magnetoplasma railgun, during the laser and laser plasma interaction with the target, as well the arcing mechanism by metallic wall interaction with plasma flow containing dopant atoms.

It should be noted that the cathode spot is a confusing object that was open for modelling and discussion mostly about two centuries. Different reasons contributed to this. Primarily, it is because the cathode spots appear in various forms depending on unpredictable conditions, which are also produced (surface modification, outgassing and others) during the arc burning. Therefore, an attempt to complete the spot description can meet unpredictable difficulties. As a result, probably some general theory produced on base of unitary principles, in essence, cannot be constructed to describe any observed arc discharge phenomena. Consequently, the relevant and correct theory has to contain non-contradictory assumptions. However, due to uncertainty in cathode spot diagnostics the agreement with experiment should be considered as necessary but not enough as some criterion for approval of the theory.

The spot theories reported in this book are self-consistent and as such allow explanation of the mechanisms of cathode plasma generation to support the current continuity, the plasma jet acceleration, the typical spot phenomena, its characteristics, behaviour as well the phenomena and the parameters of the anode spots. The author hopes that the detailed description of approaches developed over many years will be useful for students and researchers to understand the fundamentals of the electrical arcs. These tools hopefully will assist in describing new original phenomena and will serve as a springboard for further improvements in theory and understanding.

Appendix
Constants of Metals Related to Cathode Materials Used in Vacuum Arcs

Table A.1 Part 1 Electrode metals and their thermo-physical constants: T_m is the melting temperature, T_b is the boiling temperature, ρ is the electrical resistivity, φ is the work function, λ_T is the thermal conductivity, a is the thermal diffusivity, which are both presented for the indicated interval of metal temperatures, λ_s is the heat of vaporization, χ is the heat of vaporization in eV ratio to the fork function, A_i is the atom number, u_c is the cathode potential drop, u_i is the atom potential ionization. A, B, and C are constants, which are determined according to Dushman's formulas [A1] the saturating pressure p of vapor $\lg p(T) = A - \frac{B}{T}$ and rate of vaporization $\lg W(T) = C - 0.5 \lg T - \frac{B}{T}$

Number	1	2	3	4	5	6	7	8	9	10	11
Metal	Li	Be	Boron	C	Na	Mg	Al	Si	K	Ca	Ti
A_i	7	9	11	12	23	24	27	28	39	40	48
u_c, V	11.4	18.9		13.1	8.85	12.3	17.9		7.05	11.1	17.2
u_i, eV	5.4	9.332	8.3	11.3	5.14	7.64	6	8.15	4.34	6.11	6.8
T_m, °C	180.5	1283	2030	–	97.82	649.5	660.1	1432	63.4	850	1668
T_bC	1317	2477	3900	4502	890	1120	2447	2355	753.8	1487	3280
$\rho \times 10^{-6}$ Ω cm	8.5–45.2	2.78	20	800–103	4.27–9.6	3.94–4.4	2.5–8		6.15–31.4	3.6–4.1	42–55
°C	0–230 °C	0 °C	873 °C	0–2500 °C	0–98 °C	0–20 °C	0–400 °C		0–350 °C	0–20 °C	0–20 °C
φ (eV)	2.28	3.92	4.5	4.34	2.29	3.4	4.25	4.8	2.26	2.24	4
λ_T, cal/cm s°	0.103–0.264	0.44–0.151	0.064–0.023	0.274–0.06	0.32–0.103	0.396–0.235	0.497–0.15	0.4–0.062	0.24–0.058	0.235	0.0372–0.021
	182–1727 °C	27–1500 °C8	300–700 °C	20–1800 °C	7–927 °C	0–900 °C	27–927 °C	0–1414 °C	7–1227 °C	0 °C	20–1660 °C
λ_T, W/m°	42.8–110	182–62.8	26.6–9.5	114–25	133–43	165–98	207–62	167–25.9	100–24.3	98	15.5–8.8
$\lambda_s \times 10^{-3}$ cal/g	4.64	8.211	6.82	15	1.005	1.311	2.516	3.38	0.48	0.92	1.75
γ, g/cm³	0.53	1.8	2.34	2.3/0.163	0.97	17.4	2.7	2.42	0.86	1.55	4.5
a, cm²/s	0.245–0.63	0.505–0.174	0.11–0.04	0.741–0.163	0.997–0.322	0.095–0.056	0.893–0.268	0.743–0.115	1.55–0.377	1.011	0.065–0.037
c, cal/g grad	0.79	0.481	0.245	0.1604	0.331	0.24	0.206	0.223	0.18	0.15	0.126
Ws, eV	1.353	3.079	5.4	7.5	0.963	1.311	2.83	3.94	0.78	1.53	3.5
$\chi = \varphi/Ws$	1.685	0.974	0.833	0.579	2.378	2.593	1.237	1.22	2.897	1.46	1.14
A	10.99	12.01	13.07	15.73	10.72	11.64	11.79	12.72	10.28	11.22	12.5
$B \times 10^{-3}$	8.07	16.47	29.62	40.03	5.49	7.65	15.94	21.3	4.48	8.94	23.23
C	7.18	8.25	9.36	12.04	7.17	8.1	8.27	9.21	6.84	7.79	9.11

[A1] S. Dushman, Scientific foundations of vacuum technique, J. M. Lafferty, Ed., J. Wiley & Sons, NY-London, 1962

Appendix: Constants of Metals Related to Cathode Materials Used in Vacuum Arcs

Table A.1 Part 2

Number	12	13	14	15	16	17	18	19	20	21	22
Metal	V	Cr	Fe	Co	Ni	Cu	Zn	Ga	Rb	Sr	Zr
A_i	51	52	56	59	59	64	65	70	85	88	91
u_c, V	18.1	17	18.2	17.77	18	15.05	10	12.6		8.8	18.1
u_i, eV	6.74	6.8	7.9	7.9	7.6	7.72	9.4	6	4.18	5.7	7
T_m, °C	1730	1903	2	1492	1453	1083	419.51	29.78	327.3	770	1855
T_bC	3380	2642	1535	2255	2800	2595	907	2227	1751	1367	4380
$\rho \times 10^{-6}$ Ω cm	18.2–26	1.5–18.9	9.71–24.5	6.24–43.3	6.14–6.02	1.55–24.6	5.45–35.74	13.7–54.3	11.6–19.6	20	40.5
°C	0–20 °C	0–20 °C	0–200 °C	20–400 °C	0–400 °C	0–1500 °C	0–850 °C	0–20 °C	0–48 °C	0 °C	0 °C
φ (eV)	4	4.7	4.5	4.4	4.8	4.5	4.26	4	2.1	2.6	4.15
λ_T, cal/cm s°	0.08–0.11	0.161–0.07	0.18–0.085	0.17–0.026	0.22–0.132	0.95–0.722	0.266–0.137	0.098–0.065	0.852–3.39×10^{-5}		0.0514–0.047
	20–1500 °C	27–1127 °C	0–1530 °C	17–1397 °C	20–1300 °C	20–1083 °C	20–619 °C	0–100 °C	20–1227 °C	20 °C	20–1600 °C
λ_T, W/m°	33.2–45.6	67–29	75–35.6	70.9–11	92–55	395–301	111–57	41–27	35.5–0.014	30.7	21.4–19.6
$\lambda_s \times 10^{-3}$ cal/g	2.08	1.4	1.51	1.576	1.516	1.14	0.43	0.857	0.213	0.38	1.318
γ, g/cm³	6.1	7.1	7.9	8.9	8.9	8.94–8.2	7.14	5.97	1.5	2.6	6.5
a, cm²/s	0.11–0.15	0.206–0.089	0.217–0.103	0.18–0.028	0.236–0.141	1.14–0.75	0.405–0.208	0.183–0.121	0.71–0.00028		0.12–0.11
c, cal/g grad	0.12	0.11	0.105	0.1056	0.105	0.0914–0.117	0.092	0.09	0.08	0.074	0.066
W_s, eV	4.42	3.033	3.523	3.874	3.727	3.04	1.164	2.5	0.754	1.393	4.997
$\chi = \varphi/W_s$	0.905	1.498	1.277	1.136	1.288	1.48	3	1.6	2.78	1.866	0.89
A	13.07	12.94	12.44	12.7	12.75	11.96	11.63	11.41	10.11	10.71	12.33
$B \times 10^{-3}$	25.72	20	19.97	21.11	20.96	16.98	6.56	13.84	4.08	7.83	30.26
C	9.69	9.56	9.08	9.35	9.4	8.63	8.3	8.09	6.84	7.45	9.08

1120 Appendix: Constants of Metals Related to Cathode Materials Used in Vacuum Arcs

Table A.1 Part 3

Number	23	24	25	26	27	28	29	30	31	32	33
Metal	Nb	Mo	Ag	Cd	In	Sn	Sb	Te	Cs	Ba	Gd
A_i	93	96	108	112	115	118	122	128	133	137	157
u_c, V	18.2	22.5	12.85	9.4	10.7	11.8	9.6	11.85	6.2	7.8	
u_i, eV	6.8	7.2	7.6	9	5.8	7.33	8.64	9	3.9	5.2	
T_m, °C	2487	2625	960.8	321.03	156.6	231.9	630.5	449.5	28.64	710	1312
T_bC	4900	4800	2212	765	2075	2687	1637	989.8	685	1637	3230
$\rho \times 10^{-6}$ Ω cm	23.3–13.1	5.03–91.8	1.47–4	6.73–35.78	8.2	10.1–61.2	41.7–120	2.105	19–36.6	36–50	140.5
°C	18 °C	0–2327 °C	0–400 °C	0–700 °C	0 °C	0–750 °C	20–860 °C	19.6 °C	0–30 °C	0–20 °C	25 °C
φ (eV)	4	4.3	4.3	3.7	3.8	4	4	4.73	1.9	2.49	3.1
λ_T, cal/cm s°	0.127–0.245	0.39–0.176	1.003–0.014	0.223–0.1	0.211–0.107	0.156–0.078	0.055–0.05	0.139–0.003	0.057–0.024		25–40 °C
	27–2427 °C	27–2620 °C	27–960 °C	20–700 °C	20–140 °C	20–500 °C	20–730 °C	20–1180 °C	20–1227 °C	20 °C	10.4–11.2
λ_T, W/m°	53–102	162–73.3	418–5.82	93–44	88–71	65–32.5	23–20.9	58.15–1.38	23.8–10.1	18.4	0.608
$\lambda_s \times 10^{-3}$ cal/g	1.67	1.33	0.562	0.213	0.467	0.592	0.33	0.0934	0.122	0.26	7.87
γ, g/cm³	8.6	10.2	10.5	8.65	7.3	7.3	16.7	6.24	1.9	3.5	
a, cm²/s	0.164–0.31	0.586–0.265	1.737–0.024	0.408–0.193	0.508–0.41	0.4–0.2	0.056–0.051	0.479–0.011	0.577–0.243		
c, cal/g grad	0.09	0.065	0.055	0.0632	0.057	0.0534	0.0589	0.047	0.052	0.05	
W_s, eV	6.471	5.32	2.529	0.994	2.238	2.91	1.677	0.498	0.676	1.484	19.2
$\chi = \varphi/W_s$	0.6181	0.874	1.7	3.72	1.7	1.374	2.385	9.495	2.8	1.678	0.16
A	14.37	11.64	11.85	11.56	11.23	10.88	11.15		9.91	10.7	
$B \times 10^{-3}$	40.4	30.85	14.27	5.72	12.48	14.87	8.63		3.8	8.76	
C	11.12	8.4	8.63	8.35	8.03	9.71	7.96		6.74	7.54	

Table A.1 Part 4

Number	34	35	36	37	38	39	40	41	42	43
Metal	Hf	Ta	W	Pt	Au	Hg	Tl	Pb	Bi	Ur
A_i	178	181	184	195	197	200	204	207	209	238
u_c, V	17.9	19.1	19.4	16.85	14	9.5	8.9	9.5	8.55	
u_i, eV	7	7.9	7.98	9	9.22	10.4	6.1	7.42	7.3	6.2
T_m, °C	2220	2996	3380	1796	1063	−38.9	303.5	327.3	271.3	1133C
T_bC	5200	5400	5530	4310	2700	356.73	1457	1751	1559	3900C
$\rho \times 10^{-6}$ Ω cm	26.5	12.4	4.89–131.4	9.81–55.4	2.065–37	96–135.5	17.6	19.3–116.2	116–151	29–45.41
°C	0 °C	0 °C	0–3500 °C	0–1500 °C	0–1500 °C	20–350 °C	0 °C	20–800 °C	20–700 °C	20–1000 °C
φ (eV)	3.6	4.2	4.5	5.32	4.4	4.5	3.7	4	4.5	3.3
λ_T, cal/cm s°	0.05–0.04	0.15–0.264	0.312–0.21	0.178–0.24	0.744	0.02–0.036	0.113–0.086	0.084–0.0475	0.019–0.0456	0.054–0.033
	20–2300 °C	27–3000 °C	27–3650 °C	20–3000 °C	0–100 °C	20–800 °C	27–227 °C	20–800 °C	20–800 °C	27–1227 °C
λ_T, W/m°	20.9–16.7	63–110	130–88	74.1–100	310	8.45–15	47–36	34.9–19.8	8–19	22.5–13.8
$\lambda_s \times 10^{-3}$ cal/g	0.874	1.01	1.003	0.625	0.41	0.069	0.19	0.204	0.2	0.462
γ, g/cm³	13.31–12	16.6	19.9	21.5	19.3	13.6	11.8	11.34	9.8	18.6
a, cm²/s	0.108–0,087	0.251–0.442	0.461–0.312	0.263–0.354	1.24	0.0445–0.79	0.308–0.236	0.246–0.138	0.067–0.16	0.1–0.062
c, cal/g grad	0.0347	0.036	0.0343	0.0315	0.031	0.0335	0.031	0.03	0.0292	0.0284
W_s, eV	6.808	7.617	7.69	5.078	3.365	0.585	1.615	1.76	1.741	4.12
$\chi = \varphi/W_s$	0.529	0.551	0.585	1.048	1.307	7.826	2.291	2.273	1.27	0.801
A		13.04	12.4	12.53	11.89		11.07	10.77	11.18	11.59
$B \times 10^{-3}$		40.21	40.68	27.28	17.58		8.36	9.71	9.53	23.31
C		9.93	9.3	9.44	8.8		7.99	7.69	8.1	8.54

Index

A

Acute angle effect, 769, 795, 807, 809
Ambipolar mechanism, 730, 731, 761
Amperian spot motion, 806
Anode erosion, 531, 534, 535, 537, 539
Anode erosion rate, 215, 221, 222, 243
Anode modes measurements, 506
Anode phenomena, 493–495, 501, 507, 509, 517, 518
Anode plasma density, 493, 503, 527, 529
Anode plasma flow, 849, 867, 871, 878
Anode region, 829, 832, 833, 835, 838–840, 843–845, 848–851, 854, 858, 860–865, 871, 874, 875, 878, 880, 883, 886
Anode spot, 493–495, 497–499, 501–504, 506–515, 518–527, 529–531, 533, 536–538
Anode spots theory, 829, 842, 858
Anode temperature, 503, 506, 523–527, 529–531, 537, 538
Anode thermal instability, 154, 155
Anode vaporization, 854, 862, 864, 867, 871, 883
Anomalous electron emission, 1095, 1096, 1107
Arc column constriction, 1087, 1108
Arc deposition techniques, 933
Arc ignition phenomena, 148
Arcing film cathode, 925
Arc plasma sources, 933, 937, 988
Arc refractory anode, 933, 938, 945, 952, 983, 992, 993
Arc thrust source, 1005
Arc triggering, 689, 698, 717, 718
Atom ionization, 8, 10, 11, 14–18, 20

B

Black body assembly, 933, 940
Black body electrodes, 296, 300
Breakdown insulator surface, 148, 157, 159, 160

C

Cathode boundary condition, 737–740, 744, 747, 748, 762
Cathode erosion, 1004, 1006, 1007, 1011, 1014–1016, 1019, 1020
Cathode erosion rate, 215, 218, 220–222, 224, 228, 229, 231, 232, 234, 237–240, 242, 245, 250
Cathode force, 1005, 1006, 1016–1021
Cathode ion flux, 548, 557, 568, 572, 573, 575, 584, 589, 590
Cathode plasma jet-theory, 734, 742, 743, 750, 752, 760
Cathode sheath, 113, 116–118, 125, 127, 129
Cathode spot, 545–547, 549–551, 553, 555–557, 562–564, 566, 567, 569, 573, 576, 577, 579, 580, 582, 584–586, 588–592, 1067, 1080–1082, 1084, 1096, 1107, 1108
Cathode spot theory, 599, 627, 628, 634, 635, 670
Cathode trace model, 564, 566, 567
Cathode vaporization, 555, 582, 585, 588, 589
Channel model, 1087, 1088, 1093, 1094
Characteristic physical zones, 605
Classical approach, 14, 17
Collisionless approach, 127
Collisions cross-section, 8, 10

© Springer Nature Switzerland AG 2020
I. Beilis, *Plasma and Spot Phenomena in Electrical Arcs*, Springer Series on Atomic, Optical, and Plasma Physics 113, https://doi.org/10.1007/978-3-030-44747-2

Contact electroerosion phenomena, 216, 219, 221
Contact phenomena, 144
Current constriction, 1067, 1076, 1080

D
Debye radius, 2, 8
Deposition systems, 934, 942, 976, 979, 988, 989
Developed spot models, 849, 855
Double layer electrode, 1071, 1107
Double sheath model, 633, 638, 639
Double-valued current density, 562
Dropping volt-current characteristic, 1069, 1075

E
Effective voltage measurements, 285, 303
Electrical breakdown, 143, 148, 149, 152, 154, 157
Electric field, 113, 114, 117, 120–128, 130, 131, 134, 135
Electrode erosion measurements, 241, 249
Electrode repulsive effect, 308, 316, 319
Electron beam, 3, 4
Electron emission, 37, 44–51, 53, 54, 56–58, 545–547, 549, 550, 553–556, 562–565, 568, 569, 571–575, 577–580, 582, 584, 588–590
Electron pressure gradient, 726, 729, 731, 733, 741, 761
Electron relaxation length, 604, 639
Energetic and momentum length, 604
Equations of conservation, 672, 676, 677, 685
Erosion-air, 243
Erosion-vacuum, 216, 226, 227, 237, 240, 241
Evolution models, 829, 858
Excite atoms interaction, 552
Experiment & electrode energy losses, 285–287, 302
Explosive electron emission, 56, 58
Explosive electron emission center, 563, 578, 579, 582, 586, 587
Explosive model, 733

F
Fermi-Dirac distribution, 47
Filling trenches, 936, 975, 976, 991, 994
Five-component diffusion, 109, 110

Force measurements, 312, 317, 320, 321, 327, 332
Fusion devices, 895, 896, 898, 903–905, 908, 911, 912, 916, 924, 926, 927

G
Gap distance, 1008–1010
Gasdynamic acceleration, 735, 753, 761, 762
Gasdynamic approximation, 864
Gasdynamic-film cathode, 650
Gasdynamic model, 624, 625, 627, 635, 650, 658, 661, 663
Gas pressure arc measurements, 292

H
Heat conduction energy, 286, 292, 304
Heat conduction in a body, 69
Helium plasma, 895, 903–906
High current erosion, 223, 227, 242, 243, 248
High field emitters, 245, 246
Highly ionized plasma, 604
Hot anode macroparticle generation, 277
Hot refractory anode, 295, 296, 298
Hump potential, 731, 732, 739, 740, 742, 743, 761, 763
Hydrostatic pressure, 308, 314, 316, 319, 329

I
Impeded plasma, 708, 713, 715
Impeded plasma flow mechanism, 795, 796, 822, 823
Initial specific conditions, 718
Intermediate materials, 962, 967, 993
Ion and magnetic field, 352, 393–396, 403, 415
Ion-atom, 601, 602, 611
Ion charge state, 347, 354, 355, 358, 360, 363, 369, 373, 375, 377, 396–398, 400, 402–404, 412–415
Ion current, 347, 352–354, 359, 360, 363–368, 370–373, 375, 376, 393–396, 402, 404–415
Ion current & refractory anode, 368
Ion diffusion, 602, 611, 661
Ion energy, 347, 351–365, 373, 375–377, 397, 401, 412–414
Ion fraction, 354, 356, 357, 365, 367–369, 371, 373, 376, 394, 395, 398, 400, 402, 404–407, 412–415

Index   1125

K
Kinetic-film cathode, 709
Kinetic model, 670, 673, 685, 690, 696, 697, 709, 713, 718, 719
Kinetics anode region, 862, 865
Kinetics cathode vaporization, 671, 674, 685

L
Laser breakdown, 1038
Laser electron emission, 1057
Laser energy absorption, 1039, 1044, 1054
Laser energy converting, 1056
Laser irradiation, 1028, 1029, 1031, 1037–1039, 1042, 1046, 1050, 1051, 1054, 1056–1059, 1061
Laser plasma, 1027, 1029, 1030, 1035–1037, 1039, 1042, 1043, 1046, 1052, 1057–1059, 1061
Laser plasma acceleration, 1058, 1059
Laser plasma heating, 1052
Light-ionization dopant, 1080
Local explosion, 551, 563, 575, 588, 589, 591

M
Macrodroplets formation, 656
Macroparticle charging, 264–267, 271, 273–276
Macroparticle contamination measurements, 255, 275, 277, 278
Macroparticle generation, 255, 277
Magnetic field spot motion theory, 769, 805
Magneto plasma acceleration, 1104
Mathematically closed approach, 612
Mechanism "hot electrons", 548
Mechanism of thermal plasma, 547
Metallic deposition, 933, 935, 938, 992, 993
Metallic tip erosion, 245
Metal vaporization, 40
Methodology of heating calculations point heat source, 93
Microplasma, 1007, 1012, 1021
Microscale, 1007, 1008, 1012, 1015, 1020, 1021
Modes high current, 497, 499, 503, 505, 506, 508, 509, 514, 518–523, 528, 536, 538
Modes moderate current, 506
Moving heat source, 73, 74, 76–79, 82–84, 93
Multicomponent plasma, 101
Multiple charged ions, 130

N
Nanosecond cathode spots, 620, 661
Nanostructured surfaces, 895, 903, 906, 907, 927
Non-equilibrium cathode region, 670
Non equilibrium (Knudsen) layer, 38, 40
Nonlinear heat conduction, 94, 98
Normally distributed heat source, 77, 79, 83
Normal-strip heat source, 80–83

O
Oblique magnetic field, 769, 807, 809, 813–818, 820
Overheating instability, 1068, 1076

P
Particle recombination, 25
Particle transport phenomena, 101
Physically closed approach, 669, 685, 719
Plasma acceleration-mechanism, 734–736, 753, 759, 761, 763
Plasma boundary layer, 1067, 1073, 1076, 1078, 1107
Plasma density gradient, 601, 611
Plasma double layer, 132, 133, 136
Plasma instabilities, 734
Plasma-macroparticle interaction, 269–271
Plasma oscillations, 2–4
Plasma particle, 1, 2, 4, 8, 30
Plasma polarization, 729, 730, 761
Plasma-substrate interaction, 269, 271
Plasma velocity, 351, 352
Positive space charge, 546, 552
Primary hypothesis, 770
Protrusion cathode, 690–692, 713

Q
Quantum mechanical approach, 13, 17

R
Rail gun, 1067, 1098–1100, 1104, 1108
Rate of current rise, 620–623, 661, 662
Refractory cathode sheath, 130
Refractory metals deposition, 993
Resonance charge-exchange, 599, 661
Retrograde motion, 769–787, 789, 790, 792, 794, 795, 801, 804, 805, 807, 810, 811, 819–822

S

Saturation vapor, 38
Schottky effect, 48, 52, 54
Self-consistent study, 744, 748
Sheath stability, 116, 118, 119
Short arc, 1007, 1008, 1011, 1012, 1015, 1021
Silicon wafer, 975, 976, 990
Sound speed, 733, 739–741, 743, 750, 761
Sound speed plasma, 671, 695, 701, 702
Space charge, 113–118, 120, 121, 123, 127–129, 132
Spot formation mechanism, 842
Spot grouping, 769, 798, 801, 819
Spot-jet transition, 744
Spotless mode, 504–506
Spot-refractory cathode, 627
Spot splitting, 769, 780, 787, 795, 799, 815, 816, 818–820
Spot types-mechanisms, 714, 718, 719
Spot-volatile cathode, 601, 624
Strong magnetic field, 775, 783, 804–807, 810, 822
Structured spot, 562
Surface relief, 920
Surrounding gas, 796, 801, 804, 805
System of equations approach, 557, 582

T

Target electrical sheath, 1039, 1042
Target vaporization, 1036, 1037, 1042, 1057
Temperature field, 80–88, 93, 96
Thermionic, 44, 45, 47, 50, 53
Thin film deposition, 933–935, 938, 952, 958, 967, 969, 972, 975, 993
Thin plate heating, 75, 79, 91, 93, 94
Three-component diffusion, 103, 104
Threshold current, 507–511, 513, 514, 516, 521, 529, 536, 537, 539
Thrust, 1003–1006, 1016, 1018–1021
Thruster efficiency, 1003, 1004
Tokamaks, 895, 897–903, 908, 909, 912, 914–919, 926
Transient heating, 95
Transient model, 695, 696
Transport coefficients, 101
Triggering mechanisms, 149, 154
Tunneling electrons, 50, 51, 53–55

U

Unipolar arcs, 895–898, 901–903, 905, 906, 912–921, 923, 924, 926, 927
Unipolar arcs mechanisms, 913, 914
Unipolar spots, 927

V

Vacuum arc, 1003–1005, 1007, 1008, 1012, 1015, 1016, 1020, 1021
Vacuum arc measurements, 285
Vacuum spot grouping, 787
Velocity function distribution, 672, 675
Virtual cathode, 115, 128, 129
Voltage oscillations-mechanism, 715

W

Wall-plasma transition, 113, 121
Weakly ionized plasma, 601